Dedicated to Alexis Warnes

Electronic and Electrical Engineering

Principles and Practice

Third edition

Lionel Warnes

palgrave
macmillan

First edition 1994
Second edition 1998
Third edition 2003
Published by
PALGRAVE MACMILLAN
Houndmills, Basingstoke, Hampshire RG21 6XS and
175 Fifth Avenue, New York, N. Y. 10010
Companies and representatives throughout the world

PALGRAVE MACMILLAN is the global academic imprint of the Palgrave
Macmillan division of St. Martin's Press, LLC and of Palgrave Macmillan Ltd.
Macmillan® is a registered trademark in the United States, United Kingdom
and other countries. Palgrave is a registered trademark in the European
Union and other countries.

ISBN 978–0–333–99040–7

This book is printed on paper suitable for recycling and made from
fully managed and sustained forest sources. Logging, pulping and
manufacturing processes are expected to conform to the environmental
regulations of the country of origin.

A catalogue record for this book is available from the British Library.

10   9   8   7   6
12 11 10 09 08

Printed in China

Contents

Physical constants

The speed of light in a vacuum, c	3×10^8 m/s
The permeability of free space or the magnetic constant, μ_0	$4\pi \times 10^{-7}$ H/m
The permittivity of free space or the electric constant, ϵ_0	8.85×10^{-12} F/m
Planck's constant, h	6.626×10^{-34} J.s
Boltzmann's constant, k	1.38×10^{-23} J/K
The magnitude of the electronic charge, q	1.6×10^{-19} C
The ice point	273.2 K

Notes: $c = 1/\surd(\mu_0\epsilon_0)$. Impedance of free space, $Z_0 = \surd(\mu_0/\epsilon_0) = 377\ \Omega$. The above, where approximated, are accurate to ±0.14%.

Unit prefixes

a	atto	10^{-18}
f	femto	10^{-15}
p	pico	10^{-12}
n	nano	10^{-9}
μ	micro	10^{-6}
m	milli	10^{-3}
k	kilo	10^{3}
M	mega	10^{6}
G	giga	10^{9}
T	tera	10^{12}
P	peta	10^{15}
E	exa	10^{18}

Notes: Order is important, so that 1 ms.V = 1 millisecond-volt while 1 V.m.s = 1 volt-metre-second; for clarity a dot may be used to separate the different units. Lower and upper case styles for units and prefixes must be carefully followed: 1 K.gm = 1 Kelvin-gram-metre, while 1 kg.m = 1 kilogram-metre and 1 kg = 1 kilogram; 1 mS = 1 millisiemens, 1 ms = 1 millisecond. Note also that $1\ mm^2 = 1$ square mm $= 10^{-6}\ m^2$. Submultiples in denominators should be removed to the numerator, so that 1 pF/mm^2 becomes 1 μF/m^2.

Units

Base units

quantity	unit	symbol	dimension
Mass	kilogram	kg	M
Length	metre	m	L
Time	second	s	T
Current ampere	A	I	

Derived units

quantity	unit	symbol	dimensions
frequency	hertz	Hz	T^{-1}
speed	metre/second	m/s	LT^{-1}
acceleration	metre/sec/sec	m/s^2	LT^{-2}
force	newton	N	MLT^{-2}
work, energy	joule	J	ML^2T^{-2}
power	watt	W	ML^2T^{-3}
charge	coulomb	C	TI
potential	volt	V	$ML^2T^{-3}I^{-1}$
resistance	ohm	Ω	$ML^2T^{-3}I^{-2}$
conductance	siemens	S	$M^{-1}L^{-2}T^3I^2$
capacitance	farad	F	$M^{-1}L^{-2}T^4I^2$
inductance	henry	H	$ML^2T^{-2}I^{-2}$
magnetic field	ampere/metre	A/m	$L^{-1}I$
flux density	tesla	T	$MT^{-2}I^{-1}$
magnetic flux	weber	Wb	$ML^2T^{-2}I^{-1}$

1   Circuit analysis

BEFORE any use can be made of electricity or of any electrical machine or device, it must form part of an electrical circuit. Even complex machines may be modelled by simple elements that, when assembled into a circuit in the right way, can be analysed and so predict the machine's behaviour. Accordingly, circuits are the foundation of any study of electrical or electronic engineering. We begin by defining simple circuit elements, then we shall incorporate them into circuits for analysis with the help of a number of laws and theorems. There are not many laws to remember – Ohm's law and Kirchhoff's laws are almost the only ones – but from these a number of theorems have been deduced to assist in circuit analysis. In this chapter we shall primarily be concerned with direct currents and voltages (DC for short), but the principles developed will serve for analysing circuit behaviour with alternating currents and voltages (or AC).

The five circuit elements considered initially are voltage and current sources and three passive elements: resistance, inductance and capacitance. Though simple these can be combined to form powerful equivalent circuits.

1.1 Sources

No electricity can flow continuously in a circuit lacking a source, because sources are essential for supplying power. Practical sources include batteries, radio antennas and electromechanical generators, all of which can be modelled by ideal sources in combination with other circuit elements. Two sources are defined: the voltage source and the current source, the symbols for which are shown in Figure 1.1; the small circles, A and B, represent terminals.

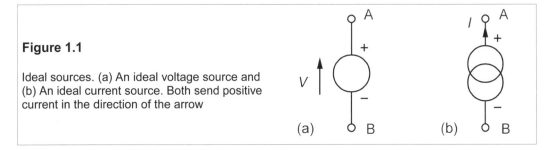

Figure 1.1

Ideal sources. (a) An ideal voltage source and (b) An ideal current source. Both send positive current in the direction of the arrow

Polarity signs are usually omitted and the strength of the source is written next to the arrow. An ideal voltage source always maintains the voltages across its terminals at the value indicated. An ideal current source always drives the stated current out of the positive terminal (A) and it returns to the source at the negative terminal (B). A car battery makes a good approximation to an ideal voltage source. Constant-current sources – within limits – can be made from transistors or operational amplifiers, but the simplest form when small currents are required is a voltage source in series with a large resistance.

1.1.1 Sign conventions

The head of a voltage arrow points to the positive terminal when the voltage is positive. A current arrow on the black line representing a resistanceless conductor indicates the direction of positive current flow. With the polarities as in Figure 1.1, the positive (conventional) current from each source will flow from the positively marked terminal. In reality electrons, being negatively-charged particles, will flow the other way. If the arrows on the sources are reversed, the voltage and current values become negative. Thus the voltage sources in Figure 1.2a and the current sources in Figure 1.2b are identical.

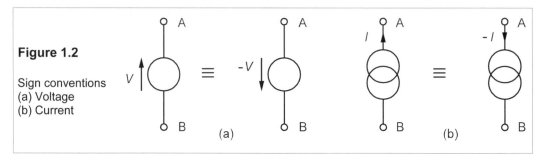

Figure 1.2

Sign conventions
(a) Voltage
(b) Current

1.1.2 Power

When the current and voltage arrows of a source point in the same direction the source is delivering power (VI) to the circuit[1]; if they are opposed, the source is consuming power. For example, in Figure 1.3, the 3V source is sending out a current of 2 A, so delivers $2 \times 3 = 6$ watts (W) to the rest of the circuit. The 2 V source receives a current of 2 A (the 2 A arrow opposes the 2 V arrow), so the power it consumes is $2 \times 2 = 4$ W: the rest of the power is consumed by the resistance. If power delivered has a positive sign and power consumed a negative sign, the algebraic sum of all the power in any circuit must be zero.

Figure 1.3

The 3V source is delivering power, while the 2V source is consuming it. Resistances, however, *always* consume power

---

[1] Direct (unidirectional) voltages and current are written in upper-case (capitals) italic: *V, I*. Instantaneous values are written in lower-case (small) italic: *v, i*. Sometimes the distinction is blurred.

1.2 **Passive circuit elements**

In this chapter we shall look at only three ideal passive circuit elements: resistance, capacitance and inductance. When combined with ideal sources these are sufficient to enable us to model the behaviour of many practical systems, once we know the appropriate voltage-current relationships.

1.2.1 Resistance[2]

A resistor is a practical device that ideally possesses only resistance. An ideal resistor obeys Ohm's[3] law:

$$V = IR \tag{1.1}$$

That is, the voltage across a resistance is directly proportional to the current through it, resistance being the constant of proportionality. When I is in amperes (A) and V in volts (V), then R is in ohms (Ω).

Figure 1.4

Ohm's law: current and voltage arrows are opposed when both current and voltage are positive

In Figure 1.4, if the current through the resistance is in the direction shown, the voltage arrow on the resistance must be opposed to it as indicated. If the arrow is reversed the voltage must be negative, following the sign convention previously described for sources. Ohm's law has been generalised and extended far beyond the scope originally intended, making it one of the most important laws – if not *the* most important – in electrical engineering.

1.2.2 Capacitance

A capacitor is said to be a device for storing electric charge. Ideally it possesses only capacitance and no resistance or inductance. In Figure 1.5, the current, I, flows into the positive plate and charges up the capacitance to a potential, V. The charge on the capacitance is said to be Q coulombs (C). In reality the total charge is zero as positive and negative charges are equal.

---

[2] The terms resistor, inductor and capacitor will be used to describe the actual device and the terms resistance, etc. to mean either the ideal device, or its characteristic property. In a circuit diagram all the elements are assumed to be ideal, and are described as resistances, etc.

[3] Georg Simon Ohm (1787–1854). A schoolteacher of modest means, Ohm published his law in its modern form in 1827, after suggesting several more complicated formulations of 1825 and 1826. It was greeted with near-universal scepticism. Not until 1849 was Ohm rewarded with a professorial chair at Munich.

Figure 1.5

A capacitance with charge +Q on the positive
plate and −Q on the negative. The *total* charge on
the plates is zero. Charging results in the separation of positive and negative charges

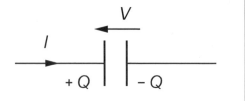

Capacitance is the ratio of the charge on the positive plate to the potential across the capacitance:

$$C = Q/V \quad \text{or} \quad Q = CV \tag{1.2}$$

When Q is in coulombs (C) and V in volts, C is in *farads* (F). (Note that the unit of charge,
the coulomb, is abbreviated to upper-case C, while the symbol for the value of a capacitance
is upper-case italic: C.) A 1 F capacitor was once thought large, but new technology has
produced cylindrical 1 F capacitors rated at 5 V that are 20 mm in diameter and 10 mm high.
    Current is the rate of flow of charge past a point in a circuit:

$$i = \mathrm{d}q/\mathrm{d}t \tag{1.3}$$

Here i and q are the instantaneous current and charge respectively. Substituting Cv for q using
Equation 1.2 yields

$$i = C\mathrm{d}v/\mathrm{d}t \tag{1.4}$$

which may be integrated to give the voltage across the capacitance:

$$v = C^{-1}\int i\,\mathrm{d}t \tag{1.5}$$

Equation 1.5 shows the voltage across a capacitance is the area under its current-time (i-t)
graph divided by C. The area under the i-t graph is the charge, Q.

Example 1.1

If the i-t graph for a capacitance of 1 μF looks like that of Figure 1.6a, what will the V-t graph look like?
(Assume that the capacitance was uncharged at $t = 0$.)
    The current is at first positive (the capacitance is therefore being charged positively) and constant
at 5 A for 1.5 μs, so the area under the i-t graph is 5 × 1.5 = 7.5 μC at $t = 1.5$ μs, and the potential drop
across the capacitance is

$$V = Q/C = 7.5 \times 10^{-6} \div 10^{-6} = 7.5 \text{ V}$$

The voltage increases linearly with time from 0 to 1.5 μs because the current is constant. After 1.5 μs
the current becomes negative and constant at −3 A until $t = 4$ μs, consequently the capacitance is
discharging and the voltage across it is declining at a constant rate. The area under the i-t graph for
$1.5 \le t \le 4$ μs is −2.5 × 3 = −7.5 μC, so the change in the capacitance's voltage during this time must
be −7.5 V. Since the voltage was +7.5 V at $t = 1.5$ μs, it must be zero at $t = 4$ μs. The areas above and
below the time axis are equal so the charge on the capacitance must be zero at $t = 4$ μs. The v-t graph
must appear like Figure 1.6b.

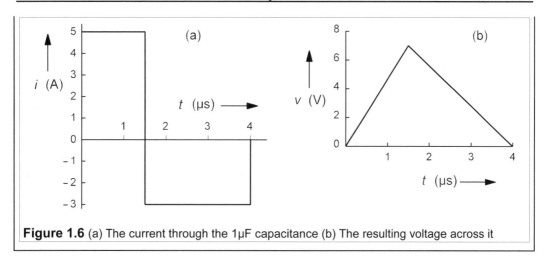

Figure 1.6 (a) The current through the 1µF capacitance (b) The resulting voltage across it

The calculation of capacitance

The capacitance of a parallel plate capacitor can be found from the formula

$$C = \varepsilon A/d \tag{1.6}$$

where A is the area of one plate (the other is assumed to be identical), d is the plate separation (assumed uniform) and ε ($= \varepsilon_r \varepsilon_0$) is the *permittivity* of the medium between the plates. ε_r is the relative permittivity (or dielectric constant) of the medium, being unity for a vacuum (or air), about 2.3 for polystyrene and 1000 or more for PZT. The permittivity of free space (or a vacuum or air), ε_0, is numerically 8.85×10^{-12} F/m. It arises from the choice of the SI unit for capacitance. In some books it is called the *electric constant*. In many cases the capacitance of a system can be calculated if its geometry (layout and dimensions) is known.

1.2.3 Inductance

An ideal inductor possesses only inductance, which is sometimes called self inductance to distinguish it from mutual inductance. Figure 1.7 shows the circuit symbol for an ideal inductor carrying a current, i, and with a voltage across it of v.

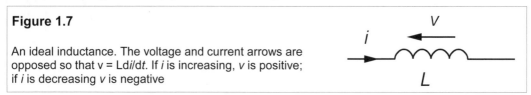

Figure 1.7

An ideal inductance. The voltage and current arrows are opposed so that $v = Ldi/dt$. If i is increasing, v is positive; if i is decreasing v is negative

The voltage and current arrows are opposed, and the voltage is given by

$$v = L\,di/dt \tag{1.7}$$

Equation 1.7 indicates that there is no voltage across an inductance *unless the current through it is changing*. It also shows that if the current into the inductance is increasing, then the opposing voltage increases also. If the current is decreasing, the voltage is negative – in other words, the voltage arrow is reversed.

The calculation of inductance

The inductance of a long solenoid (coil of uniformly wound wire of circular cross-section) is given by

$$L = \mu N^2 A / \ell \tag{1.8}$$

where N is the number of turns or wire, A is the cross-sectional area of the solenoid and ℓ is its length. μ $(= \mu_r\mu_0)$ is the permeability of the medium inside the solenoid. μ_r is the relative permeability of the medium (unity for a vacuum or air) and μ_0 has the numerical value $4\pi \times 10^{-7}$ H/m and is known as the permeability of free space (or a vacuum or air), sometimes as the *magnetic constant*. It arises because of the choice of the SI unit for inductance. Inductance, like capacitance, can be calculated from the geometry of the system.

Example 1.2

The current through an inductance of 5 mH is given in Figure 1.8. Derive the *v-t* graph. We must use Equation 1.7. From Figure 1.9 we can see that the current at first goes linearly from 0 to -5 A in 2 ms, so di/dt is $-5/2.5 \times 10^{-3} = -2000$ A/s. As $L = 5 \times 10^{-3}$ H,

$$v = L\,di/dt = 5 \times 10^{-3} \times -2000 = -10 \ \text{V}$$

Because di/dt is constant, v must be constant also during this period. From $t = 2$ ms to 3.5 ms, the current is constant, meaning both di/dt and v are zero. From 3.5 ms to 6 ms, the current increases linearly from -5 A to 5 A, a rate of change of current of 10 A in 2.5 ms, or 4000 A/s. Therefore

$$v = 5 \times 10^{-3} \times 4000 = +20 \ \text{V} \tag{1.9}$$

From then on *i* is constant, and the *v-t* graph is as in Figure 1.9.

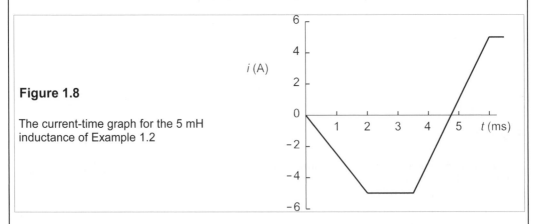

Figure 1.8

The current-time graph for the 5 mH inductance of Example 1.2

The current-voltage relations for capacitances and inductances involve time differentials, and in consequence *neither the voltage across a capacitor nor the current through an inductor can change instantaneously*: the capacitor current or the inductor voltage would have to be infinite for that to happen. This is not true for a resistance because it stores no energy, whereas a capacitor stores energy in its electric field and an inductor in its magnetic field.

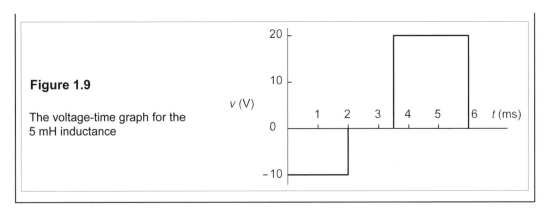

Figure 1.9

The voltage-time graph for the 5 mH inductance

1.2.4 Mutual inductance

When two coils are wound onto a ferromagnetic core, or are in close proximity to each other, some of the magnetic flux produced by a one coil will link with the other coil, which is called *coupling*. In Figure 1.10 a changing current, i_1, in coil 1 induces a voltage, v_2, in coil 2. The current and voltage are related by

$$v_2 = M\frac{di_1}{dt} \tag{1.10}$$

where M is the *mutual inductance* of the coils. A changing current, i_2, in coil 2 would similarly produce a voltage, v_1, in coil 1 of magnitude Mdi_2/dt. The dots indicate the winding direction, which is such that a positive, increasing current going into the dotted terminal will produce a positive voltage at the other dotted terminal.

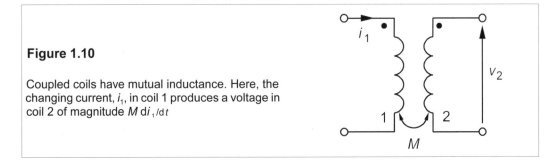

Figure 1.10

Coupled coils have mutual inductance. Here, the changing current, i_1, in coil 1 produces a voltage in coil 2 of magnitude $M \, di_1/dt$

When all the magnetic flux of one coil links with the other coil, it can be shown that

$$M = \sqrt{L_1 L_2} \tag{1.11}$$

where L_1, L_2 are the self inductances of the coils. The coils are then said to be perfectly coupled, which is very nearly true in large transformers (see Chapter 15). If the coupling is not perfect then

$$M = k \sqrt{L_1 L_2} \tag{1.12}$$

k, the *coefficient of coupling*, lies between 0 (no coupling) and 1 (perfect coupling).

1.3 **Practical circuit elements**

The circuit elements discussed so far have been ideal but practical circuit elements can be modelled from them. Some circuit elements are more nearly ideal than others; ideal sources, for example, cannot exist. Were we to short-circuit the terminals of an ideal voltage source, the current flowing would be infinite; and were we to open circuit the terminals of an ideal current source, the voltage across its terminals would be infinite. But at comparatively low frequencies capacitors and resistors are for practical purposes ideal.

1.3.1 The practical voltage source

With a practical voltage source such as a torch battery, the terminal voltage varies according to the current drawn from it, as in Figure 1.11.

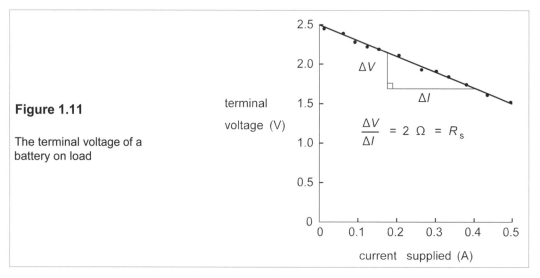

Figure 1.11

The terminal voltage of a battery on load

What can be deduced from this? Firstly, when the current is zero the terminal voltage is 2.5 V, which is the *open-circuit voltage*. However, when the current drawn is 0.5 A, the terminal voltage has dropped to 1.5 V. The *V-I* graph is also a straight line, within experimental error, having a slope with dimensions of ohms. These observations suggest that the model of a practical voltage source should be a resistance in series with an ideal voltage source as in Figure 1.12.

Figure 1.12

A model of a practical voltage source. If the resistance were zero the source would be ideal

The ideal source's e.m.f. must be 2.5 V – the same as the open-circuit voltage of the battery – while the resistance value must be given by the slope of the *V-I* graph (with the sign changed), or 2 Ω. This resistance is known as the *internal resistance* of the source. Then when the current drawn is 0.5 A, the internal resistance has a voltage across it of

$$V = IR_s = 0.5 \times 2 = 1 \text{ V} \tag{1.13}$$

The terminal voltage differs from the open-circuit voltage by this much when the battery is delivering 0.5 A.

1.3.2 The practical current source

Figure 1.13 shows the results of current and voltage measurements made at the terminals of a practical current source. The experimental data suggest that some of the source current is being diverted through an internal resistance in parallel with the load across the terminal.

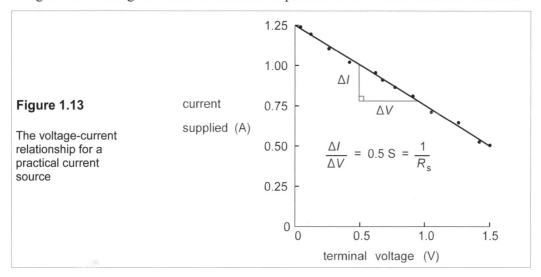

Figure 1.13

The voltage-current relationship for a practical current source

Here it can be seen that the current supplied when the terminal voltage is zero – the *short-circuit current* – is 1.25 A, but when a load is connected across the terminals, the current falls and as it does so the voltage across the terminals rises in step apart from experimental error. The slope of the graph has dimensions of conductance, $Ω^{-1}$ or *siemens*, S.

The value of this resistance is the reciprocal of the slope of the graph (with change of sign). When the terminal voltage is 1.5 V, the terminal current is 0.5 A, so the graph's slope is $(1.25 - 0.5)/1.5 = 0.5$ S. Taking the reciprocal gives a resistance of 2 Ω. The value of the ideal current source must be the short-circuit current of 1.25 A, so the model developed for this practical current source is that shown in Figure 1.14.

Figure 1.14

A practical current source. When the resistance is infinite (or the conductance is zero), the current source is ideal

We shall see later when discussing circuit theorems that practical voltage and current sources are just two different ways of describing the same observations. Were someone to be given a 'black box' with two terminals at which currents and voltages could be measured, he is at liberty to choose whether he wishes to model it as a practical voltage source or a practical current source. The two examples given are different sets of measurements on the *same* source.

1.3.3 Practical resistors

Table 1.1 gives a few of the readily-available, low-wattage, resistor types. Metal-film or carbon-film resistors commonly used in circuits and rated from 0.125 W to 2 W have very little capacitance or inductance – virtually negligible up to 100 kHz; wirewound resistors of higher wattage generally have a few µH of inductance and may be modelled as a small inductor in series with a resistance.

Table 1.1 *Low-wattage resistor types*

Technology	P[A]	Range[B]	Tol.[C]	Tco.[D]	V[E]	Ins. R.[F]	Size[G]	Price[H]
carbon film[J]	0.25	2R2-4M7	±5	−400	250	1 GΩ	3.2L × 1.7D	4
	0.33	1R-10M	±5	−500	250	1 GΩ	5.6L × 2.3D	2.9
	0.66	10R-10M	±5	−500	350	1 GΩ	8L × 3D	3.9
	1	10R-1M	±5	−500	500	1 GΩ	12.1L × 4.2D	5.7
metal film	0.125	10R-1M	±1	+50	200	10 GΩ	4.1L × 1.8D	5.5
	0.25	R10-R15	±5	+200	250	10 GΩ	6.4L × 2.4D	34
	0.25	R22-R68	±5	+200	250	10 GΩ	6.4L × 2.4D	22
	0.25	1R-9R1	±1	+200	250	10 GΩ	6.4L Γ 2.4D	8.7
	0.25	10R-1M	±1	+100	250	10 GΩ	6.4L × 2.4D	3.4
	0.25	51R1-1M	±0.1	+15	200	10 GΩ	7.2L × 2.5D	7.9
	0.5	10R-1M	±1	+100	350	10 GΩ	10L × 3.4D	4.8
SMT[K]	0.063	1R-1M	±1	±100	50		1.6 × 0.8 × 0.45	3.1
	0.1	100R-100k	±0.1	±10	100		2 × 1.25 × 0.5	85
	0.125	1R-1M	±1	+50	100		2 × 1.25 × 0.55	3.1
	0.125	15M-100M	±20	+500	100		2 × 1.2 × 0.4	50
	0.25	1R-1M	±1	±100	200		3 × 1.55 × 0.55	2.5
	0.25	25R-10k	±0.01	< 0.6	54		6 × 3.2 × 2.5	750
	1	1R-1M	±5	±200	250		6.3 × 3.1 × 1.1	19.5
metal oxide	1	1R-1M	±5	+300	350	1 GΩ	9L × 3D	14.5
cermet	0.4	10M-100M	±1	±100	1000		8 × 12.5 × 2	126
wirewound	0.25	10R-100k	±0.1	±5	300		9.5L × 4.75D	328
	0.5	10R-100k	±0.1	±5	400		12.7L × 6.35D	300

Notes: Working temperature range from −55°C to 125°C mostly. [A] Max power in W at 70°C. [B] R = Ω, k = kΩ, M = MΩ, placed at decimal point, so that R10 = 0.10 Ω and 2M2 = 2.2 MΩ. [C] Tolerance in %. [D] Tempco in ppm/°C. [E] Max. DC voltage. [F] Insulation resistance [G] In mm, cylindrical: L = length, D = diameter; otherwise rectangular: height × width × thickness. [H] In pence per piece in small quantities   (1-10). [J] Temperature coefficient (tempco) varies from −100 to −900 ppm/°C, value given is average. [K] Thick-film or thin-film

Often of more serious concern is the temperature coefficient of resistance (sometimes known as the *tempco*) and the *tolerance*, that is the most the actual value of the resistance can differ from the nominal value given by the colour-band code on the resistor. Practical resistors also have rated voltages and temperatures that must not be exceeded. Quite recently surface mount technology (SMT) resistors – chips without leads that can be soldered directly onto printed-circuit boards (PCBs) or thin-film circuits (see Chapter 10) – have become widely used, though they have lower voltage ratings than conventional wire-lead resistors.

Preferred values

In order to improve equipment maintenance and reduce costs, manufacturers use only *preferred values* wherever possible. There were six of these per decade[4] of resistance, spaced in geometric progression, for example: 10, 15, 22, 33, 47, 68 and 100 Ω. Increasingly, interpolated values are as readily available also, for example: 12, 18, 27, 39, 56 and 82 Ω. Most in-house stores will stock all of these.

Example 1.3

A ¼W resistor has a nominal resistance of 2.2 kΩ at 20°C. If it has a tolerance of 5% and must work at 100°C, what is the maximum percentage deviation from the nominal resistance for those resistor types listed in Table 1.1? What is the maximum current and voltage for the nominal-value resistor?

The only ¼W resistors with this tolerance listed in Table 1.1 are carbon-film and SMT types. These have tempcos of –400 and +200 ppm/°C respectively. At 100°C the temperature is 80°C above nominal, so that the resistance changes are 400 × 80 = –32000 ppm = **–3.2%**, and +200 × 80 = +16000 ppm = **+1.6%**. If the carbon-film type is 5% below its nominal value then the total deviation is –**8.2%**, that is the resistance is (1 – 0.082) × 2.2 = 2.02 kΩ. If the metal-film resistor is +5% from nominal then the total is **+6.6%** at 100°C, making the resistance (1 + 0.066) × 2.2 = 2.345 kΩ.

The power dissipation (see Sections 1.1.2 and 1.6) is 0.24 W, and that is given by

$$P = VI = I^2R = V^2/R = 0.25 \text{ W}$$

As R = 2.2 kΩ, we find V^2 = 0.25 × 2200, so **V(max) = 23.5 V**, and I^2 = 0.25/2200, making **I(max) = 10.7 mA**. This assumes that the ambient temperature is not higher than it should be.

1.3.4 Practical capacitors

A practical capacitor has some leakage resistance which can be modelled as a resistance in parallel with a capacitance, as in Figure 1.15.

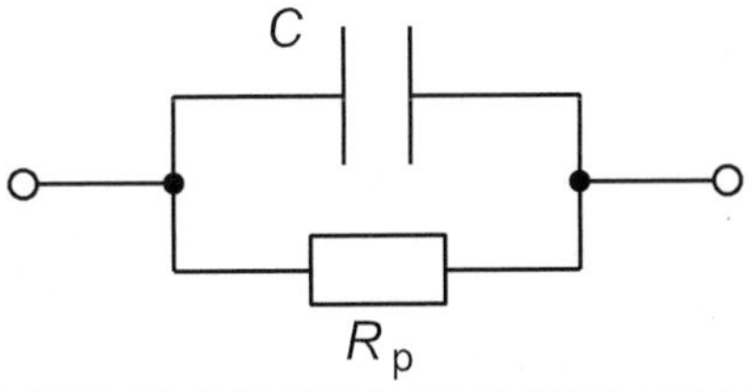

Figure 1.15

A model of a practical capacitor. In a low-value, high-quality capacitor at low frequencies, R_p could be as much as a GΩ

---

[4] A *decade* means over a range of 1:10, such as from 3 Ω to 30 Ω.

Leakage resistance is of two origins: surface leakage and internal (dielectric leakage); the latter is more fundamental but may be less important that the former: both are dependent on frequency. The capacitor's inductance must also be considered at frequencies above about a MHz. The physical size of a capacitor depends on the type of construction, the capacitance and the working voltage. Electrolytic and tantalum capacitors are polar (the terminal marked + must be connected to the more positive voltage, else the capacitor can explode) and have relatively high leakage (that is a *lower* parallel leakage resistance); however, they are small and cheap for high capacitance values. They are usually restricted to low-frequency operation with DC bias. R_p is not always given by manufacturers but can be found from $R_p = 1/(2\pi f C \tan\delta)$, where f is the frequency and $\tan\delta$ is the *loss tangent* of the capacitor. For a polystyrene capacitor with $C = 1$ nF, $f = 1$ kHz and $\tan\delta = 0.0005$, $R_p = 318$ MΩ.

Table 1.2 *A selection of capacitors*

Type[A]	μF	Tolerance	V_{max} [B]	R_p [C]	Size [D]	Frequency	Price
E	22	±20%	6.3	5	7L × 4D	DC + low	5p
E	10^4	±20%	100	0.01	105L × 40D	DC + low	£18
SMTE	1	±20%	50	100	5.4L × 4D	DC + low	16p
EM	10^6	−20% +80%	5.5	0.02	6L × 20D	DC + low	£2.79
TA	68	±20%	16	0.7	15.5 × 9	DC + low	£1.57
SMTA	10	±10	25	10	6 × 3.2 ×2.6	DC + low	£1.61
CD	10^{-2}	±20%	750	10^4 [E]	5L × 16D	< 1 MHz	41p
PP	10^{-3}	±20%	1000	20	20L × 9D	< 100 kHz	28p
PC	1	±5%	100	0.1	18L × 7.5D	< 10 kHz	68p
PE	1	±10%	400	10^4 [E]	32L × 13.5D	< 100 kHz	£1.17
PS	10^{-2}	±1%	63	10^6 [E]	13L × 7.5D	< 10 MHz	54p

Notes: [A] E = electrolytic, SMTE = surface-mount technology electrolytic, EM = electrolytic, memory back-up,   TA = tantalum, SMTA = surface-mount tantalum, CD = ceramic disc, PP = polypropylene, PC = polycarbonate, PE = polyethylene, PS = polystyrene. [B] Maximum DC voltage. For non-electrolytic types the maximum working r.m.s. alternating voltage is approximately one third of this.[C] Leakage resistance in MΩ.[D] Cylindrical capacitors, L = length, D = diameter. [E] Insulation resistance: internal leakage resistance will be much less.

Table 1.2 gives some specifications for commonly obtainable capacitors. Prices do not fall much with increasing quantities ordered. Temperature coefficients of capacitance are usually in the range − 100/+200 ppm. More costly than resistors, capacitors are available in preferred values, usually six per decade of capacitance, such as 100, 150, 220, 330, 470 and 680 pF.

1.3.5 *Practical inductors (see also Section 13.6)*

For many purposes it is possible to treat capacitors and resistors as ideal, but one can seldom do so with inductors: they *must* be modelled as an ideal inductor in series with a resistance as in Figure 1.16. R_c is the resistance of the inductor to DC. Specifications for some commercial small inductors (all of which are made from a coil of insulated copper wire wound round a ferrite core) are given in Table 1.3.

Figure 1.16

A practical inductor. R_C cannot always be ignored in circuit analysis

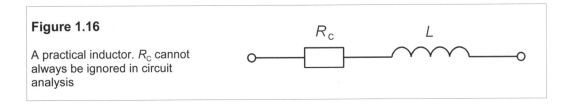

Table 1.3 *Some small, ferrite-cored, commercial inductors*

L [A]	tol.	R_c (Ω)	I [B]	f [C]	Size [D]	Price [E]
0.0001	±20%	0.44	0.45	100 MHz	4.5 × 3 × 3[F]	37p
0.001	±20%	0.5	0.45	8 MHz	4.5 × 3 × 3[F]	25p
0.1	±10%	40	0.03	250 kHz	4.5 × 3 × 3[F]	23p
1	±10	30	0.1	800 kHz	11.4L × 5.4D	£2.57
10	±10	105	0.04	250 kHz	10L × 5D	32p
100	±5	1200	0.02	80 kHz	12L × 5.2D	40p

Notes: [A] In mH. Temperature range –20°C to 70°C. [B] Maximum DC in A. [C] Frequency for max Q (approximates to optimal frequency of operation). [D] In mm, for a cylindrical shape, L = length, D = diameter. [E] Price for one. Unit price falls little with increasing quantity. [F] Rectangular, surface-mount technology.

     Generally speaking, inductors are used as little as possible; though they are often found in passive filters. They are almost impossible to incorporate into integrated circuits (unlike resistors and capacitors), consequently they are habitually 'designed out', even at the expense of component count. Large inductors (sometimes called *chokes*) are normally constructed as required from a ferromagnetic core, coil former and insulated copper wire.

1.4 Circuit analysis and Kirchhoff's laws[5]

Circuit analysis is designed to find the voltage across or the current through any circuit element, and if need be the power consumed by its resistances or supplied by its sources. By far the most important aids to this analysis are Kirchhoff's two laws. They are fundamental to circuit analysis, and with them, and the current and voltage relations for circuit elements given in the previous sections, it is possible in principle to deduce the current in any branch[6] of a circuit, or the voltage at any point. The two laws are:

1. Kirchhoff's voltage law (KVL)

The algebraic sum of the voltages at any instant around any loop in a circuit is zero

---

[5] Gustav Robert Kirchhoff (1824–87) was Professor of Physics at Berlin. He enunciated his laws c. 1848.

[6] A *branch* of a circuit is a path containing one circuit element. In Figure 1.17 AB is a branch, EF is not.

Symbolically,

$$\sum v = 0 \tag{1.14}$$

Kirchhoff's voltage law is a consequence of the principle of energy conservation; so frequent is its use in circuit theory that it is usually abbreviated to KVL.

2. Kirchhoff's current law (KCL)

The algebraic sum of the currents at any instant at any node[7] in a circuit is zero.

Symbolically,

$$\sum i = 0 \tag{1.15}$$

KCL – its usual abbreviation – is a result of the principle of charge conservation. An example will show how the laws can be applied: let us find all the currents and voltages in the circuit of Figure 1.17.

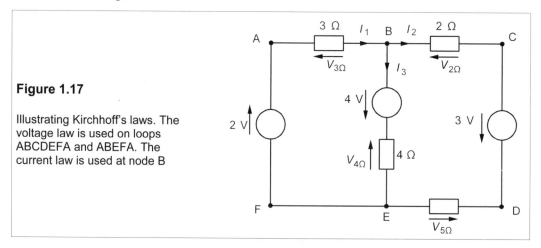

Figure 1.17

Illustrating Kirchhoff's laws. The voltage law is used on loops ABCDEFA and ABEFA. The current law is used at node B

Selecting loop ABEFA, and taking clockwise-pointing voltages as positive and anticlockwise voltages as negative, we find:

$$- V_{3\Omega} + 4 - V_{4\Omega} + 2 = 0 \tag{1.16}$$

The choice of positive for clockwise voltages is arbitrary. Once you have assigned voltage arrows to the circuit elements (with voltage arrows on passive elements opposing the currents through them) you must wait until the analysis is complete to discover whether the polarity is 'right' (in which case the voltage will turn out to be positive) or 'wrong' (in which case the voltage will turn out to be negative). For loop ABCDEFA, KVL yields

$$- V_{3\Omega} - V_{2\Omega} + 3 - V_{5\Omega} + 2 = 0 \tag{1.17}$$

and loop BCDEB produces

---

[7] A *node* is a point in a circuit where branches join. In Figure 1.17, A and B are nodes, but F is not. There is also a node between the 4Ω resistance and the 4V source.

$$-V_{2\Omega} + 3 - V_{5\Omega} + V_{4\Omega} - 4 = 0 \tag{1.18}$$

Note that subtracting Equation 1.16 from Equation 1.17 gives Equation 1.18 – the application of KVL to this circuit produces only two independent equations. Voltages across the resistances can be found from Ohm's law; $V_{3\Omega}$, for example, is $3I_1$. Using this method Equations 1.16 and 1.18 become

$$-3I_1 - 4I_3 + 6 = 0 \tag{1.19}$$

$$-2I_2 - 5I_2 + 4I_3 - 1 = 0 \quad \Rightarrow \quad -7I_2 + 4I_3 = 0 \tag{1.20}$$

KCL is now needed to find a third equation for the unknown currents. If in Figure 1.17 we consider node B and take currents into the node as positive and currents out of the node as negative, then

$$I_1 - I_2 - I_3 = 0 \tag{1.21}$$

The three independent equations produced by application of Kirchhoff's and Ohm's laws to the circuit of Figure 1.18 are numbers 1.19, 1.20 and 1.21, and solving these gives $I_1 =$ **1.017 A, I_2 = 0.279 A, I_3 = 0.738 A**. The voltages across the resistances may be found from these currents with Ohm's law.

If inductances and capacitances are present Kirchhoff's laws can still be applied using the appropriate *v-i* relationships ($v_L = L\,di/dt$ and $v_C = C^{-1}\int i\,dt$), which yield instantaneous values. (Note that Kirchhoff's laws apply to direct (DC), sinusoidal (AC) and instantaneous voltages and currents.) Many useful results are soon deduced from Kirchhoff's laws, starting with the addition of resistances, capacitances and inductances in series and in parallel.

1.4.1 Resistances in series

The vital point about resistances in series is that *the same current passes through each resistance.*

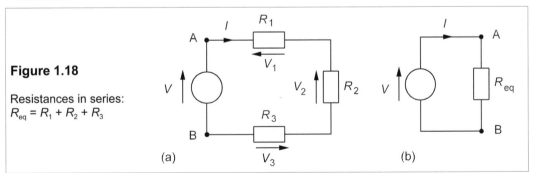

Figure 1.18

Resistances in series:
$R_{eq} = R_1 + R_2 + R_3$

(a)   (b)

Examining Figure 1.18a – three resistances in series – and applying KVL we find

$$V - V_1 - V_2 - V_3 = 0 \tag{1.22}$$

But $V_1 = IR_1$ and $V_2 = IR_2$ and $V_3 = IR_3$, and substitution into Equation 1.22 yields

$$V = IR_1 + IR_2 + IR_3 = I(R_1 + R_2 + R_3) = IR_{eq} \tag{1.23}$$

where $R_{eq} = R_1 + R_2 + R_3$. The circuit simplifies to that of Figure 1.18b. Since $I = V/R_{eq}$, then

$$V_1 = IR_1 = V \times R_1 / R_{eq} \qquad (1.24)$$

which illustrates the *voltage-divider rule*: the voltage across resistances in series divides in proportion to the value of each.

1.4.2 Conductances in parallel

The vital point about parallel circuit elements is that *the voltage is the same across each one*. *Conductance* is the reciprocal of resistance:

$$G = 1/R = R^{-1} \qquad (1.25)$$

The units of conductance are siemens (S). A conductance of 0.1 S is exactly the same as a resistance of 10 Ω.

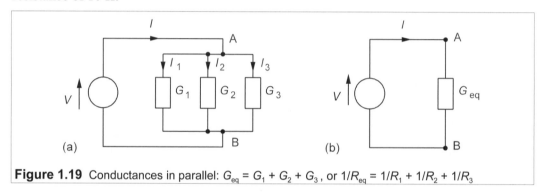

Figure 1.19 Conductances in parallel: $G_{eq} = G_1 + G_2 + G_3$, or $1/R_{eq} = 1/R_1 + 1/R_2 + 1/R_3$

Taking the circuit in Figure 1.19, which has three conductances in parallel, we can apply KCL at node A to obtain

$$I = I_1 + I_2 + I_3 \qquad (1.26)$$

But the voltage across each conductance is the same, and Ohm's law gives

$$V = I_1 R_1 = I_2 R_2 = I_3 R_3 \qquad (1.27)$$

Thus $I_1 = V/R_1 = VG_1$, $I_2 = V/R_2$ etc., and substitution into Equation 1.26 leads to

$$I = I_1 + I_2 + I_3 = V/R_1 + V/R_2 + V/R_3$$
$$= VG_1 + VG_2 + VG_3 = VG_{eq} \qquad (1.28)$$

where $G_{eq} = G_1 + G_2 + G_3$. We can add parallel conductances.

If we let the equivalent resistance be R_{eq} ($= 1/G_{eq}$), then Equation 1.28 gives

$$1/R_{eq} = 1/R_1 + 1/R_2 + 1/R_3 \qquad (1.29)$$

To combine parallel resistances you *add the reciprocal resistances and take the reciprocal of the result*.

In the special case where only two resistances are in parallel, $1/R_{eq} = 1/R_1 + 1/R_2$, so that

$$R_{eq} = \frac{R_1 R_2}{R_1 + R_2} \qquad (1.30)$$

This is sometimes called the *product-over-sum rule*.

It is a straightforward matter to show that the division of current between parallel resistances is given by

$$I_n = I \times G_n/\Sigma G \qquad (1.31)$$

where I_n is the current through G_n, and ΣG is the sum of the parallel conductances. The current through parallel conductances divides in proportion to each conductance.

1.4.3 Capacitances in parallel

In Figure 1.20 the voltage is the same across each capacitance and KCL produces

$$i = i_1 + i_2 + i_3 \qquad (1.32)$$

Then using the *i-v* relation for capacitances, $i = C dv/dt$, in Equation 1.32 yields

$$i = C_1\frac{dv}{dt} + C_2\frac{dv}{dt} + C_3\frac{dv}{dt} = (C_1 + C_2 + C_3)\frac{dv}{dt} = C_{eq}\frac{dv}{dt} \qquad (1.33)$$

where $C_{eq} \equiv C_1 + C_2 + C_3$, and it can be seen that capacitances in parallel add.

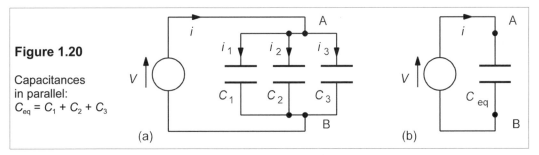

Figure 1.20

Capacitances
in parallel:
$C_{eq} = C_1 + C_2 + C_3$

(a) (b)

1.4.4 Capacitances in series

In Figure 1.21a KVL can be used to obtain $v = v_1 + v_2 + v_3$ which, differentiated with respect to t, gives

$$dv/dt = dv_1/dt + dv_2/dt + dv_3/dt \qquad (1.34)$$

Now the current is the same in all three capacitors, thus

$$i = C_1 \, dv_1/dt = C_2 \, dv_2/dt = C_3 dv_3/dt \qquad (1.35)$$

so that $dv_1/dt = i/C_1$ etc., and Equation 1.34 becomes

$$dv/dt = i/C_1 + i/C_2 + i/C_3 = i/C_{eq} \qquad (1.36)$$

where

$$1/C_{eq} = 1/C_1 + 1/C_2 + 1/C_3 \tag{1.37}$$

The rules for adding inductances are left as exercises (see also Problem 1.5). We shall see later how to use Kirchhoff's laws systematically to minimise the likelihood of errors.

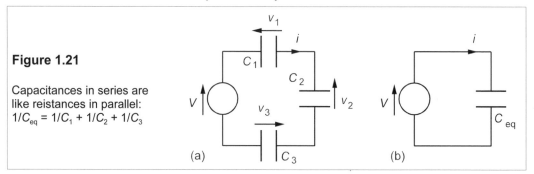

Figure 1.21

Capacitances in series are like reistances in parallel:
$1/C_{eq} = 1/C_1 + 1/C_2 + 1/C_3$

1.5 Circuit theorems and transformations

Many theorems and transformations (all derived from Kirchhoff's laws) can be used to assist with circuit analysis, but we shall use only a few, the first being the most useful:

1.5.1 Thévenin's theorem[8]

Thévenin's theorem states

Any two-terminal, linear network of sources and resistances may be replaced by a single voltage source in series with a resistance. The voltage source has a value equal to the open-circuit voltage appearing at the terminals of the network. The resistance value is the resistance that would be measured at the network's terminals when all the sources have been replaced by their internal resistances.

Thévenin's equivalent circuit is shown in Figure 1.22. To replace complex networks by such a simple equivalent circuit can be highly convenient.

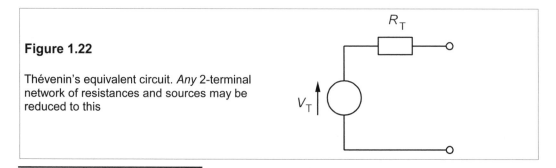

Figure 1.22

Thévenin's equivalent circuit. *Any* 2-terminal network of resistances and sources may be reduced to this

---

[8] L Thévenin was a French telegraphist, as was the eminent English electrical engineer, Oliver Heaviside. He published his theorem in 1883, though Helmholtz's work in 1853 foreshadowed both it and Norton's theorem.

For example, let us find the Thévenin equivalent of the two-terminal network of Figure 1.23a. There are three steps in essence. First, what is the voltage across AB? – just the voltage across the 4Ω resistance, which may be found by the voltage-divider rule as the resistances are in series with the voltage source:

$$V_{AB} = V_{4\Omega} = 4 \times 3/6 = 2 \text{ V} \qquad (1.38)$$

This is the open-circuit voltage across AB[9], and consequently the source voltage, V_T, in the Thévenin equivalent circuit.

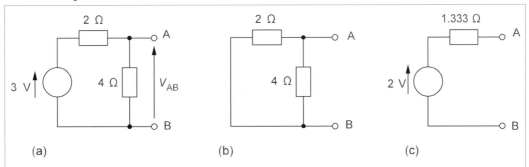

Figure 1.23 (a) The original circuit. V_{AB} is the Thévenin source voltage (b) The voltage source is 'killed' by short-circuiting it (c) The Thévenin equivalent of Figure 1.23(a)

Secondly, the voltage source must be replaced by its internal resistance. Because the voltage source is ideal and thus has zero internal resistance by definition, all this means is replace the *voltage source* by a *short circuit*. The circuit then becomes that of Figure 1.24b – simply two resistances in parallel when looked at from AB. So, thirdly, we combine them by the product-over-sum rule:

$$R_T = (2 \times 4)/(2 + 4) = 1.333 \text{ } \Omega \qquad (1.39)$$

And so the Thévenin equivalent of the network in Figure 1.23a is that of Figure 1.23c.

It is important to realise that the circuits of Figures 1.23a and 1.23c are equivalent only for measurements of current and voltage made at *terminals A and B*, and nowhere else. Having found the Thévenin equivalent we can put it in the network's place, connect any other network across AB, then go ahead and calculate all the currents and voltages in that network. However, we cannot calculate the power developed *within* a network by considering the power developed in its Thévenin equivalent (see Problem 1.11).

Example 1.4

A second 4Ω resistance is connected across AB in Figure 1.23a. What is the current in it?
 We can calculate the current through it by using the Thévenin circuit of Figure 1.23c, as in Figure 1.24a. The 4Ω resistance is in series with the 1.333Ω Thévenin resistance, so the current through them is 2/(4 + 1.333) = 0.375 A, while the voltage across AB will be 4 × 0.375 = 1.5 V (by the voltage-divider rule it will be 2 × 4/(4 + 1.333) = 1.5 V too).

---

[9] The voltage across AB is written V_{AB} in *double-suffix notation*, implying that the head of the arrow points to A and the tail to B. By reversing the arrow we get V_{BA}, so that $V_{AB} = -V_{BA}$.

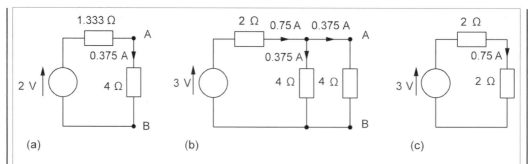

Figure 1.24 (a) Using Thévenin's equivalent circuit. The 4Ω resistance across AB is the same as (b) Connecting the 4Ω resistance across the original circuit

Looking at the original circuit in Figure 1.23a, we can connect the 4Ω resistance across AB (Figure 1.24b), and find the current from the source by combining the two parallel 4Ω resistances using the product-over-sum rule, making 2 Ω (Figure 1.24c). This result can be added to the 2Ω resistance in series to give 4 Ω, so the current from the 3V source is 3/4 = 0.75 A. This current passes through the 2Ω resistance and divides equally between the 4Ω resistances, **0.375 A** each. The voltage across the 4Ω resistances is 4 × 0.375 = 1.5 V, as found previously using the Thévenin circuit.

1.5.2 Norton's theorem[10]

Norton's theorem is a useful complement to Thévenin's:

Any two-terminal, linear network of sources and resistances may be replaced by a single current source in parallel with a resistance. The value of the current source is the current flowing between the terminals of the network when they are short-circuited. The value of the resistance is the resistance measured at the terminals of the network when all sources have been replaced by their internal resistances.

Norton's equivalent circuit is shown in Figure 1.25.

Figure 1.25

Norton's equivalent circuit. *Any* two-terminal network of sources and resistances may be replaced by this

Since Thévenin's theorem applies to *any* two-terminal network, it must apply to Norton's equivalent circuit too. The Thévenin equivalent of Norton's circuit can be derived as follows: In Figure 1.25 the open-circuit voltage is that across R_N, which must be $I_N R_N$, since all the current from the source flows through it. Therefore

$$V_T = I_N R_N \tag{1.40}$$

---

[10] E L Norton worked for Bell Telephone Laboratories in the USA. He published his theorem in 1926.

The next step is to replace the source by its internal resistance – infinity for an ideal current source by definition – which amounts to *open-circuiting* an ideal *current* source. On doing this we are left with R_N alone connected across AB, so the Thévenin resistance is identical to the Norton resistance:

$$R_T = R_N \qquad\qquad (1.41)$$

Figure 1.26 shows the transformation, which requires only the calculation of $V_T (= I_N R_N)$, the open-circuit voltage of Norton's equivalent circuit. Whether to use the Norton or the Thévenin form of equivalent circuit is for each to decide. As a general rule one should use the *Norton* circuit to combine *parallel* sources and the *Thévenin* circuit to combine *series* sources.

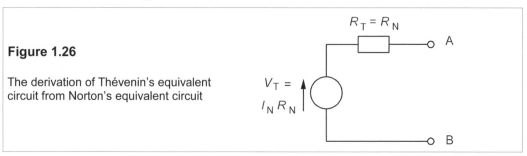

Figure 1.26

The derivation of Thévenin's equivalent circuit from Norton's equivalent circuit

---

Example 1.5

For example, consider the three generators in parallel in Figure 1.27a. What is the open-circuit voltage and the source resistance? In other words, what is the Thévenin equivalent?

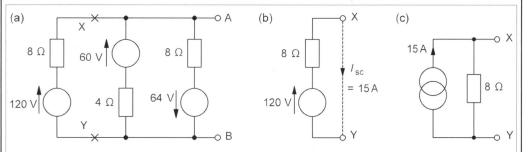

Figure 1.27 To transform a source (a) Cut the circuit at XY (b) Isolate the voltage source (c) Transform the voltage source into a current source

We proceed by cutting the circuit at XY and looking at the source on the left by itself, as in Figure 1.27b. This source is in Thévenin form and must be transformed to the Norton form to enable us to add the parallel sources. The Norton source current is the current flowing through a short circuit across XY, which is 120/8 = 15 A, by Ohm's law; and the Norton resistance is equal to the Thévenin resistance, 8 Ω. So the Norton form of the 120V source is as in Figure 1.27c. Note that the current arrow in Figure 1.27c has the *same direction* as the voltage arrow of the original source. It is a good idea if in doubt to check this by examining which way current would flow through the short-circuited terminals. Having transformed the source we can re-attach it to the network at XY as in Figure 1.28a.

The next voltage source is then transformed (it makes no difference which way round the voltage source and its series resistance are placed), and finally the one nearest to AB. In the latter case the Norton equivalent must have its current arrow pointing down, like the voltage source it replaced. Figure

1.28b shows the circuit at this stage.

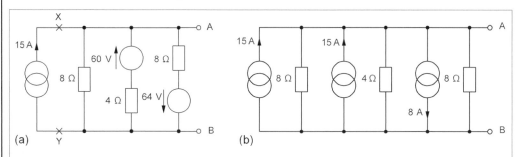

Figure 1.28 (a) Current source re-attached at XY (b) The other voltage sources are turned into current sources

The three parallel current sources can now be added algebraically to give a single source of 15 + 15 – 8 = 22 A. The third current source's value is subtracted from the other two as its direction is down not up. Then the three parallel resistances in Figure 1.28b are combined to give a single 2Ω resistance (by adding the reciprocals and taking the reciprocal of the result), giving the Norton circuit of Figure 1.29a. Finally, the Norton circuit is turned into the Thévenin circuit of Figure 1.29b, using V_T = $I_N R_N$.

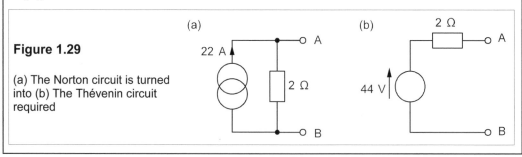

Figure 1.29

(a) The Norton circuit is turned into (b) The Thévenin circuit required

Thévenin's and Norton's theorems enable us to take any circuit inside a 'black box', make a measurement of the open-circuit voltage at its terminals, then the short-circuit current between them and from this represent the unknown contents of the box in the form of either equivalent circuit. It is not necessary to take a look inside the 'box' – which might be neither possible nor desirable – to deduce its effect on an external circuit.

1.5.3 The superposition theorem

Sometimes it is helpful to consider separately the effects of sources on a particular part of the circuit; the superposition theorem lets us do that. It states

The current in any branch of a circuit, or the voltage at any node, may be found by the algebraic addition of the currents or voltages produced by each source separately. When the effect of one source is considered, the other sources are replaced by their internal resistances.

Superposition is most useful when the sources are alternating sources of differing frequency, or a combination of direct and alternating sources, but we can illustrate the use of superposition with direct sources alone.

Example 1.6

Consider once more the circuit of Figure 1.27a, in which we wish to find V_{AB} using superposition. For this we must add the voltages produced across AB by each source in turn. Taking the 120V source first, we replace the other voltage sources by short circuits to obtain the circuit in Figure 1.30a.

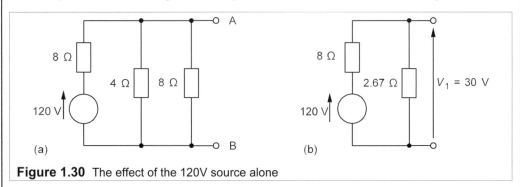

Figure 1.30 The effect of the 120V source alone

The parallel 4Ω and 8Ω resistances may be combined to give one of 2.67 Ω, which is in series with the remaining 8Ω resistance, for a total resistance of 10.67 Ω. By the voltage-divider rule therefore, V_1 must be given by $V_1 = (2.67/10.67) \times 120 = 30$ V, as in Figure 1.30b.

Proceeding in the same way to examine the effect of the 60V source we 'kill' the other two sources by replacing them with short circuits as in Figure 1.31a. The two 8Ω resistances are in parallel as far as the 60V source is concerned, so they may be combined into a 4Ω resistance. This is in series with the 4Ω resistance next to the source, as in Figure 1.31b, so the voltage divider rule gives V_2 as 30 V.

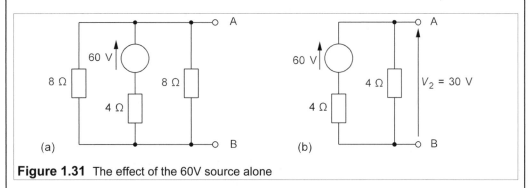

Figure 1.31 The effect of the 60V source alone

The last step is to examine the effect of the 64V source, as in Figure 1.32. V_3, the required voltage, is that across the parallel combination of 8Ω and 4Ω resistances, but as the 64V source points to B and not A like the other two, V_3 must be *negative*. Once again the parallel resistances combine to give one of 2.67 Ω, so V_3, calculated by the voltage divider rule, is $(2.67/10.67) \times (-64)$ V, or -16 V. Summing:

$$V_{AB} = V_1 + V_2 + V_3 = 30 + 30 - 16 = 44 \text{ V} \qquad (1.42)$$

The result agrees with that found by source transformations. In general the superposition theorem is not very useful for solving many problems; it is too lengthy when three or more sources are present.

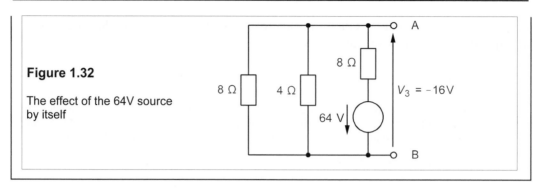

Figure 1.32

The effect of the 64V source
by itself

1.5.4 The star-delta transformation

Networks containing components that are neither in series nor in parallel are frequently encountered; the star-delta transformation enables us to change the form of networks like these so that they may be more readily analysed. In Figure 1.33a is shown a 3-terminal network in delta form, while the star form is in Figure 1.33b.

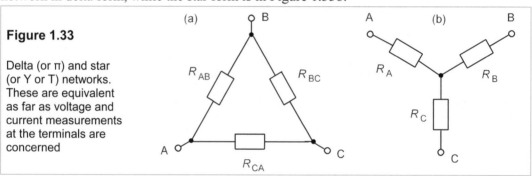

Figure 1.33

Delta (or π) and star
(or Y or T) networks.
These are equivalent
as far as voltage and
current measurements
at the terminals are
concerned

In North America, principally, this transformation is called the T-π (Tee-Pi), or Y-π (Wye-Pi) since it can be drawn in these shapes too, as in Figure 1.34. Although this form looks like a four-terminal network, terminals C and C′ are common. The networks are equivalent if we can find the values of one set of resistances in terms of those of the other.

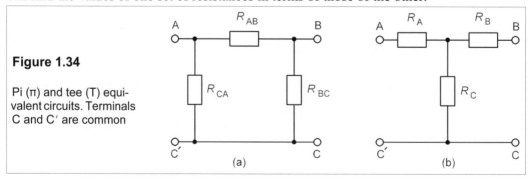

Figure 1.34

Pi (π) and tee (T) equi-
valent circuits. Terminals
C and C′ are common

To calculate the resistance of the network of Figure 1.34a between terminals A and C′, we see that it comprises R_{CA} in parallel with the series resistances, R_{AB} and R_{BC}. In symbols, R_{CA} // $(R_{AB} + R_{BC})$, to give (by the product-over-sum rule)

$$\frac{R_{CA}(R_{AB} + R_{BC})}{R_{AB} + R_{BC} + R_{CA}} = \frac{R_{CA}(R_{AB} + R_{BC})}{\Sigma R} \tag{1.43}$$

where $\Sigma R = R_{AB} + R_{BC} + R_{CA}$.

Looking into the same terminals in Figure 1.34b we see the resistance is $R_A + R_C$ and this must be the same for both networks, that is
Then the resistance between terminals B and C is

$$R_B + R_C = \frac{R_{BC}(R_{AB} + R_{CA})}{\Sigma R} \tag{1.45}$$

and looking into terminals A and B the final equation is obtained

$$R_A + R_B = \frac{R_{AB}(R_{BC} + R_{CA})}{\Sigma R} \tag{1.46}$$

Subtracting Equation 1.45 from 1.44 gives

$$R_A - R_B = \frac{R_{AB}R_{CA} - R_{AB}R_{BC}}{\Sigma R} \tag{1.47}$$

Adding this to Equation 1.46 yields

$$2R_A = 2R_{AB}R_{CA}/\Sigma R \quad \text{or} \quad R_A = R_{AB}R_{CA}/\Sigma R \tag{1.48}$$

Solving for R_B and R_C similarly leads to

$$R_B = R_{AB}R_{BC}/\Sigma R \quad \text{and} \quad R_C = R_{BC}R_{CA}/\Sigma R \tag{1.49}$$

These transformations are easier to remember than it may appear. For instance the formula for R_A (the resistance attached to terminal A of the star form) is just the *product* of the two resistances attached to terminal A of the delta form, divided by the *sum* of all three resistances. R_B and R_C are found the same way.

It is left as an exercise (Problem 1.29) to show that the reverse transformation from star to delta requires

$$G_{AB} = \frac{G_A G_B}{\Sigma G} \qquad G_{BC} = \frac{G_B G_C}{\Sigma G} \qquad G_{CA} = \frac{G_C G_A}{\Sigma G} \tag{1.50}$$

Remembering this is also easy, since each formula involves just the conductances between a pair of terminals. The letters DSR are a helpful mnemonic – **D**elta to **S**tar: use **R**esistances – the reverse transformation must then employ conductances.

The star-delta transformation is very useful when it is necessary to analyse networks such as that in Figure 1.35a, an unbalanced bridge circuit.

Example 1.7

If the galvanometer in Figure 1.35a has a resistance of 1 Ω, what is the current through it?

The redrawn circuit is as shown in Figure 1.35b, where the galvanometer has been replaced by its internal resistance and the delta networks have been replaced by π networks. We then transform the π network between A, B and D of Figure 1.35a into the inverted Y network in Figure 1.35c. It is important not to lose sight of which resistance is attached to which terminal at this stage: make a sketch and carefully label the terminals when you do this! The network in Figure 1.35c now has no bridging resistance and hence we have lost sight of I_G; notice also that the 6Ω and 3.1Ω resistances have retained their identities. We next find the current in the 3.1Ω resistance, then the current in the 3Ω resistance between A and B, and thence I_G.

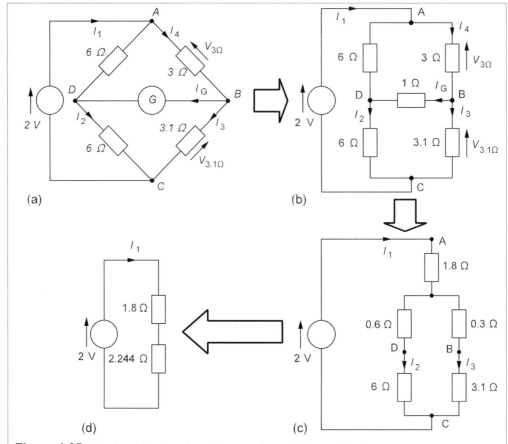

Figure 1.35 The star-delta transformation used on a bridge network

The resistance attached to A in Figure 1.35c is found by taking products of resistances attached to A in Figure 1.35b (3 × 6 = 18), and dividing by the sum of the resistances in the delta network between A, B and C (6 + 3 + 1 = 10), to give 1.8 Ω attached to A in Figure 1.35c; then likewise for those attached to B and D. Now the 6Ω and 0.6Ω resistances can be added to make 6.6 Ω, and the 3.1 Ω added to the 0.3 Ω to make 3.4 Ω. The 6.6 Ω and 3.4 Ω are in parallel so may be combined (product/sum) to give 2.244 Ω as in Figure 1.35d. So the total resistance in Figure 1.35d is 4.044 Ω, and the current, I_1, is 2/4.044 = 0.4946 A.

Going back to Figure 1.35c, we see that I_1 splits to form I_2 and I_3. We can use the current divider rule to obtain I_3: 0.3264 A. I_3 must flow through the 3.1Ω resistance in Figure 1.35a, since it is unaffected by the transformation. Thus the voltage across it by Ohm's law is $V_{3.1\Omega} = 3.1 \times 0.3264 = 1.0118$ V. Looking at the loop ABCA in Figure 1.35a that includes the 2V source, we can use KVL:

$$2 - V_{3\Omega} - V_{3.1\Omega} = 2 - V_{3\Omega} - 1.0118 = 0 \qquad (1.51)$$

to find $V_{3\Omega} = 0.9882$ V and the current through it, I_4, as 0.9882/3 = 0.3294 A. By KCL at B (in Figure 1.35a), $I_4 = I_3 + I_G$, so $I_G = I_4 - I_3 = 0.3294$ A – 0.3264 A = 3 mA.

1.6 **Power and energy**

The instantaneous electric power is given by

$$p = vi \qquad (1.52)$$

p is in watts (W) if v is in volts and i in amperes. This enables us to find the power sent out by a voltage or current source, or the power dissipated in a resistance.

Energy is given by

$$E = \int p\,dt = \int vi\,dt \qquad (1.53)$$

In electrical engineering v and i are the quantities usually measured, calculated or otherwise known, so Equation 1.53 is more useful than the converse relation $p = dE/dt$. From Equations 1.52 and 1.53 we can derive others: for example, the voltage across a resistance is iR by Ohm's law, so the power dissipated in it as heat will be

$$p = vi = i^2R = v^2/R \qquad (1.54)$$

Only resistances dissipate energy, while inductances and capacitances store it: Equation 1.53 can be used to find the amount stored. In the case of a capacitance, C, the current is given by $i = C\,dv/dt$, therefore

$$E = \int vi\,dt = \int vC\frac{dv}{dt}\,dt = \int_0^V Cv\,dv = \tfrac{1}{2}CV^2 \qquad (1.55)$$

$$E = \tfrac{1}{2}CV^2 \qquad (1.56)$$

In a like manner we may deduce that energy stored by an inductance of L henries carrying a current, I amps, is

$$E = \tfrac{1}{2}LI^2 \qquad (1.57)$$

⌐ **Example 1.8**

For example, suppose in the circuit of Figure 1.36a that the current waveform is as shown in Figure 1.36b. What is the maximum energy stored in the capacitor, the inductor and the whole circuit, if the capacitor was initially uncharged? What energy is dissipated by the circuit?

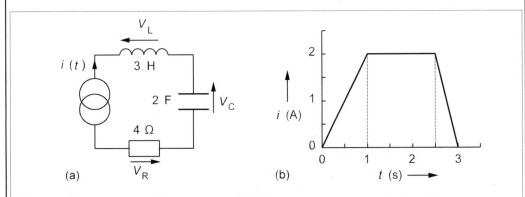

Figure 1.36 (a) The circuit for example 1.8 (b) The current-time graph for the source

To find the maximum energy stored in the capacitance using $E = \frac{1}{2}CV^2$ we must find the maximum voltage across it. The voltage across the capacitance is given by

$$V_C = C^{-1}\int i\,dt \qquad (1.58)$$

which is simply the area under the i-t graph, and this will be a maximum at $t = 3$ s. The area is made up of a triangle of area $\frac{1}{2} \times 2$ A $\times 1$ s ($= 1$ C), then a rectangle of area 2 A $\times 1.5$ s ($= 3$ C), and finally a triangle of area $\frac{1}{2} \times 2$ A $\times 0.5$ s ($= 0.5$ C). These add up to 4.5 C, so the maximum voltage across C is 2.25 V, and the energy stored is then $\frac{1}{2} \times 2 \times 2.25^2 = 5.0625$ **J**.

The energy stored in the inductance is maximum when the current is maximum, that is 2 A, then

$$E_{max} = 0.5 \times 3 \times 2^2 = 6\ \mathbf{J} \qquad (1.59)$$

There is no energy stored in the resistance, so to find the maximum energy stored in the circuit we need only consider the inductor and the capacitor. However, we cannot just add the two maxima together as they occur at different times. After 2 s the inductor carries less than maximum current, so its stored energy declines, while that of the capacitor increases.

The analytical solution is complicated and it is best to use a numerical approach. What is the circuit's stored energy at $t = 2.5$ s? The voltage across the capacitor is 2 V, so its stored energy is $\frac{1}{2} \times 2 \times 2^2 = 4$ J, while the inductor stores 6 J, for a total of 10 J. At $t = 3$ s the stored energy is just in the capacitor: 5.0625 J. What happens at $t = 2.501$ s? The inductor's current is $2 - 0.004$ A, or 1.996 A, so its stored energy is

$$E_L = 0.5 \times 3 \times 1.996^2 = 5.976\ \mathbf{J} \qquad (1.60)$$

The capacitor's voltage increases by 1 mV (nearly 2 A for 1 ms), so its stored energy is

$$E_C = 0.5 \times 2 \times 2.001^2 = 4.004\ \mathbf{J} \qquad (1.61)$$

The total is 9.98 J, hence for $2.5 < t < 3$ s the inductor is losing energy faster than the capacitor gains it; we therefore conclude that **the maximum stored energy is 10 J at 2.5 s**.

The resistance is the only element that dissipates energy, the quantity being given by $E = \int i^2 R\,dt$. The i-t graph has three parts which require individual integrations. From $t = 0$ to $t = 1$ s, the current is given by $i = 2t$ A. When $1 < t < 2.5$ s, i is constant at 2 A. And for 2.5 s $< t < 3$ s, the current is $2 - 4t_1$ A, where $t_1 = t - 2.5$ s. This substitution means that the last integral is for the time interval $0 < t_1 < 0.5$ s, and the limits change accordingly. The energy dissipated is therefore

$$E = 4\left(\int_0^1 (2t)^2\,dt + \int_1^{2.5} 2^2\,dt + \int_0^{0.5} (2 - 4t_1)^2\,dt_1 \right) \qquad (1.62)$$

$$E = 4\,[4t^3/3]_0^1 + 4\,[4t]_1^{2.5} + 4\,[4t_1 - 8t_1^2 + 16t_1^3/3]_0^{0.5} \qquad (1.63)$$

that is, $E = 4[4/3] + 4[10 - 4] + 4[2 - 2 + 2/3] = \mathbf{32\ J}$

1.6.1 The maximum-power-transfer theorem

This theorem shows how to match networks to loads for maximum power output. It states:

Maximum power is transferred to a load resistance connected between two terminals of a network when the load resistance is equal to the resistance of the network. The network resistance is measured between the same two terminals with the sources of the network being replaced by their internal resistances.

The proof is as follows. Any two-terminal network can be reduced to a single voltage source in series with a resistance (Thévenin's theorem) as in Figure 1.37.

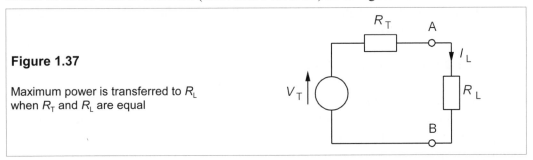

Figure 1.37

Maximum power is transferred to R_L when R_T and R_L are equal

If a resistance, R_L, is attached across the network's terminals, AB, the current through it by Ohm's law is

$$I_L = V_T/(R_T + R_L) \qquad (1.64)$$

The power in R_L is

$$P_L = I_L^2 R_L = \left(\frac{V_T}{R_T + R_L} \right)^2 R_L \qquad (1.65)$$

Differentiating with respect to R_L and equating to zero for turning points in P_L,

$$\frac{dP_L}{dR_L} = \frac{V_T^2(R_T + R_L)^2 - 2(R_T + R_L)V_T^2 R_L}{(R_T + R_L)^4} = 0 \qquad (1.66)$$

so that

$$V_T^2(R_T + R_L)^2 = 2(R_T + R_L)V_T^2 R_L \qquad (1.67)$$

$$R_T + R_L = 2R_L$$

That is
$$R_T = R_L \qquad (1.68)$$

It is left as an exercise to check that this condition leads to maximum power in the load by deciding if d^2P_L/dR_L^2 is negative. The maximum power transfer theorem reduces to

Maximum power is transferred from a resistive network when the load resistance is equal to the Thévenin resistance of the network.

When $R_T = R_L$, the current through the load must be $V_T/2R_L$, so the maximum power developed will be $I_{max}^2 R_L$, or

$$P_{max} = \frac{V_T^2 R_L}{(2R_L)^2} = \frac{V_T^2}{4R_L} \qquad (1.69)$$

Later, we shall see how transformers can be used to achieve load matching in AC circuits. It is worth emphasising here that transferring maximum power to a load will *not* result in maximum efficiency of power use, since at least half of the power consumed by the whole system must be lost in the source network (see Problem 1.11). Maximum power transfer is most needed when the source is weak, such as the signal from a radio or TV aerial, as it allows maximisation of the signal-to-noise ratio. Other theorems will be introduced as required.

1.7 Mesh analysis and nodal analysis

Until now we have made use of Kirchhoff's laws only in very simple circuits, for a good reason: without systematic use they rapidly generate a large number of unknown quantities (though also equations by which they may be found). Mesh and nodal analyses are complementary ways of making systematic the application of Kirchhoff's laws to circuits, so that fewer errors are likely.

1.7.1 Mesh analysis

Mesh analysis applies KVL systematically to a circuit and produces simultaneous equations which may be solved to give the mesh currents. A *mesh* is a loop in a circuit with no loops inside it. Consider the circuit of Figure 1.38, which represents a domestic 2-phase supply common in North America.

It contains three meshes: ABCDA, ADEFA and CEDC; numbered 1, 2 and 3: the large loop ABCDEFA is not a mesh as it can be divided into smaller loops. Meshes one and two represent low-voltage circuits for lighting and appliances such as vacuum cleaners, hair-driers, etc., while mesh three represents a two-phase circuit for supplying relatively large loads such as a cooker or a heater. The 0.1Ω resistances represent the wiring resistance.

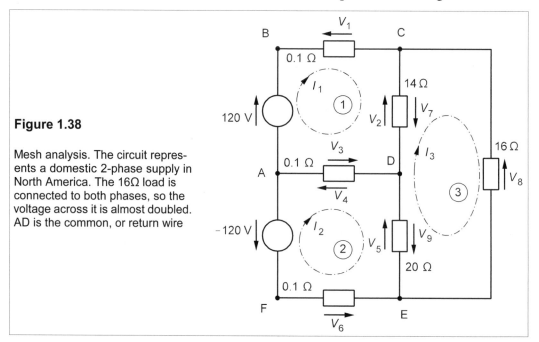

Figure 1.38

Mesh analysis. The circuit repres-ents a domestic 2-phase supply in North America. The 16Ω load is connected to both phases, so the voltage across it is almost doubled. AD is the common, or return wire

The first step in the analysis is to assign clockwise currents to each mesh: in this case I_1, I_2

and I_3 to meshes 1, 2 and 3 respectively. The next step is to use KVL to find these currents. Around mesh 1 KVL gives, taking clockwise voltages as positive and anticlockwise negative:

$$120 - V_1 - V_2 - V_3 = 0 \qquad (1.70)$$

V_1 is the voltage across the 0.1Ω resistance, which is $0.1I_1$ by Ohm's law. V_2 is the voltage across the 14Ω resistance, which is *not* $14I_1$, since the 14Ω resistance is *shared* between meshes 1 and 3, so that the resultant current flow from C to D is $I_1 - I_3$. Therefore V_2 must be $14(I_1 - I_3)$. V_3 is the voltage across the middle 0.1Ω resistance (the common wire for the two phases), which is shared between meshes 1 and 2. Therefore V_3 is $0.1(I_1 - I_2)$ and Equation 1.70 becomes

$$120 - 0.1I_1 - 14(I_1 - I_3) - 0.1(I_1 - I_2) = 0 \qquad (1.71)$$

$$\Rightarrow \quad 14.2I_1 - 0.1I_2 - 14I_3 = 120 \qquad (1.72)$$

after rearranging the terms.

The use of KVL in mesh 2 yields

$$-(-120) - V_4 - V_5 - V_6 = 0 \qquad (1.73)$$

The -120V source points anticlockwise and is written $-(-120)$ in Equation 1.73 (if in doubt, reverse the arrow *and* the sign on this source, which leaves its value unchanged). V_4, the voltage across the middle 0.1Ω resistance, is $0.1(I_2 - I_1)$ by Ohm's law (V_4 is obviously just $-V_3$, but it is best to give it a separate identity in a different mesh). V_5 is the voltage across the 20Ω resistance, and must be $20(I_2 - I_3)$ by Ohm's law, while V_6 is $0.1I_2$, so that Equation 1.73 becomes

$$120 - 0.1(I_2 - I_1) - 20(I_2 - I_3) - 0.1I_3 = 0 \qquad (1.74)$$

which rearranges to

$$-0.1I_1 + 20.2I_2 - 20I_3 = 120 \qquad (1.75)$$

Mesh 3 gives

$$V_7 + V_8 + V_9 = 0 \qquad (1.76)$$

Again the voltages are found by using Ohm's law: $V_7 = 14(I_3 - I_1)$, $V_8 = 16I_3$ and $V_9 = 20(I_3 - I_2)$. Then Equation 1.76 after substitution and rearrangement is

$$-14I_1 - 20I_2 + 50I_3 = 0 \qquad (1.77)$$

Three meshes have produced three equations, numbered 1.72, 1.75 and 1.77:

$$\begin{aligned} 14.2I_1 - 0.1I_2 - 14I_3 &= 120 \\ -0.1I_1 + 20.2I_2 - 20I_3 &= 120 \\ -14I_1 - 20I_2 + 50I_3 &= 0 \end{aligned} \qquad (1.78)$$

Solving gives the three unknown mesh currents: $I_1 = 23.1$ **A**, $I_2 = 20.6$ **A** and $I_3 = 14.7$ **A**. The currents through the 14Ω and 20Ω loads are $I_1 - I_3$ ($= 8.4$ A) and $I_2 - I_3$ ($= 5.9$ A), while the common wire carries the relatively small current $I_1 - I_2$ ($= 2.5$ A). The power consumed by the 14Ω load is $14 \times 8.4^2 = 990$ W, by the 20Ω load is $20 \times 5.9^2 = 700$ W, and the 16Ω load consumes $16 \times 14.7^2 = 3.5$ kW.

Take note of the form of Equations 1.78: the first comes from applying KVL to mesh 1, so the coefficient of I_1 is *opposite in sign* to those of the other two currents. The second equation arises from application of KVL to mesh 2 so the coefficient of I_2 is positive and the rest negative. And the third equation comes from mesh 3, so the coefficient of I_3 is positive, the others negative. If the unknown currents and voltages are assigned in the systematic way described, the signs of the current coefficients should *always* follow this pattern – check if they do not! With a little practice the student will find it superfluous to put voltages across resistances and will be able to write down the current equations directly. Aided by computer solution of simultaneous equations, an experienced student can solve a 7-mesh problem in two to three minutes, and get it right first time.

Current sources may cause trouble in mesh analysis. Sometimes they can be turned into voltage sources and dealt with as such, but there are circumstances when one cannot do so. Take for example the circuit of Figure 1.39: there is no resistance in parallel with the 2A source, making its transformation into a voltage source impossible.

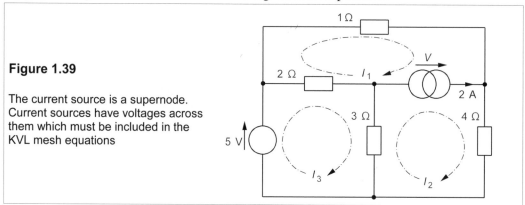

Figure 1.39

The current source is a supernode. Current sources have voltages across them which must be included in the KVL mesh equations

We assign currents to the numbered meshes as before, and also a voltage, V, to the current source, to generate the mesh equations

$$-3I_1 + 2I_2 - V = 0 \qquad (1.79)$$
$$-7I_2 + 3I_3 + V = 0 \qquad (1.80)$$
$$2I_1 + 3I_2 - 5I_3 + 5 = 0 \qquad (1.81)$$

Equations 1.79 and 1.80 can immediately be added to give

$$-3I_1 - 7I_2 + 5I_3 = 0 \qquad (1.82)$$

The voltage across the current source has been eliminated (such a source is called a *super-node*), leaving two equations and three unknown currents. The third equation comes from the fact that 2 A must flow through the current source, making $I_2 - I_1 = 2$ (make sure the sign is correct here). The solution of the three simultaneous equations

$$
\begin{aligned}
2I_1 + 3I_2 - 5I_3 &= -5 \\
-3I_1 - 7I_2 + 5I_3 &= 0 \\
-1I_1 + 1I_2 + 0I_3 &= 2
\end{aligned}
\qquad (1.83)
$$

then yields $I_1 = -0.6$ A, $I_2 = 1.4$ A and $I_3 = 1.6$ A. From these currents we can find any current or voltage readily, for example the current through the 3Ω resistance is $I_3 - I_2 = 0.2$ A, and the voltage across it must be $3 \times 0.2 = 0.6$ V.

1.7.2 Nodal analysis

While mesh analysis employs KVL systematically, nodal analysis uses KCL to derive a set of simultaneous equations from which all the nodal voltages may be found. Voltages are assigned to each principal node of a network, principal nodes being points where three or more branches of a circuit join.

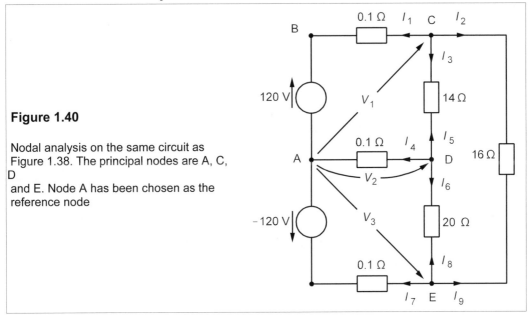

Figure 1.40

Nodal analysis on the same circuit as Figure 1.38. The principal nodes are A, C, D and E. Node A has been chosen as the reference node

Consider the circuit of Figure 1.40 (which is the same circuit as Figure 1.38). The (principal) nodes are A, C, D and E, node A being chosen as the reference node. (The reference node need not be at ground potential, nor at any particular place in the circuit: the choice is arbitrary.) Voltages are then assigned to the other nodes[11] representing the potential differences between them and the reference node: these are V_1, V_2 and V_3 in Figure 1.40. We then apply KCL at nodes C, D and E. At node C, taking current *out* of the node to be *positive*, we see that

$$I_1 + I_2 + I_3 = 0 \qquad (1.84)$$

Now I_1 is the current through the uppermost 0.1Ω resistance, which is given by Ohm's law as $(V_1 - 120)/0.1$ (make sure the sign is correct here!), since the voltage across the resistance is $V_1 - 120$. Similarly, I_2 is $(V_1 - V_3)/16$ and I_3 is $(V_1 - V_2)/14$. Equation 1.84 then becomes

---

[11] Strictly a *node* is the point at which *two* or more branches join and a *principal node* one where three or more branches join. In this sense A and B are both nodes but A is a principal node. However, if only two branches join, the current is the same in each and use of KCL is redundant; so principal node is implied when the word 'node' is used.

$$\frac{V_1 - 120}{0.1} + \frac{V_1 - V_2}{14} + \frac{V_1 - V_3}{16} = 0 \tag{1.85}$$

At node D,

$$I_4 + I_5 + I_6 = 0 \tag{1.86}$$

Now $I_4 = V_2/0.1$, as the middle 0.1Ω resistance lies between the reference node A and node D, then $I_5 = (V_2 - V_1)/15$ (note that $I_5 = -I_3$, but one should always give the currents at each node new identities, and work out afresh what they are) and $I_6 = (V_2 - V_3)/20$. Thus Equation 1.86 is

$$\frac{V_2}{0.1} + \frac{V_2 - V_1}{14} + \frac{V_2 - V_3}{20} = 0 \tag{1.87}$$

Finally, at node E

$$I_7 + I_8 + I_9 = 0 \tag{1.88}$$

where $I_7 = (V_3 - (-120))/0.1$ (take great care here!!), $I_8 = (V_3 - V_2)/20$ and $I_9 = (V_3 - V_1)/16$. Substituting these values into Equation 1.88 gives

$$\frac{V_3 + 120}{0.1} + \frac{V_3 - V_2}{20} + \frac{V_3 - V_1}{16} = 0 \tag{1.89}$$

Rearranging Equations 1.85, 1.87 and 1.89 produces the three simultaneous equations

$$\begin{aligned}
10.13V_1 - 0.071V_2 - 0.0625V_3 &= 1200 \\
-0.071V_1 + 10.12V_2 - 0.05V_3 &= 0 \\
-0.0625V_1 - 0.05V_2 + 10.11V_3 &= -1200
\end{aligned} \tag{1.90}$$

The form of these equations is worth remarking: the first comes from using KCL at node C, where the voltage relative to node A is V_1, and the only positive coefficient is that of V_1. The same pattern is followed in the other two equations. Any mistakes in sign should soon be noticed and put right if the equations are written in this systematic way. The solution to the equations is $V_1 = 117.7$ V, $V_2 = 0.24$ V and $V_3 = -118$ V. The voltage across the 14Ω load is $117.7 - 0.24 = 117.5$ V, and the current through it is $117.5/14 = 8.4$ A, as before. The voltage across the 20Ω resistance is $0.24 - (-118) = 118.2$ V, so the current through it is 5.9 A, and finally the voltage across the 16Ω resistance is $117.7 - (-118) = 235.7$ V and the current through it is 14.7 A. The solution is, as it must be, the same as that found by mesh analysis earlier.

Whether in a particular case nodal or mesh analysis is preferable is largely a matter of individual taste; in the example given above, neither method has a clear advantage. However, the topology of most networks is such that nodal analysis produces fewer unknowns, and may be preferred for that reason. In simple examples there is seldom a great difference in complexity. Though we have dwelt on DC analysis here, we more frequently require AC analysis; however, all of the DC techniques and circuit theorems may be applied to AC circuits, simply by using impedances instead of resistances.

Problems

1. In the circuits of Figure P1.1 what power is delivered (+), or consumed (−) by each source?
   [(a) +8 W (b) + 24 µW (c) +8 kW (d) −12 mW (e) −99 MW (f) −800 W]

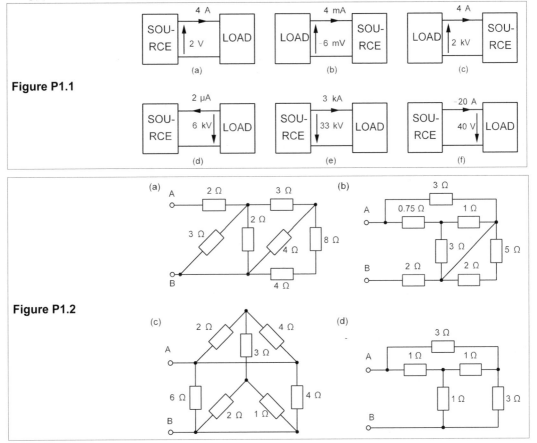

Figure P1.1

Figure P1.2

2. Find the equivalent resistance between the terminals of the networks of Figure P1.2.
   [(a) 3 Ω (b) 3 Ω (c) 1.62 Ω (d) 1.5 Ω]
3. In the circuit of Figure P1.3 the battery has an internal resistance of 2 Ω, the ammeter one of 0.1 Ω and the voltmeter one of 1 kΩ. What are the readings of the ammeter and voltmeter? What would the voltmeter read if it were to be reconnected across XY? *[0.897 A, 7.117 V, 7.206 V]*

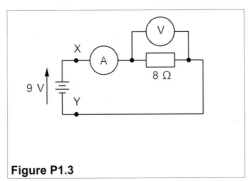

Figure P1.3

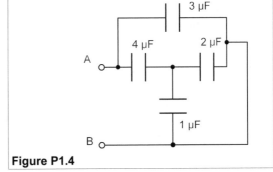

Figure P1.4

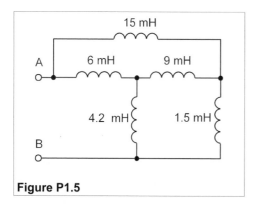

Figure P1.5

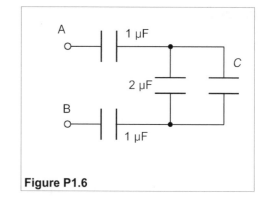

Figure P1.6

4. Find the equivalent capacitance between AB of the circuit of Figure P1.4. *[4.714 μF]*
5. Find the equivalent inductance between AB of the circuit of Figure P1.5. *[6 mH]*
6. What must C be for the equivalent capacitance between the terminals AB in Figure P1.6 to be equal to C? *[0.414 μF]*
7. You are given an ohmmeter of infinite accuracy and a large number of resistors of which all but one have exactly the same (given) resistance. How can the odd resistor (also of known resistance) be found with just two measurements?*[Hint: connect in series/parallel twice]*
8. Use Kirchhoff's laws and the *v-i* relationship for inductors to show that inductors in series may be added like resistors to find the equivalent inductance. Also show that inductors in parallel behave like resistors in parallel when their equivalent is required.
9. What is the resistance of the circuit formed by placing a resistance of 6 Ω along each edge of a cube and joining them at all eight corners? The resistance is measured between two corners at the opposite ends of the longest (body) diagonal. *[5 Ω]*
10. A capacitance of 1 F is charged up to 5 V and then is connected to an initially uncharged capacitance of 2 F in series with a 1Ω resistance. (a) What is the voltage across each capacitance after the connection is made? (b) What energy is stored in the circuit before and after the connection is made? (c) What energy is lost in the resistance?
   (d) Does the resistance affect the answer to (c)? (e) What happens to the energy if the resistance is zero? *[(a) 1.67 V and –1.67 V (b) 12.5 J and 4.17 J (c) 8.33 J (d) No]*
11. Show using a Thévenin equivalent circuit and a resistive load, that maximum power transfer does not result in maximum efficiency for energy transfer, and that for the latter the load resistance should be infinite.
12. If all the energy stored in a 32μF capacitor charged to 24 kV is transferred to an inductance of 0.5 μH, what is the current? (This is an exploding-wire experiment, but in practice the resistance slightly reduces the current.) *[192 kA]*
13. A superconducting magnet has been proposed for storing energy. If the current in the magnet is to be 200 kA when the stored energy is 5 GWhr, what is its inductance? If the maximum allowable voltage across its terminals is 5 kV, what is the maximum rate of change of current? What is the minimum time needed to discharge all the stored energy? What is the average power delivered in this time? What is the maximum output power? How practical is this proposal?
   [900 H, 5.56 A/s, 10 hrs, 500 MW, 1 GW]
14. Find the Thévenin and the Norton equivalents of the circuit between AB in Figure P1.14. Find the maximum power transferable from these terminals. What power is consumed internally by the circuit when delivering maximum power (that is excluding the load) and when unloaded?
   [V_T = 2 V, R_T = 1.11 Ω, I_N = 1.8 A. 0.9 W. 6.9 W and 12 W]
15. Use the star-delta transformation to find *I* in Figure P1.15. *[I = 0.7 A]*
16. Use mesh analysis to find the current from the 18V source in Figure P1.15. *[0.815 A]*

17. Use nodal analysis to find the voltage across the 2A source in Figure P1.15. *[15.6 V]*
18. Use Kirchhoff's laws and a full branch-current analysis to confirm the answers to Problems 1.15, 1.16 and 1.17.
19. Use the superposition theorem to find all the currents in Figure P1.14. Hence show that the power dissipated in the resistances is equal to the power from the sources.
$[I_{1\Omega} = 2\ A;\ P_{5V} = 10\ W;\ P_{2V} = 2\ W;\ P_{3A} = 0\ W]$

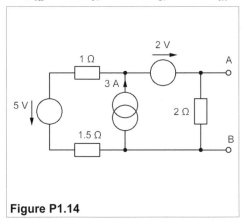

Figure P1.14

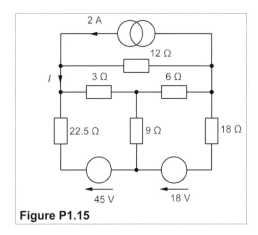

Figure P1.15

20. Show that the power consumed by the resistances in Figure P1.15 is equal to the power from the sources. $[P_{2A} = 31.2\ W,\ P_{45V} = 65.475\ W;\ P_{18V} = 14.67\ W]$
21. Use successively Norton's and Thévenin's theorems to find the voltage across AB in the ladder network of Figure P1.21. $[V_T = -4\ V]$
22. In the infinite network of Figure P1.22 what must R and R_L be for the resistance to be 2 Ω between terminals AB? *[1 Ω and any resistance]*

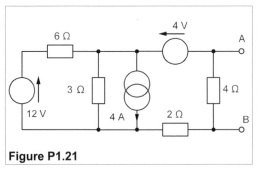

Figure P1.21

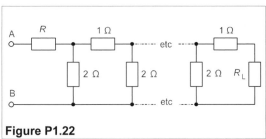

Figure P1.22

23. Find the power delivered by (a) the 121V (b) the 120V and (c) the 119V source in Figure P1.23. Repeat this problem when the source resistances are 0.3 Ω for the 121V, 0.2 Ω for the 120V and 0.1 Ω for the 119V source. *[(a) 1.692 kW (b) 478 W (c) –716 W, 812 W, 608 W and 15 W]*
24. In the circuits of Figures P1.21 and P1.24 use the superposition principle to find V_{AB} and V_T $[V_{AB} = -4\ V,\ V_T = -6\ V]$
25. A capacitance of 200 μF carries a current of $(2 + 3\sin 2t)$ mA. What is the voltage across it as a function of time, t, if it was initially uncharged? $[(10t - 7.5\cos 2t)\ V]$
26. What is the energy consumed by a resistance of 583 Ω from time $t = 0$ to $t = 10$ ms if the current through it is $0.63\exp(-100t)$ A? What are the average power and r.m.s. current during this time? *[1 J, 100 W, 0.414 A]*
27. The bridge circuit of Figure P1.27 is used to measure the value of R using a resistance-less galva-

nometer to detect imbalance. If the galvanometer's minimum detection current, I_G, is 0.1 μA, what is the percentage error in R? *[0.05 %]*

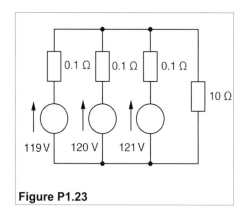

Figure P1.23

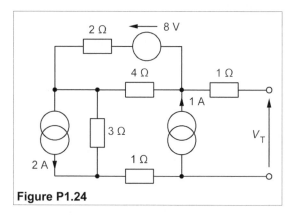

Figure P1.24

28. In the circuit of Figure P1.28 no energy is stored at time $t = 0$. I_C is $20t$ mA and starts from zero at $t = 0$. What is the voltage across AB at $t = 0.3$ s? *[0.42 V]*

29. Show that the transformation from star form to delta form requires that the equations numbered 1.50 be satisfied.

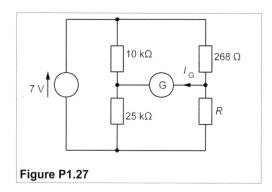

Figure P1.27

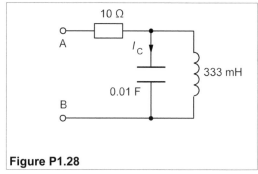

Figure P1.28

2 Sinusoidally-excited circuits

ITHERTO techniques for circuit analysis – and various useful theorems and transformations – have been illustrated using only direct voltages and currents. However, nearly all electricity is generated and consumed in the form of AC. In order to continue to use the methods devised originally for DC circuits, Heaviside[1] developed the use of complex numbers for currents, voltages and impedances. This has proved to be the most powerful tool ever put into the hands of electrical engineers.

2.1 Sinusoidal excitation

Conventional, rotating generators naturally produce sinusoidal voltage waveforms so that the most important circuits in practice are those which are sinusoidally excited (sometimes called 'AC circuits'). Besides, the Fourier theorem states that any repetitive waveform may be synthesised from purely sinusoidal components. A study of circuit behaviour with sinusoidal excitation at a single frequency can then be extended to complex waveforms, though in many instances only a single frequency may be of interest.

Figure 2.1 shows a graph of the voltage sinusoid, $v(t) = V_m\sin(\omega t + \phi)$ V.

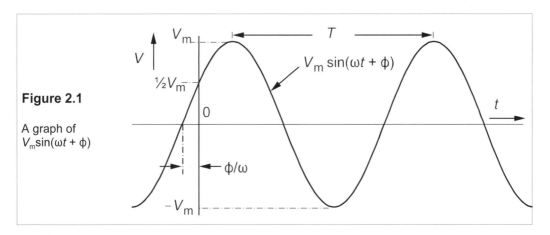

Figure 2.1

A graph of $V_m\sin(\omega t + \phi)$

---

[1] Oliver Heaviside (1850–1925) was a self-taught telegraphic engineer until he gave up work in his twenties and devoted himself to research, making outstanding contributions to mathematics, physics and above all electrical engineering. He coined the terms resistivity, impedance and inductance among others in the 1880s; his complex notation for impedance, $Z = R + jX$, became standard in 1911.

Its peak amplitude is V_m and the time between successive peaks, the period, is T, making its frequency, f, $1/T$. When T is in seconds, f is in hertz (Hz). Its angular frequency is ω, where

$$\omega = 2\pi f = 2\pi / T \tag{2.1}$$

ω is in radians per second when f is in Hz. At the origin, $v(0) = 0.5V_m$, but when $t = 0$, $v = V_m \sin\phi$, so according to Equation 2.1, ϕ, the *phase angle* is 30° (or $\pi/6$ rad) in this case. In terms of time the sinewave here is *advanced* by ϕ/ω seconds, if ϕ is in rads. When this voltage appears across a resistance, R, by Ohm's law the instantaneous current is

$$i = \frac{v}{R} = \frac{V_m \sin(\omega t + \phi)}{R} = I_m \sin(\omega t + \phi) \tag{2.2}$$

where $I_m = V_m/R$. The current waveform is precisely the same as the voltage waveform – same frequency, ω, same phase, ϕ – only its amplitude is scaled by a factor of $1/R$.

2.1.1 The effect of inductance

When a sinusoidal current, $i = I_m \sin\omega t$, flows in an inductance, L, the *v-i* relation $v = L\,di/dt$ can be used to find v, that is

$$v = L\frac{d(I_m \sin\omega t)}{dt} = \omega L I_m \cos\omega t \tag{2.3}$$

$$= \omega L I_m \sin(\omega t + 90°)$$

Writing v in the form $V_m \sin(\omega t + \phi)$ we obtain

$$v = V_m \sin(\omega t + \phi) = \omega L I_m \sin(\omega t + 90°) \tag{2.4}$$

where $\phi = 90°$ and $V_m = \omega L I_m$. The voltage and current are no longer in phase; in fact the phase of the current is 0°, while the phase of the voltage is +90°. The current through an inductance is said to *lag* the voltage by 90°, or the voltage to *lead* the current by 90°.

Equation 2.4 shows that $V_m = \omega L I_m$ for an inductance, which is a form of Ohm's law, with ωL replacing R. The inductance behaves as if it had a resistance ωL, called the *reactance* of the inductance and given the symbol X_L. This is an important finding:

$$V_m = I_m X_L \tag{2.5}$$

where $X_L \equiv \omega L$. If ω is in rad/s and L is in henries (H), X_L is in Ω.

There are two principal consequences of applying sinusoidal excitation to an inductance:

- it behaves as if it had a resistance of ωL
- the current *lags* the voltage by 90°

2.1.2 The effect of capacitance

What happens to the current when a sinusoidal voltage is applied to a capacitance? Using $i = C\,dv/dt$ for a capacitance, C, and taking $v = V_m \sin\omega t$ we find

$$i = C d(V_m \sin\omega t)/dt = \omega C V_m \cos\omega t$$

which can be written

$$i = \omega C V_m \sin(\omega t + 90°) \tag{2.6}$$

Letting $I_m = \omega C V_m$, it follows that $i = I_m \sin(\omega t + 90°)$; the current *leads* the voltage by 90°. Rearranging $I_m = \omega C V_m$ in the form $V = IR$ gives

$$V_m = I_m / \omega C = I_m X_C \tag{2.7}$$

where $X_C (= 1/\omega C)$ is the reactance of the capacitance, C. If C is in farads and ω in rad/s, then X_C is in Ω. Again there are two main consequences of applying sinusoidal excitation to a capacitance:

- it behaves as if it had a resistance of $1/\omega C$
- the current *leads* the voltage by 90°

A useful mnemonic for remembering whether voltage or current leads or lags in inductive or capacitive circuits is

$$C \ I \ V \ I \ L$$

With C, I leads V; V leads I with L.

2.1.3 Power

Instantaneous power is given by

$$p = vi \tag{2.8}$$

If v is in volts (V) and i in amperes (A), p is in watts (W). For a resistance subject to sinusoidal excitation, v and i are in phase, so that

$$p_R = V_m \sin \omega t \times I_m \sin \omega t = V_m I_m \sin^2 \omega t$$

$$= 0.5 V_m I_m (1 - \cos 2\omega t) \tag{2.9}$$

Equation 2.9 shows that p_R has a minimum value of zero: the power (which is lost as heat) produced in a resistance carrying a sinusoidal current is always positive. It has a direct component of $0.5 V_m I_m$ and a sinusoidal component of twice the current's frequency. Figure 2.2 shows the relationships of the waveforms.

The *average* power is seen from Equation 2.9 to be

$$P_{av} = 0.5 V_m I_m \tag{2.10}$$

since the average value of $\cos 2\omega t$ is zero over a cycle (See the next section on root-mean-square values).

As $V_m = I_m R$, Equation 2.10 can be written, dropping the subscript on P,

$$P = 0.5 I_m^2 R \tag{2.11}$$

This form should be compared to the power, $I^2 R$, developed by a direct current, I.

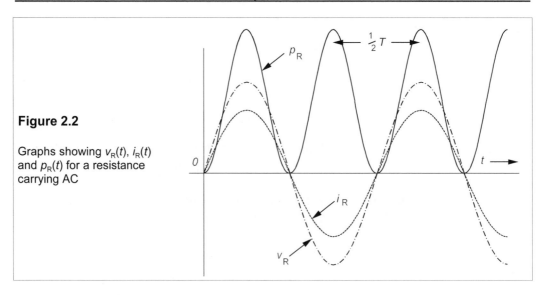

Figure 2.2

Graphs showing $v_R(t)$, $i_R(t)$ and $p_R(t)$ for a resistance carrying AC

Root-mean-square (r.m.s.) values

An r.m.s. quantity, q_{rms}, is defined by

$$q_{rms} = \sqrt{\frac{1}{T} \int_0^T q^2(t)\,dt} \tag{2.12}$$

For example if $i(t) = I_m \sin(\omega t + \phi)$, then

$$I_{rms}^2 = \frac{I_m^2}{T} \int_0^T \sin^2(\omega t + \phi)\,dt = \frac{I_m^2}{2\pi} \int_0^{2\pi} \sin^2(x + \phi)\,dx \tag{2.13}$$

making the usual substitution of x for ωt. From this

$$I_{rms}^2 = \frac{I_m^2}{4\pi} \int_0^{2\pi} [1 - \cos(2x + 2\phi)]\,dx = \frac{I_m^2}{4\pi}\left[x - \frac{\sin(2x+2\phi)}{2}\right]_0^{2\pi} \tag{2.14}$$

Evaluating the term in square brackets gives

$$I_{rms}^2 = \frac{I_m^2}{4\pi}\left[2\pi - \frac{\sin(4\pi+2\phi)}{2} + \frac{\sin 2\phi}{2}\right] = \frac{I_m^2}{2} \tag{2.15}$$

since $\sin(4\pi + 2\phi) = \sin 2\phi$. Equation 2.15 shows that any sinusoid will have an r.m.s. value equal to $1/\sqrt{2}$ of the peak amplitude:

$$I_{rms} = I_m/\sqrt{2} \quad \text{and} \quad V_{rms} = V_m/\sqrt{2} \tag{2.16}$$

Examining Equation 2.11 we see that replacing $0.5I_m^2$ by I_{rms}^2 preserves the DC form,

$$P = I^2R \qquad (2.17)$$

provided $I \equiv I_{rms}$.

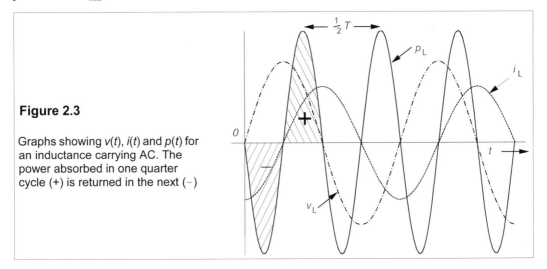

Figure 2.3

Graphs showing $v(t)$, $i(t)$ and $p(t)$ for an inductance carrying AC. The power absorbed in one quarter cycle (+) is returned in the next (−)

For an inductance the current lags the voltage by 90°, so the instantaneous power is

$$p_L = v_L i_L = V_m \sin \omega t \times I_m \sin(\omega t - 90°)$$

$$= 0.5 V_m I_m [\cos 90° - \cos(2\omega t - 90°)] = -0.5 V_m I_m \sin 2\omega t$$

$$(2.18)$$

This is just a sinewave of twice the frequency of current and voltage, and over a time of $T/2$ it will average to zero. Power absorbed in one quarter of a cycle is returned during the next quarter cycle, as Figure 2.3 shows. Compare this with Figure 2.2, which shows the voltage, current and power waveforms for a resistance.

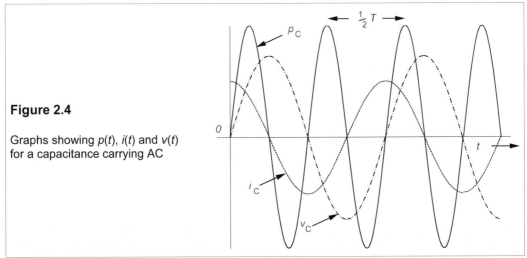

Figure 2.4

Graphs showing $p(t)$, $i(t)$ and $v(t)$ for a capacitance carrying AC

The instantaneous power in a capacitance carrying a current $i_C = I_m \sin(\omega t + 90°)$ can be derived similarly:

$$p_C = 0.5 V_m I_m \sin 2\omega t \tag{2.19}$$

Once again the average power over half a cycle is zero. The relationships between v_C, i_C and p_C are illustrated in Figure 2.4.

To summarise:

- The average power consumed by reactances is 0
- The average power consumed by resistances is $I_{rms}^2 R$

2.1.4 Form factor and crest factor

The *form factor* of a periodic voltage waveform is defined as

$$FF = V_{rms} / V_{av} \tag{2.20}$$

where V_{av}, the time-averaged value, is given by

$$V_{av} = \frac{1}{T} \int_0^T |v(t)| \, dt = \frac{\omega}{2\pi} \int_0^{2\pi/\omega} |v(t)| \, dt \tag{2.21}$$

and $|v(t)|$ means the absolute value or magnitude – the *rectified* value.

The *crest factor* is the ratio of the peak to the r.m.s. value:

$$CF = V_m / V_{rms} \tag{2.22}$$

Example 2.1

What are V_{av}, V_{rms}, the FF and the CF for $v(t) = V_m \sin \omega t$?
   The voltage is sinusoidal so $V_{rms} = V_m/\sqrt{2}$, while

$$V_{av} = \frac{2}{T} \int_0^{T/2} V_m \sin \omega t \, dt = \frac{2V_m}{\omega T} \int_0^{\pi} \sin x \, dx = \frac{V_m}{\pi} [-\cos x]_0^{\pi} = \frac{2V_m}{\pi} \tag{2.23}$$

We integrate over only half the period because the rectified sinewave is used (integration over the whole period gives zero of course). Equation 2.23 shows that the time average of a rectified sinewave is $2/\pi$ of its peak amplitude.
   The form factor for $v(t)$ is

$$FF = \frac{V_{rms}}{V_{av}} = \frac{V_m/\sqrt{2}}{2V_m/\pi} = \frac{\pi}{2\sqrt{2}} = 1.111 \tag{2.24}$$

while the crest factor is $V_m/V_{rms} = V_m/(V_m/\sqrt{2}) = \sqrt{2}$. A square wave obviously has $FF = CF = 1$, the lowest possible value for practical waveforms. The peakier the waveform the higher are the crest and form factors.

2.1.5 Waveforms with DC components

Waveforms are often encountered which have DC components. A DC component is easily identified with a CRO by switching from DC-coupled input to AC coupled (see also Section 29.3.4). A DC-coupled input is one fed directly into the signal amplifiers and thence to the Y-deflection plates, so that any DC component is unaffected. An AC-coupled input is connected to a small ($\approx$20 pF) capacitor in series with the signal source and the channel amplifier. It is best to remember these facts by the mnemonic: *DC = direct coupled, AC = after capacitor.*

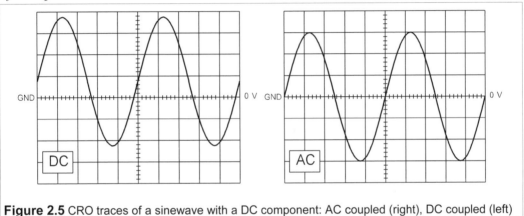

Figure 2.5 CRO traces of a sinewave with a DC component: AC coupled (right), DC coupled (left)

The capacitor removes any DC component and displays a waveform which has equal areas above and below the 0 V line, that is its time average is zero. Figure 2.5a shows a sinusoidal waveform with a DC component which is DC coupled to a CRO and in Figure 2.5b the same signal is AC coupled. In the latter case the trace is symmetrical about the 0V line, that is it has no DC component. If the CRO's vertical scale is 1 V/div, then we can see that the signal is 6 V peak-to-peak, with a DC component of 0.8 V, that is we can write the signal as

$$v(t) = 0.8 + 3\sin\omega t \text{ V} \tag{2.25}$$

The average power developed in a 1Ω resistance over a cycle by this voltage can be found by considering the DC and AC parts separately. Thus the AC part has an r.m.s. value of $3/\sqrt{2}$ = 2.12 V (see Section 2.2 for why this is so) and the power developed from it is $2.12^2 = 4.5$ W, while the DC component develops $0.8^2 = 0.64$ W, for a total of 5.14 W. It is left as an exercise to show that taking the average of $[v(t)]^2$ produces the same result.

2.1.6 The series RL circuit

Let a sinusoidal voltage source be connected to a series RL circuit as in Figure 2.6. By KVL,

$$V_m \sin(\omega t + \phi) = v_R + v_L \tag{2.26}$$

where $v_R = iR$ and $v_L = L\,di/dt$. The current waveform will also be sinusoidal with the same frequency as the source:

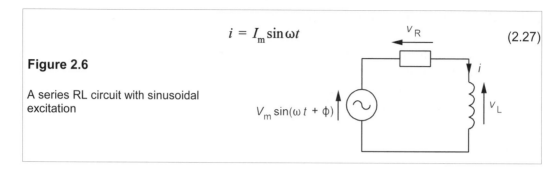

$$i = I_\text{m} \sin \omega t \tag{2.27}$$

Figure 2.6

A series RL circuit with sinusoidal excitation

Then
$$v_\text{R} = I_\text{m} R \sin \omega t \tag{2.28}$$

and
$$v_\text{L} = L d/dt \ (I_\text{m} \sin \omega t) = \omega L I_\text{m} \cos \omega t \tag{2.29}$$

Substituting for v_R and v_L in Equation 2.26 leads to

$$V_\text{m} \sin (\omega t + \phi) = I_\text{m} R \sin \omega t + \omega L I_\text{m} \cos \omega t \tag{2.30}$$

The left-hand side of Equation 2.30 can be written in the form

$$V_\text{m} \cos \phi \sin \omega t + V_\text{m} \sin \phi \cos \omega t = I_\text{m} R \sin \omega t + \omega L I_\text{m} \cos \omega t \tag{2.31}$$

Comparing terms on either side we see that $V_\text{m} \cos \phi = I_\text{m} R$ and $V_\text{m} \sin \phi = \omega L I_\text{m}$, implying that $\tan \phi = \omega L / R$ and that $V_\text{m} = I_\text{m} \sqrt{(R^2 + \omega^2 L^2)}$. The term $\sqrt{(R^2 + \omega^2 L^2)}$ has dimensions of ohms and is called the *impedance*, Z, of the circuit (as we shall see the impedance can be expressed as a complex number, in which case it is written in bold: **Z**). Then we can write

$$V_\text{m} = I_\text{m} \sqrt{(R^2 + \omega^2 L^2)} = I_\text{m} Z \tag{2.32}$$

Equation 2.32 shows that voltage and current in the circuit are now related by a modified form of Ohm's law. It will help us to display diagrammatically the relationships between voltage and current in the series RL circuit, as in Figure 2.7. On the left is the impedance triangle and on the right the phase and magnitude relationships between V_R, V_L, V_m and I.

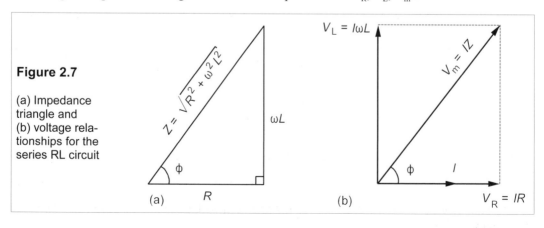

Figure 2.7

(a) Impedance triangle and (b) voltage relationships for the series RL circuit

2.2 **Phasors**

Figure 2.7b shows that the voltage across and current though the resistance are in phase, while the voltage across the inductance leads by 90°. Rotation of V_L by 90° suggests the use of complex numbers, since in the Argand diagram representation, the imaginary axis is perpendicular to the real number axis and multiplying by j (in electrical engineering $\sqrt{(-1)}$ is represented by j) is the same as rotating the line representing it through +90°. In Figure 2.7b therefore, if I and V_R lie on the real axis, V_L would lie along the imaginary ($+j$) axis. Now we can represent voltages and currents as complex quantities which encapsulate the phase relationships between them. For this reason they are called *phasors*.

The effects of inductance and capacitance in a circuit can be taken into account by expressing them as imaginary numbers. Instead of writing $V_L = IX_L$, with V_L leading I by +90°, we can write

$$V_L = IjX_L = jI\omega L \qquad (2.33)$$

where the impedance of L is jX_L or $j\omega L$. Similarly the impedance of a capacitance, which phase shifts V_C by −90° with respect to I, can be written as

$$-jX_C = -j/\omega C = 1/j\omega C \qquad (2.34)$$

Impedances then consist of real and imaginary parts:

$$\mathbf{Z} = R + jX \qquad (2.35)$$

Phasors and impedances are written in bold capitals: $\mathbf{V}$, $\mathbf{I}$, $\mathbf{Z}$; their magnitudes are written as italic capitals: V, I, Z; instantaneous values are written in lower case italics: v, i.

Complex numbers can be expressed in two forms: *polar* and *rectangular*. An impedance of $10 - j6\ \Omega$ is in rectangular form, drawn on an Argand diagram as in Figure 2.8.

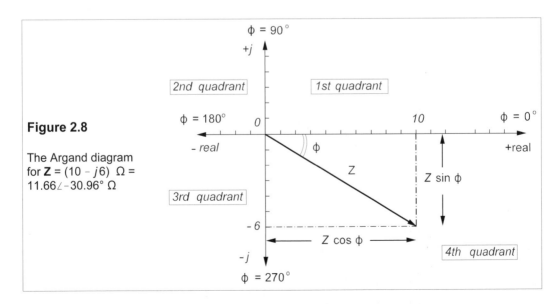

Figure 2.8

The Argand diagram for $\mathbf{Z} = (10 - j6)\ \Omega = 11.66\angle{-30.96°}\ \Omega$

The projection of $\mathbf{Z}$ on the real axis is $Z\cos\phi$, and its projection on the j-axis is $Z\sin\phi$, implying

$$Z\cos\phi = 10 \quad \text{and} \quad Z\sin\phi = -6 \tag{2.36}$$

Therefore $\qquad Z = \sqrt{10^2 + (-6)^2} \quad \text{and} \quad \tan\phi = -0.6 \tag{2.37}$

Thus the magnitude of $\mathbf{Z}$ (Z or $|\mathbf{Z}|$) is $\sqrt{136} = 11.66\ \Omega$, and $\phi = \tan^{-1}(-0.6) = -30.96°$. In polar form $\mathbf{Z} = 11.66\angle-30.96°\ \Omega$. The polar form is more suitable for multiplication and division, while the rectangular form must be used for addition and subtraction.

In the previous section we analysed the behaviour of simple AC circuits using expressions for currents and voltage such as $v = V_m\sin(\omega t + \phi)$. Mathematically it was quite a tedious task even for a simple series RL circuit. AC analysis would be a daunting task but for the powerful aid of complex numbers. Alternating voltages and currents are characterised by their magnitudes and phases and the frequency is understood. Thus a voltage such as $10\sin(100t - 30°)$ V could be written as $10\angle-30°$ V, $\omega = 100$ rad/s understood and omitted. To write power as VI, instead of using the peak amplitude of phasor we use its r.m.s. value, which for a sinewave means

$$V = V_m/\sqrt{2} \quad \text{and} \quad I = I_m/\sqrt{2} \tag{2.38}$$

and $10\sin(100t - 30°)$ V becomes $7.07\angle-30°$ V.

2.2.1 Complex arithmetic

Addition (or subtraction) of complex numbers requires the real parts to be added (or subtracted) separately from the imaginary parts. Thus

$$(a + jb) + (c + jd) = (a + c) + j(b + d) \tag{2.39}$$

This implies that $a + jb = c + jd$ only if $a = c$ *and* $b = d$: a complex equality gives us two for the price of one.

It is a *Golden Rule* always to use the polar form for multiplication and division, for which we must make use of *de Moivre's theorem*:

$$\exp(j\theta) = \cos\theta + j\sin\theta \tag{2.40}$$

Multiplication of two complex numbers then becomes easier as

$$\alpha\exp(j\theta) \times \beta\exp(j\phi) = \alpha\beta\exp[j(\theta + \phi)] \tag{2.41}$$

And division too

$$\frac{\alpha\exp(j\theta)}{\beta\exp(j\phi)} = \frac{\alpha}{\beta}\exp[j(\theta - \phi)] \tag{2.42}$$

As an example, to divide $12 + j\,5$ by $3 - j\,4$, one first calculates the polar forms $\sqrt{(12^2 + 5^2)}\angle\tan^{-1}(5/12)$ ($= 13\angle22.62°$) and $\sqrt{(3^2 + 4^2)}\angle\tan^{-1}(-4/3)$ ($= 5\angle-53.13°$), then one divides:

$$\frac{12 + j5}{3 - j4} = \frac{13\angle22.62°}{5\angle-53.13°} = \frac{13}{5}\angle[22.62° - (-53.13°)] \tag{2.43}$$

$$= 2.6\angle75.75° = 0.64 - j2.52$$

Example 2.2

A voltage, **V_s**, of peak amplitude 339 V and frequency 50 Hz is applied to a series RC circuit in which $R = 10\ \Omega$ and $C = 470\ \mu F$ as in Figure 2.9a. Find the current and all the voltages.

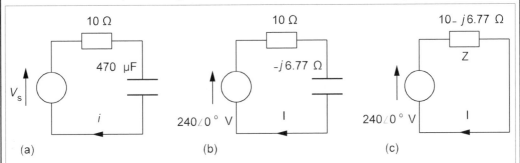

(a)   (b)   (c)

Figure 2.9 (a) The time-domain circuit (b) The phasor-domain circuit with separate resistance and reactance (c) Phasor-domain circuit with resistance and reactance combined as an impedance

The circuit of Figure 2.9a is said to be in the *time domain* and we must transform it into the *phasor domain*. The angular frequency, $\omega = 2\pi f = 100\pi$ rad/s. The reactance of the capacitance is

$$X_C = 1/\omega C = 1/(100\pi \times 470 \times 10^{-6}) = 6.77\ \Omega \tag{2.44}$$

This reactance has an imaginary impedance of $-j\,6.77\ \Omega$,, since it is capacitive. The voltage has an r.m.s. amplitude of $339/\sqrt{2} = 240$ V, and we shall make it the *reference phasor*, that is its phase angle is 0°, so the supply voltage is $240\angle 0°$. Thus the phasor-domain circuit of Figure 2.9b is obtained.

Figure 2.10

Phasor diagram for the circuit of Figure 2.9. Note that **V_c** lags **V_R** by 90°, that **I** is in phase with **V_R** and leads **V_s**. It is also clear that
$$|V_c|^2 + |V_R|^2 = |V_s|^2$$

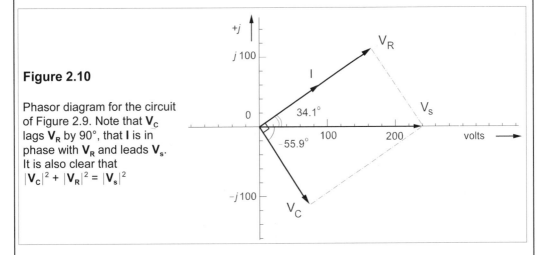

The resistance of 10 Ω and the capacitive reactance of 6.77 Ω can be combined into an impedance, **Z**, $= 10 - j\,6.77\ \Omega$ producing the much simpler circuit of 2.9c. Then Ohm's law for phasors, **V = IZ**, yields

$$\mathbf{I} = \frac{\mathbf{V}}{\mathbf{Z}} = \frac{240\angle 0°}{10 - j6.77}\ \mathbf{A} \tag{2.45}$$

The denominator is converted into polar form before dividing:

$$\mathbf{I} = 240\angle 0° \div 12.08\angle -34.1° = 19.87\angle 34.1° \text{ A} \qquad (2.46)$$

Once the current has been found the voltages follow from Ohm's law:

$$\mathbf{V_R} = \mathbf{I}R = 19.87\angle 34.1° \times 10 = 198.7\angle 34.1° \text{ V} \qquad (2.47)$$

$$\mathbf{V_C} = \mathbf{I} \times -jX_C = 19.87\angle 34.1° \times 6.77\angle -90° = 134.52\angle -55.9° \text{ V} \qquad (2.48)$$

Figure 2.10 is the phasor diagram of these voltages and currents.

2.2.2 *Determining the phase angle of a phasor*

Because $\tan^{-1}\alpha$ is restricted to $\pm 90°$ (the first and fourth quadrants of the Argand diagram), the phase angle of phasor such as $\mathbf{V} = -5 + j4$ V cannot be found by taking the inverse tangent of the imaginary part, $\mathbb{I}(\mathbf{V})$, divided by the real part, $\mathbb{R}(\mathbf{V})$, since that would give in this case $\tan^{-1}[4/(-5)] = \tan^{-1}(-0.8) = -38.6°$. This phasor lies in the second quadrant and its phase angle is $180° - 38.7° = 141.3°$, as can be seen in Figure 2.11.

Similarly for a phasor such as $\mathbf{I} = -2 - j4$ A, which lies in the third quadrant and has a phase angle of $-180° + \tan^{-1}(4/2) = -116.6°$. For this reason, care must be taken in determining phase angles from rectangular forms: *check the quadrant!*

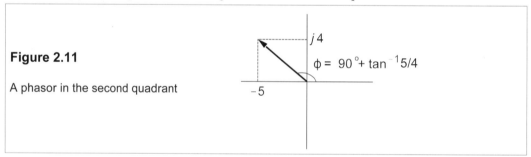

Figure 2.11

A phasor in the second quadrant

2.3 **Circuit analysis with AC**

The techniques for AC circuit analysis are essentially the same as for DC analysis, except that the circuits are in the phasor domain and complex numbers are used. With minor modifications – often just replacing *resistance* by *impedance* – all the circuit theorems discussed in the previous chapter can be used for AC circuits; in particular we shall consider first the use of Thévenin's and Norton's theorems.

2.3.1 *Deriving Thévenin and Norton equivalent circuits*

In Chapter 1, Thévenin's and Norton's theorems were couched in terms of resistance; accordingly this must be replaced by impedance, so an AC Thévenin circuit comprises a single voltage source in series with an impedance and a Norton circuit comprises a single current source in parallel with an impedance. To give an example:

Example 2.3

What are the Thévenin and Norton equivalents between the terminals A and B of the circuit in Figure 2.12, if $v(t) = 14.14 \sin(\omega t + 45°)$ V and $i(t) = 2.828 \sin(\omega t - 30°)$ A?

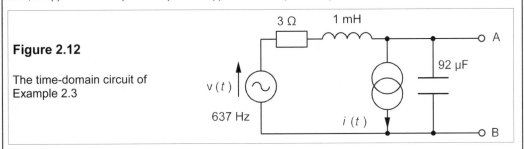

Figure 2.12

The time-domain circuit of Example 2.3

The circuit must be transformed into the phasor domain by finding ω (= $2\pi f$) = 4000 rad/s. Then $jX_L = j\omega L = j4000 \times 1 \times 10^{-3} = j4\ \Omega$, and $-jX_C = -j/\omega C = -j/(4000 \times 92 \times 10^{-6}) = -j2.72\ \Omega$. The sources must be given their r.m.s. values by dividing their amplitudes by $\sqrt{2}$, giving $\mathbf{V} = 10\angle45°$ V and $\mathbf{I} = 2\angle-30°$ A. The phasor-domain circuit is then as in Figure 2.13

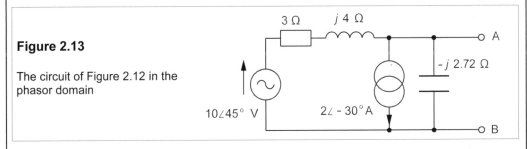

Figure 2.13

The circuit of Figure 2.12 in the phasor domain

The voltage source and the series impedance of $3 + j4\ \Omega$ can now be transformed into a current source by Norton's theorem, the source current being

$$\mathbf{I} = \frac{\mathbf{V}}{\mathbf{Z}} = \frac{10\angle45°}{3+j4} = \frac{10\angle45°}{5\angle53.13°} = 2\angle-8.13°\ \text{A} \qquad (2.49)$$

The circuit now looks like Figure 2.14.

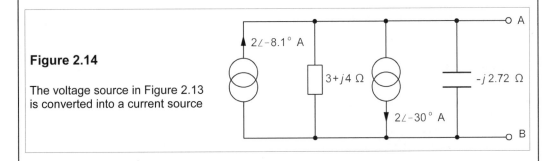

Figure 2.14

The voltage source in Figure 2.13 is converted into a current source

The impedance of $3 + j4\ \Omega$ and the capacitive impedance of $-j2.72\ \Omega$ are in parallel and can be combined by the product-over-sum rule:

$$\mathbf{Z_N} = \frac{(3+j4) \times -j2.72}{3+j4-j2.72} = \frac{5\angle 53.1° \times 2.72\angle -90°}{3+j1.28}$$

$$= \frac{13.6\angle -36.9°}{3.26\angle 23.1°} = 4.17\angle -60° \ \Omega \tag{2.50}$$

The parallel current sources may be added algebraically, which means that, since the arrows are pointing in opposite directions, one must be subtracted from the other:

$$\mathbf{I_N} = 2\angle -8.1° - 2\angle -30° = (1.98 - j0.28) - (1.73 - j1) \tag{2.51}$$

$$= 0.25 + j0.72 = 0.76\angle 70.9° \ \mathbf{A} \tag{2.52}$$

Consequently the Norton circuit is that of Figure 2.15a.

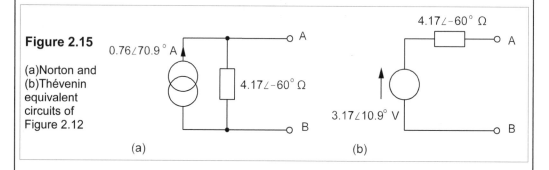

Figure 2.15

(a)Norton and
(b)Thévenin
equivalent
circuits of
Figure 2.12

(a)   (b)

The Thévenin voltage in Figure 2.14b is found from

$$\mathbf{V_T} = \mathbf{I_N Z_N} = 0.76\angle 70.9° \times 4.17\angle -60° = 3.17\angle 10.9° \ \mathbf{V} \tag{2.53}$$

The circuit derivation follows precisely the same path that would be taken for a DC circuit, apart from using impedances, phasors and complex numbers.

2.3.2 Mesh analysis with AC

The method is the same as in DC circuits, but with complex numbers. The following example demonstrates the technique.

⌐ **Example 2.4**

In the circuit of Figure 2.16, what is $\mathbf{V_{in}}/\mathbf{V_o}$? At what frequency is $\mathbf{V_o}$ 180° out of phase with $\mathbf{V_{in}}$? What is then $\mathbf{V_o}/\mathbf{V_{in}}$?

Using mesh analysis, currents $\mathbf{I_1}$, $\mathbf{I_2}$ and $\mathbf{I_3}$ are assigned to meshes 1, 2 and 3 as shown, then KVL is used on each mesh in turn, starting with mesh 1 (clockwise voltages are negative):

$$-\mathbf{V_{in}} + \mathbf{I_1}/j\omega C + (\mathbf{I_1} - \mathbf{I_2})R = 0$$

giving

$$\mathbf{V_{in}} = \mathbf{I_1}(R + 1/j\omega C) - \mathbf{I_2}R$$

Dividing this by R

$$\mathbf{V}_{in}/R = \mathbf{I}_1(1 + \beta) - \mathbf{I}_2 \qquad (2.54)$$

where $\beta \equiv 1/j\omega RC$.

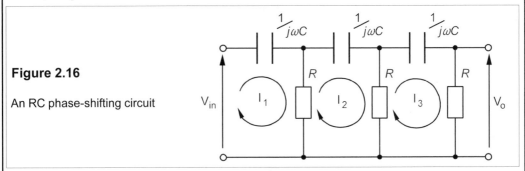

Figure 2.16

An RC phase-shifting circuit

Then in mesh 2:

$$(\mathbf{I}_2 - \mathbf{I}_1)R + \mathbf{I}_2/j\omega C + (\mathbf{I}_2 - \mathbf{I}_3)R = 0 \qquad (2.55)$$

Rearranging,

$$\mathbf{I}_2(2 + 1/j\omega CR) = \mathbf{I}_2(2 + \beta) = \mathbf{I}_1 + \mathbf{I}_3 \qquad (2.56)$$

And in mesh 3:

$$(\mathbf{I}_3 - \mathbf{I}_2)R + \mathbf{I}_3/j\omega C + \mathbf{I}_3 R = 0 \qquad (2.57)$$

which rearranges to

$$\mathbf{I}_2 = \mathbf{I}_3(2 + 1/j\omega RC) = \mathbf{I}_3(2 + \beta) \qquad (2.58)$$

Substituting into Equation 2.56,

$$\mathbf{I}_3(2 + \beta)(2 + \beta) = \mathbf{I}_1 + \mathbf{I}_3 \qquad (2.59)$$

which gives

$$\mathbf{I}_1 = \mathbf{I}_3[(2 + \beta)^2 - 1] = \mathbf{I}_3(3 + 4\beta + \beta^2) \qquad (2.60)$$

Substituting for $\mathbf{I}_2$ from Equation 2.58 and for $\mathbf{I}_1$ from Equation 2.60 into Equation 2.54 leads to

$$\mathbf{V}_{in}/R = \mathbf{I}_3(3 + 4\beta + \beta^2)(1 + \beta) - \mathbf{I}_3(2 + \beta)$$

giving

$$\mathbf{V}_{in} = \mathbf{I}_3 R(1 + 6\beta + 5\beta^2 + \beta^3) \qquad (2.61)$$

Now $\mathbf{V}_o = \mathbf{I}_3 R$, and Equation 2.61 becomes

$$\mathbf{V}_{in}/\mathbf{V}_o = 1 + 6\beta + 5\beta^2 + \beta^3$$

$$= (1 - 5/\omega^2 R^2 C^2) + j(1/\omega^3 R^3 C^3 - 6/\omega RC) \qquad (2.62)$$

since $\beta \equiv 1/j\omega RC = -j/\omega RC$.

In Equation 2.62, for $\mathbf{V}_o$ to be 180° out of phase with $\mathbf{V}_{in}$, $\mathbf{V}_{in}/\mathbf{V}_o$ must be wholly real. Looking at Equation 2.62 we see this requires that $6/\omega RC = 1/\omega^3 R^3 C^3$, that is $\omega RC = 1/\sqrt{6}$, or $\omega = \mathbf{1/RC\sqrt{6}}$. Then $\mathbf{V}_{in}/\mathbf{V}_o = 1 - 5/\omega^2 R^2 C^2 = -29$, or $\mathbf{V}_o/\mathbf{V}_{in} = -\mathbf{1/29}$. The minus sign represents a phase shift of 180°, since $V\angle 180° = V\cos 180° + jV\sin 180° = -V$.

Examining Equation 2.62 again, it can be seen that, as $1/\omega RC \rightarrow 0$ (which is the same as $\omega \rightarrow \infty$),

$\mathbb{I}(V_{in}/V_o)$, the imaginary part of $V_{in}/V_o \rightarrow 0$, while $\mathbb{R}(V_{in}/V_o)$, the real part $\rightarrow 1$. Thus the the phase angle, $\phi \rightarrow 0°$ as $\omega \rightarrow \infty$. When $\omega \rightarrow 0$, $\mathbb{R}(V_{in}/V_o) \rightarrow -5/(\omega RC)^2$ while $\mathbb{I}(V_{in}/V_o) \rightarrow +1/(\omega RC)^3$. Thus the imaginary part tends to $+\infty$ faster than the real part tends to $-\infty$, placing V_{in}/V_o in the third quadrant and making $\phi = -270°$ when $\omega = 0$. The locus of ϕ as ω goes from 0 to ∞ is shown in Figure 2.17. Though the arithmetic is more complicated because it deals with complex numbers, in no other way does the procedure differ from that for DC circuits.

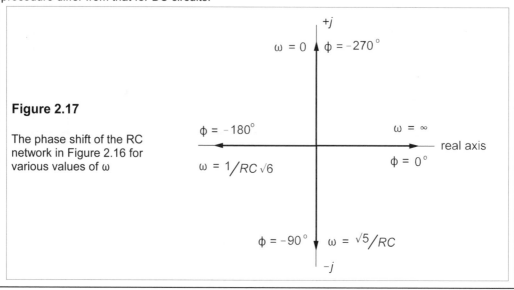

Figure 2.17

The phase shift of the RC network in Figure 2.16 for various values of ω

2.3.3 Nodal analysis with AC

Nodal analysis likewise follows the same method using complex arithmetic, as the next example demonstrates.

Example 2.5

Analyse the circuit of Figure 2.16 as in Example 2.4 but using nodal analysis instead.

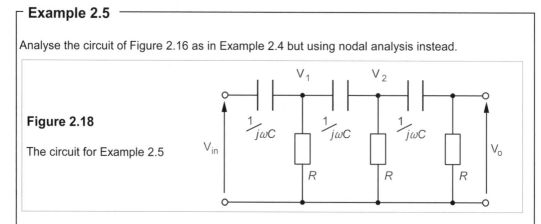

Figure 2.18

The circuit for Example 2.5

The circuit has two unknown nodal voltages, shown in Figure 2.18, V_1 and V_2. At node 1 the nodal equations by KCL are

$$(V_1 - V_{in})j\omega C + V_1 G + (V_1 - V_2)j\omega C = 0$$

where we have used G instead of $1/R$ and taken $1/(-jX_C) = 1/(1/j\omega C) = j\omega C$. This equation rearranges to

$$V_1(G+2j\omega C) - V_2 j\omega C = V_{in} j\omega C \tag{2.63}$$

Then at node 2, KCL yields

$$(V_2 - V_1)j\omega C + V_2 G + (V_2 - V_0)j\omega C = 0$$

which rearranges to

$$V_2(G+j2\omega C) - V_0 j\omega C = V_1 j\omega C \tag{2.64}$$

And at the output node

$$(V_0 - V_2)j\omega C + V_0 G = 0$$

from which we find V_2 in terms of V_0:

$$V_2 = \frac{V_0(G+j\omega C)}{j\omega C} = V_0(G/j\omega C + 1) = V_0(\beta + 1)$$

where $\beta \equiv G/j\omega C$. Substitution for V_2 in Equation 2.64 leads to

$$V_0(\beta + 1)(G+j2\omega C) - V_0 j\omega C = V_1 j\omega C$$

This can be divided throughout by $j\omega C$ to give

$$V_0(\beta + 1)(\beta + 2) - V_0 = V_0(\beta^2 + 3\beta + 1) = V_1$$

Both V_1 and V_2 can now be substituted for in Equation 2.63 to produce

$$V_0(\beta^2 + 3\beta + 1)(G + j2\omega C) - V_0(\beta + 1)j\omega C = V_{in} j\omega C$$

Dividing this throughout by $j\omega C$ gives

$$V_0(\beta^2 + 3\beta + 1)(\beta + 2) - V_0(\beta + 1) = V_{in} = V_0(\beta^3 + 5\beta^2 + 6\beta + 1)$$

This is the same as Equation 2.62 and the rest of the analysis is identical to Example 2.4.

2.3.4 Admittance

Admittance is the reciprocal of impedance: $Y = 1/Z$, as conductance is the reciprocal of resistance ($G = 1/R$). The units of both admittance and conductance are siemens (S), formerly Ω^{-1}, mhos or $\mho$. However, admittance is more widely used for parallel AC circuits than is conductance in parallel DC circuits. Remember that reciprocals of complex numbers should always be taken using the polar form. For example if $Z = 3 - j4\ \Omega$, then

$$Y = \frac{1}{3-j4} = \frac{1}{5\angle -53.13°} = 0.2\angle 53.13° = 0.12 + j0.16\ S \tag{2.65}$$

When numbers are given the rule is *avoid rationalising*. In general, though

$$Y = \frac{1}{Z} = \frac{1}{R \pm jX} = \frac{R \mp jX}{R^2 + X^2}$$

$$= \frac{R}{R^2 + X^2} \mp j\frac{X}{R^2 + X^2} = G \mp jB \qquad (2.66)$$

G is the real part of $\mathbf{Y}$, $\mathbb{R}(\mathbf{Y})$, known as the conductance, and B, the imaginary part of $\mathbf{Y}$, $\mathbb{I}(\mathbf{Y})$, is the *susceptance*.

2.3.5 Series and parallel forms

Any complex impedance, $R + jX$, comprises a resistance, R, in *series* with a reactance, X; while any complex admittance, $G + jB$, comprises a conductance, G, in *parallel* with a susceptance, B. Consequently an impedance can be expressed in series or parallel form as an admittance; we can make whichever choice is convenient, but the component values will in general be completely different for the two forms. For example, suppose a capacitance is in series with a resistance and the pair is connected to a 120V AC supply as in Figure 2.19a.

Now if, for example, the capacitance is 25 μF, the resistance is 100 Ω and the supply frequency is 60 Hz, then the capacitance has an impedance of $-j/(2\pi \times 60 \times 25 \times 10^{-6}) = -j106\,\Omega$. The series combination is $\mathbf{Z} = 100 - j106 = 146\angle -46.7°\,\Omega$, and the current flowing is $\mathbf{V_s}/\mathbf{Z} = 120\angle 0° \div 146\angle -46.7° = 0.82\angle 46.7°$ A.

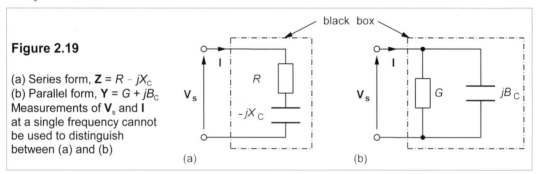

Figure 2.19

(a) Series form, $\mathbf{Z} = R - jX_C$
(b) Parallel form, $\mathbf{Y} = G + jB_C$
Measurements of $\mathbf{V_s}$ and $\mathbf{I}$ at a single frequency cannot be used to distinguish between (a) and (b)

The admittance of the combination is

$$\mathbf{Y} = \frac{1}{\mathbf{Z}} = \frac{1}{146\angle -46.7°} = 0.00685\angle 46.7°\text{ S} = 4.7 + j5.0\text{ mS} \qquad (2.67)$$

which is $G + jB_C$. But a conductance of 4.7 mS is just a resistance, R_p, of $1/(4.7 \times 10^{-3}) = 213\,\Omega$ while a susceptance of 5 mS is a reactance, X_p, of $1/(5.0 \times 10^{-3}) = 200\,\Omega$. And if $X_p = 200\,\Omega$ at 60 Hz, its capacitance is

$$C = 1/\omega X_p = 1/(2\pi \times 60 \times 200) = 13.3\ \mu\text{F} \qquad (2.68)$$

The current flowing through the 213Ω resistance would be $\mathbf{V_s}/R_p = 120\angle 0°/213 = 0.56\angle 0°$ A. The current flowing through X_p would be $120\angle 0° \div -j\,200 = 120\angle 0° \div 200\angle -90° = 0.6\angle 90°$ A. The total current flowing through the parallel combination is then

$$\mathbf{I} = 0.56\angle 0° + 0.6\angle 90° = 0.56 + j0.6 = 0.82\angle 47°\text{ A} \qquad (2.69)$$

which is the same as that flowing through the series combination.

Note that in general the components have different values in the two forms. At 60 Hz the

100Ω resistance and $25\mu F$ capacitance in series presents the same impedance to current flow as the parallel combination of a 213Ω resistance and a $13.3\mu F$ capacitance. If these components were placed in black boxes as in Figures 2.19a and 2.19b we could not tell by measuring the terminal currents and voltages at 60 Hz whether the connections were series or parallel. Of course a simple DC measurement would soon distinguish between them because the capacitor is open circuit to DC and the series combination would present infinite resistance, while the parallel combination would have a resistance of 213 Ω.

2.3.6 *The star-delta transformation*

Again the DC form is adapted to AC circuits by replacing *resistance* by *impedance* and *conductance* by *admittance*. Then the delta-to-star transformation becomes

$$\mathbf{Z_A} = \frac{\mathbf{Z_{AB}Z_{CA}}}{\Sigma\,\mathbf{Z}} \;\; ; \;\; \mathbf{Z_B} = \frac{\mathbf{Z_{AB}Z_{BC}}}{\Sigma\,\mathbf{Z}} \;\; ; \;\; \mathbf{Z_C} = \frac{\mathbf{Z_{BC}Z_{CA}}}{\Sigma\,\mathbf{Z}} \tag{2.70}$$

where $\Sigma\mathbf{Z} = \mathbf{Z_{AB}} + \mathbf{Z_{BC}} + \mathbf{Z_{CA}}$. The single-suffix impedances are those of the star form (Figure 2.20b) and the double-suffix the delta form (Figure 2.20a).

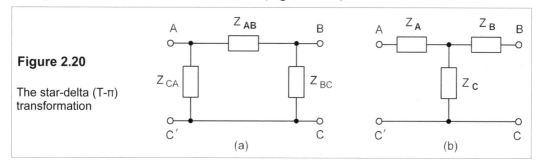

Figure 2.20

The star-delta (T-π) transformation

(a)   (b)

The reverse transformation from star to delta uses admittances:

$$\mathbf{Y_{AB}} = \frac{\mathbf{Y_A Y_B}}{\Sigma\,\mathbf{Y}} \;\; ; \;\; \mathbf{Y_{BC}} = \frac{\mathbf{Y_B Y_C}}{\Sigma\,\mathbf{Y}} \;\; ; \;\; \mathbf{Y_{CA}} = \frac{\mathbf{Y_A Y_C}}{\Sigma\,\mathbf{Y}} \tag{2.71}$$

where $\Sigma\mathbf{Y} = \mathbf{Y_A} + \mathbf{Y_B} + \mathbf{Y_C}$ and $\mathbf{Y_A} = 1/\mathbf{Z_A}$ etc. Once the correct choice of $\mathbf{Y}$ or $\mathbf{Z}$ has been made, the rule is the same: *multiply together the Zs (or Ys) attached to the terminal(s) of interest and divide by the sum of the Zs (Ys).* An example will make this clear.

Example 2.6

In the network of Figure 2.21a, $\mathbf{Z_p} = 1/\mathbf{Y_p} = R_L + j\omega L$. What must $\mathbf{Y_p}$ be for $\mathbf{V_o}$ to be zero? Show that for $\mathbf{V_o}$ to be zero, $L = 2CRR_L$ and $\omega = \sqrt{(2/LC)}$.

We recognise that the T-network centred on node B can be transformed into a π-network as in Figure 2.21b, which then can be turned into the network of Figure 2.21c. The transformation from star to delta requires the use of admittances and Equation 2.71, where $\mathbf{Y_A} = j\omega C$, $\mathbf{Y_B} = j\omega C$, $\mathbf{Y_C} = 1/R$ and $\Sigma\mathbf{Y} = 1/R + j2\omega C$, so that

$$Y_{AB} = \frac{Y_A Y_B}{\Sigma Y} = \frac{(j\omega C)^2}{1/R + j2\omega C} = \frac{-\omega^2 C^2}{1/R + j2\omega C} \tag{2.72}$$

and
$$Y_{BC} = Y_{CA} = \frac{j\omega C \times 1/R}{1/R + j2\omega C} = \frac{j\omega C}{1 + j2\omega CR} \tag{2.73}$$

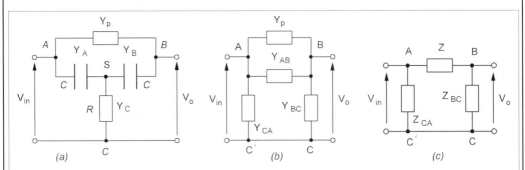

Figure 2.21 (a) The original network has star and delta parts (b) The star network radiating from B is transformed into delta (c) Y_1 and Y_p are added to give the final delta network

The output voltage is then

$$V_0 = \frac{Z_{BC}}{Z + Z_{BC}} V_{in} \tag{2.74}$$

where $Z_{BC} = 1/Y_{BC}$ and $Z = 1/(Y_p + Y_{AB})$ and $Y_p = 1/Z_p = 1/(R_L + j\omega L)$. As $|Z_{BC}|$ is always positive, V_0 can only be zero if $|Z| = \infty$, or $|Y| = |Y_p + Y_{AB}| = 0$. This means that $Y_p = -Y_1$, or

$$\frac{1}{R_L + j\omega L} = \frac{\omega^2 C^2}{1/R + j2\omega C} \tag{2.75}$$

which gives

$$1/R + j2\omega C = \omega^2 C^2 R_L + j\omega^3 C^2 L \tag{2.76}$$

For Equation 2.76 to hold, the real parts must be equal *and* the imaginary parts must also be equal, that is

$$1/R = \omega^2 C^2 R_L \quad \text{and} \quad 2\omega C = \omega^3 C^2 L \tag{2.77}$$

which leads to

$$\omega^2 C^2 = 1/RR_L \quad \text{and} \quad \omega^2 C^2 = 2C/L \tag{2.78}$$

and hence $L = 2CRR_L$ and $\omega^2 C^2 = 2C/L$, so that $\omega^2 = 2/LC$ or $\omega = \sqrt{(2/LC)}$, as required.

2.4 Power in AC circuits

Power in AC circuits is not so simply calculated as in DC circuits because voltages and currents are usually not in phase. However, if the phase angle between them is known the power is readily calculated by using the power triangle. The power calculated is an *average* power: the actual power fluctuates sinusoidally from zero to a maximum at twice the excitation frequency. Power is a *scalar* quantity, not a phasor like **V** and **I**.

2.4.1 *The power triangle*

In Section 2.1.4 we saw that pure reactances consume no power and that resistances dissipate power given by I^2R, where I is the r.m.s. current through the resistance. If the voltage across an impedance, $\mathbf{Z} = R + jX = Z\angle\phi$, is $V\angle0°$, the current though it is

$$\mathbf{I} = \frac{\mathbf{V}}{\mathbf{Z}} = \frac{V\angle0°}{Z\angle\phi} = I\angle-\phi \tag{2.79}$$

Regardless of the sign of X or ϕ, the power dissipated or *active power*, P, is given by

$$P = I^2R = I^2Z\cos\phi = VI\cos\phi \tag{2.80}$$

since $R = Z\cos\phi$ and $V = IZ$. The power dissipated is determined by the magnitude of the angle between $\mathbf{V}$ and $\mathbf{I}$, not its sign. It is the active power, P, which is the useful power, and so it is the power that is charged for by electricity suppliers.

The power delivered to the reactance during one quarter cycle is returned to the circuit in the next quarter cycle, so the net consumption is zero. However, it is convenient to define a quantity known as the *reactive power*, Q, which is given by

$$Q = I^2X = VI\sin\phi \tag{2.81}$$

The units of Q are var (reactive volt-amperes). If X is an inductive reactance, then Q is positive, while if X is a capacitive reactance Q is negative.

The *apparent power* taken by the impedance is given the symbol S, defined by

$$S = VI = I^2Z \tag{2.82}$$

S has units of VA (volt-amperes). S must be larger than (or at least equal to) P (or Q).

Looking at Equations 2.80, 2.81 and 2.82 we can see that $S^2 = P^2 + Q^2$. Thus P, Q and S can be represented by the sides of a right-angled triangle, called the *power triangle*, as in Figure 2.22. The triangle is by convention drawn this way up for an inductive load, both ϕ and Q being positive. For a capacitive load, the triangle is drawn with ϕ and Q both negative.

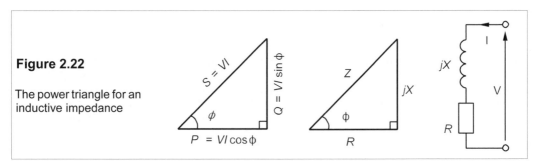

Figure 2.22

The power triangle for an inductive impedance

2.4.2 *Power factor and its correction*

The important quantity, $\cos\phi$, is given the name *power factor* (p.f.). Inductive loads are said to have lagging power factors, since current lags voltage in them, while capacitive circuits are said to have leading power factors. Generally speaking both the electricity supplier and his customer both want a power factor as close to unity as possible so that all the current does useful work, represented by P. Reactive volt-amperes increase the current needed (hence losses in wires etc.) by the load without producing any useful output. Industries often operate with a tariff which penalises power factors deviating from unity, so it is economic to *correct* the p.f. by putting capacitors or synchronous motors (see Chapter 18) in parallel with the rest of the load.

Example 2.7

A 5kVA motor operates from a 240V, 50Hz supply at a lagging p.f. of 0.6 and is in parallel with a heating and lighting load of 3 kW. What is the overall p.f. and what capacitance is required to improve this to 0.95 lagging? What is the apparent power before and after correction? What is the total current flow before and after correction?

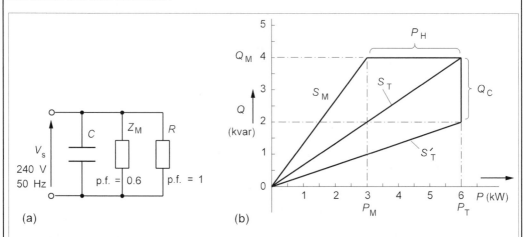

Figure 2.23 (a) Motor, heating load, correction capacitance in parallel (b) The power triangles

Figure 2.23a shows the circuit diagram, the capacitor being for p.f. correction. Problems such as these can be solved graphically or by simply calculating the Ps and Qs. The motor has $\cos\phi = 0.6$ so $P_M = 5\cos\phi = 3$ kW. The heating and lighting load has a p.f. of unity presumably, so $P_H = 3$ kW. The total kW are thus

$$P_T = P_M + P_H = 3 + 3 = 6 \text{ kW} \tag{2.83}$$

If $\cos\phi = 0.6$, $\sin\phi = 0.8$ and $Q_M = 5\sin\phi = 5 \times 0.8 = 4$ kvar. The motor is the only source of vars, so $Q_T = Q_M$ and $\tan\phi_T = Q_T/P_T = 4/6$, giving $\phi_T = 33.7°$ and $\cos\phi_T = 0.83$, the overall lagging p.f. If $\cos\phi_T$ is to be raised to 0.95, then $\tan\phi_T = 0.33$. Putting capacitance in parallel with the load does not change P_T, which stays at 6 kW, so Q_T', the reactive power after correction, should be $0.33P_T = 2$ kvar. This is a reduction of 2 kvar from the uncorrected value and these must be supplied by the capacitance:

$$Q_C = I_C^2 X_C = -2 \text{ kvar} \tag{2.84}$$

But $I_C = V_s/X_C = 240/X_C$ and Equation 2.84 becomes

$$V_s^2/X_C = 240^2/X_C = 2 \text{ kvar} \tag{2.85}$$

Solving leads to $X_C = 1/\omega C = 29 \ \Omega$, and as $\omega = 100\pi$, $C = 110 \ \mu F$. Figure 2.23b shows the power triangles.

The apparent power before correction is $S_T = \sqrt{(P_T^2 + Q_T^2)} = \sqrt{(6^2 + 4^2)} = 7.2 \text{ kVA}$. The total current, $I_T = S_T/V_s = 7200/240 = 30$ A. After p.f. correction $S_T' = \sqrt{(6^2 + 2^2)} = 6.3 \text{ kVA}$, and $I_T' = 6300/240 =$

26 A, a modest reduction. Capacitors for power-factor correction are expensive; a 250V (r.m.s.), 8μF capacitor costs £18.40. There is an optimal corrected power factor which must be determined by an economic analysis.

2.4.3 Maximum-power transfer

One exception to the rule that one replaces 'resistance' in the DC theorem with 'impedance' for the corresponding AC theorem is the maximum-power-transfer theorem.

Consider the circuit of Figure 2.24, in which the current will be

$$\mathbf{I_L} = \mathbf{V_T}/(\mathbf{Z_T} + \mathbf{Z_L}) \tag{2.86}$$

where $\mathbf{Z_T} = R_T + jX_T$ and $\mathbf{Z_L} = R_L + jX_L$. Consequently the power in the load will be

$$P_L = I_L^2 R_L = \frac{V_T^2 R_L}{|\mathbf{Z_T} + \mathbf{Z_L}|^2} = \frac{V_T^2 R_L}{(R_T + R_L)^2 + (X_T + X_L)^2} \tag{2.87}$$

Figure 2.24

Maximum power transfer from an AC source to a load impedance occurs when $\mathbf{Z_L} = \mathbf{Z_T}^*$

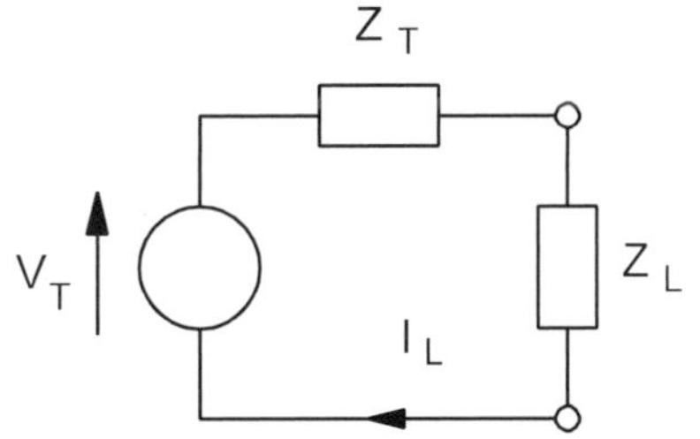

We can maximise the load power, first by varying X_L:

$$\frac{\partial P_L}{\partial X_L} = \frac{-2(X_T + X_L)V_T^2 R_L}{D^2} \tag{2.88}$$

where $D \equiv (R_T + R_L)^2 + (X_T + X_L)^2$. Setting $\partial P_L/\partial X_L = 0$ leads to $X_L = -X_T$. Then differentiating P_L with respect to R_L leads to

$$\frac{\partial P_L}{\partial R_L} = \frac{V_T^2 [(R_T + R_L)^2 + (X_T + X_L)^2 - 2R_L(R_T + R_L)]}{D^2} \tag{2.89}$$

But if $X_T = -X_L$, and $\partial P_L/\partial R_L = 0$, we find that Equation 2.89 reduces to $R_L = R_T$, and the condition for maximum power transfer is that

$$\mathbf{Z_L} = R_T - jX_T = \mathbf{Z_T}^* \tag{2.90}$$

that is the load is the complex conjugate of the source impedance. This condition is called *conjugate matching*. There are ways of arranging maximum power transfer also according to whether R_L or X_L alone can be varied (see Problems 2.19 and 2.20).

2.5 **Resonant circuits**

Capacitors and inductors store energy at different times in the cycle, and release it at different times too; a certain amount of energy is exchanged between the two elements, and it is possible to hit just the right frequency to cause this to build up to a maximum. The result is that the voltage across these storage elements is increased to a value greatly beyond that of the supply – perhaps by several hundred-fold. Because resonant circuits are frequency selective they are used in filters and demodulators, but in some circuits resonance is a nuisance.

2.5.1 The series RLC circuit

Consider the series RLC circuit of Figure 2.25.

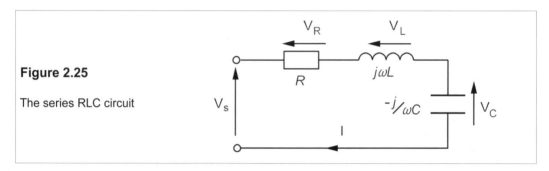

Figure 2.25

The series RLC circuit

The current flowing is

$$\mathbf{I} = \frac{\mathbf{V_s}}{\mathbf{Z}} = \frac{\mathbf{V_s}}{R + j(\omega L - 1/\omega C)} \tag{2.91}$$

I is a maximum (= V_s/R) when $\omega L = 1/\omega C$, or $\omega = 1/\sqrt{(LC)} = \omega_0$, where ω_0 is the *series resonant frequency*. What has happened is that the reactances, being equal and opposite in sign, have cancelled each other and no longer present an impediment to the flow of current. When $\omega = \omega_0$, the voltage across the inductance ($\mathbf{V_L} = -\mathbf{V_C}$) will be given by

$$\mathbf{V_L} = \mathbf{I}X_L = \mathbf{V_s}\omega_0 L/R = Q\mathbf{V_s} \tag{2.92}$$

since $\mathbf{I} = \mathbf{V_s}/R$ and $X_L = \omega_0 L$. Here Q (not to be confused with reactive power) is the *quality factor* or *Q-factor*[2] of the circuit:

$$Q = \omega_0 L/R = 1/\omega_0 CR \tag{2.93}$$

since at resonance $\omega_0 L = 1/\omega_0 C$. In a series RLC circuit at resonance the voltage across each reactance is the supply voltage amplified by Q. Now Q can easily be as high as 100, and therefore dangerous voltages can be generated from low-voltage sources.

---

[2] The Q-factor for *any* circuit is $2\pi \times$ maximum energy stored ÷ energy dissipated per cycle. See Problem 2.23

The half-power frequencies

The power consumed at resonance is just $I_{max}^2 R$. The half-power frequencies are defined as the frequencies at which the power consumption is half that at resonance, that is $0.5 I_{max}^2 R = V_s^2/2R$. Now **I** is given by Equation 2.91, therefore

$$I^2 = \frac{V_s^2}{R^2 + (\omega L - 1/\omega C)^2} \tag{2.94}$$

Setting this equal to $0.5 I_{max}^2$ or $V_s^2/2R^2$,

$$\frac{V_s^2}{R^2 + (\omega L - 1/\omega C)^2} = \frac{V_s^2}{2R^2} \tag{2.95}$$

This means that

$$(\omega L - 1/\omega C)^2 = R^2$$
$$\Rightarrow \quad \omega L - 1/\omega C = \pm R \tag{2.96}$$

yielding the quadratic equations

$$\omega^2 \mp \frac{R}{L}\omega - \frac{1}{LC} = 0 \tag{2.97}$$

The solutions to these that are positive in ω are

$$\omega_{1,2} = \mp R/2L + \sqrt{R^2 4L^2 + 1/LC} \tag{2.98}$$

where ω_1 and ω_2 are the lower and upper half-power angular frequencies respectively. Now $R^2/4L^2 = (\omega_0/2Q)^2$, and $1/LC = \omega_0^2$, and then Equation 2.98 becomes

$$\omega_{1,2} = \mp \omega_0/2Q + \sqrt{(\omega_0/2Q)^2 + \omega_0^2} \tag{2.99}$$

Thus

$$\Delta\omega = \omega_2 - \omega_1 = \omega_0/Q \tag{2.100}$$

where $\Delta\omega$ is the *bandwidth* of the circuit. Equation 2.100 can be used to define Q:

$$Q = \omega_0/\Delta\omega = \omega_0/(\omega_1 - \omega_2) \tag{2.101}$$

When $Q \geq 5$ Equation 2.99 reduces to

$$\omega_{1,2} \approx \omega_0 \mp \omega_0/2Q = \omega_0 \mp \Delta\omega/2 \tag{2.102}$$

The half-power frequencies (sometimes called the 3dB or -3dB points)[3] are equally spaced about the resonant frequency and half the bandwidth from it.

---

[3] dB stands for decibel, defined as $10\log_{10}(P/P_0)$, where P_0 is a reference power level (which must be specified or is understood). Because $P = V^2/R$ in a resistance, it is convenient to assume that voltages are impressed on the same resistance. Then voltages in dB are given by $10\log_{10}(V/V_0)^2 = 20\log_{10}(V/V_0)$ where V_0 is the reference voltage. Thus a voltage of +6 dB re 1 V (or +6 dBV) is 2 V. In the same way currents can be expressed in dB: $I_{dB} = 20\log_{10}(I/I_0)$. Voltage gains are often given in dB – a voltage gain of 1000 is 60 dB.

Example 2.8

If, in Figure 2.25, V_s = 240 V, R = 100 Ω, L = 100 mH and C = 100 nF, what are f_0, Q, Δf, f_1 and f_2? What are the maximum voltages across R, L and C? Plot a graph of I against f.
   First work out the resonant frequency:

$$\omega_0 = 1/\sqrt{LC} = (100 \times 10^{-3} \times 100 \times 10^{-9})^{-1/2} = 10^4 \text{ rad/s} \qquad (2.103)$$

or $f_0 = 10^4/2\pi$ = **1.592 kHz**. Then find $Q = \omega_0 L/R = 10^4 \times 100 \times 10^{-3}/100$ = **10**, which is substantially greater than 5, so Equation 2.100 may be used to find the half power frequencies. The bandwidth, $\Delta\omega$ = $\omega_0/Q = 10^4/10$ = 1 krad/s (or Δf = 1000/2π = **159 Hz**), giving $\omega_1 = \omega_0 - 0.5\Delta\omega = 10^4 - 500$ = 9.5 krad/s and ω_2 = 10.5 krad/s, or f_1 = 9.5/2π = **1.512 kHz** and f_2 = **1.671 kHz**. (The exact half-power frequencies are 1.514 and 1.673 kHz.)
   The maximum voltage across both the capacitance and the inductance is very nearly QV_s (see question 2.12 – for this work out $|V_C|^2$ and set $d|V_C|^2/d\omega = 0$), which is **2.4 kV**. In phasor terms these will be 2.4$\angle$–90° kV and 2.4$\angle$+90° kV for the capacitance and the inductance respectively. The current is given by Equation 2.94 and the graph of I against frequency is shown in Figure 2.26. This will also be the graph of the voltage across R against ω, when scaled up by a factor of a hundred. For comparison the curves for R = 200 Ω and R = 400 Ω (Q = 5 and Q = 2.5 respectively) are shown also.

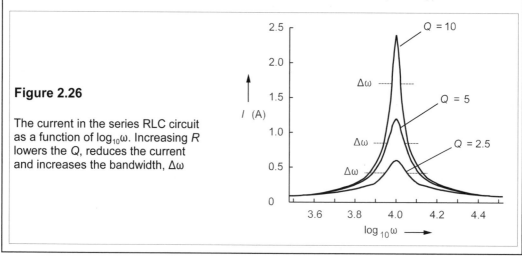

Figure 2.26

The current in the series RLC circuit as a function of $\log_{10}\omega$. Increasing R lowers the Q, reduces the current and increases the bandwidth, $\Delta\omega$

2.5.2 The parallel resonant circuit

The ideal parallel resonant circuit is shown in Figure 2.27a.
   The admittance of the circuit is

$$\mathbf{Y} = \mathbf{Y_R} + \mathbf{Y_C} + \mathbf{Y_L} = 1/R_p + j\omega C + 1/j\omega L \qquad (2.104)$$

If the circuit were excited by an ideal current source which supplies constant current at all frequencies, the voltage across it would be

$$\mathbf{V} = \frac{\mathbf{I}}{\mathbf{Y}} = \frac{\mathbf{I}}{1/R_p + j\omega C + 1/j\omega L} \qquad (2.105)$$

The magnitude of this voltage is

$$|\mathbf{V}| = V = \frac{I}{\sqrt{1/R_p^2 + (\omega C - 1/\omega L)^2}} \qquad (2.106)$$

which will be a maximum at the resonant frequency, $\omega_0 = 1/\sqrt{(LC)}$, when $V_{max} = IR_p$. The imaginary part of $\mathbf{Y}$ vanishes at resonance, when $\mathbf{I}_L = -\mathbf{I}_C$ and all the source current flows through R_p. The half-power frequencies occur when $V = V_{max}/\sqrt{2}$, for then the power consumed is $V^2/R_p = V_{max}^2/2R_p$.

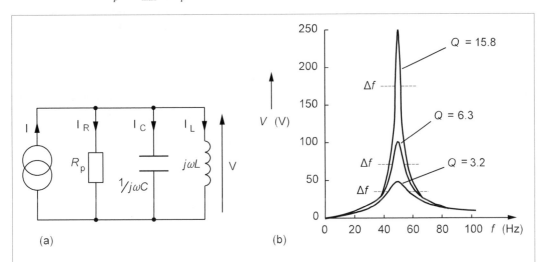

Figure 2.27 (a) The ideal parallel resonant circuit (b) V for various values of Q_p and constant source current

From Equation 2.106 we find

$$V = \frac{I}{\sqrt{1/R_p^2 + (\omega_{1,2}C - 1/\omega_{1,2}L)^2}} = \frac{V_{max}}{\sqrt{2}} = \frac{I}{\sqrt{2/R_p^2}} \qquad (2.107)$$

and therefore

$$\omega_{1,2}C - 1/\omega_{1,2}L = \mp 1/R_p \qquad (2.108)$$

leading to the quadratic equations

$$\omega_{1,2}^2 \pm \frac{1}{R_pC}\omega_{1,2} - \frac{1}{LC} \equiv \omega_{1,2}^2 \pm \beta\omega_{1,2} - \omega_0^2 = 0 \qquad (2.109)$$

(β stands for $1/R_pC$) whose real solutions are $\omega_{1,2} = \mp\beta/2 + \sqrt{(\beta^2/4 + \omega_0^2)}$. Hence the half-power bandwidth, $\omega_2 - \omega_1 = \Delta\omega = \beta = 1/R_pC$. The circuit's Q-factor is

$$Q = \frac{\omega_0}{\omega_2 - \omega_1} = \omega_0 R_p C = \frac{R_p}{\omega_0 L} \qquad (2.110)$$

For example suppose $R_p = 5$ kΩ, $L = 1$ H and $C = 10$ μF in Figure 2.27a. The resonant frequency will be $\omega_0 = 1/\sqrt{(1 \times 10 \times 10^{-6})} = 316$ rad/s, or $f_0 = \omega_0/2\pi = 50.3$ Hz. $Q_p = \omega_0 R_p C$

$= 316 \times 5 \times 1000 \times 10 \times 10^{-6} = 15.8$. The bandwidth is $\omega_0/Q_p = 316/15.8 = 20$ rad/s or 3.18 Hz. The upper half-power frequency, $\omega_2 = 10 + \sqrt{(10^2 + 316^2)} = 326$ rad/s or $f_2 = 51.9$ Hz. Then $\omega_1 = 326 - 20 = 306$ rad/s ($f_1 = 48.7$ Hz). If the supply current, $\mathbf{I} = 50\angle 0°$ mA, at resonance $\mathbf{I_R} = 50\angle 0°$ mA, and $\mathbf{V} = \mathbf{I_R}R_p = 250\angle 0°$ V. The capacitor's reactance at resonance is $1/\omega_0 C = 1/(316 \times 10 \times 10^{-6}) = 316\ \Omega$, so the current through it will be $250\angle 0° \div 316\angle -90°$ $= 0.79\angle 90°$ A. The current through the inductor at resonance will be equal and opposite (180° out of phase) to that in the capacitor: $0.79\angle -90°$ A. At resonance $|\mathbf{I_C}| = |\mathbf{I_L}| = Q|\mathbf{I}|$ – in the parallel circuit the circulating current is Q times the supply current. The *tank circuit* (an alternative name for the parallel RLC circuit) fills up with current and stores energy in capacitor and inductor though it is continually exchanged between them. Figure 2.27b shows the voltage across the circuit as a function of frequency for several values of Q, corresponding to $R_p = 1, 2$ and 5 kΩ.

Because there is always some resistance in an inductor, a purely parallel RLC circuit is not a practical possibility; instead the circuit of Figure 2.28a is used. The inductor, which is in parallel with a capacitance, C, is modelled as an inductance, L, in series with a resistance, R.

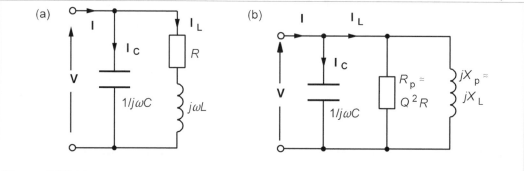

Figure 2.28 (a) Parallel resonant circuit with a practical coil (b) With coil converted to parallel form

The admittance of the combination is

$$\mathbf{Y} = \mathbf{Y_C} + \mathbf{Y_L} = j\omega C + \frac{1}{R + j\omega L} \tag{2.111}$$

Rationalising and grouping the imaginary terms together produces

$$\mathbf{Y} = \frac{R}{R^2 + \omega^2 L^2} + j\omega\left(C - \frac{L}{R^2 + \omega^2 L^2}\right) \tag{2.112}$$

The parallel resonant frequency is defined as the frequency at which $\mathbf{Y}$ is wholly real, that is when

$$C = \frac{L}{R^2 + \omega^2 L^2} \tag{2.113}$$

which leads to

$$\omega_p = \sqrt{1/LC - R^2/L^2} = \sqrt{\omega_0^2 - \omega_0^2/Q^2} \approx \omega_0\left(1 - 1/2Q^2\right) \tag{2.114}$$

where ω_p is the parallel resonant frequency, $\omega_0 = 1/\sqrt{(LC)}$ and $Q = \omega_0 L/R$, the Q-factor of the coil at its series resonant frequency. If $Q > 10$, as is almost invariably the case, then the approximation $\omega_p \approx \omega_0$ is very close.

The current drawn by the circuit is

$$\mathbf{I} = \mathbf{I_C} + \mathbf{I_L} = j\omega CV + \frac{V}{R + j\omega L} \tag{2.115}$$

If the coil had no resistance, $\mathbf{I}$ would be zero because $\mathbf{I_C}$ and $\mathbf{I_L}$ would be equal and opposite. It takes tedious, but straightforward, analysis to show that the current has a minimum at

$$\omega_m = \omega_0 \sqrt{\sqrt{1 + 2/Q^2} - 1/Q^2} \approx \omega_0(1 - 1/4Q^4) \tag{2.116}$$

For $Q \geq 3$, one can assume ω_m to be identical to ω_0 for practical purposes.

Another way of looking at the practical parallel resonant circuit is to transform the inductor into its parallel components:

$$\frac{1}{R + j\omega L} = \frac{R}{R^2 + \omega^2 L^2} - \frac{j\omega L}{R^2 + \omega^2 L^2} = G_p - jB_p \tag{2.117}$$

The resistive component, $R_p = 1/G_p = R + \omega^2 L^2/R = R(1 + Q^2)$, while the reactive component, $X_p = 1/B_p = \omega L + R^2/\omega L = X_L(1 + 1/Q^2)$, where $Q \equiv \omega L/R$ is the coil Q-factor. Then if $Q \geq 10$, $R_p \approx Q^2 R$ and $X_p \approx X_L$, to give the equivalent circuit of Figure 2.28b. At first glance it looks the same as the ideal circuit in Figure 2.27a, but it differs fundamentally because R_p is now *frequency dependent*.

Suggestions for further reading – see Chapter 5

Problems

1. Show that the form factor for the triangular waveform of Figure P2.1a is $2\sqrt{3}/3$. What is the crest factor? What are the FF and CF for the half-wave-rectified sinewave of Figure P2.1b? (Note that this is not a continuous waveform.) $[\sqrt{3}; \pi/2, 2]$
2. The circuit of Figure P2.2 is an all-pass network. Show that

$$\frac{\mathbf{V_o}}{\mathbf{V_{in}}} = \frac{1 - j\omega RC}{1 + j\omega RC}$$

What is the magnitude of $\mathbf{V_o}/\mathbf{V_{in}}$? At what frequency is the phase shift $-90°$? And when is it $-120°$? $[1, f = 1/2\pi RC, f = \sqrt{3}/2\pi RC]$

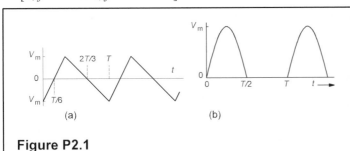

Figure P2.1

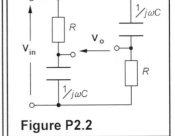

Figure P2.2

3. Find all the currents and voltages in Figure P2.3 and draw the phasor diagram.
   [$V_R = 11.985\angle32.86°$ V, $V_C = 12.045\angle27.15°$ V, $I = 11.985\angle32.86°$ mA, $I_C = 120.45\angle117.15°$ mA,
   $I_L = 119.85\angle-57.14°$ mA]
4. The circuit of Figure P2.4 is known as Owen's bridge. Show that the conditions for balancing are
   $R_1 = RC/C_2$ and $L_1 = RR_2C$, the frequency of the source being immaterial.

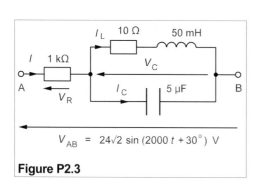

Figure P2.3

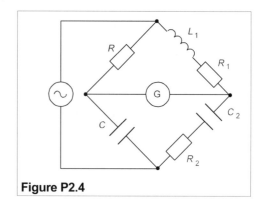

Figure P2.4

5. The circuit of Figure P2.5 is Wien's bridge. Show that it balances only when $\omega = 1/\sqrt{(R_1R_2C_1C_2)}$
   and $C_2/C_1 = R_4/R_3 - R_1/R_2$.
6. Find the Thévenin equivalent of the circuit of Figure P2.6 by source transformation and hence find
   the maximum power that can be transferred from the circuit to an external load. What is the reactive
   power in the load in this case?
   [$V_T = 5.87\angle52.8°$ V, $Z_T = 1.24\angle29.74°$ Ω, $P_L = 8.0$ W, $Q_L = 4.57$ var (lagging)]
7. Find, by using the superposition principle, the current flowing between A and B in Figure P2.6
   when the terminals are short circuited. *[$I = 4.73\angle82.5°$ A]*

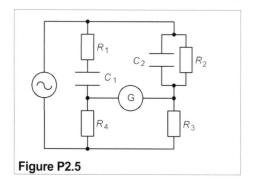

Figure P2.5

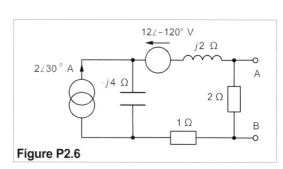

Figure P2.6

8. Transformerless matching can be achieved with an LC network as in Figure P2.8, which is the
   equivalent circuit of a dipole antenna feeding a 75Ω load. If the frequency is 1.5 MHz, what must
   L and *C* be for maximum power transfer from the antenna?
   [41.9 μH and 279 pF]
9. The circuit of Figure P2.9 represents an oscilloscope input and a ×10 probe. If the attenuation is to
   be 20 dB at all frequencies, including DC, what must *R* and *C* be?
   [R = 9 MΩ, C = 14.4 pF]
10. A synchronous motor is used to correct the p.f. of load comprising a 100 kVA motor operating at
    a lagging p.f. of 0.6 in parallel with a heater taking 25 kW. Given the synchronous motor operates
    at a leading p.f. of 0.3 and the overall p.f is 0.95 (lagging), construct the power diagram and deduce

from it the power consumed by the synchronous motor and the reactive power it supplies. If the line current was 94 A before correction, what is it after? *[P_{SM} = 14.8 kW, Q_{SM} = 47.2 kvar, 84.6 A]*

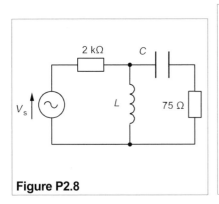

Figure P2.8

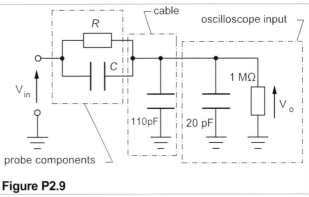

Figure P2.9

11. Mesh-connected capacitors are used to correct the p.f. of a star-connected load each impedance of which comprises an 18 mH inductance in series with a 4Ω resistance. If the p.f. is to be 0.9, what capacitance is required at 50 Hz? When the p.f. is corrected to unity, what capacitance is required? If the current taken from the supply was 100 A before correction, what is it in each case after? Comment. *[82.2 µF, 125 µF, 64.2 A, 57.7 A]*

12. Design a series RLC circuit which resonates at 300 kHz and has a bandwidth of 10 kHz, and which is to dissipate 100 W at resonance when driven by a 200V source. What is the circuit's Q? What is the magnitude of the capacitor voltage at resonance? What power is dissipated at 295 kHz? What is the current's magnitude at 305 kHz?
    [R = 400 Ω, L = 6.37 mH, C = 44.2 pF, Q = 30, V_C = 6 kV, 50 W, 0.354 A]

13. What capacitance is required for the parallel circuit of Figure 2.26a to resonate at 2 kHz, if the inductor has ainductance of 1 mH and resistance of 3 Ω. What, approximately, are the circuit's bandwidth and Q? Find the 'exact' half-power points. What currents will flow through the capacitance and the inductance at resonance if a voltage of $40\angle0°$ V is applied to the circuit? What current is then drawn from the source? *[C = 5.99 µF, Δf = 480 Hz, Q = 4.2, f_1 = 1817 Hz, f_2 = 2320 Hz, I_C = 3.01$\angle$90° A, I_L = 3.1$\angle$−76.6° A, I_s = 0.72$\angle$0° A]*

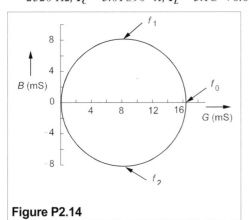

Figure P2.14

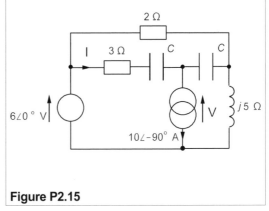

Figure P2.15

14. A high-Q, series RLC circuit has an admittance whose locus, a circle, is shown in Figure P2.14 as the frequency is varied from about 30 kHz to 60 kHz. Draw the circuit and write down an expression for its admittance and hence its conductance and susceptance. From the admittance locus determine the value of R. The points of maximum and minimum susceptance are at f_1 = 40 kHz and

$f_2 = 43$ kHz. What are these frequencies known as? What is the Q of the circuit? What are the values of the other components of the circuit? *[R = 62 Ω, Q = 13.8, L = 3.29 mH, C = 4.5 nF]*

15. Find the current, **I**, and the voltage, **V**, in the circuit of fig. P2.15 by mesh analysis, given $C = 8\ \mu F$ and the excitation frequency is 4.974 kHz. *[**I** = 4.32∠92° A, **V** = 26.7∠27.5° V]*

16. Repeat Problem 2.15 using nodal analysis.

17. Repeat Problem 2.16 using superposition (probably the easiest way of the three).

18. Calculate the r.m.s. value of the waveforms shown in Figure P2.18. Waveform (a) is equivalent to a square wave going from −3 V to +3 V plus a DC voltage of 1 V, so why is waveform (c) not equivalent to (d) plus a DC current of 2A? *[(a) √10 V (b) $V_m/2$ (c) 4√2 A (d) 6 A]*

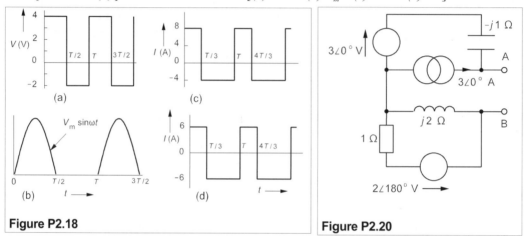

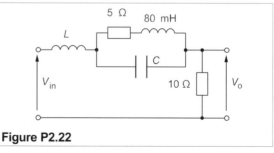

Figure P2.18   **Figure P2.20**

19. Show that the condition for maximum power transfer from a source whose internal impedance is **Z** $(= R + jX)$ to a load in which the resistance may be varied, but of constant reactance, is $R_L = \sqrt{[R^2 + (X + X_L)^2]}$.

20. Show that if, in Problem 2.19, the reactance, X_L, is variable but R_L is fixed, the condition for maximum power transfer is $X_L = -X$. Then find the maximum power that can be developed in a 2Ω resistive load across AB in the circuit of Figure P2.20, by adding reactance in series with it. *[6.64 W]*

21. The circuit of Figure P2.21 is a twin-T notch filter. Show that V_o is zero at an angular frequency, $\omega = 1/RC$. (Hint: use the star-delta transform.)

22. The circuit of Figure P2.22 is used to pass a 200Hz signal while blocking 50Hz pick-up. What must C be to block signals at 50 Hz? What must L be to pass signals at 200 Hz? What is the relative attenuation of the 50Hz signal compared to the 200Hz signal? *[C = 127 μF, L = 5.3 mH, −22.7 dB]*

23. Show that Equation 2.93 agrees with the definition of the Q-factor given in footnote 2.

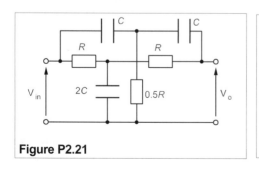

Figure P2.21   **Figure P2.22**

3 Operational amplifiers

THE OPERATIONAL amplifier is one of the most versatile devices to be found in analogue electronics as may be seen from a casual perusal of any electronics cookbook. When cheap integrated circuits began to appear, the price of op amps (the customary abbreviation for the device) plummeted coincidentally with the increase in performance, so that a high-stability, chopper-stabilised op amp costing £100+ in 1979 may be obtained for less than £2 in 2002. The ideal op amp is easy to understand for it obeys only two golden rules.

3.1 The golden rules

Figure 3.1a shows the circuit symbol for an op amp and Figure 3.1b its equivalent circuit.

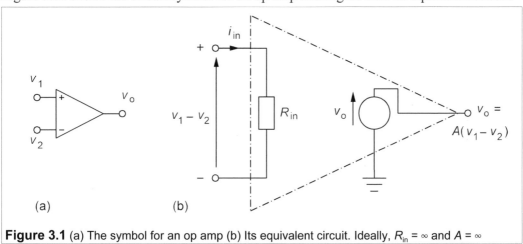

Figure 3.1 (a) The symbol for an op amp (b) Its equivalent circuit. Ideally, $R_{in} = \infty$ and $A = \infty$

For our purposes the ideal op amp has three terminals: an inverting ($-$) and a non-inverting ($+$) input terminal, and an output terminal. The input impedance of the ideal op amp is infinite (in practical op amps it is normally 100 MΩ or more), while the output voltage is a function of the difference between the voltages at the input terminals:

$$v_o = A(v_1 - v_2) \tag{3.1}$$

A is the *open-loop gain* of the amplifier, which is infinite for an ideal op amp. In consequence Equation 3.1 indicates that if v_o is finite, $v_1 - v_2 = 0$. Because of this most op-amp circuits employ *feedback*, that is the output terminal is connected, usually via a resistance or capacitance or combination of the two, to one of the input terminals. The ideal op amp's behaviour with feedback is encapsulated in the two Golden Rules:

     1. The current drawn by the op amp at the input terminals is zero.
     2. The output voltage is whatever makes the input terminal voltages equal.

The practical op amp requires two more terminals for power supply (usually bipolar, ± 15 V) and a further two for offset null compensation, which is explained later. Because the output voltage is limited to that supplied to the op amp, golden rule number 2 is broken when the amplifier is driven into *saturation*, that is when $v_1 \neq v_2$.

Figure 3.2 shows the dual-in-line (DIL) package of the popular 741 op amp together with its pin numbers and their identities. On the right is the method of offset-null compensation, which is explained later.

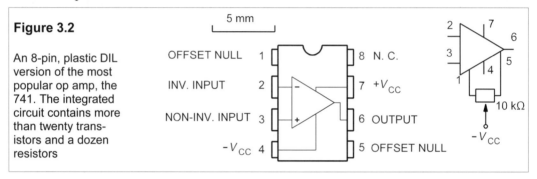

Figure 3.2

An 8-pin, plastic DIL version of the most popular op amp, the 741. The integrated circuit contains more than twenty transistors and a dozen resistors

3.2 Some common op amp circuits

The circuits in this section all treat the op amp as ideal and all employ feedback.

3.2.1 The voltage follower, or buffer

This simple – and widely used – circuit is shown in Figure 3.3.

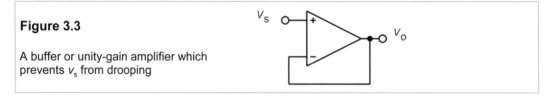

Figure 3.3

A buffer or unity-gain amplifier which prevents v_s from drooping

The output is connected directly to the inverting input, while the signal is connected to the non-inverting input. By the second golden rule $v_o = v_2 = v_1$ and by the first golden rule, the current drawn by the op amp is nil. High-impedance voltage sources must be buffered like this before being connected to other parts of a circuit with comparable or lower impedances, such as an amplifier or a meter.

3.2.2 *The inverting amplifier*

In this circuit (Figure 3.4) the output is connected via a resistance, R_f, to the inverting input, which is also connected to the signal via resistance, R_s. The non-inverting input is connected to ground (in practice this would grounded via a resistance equal to $R_f // R_s$, but we can ignore that as our op amp is ideal), so that $v_1 = 0$. Then by the second golden rule $v_A = 0$, that is point A is a *virtual ground*.

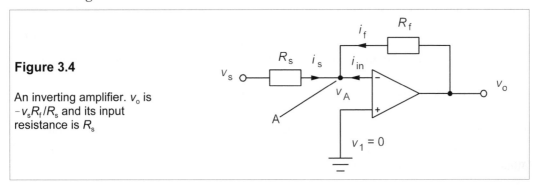

Figure 3.4

An inverting amplifier. v_o is $-v_s R_f / R_s$ and its input resistance is R_s

Using KCL at node A,

$$i_s + i_f + i_{in} = 0 \tag{3.2}$$

But $i_{in} = 0$ by golden rule 1 and Equation 3.2 implies $i_f = -i_s$. Now $i_s = (v_s - v_A)/R_s = v_s/R_s$ and $i_f = (v_o - v_A)R_f = v_o/R_f$, as $v_A = 0$, giving

$$v_o/R_f = -v_s/R_s \quad \Rightarrow \quad v_o = -v_s R_f/R_s \tag{3.3}$$

The output voltage is opposite in sign to the input voltage and scaled by R_f/R_s. For instance, to achieve a gain of 50 we could make $R_f = 100$ kΩ and $R_s = 2$ kΩ. Note that in this case the current drawn from the source is v_s/R_s, so the input resistance is only 2 kΩ – rather low. Remember that R_s is the sum of the source resistance *plus* any resistance connected to the inverting input. It is a common mistake to overlook the source resistance!

3.2.3 *The summing amplifier*

Once widely used in analogue computers, this circuit (Figure 3.5) is not used very much now.

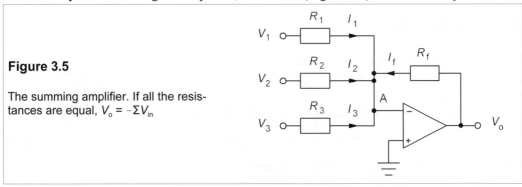

Figure 3.5

The summing amplifier. If all the resistances are equal, $V_o = -\Sigma V_{in}$

As always we proceed by using KCL at A, here called the *summing point*, and a virtual ground by golden rule 2. Ignoring I_{in} (golden rule 1), KCL gives

$$I_1 + I_2 + I_3 + I_f = 0 \tag{3.4}$$

But as A is at 0 V, $I_1 = V_1/R_1$ etc. and we obtain from Equation 3.4

$$V_1/R_1 + V_2/R_2 + V_3/R_3 + V_0/R_f = 0 \tag{3.5}$$

Making all the resistances equal yields

$$V_0 = -(V_1 + V_2 + V_3) = -\Sigma V_{in} \tag{3.6}$$

The summing point is the inverting input so the output is minus the sum of the inputs.

3.2.4 The non-inverting amplifier and constant-current source

In the circuit of Figure 3.6 the source voltage is connected to the non-inverting input and feedback is to the inverting input. The input impedance is greater than that of the amplifier alone because of feedback – effectively infinite.

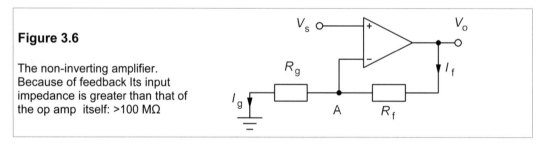

Figure 3.6

The non-inverting amplifier. Because of feedback Its input impedance is greater than that of the op amp itself: >100 MΩ

Point A must be at a potential V_s above ground, by golden rule 2. By KCL at A, $I_f = I_g$ and these currents are found by Ohm's law yielding

$$I_f = I_g = (V_0 - V_s)/R_f = V_s/R_g \tag{3.7}$$

Rearranging to find V_0:

$$V_0 = V_s(R_f/R_g + 1) \tag{3.8}$$

The current through R_f is V_s/R_g and is therefore independent of R_f; the circuit amounts to a *constant current source*, provided V_s is constant, with R_f the variable load.

3.2.5 The integrator

Figure 3.7 shows the circuit diagram for an integrator. Its output is the time integral of the input voltage, scaled by $-1/RC$. Analogue computers rely heavily on this device.

Applying KCL at A gives $i_s + i_f = 0$, and $i_s = v_R/R = v_s/R$ as A is a virtual ground. Now i_f is the current through the capacitor, $C dv_C/dt = C dv_0/dt$. Hence

$$i_s + i_f = v_s/R + C dv_0/dt = 0 \tag{3.9}$$

Rearranging and integrating Equation 3.9 leads to

$$v_o = \frac{-1}{RC} \int v_s \, dt \qquad (3.10)$$

The output is the integral of the input, inverted and scaled by a factor of $1/RC$.

Figure 3.7

The integrator. High-quality capacitors are needed, for example polystyrene, to reduce leakage and improve accuracy. Connecting the non-inverting input to ground via a resistance, R, reduces the effect of input bias current

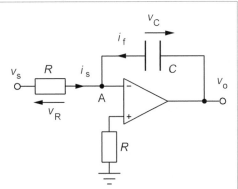

3.3 Analogue computing

Differential equations can be solved by analogue computation using little more than summers and integrators. Voltage is the analogue of the dependent variable and time the independent variable. For example the second order differential equation

$$y'' + ay' + by = c \qquad (3.11)$$

(where $y'' \equiv d^2y/dt^2$, $y' \equiv dy/dt$ and a, b and c are constants) can be solved by rewriting it in the form

$$y'' = c - ay' - by \qquad (3.12)$$

The three terms on the right – were they available – could be fed into a summer to form $y''(t)$. But then the output of the summer could be fed into an integrator to give $y'(t)$ and this could be integrated again to give $y(t)$, the desired result. Having formed $y(t)$ and $y'(t)$ in this way, all that is required is to multiply by b and a respectively before feeding into the initial summer. Since the outputs are merely voltages, they are readily scaled by using potentiometers ('coefficient pots') for values between 0 and 1, and then multiplied by ten as often as required with feedback amplifiers for values greater than unity. Figure 3.8 shows the circuit diagram derived for solving Equation 3.11.

Figure 3.8

The initial circuit derived to solve Equation 3.12. All the amplifiers are inverters with negative gains

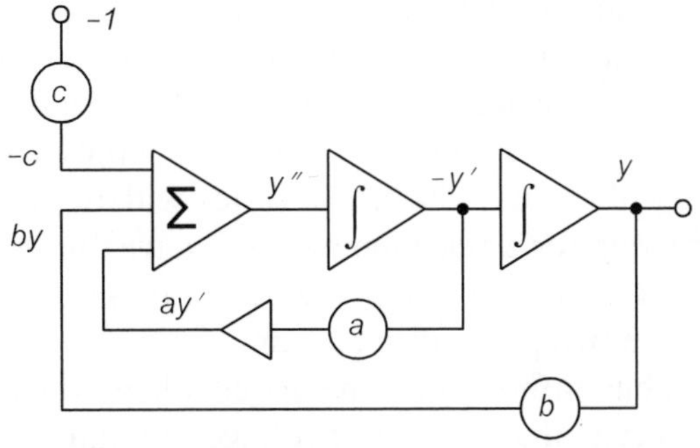

All the amplifiers in Figure 3.8 are inverting amplifiers; so for the output of the summer on the left to be y'', the inputs must be $-c + by + ay'$. Then the next amplifier integrates y'' and inverts it to $-y'$; thus after it is multiplied by a an inverter (symbol, an empty triangle) is needed to give $+ay'$. The constant term, c, is fed in as $-c$ via a coefficient pot connected to -1, which might be -10 V on a typical machine.

But the circuit diagram of Figure 3.8 is needlessly complicated, as any integrator acts as a summer and the leftmost amplifier is redundant. Figure 3.9 shows a very simple circuit that uses only two integrators – all that are necessary to solve a second order differential equation.

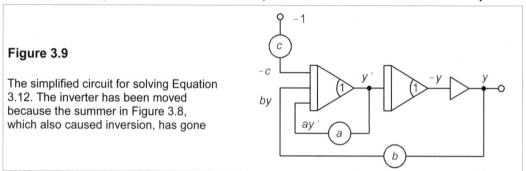

Figure 3.9

The simplified circuit for solving Equation 3.12. The inverter has been moved because the summer in Figure 3.8, which also caused inversion, has gone

Because the redundant summer also inverted, the position of the inverter has been changed from after the a coefficient pot in Figure 3.8 to after the second integrator in Figure 3.9. Initial conditions, for example $y'(0) = 3$, are easily set up by feeding the appropriate voltage into the integrator which outputs y' via an IC (initial condition) pot. Integrators are given the symbol shown in Figure 3.9 – a triangle with a rectangular box on the input side. Integrators can be given gains of 1, 10 or 100 by adjusting the feedback capacitance. This is symbolised by putting a figure into its right hand (output) corner, or *nose*: hence the term *nose gain*. In our example both nose gains are unity.

3.4 Practical op amps

3.4.1 Input offset voltage

In practice op amps are not ideal. They consist of two separate amplifier circuits – one connected to each input – which are intended to be identical, but are not, because of variations in component performance during manufacture (remember, they are ICs and the manufacturing characteristics vary a little for each component). As a result the gains of the two input channels differ slightly and there is an *input offset voltage*, V_{OS}, at the input. If the inputs are connected and grounded the output should be zero for an ideal op amp, but in practice is always either the positive supply voltage or the negative supply voltage: it is a matter of chance which. V_{OS} varies widely; from 1 μV to 3 mV in Table 3.1, which gives some data for a selection of readily-available op amps. The manufacturer usually gives a method for reducing V_{OS} to zero ('trimming'), for example by connecting a potentiometer between two pins and connecting the wiper to the negative supply as for the 741. So far so good, but unfortunately the offset voltage is prone to drift with temperature and with time, which can vitiate the offset voltage trimming.

Table 3.1 *A selection of operational amplifiers*

Device	Pack[A]	Pins	No.[B]	GB[C]	SR[D]	V_{OS}[E]	I_B[F]	V_S[G]	Price[H]	Application
TL071CN	DIP	8	1	3	8	10	0.2	±3½ - ±18	42p	general purpose, low cost
UA741CD	SO	8	1	1	0.5	5	100	±22	28p	general purpose, lowest cost
NE5534AN	DIP	8	1	10	6	5	2000	±3 - ±20	63p	general purpose
MAX473CPA	DIP	8	1	12	9	4	225	7	£2.52	general purpose, unipolar supply
AD820AR	SO	8	1	2	3.5	0.3	0.005	±15	£3.04	general purpose, low offset, low bias current
NJM2136M	SO	8	1	200	45	5	2000	±1.35 - ±6	£1.81	high frequency, low supply voltage, fast
AD648KN	DIP	8	2	1	1.8	0.5	0.01	±20	£7.11	general purpose, low offset and bias
LM358N	DIP	8	2	1	0.5	7	250	3 - ±18	29p	general purpose, low cost
OP249GP	DIP	8	2	4.7	22	2	0.04	±4½ - ±18	£2.90	general purpose, fast
TLV2463CD	SO	14	2	6.4	1.6	2.2	30	2.7 - 6	£1.87	general purpose, unipolar supply
MAX4126ES	SO	8	2	5	2	0.75	150	7.5	£2.60	general purpose, unipolar supply
AD713JN	DIP	14	4	4	20	1.5	0.15	±4½ - ±18	£6.69	general purpose, fast
KA324	DIP	14	4	1	0.5	7	250	±1½ - ±15	43p	general purpose, low voltage, cheap
TSH24IN	DIP	14	4	25	15	3.5	0.65	±5 - ±15	£1.41	general purpose, fast
LM324D	SO	14	4	1	0.5	7	250	±1½ - ±15	58p	general purpose, low voltage, cheap
OPA4132UA	SO	14	4	8	20	2	0.05	±2½ - +18	£5.97	general purpose, fast, low bias current
EL2645CN	DIP	8	1	100	275	7	–	±2 - ±5½	£2.99	fast settling (80 ns to 0.1%)
OPA4650P	DIP	14	4	160	240	5.5	–	±4½ - ±5½	£3.69	fast settling (20 ns to 0.01%)
LM7171BIM	SO	8	1	200	3100	7	–	5½ - 36	£2.14	fast settling (42 ns to 0.1%), unipolar supply
THS4062CD	SO	8	2	180	400	8	–	9 - ±16	£3.58	fast settling (40 ns to 0.1%)
AD705JN	DIP	8	1	low	0.15	0.09	–	±2 - ±18	£1.91	precision, low noise (15 nV/√Hz), low drift (1.2 μV/K)
TLC2652CN	DIP	14	1	low	2.8	0.003	–	±1.9 - ±8	£3.69	precision, low noise (23 nV/√Hz), low drift (30 nV/K)
OPA177GS	SO	8	1	low	0.3	0.06	–	±3 - ±22	£1.30	precision, low noise (10 nV/√Hz), low drift (1.2 μV/K)
OP90GP	DIP	8	1	0.02	0.012	0.45	–	±0.8 - ±18	£3.01	low voltage, low power (80 μA max supply current)
ICL7641ECPA	DIP	14	4	1.4	1.6	20	–	±2 - ±18	£2.14	low power (250 μA max current), unipolar supply
LPV324M	SO	14	4	0.15	0.1	10	–	2.7 - 5	70p	low power (46 μA max current), unipolar supply
AD620AN	DIP	8	1	1	–	0.125	–	±2.3 - ±18	£6.14	instrumentation, 73 dB CMRR, gain drift 50ppm./K
INA103KP	DIP	16	1	6	–	0.012	–	±9 - ±25	£8.79	instrumentation, 86 dB CMRR, gain drift 25ppm/K

Notes: [A] DIP = Dual in-line plastic, SO = Small outline [B] Devices/package [C] Gain-bandwidth product in MHz [D] Slew rate in V/μs [E] Input offset voltage in μV [F] Input bias current in nA [G] Supply voltage range, or preferred value if single figure [H] In single quantities

This drift is about 2% of V_{OS} per °C, so if the trimming is done at 15°C and the op amp works at 40°C, there is a drift of ½V_{OS}. With the op amp is configured as a ×100 amplifier, the output will be 50 V_{OS}, with 0 V input! The drift with time is less important except in precision circuits.

Another factor worthy considering is the output current that can be supplied by the op amp – it is no use trying to supply a load with 50 mA when the op amp can only deliver 5 mA! Though 'power' op amps are available be careful of them: they get hot, which degrades performance. It is probably best to build a separate power stage after a low-power op amp.

3.4.2 The input bias current

The *input bias current*, I_B, quoted in Table 3.1 is the current flowing into, or out of each input terminal, depending on which terminal it is. In the circuit of Figure 3.10, the bias current sees a resistance of R_2 // $R_3 = R_{eq}$ at the inverting input and this results in $V_2 = I_B R_{eq}$, if $V_s = 0$. At the non-inverting input the bias current causes V_1 to be $I_B R_1$. If $R_1 = R_{eq}$ then $V_1 = V_2$ and the effect of the input bias current will be small. If $R_1 = 0$, the output voltage will be $I_B R_{eq} × R_3$ /R_2, even if V_{os} and V_s are zero. One is therefore advised to make R_2 and R_3 as small as possible (but remember the input resistance is R_2; it may be necessary to insert a buffer). The bias current will drift with temperature causing output voltage errors in integrators even after careful adjustment for zero drift. (The output-voltage drift rate for an integrator is $C^{-1}\int I_B dt$.)

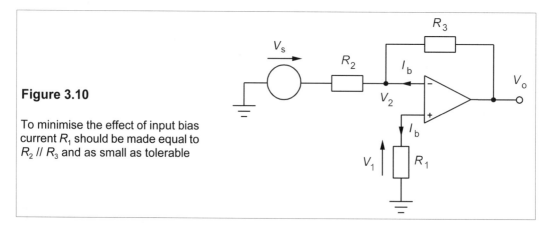

Figure 3.10

To minimise the effect of input bias current R_1 should be made equal to R_2 // R_3 and as small as tolerable

3.4.3 The input offset current

The *input offset current*, I_{OS}, is the difference in input bias currents when the inputs are connected to a common source. (The amplifier will not work unless it draws bias current, which is the base current to a bipolar transistor. FET input op amps draw very little bias current.) I_{OS} arises for the same reason as V_{OS}. Because of I_{OS} it is not possible to compensate for the effects of I_B exactly, but as I_{OS} is often only a fraction of I_B, the procedure above can remove most of the problem.

3.4.4 Frequency response

The *unity-gain frequency*, f_T, is the frequency at which the op amp's open-loop gain, A_v, is unity. The open-loop voltage gain of an op amp falls ten-fold for a ten-fold increase in frequency (that is 20 dB/decade) above a certain point (somewhere from 1–100 Hz) so that in this region the *gain-bandwidth product* (*GB*) is constant and $\approx f_T$. For a certain op amp (Figure 3.11), $A_v = 10^6$ at 5 Hz ($\log_{10} A_{v0} = 6$ and $\log_{10} f = 0.7$) before it starts to fall, and it will decline to unity at about 5 MHz: as the gain has to drop by 10^6, the frequency must increase by 10^6: $GB = f_T = 5 \times 10^6$. With feedback the closed loop gain, A_{vL}, is less and the amplifier's response drops only at about the frequency where $A_v \approx A_{vL}$. In Figure 3.11, A_{vL} has been set at 100 ($\log_{10} A_{vL} = 2$), and it starts to fall at 50 kHz ($\log_{10} f = 4.7$). An op amp with a low f_T might not achieve a voltage gain of 100 at 10 kHz.

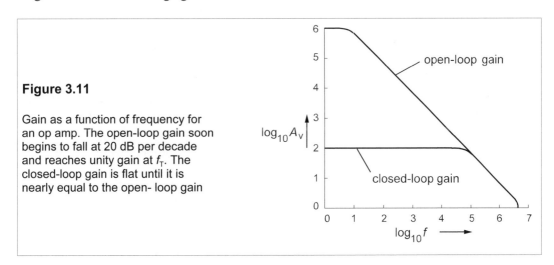

Figure 3.11

Gain as a function of frequency for an op amp. The open-loop gain soon begins to fall at 20 dB per decade and reaches unity gain at f_T. The closed-loop gain is flat until it is nearly equal to the open- loop gain

3.4.5 Slew rate

Slew rate is the rate at which the output voltage can change in V/s. If the slew rate is S, the full peak-to-peak amplitude of a sine wave, V_{pp}, is only reached when $\omega \le S/V_{pp}$. For example if an amplifier has a slew rate of 1 V/μs ($S = 1$ MV/s) and the output voltage is 10 V (r.m.s., making $V_{pp} = 20\sqrt{2}$ V), then full amplitude is obtained up to only $10^6/20\sqrt{2}$ rad/s, or 5.6 kHz.

3.4.6 CMMR

The *common-mode-rejection ratio* (CMRR) is a measure of the imbalance in the amplification of the two op amp input channels. If A_1 is the open-loop voltage gain of the non-inverting input and A_2 that of the inverting input then the CMRR in dB is

$$CMRR = 20 \log_{10} [(A_1 - A_2)/|A_1 + A_2|] \tag{3.13}$$

Ideally $A_1 = -A_2 = A_v$, but in practice it is impossible to balance the input gains exactly and A_v is defined as $(A_1 - A_2)/2$. For example, the UA741CD op amp in Table 3.1 has $A_v = 106$ dB and CMRR = 90 dB, so by Equation 3.13

$$\log_{10}[(A_1 - A_2)/|A_1 + A_2|] = 4.5 \qquad (3.14)$$

Taking antilogs gives $2A_v/|A_1 + A_2| = 3.16 \times 10^4$. But $A_v = 106$ dB $= 2 \times 10^5$, resulting in $|A_1 + A_2| = 12.6$. As $A_1 - A_2 = 2A_v = 4 \times 10^5$, we find $A_1 = 200,006.3$ or $199,993.7$ and $A_2 = -199,993.7$ or $-200,006.3$. Common mode signals – signals which are picked up by each input, such as air-borne EMI (electromagnetic interference, see Chapter 28) – are thus amplified slightly differently by each channel and will appear on the output, though attenuated by an amount depending on the gain of the feedback loop.

A word of caution about the values in Table 3.1! They are 'typical values', and the spread in parameters during manufacture can be quite large – perhaps as much as 3-fold up and down on these. The selection made is a random one from a single supplier's catalogue. There are hundreds and hundreds of different op amps.

3.5 **Comparators**

Comparators are a special type of op amp used as the name suggests to compare a signal voltage on one of the input terminals with a reference voltage on the other input terminal, which is often adjustable. When the signal voltage differs from the reference voltage the output of the comparator changes from low to high or vice versa. Negative feedback can cause oscillation, but positive feedback may be used to prevent oscillation when the inputs are almost at the same potential.

Figure 3.12 shows a comparator with the non-inverting (reference) input set to V_{ref}. The input signal here is connected to the inverting input. When $V_{in} > V_{ref}$, the comparator output is 0 V. This is because its output stage is an npn transistor (see Chapter 8) with a grounded emitter and an *open collector,* that is the collector terminal on the IC has to be connected to a positive supply by the user via a collector resistor, R_C (normally about 1 kΩ), which must also be provided. When $V_{in} > V_{ref}$ the output transistor is turned on and pulls the output down to ground. Similarly when the $V_{in} < V_{ref}$, the output transistor is switched off and the output goes to $+V_{CC}$. The comparator can drive a relay connected in place of R_C and thereby control a variety of devices such as heaters, lamps or motors. By changing the inputs around, the operation is reversed and the output is 0 V when $V_{in} < V_{ref}$. When $V_{ref} = 0$ V (grounded reference) the comparator becomes a *zero-crossing detector.*

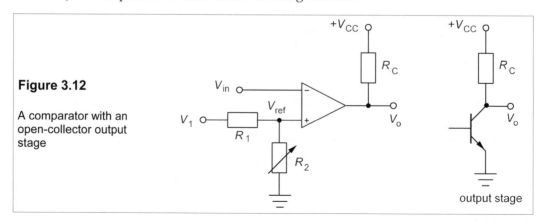

Figure 3.12

A comparator with an open-collector output stage

The chief advantage the comparator has over an ordinary op amp used for the same purpose is its very much faster slew rate. Ordinary op amps have slew rates of about 1–10 V/μs, while that of a standard comparator like the LM311 is of the order of 100 V/μs, though its slew rate is not actually on its data sheet, which specifies instead response time for a given level of overdrive, that is the amount by which V_{in} exceeds V_{ref}. At overdrives of from 2 mV to 20 mV the response time falls from 400 ns to 150 ns. At an input overdrive of 5 mV the cheap LM339 has a response time of 1.3 μs and the ultra fast LT1016 about 10 ns though for an overdrive of 0.5 V. The response time for a high-voltage to low-voltage transition is usually faster than the reverse: the data given is for the faster transition. Prices depend on response times: the LM339 is about 25p in surface-mount, quad form (four devices to a package) while the LT1016 is a single device costing £4 or so.

3.6 Schmitt triggers

Because comparators are fast acting and the input signal might be (and very often is) noisy, there is a strong possibility that the output will undergo multiple on-off transitions as the input signal crosses the reference voltage level. This undesirable behaviour can be curtailed by using positive feedback from the output as in Figure 3.13, which is called a *Schmitt trigger*.

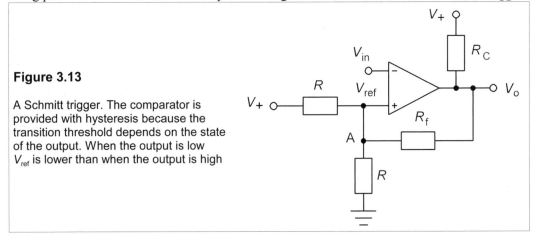

Figure 3.13

A Schmitt trigger. The comparator is provided with hysteresis because the transition threshold depends on the state of the output. When the output is low V_{ref} is lower than when the output is high

The output now has *hysteresis* because the reference voltage, V_{ref}, depends on the output level. V_{ref} can be found by nodal analysis at A, which yields

$$(V_{ref} - V_+)G + V_{ref}G + (V_{ref} - V_o)G_f = 0 \qquad (3.15)$$

where $G \equiv 1/R$ and $G_f \equiv 1/R_f$. Solving for V_{ref} gives

$$V_{ref} = \frac{V_+G + V_oG_f}{2G + G_f} \qquad (3.16)$$

For example, suppose $V_{in} < V_{ref}$, which means the output is high, $V_o = V_+ = 6$ V, say. Then if $R_f = R$, $V_{ref} = 4$ V from Equation 3.16. If V_{in} exceeds 4 V, the output will go low and $V_o = 0$ V. Equation 3.16 now gives $V_{ref} = 2$ V.

Unless the input voltage falls below 2 V, the output will stay at 0 V. The input voltage has a margin of 2 V against noise, as shown in Figure 3.14b. Without hysteresis the input and output might look like Figure 3.14a.

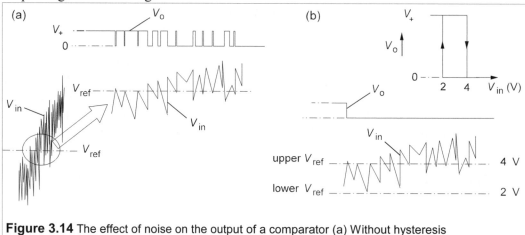

Figure 3.14 The effect of noise on the output of a comparator (a) Without hysteresis
(b) With hysteresis, which must be large enough to remove multiple output transitions

Suggestions for further reading – see Chapter 5

Problems

(In the following problems the op amps are ideal unless otherwise stated)

1. The input impedance of an op amp with feedback must be calculated from the ratio $\mathbf{V_{in}}/\mathbf{I_{in}}$. Find the input impedances of the op amps in Figure P3.1. In figure (d) what is the equivalent inductance approximately if $R = 10 \text{ k}\Omega$ and $C = 1 \text{ } \mu\text{F}$? [Figures P3.1c and P3.1d are called *negative-impedance converters*.] *[(a) R_1 (b) ∞ (c) $-R$ (d) $\mathbf{Z_{in}} = -j\omega CR^2$, $L_{eq} = -CR^2 = -100$ H]*

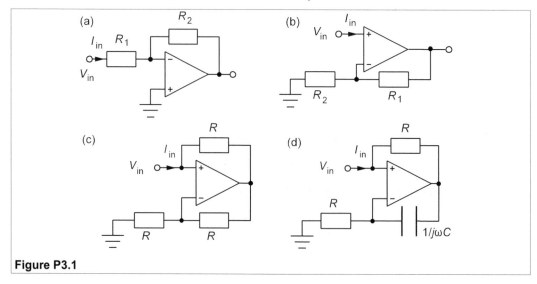

Figure P3.1

2. Find the input impedances of the circuits in Figure P3.2. In Figure P3.2a, what does Z_{in} become if $1/CR_2 \ll \omega \ll 1/CR_1$? What is the equivalent inductance then if $C = 0.1 \ \mu F$, $R_1 = 500 \ \Omega$ and $R_2 = 10 \ M\Omega$? [In Figure P3.2b the op amps are configured as in Figure P3.1c, so use that result. Figure P3.2b is the circuit of a *gyrator*.]

$[(a) \ Z_{in} = R_1(1 + j\omega CR_2)/(1 + j\omega CR_1), \ Z_{in} \approx j\omega CR_1R_2, \ L_{eq} = 500 \ H! \ (b) \ Z_{in} = R^2/Z]$

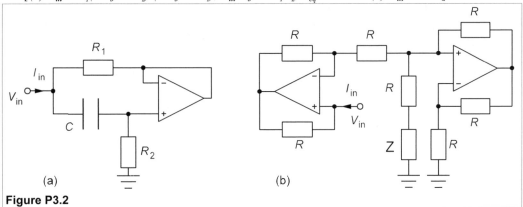

(a)   (b)

Figure P3.2

3. What is the output of the circuit of Figure P3.3 (a) at low frequencies (b) at high frequencies?

$[(a) \ V_o = -j\omega R_2C_1V_{in} \ (b) \ V_o = -V_{in}/j\omega C_2R_1]$

4. For the circuit of Figure P3.4 show that $|V_o/V_{in}| = 1/[1 + (\omega RC)^2]$. At what frequency is V_o $90°$ out of phase with V_{in}? What is V_o/V_{in} then? $[\omega = 1/RC. \ V_o/V_{in} = 0.5\angle{-90°}]$

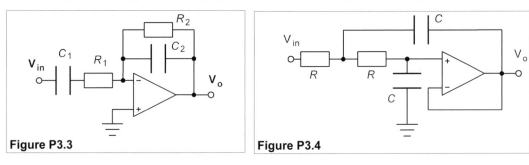

Figure P3.3   **Figure P3.4**

5. Find V_o/V_{in} for the circuit of Figure P3.5. When does V_o lead V_{in} by $90°$? What is V_o/V_{in} then? $[\alpha^2/[(\alpha^2 - 1) - j2\alpha]$ where $\alpha = \omega RC$. When $\alpha = 1; \ j/2]$

6. In the circuit of Figure P3.6, find V_o if $V_{in} = +10 \ mV$? What is the 50kΩ resistance for? $[-1.02 \ V]$

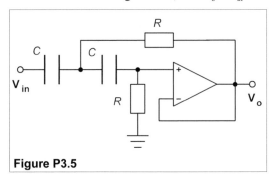

Figure P3.5

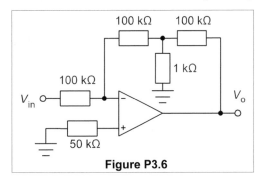

Figure P3.6

7. The circuit of Figure P3.7 is a linear differential amplifier. Show that

$$V_o = \frac{V_B(R_1 + R_2)R_4}{R_1(R_3 + R_4)} - \frac{V_A R_2}{R_1}$$

If $R_1 = R_3$ and $R_2 = R_4$, what is V_o then?

8. An RC integrator as in Figure P3.8 has $R = 147$ kΩ and $C = 1$ μF. It is constructed with an op amp whose offset-voltage drift with temperature is 1.2 μV/K and whose input bias current drift with temperature is 10 pA/K. What is the rate of drift of the output voltage if the temperature changes by 5°C after the drift has been adjusted to zero, assuming the drift voltages are additive? How can this drift be considerably reduced? In a well-made integrator the temperature changes are kept to 0.1 K. What then is the minimum time constant needed to restrict the drift at the output below 1 mV in 5 minutes, considering only the input offset-voltage? *[91 μV/s, 36 ms]*

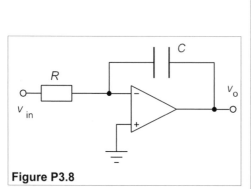

Figure P3.8

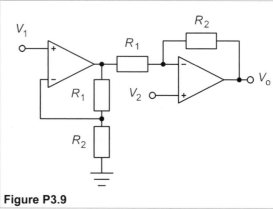

Figure P3.9

9. What is V_o in Figure P3.9 if $R_1 = 10$ kΩ, $R_2 = 90$ kΩ, $V_1 = 0.2$ V and $V_2 = 0.65$ V? *[4.5 V]*
10. Find V_o/V_{in} and the input resistance of the circuit of Figure P3.10. *[−20, 48 Ω]*
11. Find the input resistance and V_o/V_{in} for the circuit of Figure P3.11. *[1.493k Ω, −10.89]*
12. What is R_f in the circuit of Figure 3.13 if the Schmitt trigger has a hysteresis of 1 V, given $R = 20$ kΩ and $V_+ = 5$ V, assuming R_C is negligible? What is the upper value of V_{ref}? *[40 kΩ, 3 V]*

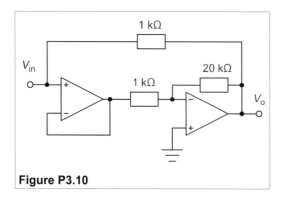

Figure P3.10

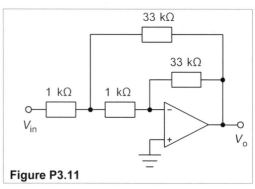

Figure P3.11

4 Transients

IN MANY instances the behaviour of a circuit or system is only of interest when it is in a steady state. There are occasions, however, when the temporary response of a circuit to a change in conditions is required. For example if a power supply in a circuit is switched on there may be a surge, possibly with oscillations, before a steady flow of current is established. One may wish to calculate the magnitude of the transient current to avoid damage to components. Circuits exhibit transients when they contain components that can store energy, such as transformers, inductors and capacitors; circuits that are purely resistive cannot display transient behaviour. Simple circuits containing only one type of storage element (that is RC and RL circuits) will be examined initially, followed by more complicated circuits that will be analysed with Laplace transforms.

4.1 Transients in RC and RL circuits

These are known as *first-order circuits* because the differential equations for the voltages and currents in them are first order – they contain only terms such as dv/dt or $\int i\,dt$ (but not both) and not d^2v/dt^2 or higher order differentials.

4.1.1 The series RC circuit

Consider the series RC circuit of Figure 4.1, in which the source has a constant voltage, V. When the switch is closed at $t = 0$, a current flows through the resistance and charges the capacitance (initially uncharged) until it eventually reaches the supply potential, V, and the current ceases.

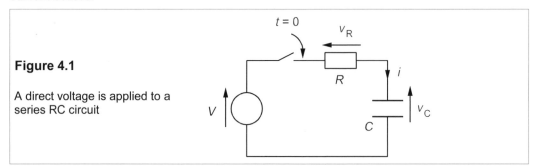

Figure 4.1

A direct voltage is applied to a series RC circuit

By KVL in Figure 4.1, for $t \geq 0$

$$V = v_R + v_C \qquad (4.1)$$

and by Ohm's law, $v_R = iR$. Now the current through a capacitor is given by $i_C = C\,dv_C/dt$, and in this case $i_C = i$, so we have

$$V = iR + v_C = RC\frac{dv_C}{dt} + v_C \tag{4.2}$$

Rearranging to separate the variables

$$dt = RC\frac{dv_C}{V - v_C} \tag{4.3}$$

And then integrating

$$\int dt = RC\int^t \frac{dv_C}{V - v_C} \tag{4.4}$$

hence $\qquad t = -RC\ln(V - v_C) + \alpha$

which yields

$$\ln(V - v_C) = \alpha/RC - t/RC \tag{4.5}$$

where α is a constant of integration. Taking antilogs produces

$$V - v_C = \beta\exp(-t/RC) \quad \Rightarrow \quad v_C = V - \beta\exp(-t/RC) \tag{4.6}$$

where β is $\exp(\alpha/RC)$. By examining the initial conditions we can find β. Initially, $v_C = 0$ when $t = 0$ and Equation 4.6 reduces to $v_C = V - \beta = 0$, as $\exp(0) = 1$ and hence $\beta = V$. The complete expression for $v_C(t)$ is

$$v_C = V - V\exp(-t/RC) = V[1 - \exp(-t/RC)] \tag{4.7}$$

when $t \geq 0$. A graph of this equation is shown in Figure 4.2.

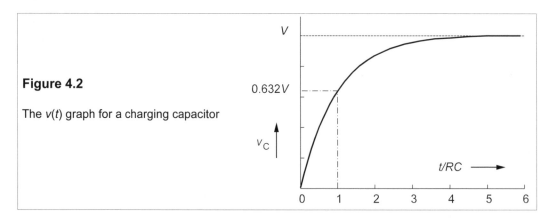

Figure 4.2

The $v(t)$ graph for a charging capacitor

The time scale is normalised to units of RC, known as the *time constant*, τ, of the circuit. After one time constant, Equation 4.7 shows that $V_C = V(1 - e^{-1}) = 0.63V$ (the capacitor is 63% charged), and after five time constants have passed $V_C = 0.99V$ – the capacitor is effectively fully charged and the 'transient' is over.

If the current-time graph for $t \geq 0$, is needed it can be found from Equation 4.7 by differentiating:

$$\frac{dv_C}{dt} = \frac{V}{RC}\exp(-t/RC) \tag{4.8}$$

and multiplying the result by C

$$C\frac{dv_C}{dt} = i = \frac{V}{R}\exp(-t/RC) \tag{4.9}$$

Equation 4.9 shows that the current decays exponentially with time as in Figure 4.3. After five time constants have passed, the current has all but ceased and a steady state has been reached.

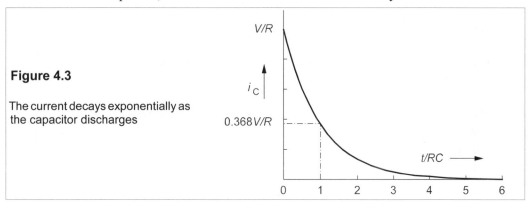

Figure 4.3

The current decays exponentially as the capacitor discharges

4.1.2 The series RL circuit

Figure 4.4 shows a series RL circuit excited by a step voltage.

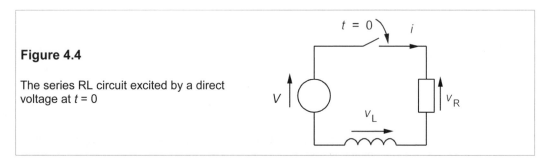

Figure 4.4

The series RL circuit excited by a direct voltage at $t = 0$

The analysis takes the same path as for the series RC circuit. After the switch is closed, by KVL

$$V = v_R + v_L \tag{4.10}$$

for $t \geq 0$. As usual, $v_R = iR$ and $v_L = L\,di/dt$, giving

$$V = iR + L\,di/dt \tag{4.11}$$

Rearranging this:

$$\frac{\mathrm{d}i}{\mathrm{d}t} = \frac{V - iR}{L} = \frac{V/R - i}{L/R} \tag{4.12}$$

The variables can be separated:

$$\frac{\mathrm{d}i}{V/R - i} = \frac{R}{L}\mathrm{d}t \tag{4.13}$$

Integrating, we find

$$\frac{R}{L}\int \mathrm{d}t = \frac{\mathrm{d}i}{V/L - i} \tag{4.14}$$

whence $$\alpha + Rt/L = -\ln(V/R - i) \tag{4.15}$$

Here α is another constant of integration. Taking antilogs produces

$$i = V/R - \beta\exp(-Rt/L) \tag{4.16}$$

where β [$= \exp(-\alpha)$] must be found by examining the initial condition, that is $i = 0$ when $t = 0$. (The initial current through the inductance must be zero as a finite current then would require an infinite voltage.) At $t = 0$, $i(0) = V/R - \beta\exp(0) = V/R - \beta = 0$. Thus $\beta = V/R$ and the current, for $t \geq 0$, is

$$i = \frac{V}{R}[1 - \exp(-Rt/L)] \tag{4.17}$$

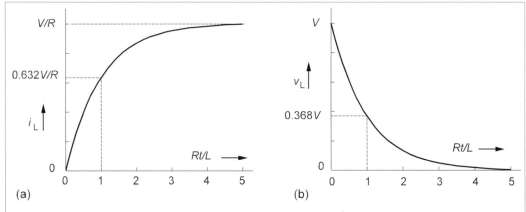

Figure 4.5 The series RL circuit's response to a step voltage input (a) The current through the inductance and (b) The voltage across it

Here the i-t graph, Figure 4.5a, is precisely the same shape as the v_C-t graph for the capacitance in the series RC circuit, and one can assume that the voltage across the inductance, v_L, will follow the same course as i_C. The only difference lies in the time constant, which is L/R for the series RL circuit and is RC for the series RC circuit. At time $t = L/R$, the current is $(1 - 1/e)V/R$ in the RL circuit. The graphs for the inductor voltage and current are shown in Figure 4.5, with time scales normalised to the time constant, L/R.

Example 4.1

In the circuit of Figure 4.6 the switch has been open a long time before being closed at $t = 0$. What are the currents through the 3Ω and 6Ω resistances and voltage across the 300mH inductance as functions of time?

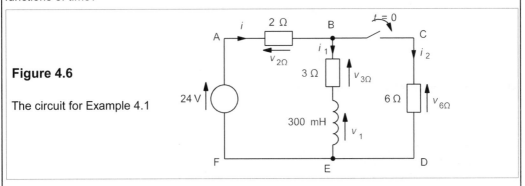

Figure 4.6

The circuit for Example 4.1

We begin by considering the initial state of the circuit, that is just before the switch is closed. The DC source will provide current through the 2Ω and 3Ω resistances and the inductance, but there will be no voltage across the inductance as the current will be constant. We find the initial current by Ohm's law to be

$$i(0) = i_1(0) = 24/(2 + 3) = 4.8 \text{ A} \tag{4.18}$$

The initial voltage across the inductance, $v_1(0) = 0$, since it carries a steady current, and $i_2(0) = 0$, since the branch is open circuit. Next we consider the circuit after the switch has been closed for a long time and the circuit is in its new steady state. Once more there can be no voltage across the inductance, and the source sees a 2Ω resistance in series with a parallel combination of 3Ω and 6Ω resistances. The 3Ω and 6Ω resistances in parallel have a combined resistance of $(3 \times 6)/(3 + 6) = 2 \text{ Ω}$ by the product-over-sum rule. Added to the 2Ω resistance this gives a total of 4 Ω presented to the source and so the current from the source in the steady state is

$$i(\infty) = 24/4 = 6 \text{ A} \tag{4.19}$$

By KCL, $i(\infty) = i_1(\infty) + i_2(\infty)$ and since $v_1(\infty) = 0$, the parallel 3Ω and 6Ω resistances act as a current divider, making

$$i_1(\infty) = 4 \text{ A and } i_2(\infty) = 2 \text{ A} \tag{4.20}$$

We can now use Kirchhoff's laws to find the time-dependent currents and voltages for $t \geq 0$. Considering the loop ABCDEF and using KVL

$$24 = v_{2\Omega} + v_{6\Omega} = 2i + 6i_2 \tag{4.21}$$

But $i = i_1 + i_2$ making Equation 4.21

$$24 = 2i_1 + 8i_2 \implies i_2 = 3 - 0.25i_1 \tag{4.22}$$

Using KVL on the loop BCDE produces

$$v_1 + v_{3\Omega} = v_{6\Omega} \implies L di_1/dt + 3i_1 = 6i_2 \tag{4.23}$$

But we can substitute for i_2 into this from Equation 4.22, and replace L by 0.3, so that we find

$$0.3 di_1/dt + 3i_1 = 6(3 - 0.25i_1) = 18 - 1.5i_1 \tag{4.24}$$

Rearranging,

$$0.3di_1/dt = 18 - 4.5i_1 \tag{4.25}$$

$$\Rightarrow \qquad di_1 = -15(i_1 - 4)dt \tag{4.26}$$

Separating variables,

$$di_1/(i_1 - 4) = -15dt \tag{4.27}$$

Integrating gives

$$\ln(i_1 - 4) = -15t + \alpha \tag{4.28}$$

where α is a constant of integration. Equation 4.28 can be expressed as

$$i_1 - 4 = \beta\exp(-15t) \tag{4.29}$$

where β (= e^{α}) is a constant, which is found by examining the initial conditions. From Equation 4.18 we have $i_1(0) = 4.8$ A, and when $t = 0$, Equation 4.29 yields $i_1(0) = 4 + \beta\exp(0) = 4 + \beta$, making the constant, $\beta = 0.8$ A. Hence the current in the inductance and 3Ω resistance is, for $t \geq 0$,

$$\mathbf{i_1(t) = 4 + 0.8exp(-15t)\ A} \tag{4.30}$$

We can find i_2, for $t \geq 0$, from Equation 4.22:

$$\mathbf{i_2 = 3 - 0.25i_1 = 2 - 0.2exp(-15t)\ A} \tag{4.31}$$

and v_1 from

$$v_1 = Ldi_1/dt = 0.3 \times [-15 \times 0.8\ \exp(-15t)] = -3.6\exp(-15t)\ V \tag{4.32}$$

when $t \geq 0$. Note that the initial current through the 6Ω resistance is not zero but 1.8 A: because it is not an energy-storing element the current can change instantaneously from 0 to 1.8 A. Also consider the time constant for the solutions, which is 1/15 seconds, implying a resistance of 0.3 × 15 = 4.5 Ω in series with the 300mH inductance. This is not what the inexperienced analyst would expect, but it arises because the resistance seen by the inductance is 3 Ω in series with the parallel combination of 2Ω and 6Ω resistances. Graphs of $i_1(t)$ and $v_1(t)$ are shown in Figure 4.7.

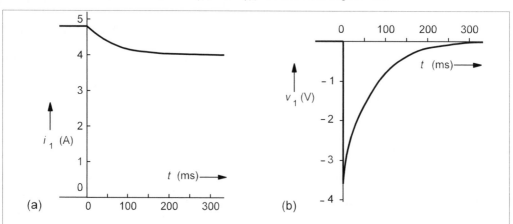

Figure 4.7 (a) Current graph (b) voltage graph for the circuit of Figure 4.6

It can be left as an exercise to show that the initial and steady state conditions are fulfilled by these solutions, and that they obey Kirchhoff's laws.

4.2 **Transient analysis by the Laplace transformation**

Heaviside was the first to apply transformations to the differential equations of electrical engineering, though in a slightly different form (for many years Heaviside transforms were used). The method is an operational one that treats the processes of differentiation and integration as inverses of each other. By using Laplace transforms we need not even derive a differential equation for a circuit; and certainly we need not trouble over the abstruse and complicated details of solving them.

The Laplace transform is defined by

$$\mathcal{L}[f(t)] = F(s) = \int_0^\infty f(t)\exp(-st)\,dt \qquad (4.33)$$

A function of time, $f(t)$, has been transformed into a function of the auxiliary variable, s. There are various restrictions placed on s and $f(t)$, essentially to ensure the convergence of the integral, but they seldom matter in problems to do with electrical circuits.

To avoid writing 'for $t \geq 0$' all the time we introduce the unit step function $u(t)$. This function has the effect of switching on any function at $t = 0$. For example the DC voltage, V, can be switched on at $t = 0$ by writing $Vu(t)$ and so forming a voltage step function as shown in Figure 4.8.

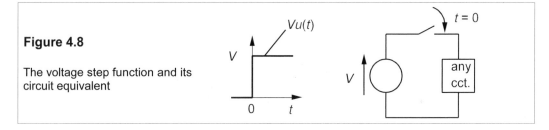

Figure 4.8

The voltage step function and its circuit equivalent

We shall transform this step voltage, $Vu(t)$, using Equation 4.33, which produces

$$\mathcal{L}[Vu(t)] = \int_0^\infty V\exp(-st)\,dt = V\int_0^\infty \exp(-st)\,dt$$

$$= V\left[\frac{\exp(-st)}{-s}\right]_0^\infty = V\left[\frac{0-1}{-s}\right] = \frac{V}{s}$$

Therefore $\qquad\qquad \mathcal{L}[Vu(t)] = V/s \qquad\qquad$ (LT2)

The inverse transform, $\mathcal{L}^{-1}[F(s)]$, is looked up in Table 4.1 when needed: for example $\mathcal{L}^{-1}[V/s] = Vu(t)$ from LT 2. (By convention lower case letters are used for functions of time and the transformed function is written in the corresponding upper case letter.) The process of taking the inverse Laplace transform is called *detransformation*.

The transform of $f'(t)$ (or df/dt) is found likewise:

$$\mathcal{L}[f'(t)] = \int_0^\infty f'(t)\exp(-st)\,dt$$

$$= [f(t)\exp(-st)]_0^\infty - \int_0^\infty f(t)[-s\exp(-st)]\,dt$$

Therefore

$$\mathcal{L}[f'(t)] = -f(0) + sF(s) = sF(s) - f(0) \tag{LT9}$$

which is our second transform pair. $F(s)$ is the Laplace transform of $f(t)$, $f(0)$ is the value of $f(t)$ at $t = 0$ and the integration by parts uses the fact that

$$\int_0^\infty w\,dv = [wv]_0^\infty - \int_0^\infty v\,dw \tag{4.34}$$

where $w \equiv \exp(-st)$ and $dv \equiv f'(t)\,dt$.

Table 4.1 *Laplace transform pairs*

$f(t)$	$F(s)$	No.
$\delta(t)$	1	LT1
$u(t)$	$1/s$	LT2
$tu(t)$	$1/s^2$	LT3
$t^n u(t)$	$1/s^{n-1}$	LT4
$\exp(-at)u(t)$	$1/(s + a)$	LT5
$f(t)\exp(-at)u(t)$	$F(s + a)$	LT6
$\sin(\omega t)u(t)$	$\omega/(s^2 + \omega^2)$	LT7
$\cos(\omega t)u(t)$	$s/(s^2 + \omega^2)$	LT8
$f'(t)$	$sF(s) - f(0)$	LT9
$f''(t)$	$s^2 F(s) - sf'(0) - f(0)$	LT10
$f(t - \tau)u(t - \tau)$	$\exp(-s\tau)F(s)$	LT11
$\int f(t)dt$	$F(s)/s$	LT12

Our third, and for the moment final transform is of $\exp(-at)$,

$$\mathcal{L}[\exp(-at)u(t)] = \int_0^\infty \exp(-at)\exp(-st)\,dt = \int_0^\infty \exp[-(a+s)t]\,dt \tag{4.35}$$

Integrating,

$$\left[\frac{\exp[-(a+s)t]}{-(a+s)}\right]_0^\infty = \frac{0-(1)}{-(a+s)} = \frac{1}{s+a}$$

Therefore $\qquad\qquad \mathcal{L}[\exp(-at)u(t)] = \dfrac{1}{s+a} \tag{LT5}$

The Laplace transform is defined only for $t \geq 0$, so transforming $f(t)$ is the same as transforming $f(t)u(t)$; however, when $F(s)$ is detransformed it is better to make it clear that $f(t)$ is zero for $t < 0$ by writing $f(t)u(t)$.

There are two ways in which the transform may be applied: either to the differential equation for the circuit (derived by using Kirchhoff's laws as above) or by transforming the circuit first and then applying Kirchhoff's laws to that. The second approach is better because *no differential equations are explicitly stated*, though they are implicitly solved.

4.2.1 Circuit elements in the s-domain

Circuit-element transformations into the *s*-domain are now derived, starting with resistance.

Resistance in the s-domain

The current-voltage relation for a resistance is given by Ohm's law and applying the Laplace transform to it produces

$$\mathcal{L}[v] = \mathcal{L}[iR] \quad \rightarrow \quad V(s) = RI(s) \tag{4.36}$$

Because R is not a function of time it remains *unchanged* after transformation. The voltage and current, however, turn into functions of s: they are said to have been transformed from the *time domain* to the *s-domain*. A resistance of $10\,\Omega$, however, stays $10\,\Omega$ in the *s*-domain. It is a little cumbersome to write $I(s)$ and $V(s)$, so we shall adopt the neater symbols $\tilde{\mathbf{i}}$ and $\tilde{\mathbf{v}}$ for current and voltage in the *s*-domain.

Inductance in the s-domain

Consider the case of an inductance connected to a step-voltage source, carrying a current of I_0 at $t = 0$, as in Figure 4.9a.

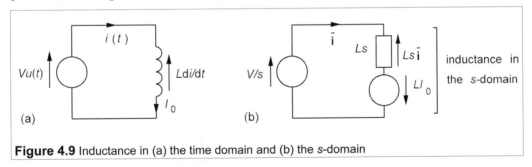

Figure 4.9 Inductance in (a) the time domain and (b) the *s*-domain

The voltage across the inductor in the time domain is given by

$$Vu(t) = L\,di/dt \tag{4.37}$$

Taking the transform of the left-hand side, $\mathcal{L}[Vu(t)] = V/s$ (by LT2); and the transform of the right-hand side is, by LT9,

$$\mathcal{L}[L\,di/dt] = Ls\tilde{\mathbf{i}} - LI_0 \tag{4.38}$$

Thus the transform of the whole of Equation 4.37 is

$$V/s = Ls\tilde{i} - LI_0 \tag{4.39}$$

And this can be rearranged in the form $V = IR$:

$$V/s + LI_0 = \tilde{i}Ls \tag{4.40}$$

The terms on the left-hand side are voltage terms, while the Ls term multiplying $\tilde{i}$ is a resistive term in the s-domain, the reactance of the inductance. An inductance therefore transforms into a reactance of Ls *together with* a series voltage source of value LI_0 in the s-domain, as in Figure 4.9b. The arrow on this voltage source points in the direction of flow of the initial time-domain current, I_0. Of course if I_0 is zero the voltage source vanishes leaving just the reactance.

Capacitance in the s-domain

Next we must find out what a capacitance becomes in the s-domain. The relationship between voltage and current for a capacitance is $v_C = \int i\,dt$, implying the need for another transform pair, $\mathcal{L}[\int f(t)\,dt]$, LT12 in Table 4.1. This transform can be deduced by setting

$$g(t) = \int_0^\infty f(\tau)\,d\tau \tag{4.41}$$

Then
$$\mathcal{L}\left[\int_0^\infty f(t)\,dt\right] = \mathcal{L}\left[g(t)\right] = G(s) \tag{4.42}$$

But $f(t) = g'(t)$ which gives

$$\mathcal{L}[f(t)] = \mathcal{L}[g'(t)] = sG(s) - g(0) \tag{4.43}$$

However $g(0) = 0$, from the definition of $g(t)$. Thus

$$F(s) = sG(s) \tag{4.44}$$

Therefore
$$G(s) = \mathcal{L}\left[\int_0^\infty f(t)\,dt\right] = \frac{F(s)}{s} \tag{4.45}$$

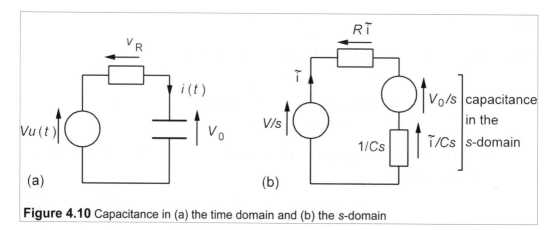

Figure 4.10 Capacitance in (a) the time domain and (b) the s-domain

Look at the capacitance in Figure 4.10a; it is initially charged to V_0 volts and then a step voltage is turned on at $t = 0$. By KVL

$$Vu(t) = v_R + V_0 + v_C = iR + V_0 + C^{-1}\int_0^t i\,dt \tag{4.46}$$

Transforming,

$$\mathcal{L}[Vu(t)] = R\mathcal{L}[i] + \mathcal{L}[V_0] + \mathcal{L}\left[C^{-1}\int_0^t i\,dt\right]$$

$$\Rightarrow \quad V/s = R\tilde{i} + V_0/s + \tilde{i}/Cs \tag{4.47}$$

In this equation there are two terms – $\tilde{i}R$ and $\tilde{i}/Cs$ – containing $\tilde{i}$ which can be combined, suggesting that $1/Cs$ is the reactance of the capacitance in the s-domain. The initially-charged capacitance in the time domain is transformed into a reactance of $1/Cs$ ohms, in series with a voltage source of value V_0/s. This is shown in Figure 4.10b.

Equation 4.47 can be rearranged to give $\tilde{i}$ as an explicit function of s:

$$\tilde{i} = \frac{(V - V_0)/s}{R + 1/Cs} = \frac{(V - V_0)/R}{s + 1/CR} \tag{4.48}$$

This expression for $\tilde{i}$ is of the form $\alpha/(s + a)$, which by LT5 has the inverse transform $\alpha\exp(-at)$ with $\alpha \equiv (V - V_0)/R$ and $a \equiv 1/CR$, resulting in

$$i(t) = \left(\frac{V - V_0}{R}\right)\exp(-t/CR)u(t) \tag{4.49}$$

When $V_0 = 0$ the current through the capacitance is seen to be identical to that found previously by solving the differential equation (see Equation 4.9).

4.2.2 Partial fraction expansions

Quite often we obtain an expression for voltage or current in the s-domain which we cannot see in the list of transforms in Table 4.1. It is then necessary to manipulate the expression algebraically to arrive at one that is detransformable. To do this we use partial fractions. Consider for example the expression

$$F(s) = \frac{2s + 3}{s^2 + 4s + 3} \tag{4.50}$$

which is not in Table 4.1. We can factorise the denominator, however, and then expand it into partial fractions, thus

$$\frac{2s + 3}{s^2 + 4s + 3} \equiv \frac{2s + 3}{(s + 1)(s + 3)} \equiv \frac{A}{s + 1} + \frac{B}{s + 3} \tag{4.51}$$

where A and B are numbers or functions of s to be determined. Notice that the equality sign ($=$) is not used here, but the identity sign ($\equiv$) instead. The distinction is necessary, because Equation 4.51 is *true for all values of s*. The partial fraction expansion on the right-hand side of Equation 4.51 can be recombined:

$$\frac{A}{s+1} + \frac{B}{s+3} \equiv \frac{A(s+3) + B(s+1)}{(s+1)(s+3)} \equiv \frac{2s+3}{s^2+4s+3} \tag{4.52}$$

Comparing the numerators we have the identity

$$A(s+3) + B(s+1) \equiv 2s+3$$

which must be satisfied by all values of s. If we choose $s = -1$, we can see that $2A = 1$, that is $A = 0.5$, since the term with B vanishes. Choosing $s = -3$ makes the term in A disappear and leaves us with $-2B = -3$, or $B = 1.5$. Thus Equation 4.52 can be rewritten as

$$\frac{2s+3}{s^2+4s+3} \equiv \frac{A}{s+1} + \frac{B}{s+3} \equiv \frac{-0.5}{s+1} + \frac{1.5}{s+3}$$

The terms now found on the right-hand side are both in Table 4.1 (LT5) and can be detransformed individually as $-0.5\exp(-t)u(t)$ and $1.5\exp(-3t)u(t)$. In symbols

$$\mathcal{L}^{-1}[F(s)] = f(t) = [1.5\exp(-3t) - 0.5\exp(-t)]u(t)$$

A different way of finding the unknowns A and Bi in Equation 4.52 is to equate coefficients of the powers of s in the numerator. Thus the identity

$$A(s+3) + B(s+1) \equiv 2s+3$$

contains s to the powers of 1 and 0 (the constant term). Equating coefficients of s^1 gives $A + B = 2$, and equating constant terms gives $3A + B = 3$. These two simultaneous equations give the same solutions as before.

It happens from time to time that the denominator of $F(s)$ has a repeated factor, for example

$$F(s) = \frac{3s+2}{2s^2+4s+2} \equiv \frac{3s+2}{2(s+1)^2}$$

First we make the coefficient highest power of s in the denominator equal to one by dividing both numerator and denominator by two, and then we divide:

$$\frac{3s+2}{2(s+1)^2} \equiv \frac{1.5s+1}{(s+1)^2} \equiv \frac{1.5(s+1) - 0.5}{(s+1)^2} \equiv \frac{1.5}{s+1} - \frac{0.5}{(s+1)^2}$$

As we shall see, the terms on the right can be detransformed using Table 4.1 into $[1.5\exp(-t) - 0.5t\exp(-t)]u(t)$.

Having shown how circuit elements can be transformed into the s-domain, we can formulate $\tilde{v}$-$\tilde{i}$ relationships by application of Ohm's law and Kirchhoff's laws to the transformed circuits, as in the next pair of examples.

Example 4.2

An inductor of inductance 0.2 H and resistance 5 Ω is placed in series with an ideal 5 Ω resistor and a voltage, $v = 6\exp(-50t)u(t)$ V, is applied to the combination as in Figure 4.11a. If the initial current through the inductor zero, find the current through the inductor and the voltage across it as functions of time.

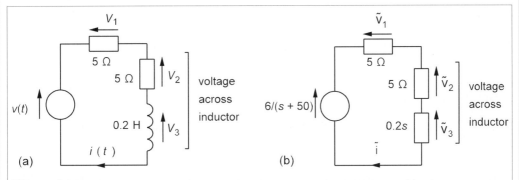

Figure 4.11 The circuit for Example 4.2 (a) In the time domain (b) In the s-domain

The coil resistance is in series with its inductance, which has a reactance of $0.2s$ in the s-domain, while the resistances are unchanged. The voltage source transforms to $6/(s + 50)$, producing the transformed circuit of Figure 4.11b. By KVL in Figure 4.11b

$$\frac{6}{s + 50} = \tilde{\mathbf{v}}_1 + \tilde{\mathbf{v}}_2 + \tilde{\mathbf{v}}_3$$

Applying Ohm's law to each circuit element in turn yields $\tilde{\mathbf{v}}_1 = 5\tilde{\mathbf{i}}$, $\tilde{\mathbf{v}}_2 = 5\tilde{\mathbf{i}}$ and $\tilde{\mathbf{v}}_3 = 0.2s\tilde{\mathbf{i}}$, then

$$\frac{6}{s + 50} = 5\tilde{\mathbf{i}} + 5\tilde{\mathbf{i}} + 0.2s\tilde{\mathbf{i}} \qquad (4.53)$$

that is,
$$\frac{6}{s + 50} = \tilde{\mathbf{i}}(10 + 0.2s) \qquad (4.54)$$

The right-hand side of Equation 4.54 shows that we can add up series resistances and reactances to form an impedance, $\mathbf{Z(s)}$ or $\mathbf{Z}$, in the s-domain. Ohm's law can then be used in the form

$$\tilde{\mathbf{v}} = \tilde{\mathbf{i}}\tilde{\mathbf{Z}} \qquad (4.55)$$

Equation 4.54 enables $\tilde{\mathbf{i}}$ to be expressed as an explicit function of s:

$$\tilde{\mathbf{i}} = \frac{6}{(s + 50)(0.2s + 10)} = \frac{30}{(s + 50)^2} \qquad (4.56)$$

Unfortunately, this is not a function that can be detransformed with any of the four transforms derived so far. Further transform pairs must be established.

Let us find $\mathscr{L}[t]$ (this is the *unit ramp function*, sometimes written $r(t) = tu(t)$ to start it at $t = 0$) using integration by parts:

$$\mathscr{L}[tu(t)] = \int_0^\infty t\exp(-st)\,\mathrm{d}t = \left[t\,\frac{\exp(-st)}{-s}\right]_0^\infty - \int_0^\infty \frac{\exp(-st)}{-s}\,\mathrm{d}t \qquad (4.57)$$

Integrating produces

$$[0 - (-0)] - \left[\frac{\exp(-st)}{s^2}\right]_0^\infty = -\left[\frac{0 - (1)}{s^2}\right] = \frac{1}{s^2} \qquad (4.58)$$

And $\mathscr{L}[tu(t)] = 1/s^2$, LT3 in Table 4.1. Next we require $\mathscr{L}[f(t)\exp(-at)]$:

$$\mathscr{L}[f(t)\exp(-at)] = \int_0^\infty f(t)\exp(-at)\exp(-st)\,\mathrm{d}t = \int_0^\infty f(t)\exp[-(s + a)t]\,\mathrm{d}t \qquad (4.59)$$

Replacing $s + a$ by α leads to

$$\mathscr{L}[f(t)\exp(-at)] = \int_0^\infty f(t)\exp(-\alpha t)\mathrm{d}t = F(\alpha) = F(s + a) \tag{LT6}$$

Therefore multiplying any function of time by $\exp(-at)$ has the effect of replacing $F(s)$ by $F(s + a)$, called *translation in the s-domain*.

Returning now to the current through the inductor of Figure 4.11, Equation 4.56 shows that $\tilde{i} = 30/(s + 50)^2 = F(s + a)$ with $a = 50$, and its detransformation by LT3 and LT6 must be

$$i(t) = 30t\exp(-50t)u(t) \text{ A} \tag{4.60}$$

The voltages can be derived either in the time domain from $i(t)$ or in the s-domain. Taking the latter route: the voltage across the inductor is the sum of $\tilde{v}_2$ and $\tilde{v}_3$, that is the voltage across the inductance, $Ls\tilde{i}$, and its coil resistance, $5\tilde{i}$, as shown in Figure 4.11b. Therefore

$$\tilde{v}_L = \tilde{v}_2 + \tilde{v}_3 = 5\tilde{i} + 0.2s\tilde{i} = (0.2s + 5)\tilde{i}$$

$$= \frac{30(0.2s + 5)}{(s + 50)^2} = \frac{6s + 150}{(s + 50)^2} = \frac{6(s + 50) - 150}{(s + 50)^2} \tag{4.61}$$

$$\Rightarrow \qquad \tilde{v}_L = \frac{6}{s + 50} - \frac{150}{(s + 50)^2} \tag{4.62}$$

which can now be detransformed using LT2, LT3 and LT6 to

$$v_L(t) = [6\exp(-50t) - 150t\exp(-50t)]u(t) \text{ V} \tag{4.63}$$

The time constant is 1/50 = 20 ms and the current and voltage are practically zero after five time constants, 100 ms. A graph of the inductor voltage is shown in Figure 4.12. The inductor voltage is zero when t = 6/150 = 40 ms, then goes negative, reaching a minimum at t = 60 ms.

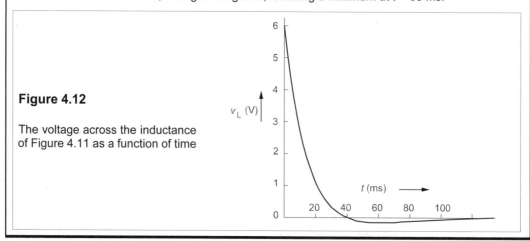

Figure 4.12

The voltage across the inductance of Figure 4.11 as a function of time

A more complicated example with two time constants will be studied next.

Example 4.3

In the circuit of Figure 4.13a the 0.5F capacitance is charged up to 3 V, then the switch is closed. Find all the currents flowing as functions of time.

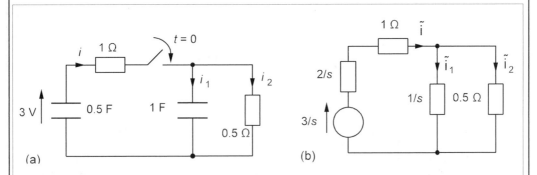

Figure 4.13 The circuit for Example 4.3 (a) In the time domain and (b) In the *s*-domain

The circuit is first transformed to that of Figure 4.13b. The impedance of the whole circuit must be found next and then Ohm's law can be used to find $\tilde{\imath}$.

The impedance of the $1/s$ reactance and the parallel $0.5\ \Omega$ resistance is given by the product-over-sum rule:

$$\frac{0.5 \times 1/s}{0.5 + 1/s} = \frac{1}{s + 2} \tag{4.64}$$

This is in series with $(1 + 2/s)\ \Omega$ for a total impedance of

$$\tilde{\mathbf{Z}} = 1 + \frac{2}{s} + \frac{1}{s+2} = \frac{s^2 + 5s + 4}{s(s+2)} \tag{4.65}$$

The current is

$$\tilde{\mathbf{i}} = \tilde{\mathbf{v}} \div \tilde{\mathbf{Z}} = \frac{3}{s} \div \frac{s^2 + 5s + 4}{s(s+2)} = \frac{3(s+2)}{s^2 + 5s + 4} \tag{4.66}$$

$\tilde{\imath}_1$ can be found by the voltage-divider rule:

$$\tilde{\mathbf{i}}_1 = \frac{0.5}{0.5 + 1/s}\,\tilde{\mathbf{i}} = \frac{s}{s + 2}\,\tilde{\mathbf{i}} = \frac{s}{(s + 2)}\,\frac{3(s + 2)}{(s^2 + 5s + 4)} \tag{4.67}$$

$$\Rightarrow \qquad \tilde{\mathbf{i}}_1 = \frac{3s}{s^2 + 5s + 4} = \frac{3s}{(s + 4)(s + 1)} \tag{4.68}$$

Expanding into partial fractions:

$$\tilde{\mathbf{i}}_1 = \frac{4}{s + 4} - \frac{1}{s + 1} \tag{4.69}$$

Detransforming:

$$i_1(t) = [4\exp(-4t) - \exp(-t)]u(t)\ \text{A} \tag{4.70}$$

$\tilde{\imath}_2$ is best found by KCL:

$$\tilde{\mathbf{i}}_2 = \tilde{\mathbf{i}} - \tilde{\mathbf{i}}_1 = 3(s + 2)/D - 3s/D = 6/D \tag{4.71}$$

where $D \equiv s^2 + 5s + 4$. Expanding into partial fractions and detransforming produces

$$\tilde{i}_2 = \frac{6}{(s+1)(s+4)} = \frac{2}{s+1} - \frac{2}{s+4} \tag{4.72}$$

$$i_2(t) = 2[\exp(-t) - \exp(-4t)]u(t) \text{ A} \tag{4.73}$$

The total current, $i(t)$, is

$$i(t) = i_1(t) + i_2(t) = [2\exp(-4t) + \exp(-t)]u(t) \text{ A} \tag{4.74}$$

Are the answers reasonable? One must check for the initial and the final (steady-state) currents. The initial current through the 1F capacitance must be $i_1(0) = V_0/R_1$, where V_0 is the initial voltage on the 0.5F capacitance and $R_1 = 1 \ \Omega$. Thus $i_1(0) = 3/1 = 3$ A, which is given correctly by Equation 4.71. The initial current through the 0.5Ω resistance, i_2, must be zero as the initial voltage across the 1F capacitance in parallel with it is zero, as Equation 4.74 correctly gives. The initial current from the 0.5F capacitance, $i(0)$, is also 3 A, as $i(0) = i_1(0)$. The final current must be zero in all cases as there is no source in the circuit and the capacitances end up fully discharged. All the solutions give a steady-state current of zero as shown in Figure 4.14.

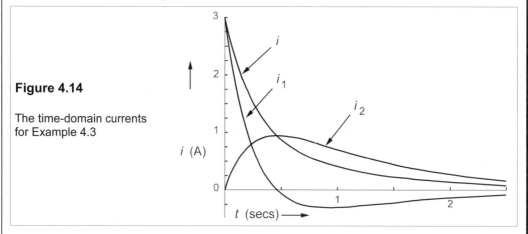

Figure 4.14

The time-domain currents for Example 4.3

The voltages across the components can be found from the currents. Looking at the solution we can see what has happened in the circuit: the 0.5F capacitance discharges into the 1F at first via the 1Ω resistance, then both capacitances discharge through the 0.5Ω resistance until the charge is zero.

The example illustrates the principles and strengths of the Laplace transform technique. Neither the time constants nor the differential equations were easy to write down by inspection, yet the transformed circuit readily yielded the solution – the time constants were generated by the algebra in which the differential equations were implicitly contained.

The rules for using Laplace transforms are:

● First transform the circuit into the s-domain
● Next use Kirchhoff's laws and Ohm's law to give equations for the unknowns
● Obtain an s-domain expression for the desired unknowns
● If necessary expand into partial fractions and manipulate into a recognisable form
● Detransform using Table 4.1

Other transforms in Table 4.1, such as those of $\sin\omega t$, $\cos\omega t$ and $f^n(t)$, are left as exercises (see Problem 4.15), but we shall derive three further transforms.

4.2.3 The delayed function, f(t − T) u(t − T)

This function is zero for $t < T$, and turns on at $t = T$, as in Figure 4.15.

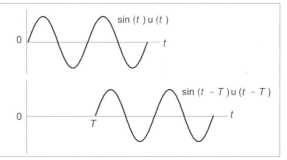

Figure 4.15

The upper graph is a sinewave which is switched on at $t = 0$, while the lower graph is a sinewave switched on T seconds later

Its Laplace transform is

$$\mathscr{L}[f(t-T)u(t-T)] = \int_0^\infty f(t-T)u(t-T)\exp(-st)\,dt \qquad (4.75)$$

Making the substitution $t - T \equiv \tau$ leads to

$$\mathscr{L}[f(t-T)u(t-T)] = \int_{-T}^\infty f(\tau)u(\tau)\exp[-s(T+\tau)]\,d\tau$$

$$= \exp(-sT)\int_0^\infty f(\tau)u(\tau)\exp(-s\tau)\,d\tau = \exp(-sT)F(s) \qquad (4.76)$$

This is LT11 in Table 4.1, and is translation in the time domain (cf translation in the s-domain in Example 4.2, LT6). The lower limit can be changed to zero as $f(\tau)u(\tau) = 0$ when $\tau < 0$.

4.2.4 The unit pulse function

Figure 4.16a shows the unit pulse function, $p(1)$, which can be written in terms of unit step functions that can then be transformed.

The unit pulse function is

$$p(1) = u(t) - u(t-1) \qquad (4.77)$$

$u(t)$ is switched off by the term $-u(t-1)$ at $t = 1$, to give the unit pulse shape of Figure 4.16a. The Laplace transform of $p(1)$ must therefore by LT2 and LT11 be

$$\mathscr{L}[p(1)] = \mathscr{L}[u(t) - u(t-1)] = 1/s - \exp(-s)/s \qquad (4.78)$$

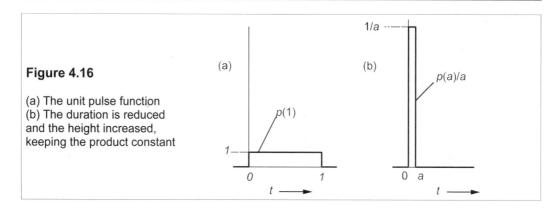

Figure 4.16

(a) The unit pulse function
(b) The duration is reduced
and the height increased,
keeping the product constant

4.2.5 The impulse function (or δ-function)

The unit pulse function can be made of briefer duration by a factor of a, where $a \ll 1$, but keeping its area the same by making its height $1/a$ times greater, as in Figure 4.16b, then

$$p(a)/a = [u(t) - u(t - a)]/a \tag{4.79}$$

Now the Laplace transform of this function is

$$\mathcal{L}\left[\frac{u(t) - u(t - a)}{a}\right] = \frac{1}{sa} - \frac{\exp(-sa)}{sa} \tag{4.80}$$

And if $a \ll 1$, this is nearly

$$\frac{1}{sa} - \frac{(1 - sa)}{sa} = 1 \tag{4.81}$$

The δ-function[1] is defined therefore as

$$\delta(t) = \lim_{a \to 0}\left[p(a)/a\right] \text{ and } \mathcal{L}[\delta(t)] = 1 \tag{LT1}$$

The δ-function aids mathematical analysis, but in physical systems it can only be approximated.

Example 4.4

Consider the circuit in Figure 4.17a, in which the 200μF capacitance is charged to 50 V and then at $t = 0$ the switch is closed. What is the current through the 300μF capacitance and the voltage across it?

   The transformed circuit is shown in Figure 4.17b. The simplest way to proceed is to transform the voltage source into a current source, whose value will be $50/s \div 5000/s$, or 0.01. Then the circuit becomes the set of parallel elements shown in Figure 4.17c. By the current-divider rule

---

[1] Sometimes known as the Dirac δ-function, after the physicist P A M Dirac (1902–90). Dirac won the Nobel prize for physics in 1933 while occupying the Lucasian chair for mathematics at Cambridge University.

$$\tilde{i} = \frac{\dfrac{s}{3333} \times 0.01}{\dfrac{s}{3333} + \dfrac{s}{5000} + \dfrac{1}{2000}} = \frac{0.006s}{s+1} = 0.006 - \frac{0.006}{s+1} \qquad (4.82)$$

Detransformation gives

$$i(t) = 6\delta(t) - 6\exp(-t)u(t) \text{ mA} \qquad (4.83)$$

$6\delta(t)$ mA seems to imply that the impulse has a magnitude of 6 times infinity at the origin – a concept difficult to grasp – but all it means is that the area under the impulse-time graph is 6 mC. This current-time graph is drawn in Figure 4.17d.

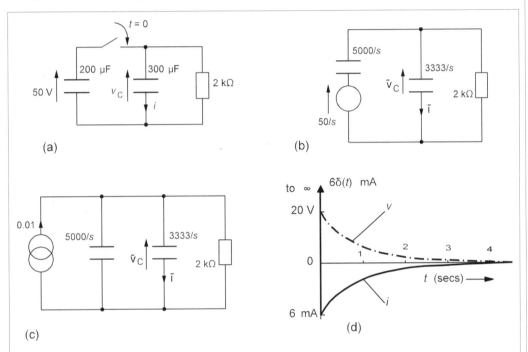

Figure 4.17 (a) The *t*-domain circuit (b) The *s*-domain circuit (c) The voltage source transformed into a current source in the *s*-domain circuit (d) The *t*-domain solutions

The capacitor voltage is $\tilde{i}\mathbf{X}_c$, that is

$$\tilde{i}\tilde{\mathbf{X}}_C = 0.006s/(s+1) \times 3333/s = 20/(s+1) \qquad (4.84)$$

whose inverse transform is

$$v_C(t) = 20\exp(-t)u(t) \text{ V} \qquad (4.85)$$

This implies that at the moment the switch is closed, the voltage across the 300μF capacitance (shown in Figure 4.17d also) changes *instantaneously* to 20 V, an impossibility in practice because there is always resistance in the connecting wires and switch contact.

4.3 **Transients in RLC circuits**

With Laplace transforms we can analyse the transient response of circuits containing inductance *and* capacitance, that is second (and higher) order circuits, with facility. The formal absence of differential equations and their mathematical jargon makes the generation of an *s*-domain solution and its detransformation seem merely workmanlike by comparison and lacking sophistication – precisely what Heaviside intended when he devised the method.

4.3.1 *The series RLC circuit*

In the series RLC circuit of Figure 4.18a suppose that $v(t)$ is a 3V, DC source switched on at $t = 0$, and that no energy is stored initially, giving the *s*-domain circuit in Figure 4.18b.

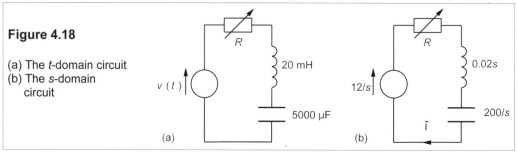

Figure 4.18

(a) The *t*-domain circuit
(b) The *s*-domain circuit

$v(t)$ R 20 mH   5000 μF   (a)

$12/s$ R 0.02s   200/s $\tilde{i}$ (b)

The resistance in the circuit can be varied, and is first set to 14 Ω, so that the impedance presented to the source in the *s*-domain is

$$\tilde{\mathbf{Z}} = R + Ls + 1/Cs = 14 + 0.02s + 200/s \qquad (4.86)$$

Hence the current is

$$\tilde{\mathbf{i}} = \tilde{\mathbf{v}}/\tilde{\mathbf{Z}} = 12/s \div (5 + 0.02s + 200/s) \qquad (4.87)$$

which can be rearranged as

$$\tilde{\mathbf{i}} = \frac{600}{s^2 + 250s + 10^4} \qquad (4.88)$$

The denominator is a quadratic with two unequal real roots found by completing the square:

$$\tilde{\mathbf{i}} = \frac{600}{(s+125)^2 - (125^2 - 10^4)} = \frac{600}{(s+125)^2 - (75)^2} \qquad (4.89)$$

which is

$$\tilde{\mathbf{i}} = \frac{600}{(s+200)(s+50)} \qquad (4.90)$$

Expanding into partial fractions gives

$$\tilde{\mathbf{i}} = \frac{4}{s+50} - \frac{4}{s+200} \qquad (4.91)$$

This is shown as trace 1 in Figure 4.19.

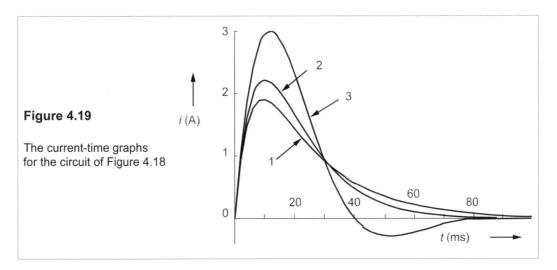

Figure 4.19

The current-time graphs
for the circuit of Figure 4.18

i (A)

Now suppose R were only 4 Ω instead of 14 Ω; $\tilde{\mathbf{i}}$ is given by

$$\tilde{\mathbf{i}} = \frac{12/s}{\tilde{\mathbf{Z}}} = \frac{12/s}{4 + 0.02s + 200/s} = \frac{600}{s^2 + 200s + 10^4} \tag{4.93}$$

This time the denominator is a perfect square (it has two equal real roots) that can be detransformed at once to

$$i(t) = 600\,t\exp(-100t)u(t) \text{ A} \tag{4.94}$$

A graph of $i(t)$ is shown as trace 2 in Figure 4.19. The peak current is about 2.2 A, whereas before it was less – about 1.9 A – something one might have expected considering that the resistance has fallen from 14 Ω to 4 Ω.

What happens when R is further decreased, say to 2.4 Ω? Once more $\tilde{\mathbf{i}}$ can be found:

$$\tilde{\mathbf{i}} = \frac{600}{s^2 + 120s + 10^4} = \frac{600}{(s+60)^2 + 80^2} \tag{4.95}$$

Looking up the table of transform pairs we see that $\tilde{\mathbf{i}}$ is of the form $\omega/[(s + a)^2 + \omega_n^2]$, the inverse transform of which is $\sin(\omega_n t)\exp(-at)u(t)$. Thus

$$i(t) = 7.5\sin(80t)\exp(-60t)u(t) \text{ A} \tag{4.96}$$

This time $i(t)$ is oscillatory, a damped sinusoid (trace 3 in Figure 4.19), with a peak value of 3 A. ω_n is 80 rad/s, which is $80/2\pi = 12.7$ Hz. The damping term, $\exp(-60t)$ has a time constant of 1/60 s, or 16.7 ms, so the current will cease after about $5 \times 16.7 = 83$ ms.

In general, a series RLC circuit in the s-domain has an impedance of $(R + Ls + 1/Cs)$. If a step voltage, V/s, is applied, then

$$\tilde{\mathbf{i}} = \frac{V/s}{R + Ls + 1/Cs} = \frac{V/L}{s^2 + Rs/L + 1/LC} \tag{4.97}$$

Completing the square of the denominator:

$$i(t) = 4[\exp(-50t) - \exp(-200t)]u(t) \text{ A} \tag{4.92}$$

$$s^2 + \frac{R}{L}s + \frac{1}{LC} = \left(s + \frac{R}{2L}\right)^2 - \left(\frac{R^2}{4L^2} - \frac{1}{LC}\right) \qquad (4.98)$$

Three cases may be considered:

1. $R^2/4L^2 > 1/LC$, or $R > 2\sqrt{(L/C)}$

The denominator has two unequal real roots leading to a current in a form like Equation 4.92, trace 1 in Figure 4.19. The circuit is said to be *overdamped* and the current flow is unidirectional – there are no oscillations.

2. $R^2/4L^2 = 1/LC$, or $R = 2\sqrt{(L/C)}$

The denominator has two equal, real roots and the form of $i(t)$ is like Equation 4.94, trace 2 in Figure 4.19. There are no oscillations but the current falls to zero faster than in the underdamped case. The circuit is *critically damped*.

3. $R^2/4L^2 < 1/LC$, or $R < 2\sqrt{(L/C)}$

The denominator has two complex, conjugate roots. The current is oscillatory, as in Equation 4.96, with a *natural frequency*, ω_n, of $\sqrt{(1/LC - R^2/4L^2)}$, trace 3 in Figure 4.19. These oscillations die away at a rate determined by the term $\exp(-t/\tau)$, where τ, the time constant of the exponential decay is $2L/R$. The circuit is *underdamped*.

When R is zero the circuit is undamped and then it oscillates at the *resonant frequency*, $\omega_0 = 1/\sqrt{(LC)}$ – the excitation frequency that results in maximum stored energy.

Example 4.5

A solenoid magnet has a DC resistance of 0.8 Ω and an inductance of 0.2 H. It is operated from a 48V battery of negligible internal resistance. In order to prevent arcing at the switch when it is opened a shunt capacitor of 20000 µF is placed across it as in Figure 4.20a. If the switch has been closed for

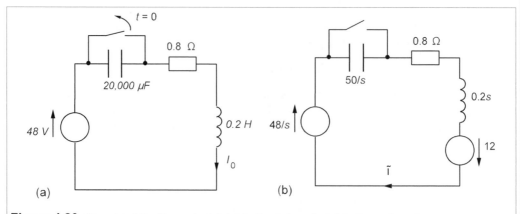

Figure 4.20 The circuit for Example 4.5 (a) In the *t*-domain (b) In the *s*-domain

a long time before being opened at $t = 0$, what is the maximum voltage across the switch? What is the maximum energy stored in (a) the capacitor and (b) the solenoid?

The steady current through the inductor before the switch is opened is $48/0.8 = 60$ A, since the capacitor is short-circuited. When the switch is opened the inductance becomes a reactance of $0.2s$ in series with a voltage source of $LI_0 = 12$. The 48V source is switched on at $t = 0$ as far as the capacitor is concerned, so becomes $48/s$ in the s-domain, giving the transformed circuit of Figure 4.20b. We can use the voltage-divider rule to find the capacitor voltage as follows:

$$\tilde{\mathbf{v}}_C = \tilde{\mathbf{v}}_T \tilde{\mathbf{X}}_C / \tilde{\mathbf{Z}} \tag{4.99}$$

where $\mathbf{X}_C = 50/s$, $\mathbf{Z} = 0.2s + 0.8 + 50/s$, and we add voltage sources giving $\tilde{\mathbf{v}}_T = 12 + 48/s$. Substituting these into Equation 4.99 leads to

$$\tilde{\mathbf{v}}_C = \frac{(12 + 48/s)(50/s)}{0.2s + 0.8 + 50/s} = 3\,000s + 12\,000 overs(s^2 + 4s + 250) \tag{4.100}$$

Expanding into partial fractions produces

$$\tilde{\mathbf{v}}_C = \frac{3\,000s + 12\,000}{s(s^2 + 4s + 250)} \equiv \frac{K_1}{s} + \frac{K_2 + K_3}{s^2 + 4s + 250} \tag{4.101}$$

and this gives the identity

$$K_1(s^2 + 4s + 250) + K_2 s^2 + K_3 s \equiv 3000s + 12000 \tag{4.102}$$

Equating coefficients on both sides:

$$\begin{aligned}
K_1 + K_2 &= 0 &&\text{(coeff. of } s^2) \\
4K_1 + K_3 &= 3000 &&\text{(coeff. of } s) \\
250K_1 &= 12000 &&\text{(constant term)}
\end{aligned} \tag{4.103}$$

Solving these produces $K_1 = 48$, $K_2 = -48$ and $K_3 = 2808$. Then

$$\tilde{\mathbf{v}}_C = \frac{48}{s} - \frac{48s - 2808}{(s+2)^2 + (250 - 4)} = \frac{48}{s} - \frac{48(s+2) - 2904}{(s+2)^2 + 15.7^2}$$

$$= \frac{48}{s} - \frac{48(s+2)}{(s+2)^2 + 15.7^2} + \frac{185(15.7)}{(s+2)^2 + 15.7^2} \tag{4.104}$$

which detransforms to

$$v_C(t) = [48 - 48\exp(-2t)\cos(15.7t) + 185\exp(-2t)\sin(15.7t)]u(t) \text{ V} \tag{4.105}$$

The sine and cosine terms can be combined to give

$$v_C(t) = [48 + \alpha\exp(-2t)\sin(15.7t + \phi)]u(t) \text{ V} \tag{4.106}$$

where $\alpha \sin\phi = -48$ and $\alpha \cos\phi = 185$, so that $\alpha = \surd(48^2 + 185^2) = 191$, and $\tan\phi = -48/185 = -0.2595$, or $\phi = -0.255$ rad. The final expression for $v_C(t)$ is

$$v_C(t) = [48 + 191\exp(-2t)\sin(15.7t - 0.255)]u(t) \text{ V} \tag{4.107}$$

A plot of Equation 4.107 is shown in Figure 4.21; it ends up as the battery voltage, 48 V.

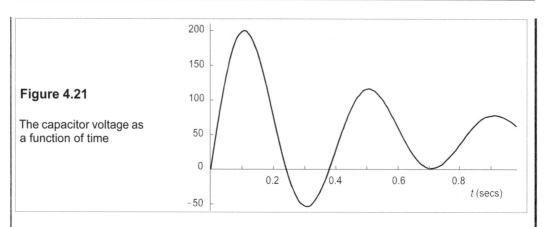

Figure 4.21

The capacitor voltage as
a function of time

$v_{C\,max}$ is found by differentiating v_C with respect to t:

$$\frac{dv_C}{dt} = [3000\cos(15.7t - 0.255) - 382\sin(15.7t - 0.255)]\exp(-2t) \qquad (4.108)$$

setting dv_C/dt to zero for a maximum produces

$$\tan(15.7t - 0.255) = 3000/382 = 7.853 \qquad (4.109)$$

which leads to $(15.7t - 0.255) = \tan^{-1}(7.853) = 1.444$ radians. Thus $15.7t = 1.444 + 0.255$, and then $t = 0.108$ s. Substituting $t = 0.108$ s into Equation 4.107 gives $v_{C\,max}$ = **201 V**. The maximum energy stored in the capacitor is $\frac{1}{2}Cv_{C\,max}^2$ = **0.5 × 20 × 10⁻³ × 201² = 404 J**.

The energy stored in the solenoid is $\frac{1}{2}Li^2$, and maximum energy is stored when the current is a maximum, so we must find $i(t)$. The transformed current is given by

$$\tilde{\mathbf{i}} = \tilde{\mathbf{v}}_T/\tilde{\mathbf{Z}} = \frac{48/s + 12}{0.2s + 0.8 + 50/s} = \frac{60s + 240}{s^2 + 4s + 250}$$

$$= \frac{60(s+2) + 120}{(s+2)^2 + (15.7)^2} \qquad (4.110)$$

which detransforms, after combining sine and cosine terms, to

$$i(t) = 60.5\exp(-2t)\sin(15.7t + 1.443)\ \text{A} \qquad (4.111)$$

This is a maximum at $t = 0$, at which time $i = 60$ A and the energy stored in the solenoid is 360 J. A 360J spark is a fairly substantial one – enough to melt about 1 g of copper.

The current must be calculated in this way, and not simply assumed to be maximal at $t = 0$, because the 48V supply has not been disconnected from the circuit and still supplies energy. This fact explains why the maximum energy stored in the capacitor is greater than the maximum energy stored in the inductor, even though there would have been significant energy losses in the resistance. (It is a tedious exercise to find $\int i^2R\,dt$, which is 150 J between $t = 0$ and $t = 0.108$ s. The battery supplies energy equal to $\int 48i\,dt$, or 193 J, for $0 \le t \le 0.108$ s.) In practice a relatively large shunt resistance (about 10 kΩ) is placed across the capacitance to discharge it after the switch is opened.

Suggestions for further reading – see Chapter 5

Problems

1. The capacitor in Figure P4.1 is initially uncharged. Transform the circuit into the *s* domain and hence find *i(t)* and *v(t)*. *[24 exp(-6t)u(t) mA, 2u(t) - 2 exp(-6t)u(t) V]*

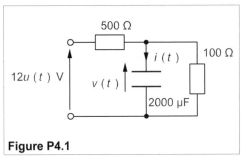

Figure P4.1

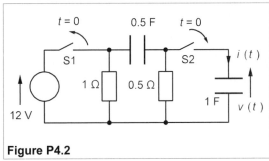

Figure P4.2

2. In the circuit of Figure P4.2, S1 has been closed for a long time before being opened at *t* = 0, when S2 is simultaneously closed. By inspection write down *i(0)*, *v(0)*, *i(∞)* and *v(∞)*. Obtain *i(t)* and *v(t)* using the Laplace transformation. When is |*v(t)*| greatest?
    [i(t) = [4 exp(-t) - 16 exp(-4t)]u(t) A, v(t) = 4[exp(-4t) - exp(-t)]u(t) V. At t = 0.462 s]

3. From Figure P4.3 write down *i(0)* and *i(∞)* by inspection. Find *i(t)*. What is the voltage across the inductance at *t* = 10 ms? *[i(t) = {0.2 - 0.1 exp(-1000t)}u(t) A. 1.362 mV]*

4. In the circuit of Figure P4.4 the current in the 2H inductance is at its steady-state value when the switch is closed at *t* = 0. What are the currents through the inductances as functions of time? *[i_{2H} (t) = {15exp(-0.5t) - 10exp(-t)}u(t) A , i_{1H} (t) = {10 - 10exp(-t)}u(t) A]*

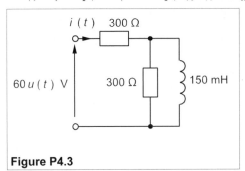

Figure P4.3

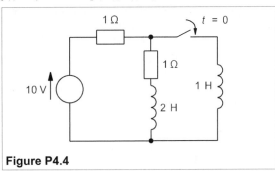

Figure P4.4

5. For the circuit of Figure P4.5, what is $v_o(t)$ if $v_{in}(t) = 5\exp(-100t)u(t)$ V? What is the minimum of $v_o(t)$? *[v_o(t) = {5 exp(-100t) - 500t exp(-100t)}u(t) V, -0.677 V]*

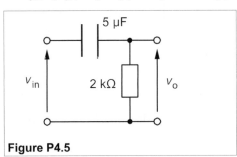

Figure P4.5

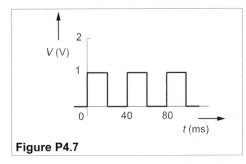

Figure P4.7

6. For the circuit of Figure P4.5, what is $v_o(t)$ if $v_{in}(t) = V\exp(-at)u(t)$? Does this answer agree with that of Problem 4.5? *[{aV/(a - 100)}exp(-at) - {100V/(a - 100)}exp(-100t)]*

7. The continuous waveform of Figure P4.7 is applied to the circuit in Figure P4.5 in which $C = 35$ μF and $R = 1$ kΩ. After what time will the resistor voltage settle to within 1% of its steady state value? Graph the resistor waveform for the first 100 ms. *[About 175 ms]*

8. Find $v_o(t)$ for the circuit of Figure P4.8 when $v_{in}(t) = 0.2u(t)$ V and when $v_{in}(t) = 10tu(t)$ mV, given $R_1 = 80$ kΩ, $R_2 = 2$ kΩ and $C = 4$ μF.
[$v_o(t) = \{8.2 - 8\exp(-3.125t)\}u(t)$ V, $v_o(t) = \{-128 + 410t + 128\exp(-3.125t)\}u(t)$ mV]

9. Find $v_o(t)$ for the circuit of Figure P4.9 if $v_{in}(t) = 10\delta(t)$ V, $R_1 = 10$ kΩ, $R_2 = 1$ MΩ, $C_1 = 100$ μF and $C_2 = 1$ μF. *[$v_o(t) = -4.975\{1 - \exp(-2.01t)\}u(t)$ V]*

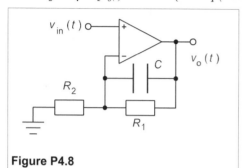

Figure P4.8

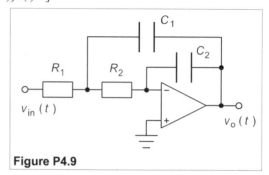

Figure P4.9

10. Repeat Problem 4.9 with $v_{in}(t) = 2.5u(t)$ V. If the supply voltages are ±15 V, how long does it take to drive the op amp into saturation? *[$v_o(t) = \{0.619 - 1.244t - 0.619\exp(-2.01t)\}u(t)$ V, 12.6 s]*

11. For the circuit of Figure P4.11 find $v_o(t)$ if $v_{in}(t) = 1.4u(t)$ V and $R_1 = 250$ Ω, $R_2 = 12.5$ kΩ, $C_1 = C_2 = 100$ nF. Find the Q-factor of the circuit. Draw a graph of the solution.
[$v_o(t) = -10\exp(-800t)\sin(5600t)u(t)$ V, $Q = 3.54$]

12. Repeat Problem 4.11 but with $C_1 = 100$ nF, $C_2 = 2.5$ μF, $R_1 = 1$ kΩ and $R_2 = 6.76$ kΩ. What is the greatest value of $|v_o(t)|$? *[$v_o(t) = -1.4 \times 10^4 t\exp(-769t)u(t)$ V, 6.7 V]*

13. The series resonant circuit of Figure P4.13 represents a circuit for exploding a wire. What is $i(t)$? What is the peak power in the resistance? At what time is the current a maximum?
[$i(t) = 165\exp(-1.667 \times 10^4 t)\sin(5.061 \times 10^5 t)u(t)$ kA, 245 MW, $t = 3.04$ μs]

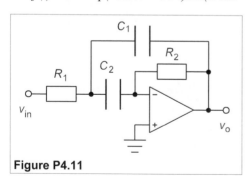

Figure P4.11

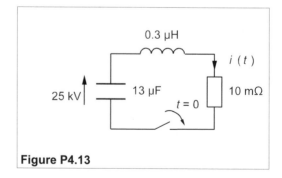

Figure P4.13

14. Repeat Problem 4.13 with the resistance changed to produce critical damping. Draw graphs of this solution and that of Problem 4.13.
[$i(t) = 83.3t\exp(-5.064 \times 10^5 t)u(t)$ GA, 1.11 GW, $t = 1.975$ μs]

15. Use the method of integration by parts to show that the Laplace transforms of $\cos\omega t$, $\sin\omega t$ and $f''(t)$ are as given in Table 4.1.

5 Bode diagrams and 2-port networks

5.1 The steady-state frequency response of circuits

OFTEN WE want to know how a circuit will behave as the frequency changes, apart from any transient response, particularly with devices such as filters. (To some extent any circuit containing reactive components acts as a filter, that is it attenuates more at some frequencies than at others.) Any network may be so arranged as to have a pair of input terminals (the input *port*) and another pair for output (the output port) as in Figure 5.1a. The ratio of output voltage to input voltage is written

$$\mathbf{V_o}/\mathbf{V_{in}} = A_v \angle \phi \tag{5.1}$$

where A_v $(= |\mathbf{V_o}/\mathbf{V_{in}}|)$ is the *amplitude response* and ϕ the *phase response* of the circuit. Conventionally these are displayed together on a *Bode diagram*[1], which is a plot of voltage gain (in dB, or $20\log_{10}A_v$) and phase against log frequency.

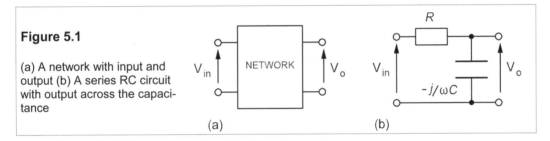

Figure 5.1

(a) A network with input and output (b) A series RC circuit with output across the capacitance

(a)   (b)

5.1.1 The series RC circuit

This circuit is a simple first-order filter (filters are dealt with more fully in Chapter 11), the Bode plot of which depends on whether the output is taken as the voltage across the resistance or the voltage across the capacitance. Consider the series RC circuit of Figure 5.1b in which the output is the voltage across the capacitance, $\mathbf{V_o}$, and the excitation, $\mathbf{V_{in}}$, is a sinusoidal voltage of angular frequency, ω. By the voltage-divider rule

$$\frac{\mathbf{V_o}}{\mathbf{V_{in}}} = \frac{-j/\omega C}{R - j/\omega C} = A_v \angle \phi \tag{5.2}$$

The voltage gain, A_v, is

---

[1] Named after H W Bode who invented them in 1940.

111

$$A_v = |\mathbf{V}_o/\mathbf{V}_{in}| = \frac{1/\omega C}{\sqrt{R^2 + 1/\omega^2 C^2}} = \frac{1}{\sqrt{\omega^2 R^2 C^2 + 1}} \tag{5.3}$$

RC is the time constant, τ, and it is useful to define the characteristic frequency, ω_c, as $1/\tau$, so that $\omega_c = 1/RC$, and then Equation 5.3 becomes

$$A_v = \frac{1}{\sqrt{(\omega/\omega_c)^2 + 1}} \tag{5.4}$$

This is a first-order *lowpass* response. The gain response in dB is

$$A_{dB} = 20\log_{10} A_v = 20\log_{10}\left[\frac{1}{\sqrt{(\omega/\omega_c)^2 + 1}}\right] \tag{5.5}$$

When $\omega \ll \omega_c$, the term under the square root sign approximates to unity, so does A_v, and $A_{dB} = 0$. When $\omega \gg \omega_c$ the term under the square root sign is approximately $(\omega/\omega_c)^2$ and $A_v = \omega_c/\omega$; leading to

$$A_{dB} = 20\log A_v \approx 20\log\omega_c - 20\log\omega \tag{5.6}$$

where log stands for $\log_{10}$. The first term on the left is a constant while the second term represents a rate of fall in gain of 20 dB per frequency decade ($= 6$ dB per octave, or doubling of frequency). As an approximation we can take A_{dB} to be zero up to $\omega = \omega_c$ and to fall thereafter at 20 dB/decade. The approximate graph of A_{dB} against $\log\omega$ has the form of the dotted line in Figure 5.2a. Because of the shape of this approximate plot, ω_c is called the *corner frequency* (sometimes the *break point*). In Figure 5.2a, the frequency has been normalised to ω_c, so the corner frequency is zero on the $\log(\omega/\omega_c)$ axis. We could write $\log(f/f_c)$ for $\log(\omega/\omega_c)$ in this diagram since $\omega = 2\pi f$ and the 2π cancels.

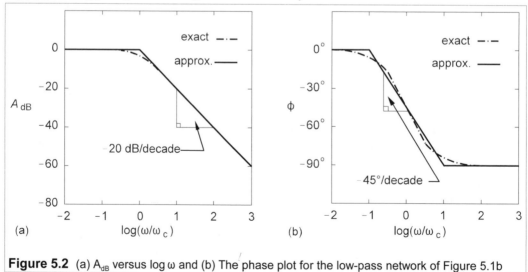

Figure 5.2 (a) A_{dB} versus $\log\omega$ and (b) The phase plot for the low-pass network of Figure 5.1b

Looking back to Equation 5.4, the exact expression for A_{dB}, one sees that when $\omega = \omega_c$, $A_{dB} = 20\log_{10}(1/\sqrt{2}) = -3$ dB, whereas the approximate expression in Equation 5.5 gives $A_{dB} = 0$. This 3dB difference at the corner frequency is the *largest* error in the straight line approximation for A_{dB}. When the frequency is half a decade up or down from the corner frequency (that is when $\omega = \sqrt{10}\omega_c$ or $\omega_c/\sqrt{10}$) the error is only 0.4 dB, almost negligible.

The graph of A_{dB} is only half the Bode diagram; the other half is a phase plot against log ω. To find the phase shift we must go back to Equation 5.1 which gives $\mathbf{V}_o/\mathbf{V}_{in}$. The numerator is wholly imaginary and negative, corresponding to a phase shift of $-90°$, while the denominator has a phase of $\tan^{-1}(-1/\omega RC) = \tan^{-1}(-\omega_c/\omega)$. The overall phase shift is therefore given by

$$\phi = \tan^{-1}(\omega_c/\omega) - 90° \qquad (5.7)$$

When $\omega \ll \omega_c$, $\tan^{-1}(\omega_c/\omega) \approx 90°$ and $\phi \approx 0°$, and when $\omega \gg \omega_c$, $\tan^{-1}(\omega_c/\omega) \approx 0°$ and $\phi \approx -90°$. When $\omega = \omega_c$ Equation 5.7 yields $\phi = \tan^{-1}(1) - 90° = -45°$: at the corner frequency the phase shift is halfway between its maximum and minimum values. A decade from ω_c the phase shifts are close to the asymptotic values $0°$ and $-90°$ (the exact values are $-6°$ and $-84°$), so the straight-line approximation to the phase shift is as shown by the unbroken line in Figure 5.2b. The phase shift falls approximately linearly with log ω from $0°$ at $\omega = 0.1\omega_c$ to $-90°$ at $\omega = 10\omega_c$, that is at $-45°$/decade. To fit an 'exact' curve using the approximation, all we have to do is find ϕ at $\omega = \sqrt{10}\omega_c$ and $\omega = \omega_c/\sqrt{10}$ from Equation 5.6, which gives $\phi = -17.5°$ and $\phi = -72.5°$ respectively, compared to $-22.5°$ and $-67.5°$ from the approximation. A smooth curve may be drawn by eye through these points to give the dotted ('exact') line of Figure 5.2b. Thus the *maximum* error of the straight-line approximation is $6°$ at a decade up and down from the corner frequency. For most purposes this error may be ignored.

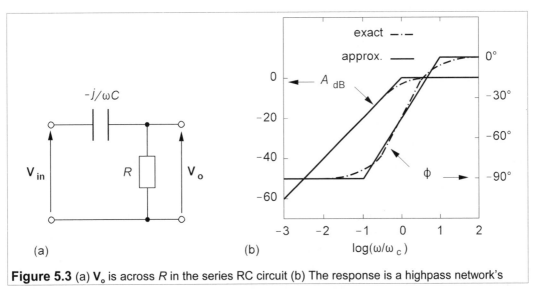

Figure 5.3 (a) $\mathbf{V}_o$ is across R in the series RC circuit (b) The response is a highpass network's

Taking the output to be the voltage across the resistance, as in Figure 5.3a, instead of the capacitance we obtain a Bode diagram that is the reverse of Figure 5.2; in other words a *highpass network* with the frequency response of Figure 5.3b.

Always ask the question: 'Does the Bode plot make sense?' The series RC circuit contains a capacitance and a resistance, so we expect no resonant effects (which require both capacitance *and* inductance) and A_{dB} must always be zero or negative. Capacitors block DC and pass AC, that is they will tend to look more like a short circuit as the frequency goes up.

In Figure 5.3a the output voltage is taken across the resistance, therefore we should expect the response to grow larger as the frequency goes up, that is the circuit is a highpass network. The phase shift should start at $-90°$, since at low frequencies the capacitance dominates and at high frequencies it will approach $0°$ as $R \gg X_C$.

5.1.2 Circuits with more than one corner frequency

Even simple RL or RC circuits can have more than one break point as in the circuit of Figure 5.4, in which an inductance is in series with a resistance.

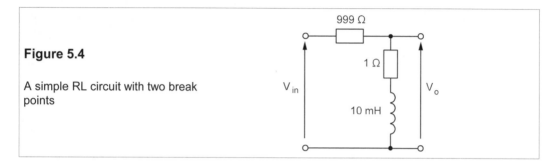

Figure 5.4

A simple RL circuit with two break points

The coil resistance is $1\ \Omega$ and has an inductance of 10 mH; the output voltage is measured across the coil, that is both of these. The response is given by the voltage divider rule as

$$\mathbf{V_o}/\mathbf{V_{in}} = \frac{1 + j0.01\omega}{1000 + j0.01\omega} = \frac{100 + j\omega}{10^5 + j\omega} \tag{5.8}$$

We see immediately that there are two break points, ω_1 and ω_2, at 100 rad/s and 100 krad/s. The amplitude response is

$$A_v = \frac{|j\omega + 100|}{|j\omega + 10^5|} = \frac{100|j(\omega/10^2) + 1|}{10^5|(j\omega/10^5) + 1|} = 10^{-3}\sqrt{\frac{(\omega/10^2)^2 + 1}{(\omega/10^5)^2 + 1}} \tag{5.9}$$

In this form the response is more readily drawn on a Bode plot. In dB,

$$A_{dB} = 20\log A_v = -60 + 10\log[(\omega/10^2)^2 + 1] - 10\log[(\omega/10^5)^2 + 1] \tag{5.10}$$

A consequence of the inductor's finite resistance is that the low-frequency roll-off is limited to three decades. Below 100 rad/s (the first break point) the second and third terms of the right-hand side of Equation 5.10 are zero, but the first term is a constant which shifts the whole plot -60 dB, and so the graphs start off at this point. Above 100 rad/s the second term on the right (which comes from the numerator of Equation 5.9) increases at 20 dB/decade. The last term on the right of Equation 5.10 (which originates from the denominator of Equation 5.9) starts to decrease at 20 dB/decade when $\omega > 100$ krad/s, and thereby cancels the $+20$ dB/decade of the second term. Consequently the overall response is 0 dB when $\omega > 100$ krad/s. The separate responses of the terms are shown in Figure 5.5a. They may then be combined to give the overall amplitude response of the circuit in Figure 5.5b.

Again, is this plot as expected? At low frequencies the inductance is a short circuit and A_v is just $R_2/(R_1 + R_2)$, or $1/1000$ ($A_{dB} = -60$ dB). At high frequencies the inductance will block the signal making $V_o = V_{in}$ and $A_v = 1$, $A_{dB} = 0$. Between the two there will be an increase of

+20 dB/decade. Figure 5.5b is in accordance with these deductions.

The phase response is given by

$$\phi = \phi_N - \phi_D = \tan^{-1}(\omega/100) - \tan^{-1}(\omega/10^5) \tag{5.11}$$

Here ϕ_N and ϕ_D are the respective phases of the numerator and denominator of Equation 5.8.

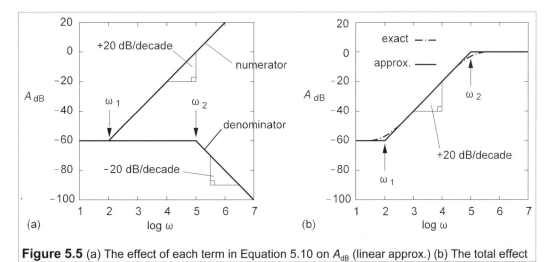

Figure 5.5 (a) The effect of each term in Equation 5.10 on A_{dB} (linear approx.) (b) The total effect

ϕ_N comes into play only when $\omega = \omega_1/10$ (= 10 rad/s) and ϕ_N at $\omega_2/10$ (= 10 krad/s). Therefore when $\omega < \omega_1/10$, $\phi = 0°$, and at $\omega = 10$ rad/s starts to increase at 45°/decade by the straight-line approximation, reaching 45° at 100 rad/s (ω_1) and 90° at 1 krad/s ($10\omega_1$). ϕ remains at 90° as ϕ_D is zero up to $\omega_2/10$ (10 krad/s). At this point ϕ starts to fall linearly back to zero again at $\omega = 1$ Mrad/s ($10\omega_2$). The corrections to the straight-line approximation are +6° at $0.1\omega_1$ and $10\omega_2$, −6° at $10\omega_1$ and $0.1\omega_2$, −5° at $\omega_1/\sqrt{10}$ and $\sqrt{10}\omega_2$ and +5° at $\sqrt{10}\omega_1$ and $\omega_2/\sqrt{10}$ as shown in Figure 5.6.

Figure 5.6

The phase response of the circuit of Figure 5.4

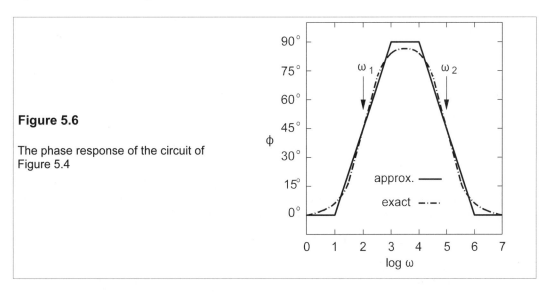

5.1.3 The general case (real poles and zeros)

The general expression for A_v is of the form

$$A_v = \frac{(j\omega)^n (j\omega + z_1)(j\omega + z_2)...}{(j\omega + p_1)(j\omega + p_2)...} \tag{5.12}$$

The corner frequencies are $z_1, z_2...$ and $p_1, p_2...$ in numerator and denominator respectively. In the numerator these are called *zeros* and in the denominator *poles*. Each factor, $(j\omega + z_n)$, in the numerator of A_v gives rise to an increase of 20 dB/decade starting at its 'zero', z_n, while each factor, $(j\omega + p_n)$ in the denominator causes a decrease of 20 dB/decade commencing at p_n. The term $(j\omega)^n$ contributes $20n$ dB/decade (n may be negative) from $\omega = 0$ to $\omega = \infty$.

In the same way each $(j\omega + z_n)$ term causes a change in phase of $+45°$/decade for $0.1z_n \le \omega \le 10z_n$, while each term in the denominator of A_v causes a phase change of $-45°$/decade for $0.1p_n \le \omega \le 10p_n$. The $(j\omega)^n$ term contributes $90n°$, at all frequencies.

5.1.4 An example with operational amplifiers

The determination of the frequency response of real circuits containing op amps is fraught with pitfalls, which can be avoided by using a sophisticated circuit analysis programme such as SPICE[2], several versions of which are available. However, over a restricted range of (low) frequencies, many op amps are reasonably close to ideal and the following simple analysis yields valuable insights.

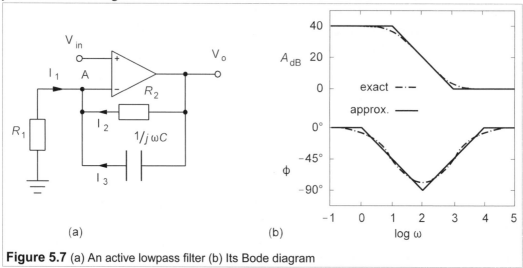

Figure 5.7 (a) An active lowpass filter (b) Its Bode diagram

Figure 5.7a is a lowpass filter circuit whose frequency response is sought. We first use nodal analysis at A, that is we sum the current into the node and set them equal to zero:

$$\mathbf{I_1} + \mathbf{I_2} + \mathbf{I_3} = 0 \tag{5.13}$$

---

[2] SPICE – Simulation Programme with Integrated Circuit Emphasis. Op amps and other circuit elements can be modelled fairly accurately by Spice and the models can be tailored by the user to fit almost any device.

Then we note that by the first golden rule the voltage at A is $\mathbf{V_{in}}$ and by Ohm's law $\mathbf{I_1} = -\mathbf{V_{in}}G_1$, $\mathbf{I_2} = (\mathbf{V_o} - \mathbf{V_{in}})G_2$ and $\mathbf{I_3} = (\mathbf{V_o} - \mathbf{V_{in}}) \times j\omega C$, where $G_1 = 1/R_1$ etc. Substituting for the currents in Equation 5.13 gives

$$-\mathbf{V_{in}}G_1 + (\mathbf{V_o} - \mathbf{V_{in}})G_2 + (\mathbf{V_o} - \mathbf{V_{in}})j\omega C = 0 \qquad (5.14)$$

Rearranging this we find

$$\mathbf{V_o}/\mathbf{V_{in}} = \frac{j\omega C + G_1 + G_2}{j\omega C + G_1} = \frac{j\omega + \omega_1}{j\omega + \omega_2} \qquad (5.15)$$

where $\omega_1 \equiv 1/R_1C + 1/R_2C$ and $\omega_2 \equiv 1/R_2C$. Suppose $R_1 = 1.01$ kΩ, $R_2 = 100$ kΩ and $C = 1$ μF, leading to $\omega_1 = 10$ rad/s and $\omega_2 = 1$ krad/s so that Equation 5.15 becomes

$$\mathbf{V_o}/\mathbf{V_{in}} = \frac{j\omega + 1000}{j\omega + 10} = \frac{100[j(\omega/1000) + 1]}{j(\omega/10) + 1} \qquad (5.16)$$

This yields A_{dB}:

$$A_{dB} = 20\log\left|\frac{\mathbf{V_o}}{\mathbf{V_{in}}}\right| = 40 + 10\log[(\omega/1000)^2 + 1] - 10\log[(\omega/10)^2 + 1] \qquad (5.17)$$

And the amplitude response is readily drawn, as in Figure 5.7b.

From Equation 5.16 the phase is

$$\phi = \tan^{-1}(\omega/1000) - \tan^{-1}(\omega/10) \qquad (5.18)$$

The approximate plot starts at zero up to $\omega_1/10$ (1 rad/s) and then falls at 45°/decade to $-90°$ at $10\,\omega_1$ (100 rad/s). Here the second break point comes into play and ϕ increases at 45°/decade up to 0° at $10\omega_2$ (1 Mrad/s). Because the break points are relatively close this time ($\omega_2/\omega_1 = 100$), the correction at 100 rad/s – midway between break points on a log scale – is double the usual 6°. Whenever corner frequencies are close together the approximate plots are less accurate at the frequencies between them.

5.1.5 Resonant circuits

Resonant circuits, in the form of the series RLC circuit, have been discussed in the previous chapter in Section 4.3; they present problems for those seeking to draw Bode diagrams because, in the region of resonance, the amplitude and phase of the output may rapidly change by a huge amount, especially in high-Q circuits. And because resonant circuits produce complex poles in their transfer functions, the method of analysis in Section 5.1.4 is inapplicable. We shall look at only one example – the series RLC circuit once more, of Figure 5.8, where the output is the voltage across the capacitance.

By the voltage-divider rule, the response is

$$\frac{\mathbf{V_o}}{\mathbf{V_{in}}} = \frac{1/j\omega C}{R + j\omega L + 1/j\omega C} = \frac{1/LC}{1/LC - \omega^2 + j\omega R/L} \qquad (5.19)$$

Replacing $\sqrt{(LC)}$ by ω_0 and R/L by B leads to

$$\frac{\mathbf{V_o}}{\mathbf{V_{in}}} = \frac{\omega_0^2}{\omega_0^2 - \omega^2 + jB\omega} = \frac{1}{1 - (\omega/\omega_0)^2 + j(B/\omega_0)(\omega/\omega_0)} = \frac{1}{1 - \Omega^2 + j\Omega/Q} \qquad (5.20)$$

where $\Omega \equiv \omega/\omega_0$ and $Q \equiv \omega_0/B$. If the denominator is put into the form $(j\omega + a)(j\omega + b)$, a and b will be complex poles.

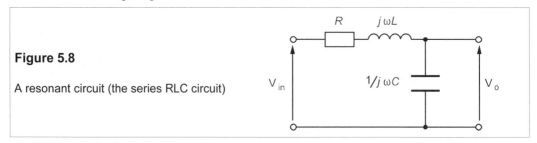

Figure 5.8

A resonant circuit (the series RLC circuit)

From Equation 5.20 A_v is found:

$$A_v = \frac{1}{|1 - \Omega^2 + j\Omega/Q|} = \frac{1}{\sqrt{(1 - \Omega^2)^2 + (\Omega/Q)^2}} \qquad (5.21)$$

Hence

$$A_{dB} = -10\log[(1 - \Omega^2)^2 + (\Omega/Q)^2] \qquad (5.22)$$

When $\Omega \ll 1$, that is $\omega \ll \omega_0$, $A_{dB} = 0$; and when $\Omega \gg 1$, $A_{dB} \rightarrow -10\log\Omega^4 = -40\log\Omega$: the amplitude response at high frequencies falls by 40 dB/decade. The approximate amplitude graph comprises two straight lines intersecting at $\Omega = 1$ ($\omega = \omega_0$). But this analysis overlooks the important contribution of the $(\Omega/Q)^2$ term at frequencies close to resonance, where Ω is close to unity. Putting $\Omega = 1$ into Equation 5.22 yields $A_{dB} = -10\log(1/Q)^2 = +20\log Q$. For $Q = 10$, $A_{dB} = +20$ dB at $\omega = \omega_0$, and this is the error in the straight-line approximation! It is left as an exercise to show that, if Q is large, $A_{dB} = +3$ dB when $\Omega = 0.54$ and 1.31. The amplitude plot goes as Figure 5.9a.

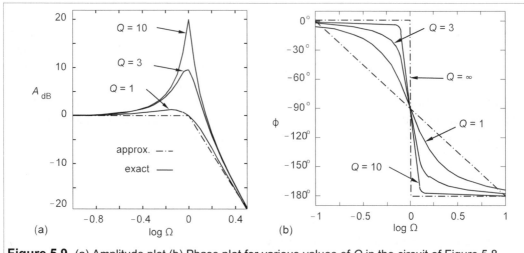

Figure 5.9 (a) Amplitude plot (b) Phase plot for various values of Q in the circuit of Figure 5.8

From Equation 5.20 the phase response is $-\tan^{-1}[\Omega Q^{-1}/(1 - \Omega^2)]$; when $\Omega = 0$ the phase is $0°$. When Ω is small, $\phi \approx -\tan^{-1}(\Omega/Q) \approx \Omega/Q$; thus for large Q-values the phase angle stays small until Ω approaches unity (that is $\omega \approx \omega_0$). When $\Omega = 1$, $\phi = -90°$ and for large Ω, $\phi \approx -\tan^{-1}[[1/(-\Omega Q)]$. This expression requires care in evaluation. The numerator inside the square bracket corresponds to positive j, while the denominator is negative real, so the angle lies in the second quadrant, but the minus sign in front of the $\tan^{-1}$ puts ϕ into the third quadrant, making $\phi \approx \tan^{-1}(1/\Omega Q) - 180°$. For instance, if $\Omega = 10$ and $Q = 5$, $\phi \approx -179°$. The phase angle goes from $0°$ to $-90°$ then to $-180°$ as Ω goes from 0 to 1 and then to infinity. When Q is high the phase change occurs near ω_0. Figure 5.9b shows the phase plot for $Q = 1, 3, 10$ and infinity. When Q is not large the phase change can be approximated by a straight line of slope $-90°$/decade, as shown by the dashed line in Figure 5.9b, but the errors are still large close to resonance.

5.2 **Two-port networks**

A *port* is a pair of network terminals which are used as either an input or an output – many circuits already discussed happen to have been 2-port networks. 2-port network analysis is useful when one wishes to combine networks, such as amplifiers, filters, transmission lines, matching circuits etc. Figure 5.10 shows the conventional representation of a 2-port network.

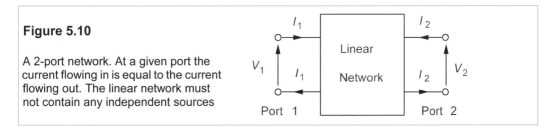

Figure 5.10

A 2-port network. At a given port the current flowing in is equal to the current flowing out. The linear network must not contain any independent sources

There is a restriction on the currents flowing in and out of the four terminals: the current into the upper terminal, I_1, is equal to the current flowing out of its lower terminal, while the current into the upper terminal of port 2, I_2, also equals that flowing out of its bottom terminal. Each 2-port network is described by a set of four parameters by which input and output voltages and currents are related. Altogether there are six such sets of parameters, but once one set of four has been determined, all the other five may be derived from it. Table 5.1 gives the conversions between sets of parameters.

5.2.1 *The impedance or z-parameters*

These parameters are used to relate the voltages to the currents:

$$\begin{aligned} V_1 &= z_{11}I_1 + z_{12}I_2 \\ V_2 &= z_{21}I_1 + z_{22}I_2 \end{aligned} \tag{5.23}$$

Looking at Equation 5.23 we see that $z_{11} = V_1/I_1$, when $I_2 = 0$: z_{11} is the input impedance of port 1 while port 2 is open-circuited. And similarly, $z_{12} = V_1/I_2$, when $I_1 = 0$, that is when port 1 is

Table 5.1 *2-port parameter conversions (note: Δh = h₁₁h₂₂ − h₁₂h₂₁ etc.)*

$z_{11} = \dfrac{y_{22}}{\Delta y} = \dfrac{\Delta h}{h_{22}} = \dfrac{1}{g_{11}} = \dfrac{a_{11}}{a_{21}} = \dfrac{b_{22}}{b_{21}}$	$g_{11} = \dfrac{1}{z_{11}} = \dfrac{\Delta y}{y_{22}} = \dfrac{h_{22}}{\Delta h} = \dfrac{a_{21}}{a_{11}} = \dfrac{b_{21}}{b_{22}}$
$z_{12} = \dfrac{-y_{12}}{\Delta y} = \dfrac{h_{12}}{h_{22}} = \dfrac{-g_{12}}{g_{11}} = \dfrac{\Delta a}{a_{21}} = \dfrac{1}{b_{21}}$	$g_{12} = \dfrac{-z_{12}}{z_{11}} = \dfrac{y_{12}}{y_{22}} = \dfrac{-h_{12}}{\Delta h} = \dfrac{\Delta a}{a_{11}} = \dfrac{-1}{b_{22}}$
$z_{21} = \dfrac{-y_{21}}{\Delta y} = \dfrac{-h_{21}}{h_{22}} = \dfrac{g_{21}}{g_{11}} = \dfrac{1}{a_{21}} = \dfrac{\Delta b}{b_{21}}$	$g_{21} = \dfrac{z_{21}}{z_{11}} = \dfrac{-y_{21}}{y_{22}} = \dfrac{h_{21}}{\Delta h} = \dfrac{1}{a_{11}} = \dfrac{\Delta b}{b_{22}}$
$z_{22} = \dfrac{y_{11}}{\Delta y} = \dfrac{1}{h_{22}} = \dfrac{\Delta g}{g_{11}} = \dfrac{a_{22}}{a_{21}} = \dfrac{b_{11}}{b_{21}}$	$g_{22} = \dfrac{\Delta z}{z_{11}} = \dfrac{1}{y_{22}} = \dfrac{h_{11}}{\Delta h} = \dfrac{a_{12}}{a_{11}} = \dfrac{b_{12}}{b_{22}}$
$y_{11} = \dfrac{z_{22}}{\Delta z} = \dfrac{1}{h_{11}} = \dfrac{\Delta g}{g_{22}} = \dfrac{a_{22}}{a_{12}} = \dfrac{b_{11}}{b_{12}}$	$a_{11} = \dfrac{z_{11}}{z_{21}} = \dfrac{-y_{22}}{y_{21}} = \dfrac{-\Delta h}{h_{21}} = \dfrac{1}{g_{21}} = \dfrac{b_{22}}{\Delta b}$
$y_{12} = \dfrac{-z_{12}}{\Delta z} = \dfrac{-h_{12}}{h_{11}} = \dfrac{g_{12}}{g_{22}} = \dfrac{-\Delta a}{a_{12}} = \dfrac{-1}{b_{12}}$	$a_{12} = \dfrac{\Delta z}{z_{21}} = \dfrac{-1}{y_{21}} = \dfrac{-h_{11}}{h_{21}} = \dfrac{g_{22}}{g_{21}} = \dfrac{b_{12}}{\Delta b}$
$y_{21} = \dfrac{-z_{21}}{\Delta z} = \dfrac{h_{21}}{h_{11}} = \dfrac{-g_{21}}{g_{22}} = \dfrac{-1}{a_{12}} = \dfrac{-\Delta b}{b_{12}}$	$a_{21} = \dfrac{1}{z_{21}} = \dfrac{-\Delta y}{y_{21}} = \dfrac{-h_{22}}{h_{21}} = \dfrac{g_{11}}{g_{21}} = \dfrac{b_{21}}{\Delta b}$
$y_{22} = \dfrac{z_{11}}{\Delta z} = \dfrac{\Delta h}{h_{11}} = \dfrac{1}{g_{22}} = \dfrac{a_{11}}{a_{12}} = \dfrac{b_{22}}{b_{12}}$	$a_{22} = \dfrac{z_{22}}{z_{21}} = \dfrac{-y_{11}}{y_{21}} = \dfrac{-1}{h_{21}} = \dfrac{\Delta g}{g_{21}} = \dfrac{b_{11}}{\Delta b}$
$h_{11} = \dfrac{\Delta z}{z_{22}} = \dfrac{1}{y_{11}} = \dfrac{g_{22}}{\Delta g} = \dfrac{a_{12}}{a_{22}} = \dfrac{b_{12}}{b_{11}}$	$b_{11} = \dfrac{z_{22}}{z_{12}} = \dfrac{-y_{11}}{y_{12}} = \dfrac{1}{h_{12}} = \dfrac{-\Delta g}{g_{12}} = \dfrac{a_{22}}{\Delta a}$
$h_{12} = \dfrac{z_{12}}{z_{22}} = \dfrac{-y_{12}}{y_{11}} = \dfrac{-g_{12}}{\Delta g} = \dfrac{\Delta a}{a_{22}} = \dfrac{1}{b_{11}}$	$b_{12} = \dfrac{\Delta z}{z_{12}} = \dfrac{-1}{y_{12}} = \dfrac{h_{11}}{h_{12}} = \dfrac{-g_{22}}{g_{12}} = \dfrac{a_{12}}{\Delta a}$
$h_{21} = \dfrac{-z_{21}}{z_{22}} = \dfrac{y_{21}}{y_{11}} = \dfrac{-g_{21}}{\Delta g} = \dfrac{-1}{a_{22}} = \dfrac{-\Delta b}{b_{11}}$	$b_{21} = \dfrac{1}{z_{12}} = \dfrac{-\Delta y}{y_{12}} = \dfrac{h_{22}}{h_{12}} = \dfrac{-g_{11}}{g_{12}} = \dfrac{a_{21}}{\Delta a}$
$h_{22} = \dfrac{1}{z_{22}} = \dfrac{\Delta y}{y_{11}} = \dfrac{g_{11}}{\Delta g} = \dfrac{a_{21}}{a_{22}} = \dfrac{b_{21}}{b_{11}}$	$b_{22} = \dfrac{z_{11}}{z_{12}} = \dfrac{-y_{22}}{y_{12}} = \dfrac{\Delta h}{h_{12}} = \dfrac{-1}{g_{12}} = \dfrac{a_{11}}{\Delta a}$

open circuit; $z_{21} = V_2/I_1$, when $I_2 = 0$ and $z_{22} = V_2/I_2$, when $I_1 = 0$: z_{22} is the input impedance of port 2 when port 1 is open circuit. In matrix form Equation 5.23 is:

$$\begin{bmatrix} V_1 \\ V_2 \end{bmatrix} = \begin{bmatrix} z_{11} & z_{12} \\ z_{21} & z_{22} \end{bmatrix} \begin{bmatrix} I_1 \\ I_2 \end{bmatrix} \tag{5.24}$$

Or,

$$\mathbf{V} = \mathbf{zI} \tag{5.25}$$

where $\mathbf{V}$, $\mathbf{I}$ and $\mathbf{z}$ are the voltage, current and impedance *matrices*, which is the reason for writing Equation 5.25 in this manner.

Example 5.1

What is the *z*-matrix for the circuit of Figure 5.11?

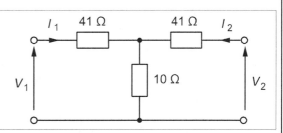

Figure 5.11

A symmetrical, 20dB, T-attenuator for a 50Ω line

This circuit is called a *symmetrical T-attenuator*. When a voltage source is connected via a transmission line to a load, it can be reduced in amplitude by inserting a resistive attenuator in the line. However, unless the attenuator has certain properties, it will cause problems with signal transmission. One property the attenuator should have is that when the correct load resistance is connected across one set of terminals, the input resistance measured at the other set is the same as the load resistance. This is the *characteristic impedance* of the line. 50 Ω and 600 Ω are frequently-encountered values for the characteristic impedance.

The input resistance of port 1 with port 2 open-circuit is clearly 41 + 10 = 51 Ω, which is z_{11}. By the symmetry of the network, $z_{22} = z_{11}$ and z_{12} is V_1/I_2 with port 1 open-circuit, so that I_2 flows through the right-hand 41Ω resistance and then the 10Ω resistance, no current flowing through the left-hand 41 Ω. Thus V_1 (= $10I_2$) is the voltage across the 10Ω resistance and $z_{12} = 10I_2/I_2 = 10$ Ω. Again by symmetry $z_{12} = z_{21}$ and the *z*-matrix is

$$\mathbf{z} = \begin{bmatrix} 51 & 10 \\ 10 & 51 \end{bmatrix} \Omega \qquad (5.26)$$

The *z*-parameters are not often used, but the *y*-parameters are.

5.2.2 *The admittance or y-parameters*

As the name implies, these relate currents to voltages. The defining relationships are

$$\begin{aligned} I_1 &= y_{11}V_1 + y_{12}V_2 \\ I_2 &= y_{21}V_1 + y_{22}V_2 \end{aligned} \qquad (5.27)$$

This can be written as $\mathbf{I} = \mathbf{yV}$, where $\mathbf{I}$, $\mathbf{y}$ and $\mathbf{V}$ are the appropriate matrices. We can easily find the *y*-parameters of the symmetrical T-attenuator of the previous example. y_{22} is the input admittance of port 2 when $V_1 = 0$, that is when port 1 is *short*-circuited (as opposed to being open-circuit for the *z*-parameters), as in Figure 5.12, where the resistances have been replaced by their equivalent conductances (1/41 = 24.4 mS etc.)

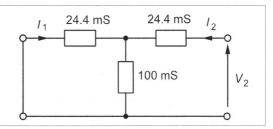

Figure 5.12

Port 1 is short-circuited to find y_{22} and y_{12}

Short-circuiting port 1 places the left-hand 24.4mS conductance in parallel with the 100mS conductance, for a combined conductance of 124.4 mS which is in series with the left-hand 24.4mS conductance making $(124.4 \times 24.4)/(124.4 + 24.4) = 20.4$ mS in total. This is the input admittance, $y_{22} = 20.4$ mS. Since the network is symmetrical $y_{22} = y_{11}$. To find y_{12} requires I_1/V_2 when $V_1 = 0$, that is when port 1 is short-circuited. I_2 is the input current at port 2, that is $y_{22}V_2$, and it flows through the right-hand 24.4mS conductance and then the parallel combination. The current flowing from right to left through the left-hand 24.4mS conductance is $24.4I_2/124.4$, by the current divider rule. Now I_1 is the current flowing through the left-hand 24.4mS conductance from left to right, $I_1 -24.4I_2/124.4 = -24.4y_{22}V_2/124.4$. Therefore

$$y_{12} = \frac{I_1}{V_2} = \frac{-24.4y_{22}V_2}{124.4V_2} = \frac{-24.4 \times 20.4}{124.4} = -4 \text{ mS} \tag{5.28}$$

As the network is symmetrical, so is the y-matrix:

$$\mathbf{y} = \begin{bmatrix} 20.4 & -4 \\ -4 & 20.4 \end{bmatrix} \text{mS} = \begin{bmatrix} 0.0204 & -0.004 \\ -0.004 & 0.0204 \end{bmatrix} \text{S} \tag{5.29}$$

The $\mathbf{y}$ and $\mathbf{z}$ matrices are *inverses*: $\mathbf{y} = \mathbf{z}^{-1}$ (this relationship does *not* apply to the individual matrix elements), or $\mathbf{yz} = \mathbf{u}$, where $\mathbf{u}$ is the unit 2×2 matrix consisting of just ones on the leading diagonal and zeros off it. It is a worthwhile exercise to verify this relationship; the multiplication goes as follows:

$$\begin{bmatrix} 0.0204 & -0.004 \\ -0.004 & 0.0204 \end{bmatrix}\begin{bmatrix} 51 & 10 \\ 10 & 51 \end{bmatrix} = \begin{bmatrix} 0.0204 \times 51 - 0.004 \times 10 & 0.0204 \times 10 - 0.004 \times 51 \\ -0.004 \times 51 + 0.0204 \times 10 & -0.004 \times 10 + 0.0204 \times 51 \end{bmatrix}$$

$$= \begin{bmatrix} 1 & 0 \\ 0 & 1 \end{bmatrix} = \mathbf{u} \tag{5.30}$$

The y-parameters are used in describing the properties of field-effect transistors and amplifiers (see Chapter 9), for which we need the y-equivalent circuit. Looking at the defining equations above we see that the first, $I_1 = y_{11}V_1 + y_{12}V_2$ relates the current into port 1 to the voltages at both ports. The first term on the right is just an admittance, y_{11}, with the voltage at port 1, V_1, across it. The second term on the right involves the voltage at the other port and can be modelled as a *dependent* current source of value $y_{12}V_2$. The second defining equation yields a similar pair of components at port 2, to make the complete y-equivalent circuit of Figure 5.13.

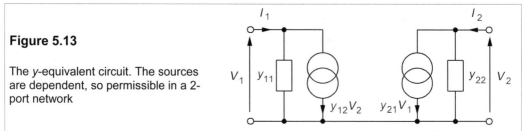

Figure 5.13

The y-equivalent circuit. The sources are dependent, so permissible in a 2-port network

The y-equivalent circuit can be used to derive any of the other sets of parameters, for example the z-parameters, in terms of the y-parameters. From the definition, Equation 5.23, $z_{11} = V_1/I_1$, $I_2 = 0$, that is port 2 is open circuit and the current from the right-hand source passes through the admittance y_{22}. Thus $V_2 = -y_{21}V_1/y_{22}$ (note the minus sign). Now by KCL, the current through y_{11} is $I_1 - y_{12}V_2$, so the voltage across it, $V_1 = (I_1 - y_{12}V_2)/y_{11}$. Substituting into this expression for V_2 leads to

$$V_1 = \frac{I_1}{y_{11}} + \frac{y_{12}y_{21}V_1}{y_{11}y_{22}} \tag{5.31}$$

Rearranging,
$$V_1\left(1 - \frac{y_{12}y_{21}}{y_{11}y_{22}}\right) = \frac{I_1}{y_{11}} \tag{5.32}$$

This produces

$$\frac{V_1}{I_1} = z_{11} = \frac{y_{22}}{y_{11}y_{22} - y_{12}y_{21}} = \frac{y_{22}}{\Delta y} \tag{5.33}$$

where Δy stands for the determinant of the y-matrix; the other parameters are found similarly.

5.2.3 *The hybrid or h-parameters*

The z- and y-parameters (or *immittance* parameters) are dimensionally homogeneous, but there are two sets of parameters which are mixed dimensionally – the h-parameters and the inverse hybrid or g-parameters. The h-parameters are frequently used in modelling bipolar transistors (see Chapter 8). They are defined by the equations

$$\begin{aligned} V_1 &= h_{11}I_1 + h_{12}V_2 \\ I_2 &= h_{21}I_1 + h_{22}V_2 \end{aligned} \tag{5.34}$$

while the g-parameters are defined by

$$\begin{aligned} I_1 &= g_{11}V_1 + g_{12}I_2 \\ V_2 &= g_{21}V_1 + g_{22}I_2 \end{aligned} \tag{5.35}$$

The g- and h-matrices form an inverse pair like the z- and y-matrices. We shall use these parameters when transistor amplifiers are described.

5.2.4 *The transmission or a-parameters*

The transmission parameters are especially useful for dealing with 2-port networks joined in cascade, as in Figure 5.14. The a-parameters relate voltage and current at the output (port 2) to voltage and current at the input (port 1) and are defined by

$$\begin{aligned} V_1 &= a_{11}V_2 + a_{12}I_2 \\ I_1 &= a_{21}V_2 + a_{22}I_2 \end{aligned} \tag{5.36}$$

Note that now I_2 flows *out* of port 2. In power transmission the a-parameters are known as the

ABCD-parameters, where, $A = a_{11}$, $B = a_{12}$, $C = a_{21}$ and $D = a_{22}$.

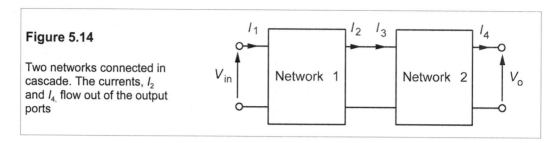

Figure 5.14

Two networks connected in cascade. The currents, I_2 and I_4, flow out of the output ports

There is a sixth and final set of parameters known as the inverse transmission or *b*-parameters which relate current and voltage at the output (port 2) to those at the input:

$$V_2 = b_{11}V_1 + b_{12}I_1$$
$$I_2 = b_{21}V_1 + b_{22}I_1$$

(5.37)

with I_1 flowing *out* of port 1. In power transmission the *b*-matrix is known as the *abcd*-matrix, with $a = b_{11}$, $b = b_{12}$, $c = b_{21}$ and $d = b_{22}$. The *a*- and *b*-matrices are also inverses of each other. A most important property of the cascade connection is that the *a*-parameters of two cascaded 2-ports are found by multiplying the *a*-matrices of the constituent networks and the same for the *b*-matrices.

5.2.5 *The reciprocity theorem*

The reciprocity theorem states:

If, in a linear network containing only passive bilateral impedances, the application of an ideal voltage source at a pair of terminals produces a current, I, in an ideal ammeter in any branch of that network, then the interchange of ammeter and voltage source will make no difference to the ammeter reading.

'Bilateral' simply means that it does not matter which way round the impedance is connected. The network must not contain any sources, whether dependent or independent. When applied to a 2-port network, the reciprocity theorem simply means that if the voltage source is connected across port 1 and the ammeter across port 2, as in Figure 5.15, the ammeter deflexion will be the same when they are interchanged.

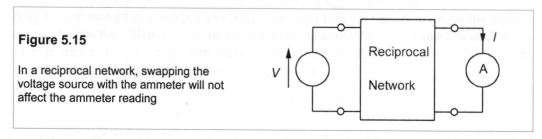

Figure 5.15

In a reciprocal network, swapping the voltage source with the ammeter will not affect the ammeter reading

Example 5.2

As a demonstration of the reciprocity theorem, look at the circuit of Figure 5.16 in which a voltage source is connected to three impedances with an ammeter in series with the 2Ω resistance. The truth of the reciprocity theorem in this case is readily shown by transforming the voltage source into a current source, but in the following method there is less inevitability in the final result.

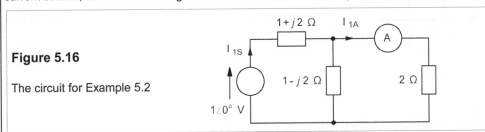

Figure 5.16

The circuit for Example 5.2

I_{1S} is the total current supplied by the source which divides, part through the $(1 - j2)\Omega$ branch and part through the 2Ω branch and ammeter. The total impedance seen by the source is $(1 + j2)$ Ω in series with [2 Ω // $(1 - j2)$ Ω]. The latter is, by the product-over-sum rule

$$\frac{2(1 - j2)}{2 + 1 - j2} = \frac{4.472\angle-63.4°}{3.606\angle-33.7°} = 1.24\angle-29.7° = 1.077 - j0.614 \text{ A} \qquad (5.38)$$

The total impedance is then

$$(1.077 - j0.614) + (1 + j2) = 2.077 + j1.386 = 2.497\angle33.7° \text{ Ω} \qquad (5.39)$$

and $I_{1S} = 1\angle0° \div 2.497\angle33.7° = 0.4\angle-33.7°$ A. The current-divider rule gives I_{1A} as

$$\mathbf{I_{1A}} = \frac{(1 - j2)\mathbf{I_{1S}}}{2 + (1 - j2)} = \frac{2.236\angle-63.4° \times 0.4\angle-33.7°}{3.606\angle-33.7°} = 0.248\angle-63.4° \text{ A} \qquad (5.40)$$

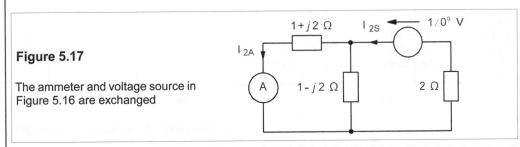

Figure 5.17

The ammeter and voltage source in Figure 5.16 are exchanged

Now the voltage source and ammeter are interchanged as in Figure 5.17. This time the source current, I_{2S}, divides between the $(1 + j2)\Omega$ and the $(1 - j2)\Omega$ impedances, which in parallel amount to an impedance of $2.5\angle0°$ Ω. The source therefore sees an impedance of 4.5 Ω, making $I_{2S} = 0.222\angle0°$ A. I_{2A} becomes

$$\mathbf{I_{2A}} = \frac{1 - j2}{(1 - j2) + (1 + j2)} \times \mathbf{I_{2S}} = \frac{2.236\angle-63.4°}{2} \times 0.222\angle0° \qquad (5.41)$$

$$= 0.248\angle-63.4° \text{ A}$$

As expected from the reciprocity theorem the answer agrees with that of Equation 5.40.

Suggestions for further reading

The following books cover all the topics in Chapters 1–5:

Introduction to electric circuits by R Dorf and J A Svoboda (Wiley, 5th ed., 2000). Very lengthy after the American fashion.
Electric circuits by J W Nilsson (Addison-Wesley, 6th ed., 2000). Thorough, popular, but long-winded.
Introduction to electric circuits by R Powell (Arnold, 1995). Succinct.
Schaum's electronic tutor of electric circuits by J Edminister and M Nahvi (McGraw-Hill, 3rd ed., 1996). Good value for money.
Introductory circuit analysis by R Boylestad (Prentice Hall, 9th ed. 1999). Bulky.

Problems

(The problems on Bode diagrams assume sinusoidal excitation)

1. For the circuit of Figure P5.1 find V_o/V_{in} as a function of frequency then draw the Bode diagram.
   $[0.9/\{1 + j(2\pi f/10^4)\}]$
2. Repeat Problem P5.1 for the circuit of Figure P5.2. $[jf/(5 \times 10^5 + j2f)]$

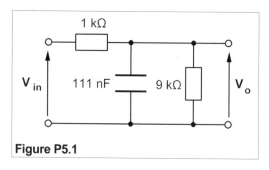

Figure P5.1

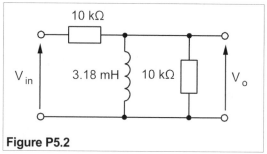

Figure P5.2

3. What will the circuit of Figure P5.3 approximate to at low and at high frequencies? Find V_o/V_{in} as a function of ω and confirm the low and high frequency responses. Draw the Bode plot when $C_1 = 100$ nF, $C_2 = 1$ μF, $R_1 = 1$ kΩ and $R_2 = 10$ kΩ.
   $[j\omega R_2 C_1/\{1 - \omega^2 R_1 R_2 C_1 C_2 + j\omega(R_2 C_1 + R_1 C_1 + R_2 C_2)\}]$
4. Find V_o/V_{in} as a function of ω for the circuit of Figure P5.4 and draw the Bode diagram when $R_1 = 5$ kΩ, $R_2 = 500$ kΩ and $C = 20$ nF. $[-R_2/(R_1 + j\omega CR_1R_2)]$

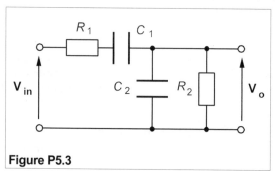

Figure P5.3

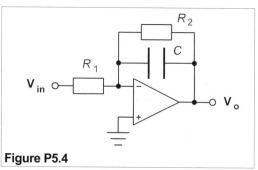

Figure P5.4

5. To what does the response of the circuit of Figure P5.5 approximate at high and at low frequencies?

Find V_o/V_{in} as a function of ω to confirm your predictions. Draw the Bode diagram when $R = 10$ kΩ and $C = 30$ nF. $[(1 + j\omega RC)/(1 - j\omega RC)]$

6. To what does the circuit of Figure P5.6 approximate at high and low frequencies? Find V_o/V_{in} and confirm this. Draw the Bode diagram for $R_1 = R_2 = 33$ kΩ, $C_1 = 30$ nF and $C_2 = 1$ nF.
    $[-j\omega R_2 C_1/\{(1 - \omega^2 R_1 R_2 C_1 C_2) + j\omega C_2 (R_1 + R_2)\}]$

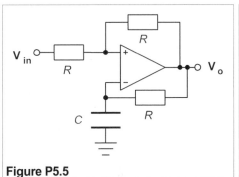

Figure P5.5

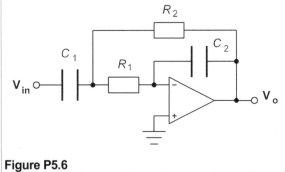

Figure P5.6

7. Repeat the first two parts of Problem 5.6 for the circuit of Figure P5.7. Draw the Bode diagram for $R = 22$ kΩ and $C = 455$ pF. $[\{1 - (\omega RC)^2 + j3\omega RC\}^{-1}]$

8. Repeat Problem 5.6 for the circuit of Figure P5.8. $[(1 + j\omega RC)^{-2}]$

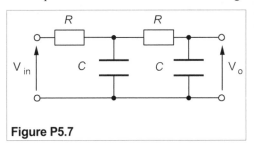

Figure P5.7

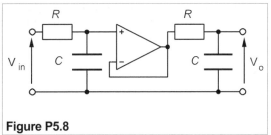

Figure P5.8

9. How would you expect the circuit of Figure P5.9 to behave at high and low frequencies? Find V_o/V_{in} and draw the Bode diagram when $R = 10$ kΩ, $C = 3$ nF and $L = 3$ mH.
    $[-\omega^2 LC/(1 - \omega^2 LC + j\omega L/R)]$

10. For the two-port network of Figure P5.10, find the y-parameters and draw the y-parameter equivalent circuit. From the y-equivalent circuit find V_2 if $V_1 = 15\angle-60°$ V.
    $[y_{11} = y_{22} = 120\angle17.2°$ mS, $y_{12} = y_{21} = 92\angle157.4°$ mS. $V_2 = 11.5\angle-99.8°$ V]

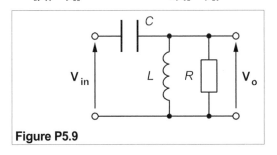

Figure P5.9

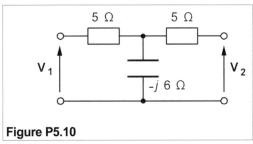

Figure P5.10

11. For the circuit of Figure P5.10 find the z-parameters and draw the z-parameter equivalent circuit. From this find I_2 if $V_1 = 120\angle60°$ V and a reactance of $j6$ Ω is connected across the output terminals. $[z_{11} = z_{22} = 5 - j6$ Ω, $z_{12} = z_{21} = -j6$ Ω. $I_2 = 10.6\angle176.2°$ A]

12. In the network of Figure P5.12 the short-circuit current though XY is $0.662\angle 70.7°$ A when $\mathbf{V_{AB}} = 10\angle 0°$ V. Show by the reciprocity theorem that the current though the inductance is $69\angle -26°$ mA, when $\mathbf{V_{XY}} = 1.34\angle -58°$ V and AB is short circuited.

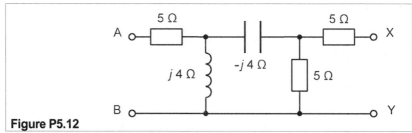

Figure P5.12

13. A transistor has $h_{11} = 1$ kΩ, $h_{12} = 0.001$, $h_{21} = 100$ and $h_{22} = 100$ μS. Derive the *h*-parameter equivalent circuit using Equation 5.34 and hence find the voltage and current gains (V_2/V_1 and I_2/I_1) if the load across the output terminals is 2 kΩ. *[−200 and −83]*

14. Two identical amplifiers, with the same *h*-parameters as in Problem 13, are connected in cascade. Find the *a*-parameters from the *h*-parameters using Table 5.1, and hence the *a*-parameters of the combination. What is the voltage gain of the combination (a) with the output open-circuit and (b) with a 10 kΩ resistance connected across the output?
[$a_{11} = 10^{-5}$, $a_{12} = 0.1$, $a_{21} = 10^{-8}$ and $a_{22} = 1.1 \times 10^{-4}$ (a) 1.22×10^5 (b) 4.76×10^4]

15. Find the *y*-parameters of the low-frequency equivalent circuit of the FET common-source amplifier shown in Figure P5.15, and hence find the voltage gain, v_o/v_{in}. What are the *h*-parameters of this circuit? *[$y_{11} = 1$ μS, $y_{12} = 0$, $y_{21} = 5$ mS, $y_{22} = 1$ mS. Voltage gain = −5. $h_{11} = 1$ MΩ, $h_{12} = 0$, $h_{21} = 5000$, $h_{22} = 1$ mS.]*

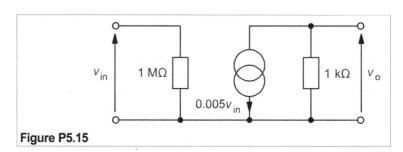

Figure P5.15

6 Semiconductors

SEMICONDUCTORS are the materials at the heart of many electronic devices. The elemental semiconductors are few, the only ones of any practical use being germanium (Ge) and silicon (Si) and only silicon is widely used nowadays. However, there are some compound semiconductors which find applications in light-emitting diodes, semiconductor lasers, microwave devices and other specialised areas. Most of these are based on elements from groups III and V of the periodic table (see Table 6.1) and occasionally from groups II and VI. These are known respectively as III–V and II–VI compounds, examples being gallium arsenide (GaAs), gallium phosphide (GaP) and cadmium sulphide (CdS).

Table 6.1 *The Periodic Table (A-group elements only)*

I	II	III	IV	V	VI	VII	VIII
H							He
Li	Be	**B̲**	**C̲**	N	O	F	Ne
Na	Mg	Al	**S̲i̲**	P	S	Cl	Ar
K	Ca	Ga	**G̲e̲**	As	**S̲e̲**	Br	Kr
Rb	Sr	In	Sn	Sb	**T̲e̲**	I	Xe
Cs	Ba	Tl	Pb	Bi	Po	At	Rn

Notes: Non-metallic elements which are solid at 300 K are in boldface type, such as **S**. Semiconducting elements are underlined, such as **B̲**.

6.1 Electrons and holes in semiconductors

Semiconductors are only different from insulators because of conduction brought about by thermally generated charge carriers (*intrinsic* conduction) or by the addition of controlled amounts of impurities (*extrinsic* conduction), called *dopants*. In most semiconductor devices only extrinsic conduction is desirable. The charge carriers are electrons and holes, which act just like positively-charged electrons. By adding the right kind of dopants it is possible to make semiconductor material which conducts either with electrons alone (n-type material) or with holes alone (p-type material). But first we must consider the ordinary processes of electrical conduction.

6.2 **Electrical conductivity**

The resistance, R, of a bar of material of length, ℓ, and uniform cross-sectional area, A, is

$$R = \ell/\sigma A \qquad (6.1)$$

where σ is the *electrical conductivity* of the material. If ℓ is in m, A in m^2 and R in Ω, then σ is in S/m. The reciprocal of the conductivity is the *resistivity*, ρ, which has units of Ωm. The electrical conductivity of materials is found to vary over a very wide range – wider than any other material property.

Materials came to be classified according to the behaviour and magnitude of their conductivities. Some, mainly metals and alloys, were found to be good conductors at room temperature (σ ranging from about 10^8 S/m for silver to 10^6 S/m for nichrome resistance wire), with a conductivity which decreased with increasing temperature and did not change much when the metal was impure. Other materials, mainly non-metals such as sulphur, polystyrene and silica, were found to have a very low conductivity at room temperature ($\sigma = 10^{-10}$ to 10^{-16} S/m) which increased very rapidly with temperature. Again it was found that the purity of the material had relatively little effect on the conductivity or its temperature coefficient. Then there was another class of materials, such as germanium, silicon and silicon carbide, which had very variable conductivities ($\sigma \approx 10^6$ to 10^{-2} S/m) and temperature coefficients that could be positive or negative and depended, it was eventually discovered, on the kind of impurities present, sometimes only in tiny amounts. Though these materials came to be called semiconductors, there is no fundamental difference between insulators and semiconductors: diamond is widely considered to be an insulator, and mostly it is, but semiconducting diamonds are known. Finally a fourth type of material was discovered (by Kammerlingh Onnes in 1911) that had no resistance at all below a critical temperature: it was a *superconductor*.

6.2.1 *Intrinsic conduction*

Generally speaking the conductivity of metals is due to free electrons in them and these are largely absent from semiconductors. There does, however, appear to be a characteristic energy associated with the conduction process. If sufficiently energetic photons (quanta of e.m. radiation) are incident on a semiconductor its conductivity rises enormously. The photon's energy can be equated with the energy needed to transfer an electron from a bound state (the electron is said to be in the *valence band*) to a free or conducting state (then it is said to be in the *conduction band*). At 0 K the conduction band is empty and the valence band full of electrons. The energy difference between these bands is known as the *bandgap energy* or just the *bandgap*.

Measurements in magnetic fields (Hall-effect measurements) show that there are two types of charge carrier: a negatively charged one identified with the electron and a positively charged one called a hole. When a photon causes an electron to transfer from valence band to conduction band it leaves behind a hole in the valence band. This hole can move under the influence of an electric field; it is a *charge carrier* that acts like a positively charged electron.

Silicon has a band gap of about 1.1 eV (electron volts, 1 eV = 1.6×10^{-19} J), meaning that an electron (charge, $-q = -1.6 \times 10^{-19}$ C) changes its potential by 1.1 V on crossing the band gap. Germanium has a bandgap of 0.7 eV and diamond has one of 5.4 eV. Thermal energy can also create electron-hole pairs that cause conduction. In a pure semiconductor kept in the dark, thermally generated charge carriers are the sole means of conduction. The numbers of them per unit volume (the *intrinsic carrier concentration*, n_i) are given by

$$n_i = N \exp(-E_g/2kT) \tag{6.2}$$

where N is a constant for a given semiconductor, E_g is the bandgap energy *in joules*, k is Boltzmann's constant (1.38×10^{-23} J/K) and T is the temperature in K. The factor of 2 in the denominator of the exponent means that in classical terms, the electron starts off in the middle of the band gap. This is because electrons are not classical particles. Be that as it may, no matter how carefully prepared the material is, there will always be thermal carrier generation and associated conductivity.

The conductivity in any material is given by $\sigma = nq\mu$, where μ is the *mobility* of the charge carrier. The mobility is the speed (v) that a charge carrier acquires per unit electric field ($\mathscr{E}$), that is $\mu = v/\mathscr{E}$. q is the magnitude of the electronic charge and n the number of charge carriers per unit volume. For an intrinsic semiconductor

$$\sigma = nq(\mu_e + \mu_h) \tag{6.3}$$

where n is the number of electron-hole pairs and μ_e and μ_h are the respective electron and hole mobilities. It is worthwhile putting some numbers into Equations 6.2 and 6.3. The material constants for silicon are $N = 3 \times 10^{25}$ m^{-3}, $E_g = 1.1$ eV $= 1.76 \times 10^{-19}$ J, $\mu_e = 0.14$ m^{2}V^{-1}s^{-1} and $\mu_h = 0.05$ m^{2}V^{-1}s^{-1}, so at 300 K, Equation 6.2 yields $n_i \approx 2 \times 10^{16}$ m^{-3} and Equation 6.3 gives $\sigma = 6 \times 10^{-4}$ S/m or $\rho = 1/\sigma = 1650$ Ωm. A bar of intrinsic silicon 100 mm long and of square cross-section, 10 mm by 10 mm, would then have, by Equation 6.1, a resistance of 1.7 MΩ. Compare this with a copper bar of the same size, which would have a resistance of 17 $\mu\Omega$. Equation 6.2 also shows that the carrier density and hence σ will rapidly increase with temperature (by 7%/K – see Problem 6.2).

6.2.2 Extrinsic conduction

At 300 K the intrinsic carrier concentration is very small in silicon and the electrons and holes are equal in number. Silicon is a covalently-bonded element with the same structure as diamond and germanium. Each silicon atom has four outer, bonding electrons and with them it forms four covalent bonds by which it is joined to four other silicon atoms arranged at the corners of a regular tetrahedron as in Figure 6.1a.

If an atom of silicon is replaced by one of phosphorus, which has five electrons available for bonding, then it can form four bonds with neighbouring silicon atoms, using up four electrons and having one left over. This spare electron is readily detached from its parent atom (the ionisation energy is of the order of 10 meV) to move into the nearby conduction band; at 300 K practically all the phosphorus atoms are ionised. The concentration of phosphorus atoms, $N_D \approx n$, the conduction-electron concentration. Phosphorus is known as a *donor* atom because it donates electrons to the conduction band. Arsenic and antimony behave as donors in silicon too.

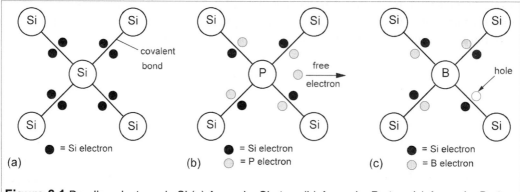

Figure 6.1 Bonding electrons in Si (a) Around a Si atom (b) Around a P atom (c) Around a B atom

When an atom of boron is substituted for one of silicon the position is different, as boron has only three electrons available for bonding, as a result, when it forms four bonds with neighbouring silicon atom as in Figure 6.1c, there is a deficiency of an electron – a hole – in one of the bonds. This hole can also move about freely in the valence band and acts as a mobile positive charge carrier, though really the electrons move the other way. The boron atoms are also effectively 100% ionized in silicon (though negatively, not positively like phosphorus), so the concentration of boron atoms, $N_A \approx p$, the hole concentration. Boron is termed an *acceptor* because it accepts an electron from a silicon atom. In semiconductor devices, the dopant atom concentration far exceeds the intrinsic carrier density; conduction is thus said to be *extrinsic*. Figure 6.2 shows the energy levels in extrinsic silicon. The higher up an electron is in Figure 6.2, the higher is its energy. However, for holes, the *lower* it is, the higher is its energy.

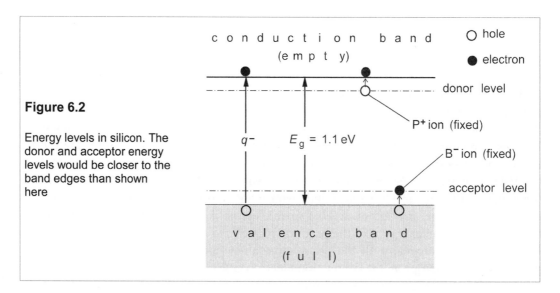

Figure 6.2

Energy levels in silicon. The donor and acceptor energy levels would be closer to the band edges than shown here

The law of mass action applies to the charge carrier concentration in a semiconductor, that is the product of electron and hole concentration, *pn*, is constant at constant temperature. If electrons predominate, the material is said to be *n-type* and the electrons are the *majority* carriers, holes the *minority* carriers. If holes predominate, the material is *p-type*, holes are the majority and electrons the minority carriers.

6.3 **The p-n junction**

Whether a particular piece of semiconductor is p-type, n-type or intrinsic makes no difference to the direction of (conventional) current flow when a voltage is applied to its ends, though the current's magnitude will depend on the doping level. But when p- and n-type materials are joined, or a *p-n junction* is formed, matters undergo a great transformation: current is now found to flow much more easily in one direction than in the other. The junction is a *rectifier*, or diode. P-n junctions, singly or in combinations of two or more, are essential for the operation of many semiconductor devices. Figure 6.3 shows a cross-section of a p-n junction, with no externally applied potential.

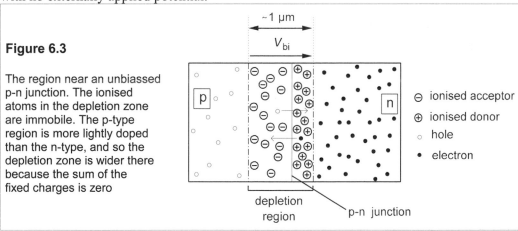

Figure 6.3

The region near an unbiassed p-n junction. The ionised atoms in the depletion zone are immobile. The p-type region is more lightly doped than the n-type, and so the depletion zone is wider there because the sum of the fixed charges is zero

Before the junction is formed the n-type material consists of fixed, positively-charged donor ions and an equal number of conduction electrons, while the p-type material consists of fixed, negatively-charged acceptor ions and an equal number of holes. When the junction is formed, electrons from the n-type region diffuse into the p-type region (leaving a fixed positive charge behind) and combine with holes. Simultaneously, holes from the p-type region diffuse into the n-type region (leaving a fixed negative charge behind) and combine with free electrons, thus causing an absence of mobile charge carriers in the junction region. In the *depletion region* (or zone) in the vicinity of the junction a voltage called the *built-in potential*, V_{bi}, is thus set up which opposes the diffusion of charge carriers across the junction. V_{bi} is about 0.6 V in silicon, about 1.2 V in GaAs and about 0.3 V in germanium. Thus with no bias applied, the diffusion current for the electrons (from the n-region), I_{er}, is exactly equal to the current (due to thermally generated electrons in the p-region), I_{eg}, produced by V_{bi}. The same is true for the hole currents, I_{hg} and I_{hr}. The holes and electrons move in opposite directions, though the conventional currents are in the same direction because of the negative electronic charge: thus $I_g = I_{eg} + I_{hg}$.

When *reverse bias* is applied to the junction, that is the p-type side is made more negative than the n-type, the depletion region becomes larger as the bias voltage adds to V_{bi}. The diffusion current, I_r, that flows across the junction is very small, since only carriers with enough energy can surmount the potential barrier. But the current, I_g, due to thermally generated minority carriers, remains unaffected by the bias field, resulting in a net reverse current, $I_g - I_r$. When the junction is *forward biassed* (p-type side positive with respect to the n-type) the depletion zone is narrower than under zero bias and I_r is now very much greater than I_g. Figure 6.4 shows the effect of bias polarity on the depletion zone widths and the forward current.

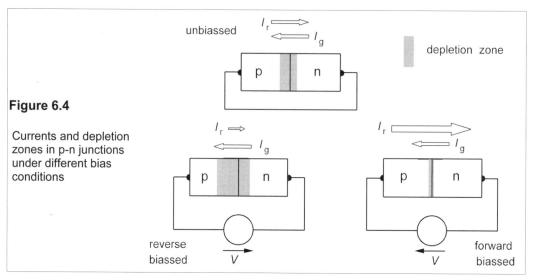

Figure 6.4

Currents and depletion zones in p-n junctions under different bias conditions

Reverse current is often called *leakage* current and its magnitude depends on the material and the size of the diode. Germanium diodes have high leakage currents – about 1,000 times that of silicon diodes (a silicon signal diode at room temperature typically has a leakage of around 1 nA). Light causes the generation of additional electron-hole pairs in semiconductors thereby increasing the leakage current. The so-called *dark current* is the leakage current when no light falls on the semiconductor material.

The behaviour of p-n junctions in semiconductor devices can now be discussed, starting with the simplest – diodes – in Chapter 7. The next section sets out some of the new terms this chapter has introduced.

6.4 A glossary of terms

acceptor   a dopant atom that has fewer valence electrons than the host semicon-
                      ductor material, so that it can accept one or more electrons from the
                      valence band of the host and gives rise to a corresponding number of
                      holes in the valence band. Boron is an acceptor atom in silicon

bandgap   the energy difference, E_g, between the top of the valence band and the
                      bottom of the conduction band, usually given in electron volts (eV),
                      where 1 eV = 1.6×10^{-19} J. In silicon the band gap is about 1.1 eV

bias	The potential applied to a p-n junction. Positive bias means that the p-type side of the junction is at a higher potential than the n-type side, negative bias means vice versa
charge carrier	Either a negatively-charged electron or a positively charged hole
compensation	The reduction in charge-carrier concentration in a semiconductor that is brought about by incorporating both donors and acceptors into the semiconductor material. Hence reduced conductivity
conduction band	A region of energy states lying above the valence band and separated from it by an energy gap in which electrons are not bound to their parent atoms and can move under the influence of applied fields
conductivity	Defined by $\sigma = nq\mu$, where n = charge carriers/m^3, q, the electronic charge = 1.6×10^{-19} C, μ = mobility (q.v.) in m^2/V/s. The reciprocal of the resistivity (q.v.)
conductivity type	Determined by the majority charge carriers. When these are electrons the material is n-type and when holes it is p-type
depletion region	A volume of material near a p-n junction in which oppositely charged carriers have combined so that the mobile charge carrier concentration is very small
donor	An dopant atom which has more valence electrons than its host semiconductor, so that it tends to give up electrons to the conduction band, so making the material n-type. Examples are phosphorus and arsenic in silicon
dopant	An element which is incorporated in controlled amounts into a semiconductor during crystal growth. At working temperatures the dopant is ionised and supplies electrons or holes to the host to produce a vast excess of one or the other, so making the material p-type or n-type
extrinsic conduction	Conduction brought about by charge carriers from ionised dopants
hole	The absence of an electron in the valence band, which acts like a positively charged electron
intrinsic conduction	Conduction by hole-electron pairs produced by ionisation of the semiconductor material. Ionisation means that electrons in the valence band must cross the band gap to reach the conduction band. At normal temperatures there are few carriers produced like this and conductivity is low
majority carrier	The more-numerous charge-carrier. Electrons in n-type material, holes in p-type
minority carrier	The less-numerous charge-carrier. Holes in n-type material, electrons in p-type
mobility	The speed acquired by a charge carrier in unit electric field, $\mu = v/\mathscr{E} = \sigma/nq$
p-n junction	The plane dividing p-type from n-type material, which lies within the depletion region
resistivity	Given by $\rho = RA/l$, where R = resistance in Ω of a bar of material, of uniform cross-sectional area A m^2 and length l m
valence band	The region of energy states lying below the conduction band by E_g,

the gap energy. Holes are free to move in the valence band like electrons in the conduction band

Suggestions for further reading

Electrical engineering semiconductor physics and devices by D A Neaman (Irwin, 2nd ed., 1997)
Band theory and electronic properties of solids by J Singleton (Oxford University Press, 2001)

Problems

(Note: 'conductivity' means 'electrical conductivity')

1. (a) A length of round copper wire has a diameter of 1 mm, calculate what length of wire will have a resistance of 1 kΩ if $\sigma = 60$ MS/m. (b) If a cylindrical piece of silicon, 1 mm in diameter, is doped with 10^{20} m^{-3} atoms of phosphorus, what length is required to give a resistance of 1 kΩ, assuming the phosphorus is fully ionised and that the electronic mobility in silicon is 0.1 m^{2}V^{-1}s^{-1}. (c) If the free electron concentration in copper is 8.5×10^{28} m^{-3}, what is the electronic mobility in copper? What is the ratio of electronic mobilities in silicon and copper?
    [(a) 47.1 km (b) 1.26 mm (c) 0.0044 m^2/V/s, $\mu_{Si}/\mu_{Cu} = 22.7$]

2. Find the relative rate of change of electrical conductivity with temperature in an intrinsic semiconductor from Equation 6.2, assuming that the mobility of the charge carriers is not dependent on temperature. Given that the bandgaps of germanium, silicon and gallium arsenide are 0.66, 1.1 and 1.42 eV respectively, what will be the percentage change in their intrinsic conductivity when the temperature is increased from 300 K to 302 K?
    [$d\sigma/\sigma dT = E_g/2kT^2$. 8.5%, 14.2% and 18.3%]

3. In an extrinsic semiconductor the carrier concentration is found to be constant above 100 K, but the carrier mobility goes as $T^{-1.5}$; how will the relative conductivity change with temperature at 300 K? What will the percentage change in conductivity be as the temperature increases from 300 K to 302 K? *[$d\sigma/\sigma dT = -1.5/T$, -1%]*

4. At a certain temperature, the carrier concentration in a semiconductor is proportional to $\exp(-E_c/2kT)$, where E_c is the carrier's ionisation energy. The mobility of the carriers is proportional to $T^{-1.5}$. If $E_c = 0.05$ eV, at what temperature will the temperature coefficient of the conductivity ($d\sigma/\sigma dT$) be zero? *[193 K]*

5. The mobility of electrons is five times the mobility of holes in a certain semiconductor. When doped with 10^{22} atoms of phosphorus per m^3 it is found to have a conductivity of 10 S/m. What are the carrier mobilities? What will the conductivity of a p-type sample be if the carrier density is 10^{21} m^{-3}? (Assume full ionisation of dopants.)*[6.25×10^{-3} and 1.25×10^{-3} m^2/V/s. 0.2 S/m]*

6. In a sample of semiconducting material the electronic mobility is α times the hole mobility and the intrinsic hole concentration is p per unit volume. Derive an expression for the minimum conductivity. If $\mu_h = 0.009$ m^2/V/s, $\alpha = 3$ and $p = 10^{16}$ m^{-3}, what is the minimum conductivity and what are the corresponding electron and hole concentrations? *[$\sigma_{min} = 2pq\mu_h\sqrt{\alpha}$ ($q = 1.6 \times 10^{-19}$ C), 50 μS/m, 5.8×10^{15} m^{-3}, 1.73×10^{16} m^{-3}]*

7. A certain intrinsic semiconductor is found to have doubled its conductivity when the temperature is increased by 4%, and then doubles again when the temperature rises a further 10 K. What is its bandgap in eV? At what temperatures are the measurements taken? (Assume carrier mobility is independent of temperature.) *[$E_g = 0.72$ eV, 231 - 250 K]*

7   Diodes

A
DIODE is a two-terminal, passive, non-linear device that can be used to control voltage and current in a circuit. Some diodes are used primarily to rectify alternating current, some are used as signal detectors and others are used as voltage references or voltage regulators. There are also optical diodes which are used as indicators (light-emitting diodes, or LEDs), signal sources (LEDs and laser diodes) or optical detectors (avalanche photodiodes and PIN diodes). The solar cell is a special type of optical diode which converts light energy directly into electricity and by tonnage is probably the most important use of diodes today. The shape of a diode's current-voltage relationship determines its specific application, which in today's solid-state devices depends on the way it is doped during manufacture. There are several classes of diode besides the 'ordinary' rectifier, but of these we shall examine only Schottky diodes, Zener diodes and light-emitting diodes. Optical signal source and detector diodes are discussed in Chapter 25.

7.1 Junction diodes

Junction diodes depend on p-n junctions in semiconductors for their operation. They are primarily used for rectification and signal detection, though they can also be used for voltage limitation or regulation. The symbol for an 'ordinary' diode is shown in Figure 7.1. The n-type side of the p-n junction is the called the *cathode* (marked on the device often by a coloured band) and the p-type side is the *anode*. Forward current is said to flow when the direction of positive (conventional) current flow is from anode to cathode.

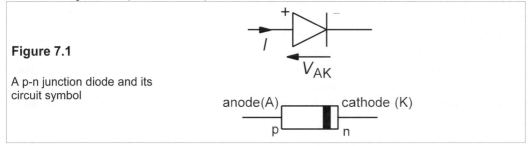

Figure 7.1

A p-n junction diode and its circuit symbol

The current-voltage relation for a junction diode is given by the rectifier equation[1]

$$I = I_s[\exp(qV_{AK}/kT) - 1]\qquad(7.1)$$

where V_{AK} is the potential across the diode, positive if forward biassed and negative if reverse biassed. I_s is the reverse saturation current (a very small quantity), q is the magnitude of the

---

[1] The departure of the current-voltage relationship from Equation 7.1 can be taken care of with an ideality factor, η, in the exponent; then $I = I_s[\exp(qV_{AK}/\eta kT) - 1]$, where $1 \le \eta \le 2$.

electronic charge (1.6×10^{-19} C), k is Boltzmann's constant (1.38×10^{-23} J/K) and T the absolute temperature in K. Putting in the values of q, k and T (290 K), Equation 7.1 is approximately, for forward currents

$$I = I_s \exp(40V_{AK}) \tag{7.2}$$

The −1 in Equation 7.1 can be neglected at normal forward currents. When the diode is reverse biassed, V is negative and Equation 7.1 reduces to $I \approx -I_s \approx 0$. Because the V-I relationship in the rectifier equation is not linear (that is, not of the form $I = \alpha V + \beta$, where α and β are constants) diodes are non-linear devices.

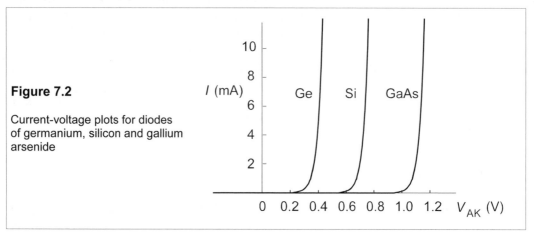

Figure 7.2

Current-voltage plots for diodes of germanium, silicon and gallium arsenide

A graph of Equation 7.1 at 300 K is drawn in Figure 7.2 for different values of I_s, corresponding to germanium, silicon and gallium arsenide signal (low current) diodes. On the scale of the graph the reverse current flow is negligible; it is also very dependent on temperature as well as the material of which the diode is made, and roughly doubles when the temperature rises by 10 K. We see from Figure 7.2 that the diodes conduct effectively only at a certain forward bias: 0.3 V for germanium, 0.6 V for silicon and 1 V for gallium arsenide diodes. Most common diodes have a forward drop of about 0.6 V and must therefore be made from silicon. The increase in current when the forward voltage rises above 0.6 V is very steep as expected from Equation 7.2, though in an ideal silicon diode it would be vertical as shown in Figure 7.3.

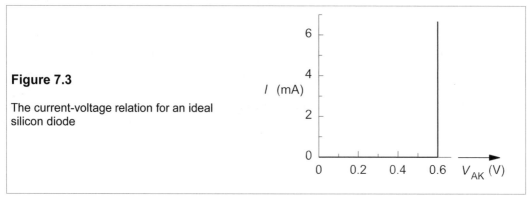

Figure 7.3

The current-voltage relation for an ideal silicon diode

Example 7.1

In the circuit of Figure 7.4, what will be the current flowing through the 1kΩ resistance, if the diodes are ideal and all have forward drops of 0.6 V?

Figure 7.4

The circuit for Example 7.1

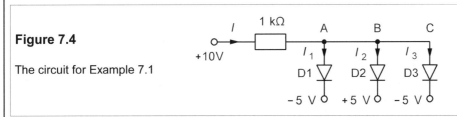

The first thing to note is that the +10V source is the highest potential in the circuit and that all three diodes have lower voltages at their cathodes than this. They could therefore all be conducting at first sight. But D1 will hold the potential at A down to +0.6 V above its cathode, which is at -5 V, that is V_A = -5 + 0.6 = -4.4 V. Since A, B and C are all wired together (a common-anode connection) they are all at the same potential, and D2 must be reverse biassed, so that $I_2 = 0$ and D2 is not conducting. What about I_1 and I_3? All we can say is that by KCL, $I_1 + I_3 = I$, the only way to calculate them would be from the exact characteristics of each diode. We *can* calculate I, however, from KVL, because the potential at one end of the 1kΩ resistor is +10 V and at the other it is -4.4 V, hence

$$I = [10 - (-4.4)]/1000 = 0.0144 \text{ A} = 14.4 \text{ mA} \tag{7.3}$$

Diode arrays, such as LEDs, fed from a common source require current-sharing resistors in series with them so that they all carry much the same current. The next example shows how currents are found when resistors are in series with the diodes (see also Problem 7.2).

Example 7.2

In the circuit of Figure 7.5, find the currents through the diodes, assuming they are both ideal and have forward voltage drops of 0.6 V, $R_1 = 5$ kΩ and $R_2 = 1$ kΩ. Repeat the analysis with all resistances equal to 1 kΩ.

Figure 7.5

The circuit for Example 7.2

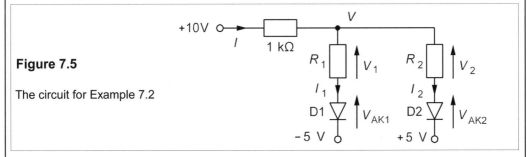

We cannot tell by inspection whether the diodes are both conducting or not, but must carry out the analysis and determine the anode potentials. Using nodal analysis and assigning voltage, V, to the only node, the current in branch 1 is

$$I_1 = \frac{V - V_{AK1} - (-5)}{5} = \frac{V - 0.6 + 5}{5} = 0.2V + 0.88 \tag{7.4}$$

working in volts, kΩ and mA. And in the branch 2 the current is

$$I_2 = (V - V_{AK2} - 5)/1 = V - 5.6 \qquad (7.5)$$

And the current through the 1kΩ resistance is

$$I = (10 - V)/1 = 10 - V \qquad (7.6)$$

By KCL $I = I_1 + I_2$ so that

$$I_1 + I_2 = 10 - V \qquad (7.7)$$

Equations 7.4, 7.5 and 7.7 contain three unknowns and can be solved to give

$$V = 6.691 \text{ V}; \qquad I_1 = 2.218 \text{ mA}; \qquad I_2 = 1.091 \text{ mA} \qquad (7.8)$$

Both diodes conduct despite their very different cathode potentials.
Repeating the analysis with $R_1 = 1$ kΩ requires only that Equation 7.3 be changed to

$$I_1 = (V - 0.6 + 5)/1 = V + 4.4 \qquad (7.9)$$

The solution is then found to be

$$V = 3.733 \text{ V}; \qquad I_1 = 8.133 \text{ mA}; \qquad I_2 = -1.867 \text{ mA} \qquad (7.10)$$

This solution cannot be correct because it shows reverse (negative) current through D2. We therefore assume D2 is not conducting, that is $I_2 = 0$, the other equations being the same. These equations are easily solved to give $V = 2.8$ V and $I_1 = I = 7.2$ mA.
One may verify that to satisfy KVL, $V_{AK2} = -2.2$ V, and D2 is indeed reverse biassed.

7.1.1 The dynamic resistance

The rectifier equation enables us to calculate the *dynamic resistance* (or AC resistance), r_d, of *all* (be they made from silicon, germanium, gallium phosphide or any other material) junction diodes when forward biassed. Differentiating Equation 7.1,

$$\frac{dI}{dV_{AK}} = \frac{q}{kT} I_s \exp\left(\frac{qV_{AK}}{kT}\right) = \frac{q}{kT} I \approx 40I \qquad (7.11)$$

$$\Rightarrow \qquad r_d = \frac{dV_{AK}}{dI} \approx \frac{1}{40I} = \frac{0.025}{I} \qquad (7.12)$$

This is at room temperature (300 K), so that $r_d = 0.025/I = 1$ Ω, when the forward current is 25 mA. Such a small value can safely be neglected, but at low diode currents r_d is significant, especially in bipolar transistors (see Chapter 8).

For a forward current of 25 mA, the voltage drop, V_{AK}, according to Equation 7.2 (taking I_s as 0.01 pA) will be 0.71 V. For $I = 1$ mA we find V_{AK} is 0.63 V and if $I = 1$ A, V_{AK} is 0.81 V; a forty-fold increase in current has resulted in only a 14% increase in the voltage across the junction. Thus V_{AK} can be assumed constant and the diode can be modelled as a voltage source, V_{AK0}, in series with a small resistance, r_d, as shown in Figure 7.6a, with the corresponding *V-I* characteristic shown in Figure 7.6b.

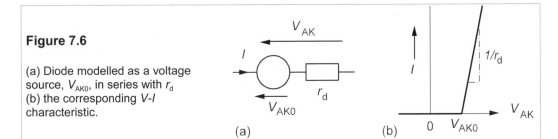

Figure 7.6

(a) Diode modelled as a voltage source, V_{AK0}, in series with r_d
(b) the corresponding *V-I* characteristic.

(a)   (b)

Example 7.3

In the circuit of Figure 7.7a, what will be the current flowing if the diode's forward bias is 0.7 V when its forward current is 25 mA, $r_d = 1\ \Omega$ and $V_s = 1 + 0.01\sin \omega t$ V?

The voltage source can be split into AC and DC parts and then we can use the superposition theorem, provided that the AC component is small compared to the DC. Taking the DC part first and using the diode model of Figure 7.6a we derive the circuit of Figure 7.7b. We must first deduce V_{AK0} from the data given, namely that the diode's voltage is 0.7 V when the current is 25 mA and $r_d = 1\ \Omega$. The voltage across r_d when $I = 25$ mA is $Ir_d = 25$ mV, making $V_{AK0} = V_{AK} - Ir_d = 0.7 - 0.025 = 0.675$ V. Then the DC current flowing in Figure 7.7b is

$$I = (V_s - V_{AK0})/(R + r_d) = (1 - 0.675)/(10 + 1) = 0.325/11 = 0.0295\ \text{A}$$

At this current the voltage drop across the diode is

$$V_{AK} = V_{AK0} + Ir_d = 0.675 + 0.0295 \times 1 = 0.7045\ \text{V} \qquad (7.13)$$

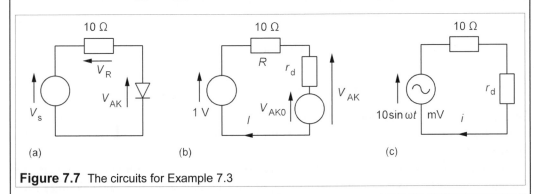

(a)   (b)   (c)

Figure 7.7 The circuits for Example 7.3

7.1.2 Load lines

By Kirchhoff's voltage law in the circuit of Figure 7.7a

$$V_s = V_R + V_{AK} = IR + V_{AK} \qquad (7.14)$$

Which rearranges to

$$I = -V_{AK}/R + V_s/R \qquad (7.15)$$

where V_s is the source voltage. If this equation is plotted on a graph of current as a function of V_{AK}, it results in a straight line of slope $-1/R$ and intercept on the *I*-axis of V_s/R, as in Figure 7.8. This line, the *load line*, is the locus of all possible values for the diode forward drop, V_{AK}, no matter what the diode. If the diode characteristic is plotted too, the intersection of the load line with it occurs at the *quiescent point* or *Q-point*, so that the diode's quiescent voltage (V_{AKQ}) and current, I_Q, can be read off the axes. We shall find load lines useful in transistor circuits (Chapters 8 and 9).

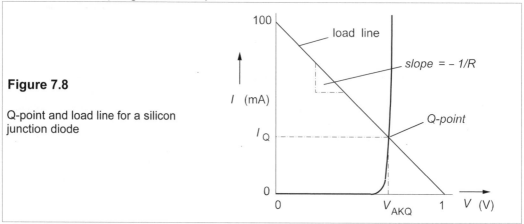

Figure 7.8

Q-point and load line for a silicon junction diode

We can now take the $10\sin\omega t$ mV AC source on its own. The diode is operating at a bias of $V_{AKQ} = 0.7035$ V, and $I_{DQ} = 28.5$ mA. The alternating source will induce small changes in current and voltage about the Q-point, which we can find by a linear approximation using the diode's dynamic resistance, r_d. The value of r_d at the Q-point must be calculated from

$$r_d = 0.025/I_Q = 0.025/0.0285 = 0.88 \ \Omega \tag{7.16}$$

and the AC circuit is as in Figure 7.6c. The current is given by Ohm's law

$$i = \frac{10\sin\omega t}{10 + 0.88} = \frac{10\sin\omega t}{10.88} = 0.92\sin\omega t \ \text{mA} \tag{7.17}$$

Direct and alternating currents can be added for a total current of $(28.5 + 0.92\sin\omega t)$ mA. Using Equation 7.1 to give an 'exact' solution is pointless since the accuracy of the approximation used here is about $\pm 1\%$.

Were the temperature of the diode to increase to 100°C, a not uncommon occurrence, then I_s would increase by about $2^{7.3}$ (because the temperature has increased 73 K and I_s doubles every 10 K) or about 160 times, while q/kT would decrease to 32. r_d would increase to $0.031/I$, or about $1.24 \ \Omega$ at 25 mA. The increase of $0.24 \ \Omega$ is only 2% of the circuit's total resistance, thus the solution is little affected even though I_s increased dramatically and r_d significantly.

We shall look at some important diode circuits later, when we have examined further properties of p-n junction diodes and discussed other types of diode.

7.1.3 *Further properties of p-n junction diodes*

Although the rectifier equation does not indicate the fact, there is a reverse voltage, V_b, dependent principally on the dopant concentration, where a diode will break down. Reverse breakdown in normal p-n junction diodes is usually irreversible, and damages the diode.

Roughly speaking, the cost of a junction diode depends on how much current it can carry (which depends on the size of the active semiconductor material) and the voltage it can withstand in reverse bias: low-current, low-reverse-voltage diodes are very cheap and high-current, high-reverse-voltage diodes are dear. For example, the 1N4148 signal diode is rated at 75 V maximum reverse voltage and 150 mA maximum forward current, and costs only 0.8p per diode for 10+ quantities, and 0.6p per diode in quantities of 500+; while the SW15PHR400 power diode rated at 1.2 kV and 400 A costs £31 for 1+ and £23 for 100+.

There are, however, other parameters of interest in diodes than V_{RRM} (the reverse, repetitive, maximum voltage rating) and $I_{F(AV)}$ (the average maximum forward current rating), such as the junction capacitance, the reverse-recovery time, the surge current rating and the leakage current. A p-n junction diode stores charge in the depletion region near the junction and this charge must be removed rapidly at high frequencies, or the diode does not work properly. The time taken for the charge to dissipate when the reverse voltage is reduced to zero is called the reverse-recovery time, t_{rr} and can vary from 10 µs down to 1 ns. Roughly, the maximum operating frequency for a diode is $1/10t_{rr}$. Diodes that operate in the UHF and VHF regions (50 MHz up) are usually Schottky-barrier devices that recover instantly.

The junction capacitance depends on the area of the junction and the width of the depletion region:

$$C_j = \varepsilon_r \varepsilon_0 A / x_j \qquad (7.18)$$

which is the same as Equation 1.6, giving the capacitance of a parallel-plate capacitor. For silicon $\varepsilon_r = 12$, $\epsilon_0 = 8.85 \times 10^{-12}$ F/m, A is the junction area and x_j is the width of the depletion region. x_j is proportional to $\sqrt{V_j}$, where V_j is the reverse voltage across the junction, so the capacitance decreases as the bias becomes more negative. Taking a small signal diode, where $x_j \approx 10$ µm and $A \approx 2 \times 10^{-7}$ m², we find $C_j \approx 2$ pF, about the right order of magnitude. *Varactor* and *varicap* diodes have a relatively large capacitance (100 pF at $V_j = 0$) that can be varied (as much as tenfold) by altering the reverse bias. They are used in circuits whose frequency of oscillation or resonance is voltage controlled (see Problem 7.9). In forward bias the diode's capacitance is largely *diffusion capacitance*, which depends principally on the current flowing.

7.2 Schottky diodes

A metal-semiconductor contact can be non-rectifying (or *ohmic*) as well as rectifying: Schottky diodes (sometimes called Schottky-barrier diodes) make use of the rectifying properties. Usually the metal is aluminium (or, rarely, gold) and the semiconductor silicon (occasionally germanium). Figure 7.9 shows a Schottky diode and its circuit symbol.

Schottky diodes rely on majority-carrier conduction – unlike p-n junction diodes which rely on minority-carrier conduction – so that they do not store charge at the junction. In consequence they are able to change very fast from conducting to non-conducting states and are used in circuits which must switch on and off rapidly, such as switch-mode power supplies (see Section 12.6). The original cat's whisker detector of the early radio years depended on the rectifying properties of a metal-semiconductor contact. They were notoriously unreliable (though point-contact diodes continued to be manufactured until quite recently) and so came to be replaced by vacuum-tube devices, which in turn were replaced by semiconductor devices once more, based on the new technology and understanding of solid-state physics brought about in the 1930s and 1940s.

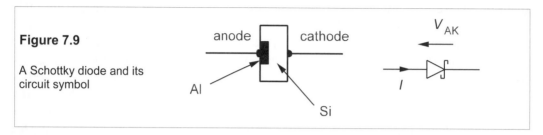

Figure 7.9

A Schottky diode and its circuit symbol

When a metal forms a contact with a semiconductor, a potential barrier is set up at the interface. At moderate to low dopant densities in n-type silicon ($\leq 10^{23}$ atoms per m^3) electrons pass from semiconductor to metal over this barrier by a process called thermionic emission. If the metal is made positive with respect to the semiconductor (forward bias), the barrier height is reduced and if reverse bias is applied it is increased. The current flow follows Equation 7.1 again. However, I_s is now given by

$$I_s = A_j A^* T^2 \exp(-q\phi_B / kT) \qquad (7.19)$$

which is known as Richardson's equation. A_j is the area of the contact, A^* is the effective Richardson constant ($\approx 10^8$ A/K^2/m^2 for n-type silicon), ϕ_B is the barrier height in volts and the other symbols have their usual significance. I_s works out to about 16 μA for a 1 mm^2 contact when ϕ_B is 0.7 V as it is for aluminium and n-type silicon. As a result of this high value of I_s (about 2×10^6 times as large as for a p-n junction diode), the forward drop across a conducting Schottky diode can be much smaller than for a p-n junction diode.

For example, a Schottky diode having a 1mm^2 Al-Si contact with $I_s = 16$ μA will conduct 25 mA at a forward bias of only 0.2 V. Of course the penalty to be paid is the relatively large reverse (leakage) current of 16 μA at 0 V. In practice the requirements for rectifying power Schottky diodes are different from those of fast-switching signal diodes, so the contact metal may be different: chromium, platinum and tungsten are all used. In integrated circuit manufacture, aluminium is preferred as it is easily evaporated.

The chief advantage of Schottky diodes is not, however, their potentially-lower forward voltages, but their fast recovery times caused by low charge storage at the junction, even when the device is carrying large currents. They are therefore used in high-speed switching circuits as well as in low-loss power supplies. The 1N6263 RF Schottky diode, for example, has a reverse breakdown voltage of 60 V (at 10 μA), a maximum capacitance of only 2.2 pF and a switching time of a few picoseconds, with a forward voltage drop of only 0.3 V at $I_F = 1$ mA. The capacitance of the diode is in parallel with it, so will short it out at high frequencies. It costs only 19p (for 1+) and as little as 10p for 5k+. High-power Schottky diodes, such as the MBR3545 (which has $V_{RRM} = 45$ V, $I_{F(AV)} = 30$ A, $V_{AK} = 0.64$ V at $I_F = 30$ A) cost a good deal more – from £8 (for 1+) to £5 (100+).

To summarise, Schottky diodes have similar current-voltage characteristics to p-n junction diodes: the difference lies chiefly in their faster switching times and in their reduced forward voltage drops, advantages that must be set against their much larger reverse (leakage) currents.

7.3 **Zener diodes**

The symbol for a Zener diode and its current-voltage relationship are shown in Figure 7.10.

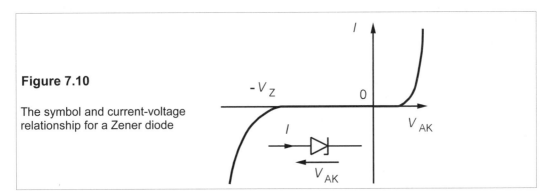

Figure 7.10

The symbol and current-voltage relationship for a Zener diode

Unlike the previous two types of diode, the Zener diode is designed to be used and to break down in reverse bias, but in a controlled fashion so that the process is reversible and not destructive. Some manufacturers give them numbers indicating the reverse breakdown voltage (or Zener voltage, V_Z), such as the BZX85C5V1, made by SGS-Thomson, which has $V_Z =$ 5.1 V. Zeners can be readily bought with Zener voltages from 2.4 V to 270 V, but the spread in breakdown voltage from the nominal values is usually about ±5%.

In forward bias silicon Zeners conduct at 0.6 V, just the same as a normal silicon diode. The breakdown process in reverse bias in a 'normal' p-n junction diode is caused by an avalanche effect: electrons are accelerated in the junction region by the very high fields (as much as 1 GV/m) to such a high speed that they can remove bound electrons. In turn these electrons will be accelerated by the applied field to produce more electrons by the same process. The result is an avalanche of current at a well-defined reverse voltage, V_Z.

In low-voltage Zener diodes, this is not the process by which breakdown occurs, though it *is* in 'Zeners' with breakdown voltages above about 6 V. Low-voltage Zeners breakdown by a tunnelling process called *Zener breakdown*, in which electrons in the valence band find themselves close to empty states in the conduction band, so that they are able to tunnel through the bandgap between them, though they do not have the necessary energy, classically.

The dynamic resistance of Zeners in reverse breakdown is at a minimum around a Zener voltage of 7 V (where it may be about 10 Ω for a current of 5 mA), rising substantially as the Zener voltage drops, to as much as 100 Ω for a 5mA current in a 3V Zener, and also as the Zener voltage rises to about 100 Ω again for a 5mA current and $V_Z= 270$ V. In addition to this drawback, when V_Z is greater than about 7 V it has a large temperature coefficient, θ_{VZ} ($\equiv \Delta V_Z/V_Z\Delta T$), of about +0.1%/K. To some extent this can be offset by connecting forward biassed p-n junction diodes in series with the Zener, as these have negative temperature coefficients. When V_Z falls below about 5 V, θ_{VZ} becomes negative. Nevertheless, even with these deficiencies Zeners are a cheap and reasonably effective way of achieving voltage regulation and voltage reference.

Example 7.4

An example of a simple voltage-regulating circuit is shown in Figure 7.11a. In this circuit $V_Z = 20$ V at a current of 10 mA, when the dynamic resistance, $r_z = 20$ Ω. The load current varies from 20–30 mA, while the supply voltage varies independently from 40–45 V. Given that $R_s = 560$ Ω, calculate the percentage regulation of the load voltage. What power rating is required for the Zener?

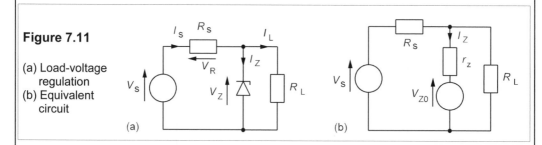

Figure 7.11

(a) Load-voltage regulation
(b) Equivalent circuit

(a)   (b)

The equivalent circuit of the Zener is used in Figure 7.11b, which we can analyse to find the Zener current and hence the Zener and load voltage. By KVL

$$V_s = V_R + V_Z = I_s R_s + I_Z r_z + V_{Z0} \tag{7.20}$$

We find V_{Z0} from the data given for the diode and the equation

$$V_{Z0} = V_Z - I_Z r_z = 20 - 10 \times 10^{-3} \times 20 = 19.8 \text{ V} \tag{7.21}$$

And since by KCL $I_s = I_L + I_Z$, Equation 7.20 becomes

$$V_s = (I_Z + I_L)R_s + I_Z r_z + V_{Z0} \tag{7.22}$$

Rearranging this gives I_Z as

$$I_Z = \frac{V_s - I_L R_s - V_{Z0}}{R_s + r_z} \tag{7.23}$$

We see now that I_Z will be a maximum when V_s is maximum (= 45 V) and I_L is minimum (= 20 mA), and this will give the maximum load voltage:

$$I_{Zmax} = \frac{V_{smax} - I_{Lmin}R_s - V_{Z0}}{R_s + r_z} = \frac{45 - 20 \times 10^{-3} \times 560 - 19.8}{560 + 20} = 24.1 \text{ mA} \tag{7.24}$$

And then $V_{Zmax} = V_{Z0} + I_{Zmax}r_z = 19.8 + 24.1 \times 10^{-3} \times 20 = 20.28$ V. Similarly I_Z will be at a minimum when V_s is minimum (= 40 V) and I_L maximum (= 30 mA). Substituting these values into Equation 7.23 gives $I_{Zmin} = 5.86$ mA and then $V_{Zmin} = 19.92$ V. The load-voltage regulation is

$$(V_{Zmax} - V_{Zmin})/V_{Zmax} = (20.28 - 19.92)/20.28 = 0.0178 \tag{7.25}$$

In percentage terms %reg = 1.78%. Had we also considered the effect of temperature variations, the result would have been worse; still, the performance of the Zener is creditable.

The maximum power dissipation in the Zener occurs when I_Z is max and then $P_{Zmax} = I_{Zmax}V_{Zmax} = 24.1 \times 20.28 = 489$ mW. A 500mW device might just do, allowing virtually no safety margin, but it is best to use a 1.3W device such as the BZX85C20, costing only 13p.

7.3.1 Bandgap references ('Voltage references')

Because the Zener-breakdown voltage has a poor temperature coefficient, particularly for low values of V_Z, a class of voltage-reference ICs known as *bandgap references* has become popular. The reference voltage they provide is the bandgap of silicon at 0 K, which is 1.24 V, hence the name. For example two-terminal devices LT1004CZ-1.2 has $V_Z = 1.235$ V and the LT1004CZ-2.5 has $V_Z = 2.500$ V, while their temperature coefficient is only 20 ppm (parts per million) or 0.002%/°C. They cost £2.15 in single quantities. The manufacturing spread of breakdown voltages is also small, about ±1.5%. In addition r_d is only 0.2 Ω, typically, when the reverse current is from 1 to 10 mA. There are also trimmable bandgap-reference ICs available such as the LM385Z-1.26, whose output voltage can be adjusted between 1.24 V and 5.3 V. Though less stable ($dV/dT = 0.015\%/°C$) they are cheap (39p for 1+) and versatile. The more stable and expensive types (> £2) are used in digital-to-analogue converters and DVMs. Other applications include cold-junction compensation for thermocouples (as shown in Figure 7.12), temperature-stabilised amplifiers, linear-voltage-output thermometers, highly stable voltage regulators, etc.

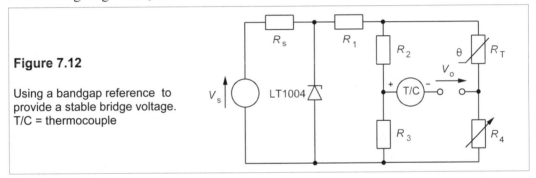

Figure 7.12

Using a bandgap reference to provide a stable bridge voltage.
T/C = thermocouple

In Figure 7.12, V_s can be a 3V battery, with $R_S = 100$ kΩ, so that the supply current will be about 18 µA if $V_Z = 1.235$ V. The battery can thus last for a considerable time. Bandgap references offer stable reference voltages with low operating currents which is often a boon. The value of R_1 (also about 100 kΩ) sets the current through the bridge network, typically 10 µA, while R_T is an NTC (negative temperature coefficient) thermistor. R_4 can be adjusted to give the correct output for the thermocouple in contact with a body at a known temperature, with the thermistor at ambient temperature.

7.4 Light-emitting diodes (LEDs)

In a forward-biassed p-n junction diode, majority carriers either side of the junction move in opposite directions and cross the junction, becoming minority carriers on the other side. The process is called *minority-carrier injection* and the density of minority carriers is thereby increased near the junction. However, majority carriers will combine with the injected excess minority carriers and re-establish equilibrium, possibly producing a photon, a *radiative recom-bination*. Ideally one would like all such recombinations in a LED to be radiative, that is for the *quantum efficiency* to be 100%, but in fact quantum efficiencies are much lower, ranging from about 0.01% to 10%, depending on the structure of the valence and conduction bands. In materials where the bandgap is *direct*, a charge carrier can recombine without changing its momentum, but in *indirect-gap* materials a change of carrier momentum requires a *phonon* (quantum of lattice vibrational energy) to participate. The probability of a radiative recombination is thereby much reduced in indirect-gap materials.

Figure 7.13 shows radiative recombination of an electron with a hole and a hole with an electron in a LED.

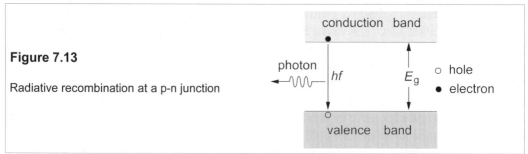

Figure 7.13

Radiative recombination at a p-n junction

The photon (quantum of light energy) emitted has an energy very close to the gap energy, E_g, so that the frequency of the emitted light is given by

$$E_g = hf = hc/\lambda \tag{7.26}$$

where h is Planck's constant (6.627×10^{-34} Js), c the speed of light in a vacuum (3×10^8 m/s) and λ is the wavelength in a vacuum of the light emitted. In fact the most probable energy of the photon is a little larger than this, by about kT:

$$hc/\lambda = E_g + kT \;\Rightarrow\; \lambda = hc/(E_g + kT) \tag{7.27}$$

Thus the wavelength of the light is largely determined by the bandgap of the material.

The photons emitted range over about $3kT$ in energy, and it can be shown (see Problem 7.16) that the *linewidth* (the range of wavelengths emitted) of the LED, $\Delta\lambda$, is given by

$$\Delta\lambda \approx 2.4\,kT\lambda^2 \tag{7.28}$$

where $\Delta\lambda$ and λ are in µm and kT is in eV. For example in a red LED the emission is at 0.65 µm, so at 300 K where kT is about 0.025 eV, the linewidth is about $2.24 \times 0.65^2 \times 0.025$ = 0.024 µm (24 nm). The linewidth does not matter if the LED is an indicator lamp, but has an important bearing on their use as fibre-optic communications sources. The narrower linewidths of semiconductor lasers makes then much more suitable for this purpose. LEDs are designed to emit light from the top surface, while signal LEDs are designed to emit in the plane of the p-n junction as shown in Figure 7.14. The latter mode of emission is known as *edge emission*. It results in a greater concentration of the light which increases the amount coupled into an optical fibre.

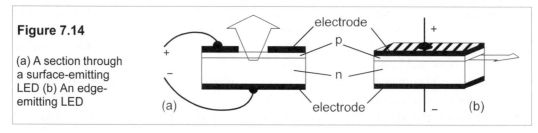

Figure 7.14

(a) A section through a surface-emitting LED (b) An edge-emitting LED

7.5 **Solar cells**

Solar cells have now assumed considerable importance for generating electricity, and not only in remote, hot regions. Unfortunately their overall efficiency is still only 10–15%, but if one considers that on a sunny summer day in Britain the amount of solar energy reaching the ground at noon is about 1.2 kW/m², the opportunity for substantial electrical power production is clear. Of course solar-cell-powered equipment must have battery backup, even in deserts, and in Britain its economic operation must be considered doubtful. In the USA there are large arrays of solar cells generating many megawatts. The key to further development is the production of cheaper silicon cells, though these are already relatively cheap compared to other semiconductor devices, because they need not be made from expensive single-crystal material. They cost, in single quantities, about £700/m², or 1% of the price of a power rectifier of the same area. For a capital outlay of about £700 one can therefore hope to produce about 100 W of electricity on a sunny day, which is an order of magnitude greater than the capital cost per watt of commercial generators.

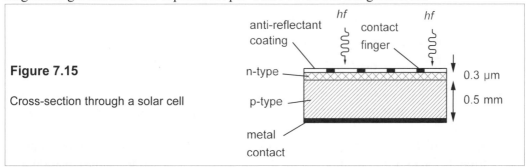

Figure 7.15

Cross-section through a solar cell

Figure 7.15 shows a cross-sectional diagram of a typical solar cell. The drawing is not to scale, since the metal contacts, the antireflectant layer and the uppermost n-type silicon layer are all much thinner than the p-type material which makes up virtually all of the volume. Photons enter the top surface through the antireflectant window and create electron-hole pairs in the junction region. The thin n-type top layer ensures that relatively few charge carriers can recombine before crossing the junction. The equivalent circuit of a solar cell (and a photodiode) is shown in Figure 7.16a and its *V-I* characteristics in Figure 7.16b, which are seen to be that of a p-n junction diode, but displaced down the current-axis by an amount I_{ph}, which is the short-circuit current of the cell.

Examining Figure 7.16a we see that by KCL

$$I_L = I_D - I_{ph} = I_s[\exp(qV_{AK}/kT) - 1] - I_{ph} \tag{7.29}$$

since the diode current is given by the rectifier equation (Equation 7.1). If the cell is open circuit $I_L = 0$ and $V_{AK} = V_{oc}$, so that Equation 7.29 becomes

$$I_{ph} = I_s[\exp(qV_{oc}/kT) - 1] \tag{7.30}$$

In silicon solar cells I_{ph} is about $5 \times 10^8 I_s$ with a solar radiation intensity of about 1 kW/m² (= 1 Sun), and is about a hundred times smaller than the typical current in an ordinary diode. Thus at 27°C or 300 K, where $q/kT \approx 40$, Equation 7.30 leads to $V_{oc} = 0.5$ V, as indicated in Figure 7.16b.

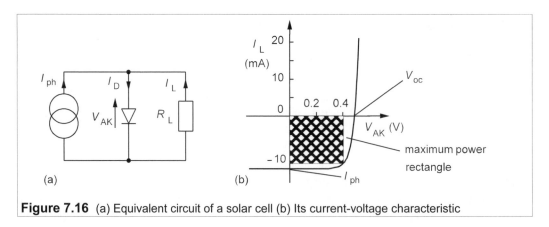

Figure 7.16 (a) Equivalent circuit of a solar cell (b) Its current-voltage characteristic

When on load the power supplied to the load is $-V_{AK}I_L$ and using Equation 7.29 for I_L, with $q/kT = 40$, this is

$$P = -V_{AK}I_L = -V_{AK}[I_s\exp(40V_{AK}) - I_{ph}] \tag{7.31}$$

ignoring the -1. Differentiating to find P_{max} gives

$$dP/dV_{AK} = -I_s\exp(40V_{AK}) + I_{ph} - V_{AK}[40I_s\exp(40V_{AK})] = 0 \tag{7.32}$$

Hence

$$I_{ph} = 40V_{AK}I_s\exp(40V_{AK}) + I_s\exp(40V_{AK}) = 5 \times 10^8 I_s \tag{7.33}$$

taking $I_{ph} = 5 \times 10^8 I_s$. The solution to Equation 7.33 is $V_{AK} = V_m = 0.428$ V. The corresponding load current is found from Equation 7.29 to be $-0.946I_{ph} (= I_m)$ and the same equation then gives $P_{max} = -V_mI_m = 0.405I_{ph}$ watts. In a silicon solar cell with an area of 10×10 mm^2, $I_{ph} \approx$ 30 mA, and the maximum power becomes 12 mW, or 120 W/m^2.

Table 7.1 *Data for some solar panels*

P_{max} (W)	V_m (V)	V_{oc} (V)	V_m/V_{oc}	I_m (A)	I_{ph} (A)	I_m/I_{ph}	Size (mm)	Price	(£/W)
0.446	3.3	4.6	0.72	0.15	0.16	0.94	147 × 79	£18	40
1	7.5	10.3	0.73	0.15	0.16	0.94	161 × 139	£27	27
5	17.5	20.5	0.85	0.27	0.29	0.93	269 × 249	£108	22
20	17.1	20.8	0.82	1.17	1.27	0.92	502 × 421	£231	11.6
53	17.2	21.3	0.81	3.08	3.33	0.92	937 × 500	£362	6.8

Prices in the USA are about 30–50% of those in Table 7.1, and are beginning to be competitive with the price per watt of conventional power stations. Examining the data for commercial solar panels we can see that many cells must be wired in series to give the output voltages stated. The ratios of V_m/V_{oc} and I_m/I_{ph} calculated above are 0.85 and 0.946 respectively, which agree with those in the table reasonably well. The largest panel gives an output of about 100 W/m^2, again close to that expected. The maximum power is stated in the table for 1 kW/m^2 solar intensity. In order to generate maximum power the load has to be matched to the cell correctly and interfacing circuits are usually necessary, particularly if batteries are to be recharged by the cells. When the cells are not generating current they act as diodes and will discharge the battery, possibly being destroyed in the process.

7.6 Applications for ordinary p-n junction diodes

There are so many applications for diodes that we can only look at a few typical uses.

7.6.1 The half-wave rectifier

The half-wave rectifier is shown in Figure 7.17a, together with the voltage waveform across R_L, Figure 7.17b.

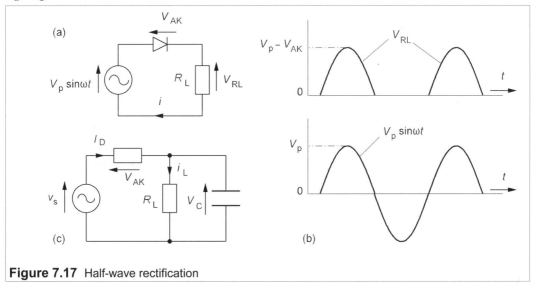

Figure 7.17 Half-wave rectification

Though the current is unidirectional, it varies from zero to a maximum of $(V_p - V_D)/R_L$, since by KVL, $v_L = V_p \sin\omega t - V_{AK}$. By placing a capacitor across the load, as in Figure 7.17c, we can smooth the current flow through R_L. During the positive half cycle, while the diode conducts, the capacitor charges up, so that it supplies current when the diode is not conducting.

Suppose $V_p \gg V_{AK}$, so that the load charges up very nearly to V_p volts. As the voltage drops from its peak value the capacitor discharges according to

$$V_C = V_p \exp(-t/R_L C) \tag{7.34}$$

Then when the applied voltage increases to $V_{C\min}$, the capacitor begins to charge again as in Figure 7.18.

The time during which the capacitor discharges is almost one period, $T\,(= 1/f)$, so if $T \ll R_L C$, the time constant, the exponential decay is approximately a straight line:

$$V_{C\min} \approx V_p(1 - T/R_L C) \approx V_p(1 - 1/R_L Cf) \tag{7.35}$$

(because $e^{-x} \approx 1 - x$ for small x). T has been replaced by $1/f$, assuming that the charging time is negligible. The capacitor now charges up almost to V_p once again. The difference between maximum (V_p) and minimum voltage is the *ripple voltage*, V_r, which is

$$V_r = V_p - V_{Cmin} = V_p - V_p(1 - 1/R_LCf) \tag{7.36}$$

Or, $$V_r \approx \frac{V_p}{R_LCf} \tag{7.37}$$

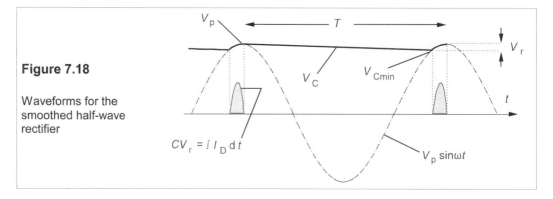

Figure 7.18

Waveforms for the smoothed half-wave rectifier

Sometimes the ripple voltage is defined as the r.m.s. value of the alternating component. A half-wave rectifier with little ripple is almost a saw-tooth wave, and the r.m.s. ripple would be $V_r/2\sqrt{3}$. But for other ripple waveforms, the conversion factor would be different. Since one normally measures peak-peak ripple, it seems preferable to define this as V_r.

Example 7.5

Suppose we wish to smooth the output of a half-wave rectifier operating at 50 Hz with $R_L = 100\ \Omega$, $V_p = 340$ V, so that the peak-to-peak ripple is 2% of V_p, what must C be? What is the peak current through the diode, approximately?

A ripple of 2% means that $V_p/V_r = 50$, and Equation 7.37 gives

$$C = (V_p/V_r)/R_Lf = 50/(100 \times 50) = 0.01\ \text{F} = 10000\ \mu\text{F} \tag{7.38}$$

It is a fairly large capacitance. Because smoothing capacitances often need to be large they are normally electrolytic types, which occupy less volume for a given capacitance. The current supplied to the capacitor must be enough to charge it up by an additional amount V_r. In the example given the capacitor charges up from $0.98V_p$ to V_p (actually a little bit longer because after the peak of the supply voltage is attained the supply voltage drops more slowly than the capacitor would discharge). Thus the phase of the voltage waveform goes from $\sin^{-1}0.98$ to $\sin^{-1}1$ (from 78.5° to 90°), which takes (90 − 78.5)/360 of one cycle, or 0.032 × 0.02 s, since one cycle is 0.02 s. This is only 0.64 ms. Let us suppose the current pulse shape is square, then

$$i = Cdv/dt \approx CV_r/\Delta t = (0.01 \times 6.8)/(6.4 \times 10^{-4}) = 106\ \text{A} \tag{7.39}$$

If we assume a form factor of $2/\pi$ for the current (taking it to be sinusoidal), the maximum current flow is magnified by $\pi/2$ to 167 A. However, this current is carried for only a very short time: the average diode current during one cycle (0.02 s) is only 3.4 A, and the diode need only have this current rating. The maximum energy stored in the capacitor is $\frac{1}{2}CV_p^2$, or 578 J, and the minimum is $\frac{1}{2}CV_{Cmin}^2 = 555$ J, so only 23 J are supplied to the load every cycle, at a rate of 23 × 50 = 1.15 kW, and this is the power dissipated in the load.

7.6.2 *The full-wave rectifier*

The half-wave rectifier in effect uses only half the supply. The full-wave rectifier uses all of the supply waveform and is therefore more readily smoothed and produces only half as much peak current flow into the smoothing capacitor as the half-wave rectifier. There are two ways principally of producing full-wave rectification: one uses a centre-tapped transformer and two diodes (Figure 7.19a), the other uses a bridge diode network (Figure 7.17b).

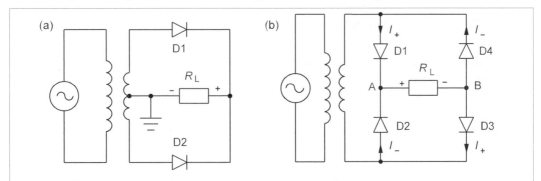

Figure 7.19 Full-wave rectifiers (a) Centre-tapped transformer (b) Bridge rectifier. Smoothing capacitors are placed in parallel with R_L in both circuits.

In Figure 7.19a, D1 conducts on the positive half-cycle, while D2 blocks. On the negative half-cycle D2 conducts while D1 blocks. Only half of the transformer secondary is used at a time, though it must still be wound for maximum current rating. A capacitor across the load must be used for smoothing, but need only be half of the capacitance of that used in a half-wave rectifier for the same ripple.

The diode-bridge network shown in Figure 7.19b works like this: on the positive-going waveform diodes D1 and D3 conduct as they are forward biassed, while diodes D2 and D4 do not as they are reverse biassed. During the negative half-cycle, D2 and D4 are forward biassed, so they conduct while D1 and D3 are reverse biassed and do not. The current flow on the positive half-cycle, I_+, is from A to B through the load, R_L, and so also is the current flow during the negative half-cycle, I_-. A smoothing capacitor across the load, as in Figure 7.20a, gives the current and voltage waveforms in Figure 7.20b. The analysis of the effect of this capacitor follows the same path as for the half-wave rectifier, and leads to $\Delta V = V_p /2R_L Cf$, so the capacitance is halved for the same ripple, as is the charging current.

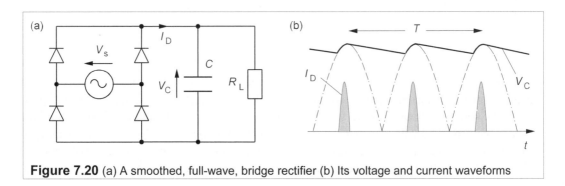

Figure 7.20 (a) A smoothed, full-wave, bridge rectifier (b) Its voltage and current waveforms

Nowadays plenty of 4-terminal diode bridges are available, ranging from the PCB-mounted type DF02M supplied by International Rectifier and rated at $V_{RRM} = 200$ V, $I_{F(AV)} = 1$ A, for 47p, through the 15A, 200V GBPC1502 from General Semiconductor at £2.58, up to the 100A, 1200V PSB105/12 from Powersem for £79, all in single quantities. In many cases it is cheaper to buy four diodes of the appropriate rating! 5-terminal diode bridges are available for rectifying 3-phase supplies. These contain six diodes with high current (13–175 A) and voltage (800–1200 V) ratings, and are discussed in Chapter 19.

7.6.3 Clipping and clamping: voltage limitation

There are occasions on which one wishes to restrict the voltage at some point in a circuit, or to clip a waveform. Consider the circuit of Figure 7.21a; because the diode must drop 0.6 V, V_o, will remain at 5.6 V. Moreover, the current through the resistor must be $(15 - 5.6)/470 = 20$ mA. The diode has clamped V_o to 0.6 V above the voltage at the diode's cathode, and it is immaterial what the voltage to the left of the 470Ω resistance is, as long as it is over 5.6 V.

The circuit of Figure 7.21b shows two diodes in use as voltage *limiters* which will clip both positive and negative parts of the input waveform to ±0.6 V as a protective measure for any sensitive components attached to the output terminal: the resistor limits the diode current.

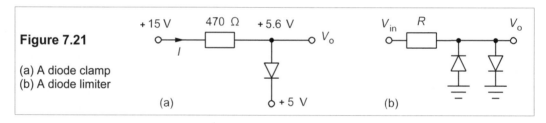

Figure 7.21

(a) A diode clamp
(b) A diode limiter

In cases where a waveform is to be clipped, two Zeners can be put in series as in Figure 7.22a, giving the waveform in Figure 7.22b. The clipping voltage in this case is 0.6 V more than the Zener breakdown because of the forward drop across the second Zener.

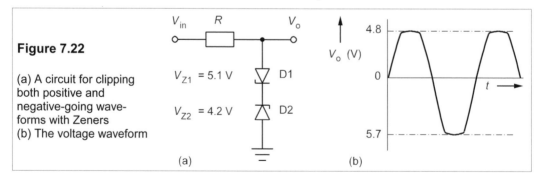

Figure 7.22

(a) A circuit for clipping both positive and negative-going waveforms with Zeners
(b) The voltage waveform

Inductive loads can cause problems when they are disconnected from their supply, as the voltages generated can damage the switch. A catch diode connected across the load as in Figure 7.23 is the solution, though it is no use if the supply is AC, since the diode will short out half of the waveform. The voltage produced by the inductor will be $V_B = L \, di/dt$, so if the supply is disconnected, di/dt is large and negative, and the voltage across the load is large and of opposite polarity to the supply. The diode then conducts and allows the load current to decay harmlessly.

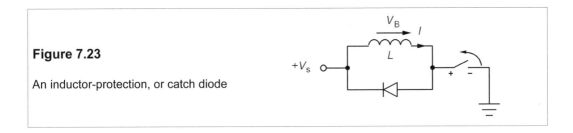

Figure 7.23

An inductor-protection, or catch diode

7.6.4 DC restoration

Capacitors are often used to couple circuits, in which case the direct component of the signal is lost, as capacitors block DC. The direct component of the signal can be restored, very nearly, by using a diode as shown in Figure 7.24a.

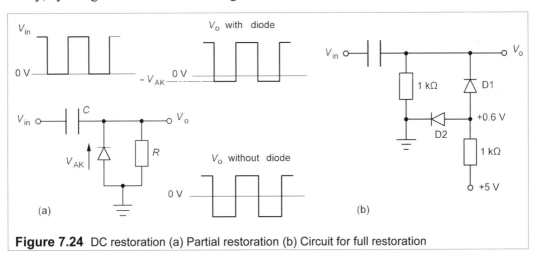

Figure 7.24 DC restoration (a) Partial restoration (b) Circuit for full restoration

The input waveform is shown and the output waveforms with and without the diode. The reverse biassed diode restores the direct component except for the 0.6 V diode forward drop. By using two diodes as in Figure 7.24b, the signal is fully restored. The second diode holds the first diode's anode at +0.6 V, so the drop across the first diode brings the signal down to 0 V. Compensation of this kind is essential of course for signals having amplitudes less than 0.6 V. The use of a second identical diode rather than a voltage divider ensures that temperature effects are fully compensated.

7.6.5 Voltage multipliers

By using ingenious combinations of diodes and capacitors it is possible to rectify and at the same time multiply the supply voltage by an integer ranging from two up. The circuit most often met is the full-wave voltage doubler that of Figure 7.25, and is the same as the full-wave bridge rectifier with diodes D2 and D3 replaced by capacitors. In this circuit the maximum voltage from the supply V_p, is doubled (but for the forward drop of two diodes). On the positive cycle D1 conducts and C1 charges up to $V_p - V_{AK}$. On the negative half-cycle D2 conducts and charges C2 up to $V_p - V_{AK}$, so the maximum output voltage is $2(V_p - V_{AK})$, with the usual ripple.

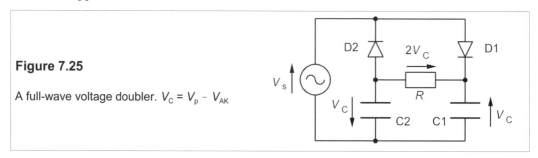

Figure 7.25

A full-wave voltage doubler. $V_C = V_p - V_{AK}$

7.6.6 Free-wheeling

The basic circuit is shown in Figure 7.26, D2 being the free-wheeling diode. D1 conducts when the supply is in its positive half cycle and so builds up current in the inductor. When D1 cuts off the current flow to the inductor, the free-wheeling diode, D2, allows the current to free-wheel through the inductor until the supply can replenish it once more. See also Section 19.2.2.

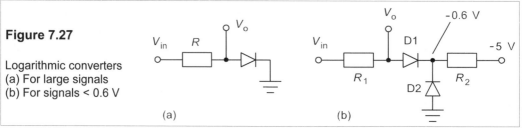

Figure 7.26

D2 is a free-wheeling diode which allows current to flow when D1 is not conducting

7.6.7 Logarithmic conversion

Since the current through a diode is an exponential function of the voltage across it, we can take the logarithm of a signal fairly easily with the circuit shown in Figure 7.27a. If the signal is smaller than 0.6 V, then compensation for diode drop such as that shown in Figure 7.27b is necessary.

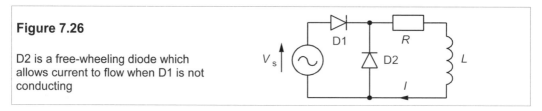

Figure 7.27

Logarithmic converters
(a) For large signals
(b) For signals < 0.6 V

(a)   (b)

In this circuit, unless $R_2 \ll R_1$, D2 will not pass enough current to keep its cathode at 0.6 V. The circuit will not work well over large input voltage ranges. Logarithmic converters are available as ICs which use operational amplifiers with a p-n junction in the feedback loop to give an output that is the logarithm of the input voltage. By suitable compensation these can be made to perform fairly well over several voltage decades.

7.6.8 AM detection

Amplitude modulation (AM) is described in Section 24.3, but here we can consider the use of an RF diode to demodulate the signal from the carrier. The diode rectifies the input signal of Figure 7.28a and the rectified signal is filtered by an RC network, Figure 7.28b, to give the demodulated signal shown in Figure 7.28c. If the carrier frequency is f_c and the modulation frequency is f_m then we require that

$$1/f_c \ll RC \ll 1/f_m \tag{7.40}$$

in order that the demodulated signal should be undistorted and have no ripple, as in Figure 7.28d. Some DC bias may be needed to offset the diode drop if the signal is not large.

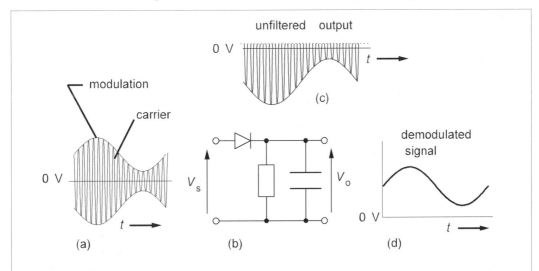

Figure 7.28 A diode used as an AM demodulator or envelope detector (a) The input waveform (b) The circuit (c) The rectified waveform (d) The demodulated waveform

Suggestions for further reading – see Chapter 9

Problems

1. A diode obeys the equation, $I \approx I_s \exp(qV/\eta kT)$, where $I_s = A\exp(-E_g/kT)$, A being constant. Show that the change in voltage with temperature, when the current is constant, is given by

$$\left.\frac{\mathrm{d}V}{\mathrm{d}T}\right|_{I=\text{const.}} = \frac{V - \eta E_g/q}{T}$$

    If $V = 0.7$ V, $E_g = 1.1$ eV, $T = 300$ K, and $\mathrm{d}V/\mathrm{d}T = -2.1$ mV/K for Si diodes, find η. *[1.21]*
2. The diodes in the circuit of Figure P7.2 have negligible resistance and forward voltage drops of 1 V. Find I_1, I_2, I_3 and I_4 . *[$I_1 = 1.143$ A, $I_2 = 0.829$ A, $I_3 = 0$, $I_4 = 0.314$ A]*

3.   Three LEDs are connected to a DC supply as shown in Figure P7.3a. If D1 and D2 have forward
     voltage drops of 2 V and D3 one of 2.02 V what is the current flowing in each, given they all have
     dynamic resistances of 1 Ω. Calculate the currents flowing in Figure P7.3b if the diodes are the
     same as in Figure P7.3a. *[$I_1 = I_2 = 26$ mA, $I_3 = 6$ mA; $I_1 = I_2 = 21.15$ mA, $I_3 = 21.09$ mA]*
4.   For the circuit of Figure P7.4, find I_1 and I_2, given that the forward-biassed diodes have a voltage
     drop of 0.7 V at all currents. *[$I_1 = 0.14$ A, $I_2 = -0.39$ A]*

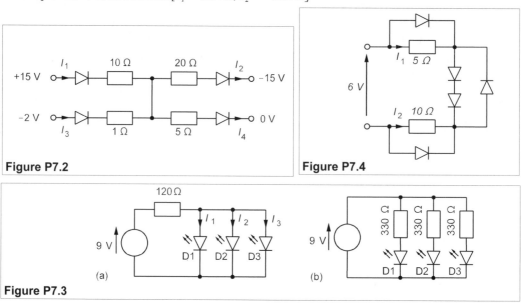

Figure P7.2   **Figure P7.4**

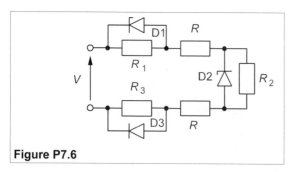

Figure P7.3

5.   Repeat the previous problem but with the source polarity reversed.
6.   Find the current through each component in Figure P7.6 assuming the diodes are ideal and D1 has
     $V_Z = 6$ V, D2 has $V_Z = 4$ V, the forward voltage drop of D3 is 0.8 V, $R = 40$ Ω, $R_1 = 300$ Ω, $R_2 = 100$
     Ω and $R_3 = 80$ Ω and (a) $V = +6$ V (b) $V = +15$ V (c) $V = +10.7$ V. What is the effective circuit
     resistance in each case? *[(a) $I_R = I_{R1} = I_{R2} = 10.83$ mA, $I_{R3} = 10$ mA, $I_{D1} = I_{D2} = 0$, $I_{D3} = 0.83$ mA, 554
     Ω (b) $I_R = 52.5$ mA, $I_{R1} = 20$ mA, $I_{R2} = 40$ mA, $I_{R3} = 10$ mA, $I_{D1} = 32.5$ mA, $I_{D2} = 12.5$ mA, $I_{D3} = 42.5$
     mA, 286 Ω (c) $I_R = 21.7$ mA, $I_{R1} = 20$ mA, $I_{R2} = 21.7$ mA, $I_{R3} = 10$ mA, $I_{D1} = 1.7$ mA, $I_{D2} = 0$, $I_{D3} =
     11.7$ mA, 494 Ω]*

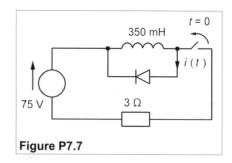

Figure P7.6   **Figure P7.7**

7.   In the circuit of Figure P7.7, steady-state conditions obtain prior to the switch's opening. Find $i(t)$
     if the diode has negligible resistance and a forward voltage drop of 0.7 V. When is $i(t) = 0$? What
     is the power consumption as a function of time? What is the average power consumption during

discharge? $[i(t) = (25 - 2t)u(t) + 2(t - 12.5)u(t - 12.5) A, t = 12.5 s, p(t) = (17.5 - 1.4t)u(t) + 1.4(t - 12.5)u(t - 12.5) W, 8.75 W]$

8. In the circuit of Figure P7.8 the switch was closed for a long time before being opened at $t = 0$. If the diode's forward drop is 1 V, what is $i(t)$? When is the current zero? Suggest ways in which the current can be reduced to zero more quickly. $[i(t) = -2 + 8\exp(-20t) A$ *(for* $0 \le t \le t_0$*).* $i(t) = 0$ when $t = t_0 = 69.3 ms]$

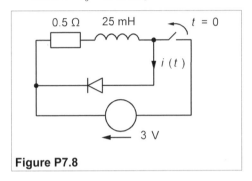

Figure P7.8

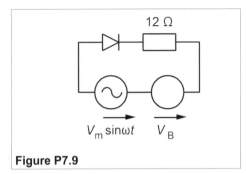

Figure P7.9

9. In the circuit of Figure P7.9, the diode's forward voltage drop is 0.7 V. What is the average power developed in the load if $V_m = 6$ V and V_B is (a) 1 V (b) 2 V and (c) 4 V?
    [(a) 322 mW (b) 165 mW (c) 15.9 mW]
10. Repeat Problem 7.9 with V_B reversed. [(a) 0.85 W (b) 1.24 W (c) 2.31 W]
11. In the circuit of Figure P7.11 the diode has a dynamic resistance of 0.9 Ω at all forward currents and a forward drop of 0.6 V at very low current. If $V_m = 0.2$ V and $V_B = 1.2$ V, find the diode current and voltage. *[18 sin ωt + 43.3 mA, 0.0162 sin ωt + 0.639 V]*

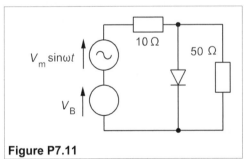

Figure P7.11

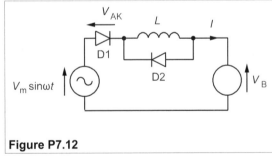

Figure P7.12

12. For the circuit of Figure P7.12 show that $I_{max} = [2V_m\cos\theta_c + (V_B + V_{AK})(2\theta_c - \pi)]/\omega L$, where θ_c, the conduction angle, is $\sin^{-1}[(V_{AK} + V_B)/V_m]$. What is I_{max} if $V_m = 10$ V, $V_B = 5$ V, $V_{AK} = 0.7$ V, $L = 15$ mH and $f = 50$ Hz? What is V_B if $I_{max} = 0.5$ A? *[$I_{max} = 1.154$ A, $V_B = 6.82$ V]*
13. A varactor diode is used in a tuned circuit that gives an output frequency of $1/2\pi\sqrt{(LC)}$, where C is the diode capacitance given by $C = 100/\sqrt{V}$ pF. V is the reverse bias voltage of the diode and can be varied from 1 V to 20 V. What is the frequency range of the tuned circuit if $L = 10$ µH? Derive an expression for the relative change of frequency with voltage. Calculate from this the change in frequency produced by a 1% change in diode voltage at 7 MHz. *[5 MHz to 10.6 MHz, 17.5 kHz]*
14. For the circuit of Figure P7.14 the Zener voltage is 10 V at a current of 10 mA. If the dynamic resistance of the diode is 15 Ω, calculate the maximum and minimum load voltages if the load current can vary from 5 mA to 25 mA (independent of the supply voltage) and the supply voltage from 20 to 25 V (independently from the load current). What is the minimum power rating required for the Zener? *[10.5 V, 9.976 V. 455 mW, but at least 600 mW is needed to give a fair safety margin]*

15. The circuit of Figure 7.15 has V_s varying from 15–18V and a load current varying independently from 2–10 mA. If D1 has $V_Z = 6.2$ V when $I_Z = 10$ mA with $r_Z = 4\,\Omega$, and D2 had $V_Z = 8.2$ V when $I_Z = 10$ mA with $r_Z = 5\,\Omega$, what will the load voltage regulation be, given $R_s = 400\,\Omega$? *[5.2%]*

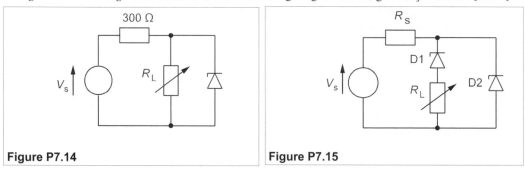

Figure P7.14   **Figure P7.15**

16. The half-wave rectifier in Figure P7.16a gives the CRO traces shown in Figure P7.16b. The voltage scales (vertical) are 0.2 V for CH1 and 0.1 V for CH2 for each of the larger divisions. The horizontal scale (time) is the same for both channels and is 50 µs per (large) division. CH1 is AC coupled and has no DC component, but CH2 is DC coupled and shows the zero voltage level. The source voltage is 12 V_{pp}. Find the values of R and C. (Take the diode's forward voltage drop to be 0.7 V.) *[4 kΩ, 0.7 µF]*

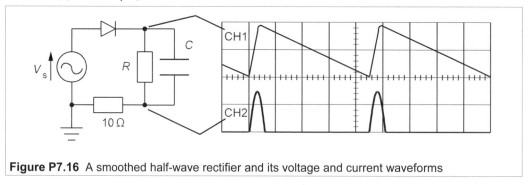

Figure P7.16 A smoothed half-wave rectifier and its voltage and current waveforms

17. The energy of a photon is given approximately by $E = 1.24/\lambda$, where E is in eV and the wavelength, λ, is in µm. If the range of photon energies in eV is $3kT$, show that the linewidth is given by $\Delta\lambda = 2.4kT\lambda^2$. What is the Q ($= \lambda/\Delta\lambda$) of a GaAs LED at 77 K (the boiling point of liquid nitrogen) and at 300 K given that its bandgap is 1.42 eV? (The Q of a LED indicates how sharp the spectral line is.) *[Q = 72 and Q = 18.4]*

18. Using the Figures given in Table 7.1, what area of solar panels would be required to deliver an average of 1 kW to a house throughout the year, given that the sun shines for 1500 h/year at sufficient intensity. Ignoring the cost of energy storage and installation, and counting only the cost of the solar cells, what would the cost be per kWh if the charge on capital is 12% per annum, including interest and depreciation? If the sun shone for twice as long, what would the price of solar panels have to be to reduce the cost/kWh to 10 p? *[51.6 m². 54.6 p/kWh. £283/m²]*

8 Bipolar junction transistors

ALTHOUGH in principle consisting merely of two p-n junction diodes back to back, in practice the bipolar junction transistor (BJT) is a completely different class of device; whereas the diode is a passive device, the transistor is *active* – it can be used to amplify voltages or signals. The BJT is termed *bipolar* because its operation – unlike the field effect transistor's – depends on both positive (holes) and negative (electrons) charge carriers. Before the transistor was invented, amplifiers used vacuum tubes ('valves') and vacuum tubes used a great deal of power, most of which was wasted as heat. Vacuum tubes were also prone to failure by filament burn out, loss of vacuum and just plain breakage; they were also difficult to miniaturise. A 'large' computer (by the standards of its day, 1950) such as EDSAC at Cambridge occupied a fair-sized room, required a large amount of power and cooling, and was prone to break down as the tubes failed, though ingenious methods were devised to minimise the stoppages. But it was not only high-technology products like computers which suffered from valve technology; radios for example were large and clumsy, and batteries for 'portable' radios were enormous, heavy and costly. Above all, the transistor could be made smaller and smaller, and cheaper and cheaper – and more and more reliable. The story of the transistor is far from finished 55 years after its invention.

8.1 Theory of operation

Figure 8.1 shows a conceptual diagram of pnp and npn transistors – the structures of practical devices are rather different (see Chapter 10) – in which a thin layer of semiconductor material, called the *base*, is sandwiched between two pieces of opposite conductivity type, known as the *emitter* and the *collector*. We shall consider briefly the behaviour of the pnp transistor: the npn behaves similarly, but with holes and electrons exchanging places; in consequence the supply polarities are different for the two types.

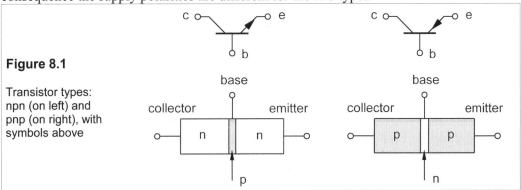

Figure 8.1

Transistor types: npn (on left) and pnp (on right), with symbols above

With the supply voltages applied as shown in Figure 8.2, the emitter-base junction is forward biassed by V_{EE}. Thus holes are emitted from the emitter region and injected into the base. The emitter is much more heavily doped than the base, so the current across the emitter-base junction, I_E, is almost entirely due to holes (majority carriers in the emitter), with only a small electron current in the opposite direction. The injected holes, being minority carriers in the base, must therefore combine with electrons (the majority carriers in the base). However, the base is very thin and most of the injected holes arrive at the base-collector junction, provided the lifetime of the holes in the base is long enough compared to their transit time across it. The base-collector junction is reverse biassed, so that holes arriving at the n-type base side of it are at a higher potential than those on the collector side, and are therefore swept across into the collector, where they are majority carriers once more.

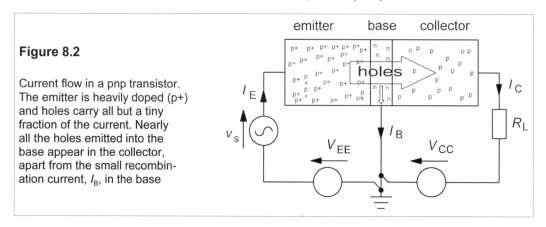

Figure 8.2

Current flow in a pnp transistor. The emitter is heavily doped (p+) and holes carry all but a tiny fraction of the current. Nearly all the holes emitted into the base appear in the collector, apart from the small recombination current, I_B, in the base

The signal voltage, v_s, in the emitter circuit causes a large change in the emitter current, since this is given by the rectifier equation

$$I_E \approx I_s \exp[q(V_{EE} + v_s)/kT] \tag{8.1}$$

The emitter current, I_E, is almost equal to the collector current, I_C, apart from a small recombination current, I_B, which is the current flowing out of the base. Thus

$$I_C = I_E - I_B = \alpha I_E \tag{8.2}$$

In well-made devices α is of the order of 0.99 to 0.999, that is $I_C \approx I_E$ and $I_B = (1 - \alpha)I_E < 0.01$ I_E. The parameter α is not often used in electronics, the parameter β is used instead, which is defined by

$$I_C = \beta I_B \tag{8.3}$$

β^1 is the forward current gain and Equation 8.3 is the fundamental equation for transistors.

Since the emitter-base junction is forward biassed, it presents a low impedance to signals, while the collector-base junction, being reverse biassed is high impedance, thus a large voltage and power gain is achieved in R_L. Next we must look at the input and output characteristics and the biassing of the transistor.

---

[1] $\beta \equiv h_{FE}$, an *h*-parameter discussed in Section 8.7; it can be shown that $\beta = \alpha/(1 - \alpha) \approx 1/(1 - \alpha)$ as $\alpha \approx 1$.

8.2 **The common-emitter amplifier**

In Figure 8.2 the base is common to both the emitter and collector circuits, giving rise to the name *common-base* amplifier. This configuration is not often used because the current gain is less than one, though the voltage gain is large. The most versatile amplifier giving both current and voltage gain is the *common-emitter* amplifier, whose input and output characteristics can be obtained using the circuit of Figure 8.3. The term 'common emitter' signifies that the emitter is common to both the input and the output circuits.

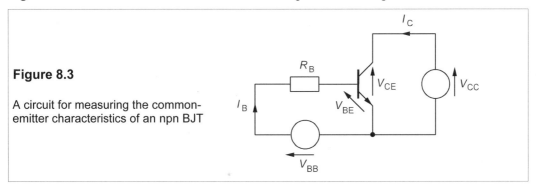

Figure 8.3

A circuit for measuring the common-emitter characteristics of an npn BJT

8.2.1 *Characteristics of the CE amplifier*

The input circuit in Figure 8.3 comprises the voltage source, V_{BB}, feeding the base with current, I_B, which combines with I_C in the base and crosses the base-emitter junction as I_E. The base-emitter junction is a forward-biassed p-n junction, so the input characteristic shown in Figure 8.4 is simply the current-voltage graph for a forward biassed p-n junction.

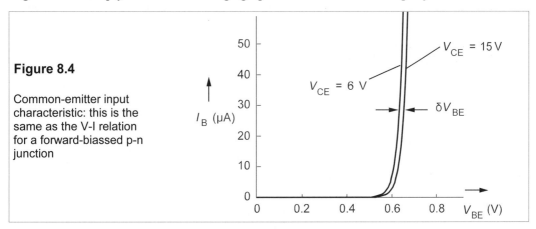

Figure 8.4

Common-emitter input characteristic: this is the same as the V-I relation for a forward-biassed p-n junction

The output circuit comprises the voltage source, V_{CC}, the collector-base junction and collector-emitter junction. The output characteristics, Figure 8.5, show a steep rise in I_C for small values of V_{CE}, followed by a levelling out to a constant value of I_C, directly proportional to I_B.

At low values of V_{CE} there is not enough voltage across the transistor to supply the collector current demanded by the base current. The transistor is said to be in *saturation* and $V_{CE\,sat} \approx 0.2$ V. The slope of the characteristic in this region is simply the resistance of the semicon-ductor material in the transistor, about 50 Ω in this case. However, when $V_{CE} > V_{CE\,sat}$, I_C is virtually constant for constant value of I_B because the proportion of charge carriers recombining in the base is constant. This is the operating region of the transistor when it is used as an amplifier. If V_{CE} is made too large there will be a steep rise in collector current due to breakdown of the collector-base junction – usually fatal because of the large amount of power dissipated in the transistor.

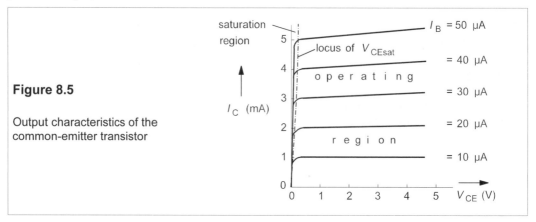

Figure 8.5

Output characteristics of the common-emitter transistor

8.2.2 Biassing the transistor: the Q-point and the load line

The purpose of an amplifier is to replicate a signal and make it larger. In the common-emitter amplifier of Figure 8.6a the output voltage, taken at the collector terminal, is derived from the DC supply voltage, V_{CC} (15 V in this case) and so must lie between 0 and 15 V. A signal, v_{in}, applied to the transistor base causes a change in base current, I_B, which is amplified by the transistor to produce a much larger change in I_C, the collector current. This produces a change in the voltage across R_C and in the collector voltage, V_{CE}.

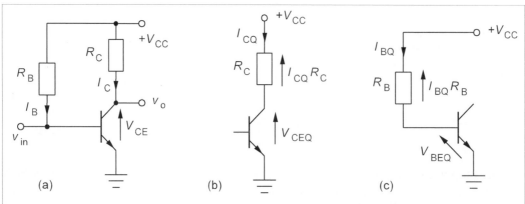

Figure 8.6 (a) The fixed-bias (sometimes called the base-bias), common-emitter (CE) amplifier
(b) the output side of the amplifier (c) the input side of the amplifier

The DC conditions in the amplifier are said to be obtained by *biassing* the transistor, which in Figure 8.6a means choosing R_B to obtain a suitable value of I_B, and choosing R_C to obtain a suitable value of V_{CE}. These are called the *quiescent* conditions because they are the unchang-ing current and voltage values when there is no signal input. Suppose the transistor in Figure 8.6a has the output characteristics of Figure 8.7 and that we want to bias it to make $V_{CEQ} = \frac{1}{2}V_{CC} = 7.5$ V and $I_{CQ} = 5$ mA, indicated on the figure by the point Q.

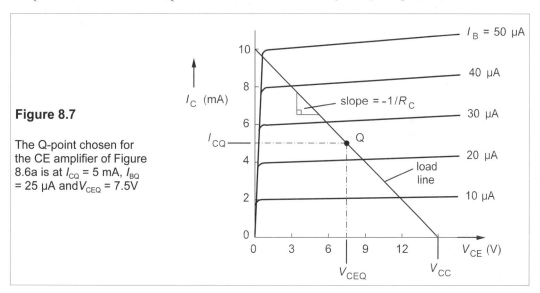

Figure 8.7

The Q-point chosen for the CE amplifier of Figure 8.6a is at $I_{CQ} = 5$ mA, $I_{BQ} = 25$ μA and $V_{CEQ} = 7.5$V

In order to operate at the desired Q-point we must insert the correct values of R_B and R_C into the circuit of Figure 8.6a; these values are found by applying KVL to the input and the output sides of the amplifier separately. In Figure 8.6b the output side is isolated and we see that KVL gives us

$$V_{CC} = I_{CQ}R_C + V_{CEQ} \tag{8.4}$$

Substituting the known values yields

$$15 = 5R_C + 7.5 \;\; \Rightarrow \;\; R_C = 7.5/5 = 1.5 \text{ k}\Omega \tag{8.5}$$

working as is customary in V, mA and kΩ.

In Figure 8.6c application of KVL gives

$$V_{CC} = I_{BQ}R_B + V_{BEQ} \tag{8.6}$$

But we know that V_{BEQ} is just the voltage across a forward-biassed p-n junction, which is always about 0.6 V in silicon devices, and we also know from Figure 8.7 that $I_{BQ} = 25$ μA. Putting these values into Equation 8.6 leads to

$$15 = 0.025R_B + 0.6 \;\; \Rightarrow \;\; R_B = 14.4/0.025 = 576 \text{ k}\Omega \tag{8.7}$$

These two resistance values are all that we need to bias the transistor to operate at the Q-point where $I_{CQ} = 5$ mA and $V_{CEQ} = 7.5$ V.

This Q-point, with $V_{CEQ} = \frac{1}{2}V_{CC}$, is optimal because it allows the largest possible output voltage swing to be obtained – from 0 to V_{CC} as shown in Figure 8.8a. If V_{CEQ} is lower than optimal, as in Figure 8.8b, then the output can be clipped at the bottom when it might have been undistorted with the optimal Q-point. Similarly if V_{CEQ} is made too high, as in Figure 8.8c, the output can be clipped at the top of the waveform. Of course, if the input signal is large enough the output will be clipped regardless of where the Q-point is placed.

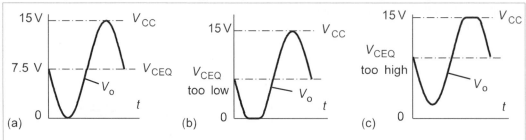

Figure 8.8 (a) V_{CEQ} is optimal (b) For the same input signal V_{CEQ} is too low; negative peaks of v_o are clipped (c) V_{CEQ} is too high; positive peaks of v_o are clipped

The collector current can be found by using KVL on the circuit of Figure 8.6b, as before

$$V_{CC} = V_{RC} + V_{CE} = I_C R_C + V_{CE} \tag{8.8}$$

which gives

Equation 8.9 is that of a straight line, called the *load line*, of slope $-1/R_C$ and intercept V_{CC}/R_C

$$I_C = (-1/R_C)V_{CE} + V_{CC}/R_C$$

$$\tag{8.9}$$

$$y = m\,x + c$$

on the I_C-axis. It is plotted on the output characteristics of Figure 8.7: by putting $V_{CC} = 15$ V and $R_C = 1.5$ kΩ the I_C-axis intercept is found to be $15/1.5 = 10$ mA, and by putting $I_C = 0$ the V_{CE}-axis intercept is found to be V_{CC} (= 15 V), while the slope is $-1/R_C$. The load line is the locus of *all* the possible values of I_C and V_{CE} for a given circuit.

The load line in Figure 8.7 can be used to derive the output voltage swing for a given base current change: for example if the minimum base current is found to be 10 μA, the load line intercepts the $I_B = 10$ μA line at $V_{CE} = 12$ V, which is the maximum output voltage. If the maximum base current is 40 μA, the intercept with the load line occurs at $V_{CE} = 3$ V, which is the minimum output voltage. As the base current goes up, the output voltage goes down and the signal is therefore amplified but inverted. We will see later on how to calculate the output voltage from the input voltage and the circuit parameters.

Example 8.1

Use the load line and Figure 8.7 to find (a) the transistor's β (b) the base current at which the transistor is driven into saturation (c) the maximum unclipped output voltage swing when $I_{CQ} = 2$ mA instead of 5 mA.

(a) The transistor's β (= h_{FE}) is the ratio I_C / I_B in the operating region[2] (where the output characteristics are nearly flat). When $I_B = 30$ μA = 0.03 mA, $I_C = 6$ mA, so β = 6/0.03 = 200.

(b) The transistor is saturated when the collector current falls below the value at the knee of the

---

[2] The convention for subscripts is that capitals are used for DC values and lower case for small-signal values.

output characteristic, that is when $I_B > \beta I_C$. Clearly when $I_B = 50\ \mu A$, the load line intercepts the output characteristic at $I_C = 9.5\ mA < \beta I_B$, so the transistor is in saturation. The base current corresponding to $I_C = 9.5\ mA$ is $I_B = 9.5/\beta = 9.5/200 = 47.5\ \mu A$.

(c) When the Q-point moves down the load line to $I_{CQ} = 2\ mA$, we see that V_{CEQ} is then 12 V and the maximum output voltage is 15 V, that is 3 V more. If the output voltage is to be unclipped it can only swing from +9 V to +15 V, that is $V_{CEQ} \pm 3$ V. This compares to $V_{CEQ} \pm 7.5$ V for the optimal Q-point. The maximum possible input signal has been reduced by a factor of 7.5/3 = 2.5.

8.2.3 The voltage-divider bias circuit

The fixed-bias circuit produces a constant value for I_{BQ}, but since $I_{CQ} = \beta I_{BQ}$ any change in β causes a corresponding change in I_{CQ}. Sadly, β is not a well-behaved parameter: it changes with temperature, with collector current, and from transistor to transistor in manufacture, even if the process is under good control. For example, the manufacturer's specification sheet for the ZTX302 npn transistor states that β can vary from 50 to 300 at 20°C and $I_C = 10\ mA$. The Q-point can therefore move from saturation almost to cut-off (that is $I_B = 0$) with different transistors. To overcome this difficulty and stabilise the collector current the voltage-divider (sometimes called the emitter-bias) circuit of Figure 8.9 is used.

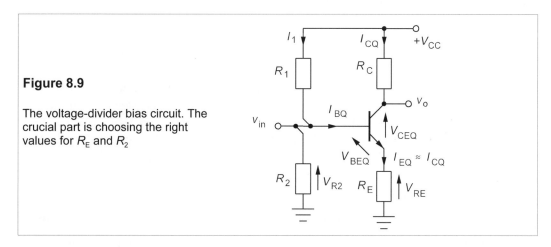

Figure 8.9

The voltage-divider bias circuit. The crucial part is choosing the right values for R_E and R_2

In this circuit the base-bias voltage is more or less fixed by the pair of resistors, R_1 and R_2, on the input side. If $I_{BQ} \gg I_1$, then R_1 and R_2 form a voltage divider that keeps the base voltage at

$$V_{R2} = R_2 V_{CC}/(R_1 + R_2) \qquad (8.10)$$

The root of the circuit's stability, however, is the emitter resistor, R_E. To see this, let us suppose β has increased so that I_C increases. In turn this increases the voltage across R_E; but if V_{R2} and V_{BEQ} are virtually fixed, I_{CQ} must fall to its former value: I_{CQ} is very nearly independent of β.

Suppose $V_{CC} = 15$ V, $\beta = 100$, and it is desired to have $I_{CQ} = 5$ mA (or $I_{BQ} = 50\ \mu A$) and $V_{CEQ} = 7.5$ V, what must R_1, R_2, R_C and R_E be in Figure 8.9? By Kirchhoff's voltage law,

$$I_{CQ}R_C + I_{EQ}R_E \approx I_{CQ}(R_C + R_E) = V_{CC} - V_{CEQ} = 7.5\ \text{V} \qquad (8.11)$$

whence $R_C + R_E = 1.5\ k\Omega$. Because the voltage drop across R_E reduces the output voltage swing, it is best to make it about $\frac{1}{4}R_C$ – 330 Ω (0.33 kΩ) is a reasonable value, with R_C = 1.2 kΩ. R_1 and R_2 must be chosen so that the current through each, I_1, is much greater than I_{BQ}. However, because they reduce the input resistance, they must be as large as possible. A reasonable compromise might be to make $I_1 \approx 10I_{BQ} = 0.5$ mA, then

$$I_1 = 0.5 \approx \frac{V_{CC}}{R_1 + R_2} = \frac{15}{R_1 + R_2} \Rightarrow R_1 + R_2 = \frac{15}{0.5} = 30 \text{ k}\Omega \qquad (8.12)$$

Now the current through R_E is 5 mA, giving a voltage drop across it, V_{RE}, of $0.330 \times 5 = 1.65$ V. But by Kirchhoff's voltage law,

$$V_{R2} = V_{BEQ} + V_{RE} = 0.7 + 1.65 = 2.35 \text{ V} \qquad (8.13)$$

Using the voltage divider rule and neglecting I_{BQ},

$$V_{R2} = V_{CC}R_2/(R_1 + R_2) = 15R_2/30 = 2.35 \text{ V} \qquad (8.14)$$

which yields $R_2 = 4.7$ kΩ and $R_1 = 30 - R_2$ (Equation 8.12) = 25.3 kΩ. Choosing a preferred value for R_1 would make it 22 kΩ. It can then be shown that the actual Q-point (as in Problem 8.5) is at $I_{CQ} = 5.2$ mA and $V_{CEQ} = 7$ V.

8.2.4 *The small-signal equivalent circuit*

In order to calculate the voltage, current and power gains of the CE amplifier we need an equivalent circuit. When using a small-signal equivalent circuit it is assumed that the non-linear characteristics of the amplifier can be considered linear if the signals are small enough. Neglecting any base biassing resistors, the input resistance presented to a small-signal source, v_{in}, in Figure 8.9, is $\delta V_{BE}/\delta I_B = r_{in}$, where δV_{BE} is the small change in V_{BE} caused by the AC signal. Conventionally, these small-signal changes are written in lower case and δ is dropped so that δV_{BE} becomes v_{be} and $r_{in} = v_{be}/i_b$. And as $i_b = i_e/(\beta + 1)$,

$$r_{in} = (\beta + 1)v_{be}/i_e = (\beta + 1)r_e \approx \beta r_e = h_{ie} \qquad (8.15)$$

h_{ie} is one of the h-parameters used to characterise the CE amplifier (h-parameters are discussed in Section 8.2.6). Now r_e ($\equiv v_{be}/i_e$) is just the dynamic resistance of a forward-biassed p-n junction, which was found in Section 7.1 to be about $0.025/I_{EQ}$ at 20°C. Taking $I_{EQ} \approx I_{CQ} = 5$ mA, as in Figure 8.7, r_e is 5 Ω; and then $r_{in} = h_{ie} = \beta r_e = 500$ Ω.

Consider the amplifier of Figure 8.6, which has no emitter resistor, R_E, and suppose a small-signal input to the base, v_{in}, causes a change in the collector current of δI_C. We find

$$V_{CE} = V_{CC} - I_C R_C \Rightarrow \delta V_{CE} = -\delta I_C R_C \qquad (8.16)$$

The AC output voltage is therefore $v_o = \delta V_{CE} = v_{ce} = -\delta I_C R_C = -i_c R_C = -\beta i_b R_C$. These considerations lead to the small-signal equivalent circuit of Figure 8.10.

Note that the transistor is modelled by only two elements in the equivalent circuit: an input resistance, βr_e, and a current source, $-\beta i_b$. The transistor's characteristics have thus been reduced to two parameters; and the only external component of the amplifier to appear is R_C, but these are sufficient to calculate the voltage, current and power gains.

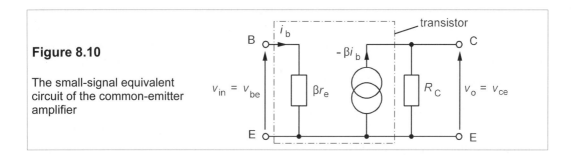

Figure 8.10

The small-signal equivalent circuit of the common-emitter amplifier

The no-load voltage gain, A_{v0}, is

$$A_{v0} = v_o/v_{in} = -\beta i_b R_C/i_b \beta r_e = -R_C/r_e \qquad (8.17)$$

The minus sign is usually retained to indicate that the output is 180° out of phase with the input. With $R_C = 1.5$ kΩ and $r_e = 5$ Ω, $A_{v0} = -1500/5 = -300$. The current gain, if we regard R_C as the load, is $A_i = \beta = 200$. The power gain is the product, $A_p = A_v A_i = 60000$. These gains in dB are $20\log_{10}|A_v| = 49.5$ dB, $20\log A_i = 46$ dB and $10\log A_p = 47.8$ dB.

8.2.5 *The practical amplifier*

Having biassed the transistor with a voltage divider as in Figure 8.9, the source and load can be connected; but if the source for example, contains a DC component it would upset the bias. This problem is solved by the use of a coupling capacitor, C_1, shown in Figure 8.11b.

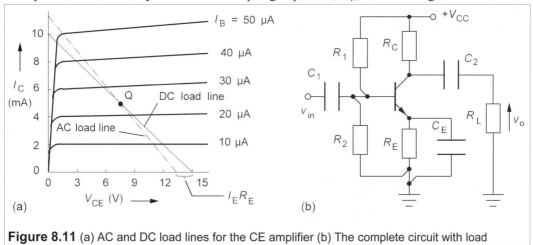

(a)

(b)

Figure 8.11 (a) AC and DC load lines for the CE amplifier (b) The complete circuit with load

The same applies to the output: the DC bias of the amplifier must be removed by a capacitor, C_2, so that only the amplified input signal remains. And there is a third difficulty which must be overcome: the presence of an emitter resistor, R_E, which would cause a great loss in gain unless a *bypass* capacitor, C_E, is placed across R_E.

We have just seen that the open-circuit voltage gain without an emitter resistor, R_E, is $-R_C/r_e$. Since R_E and r_e are in series, the analysis follows an identical path making $A_{v0} = -R_C/(r_e + R_E)$. As $R_E \gg r_e$, $A_{v0} \approx -R_C/R_E$. In the above example, with $R_C = 1.2$ kΩ and $R_E =$

330 Ω, $A_{v0} \approx -4$. Almost all the voltage gain has been lost because of R_E, a phenomenon called *gain degeneration*. The value of C_E is such that its reactance at the lowest frequencies used is negligible: a typical value might be 2200 μF. Then as far as AC is concerned, R_E is short-circuited, while still providing DC bias stabilisation. However, the output voltage swing is then reduced by the direct voltage across R_E.

The slope of the DC load line is $-1/(R_C + R_E)$ while that of the AC load line becomes $-1/R_C$. The AC load line still passes through the Q-point, which is where the instantaneous AC output voltage is zero. The DC volts dropped across R_E, $I_E R_E$, reduce the maximum signal amplitude as shown in Figure 8.11b.

8.2.6 The h-parameters

The hybrid-parameter small-signal equivalent circuit of the CE amplifier, shown in Figure 8.12, is often used.

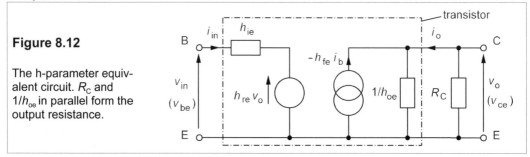

Figure 8.12

The h-parameter equivalent circuit. R_C and $1/h_{oe}$ in parallel form the output resistance.

The common-emitter h-parameters are defined by

$$v_{be} = h_{ie}i_b + h_{re}v_{ce}$$
$$i_c = h_{fe}i_b + h_{oe}v_{ce}$$

(8.18)

Hence

$$h_{ie} = \left. \frac{\delta V_{BE}}{\delta I_B} \right|_{V_{CE} = \text{const}} \qquad h_{re} = \left. \frac{\delta V_{BE}}{\delta V_{CE}} \right|_{I_B = \text{const}}$$

(8.19)

$$h_{fe} = \left. \frac{\delta I_C}{\delta I_B} \right|_{V_{CE} = \text{const}} \qquad h_{oe} = \left. \frac{\delta I_C}{\delta V_{CE}} \right|_{I_B = \text{const}}$$

The 'e' subscript indicates common-emitter configuration (other transistor amplifiers have appropriate subscripts). h_{ie} is the slope of the input characteristic, which is the input resistance or βr_e, in the order of 1 kΩ in a signal amplifier. h_{re} is the change in V_{BE} caused by changes in V_{CE}; as can be seen in Figure 8.4 it is small, even though exaggerated in the diagram for the sake of clarity. The small-signal current gain, $h_{fe} \approx h_{FE} \approx \beta$. h_{oe} is the slope of the output characteristic in the operating region and hence is the output conductance for small signals. Ideally h_{oe} is zero, but in practice it increases (see Figure 8.5) with base current. Typically h_{oe} is 10–25 μS ($\equiv$ 40–100 kΩ). As h_{oe} is in parallel with R_C and R_L it is usually neglected. If we ignore h_{re} and h_{oe}, the h-parameter equivalent circuit reduces to the small-signal equivalent circuit of Figure 8.10, with $h_{ie} = \beta r_e$ and $h_{fe} = \beta$.

Example 8.2

A transistor used in a common-emitter amplifier has $h_{ie} = 1\ k\Omega$, $h_{fe} = 100$, $h_{re} = 8 \times 10^{-4}$ and $h_{oe} = 20\ \mu S$. What will be the no-load voltage gain of a common-emitter amplifier, if $R_C = 2\ k\Omega$? What will be the current gain? The bias resistances may be neglected.

The h-parameter equivalent circuit is shown in Figure 8.12 with R_C in parallel with $1/h_{oe}$, comprising the output resistance, R_o. Now $R_C\ //\ 1/h_{oe} = 2\ k\Omega\ //\ 50\ k\Omega = 2 \times 50/52 = 1923\ \Omega = R_o$. The input current ($i_{in} \equiv i_b$) is

$$i_{in} = \frac{v_{in} - h_{re}v_o}{h_{ie}} = \frac{v_{in} - 8 \times 10^{-4}v_o}{1000} \tag{8.20}$$

Now v_o is $-h_{fe}i_{in}R_o$, so that $i_{in} = -v_o/h_{fe}R_o = -5.2 \times 10^{-6}v_o$. Comparing these two equations for i_{in} yields

$$-5.2 \times 10^{-6}v_o = \frac{v_{in} - 8 \times 10^{-4}v_o}{1000} \tag{8.21}$$

Then

$$-4.4 \times 10^{-3}v_o = v_{in} \quad \Rightarrow \quad v_o/v_{in} = A_{v0} = -227 \tag{8.22}$$

Had we neglected h_{oe}, A_{v0} would have been -238: hardly a significant difference. Neglecting both h_{oe} and h_{re} would have given $A_{v0} = -200$, which is significantly, but not greatly (about 10%), different from the 'exact' value.

The current gain is found by noting that $v_o = -i_oR_C$, and that it is also the voltage across the current source, which is $-h_{fe}i_{in}R_o$. Thus

$$i_o/i_{in} = A_i = -h_{fe}R_o/-R_C = 100 \times 1923/2000 = 97 \tag{8.23}$$

The sign is now positive because of the 2-port convention that the output current goes *into* the port. The current gain is very nearly equal to h_{fe} because $R_C \ll 1/h_{oe}$.

If the signal source, v_s, has an internal resistance, R_s, and the voltage-divider bias method is used then the input voltage to the base is reduced, and so the voltage gain is reduced in proportion. The voltage-divider bias resistances are in parallel as far as AC input are concerned (to see this, connect an ideal voltage source, V_{CC}, between the top terminal and ground in Figure 8.11b, and this will be a short-circuit to ground for AC, meaning R_2 and R_1 are both connected from the input to ground and are thus in parallel). Their combined resistance is therefore $R_B = R_1\ //\ R_2$, and $v_{in} = R_Bv_s/(R_B + R_s)$. For example, if $R_1 = 25.3\ k\Omega$, $R_2 = 4.7\ k\Omega$ and $R_s = 4\ k\Omega$, then $R_B = 3.96\ k\Omega$ and $v_{in} = 0.5v_s$, and the voltage gain is halved. For this reason the bias resistors must be kept as large as possible while still keeping the current through them much greater than the base current.

8.2.7 The low-frequency response

The low-frequency response of the common-emitter amplifier is determined by the bias-decoupling and bypass capacitors and the resistances in series with them in Figure 8.11b. The input capacitor, C_1, sees a resistance of $R_s + R_B\ //\ h_{ie}$, where R_s is the internal resistance of the input voltage source and R_B is $R_1\ //\ R_2$, so the time constant for this combination is $C_1(R_s + R_B\ //\ h_{ie})$, and the corner, or cut-off, frequency will be

$$f_{L1} = \frac{\omega_{L1}}{2\pi} = \frac{1}{2\pi C_1(R_s + R_B\ //\ h_{ie})} \tag{8.24}$$

As far as C_2 is concerned, it is in series with R_L and $R_C \, /\!/ \, 1/h_{oe}$, so the time constant is

$$f_{L2} = \frac{1}{2\pi C_2 (R_L + R_C \, /\!/ \, 1/h_{oe})} \tag{8.25}$$

Finally, C_E in combination with R_E and r_e (the resistance of the base-emitter junction seen by I_E, typically only 10 or 20 Ω) leads to a low-frequency cut-off given by

$$f_{LE} = \frac{1}{2\pi C_E (R_E \, /\!/ \, r_e)} \tag{8.26}$$

Example 8.3

In a common-emitter amplifier with voltage-divider bias (Figure 8.11), $R_1 = 25$ kΩ, $R_2 = 4.7$ kΩ, $R_s = 1$ kΩ, $R_E = 680$ Ω, $R_C = R_L = 2$ kΩ, $h_{oe} = 25$ μS, $\beta = 150$ and $I_E = 5$ mA. What is the lower cut-off frequency if $C_1 = 10$ μF, $C_2 = 10$ μF and $C_E = 100$ μF? What is the response at 10 Hz if the mid-band gain is +40 dB?

We must find the three cut-off frequencies given by Equations 8.24–8.26, to see which is the highest. First, $R_B = R_1 \, /\!/ \, R_2 = 25 \, /\!/ \, 4.7 = 3.95$ kΩ, in parallel with h_{ie}, which is given by

$$h_{ie} = V_{th}/I_B \approx \beta V_{th}/I_E = 150 \times 0.025/5 = 0.75 \text{ k}\Omega$$

taking V_{th} to be 25 mV at 300 K. Then $R_B \, /\!/ \, h_{ie} = 3.95 \, /\!/ \, 0.75 = 0.63$ kΩ. With $R_s = 1$ kΩ, substitution into equation 8.24 leads to $f_{L1} = 9.8$ Hz.

Next $1/h_{oe} = 40$ kΩ, so that $R_C \, /\!/ \, 1/h_{oe} = 1.9$ kΩ and $R_L + R_C \, /\!/ \, 1/h_{oe} = 3.9$ kΩ. Substitution into Equation 8.25 gives $f_{L2} = 4$ Hz.

We then find $r_e = V_{th}/I_E = 0.025/0.005 = 5$ Ω ($= h_{ie}/\beta$). And as $R_E = 680$ Ω, Equation 8.26 gives $f_{LE} = 321$ Hz $>> f_{L1} > f_{L2}$. The lower cut-off is therefore at 321 Hz and the response falls at 20 dB/decade as the frequency decreases to about 10 Hz, which is $\log_{10}(321/10) = 1.5$ decades below 321 Hz. Thus at 10 Hz the response should be $40 - 20 \times 1.5 = 10$ dB, but since C_1 takes effect here it might be 3 dB lower at 7 dB. Each time constant is associated with a -20 dB per decade roll-off and so the roll-off will be at up to -60 dB/decade: only if one cut-off is at a much higher frequency than the rest will the roll-off be at -20dB/decade.

8.2.8 The high-frequency response

The high-frequency response of the BJT is determined by the (very small ≈ 2 pF) p-n junction capacitances which shunt the device at high frequency, as shown in Figure 8.13.

In practice only one of these matters – the base-collector junction capacitance, C_{bc} – not because it is much larger than the other two (it is not) but because it is multiplied by the amplifier's gain, A, through the *Miller effect*. To see how this happens consider the feedback section of an amplifier, Figure 8.14a, where the output is linked to the input by an admittance, Y. The input current, I_1, is given by

$$Y_1 V_1 = Y V_1 (1 - A) \quad \rightarrow \quad Y_1 = Y(1 - A) \tag{8.27}$$

When Y is replaced by its equivalent in Figure 8.14b, then the currents and voltages must be the same, in particular $I_1 = Y_1 V_1$, which from Equation 8.26 leads to

$$I_1 = Y(V_1 - V_2) = Y V_1 (1 - V_2/V_1) = Y V_1 (1 - A) \tag{8.28}$$

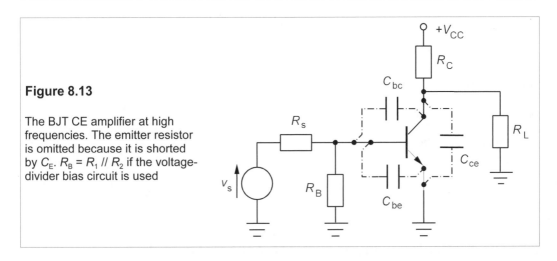

Figure 8.13

The BJT CE amplifier at high frequencies. The emitter resistor is omitted because it is shorted by C_E. $R_B = R_1 \mathbin{/\!/} R_2$ if the voltage-divider bias circuit is used

Unfortunately, C_{bc} is not a fixed quantity, being affected by the voltage across the junction and the current through it. The amplification factor, A, is best taken as the transistor β, and then the effective capacitance of the input side of the common-emitter amplifier is βC_{bc}, roughly 200 pF in a typical transistor.

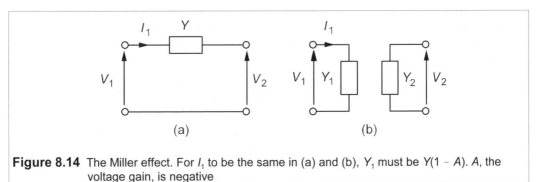

Figure 8.14 The Miller effect. For I_1 to be the same in (a) and (b), Y_1 must be $Y(1 - A)$. A, the voltage gain, is negative

If the input bias resistors are sufficiently large and the input source is low impedance, then the upper cut-off will be approximately

$$f_H = \frac{1}{2\pi\beta C_{bc}h_{ie}} \tag{8.29}$$

Taking $\beta = 100$, $C_{bc} = 2$ pF and $h_{ie} = 1$ kΩ gives $f_H = 800$ kHz, a typical value. The Miller effect on C_{bc} renders all the other capacitances (though they may be several times greater) of no account. As a result the upper frequency response rolls off at a strict -20 dB/decade, while the low-frequency response might not. If, say, $A_v = 100$ (or +40 dB) in the flat region between f_L and f_H, the unity-gain frequency would be $f_T = 100 f_H = 80$ MHz.

Figure 8.15 shows a gain-frequency plot for a common-emitter amplifier with well-separated time constants so the upper and lower roll-offs are both 20 dB/decade. The bandwidth of the amplifier is $f_H - f_L \approx f_H$, in the example above, where $f_H \gg f_L$.

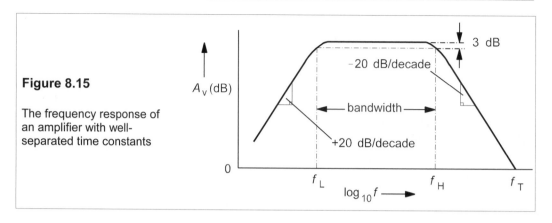

Figure 8.15

The frequency response of
an amplifier with well-
separated time constants

8.3 The emitter follower, or common-collector amplifier

We have seen that the common-emitter amplifier has moderate parameters useful for general-purpose signal amplifiers: moderate input resistance, moderate output resistance, moderate gain and moderate frequency response. The emitter follower has a better frequency response, low output resistance and high input resistance; it has however, a voltage gain less than unity. It makes a good buffer amplifier for high impedance sources. Figure 8.16a shows the emitter-follower circuit[3], also called a common-collector amplifier because the collector is grounded for AC input signals through the DC voltage source, V_{CC}; 'emitter follower' better describes its behaviour.

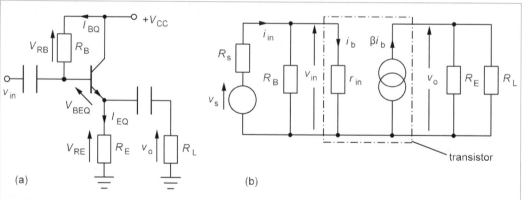

Figure 8.16 (a) The emitter-follower or common-collector circuit (b) Its approximate equivalent circuit, with a voltage source, source resistance and a load

In Figure 8.16a, the transistor $\beta = 100$, $V_{CC} = 20$ V, and I_{CQ} is to be 5 mA and $V_{CEQ} = 10$ V. Then $V_{RE} = V_{CC} - V_{CEQ} = 20 - 10 = 10$ V and $R_E = V_{RE}/I_{EQ} \approx V_{RE}/I_{CQ} = 10/5 = 2$ kΩ. The value of R_B is found from the KVL equation:

$$V_{RB} + V_{BEQ} + V_{RE} = I_{BQ}R_B + V_{BEQ} + I_E R_E = V_{CC} \qquad (8.30)$$

---

[3] For easier analysis the fixed-bias circuit is used, with the usual drawback of being dependent on transistor β. In practice a circuit independent of β would be used, such as the voltage-divider bias circuit of the CE amplifier.

But $I_{EQ}R_E = V_{RE} = 10$ V, so

$$I_{BQ}R_B = V_{CC} - V_{BEQ} - V_{RE} = 20 - 0.7 - 10 = 9.3 \text{ V} \tag{8.31}$$

We know that $I_{BQ} = I_{CQ}/\beta = 50$ μA and therefore $R_B = 9.3/50 \times 10^{-6} = 186$ kΩ (notice $R_B \approx \beta R_E$). The input resistance of the amplifier is then R_B // r_{in}, where the input resistance of the transistor, $r_{in} = \beta(r_e + R_L')$, and $R_L' = R_E$ // R_L. If there is no load, $R_L' = R_E$, and as $r_e \ll R_E$, the input resistance of the amplifier becomes very nearly R_B // $\beta R_E = 186 \times 200/(186+200) = 96$ kΩ $\approx \frac{1}{2}R_B$. The input resistance of the emitter follower amplifier is about β times that of the CE amplifier.

An equivalent circuit from which the voltage and current gains may be found is that of Figure 8.16b. From this we see that $v_o = \beta i_b R_L'$ (where $R_L' = R_E$ // R_L) and $i_b = v_{in}/r_{in}$.

Then

$$v_o = \beta v_{in} R_L'/r_{in} = \beta v_{in} R_L'/\beta(r_e + R_L') = v_{in} R_L'/(r_e + R_L') \tag{8.32}$$

And thus the voltage gain with load is

$$A_{vL} = v_o/v_{in} = R_L'/(r_e + R_L') \approx 1 \tag{8.33}$$

if $R_L' \gg r_e$, as it usually is. In the unloaded amplifier, $R_L' = R_E$, and taking the values of R_E and r_e in the example above as 2 kΩ and 5 Ω respectively leads to $A_{v0} = 0.9975$. The voltage gain of an emitter follower is *always less than one*. Equation 8.33 shows that the voltage gain will be halved when $R_L' = r_e$, which means when $R_L \approx r_e$, if $R_E \gg R_L$. In other words the output resistance of the amplifier is approximately r_e, a small value. The output resistance is not R_E because the current source's value depends on R_E – it is a *dependent* source. Its small output resistance makes the emitter follower good for driving low-resistance loads.

Example 8.4

Find the AC voltage and current gains of an emitter follower such as in Figure 8.16, where the transistor has a β of 120, R_B = 68 kΩ, R_S = 5 kΩ, R_E = 600 Ω, R_L = 100 Ω and I_{EQ} = 15 mA. Assume $r_e = 0.03/I_{EQ}$.

First to estimate r_e from $r_e = 0.03/I_{EQ} = 2$ Ω. We next find R_L' (= R_E // R_L = 86 Ω). The input resistance of the transistor is $r_{in} = \beta(R_L' + r_e) = 120 \times (86 + 2) = 10.6$ kΩ which is in parallel with R_B, making the overall input resistance $R_{in} = 10.6$ // 68 = 9.2 kΩ. v_{in} is then

$$v_{in} = v_s R_{in}/(R_{in} + R_s) = 9.2 v_s/14.2 = 0.648 v_s \tag{8.34}$$

R_L' (= 86 Ω) is substituted into Equation 8.34 to find

$$v_o/v_{in} = R_L'/(R_L' + r_e) = 86/(86 + 2) = 0.977 \tag{8.35}$$

From Equation 8.35 it follows that

$$A_{vL} = v_o/v_s = (v_o/v_{in}) \times (v_{in}/v_s) = 0.977 \times 0.648 = 0.633 \tag{8.36}$$

The voltage gain has been reduced mainly by the high source resistance and to a lesser degree by the low load resistance.

The current gain is the ratio of load current to source current. The load current is v_o/R_L, while the source current is $v_s/(R_S + R_{in})$ and the ratio is

$$\frac{v_o}{R_L} \frac{R_s + R_{in}}{v_s} = \frac{A_{vL}(R_s + R_{in})}{R_L} = \frac{0.633 \times 14.2}{0.1} = 90 \tag{8.37}$$

Thus A_{iL} = 90 and the power gain is $A_p = A_{iL}A_{vL}$ = 57, which is only 17.6 dB; fairly modest.

A summary of the salient features of the common-emitter and emitter-follower amplifiers is given in Table 8.1 with the common-base amplifier for comparison also.

Table 8.1 *Properties of BJT amplifiers*

Parameter	CE	CC	CB
A_v	high	≈ 1	high
A_i	high	high	≈ 1
A_p	high	medium	medium
R_{in}	medium	high	low
R_o	medium	low	high

CE = common emitter, CC = common collector (emitter follower), CB = common base

8.4 **BJT switches**

In logic circuits and high-power amplifiers BJTs are often used simply as switches that are either open (off) or closed (on). Consider the circuit of Figure 8.17 which is a simple transistor switch. Application of a positive voltage to the base of the transistor will cause current to flow through the collector, base and emitter to ground. The magnitude of the current flowing is nearly V_{CC}/R_C, as the transistor is in saturation and $V_{CE} \approx 0$. The switch also acts as an inverter, as v_o goes low when v_{in} goes high and vice versa. The only calculation necessary is that for R_B, which must be small enough to let the transistor operate in saturation over the full range of β.

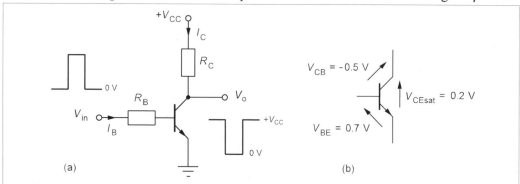

Figure 8.17 (a) A BJT used as a switch or an inverter (b) Voltages in a saturated transistor

---

Example 8.5

If β varies from 50 to 200 for a transistor switch which has to pass a current of 100 mA, what should R_C and R_B be if $V_{in} = V_{CC} = 10$ V?

With a β of 50, I_B must be 2 mA at least, so R_B has to be $(10 - V_{BE})/I_B = 4.65$ kΩ. Having $\beta = 200$ will not affect this since there will be enough base current. The voltage across R_C is $10 - V_{CEsat}$, where V_{CEsat} is the saturated collector-emitter voltage, about 0.2 V. (Note that the collector-base junction is *forward biassed* by 0.5 V in saturation, that is the n-type collector is at a lower potential than the p-type base as shown in Figure 8.17b.) Thus $R_C = (V_{CC} - V_{CEsat})/I_C = (10 - 0.2)/0.1 = 9.8/0.1 = 98$ Ω.

8.5 **BJT specifications**

Bipolar transistors come in several styles of package, some of which are shown in Figure 8.18. The E-line and TO92 packages are plastic, cheap and for power up to 1 W. The TO5 (which is almost the same as the TO18, TO39 and TO205) is a small, metal-can package for power up to 5 W. The TO220 is a flat package with the transistor mounted on a metal plate which can be bolted to heat sink. It can handle up to 100 W. The TO3 is a metal can with a large metal base for heat sinking and dissipates up to 350 W.

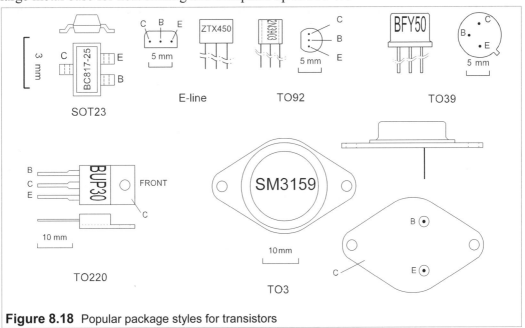

Figure 8.18 Popular package styles for transistors

Table 8.2 *A selection of low-cost BJTs*

Device	Type	Pack	I_C(A) [A]	P [B]	V_{CEO} [C]	h_{fe} [D]	f_T [E]	Price [F]	Comment [G]
BC817-25	npn	SOT23	0.5	0.25	45	160/400	>100	7 - 4	GP, SS, SMT
BC807-25	pnp	SOT23	−0.5	0.25	−45	160/400	>80	7 - 4	pnp ≡ BC817
ZTX300	npn	E-line	0.5	0.3	25	50/300	150	22 - 15	GP, SS
MPSA13	npn	TO92	0.5	0.62	30	>5000	–	20 - 18	Darlington
BFY50	npn	TO39	1	0.8	35	>30	140	48 - 34	GP, SS
BC550C	npn	TO92	0.1	1.5	45	420/800	300	4 - 2	cheap, GP, HG
BC557A	pnp	TO92	−0.1	1.5	−45	120/220	320	4 - 2	cheap, GP
FZT853	npn	SOT223	6	3	100	100/300	130	78 - 52	GP, SMT
25A1727P	pnp	SC63	0.5	10	−400	80/180	12	36 - 22	HV, SMT
BUX87P	npn	SOT82	0.5	20	450	50	20	64 - 40	HV, SMT
BD539B	npn	TO220	5	45	80	>12	–	68 - 47	med. power
2N3055	npn	TO3	15	115	60	20/70	>2.5	125 - 87	GP, hi power

Notes: [A] Maximum [B] Max. (in W) [C] Max V_{CE} (in V) [D] Min/max measured at 5% I_C(max) [E] MHz [F] In pence, for 1+ and 100+ [G] GP = general purpose, SS = small-signal, SMT = surface-mount technology, HG = high gain, HV = high voltage

Table 8.2, drawn almost at random from a single stockist's catalogue, gives some details of a range of transistors. It is difficult to believe the small, plastic-packaged devices can get any cheaper, but they almost certainly will. As usual, high-voltage, high-power devices are disproportionately costly. The specifications given are for normal ambient temperatures and must be derated in a hot environment. P_{max} is $(V_{CE}I_C)_{max}$, so if $I_C = I_{Cmax}$, it is clear that $V_{CE} \approx V_{CEsat}$. h_{fe} depends very much on the collector current and varies widely even when this is well within the normal operating region. The gain-bandwidth product, f_T, only applies when the collector current is a few percent of I_{Cmax}. It is determined by the input capacitance, but as a rule of thumb one can say that a 'normal' CE amplifier's upper frequency limit will be f_T divided by the voltage gain.

Suggestions for further reading – see Chapter 9

Problems

(In following problems take $V_{BEQ} = 0.7$ V and the reactance of the capacitors as zero.)

1. In the circuit of Figure P8.1 what is the largest value of R_B which will just drive the transistor into saturation, if $V_{CEsat} = 0.2$ V and $\beta = 150$? The circuit is to operate with any one of a particular type of transistor which can have a range of values of β from 50 to 300, what is the largest value of R_B that will guarantee saturation? In this latter case, what will be the base and collector currents if $\beta = 300$? *[279 kΩ, 93 kΩ, 71 μA, 3.55 mA]*

2. Draw AC and DC load lines for the CE amplifier of Figure P8.2. Place the Q-point optimally so that the load voltage has maximum undistorted swing. From the diagram you have drawn deduce that the optimal Q-point is at $I_{CQ} = V_{CC}/(R_{ac} + R_{dc})$, $V_{CEQ} = V_{CC} - I_{CQ}R_{dc}$, where R_{ac} is $R_C // R_L$ and $R_{dc} = R_C + R_E$.

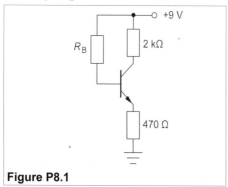

Figure P8.1

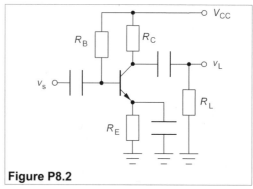

Figure P8.2

3. Figure P8.3 shows the output characteristics of an npn transistor in common-emitter configuration. What are h_{fe} and h_{oe} at the Q-point? Estimate h_{ie} at the Q-point at 300 K. Mark V_{CC} on the characteristics and label the AC and DC load lines. What are the values of R_C and R_E if the load lines are for the CE amplifier of Figure P8.2, with $R_L = \infty$? What is the maximum, unloaded, peak-to-peak output voltage approximately? Where is the optimal Q-point for a load of 2 kΩ which is capacitively coupled to the output? (Use the results of the previous problem.) *[$h_{fe} = 100$, $h_{oe} = 12$ μS, $h_{ie} = 600$ Ω. $R_C = R_E = 1$ kΩ. 8 V. $I_{CQ} = 6$ mA, $V_{CEQ} = 4$ V.]*

4. The transistor in Figure P8.4 has $\beta = 100$, $R_E = 330$ Ω and $R_C = 1.5$ kΩ. What will be the optimal values of V_{CEQ} and I_{CQ} for maximum, unloaded, output voltage? Draw the AC and DC load lines. What are R_1 and R_2 if $I_1 = 10I_{BQ}$? (Use the result of Problem 8.2.) *[$V_{CEQ} = 9$ V, $I_{CQ} = 6$ mA. $R_1 = 28.8$ kΩ, $R_2 = 5$ kΩ]*

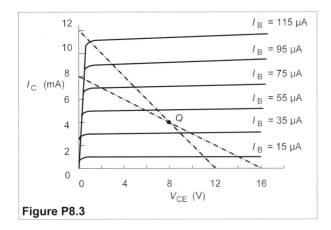

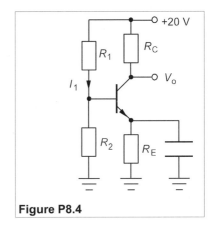

Figure P8.3   **Figure P8.4**

5. In the circuit of Figure P8.5, what is the optimal Q-point for maximum undistorted AC load voltage? What must R_B be in this case, given that $\beta = 180$? *[$V_{CEQ} = 4.5$ V, $I_{CQ} = 3.75$ mA, $R_B = 542$ kΩ]*

6. The transistor in the circuit of Figure P8.6 has $h_{fe} = 120$, $h_{ie} = 0.02/I_{BQ}$ and $h_{re} = h_{oe} = 0$. Find h_{ie} and use the approximate *h*-parameter equivalent circuit to find A_{v0} ($= v_o/v_s$). What will A_{vL} ($= v_L/v_s$) be if a capacitively-coupled load has a resistance 2 kΩ? What will A_{vL} become if the signal source has an internal resistance of 2 kΩ? If $h_{oe} = 50$ μS, recalculate A_{vL} in the latter case. *[$h_{ie} = 690$ Ω, $A_{v0} = -261$, $A_{vL} = -149$, $A_{vL} = -35$, $A_{vL} = -33.5$]*

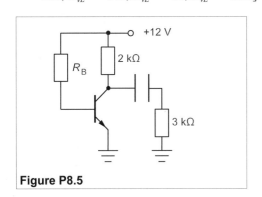

Figure P8.5

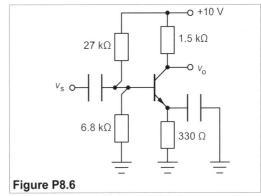

Figure P8.6

7. In the circuit of Figure P8.7 $A_{v0} = -150$, while $A_{vL} = -100$ when $R_L = 3$ kΩ. If $h_{re} = h_{oe} = 0$, what is R_C? If $V_{CC} = -12$ V and $V_{CEQ} = -6$ V, what is I_{CQ}? If $R_B = 400$ kΩ, what is h_{fe}? Find h_{ie} from A_{v0}. *[$R_C = 1.5$ kΩ, $I_{CQ} = -4$ mA, $h_{fe} = 142$, $h_{ie} = 1.42$ kΩ]*

8. The circuit of Figure P8.8 shows a Darlington pair with $\beta_1 = \beta_2 = 30$. Find R_1 if $R_1 = R_2$. *[200 kΩ]*

9. Calculate the exact no-load voltage gain of the CE amplifier of Figure P8.9, given that $V_{CC} = 9$V, $h_{fe} = \beta = 167$, $R_1 = 33$ kΩ, $R_2 = 8.2$ kΩ, $R_C = 2.2$ kΩ, $R_{E1} = 100$ Ω, $R_{E2} = 470$ Ω. Take $h_{ie} = 0.03/I_{BQ}$. What is it approximately, taking I_{BQ} to be small compared to both I_{CQ} and the current through R_1? And what is it if h_{ie} is neglected? *[-18.7, -18.9, -22]*

10. Consider the voltage-divider-biassed CE amplifier of Figure P8.4 and show that

$$I_{CQ} = \frac{\beta(V_{CC}R_B/R_1 - V_{BEQ})}{R_B + (\beta + 1)R_E}$$

where $R_B \equiv R_1 // R_2$. Hence show that $\Delta I_{CQ}/I_{CQ} \approx \Delta\beta R_B/\beta(\beta + 1)R_E$. What is the percentage change in I_{CQ} if β changes from 150 to 120 and $R_B = 10R_E$? *[-1.3%]*

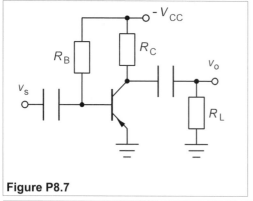

Figure P8.7

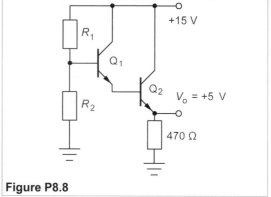

Figure P8.8

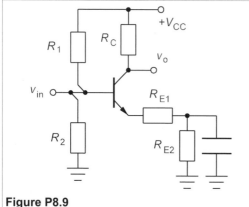

Figure P8.9

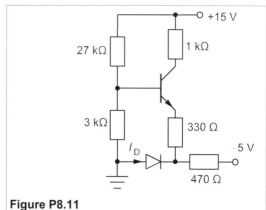

Figure P8.11

11. Figure P8.11 shows a circuit designed to stabilise I_{CQ} against temperature changes in V_{BEQ} by providing a compensating diode. Show that the circuit will do this. Calculate I_{CQ} and I_D if $\beta = 100$. What should R_D (= 470 Ω) be ideally? *[I_{CQ} = 4.163 mA, I_D = 4.944 mA, R_D = 511 Ω, ideally]*

12. For the circuit of Figure P8.12 find R_B if $V_{CEQ} = 12$ V and $\beta = 120$. If β increases by 10%, by what percentage does I_{CQ} change? *[280 kΩ, +4.6%]*

13. The emitter follower of Figure P8.13 is used as a buffer amplifier after pre-amplifier stage whose output impedance is 2 kΩ and whose no-load voltage gain is 100. What is the overall voltage gain, v_L/v_s, if $R_B = 350$ kΩ, $R_E = 2.2$ kΩ, $R_L = 20$ Ω and $\beta = 150$? If the pre-amplifier has an input resistance of 1 kΩ, what is the current gain, i_L/i_s? *[v_L/v_s = 53, i_L/i_s = 2 650]*

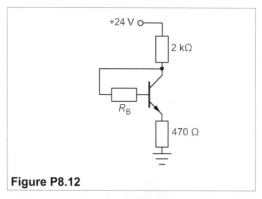

Figure P8.12

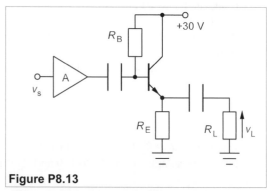

Figure P8.13

9 Field-effect transistors

FIELD-EFFECT transistors (FETs), though simpler to make than BJTs, were held back for many years by manufacturing difficulties with the gate insulation layer. In the 1970s the problems were solved and FET technology developed at a rapid pace. FETs are now more widely used than BJTs and have made an enormous impact on integrated circuit (IC) technology. Not only have very low-power CMOS (complementary metal oxide-semiconductor, a type of FET) ICs become reliable and cheap, but also devices such as HEXFETs and VMOS FETs have replaced many types of power bipolar devices, and semiconductor memories are no more than huge arrays of MOSFETs.

9.1 The junction field-effect transistor (JFET)

We shall discuss first the behaviour of JFETs (sometimes called JUGFETS) as they are typical of all FETs. Figure 9.1 shows the structure of a JFET, which consists in essence of a bar of semiconductor (n-type in Figure 9.1) sandwiched between regions of opposite type which are connected to the gate electrodes (these are electrically connected and at the same potential). The current passing through the bar of material from drain to source is controlled by the potential applied to the gate. The JFET in Figure 9.1 is an n-channel device operating at small drain-source voltages (V_{DS}) as a resistor whose value is set by the gate-source voltage (V_{GS}).

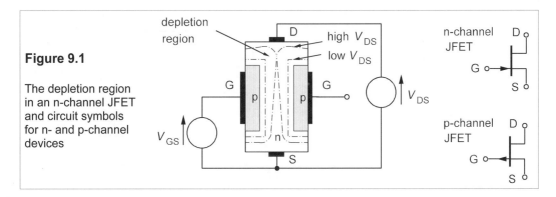

Figure 9.1

The depletion region in an n-channel JFET and circuit symbols for n- and p-channel devices

The depletion region formed around the gate electrodes constricts the channel and increases its resistance as V_{GS} is made more negative. At higher values of V_{DS}, however, the depletion regions are drawn out towards the drain and almost meet in the middle of the channel. In this region of operation (*pinch-off*) further increases in V_{DS} will not increase the current through the channel, I_D, as the rise in voltage merely serves further to constrict the channel and I_D stays constant. If V_{GS} is made sufficiently negative, no current flows and the device is said to be *cut off*. The value of V_{GS} which just cuts off the device is denoted by V_P. Because the gate-source voltage is negative, the device's p-n junctions are reverse biassed, giving in a very small gate current, roughly a nA. Thus the gate-source resistance, which is the input resistance since signals are applied to the gate, is extremely large.

9.1.1 Characteristics of an n-channel JFET

The drain, or output, characteristics of an n-channel JFET may be obtained with a circuit such as that of Figure 9.2. With V_{GS} held constant (and negative to avoid forward-biassing the gate-source p-n junction), I_D is measured as a function of V_{DS}.

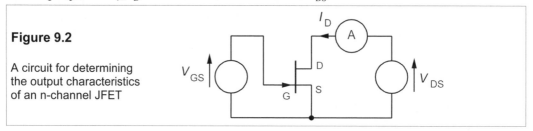

Figure 9.2

A circuit for determining the output characteristics of an n-channel JFET

On the right of Figure 9.3 are shown the results of such measurements, with V_{GS} taking values of 0, -1, -2 and -3 V. We can see that in the pinch-off region (which is also the operating region) I_D is independent of V_{DS}. The transfer characteristic, shown on the left, is a graph of I_D in the pinch-off region against V_{GS}. The maximum value of I_D is attained when $V_{GS} = 0$, and is denoted I_{DSS}. It is an important parameter of a JFET, as is V_P, the cut-off voltage.

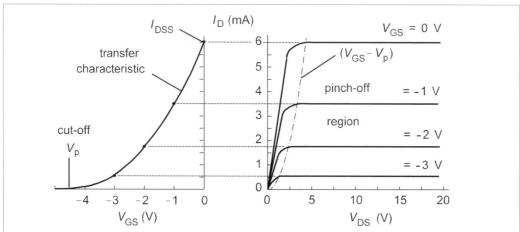

Figure 9.2 The drain characteristics of an n-channel JFET are shown on the right and the corresponding transfer characteristic on the left

In the pinch-off region I_D is given approximately by

$$I_D = I_{DSS}(1 - |V_{GS}/V_P|)^2 \tag{9.1}$$

where I_{DSS} is 6 mA and the cut-off voltage, V_P, is -4.3 V in this case. The device operates with negative values of V_{GS} from V_P to 0 V. Note that V_P can vary quite widely even for devices that are nominally the same: the BF244 JFET has a cut-off voltage ranging from -0.5 V to -8 V.

9.1.2 Biassing the JFET

The JFET requires bias like the BJT, but the operation of the devices is so different that the biassing method is sometimes dissimilar too. Figure 9.4a shows a JFET common-source amplifier circuit, the so-called self-biassed circuit, which is to operate at a quiescent point on the drain characteristics such that $V_{DSQ} = \frac{1}{2}V_{DD}$ and $I_{DQ} = \frac{1}{2}I_D(max)$.

Let us suppose that $V_{DD} = 20$ V and $I_D(max) = 5$ mA, requirements that fix the load line, which must join these points on the output characteristics as in Figure 9.4b. The optimal value of V_{DSQ} is then 10 V and $I_{DQ} = 2.5$ mA. For this value of I_D the transfer characteristic shows that V_{GSQ} is -1.5 V. Our biassing must achieve, therefore, a drain current of 2.5 mA and a gate-to-source voltage, $V_{GSQ} = -1.5$ V. The slope of the load line is $-I_{DQ}/V_{DSQ} = -1/4000$, so the resistance in the output circuit, $R_D + R_S$, must be 4 kΩ. The source resistor, R_S, fixes V_{GS} because $V_{GS} + V_{RS} = 0$, that is $V_{GS} = -V_{RS} = -I_D R_S$. Then $R_S = -V_{GSQ}/I_{DQ} = -(-1.5)/2.5 \times 10^{-3}$ $= 600$ Ω and $R_D = 3.4$ kΩ. The quiescent output voltage is given by

$$V_o = V_{DSQ} = V_{DD} - I_{DQ}R_D = 20 - 2.5 \times 3.4 = 11.5 \text{ V}.$$

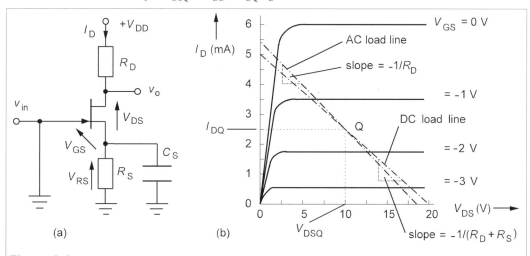

Figure 9.4 (a) JFET common-source amplifier with grounded gate
(b) The drain characteristics and load lines for the circuit

9.1.3 The equivalent circuit of the common-source amplifier

Suppose the JFET has the drain characteristics of Figure 9.4b and R_S and R_D are as before: what will the no-load voltage gain be? To find out we need an equivalent circuit for the JFET. The usual model for the common-source amplifier uses the *transconductance*, defined by

$$g_m = i_d/v_{gs} = \delta I_D/\delta V_{GS}\big|_{V_{DS} = \text{const}} \tag{9.2}$$

We recognise g_m as the slope of the transfer characteristic, taken at the Q-point. Differentiating Equation 9.1,

$$g_m = \frac{dI_D}{dV_{GS}} = \frac{-2I_{DSS}}{V_P}\left(1 - \frac{V_{GS}}{V_P}\right) = \frac{2I_{DSS}}{|V_P|}\left(1 - \frac{|V_{GS}|}{|V_P|}\right) \quad (9.3)$$

using magnitudes to avoid sign confusion. Equation 9.3 can be shown (see Problem 9.12) to be the same as

$$g_m = \frac{2\sqrt{I_D I_{DSS}}}{|V_P|} \quad (9.4)$$

The small-signal (alternating) drain current, i_d, is then $g_m v_{gs}$, where v_{gs} is the signal applied to the gate. Since the DC output voltage is given by $V_o = V_{DD} - I_D R_D$, the AC output voltage is $-i_d R_D$, and there is a 180° phase shift. The drain current increases slightly in the pinch-off region, at constant V_{GS}, so that the JFET has an output resistance, r_d, equal to the reciprocal of this slope. The output side of the FET is then a current source of value $-g_m v_{gs}$ in parallel with a resistance of r_d. The CS amplifier's equivalent circuit can therefore be represented very simply as in Figure 9.5. However, in most practical amplifiers $r_d \gg R_D$, and as they are in parallel r_d may be neglected.

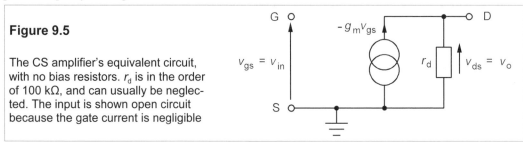

Figure 9.5

The CS amplifier's equivalent circuit, with no bias resistors. r_d is in the order of 100 kΩ, and can usually be neglected. The input is shown open circuit because the gate current is negligible

9.2 The practical common-source amplifier

The practical common-source amplifier in Figure 9.6a uses the capacitors, C_1 and C_2, to block DC from input and output. The gate resistor, R_G, provides a path to ground for the gate circuit, without which charge and hence voltage could build up on the gate producing unstable bias. R_G should be large since it is the effective input resistance of the amplifier; if, however, it is too large any gate leakage current will generate a significant voltage across it; about a MΩ is suitable. The source resistor, R_S, must be bypassed like the emitter resistor in the CE amplifier, to prevent gain degeneration caused by the input signal appearing across it. r_d has been omitted as it is much larger than R_D and in parallel with it. The equivalent circuit of this amplifier is shown in Figure 9.6b.

To estimate the amplifier's voltage gain we need to find the transconductance, g_m, using Equation 9.3 and the previously-used values for V_P (-4.2 V), V_{GSQ} (-1.5 V) and I_{DSS} (6 mA). Substituting these leads to $g_m = 1.84$ mS. Then

$$v_o = -g_m v_{gs} R_D = -1.84 \times 3.4 \times v_{in} = -6.26 v_{in} \quad (9.5)$$

taking $R_D = 3.4$ kΩ and noting that $v_{gs} = v_{in}$. Hence the no-load voltage gain is $A_{v0} = -6.26$. The minus sign again indicates inversion of the output waveform. This is typical of the modest voltage gain of FET amplifiers. The main advantage of FET amplifiers is their extremely high input resistance, so they make good buffer and first-stage amplifiers, which can then be coupled with high gain second stages, such as the CE amplifier. BIFET amplifiers are made in IC form to utilise the advantages of FET inputs and bipolar outputs.

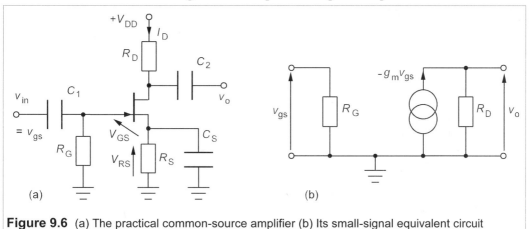

Figure 9.6 (a) The practical common-source amplifier (b) Its small-signal equivalent circuit

9.2.1 *The frequency response of the CS amplifier*

The low-frequency response of the CS amplifier is mainly determined by the smallest of the time constants of the three capacitors and associated resistances. Consider the circuit of Figure 9.7: the input resistance, R_G, is 1 MΩ and C_1, the input coupling capacitor is 10 nF giving $R_G C_1 = 0.01$ s, and the cut-off (corner) frequency of these components is $\omega_{L1} = 1/R_G C_1$ = 100 rad/s, or $f_{L1} = 16$ Hz. (Assuming that the input voltage source has an internal resistance small compared to R_G.) The output coupling capacitor, C_2, is in series with R_D (3.4 kΩ) and the load resistance, R_L.

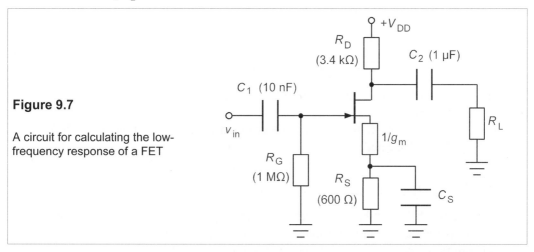

Figure 9.7

A circuit for calculating the low-frequency response of a FET

Let us suppose $R_L = 3.4$ kΩ also and as C_2 is 1 μF, then the cut-off frequency is $1/(R_D + R_L)C_2 = 147$ rad/s, or $f_{L2} = 23$ Hz. The time constant for C_S is not $C_S R_S$, because the resistance of the FET looking into the source ($1/g_m$) must be included. This resistance is effectively in parallel with R_S, making $\omega_{L3} = 1/C_S(R_S \mathbin{/\!/} 1/g_m)$. With $g_m = 1.82$ mS, $1/g_m = 550\ \Omega$, so $R_S \mathbin{/\!/} 1/g_m = 287\ \Omega$. C_S should be such that its cut-off is no higher than ω_{L2}, 147 rad/s, thus $C_S = 1/(147 \times 287) = 24\ \mu$F. The source-resistor bypass capacitor is always much the largest of the three because the source resistor is comparatively small.

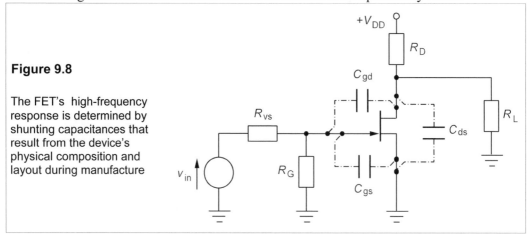

Figure 9.8

The FET's high-frequency response is determined by shunting capacitances that result from the device's physical composition and layout during manufacture

The high-frequency response is determined by very small capacitances (Figure 9.8) associated with the FET itself, which shunt the junctions and reduce the gain. The coupling and bypass capacitors all act as short-circuits at high frequencies. The most important of the device capacitances is C_{gd}, because it is amplified by a factor of $(1 - A_v)$ by the Miller effect. The input capacitance is then $C_{gs} + C_M$, where C_M is the effective capacitance of the gate-drain junction and is in parallel with C_{gs}. The parallel wiring capacitance, C_w, is far from negligible and must also be included for a total capacitance of $C_w + C_{gs} + C_M$. The input voltage-source has a resistance, R_{vs}, in parallel with the gate resistor, R_G (to see this convert v_{in} to a current source in parallel with R_{vs}). The effective capacitance of the input side is in series with $R_s \mathbin{/\!/} R_G$, so the cut-off frequency for the input side is

$$f_{H1} = \frac{1}{2\pi[C_w + C_{gs} + (1 - A_v)C_{gd}](R_{vs} \mathbin{/\!/} R_G)} \tag{9.6}$$

Taking typical values for C_w (3 pF), C_{gd} (1 pF), C_{gs} (5 pF), A_v (-6), R_{vs} (10 kΩ) and R_G (1 MΩ) gives $f_{H1} = 1.07$ MHz.

The output side's cut-off frequency is determined by $C_{ds} \mathbin{/\!/} C_w$ in series with $R_D \mathbin{/\!/} R_L$. Taking $R_D = R_L = 3.4$ kΩ and $C_{ds} = 1$ pF, $C_w = 3$ pF gives

$$f_{H2} = \frac{1}{2\pi(C_{ds} + C_w)(R_D \mathbin{/\!/} R_L)} = \frac{1}{2\pi(4 \times 10^{-12})(1700)} = 23\ \text{MHz} \tag{9.7}$$

f_{H1} is usually much lower than f_{H2} unless low-resistance voltage sources are used. If the various cut-off frequencies are well spaced, then the response drops at 20 dB/decade at both low and high frequencies like the BJT common-emitter amplifier.

9.3 **MOSFETs**

The MOSFET has become by far the most common transistor used in ICs because it is simpler to make than a BJT and occupies a smaller surface area of the chip. There are four types manufactured: the n-channel enhancement-mode (EN) MOSFET, the p-channel EN MOSFET and the n-channel and p-channel depletion-mode (also called depletion-enhancement-mode, or DE) MOSFETs. Figure 9.9 shows a cross-section of an n-channel EN MOSFET and Figure 9.10 the symbols for the n-channel devices (the p-channel device symbols have the arrow on the body terminal reversed). The p-channel EN device has the same construction as Figure 9.9, but with p-type for n-type and vice versa.

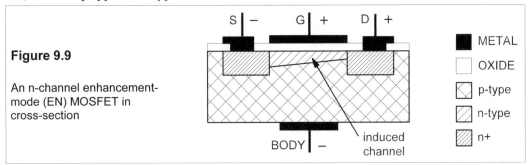

Figure 9.9

An n-channel enhancement-mode (EN) MOSFET in cross-section

When the gate electrode (G) is made positive with respect to the substrate (B, for body) electrode, some of the positive charges (holes) in the p-type silicon beneath the gate oxide insulation are driven away and electrons become the majority carriers. Thus an induced n-type channel is formed, which allows conduction between the drain (D) and source (S) electrodes. The magnitude of the voltage between gate and body determines how large is the induced channel and how readily conduction through it occurs. In consequence MOSFET characteristics are similar to those of JFETs, though there are some differences as will be seen.

Figure 9.10

n-channel MOSFET symbols
(a) enhancement mode (b) depletion mode

DE MOSFETs have a channel between drain and source which is lightly doped the same type as them, and it can be turned off by applying a negative voltage (n-channel) or positive voltage (p-channel). The transfer characteristic is therefore much like a JFET's, except that V_{GS} is not restricted to negative values for n-channel devices, or positive for p-channel.

9.3.1 Characteristics of MOSFETS

These are found with a circuit much like that used for determining JFET characteristics. The transfer characteristics of n-channel EN and DE devices are shown in Figure 9.11. The gate-source voltage required to turn off a MOSFET is denoted V_T. (For threshold voltage, rather than V_P for pinch-off voltage as in a JFET. In data sheets $V_{GS}(th)$ is used to denote both.). In the n-channel EN device V_T is negative and in the n-channel DE device, positive. The p-channel EN MOSFET has a drain characteristic which is a mirror image of the n-channel device, and so has a negative V_T. The drain characteristics are also much like those of the JFET, having a linear part up to the knee where I_D saturates, so that the device can be used as a voltage-variable resistor. There is no restriction for MOSFETs on the sign of V_{GS}, which must be negative in an n-channel JFET, and so I_{DSS} does not denote the limiting value for I_D, but signifies the value of I_D when $V_{DS} = 0$.

For all MOSFETs the transfer characteristic is parabolic in shape, so the saturation drain current is given by

$$I_D = K(V_{GS} - V_T)^2 \tag{9.8}$$

where K is a constant for a particular device with units of A/V^2 or S/V. K can usually be calculated from the manufacturer's data sheets. Equation 9.9 is more accurate than Equation 4.1 is for a JFET. The transconductance, g_m, is given by differentiating Equation 9.8:

$$g_m = dI_D/dV_{GS} = 2K(V_{GS} - V_T) \tag{9.9}$$

Example 9.1

From Figure 9.11 find K for each device, and g_m when I_D = 10 mA.

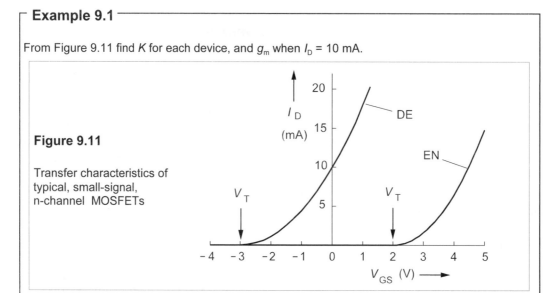

Figure 9.11

Transfer characteristics of typical, small-signal, n-channel MOSFETs

For the DE FET $V_T = -3$ V, and $V_{GS} = 0$ when $I_D = 10$ mA; then Equation 9.9 gives $K = 10/3^2 = 1.11$ mA/V^2. For the EN FET $V_T = 2$ V and $I_D = 10$ mA when $V_{GS} = 4.5$ V, making $K = 10/(4.5 - 2)^2 = 10/2.5^2 = 1.6$ mA/V^2. Equation 9.10 gives $g_m = 2 \times 1.11 \times 3 = 6.66$ mS for the DE FET and $g_m = 2 \times 1.6 \times 2.5 = 8$ mS for the EN FET.

MOSFETs are used in much the same way as JFETs, for example in the CS amplifier in the next example.

Example 9.2

The MOSFET in Figure 9.12 is the n-channel DE device whose transfer characteristic is given in Figure 9.11. What is the voltage gain? What is the power gain? What should V_{DD} be for a reasonable Q-point? What is the Q-point then? (The capacitances have negligible reactances.)

We can see that both gate and source are grounded, so $V_{GS} = 0$ V and $I_{DQ} = 10$ mA from Figure 9.11. The voltage gain is given by Equation 9.5, replacing R_D by R_D // R_L (= 0.75 kΩ), thus

$$v_L = -g_m(R_D\ /\!/\ R_L)v_{in} = -6.66 \times 0.75v_{in} = -5v_{in}$$

The voltage gain is $v_L/v_{in} = A_{vL} = -5$, or 14 dB. The input power is $v_{in}^2/1000$ mW, while the useful output power is $v_L^2/R_L = 25v_{in}^2/2$ mW, giving a power gain of 12 500 or 41 dB.

If $I_{DQ} = 10$ mA, then the voltage across R_D (1.2 kΩ) is 12 V, so a reasonable value for V_{DD} is twice this: $V_{DD} = +24$ V, making $V_{DSQ} = 12$ V. The Q-point is at $I_{DQ} = 10$ mA, $V_{DSQ} = 12$ V.

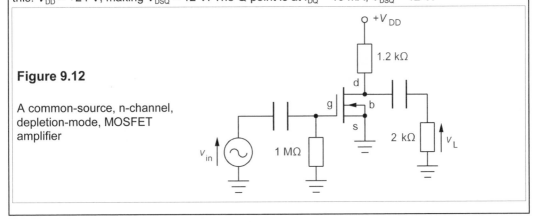

Figure 9.12

A common-source, n-channel, depletion-mode, MOSFET amplifier

9.4 **Specifications of some popular FETs**

Table 9.1 gives a very limited selection from the hundreds of FETs readily available today. Many are put into the same package styles as BJTs, though there are some peculiar to FETs. The important parameters are given in Table 9.1. For power FETs an additional parameter, R_{DS}(on) – the channel resistance, V_{DS}/I_D, when fully conducting – is given as it largely determines the power consumption for FETs used as output stages in power supplies. As a rule of thumb, $I_D(\text{max})^2 R_{DS}(\text{on}) \approx \tfrac{1}{2}P_{max}$ for most power FETs.

Table 9.1 *Specifications for some inexpensive FETs*

Device No.	Pack [A]	P_{max} [B]	R_{DS} [C]	I_{Dmax} [D]	V_{DSmax} [E]	g_m [F]	Price [G]	Comments [H]
PMBFJ177	SOT23	0.3	–	>1.5 mA	30	–	48/30	P, JFET, SMT
2N7000	TO92	0.35	5	0.2	60	0.1	13/10	N, GP, v. cheap
J113	TO92	0.36	–	>2 mA	35	–	31/20	N, JFET, GP
BSH203	SOT23	0.42	0.9	0.47	30	0.3	26/19	P, SMT, GP
BS250P	E-line	0.7	14	0.23	45	0.15	43/29	P, GP
BSS295	TO92	1	0.3	1.4	50	0.5	54/33	N, GP
VN88AFD	TO220	15	4	1.3	80	0.17	233/178	N, VMOS
IRF610	TO220	20	1.5	2.5	200	0.8	41/25	N, cheap, power
BUZ72L	TO220	40	0.2	10	100	5	89/54	N, cheap, hi g_m
PHP4N50	TO220	100	2	5	500	2	86/50	N, cheap, HV
IRFZ44A	TO220	150	0.02	50	60	15	97/76	N, cheap, HI
IRF1104	TO220	170	0.009	100	40	37	200/114	N, v. low R_{DS}

Notes: [A] See Figure 8.18 for package styles. Pin IDs vary from one type of transistor to another even if the package is the same. For example, the TO92 package comes with five different combinations for the gate, source and drain pins. [B] In W, = $(V_{DS}I_D)_{max}$ [C] Channel resistance when fully conducting [D] In A. I_{DSS} for JFETs [E] Max drain-source voltage [F] In S [G] For 1+/100+ in pence [H] N, P = channel type, HV = high voltage, HI = high current

Suggestions for further reading

The following books cover the topics in Chapters 7 to 9 inclusive.

Electronic devices by T L Floyd (Prentice Hall, 4th ed., 2001). Keeps it fairly simple.
Electronic devices and circuits by T F Bogart Jr (Prentice Hall, 5th ed., 2000). Thorough.
Principles of transistor circuits by S W Amos and M R James (Butterworth-Heinemann, 9th ed., 2000). Still going strong, now with a ghost writer!
Microelectronic circuits by A S Sedra and K C Smith (OUP USA, 4th ed., 1997). Advanced.
Microelectronic circuits and devices by M N Horenstein (Prentice Hall, 2nd ed., 1995)
Principles of electronic devices by W D Stanley (Prentice Hall, 1994). Fairly lengthy.

Problems

1.   The JFET in the circuit of Figure P9.1 has $V_p = -4$ V, $I_{DSS} = 10$ mA and it is required to operate with $I_{DQ} = 3$ mA and $V_{DSQ} = 6$ V from a +15V supply. What values should be selected for R_S and R_D?
     [$R_S = 603$ Ω, $R_D = 2.397$ kΩ]
2.   In Problem 9.1, the desired resistors are not available, but only the nearest preferred values of 560 Ω and 2.2 kΩ for R_S and R_D respectively. What will I_{DQ} and V_{DSQ} be when these are used?
     [$I_{DQ} = 3.14$ mA, $V_{DSQ} = 6.33$ V]
3.   The JFET in Problem 9.2 is defective and is replaced by another one of the same type which has $V_p = -2.8$ V and $I_{DSS} = 8$ mA, what are the values of I_{DQ} and V_{DSQ} now?
     [$I_{DQ} = 2.31$ mA, $V_{DSQ} = 8.62$ V]

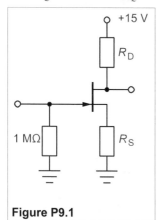

Figure P9.1

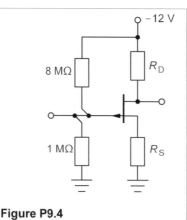

Figure P9.4

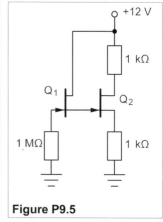

Figure P9.5

4.   In the circuit of Figure P9.4, the JFET has $V_p = +3$ V and $I_{DSS} = -6$ mA and is operated from a -12V supply with $I_{DQ} = -3$ mA and $V_{DSQ} = -5$ V. Calculate the values of R_D and R_S required.
     [$R_S = 737$ Ω, $R_D = 1.596$ kΩ]
5.   In the circuit of Figure P9.5, $R_S = R_D = 1$ kΩ, while the JFETs are identical and have $V_p = -2$ V and $I_{DSS} = 8$ mA. Find V_{GSQ1}, V_{DSQ2} and I_{DQ2}. *[$V_{GSQ1} = -2$ V, $V_{DSQ2} = 6.37$ V and $I_{DQ2} = 2.814$ mA]*

6. In the circuit of Figure P9.6 $v_g = 100$ mV and the JFET has $V_p = -3.3$ V and $I_{DSS} = 7$ mA. Find I_{DQ}, V_{GSQ} and g_m. If $R_L = 3.3$ kΩ and $r_d = \infty$, what is v_L?
   [$I_{DQ} = 3.86$ mA, $V_{GSQ} = -0.85$ V and $g_m = 3.15$ mS. $v_L = 416$ mV]

7. The JFET in the circuit of Figure P9.7 has $V_p = -5$ V and $I_{DSS} = 8$ mA, while $V_{DD} = +24$ V. If $R_D = R_S = 1.8$ kΩ, $R_1 = 6$ MΩ, $R_2 = 1$ MΩ and $R_L = 5$ kΩ, what are I_{DQ}, g_m and v_L/v_g, assuming $r_d = \infty$? What would v_L/v_g be if the signal source's internal resistance were 50 kΩ?
   [$I_{DQ} = 2.985$ mA, $g_m = 1.956$ mS, $v_L/v_g = -2.59, -2.45$]

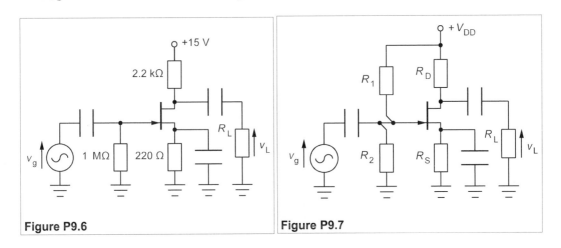

Figure P9.6   **Figure P9.7**

8. Draw a graph of the drain characteristics and load lines for Problem 9.7. Use your graph to draw the waveform of the load voltage when $v_g = 1.5 \sin \omega t$ V. What are the minimum and maximum values of V_{DS}? What is the peak-to-peak load-voltage? Does $v_L/v_g = -2.59$ as suggested by Problem 9.7? Under what conditions will v_L/v_g not be -2.59? (Assume $R_{DS}(\text{on}) = 0$ Ω)
   [$V_{DS}(min) = 8.4$ V, $V_{DS}(max) = 16.2$ V. $v_L(p\text{-}p) = 7.8$ V]

9. Repeat Problem 9.8 but with $v_g = 2 \sin \omega t$ V. *[$V_{DS}(min) = 7.2$ V, $V_{DS}(max) = 16.7$ V. $v_L(p\text{-}p) = 9.5$ V]*

10. The circuit of Figure P9.10 is a source follower (or common drain) amplifier. The output side can be modelled as a current source of value $g_m v_{gs}$ in parallel with a resistance of $R_S // R_L$. By noting that $v_{gs} = v_g - v_L$, show that the voltage gain, $A_{vL} = v_L/v_g = g_m R_o /(1 + g_m R_o)$, where $R_o \equiv R_S // R_L$. A source follower is required in which the FET has $I_{DSS} = 12$ mA and $V_p = -3$ V, with $R_2 = 1$ MΩ, $R_S = 1$ kΩ, $R_L = 5$ kΩ and $V_{DD} = +9$ V. What must R_1 be for I_{DQ} to be 4 mA? What is the voltage gain in this case? What is the input resistance? *[$R_1 = 2.294$ MΩ, $v_L/v_g = 0.794, 696$ kΩ]*

11. The circuit of Figure P9.11 is a JFET current source. With a JFET having $I_{DSS} = 12$ mA and $V_p = -2.5$ V, what are V_{GS}, V_{DS} and I_D? *[$V_{GS} = 0$ V, $V_{DS} = 6.84$ V, $I_D = 12$ mA]*

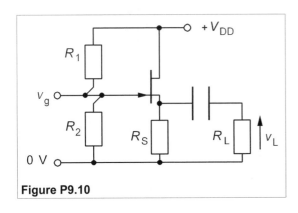

Figure P9.10

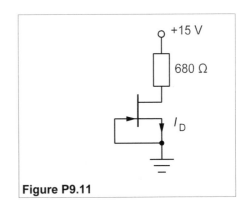

Figure P9.11

12. Derive Equation 9.4 from Equations 9.1 and 9.3.

13. By including the source resistor, R_S, into the JFET equivalent circuit of Figure 9.6b, show that the degenerated (that is without a source-resistor bypass capacitor), no-load, voltage gain of the common-source amplifier is $A_{v0} = -R_D/(R_S + 1/g_m)$. What is this gain for a self-biased CS amplifier with $I_{DSS} = 3$ mA, $V_P = -2$ V, $R_D = 3.3$ kΩ and $I_{DQ} = 1.5$ mA? What would it be with a bypass capacitor? *[-3.83, -7.01]*

14. Find the Q-point, load voltage and power gain of the CS amplifier of Figure P9.14, given that the transistor has $K = 2.2$ mA/V^2 and $V_T = +4$ V.
[$V_{GSQ} = 6.9$ V, $I_{DQ} = 18.5$ mA, $V_{DSQ} = 12.9$ V. $v_L = 0.48$ V. $A_P = 37.3$ dB]

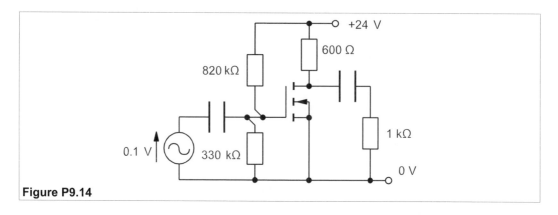

Figure P9.14

15. The circuit of Figure P9.15 is an EN FET voltage reference. If the FET has $K = 2$ mA/V^2 and $V_T = 0.9$ V, what is V_{REF}? By what percentage will V_{REF} change if the supply voltage changes by +2%? (Ignore the small gate-bias current.) *[$V_{REF} = 8.52$ V, +0.96%]*

16. What are the quiescent conditions in the circuit of Figure P9.16 if the DE FET has $I_D = 12.5$ mA when $V_{DS} = 2$ V and $I_D = 8$ mA when $V_{DS} = 1$ V? *[$I_{DQ} = 8.515$ mA, $V_{DSQ} = 1.127$ V]*

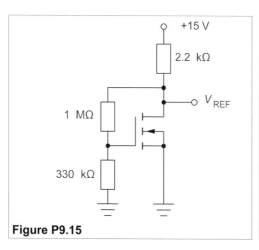

Figure P9.15

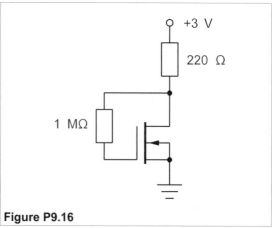

Figure P9.16

10 Integrated circuits

INTEGRATED CIRCUITS (ICs) are simply circuits whose components are formed simultaneously on a single piece of semiconducting material. Instead of wiring together the discrete components of a circuit – such as resistors, capacitors, inductors, diodes and transistors – the IC designer arranges for them to be produced and electrically interconnected as a single 'chip'. There are many advantages in so doing: the devices need no packaging, the interconnections and spacing between components can be very small (usually a few μm) and their small size means they can be mass produced in very large quantities very cheaply. In addition, the simultaneous formation of transistors in ICs means they are naturally closely matched, leading to improved circuit performance for all types of application

The development of the manufacturing process for ICs based on silicon has been rapid and continuous. Figure 10.1 shows the trend with time in the best *in-production* processes for the minimum width of 'features' such as conductors, gates and so forth, transistors/chip and microprocessor clock frequencies, all on a logarithmic scale.

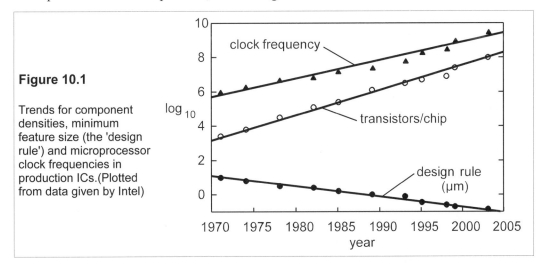

Figure 10.1

Trends for component densities, minimum feature size (the 'design rule') and microprocessor clock frequencies in production ICs.(Plotted from data given by Intel)

Because the chip size has steadily increased the number of transistors/chip has not increased as L^{-2} (L = feature size) but as $L^{-2.5}$. The number of transistors per chip has doubled every two years for more than three decades, more or less in line with the prediction {known as Moore's Law[1]) made by in a remarkably prescient article published as long ago as 1965. Other prophets have forecast the imminent or not-so-imminent end to the development of silicon ICs, ever since their invention, but the huge capital investment in silicon technology will ensure its continuation for many years – certainly until 2010. By this time the minimum feature size will be 0.05 μm giving a billion transistors in a microprocessor chip and probably

---

[1] G E Moore 'Cramming more components onto integrated circuits' in *Electronics* **38**(8) 1965

many times that number in a memory chip. At some point the number of atoms comprising a transistor will become so small that thermal fluctuations ('kT') will cause malfunction unless we are prepared to cool devices considerably below 300 K, but even cooling to liquid helium temperature (4.2 K) will produce only a four-fold reduction in minimum feature size.

10.1 Integrated circuit fabrication

Figure 10.2 shows some of the stages in making a complete bipolar IC by the planar process[2].

Figure 10.2 Steps in the planar production process

[2] So called because the devices are all in the same plane and the silicon surface is nearly flat.

10.1.1 Making silicon slices

Pure polycrystalline silicon is produced by the repeated distillation of, for example, trichlorosilane ($SiHCl_3$), followed by its reduction to silicon with hydrogen. This silicon is of almost intrinsic resistivity at 300 K, which implies that its electrically-active impurities are present in concentrations of 1 in 10^{11}. This polysilicon must be turned into single-crystal silicon of the required dopant type and concentration for use in devices. The method for doing this is to melt the silicon (it melts at about 1410°C) in a special furnace called a Czochralski grower. Boron-doped silicon is added to the melt to ensure the correct dopant concentration (about 1 atom of boron in 10 million of silicon) for a resistivity of 0.1 Ωm at 300 K. A small seed crystal of the desired crystallographic orientation is lowered into the melt and slowly withdrawn, melt and seed crystal being contra-rotated to help grow a uniform crystal. The melt freezes onto the seed forming a single crystal whose width is controlled by the withdrawal rate. Czochralski pullers are impressive machines, standing 4 m high, weighing several tonnes and costing hundreds of thousands of pounds. The crystal is about 150 mm in diameter and some 1.5 to 2 m long, weighing perhaps 50 kg. After grinding to a perfect cylinder a flat is ground along a particular crystallographic plane so that the masks can be oriented properly. The next step is to cut the crystal into 0.25mm-thick slices normal to the growth axis with a diamond saw. These slices are carefully polished to remove all traces of saw damage before starting the sequence of processes shown in Figure 10.2. The starting slice is in fact no more than a substrate or support on which the active layer is formed by *epitaxy*.

10.1.2 Oxidation, diffusion and photolithography

The crucial feature of the IC fabrication process is the periodic oxidation of the silicon surface in a furnace with an oxygen atmosphere. The oxide layer is a coherent film (meaning it sticks well to the underlying silicon, without any cracks or holes in it) about 1 μm thick and forms a barrier to diffusants such as boron, phosphorus, arsenic and antimony. It is this property of silicon which has made it especially suitable for integrated circuit manufacture. If holes are cut into the oxide film, selected areas can be exposed to gases containing the dopant atoms and circuit components – diodes, resistors and transistors – can be made. The circuit is in effect defined by the set of masks used to make patterns in the sensitive layer of photoresist which is spread over the oxide layer on the surface of the silicon slice. When the resist is exposed to ultra-violet (uv) light or an electron beam it is polymerised and becomes insoluble. On dissolving away the unpolymerized resist, silicon dioxide is exposed which can be removed by hydrofluoric acid (HF). In turn the underlying silicon is exposed and can be treated with dopant atoms that can be diffused into the slice to form the base, emitter, etc. Figure 10.3 shows the photolithographic process on a cross-section of a slice of silicon, corresponding to mask No. 1 of Figure 10.2, and the subsequent buried-layer diffusion, followed by the growth of an n-type epi-layer. For bipolar ICs the slice is oxidised prior to epitaxy, photoresisted, etched and diffused with antimony (Sb) to form the buried layer, which is a very low-resistivity n+ area designed to reduce the resistance of the collector as much as possible (the active material of the n-type collector is of relatively high resistivity, about 0.1 Ωm).

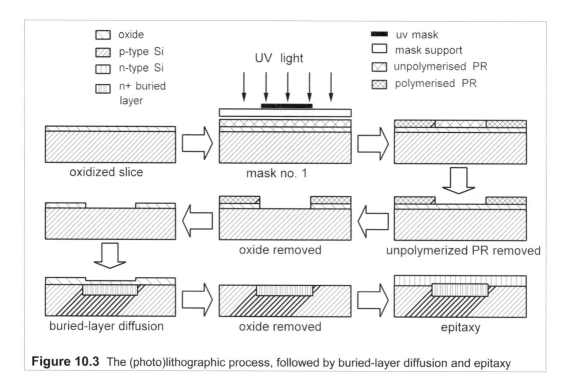

Figure 10.3 The (photo)lithographic process, followed by buried-layer diffusion and epitaxy

10.1.3 Epitaxy

After the buried-layer diffusion the slice is covered with oxide that has to be completely removed before a layer of epitaxial silicon can be deposited in a special furnace, shown in Figure 10.4. Deposition takes place at elevated temperatures which are achieved by RF-heating a susceptor made of carbon coated with silicon carbide. The incoming gases contain silicon (in the form of trichlorosilane, $SiHCl_3$, silane, SiH_4, or some other volatile compound), hydrogen and a dopant-containing species such as phosphine (PH_3). The silicon compound is reduced to silicon by the hydrogen and the high temperature of the susceptor (and slices) and deposits onto the slices, as well as the susceptor. The epi-layer takes up the same orientation as the substrate, but is of opposite conductivity type so that the components of the circuit can be isolated from each other by a p-type diffusion. Some diffusion of the buried layer into the epi-layer occurs during epitaxy, which is shown in Figure 10.3.

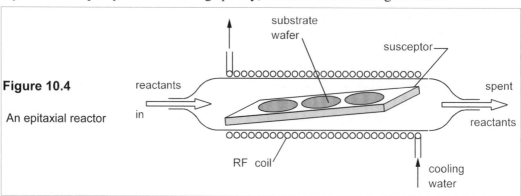

Figure 10.4

An epitaxial reactor

10.1.4 Isolation and base formation

Isolation is achieved by reverse-biassing the p-n junction formed by the isolation diffusion. The oxidation, photoresist, masking, exposure and diffusion steps are repeated to form the base and emitter of any transistors. Windows are then etched into the oxide layer for the base, emitter and collector contacts, as well as the terminals of any diodes, resistors or capacitors. Following this a layer of aluminium is evaporated over the whole of the slice. The metallization pattern is defined by photoresisting and the surplus aluminium etched away to leave the interconnected components of each circuit. The aluminium is covered by a layer of an impervious dielectric such as silicon nitride ('passivation') before pad openings for wire bonds are formed allowing each chip (or die) to be wired into its lead frame. At this stage the circuits are tested by a specially-programmed, multi-probe machine which marks the defective circuits. Figure 10.5 shows a much-condensed version of this process, with cross-sections of the slice at four stages in the formation of an npn transistor.

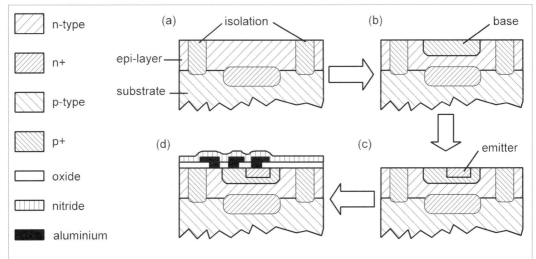

Figure 10.5 Steps in the formation of an npn transistor in an IC after the buried-layer diffusion (a) Isolation diffusion (b) Base diffusion (c) Emitter diffusion (d) Contacts and passivation

10.1.5 Die production, testing and packaging

The slice has remained in one piece, barring accidents, to this stage, but it must now be scribed like a sheet of glass with a diamond stylus and broken into its individual dice, which might be anything from 1 mm to 10 mm along a side. The optimal chip size depends on the density of defects generated by the processing, and has increased slowly but steadily in time. Chips that pass testing are fed into a machine which mounts them on a lead frame and a skilled wire bonder connects the bonding pads on each chip to the leads of the lead frame. The lead frames and attached chips are then encapsulated by plastic or other protective material before being tested for the last time and stamped with identifying code numbers.

Though there are many more steps than indicated in this brief account, it gives an idea of the complexity of the operation. The number of process steps is so large that unless the yield at each stage is virtually 100%, the overall yield will be zero! It happens when a new process is set up that the initial yield at probe-testing is zero and the entire output must be thrown away. Investigations pinpoint process deficiencies following which the yield rapidly rises to nearly 100% and the cost per circuit becomes minimal.

By similar means to those of the planar bipolar process, it is possible to make CMOS or any other ICs based on field-effect transistors; all that is needed is a change of masks and a different sequence of steps, such as polysilicon deposition for the FET gates. But the oxidation, photoresist and diffusion sequence is a common feature of all the processes, and so too are metallization, probe-testing, bonding and encapsulation.

The IC production process is well suited to p-n junction formation so that diodes and transistors are particularly favoured, but resistors much less so because they occupy more space. For the same reason capacitors can only be a few tens of picofarads and inductors are avoided altogether. Because transistors take up so little space and the marginal cost is almost nil, there is no need to economise on them and design with ICs has evolved to take this into account. Inductors are designed out completely, and capacitors as much as possible. For example, an early 741 op-amp circuit contained 24 transistors, nine resistors and one 30 pF capacitor. Despite its small value this one capacitor occupied nearly as much space as *all* the transistors!

10.2 **Hybrid circuits**

The limitations of semiconductor ICs means that it is sometimes necessary to make a circuit from one or more ICs together with other components, usually resistors and capacitors and more rarely with inductors. These are known as *hybrid* circuits because they are an amalgam of different technologies. There are two methods of making the resistors, etc. and the connections to the ICs in a hybrid circuit: thin-film and thick-film technology. Thick-film technology is quite old and relatively unsophisticated and produces relatively large circuits. It uses conductive pastes and inks, which are baked onto an insulating substrate such as alumina, to form resistors and conductors. Specially made capacitors can be stuck onto these circuits, while the IC can be wire bonded onto them just as in a lead frame. The whole circuit can then be encapsulated in plastic or hermetically sealed if required.

Thin-film technology is more recent than thick-film and uses evaporated or sputtered layers of metal or alloys to form the conductors and resistors on an insulating substrate. Thin-film circuits require more costly equipment than thick-film. They are also smaller and capable of being adjusted to finer tolerances than thick-film components. Typically thick-film conductor widths are 250 μm, and resistors are about 1 mm wide, whereas thin film conductor and resistor widths can be as little as 1 μm. Figure 10.6 shows a platinum, thin-film, resistance thermometer, which is automatically laser trimmed to a 0.1% tolerance.

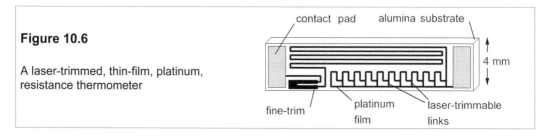

Figure 10.6

A laser-trimmed, thin-film, platinum, resistance thermometer

Suggestions for further reading

Introduction to microelectronic fabrication by R C Jaeger (Prentice Hall, 2002). Lots of examples and problems.
Semiconductor devices, physics and technology by S M Sze (Wiley, 2nd ed., 2001)
Semiconductor manufacturing technology by M Quirk and J Serda (Prentice Hall, 2000)
Technology of integrated circuits by D Widman, H Mader and H Friedrich (Springer, 1999) Advanced.
Application specific integrated circuits by M J S Smith (Addison-Wesley 1997)
Integrated circuit design, fabrication and test by P Shepherd (Macmillan, 1996). A good introduction.
ULSI technology by C Y Chang and S M Sze (McGraw-Hill, 1996)

Problems

1. A silicon wafer (or slice) is 200 mm in diameter and the cost of processing a single wafer is £1000. The yield of 'good' IC chips depends on the chip area according to the formula

$$\text{yield} = 100\exp(-A/D) \ \%$$

where A is the area of the chip and D is a failure-rate parameter equal to 120 mm². If the area of the IC chip is 100 mm², how many good chips will the wafer yield, and what is the cost of processing per good chip? What is the largest chip that can be produced at an average wafer-processing cost of £1 per good chip? What is the yield for this sized chip? What is the largest chip that can be produced at an average yield of 1 good chip per wafer? What would D have to be to enable a chip to be made at an average yield of 1 good chip per wafer if the chip area were 1200 mm²? (Ignore edge effects that would lead to incomplete chips.)
[136, £7.35, 25.4mm², 80.9%, 497.5 mm², D = 368 mm²]

2. Thin-film circuit designers often use the sheet resistance unit 'ohms per square' ($\Omega/\square$) to measure the resistivity of a conductor. The resistance is then given by $R = sl/w$, where s is the sheet resistance, l is the length and w the width of the conductor. Thus if a conductor is 8 units long and 2 units wide and has a sheet resistance of 100 $\Omega/\square$, it would present a resistance of 400 Ω. If the conductor width is 8 µm and it is 100 mm long, what must its sheet resistance be if it is to have a resistance of 100 Ω? If it has a laser trimming pad 0.5 mm long and 20 µm wide which is cut down the centre (see Figure 10.6), to what accuracy (in µm) must the laser cut be made to produce a resistor tolerance of 0.1%? What is the maximum trimming (in Ω) that is feasible?
[8 mΩ/□. 50 µm. 0.8 Ω]

3. The resistors in Figure P10.3 each have sheet resistances of 25 $\Omega/\square$. If a corner square in the right-hand pattern counts as 55% of a square in the straight part of the pattern, what are the resistances? What is the resistance ratio? (Count squares from the contact window edge nearest to the end of the resistor.) What is the simplest way of ensuring an accurate 2:1 ratio? *[225 Ω, 457.5 Ω, 2.033:1]*

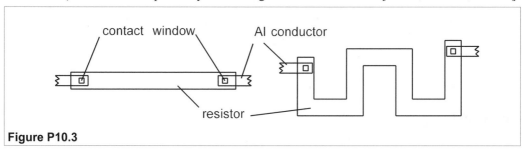

contact window   Al conductor

resistor

Figure P10.3

4. The maximum tolerable current density for an aluminium IC conductor rail is 1 GA/m². If its thickness is 0.5 µm and the maximum current it must carry is 1.2 mA, what is its minimum width?

What is the maximum voltage drop along the conductor if the conductivity of aluminium is 25 MS/m and the rail is 2 mm long? *[2.4 μm. 0.08 V]*

5. The maximum collector current in an epitaxial BJT of an IC is 1 mA. The epi-layer is 4 μm thick and of resistivity 0.1 Ωm. What is the minimum allowable cross-sectional area of the collector path if the maximum collector voltage drop is to be 0.6 V? (Assume zero resistance in the buried layer.) If the collector is effectively a hollow cylinder with inner radius equal to half its outer, what is its diameter? *[333 μm². 23.8 μm]*

6. A conductor rail has thickness, d and length, L, and its resistivity is ρ. It passes over a dielectric layer (silica) of permittivity, ϵ, and thickness, w. Considering the substrate as the other plate of a parallel-plate capacitor, and ignoring its resistance, show that the time constant for signal propagation in the conductor is

$$\tau = RC = \epsilon \rho L^2 / wd$$

(Note that the propagation time goes as the square of the conductor's length, so they must be kept as short as possible in high-speed devices.) What is the propagation time if $L = 1$ mm, $w = 0.5$ μm, $d = 0.3$ μm, $\sigma = 30$ MS/m and $\epsilon_r = 2.3$? *[4.5 ps]*

7. Epi-layer thickness is measured by infra-red reflectance-interference as shown in Figure P10.7a. The epi-layer thickness is given by

$$p\lambda = 2d\sqrt{(n^2 - \sin^2\theta)}$$

where p is the order of the extremum (integer for a maximum, half integer for a minimum), d is the layer thickness, n is the refractive index of silicon (= 3.42) and θ is the angle of incidence of the infra-red beam, which is 45°. Using the spectrophotometric output of Figure P10.7b, find the positions of the extrema, then work out the correct values for p. Calculate the thickness of the layer using an average of the values produced by all of the extrema. Find the standard deviation (σ_{n-1}, see Chapter 29) and standard error for the thickness. *[8.01 μm; 0.07 μm, 0.03 μm]*

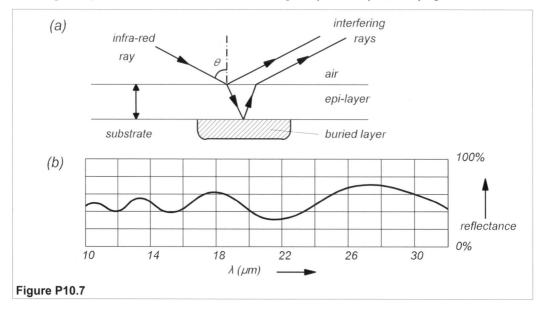

Figure P10.7

11 Analogue circuits

ELECTRONIC circuits, particularly ICs, are classified as analogue or as digital. Analogue integrated circuits of all kinds are becoming more and more diverse, but the most popular analogue IC is the operational amplifier and its close relation the comparator, both of which are discussed in Chapter 3. Of the remainder the most numerous ICs are analogue-to-digital (and digital-to-analogue) converters and voltage regulators. All of these types together account for about 80% of IC usage. In the next sections we shall look at some other classes of circuit to see how they have been adapted to IC technology. One of the most difficult to adapt has been filters, because they normally contain inductors and capacitors, although only the latter are absolutely essential.

11.1 Filters

On many occasions the signal one wishes to use is combined with noise or with some unwanted frequency components; filters may then be used in an attempt to remedy matters. If the unwanted frequencies are higher than those of interest, then one can use a *low pass* filter, whose ideal characteristic is shown in Figure 11.1a. Alternatively, one may want to remove low frequencies when the desired signal is at higher frequencies, then one uses a *highpass* filter, whose ideal response is shown in Figure 11.1b. If a particular band of frequencies is of interest one can pass these with a *bandpass* filter (Figure 11.1c) and reject all the rest. And if the interfering noise is largely confined to a narrow range of frequencies one can use a *bandstop* or *band-reject* filter as in Figure 11.1d. The characteristics shown in Figure 11.1 are known as *brickwall* responses, for obvious reasons. However, the infinite rate of change of the filter response at the cut-off frequency, f_c, (or the lower and upper pass band and band stop frequencies, f_1 and f_2) is a practical impossibility as we shall see.

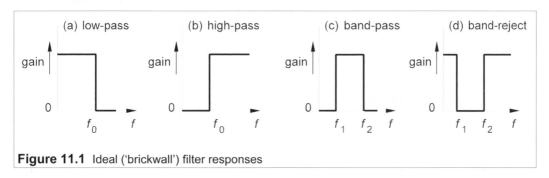

Figure 11.1 Ideal ('brickwall') filter responses

11.1.1 Passive filters

A passive filter is one made from passive components such as resistors, inductors and capacitors. The series RC circuit, whose Bode plots were derived in Section 5.1.1, can be considered to be a prototype filter. In Figure 11.2a the output is taken across the capacitor, so the response falls above $f_c = 1/2\pi\tau = 1/2\pi RC$. The characteristic is that of a low-pass filter with a roll-off (fall in response as the frequency moves out of the pass band) of -20 dB/decade. Taking the output of the circuit across the resistor causes the response to be that of Figure 11.2b, which approximates that of a high pass filter rolling off at 20 dB/decade below f_c. This is a fairly slow roll-off rate, since if one wished to attenuate frequencies below 100 Hz by 60 dB, for example, one needs f_c to be three decades higher, that is at 100 kHz! As one might wish to retain frequencies much lower than this, the roll-off is too slow. Series RL circuits will have the same response as series RC circuits of course, if we substitute components appropriately, the corner frequency then being given by $f_c = 1/2\pi\tau = R/2\pi L$.

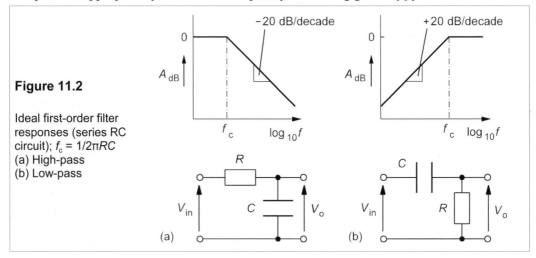

Figure 11.2

Ideal first-order filter responses (series RC circuit); $f_c = 1/2\pi RC$
(a) High-pass
(b) Low-pass

The series RC and RL circuits are both *single-pole* or *first-order* filters whose transfer functions have a zero at $\omega = \omega_c$:

$$\mathbf{H}(j\omega) = \frac{1}{1 + j(\omega/\omega_c)} \qquad \text{(lowpass)} \qquad (11.1)$$

and
$$\mathbf{H}(j\omega) = \frac{1}{1 - j(\omega_c/\omega)} \qquad \text{(highpass)} \qquad (11.2)$$

In the pass band the amplitude response, V_o/V_{in}, or voltage gain, A_v, is unity.

The next step is obviously to cascade several RC sections to improve roll-off. Putting two identical RC sections together to make a highpass filter as in Figure 11.3a, which eventually gives the required 40 dB/decade roll-off, but the -3dB point is shifted from ω_c to $2.67\omega_c$. This problem is easily solved by increasing the product of R and C by a factor of 2.67. The response is then the lower line in Figure 11.3b, which shows the single-stage RC filter's gain response for comparison. The frequency has been normalized to the respective -3dB frequencies. Unfortunately the 2-stage RC filter's response does not fall at 40 dB/decade until well below the cut-off frequency. In fact at $0.1f_c$ the response of the single-stage filter is -20 dB and the 2-stage filter's is -25 dB, though beyond this point the difference grows at the expected 20 dB/decade.

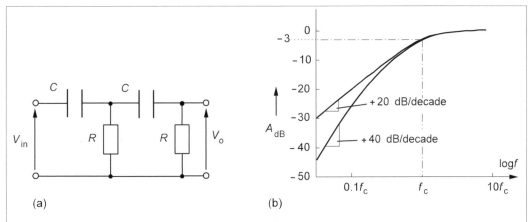

(a)   (b)

Figure 11.3 (a) A cascaded, 2-section, RC, highpass filter (b) The gain responses of single and 2-stage, RC filters, normalized to the cut-off frequencies

To get closer to the ideal the filter must be made to roll-off at the expected rate nearer to the cut-off frequency, which can be achieved with an RLC circuit as in Figure 11.4a.

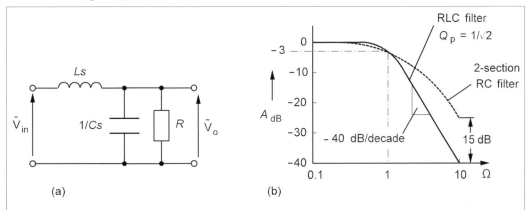

(a)   (b)

Figure 11.4 (a) A 2nd-order, lowpass, RLC filter in the *s*-domain (b) Voltage gain v. reduced frequency of the filter with optimally flat response ($Q_p = 1/\sqrt{2}$) compared to that of a 2-section, RC filter

This is intuitively seen to be a lowpass filter because it will fully pass DC (the inductance is short circuited and the capacitance open circuited for DC) and fully block high frequencies, since the inductance then acts as an open circuit and the capacitance as a short circuit. In the s-domain $R \mathbin{/\mkern-4mu/} 1/Cs$ is an impedance of

$$\check{\mathbf{Z}} = \frac{R/Cs}{R + 1/Cs} = \frac{R}{RCs + 1} \tag{11.3}$$

By the voltage-divider rule

$$\tilde{\mathbf{V}}_o = \frac{\tilde{\mathbf{V}}_{in} R/(RCs + 1)}{Ls + R/(RCs + 1)} = \frac{\tilde{\mathbf{V}}_{in} R}{R + Ls(RCs + 1)} \tag{11.4}$$

which can be rewritten:

$$\mathbf{H}(s) = \tilde{\mathbf{V}}_o/\tilde{\mathbf{V}}_{in} = \frac{1}{LCs^2 + Ls/R + 1} \tag{11.5}$$

For a sinusoidally-excited circuit we may replace s by $j\omega$ giving

$$\mathbf{H}(j\omega) = \frac{1}{(1 - \omega^2 LC) + j\omega L/R} = \frac{1}{(1 - \omega^2/\omega_0^2) + j\omega/\beta} \tag{11.6}$$

where $\omega_0^2 \equiv 1/LC$ and $\beta \equiv R/L$. Since $Q_p = \beta/\omega_0$, $\beta = \omega_0 Q_p$ and Equation 11.6 can be rewritten

$$\mathbf{H}(j\omega) = \frac{1}{(1 - \omega^2/\omega_0^2) + j\omega/\omega_0 Q_p} \tag{11.7}$$

Replacing ω/ω_0 by Ω, the reduced frequency, we find

$$\mathbf{H}(j\Omega) = \frac{1}{(1 - \Omega^2) + j\Omega/Q_p} \tag{11.8}$$

Thus the amplitude response is

$$A_{LP} = |\mathbf{H}(j\Omega)| = \frac{1}{\sqrt{(1 - \Omega^2)^2 + \Omega^2/Q_p^2}} \tag{11.9}$$

By suitably choosing components we can make $Q_p = 1/\sqrt{2}$ and then A_{LP} is

$$A_{LP,BU} = \frac{1}{\sqrt{1 + \Omega^4}} \tag{11.10}$$

This is a 2nd-order, Butterworth, lowpass filter; its response is nearly flat below $\Omega = 1$, and has no peak nearby, as it would for larger values of Q_p: it is said to be *optimally flat*. The roll-off is -40 dB/decade above $\Omega = 1$, or when $f > f_0$, the series resonant frequency. Figure 11.4b shows A_{dB} for this filter, and its superiority to the 2-section RC filter is clear: 15 dB at $\Omega = 10$. The highpass, Butterworth response of the circuit shown in Figure 11.5a, with $Q_p = 1/\sqrt{2}$, can be found likewise to be

$$A_{HP,BU} = \frac{\Omega^2}{\sqrt{1 + \Omega^4}} \tag{11.11}$$

It can soon be shown that

$$A_{HP}^2 + A_{LP}^2 = 1 \tag{11.12}$$

The normalized, bandpass response of the RLC circuit of Figure 11.5b is

$$\mathbf{H}(j\Omega) = \frac{j\Omega/Q_p}{(1 - \Omega^2) + j\Omega/Q_p} \tag{11.13}$$

This gives the Butterworth amplitude response graphed beneath the circuit when $Q_p = 1/\sqrt{2}$ and then

$$A_{\text{BP,BU}} = \frac{\Omega\sqrt{2}}{\sqrt{1 + \Omega^4}} \tag{11.14}$$

Finally the notch or band-reject response of the circuit of Figure 11.5c, when the coil's Q-factor is high and $Q_p = 1/\sqrt{2}$, is nearly

$$A_{\text{BR,BU}} = \frac{1 - \Omega^2}{\sqrt{1 + \Omega^4}} \tag{11.15}$$

This time we see that $A_{\text{BP}}^2 + A_{\text{BR}}^2 = 1$.

The performance of the bandpass and notch RLC filters is not very good; the bandpass filter's roll-offs are only at 20 dB/decade, half that of the lowpass and highpass filters, and the notch filter has a narrow bandwidth. These problems can be overcome in principle by using multi-stage RC filters, whose design parameters can all be found in standard tables.

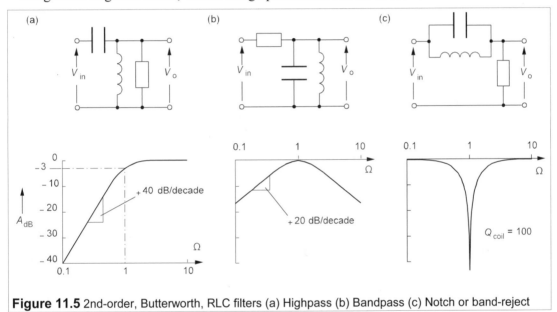

Figure 11.5 2nd-order, Butterworth, RLC filters (a) Highpass (b) Bandpass (c) Notch or band-reject

11.1.2 *Active filters*

Though passive filters can be made with very good performances, they suffer from the drawback of being bulky, especially at low frequencies, when the inductors tend to be large. They are, however, quite often used at frequencies in the MHz range and above; in fact small, fairly cheap, passive, in-line RF filters are the stock-in-trade of some factories. There are several approaches that can be used to design active filters at frequencies below the MHz region. One is based on op amps and feedback circuits such as the Sallen-and-Key family of filters, and another is based on the use of op amps as active inductors. A recent development is switched-capacitor filters which can be made as complete ICs without any need for discrete components. There are also a number of active filter ICs to be found in electronics suppliers' catalogues, but they tend to be costly. Several websites are available for designing both active and passive filters – they may save time, and are easy to use.

Sallen-and-Key filters

Named after their inventors, Sallen-and-Key filters are a means of obtaining two-pole responses without using inductors. We shall find the response of the 2-pole, lowpass, Sallen-and-Key filter in Figure 11.6a by nodal analysis.

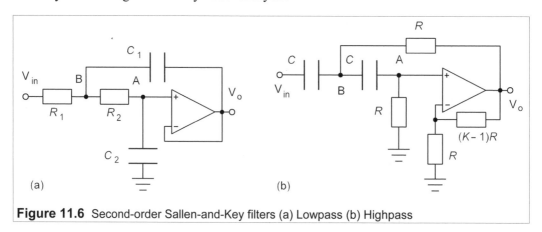

Figure 11.6 Second-order Sallen-and-Key filters (a) Lowpass (b) Highpass

We note that $V_A = V_o$, by golden rule 2 (see Section 3.1) and let V_B be the voltage at B. By Ohm's law the currents out of node B are $(V_B - V_{in})G_1$, $(V_B - V_o)j\omega C_1$ and $(V_B - V_o)G_2$, using conductances. These sum to zero by Kirchhoff's current law

$$(V_B - V_{in})G_1 + (V_B - V_o)j\omega C_1 + (V_B - V_o)G_2 = 0 \tag{11.16}$$

Hence

$$V_{in} = V_B(1 + G_2/G_1 + j\omega C_1 R_1) - V_o(G_2/G_1 + j\omega C_1 R_1) \tag{11.17}$$

Also by Ohm's law the currents out of node A are $(V_o - V_B)G_2$, $V_o j\omega C_2$ and zero into the amplifier. These also sum to zero:

$$(V_o - V_B)G_2 + V_o j\omega C_2 = 0 \tag{11.18}$$

which leads to $V_B = V_o(1 + j\omega C_2 R_2)$. Substituting for V_B in Equation 11.17 yields

$$V_{in} = V_o(1 + j\omega C_2 R_2)(1 + R_1/R_2 + j\omega C_1 R_1) - V_o(R_1/R_2 + j\omega C_1 R_1) \tag{11.19}$$

Equation 11.19 produces

$$\frac{V_o}{V_{in}} = H(j\omega) = \frac{1}{(1 - \omega^2 C_1 C_2 R_1 R_2) + j\omega C_2(R_1 + R_2)} \tag{11.20}$$

Replacing $C_1 C_2 R_1 R_2$ by $1/\omega_c^2$ in Equation 11.20 gives

$$H(j\omega) = \frac{1}{1 - (\omega/\omega_c)^2 + j\omega/\omega_c Q_p} = \frac{1}{1 - \Omega^2 + j\Omega/Q_p} \tag{11.21}$$

where $\Omega \equiv \omega/\omega_c$ and Q_p is given by

$$Q_p = \frac{\omega_c}{C_2(R_1 + R_2)} = \frac{\sqrt{C_1 C_2 R_1 R_2}}{C_2(R_1 + R_2)} \tag{11.22}$$

Equation 11.21 is exactly the same as Equation 11.8, which was derived for a second-order, RLC filter. Within reason, component values can be chosen for any value for Q_p desired.

Example 11.1

Design a second-order, Sallen-and-Key filter with a lowpass, Butterworth response and a cut-off at 2 kHz. What is its amplitude response at 10 kHz?

If we want a Butterworth response then $Q_p = 1/\sqrt{2}$, and from Equation 11.22 that is

$$Q_p^2 = 1/2 = (C_1 C_2 R_1 R_2)/[C_2(R_1 + R_2)]^2 \quad \rightarrow \quad (R_1 + R_2)^2 = 2cR_1 R_2 \tag{11.23}$$

where c is the capacitance ratio, C_1/C_2 and must be greater than 1. This is an unfortunate restriction since it is more difficult to make capacitance ratios equal to some arbitrary number than resistance ratios. Choosing $R_1 = R_2 = R$ leads to $C_1 = 2C_2$. A cut-off at 2 kHz means $\omega_c = 12.566$ krad/s, so that

$$\sqrt{R_1 R_2 C_1 C_2} = RC_2\sqrt{2} = 1/12566 = 79.6 \quad \mu s \tag{11.24}$$

Making $C_2 = 1.5$ nF and $C_1 = 3$ nF gives $R = 37.5$ kΩ. It is possible to find 1% tolerance polystyrene capacitors of these values and then the resistance value is readily achieved.

The amplitude response is $1/\sqrt{(1 + \Omega^4)}$ and at 10 kHz, $\Omega = \omega/\omega_c = f/f_c = 10/2 = 5$, so that $A_v = 1/\sqrt{(1 + 5^4)} = 0.04$ or -28 dB.

The slight difficulty with component values of this version of a Sallen-and-Key filter is overcome by the circuit of Figure 11.6b, in which the resistances are all equal and so are the capacitances. The circuit shown is a highpass filter, and the analysis takes the same path as above except that the voltage at node A is not $\mathbf{V_o}$ but $\mathbf{V_o}/K$ since the gain resistances, R and $(K-1)R$, at the inverting input form a voltage divider. Then it is straightforward to show that the response is

$$\mathbf{V_o}/\mathbf{V_{in}} = \frac{-K\omega^2 C^2 R^2}{1 - \omega^2 C^2 R^2 + j(3 - K)\omega CR} = \frac{-K\Omega^2}{(1 - \Omega^2) + j\Omega/Q_p} \tag{11.25}$$

where $\Omega = \omega/\omega_c = \omega CR$ and $Q_p = 1/(3 - K)$. If both gain and Q are to be adjusted independently then the component values cannot all be the same.

11.1.3 Types of filter response

There are many filter circuits named after their inventors but only a few principal types of filter response, which are given in Table 11.1. All of these can be constructed from 2-pole, op-amp, resistor and capacitor stages such as the Sallen-and Key-filter. The Chebyshev filter allows some ripple in the passband which can be traded off against transition rate, that is how close to the cut-off the characteristic roll-off rate is established; the more the ripple, the closer. Elliptic filters have ripple in passband and stopband (the frequencies where the response is low) and thereby have the fastest transition rates, but their performance is very sensitive to component values, and their design more complicated. Chebyshev and Bessel filters can be constructed from circuits such as the Sallen-and-Key filters shown in Figure 11.6, but more than just K must be adjusted in the circuit of Figure 11.6b.

Table 11.1 *Characteristics of some types of filter*

Type	Advantages	Disadvantages
Butterworth	Flattest pass band	Medium transition rate, medium distortion
Chebyshev	Fast transition rate	High distortion, high passband ripple
Elliptic	Fastest transition rate	High distortion, high ripple
Bessel	Least distortion	Slow transition rate
Gaussian	Low distortion	Medium transition rate

We have seen that a lowpass Butterworth filter has an amplitude response given by

$$|\mathbf{H}(j\Omega)| = \frac{1}{\sqrt{1 + \Omega^{2n}}}$$

(11.26)

where n is the number of poles and $n/2$ the number of stages. The amplitude response is shown in Figure 11.7a for Butterworth filters with various values of n: increasing n leads to a closer approach to the brickwall filter. Figure 11.7b shows the gain responses of Chebyshev, Butterworth and Bessel 4-pole, highpass filters. The Chebyshev filter has a faster roll-off for the same value of n. The Bessel filter has inferior roll-off, though still much better than cascaded RC sections. Because the component calculations are easier and less critical, Butterworth filters are often preferred, though the Bessel filter will distort less.

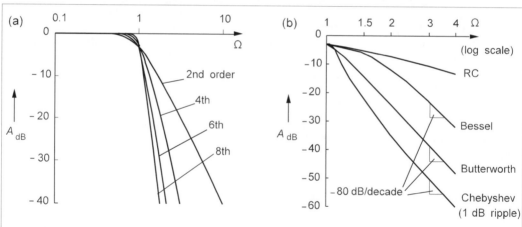

Figure 11.7 (a) Lowpass Butterworth filter responses (b) Responses of four types of 4th order filter in the transition zone

11.1.4 Active-inductor filters

Op amps, resistors and capacitors can be used to build a circuit which behaves like an inductance, and then this can be used in place of an inductor (an 'active inductor') in passive RLC or LC filter circuits. One such circuit is the Antonionou inductance simulator of Figure 11.8. By golden rule 2, all four op amp inputs are at a potential of $\mathbf{V}_{in}$. The currents out of node A are $\mathbf{V}_{in}/R_4$, $(\mathbf{V}_{in} - \mathbf{V}_1)j\omega C$ and zero into the amplifier by golden rule 1 (see Section 3.1, Chapter 3). By Kirchhoff's current law:

$$\mathbf{V}_{in}/R_4 + (\mathbf{V}_{in} - \mathbf{V}_1)j\omega C = 0 \tag{11.27}$$

which leads to

$$\mathbf{V}_{in} - \mathbf{V}_1 = -\mathbf{V}_{in}/j\omega CR_4 \tag{11.28}$$

The current, $\mathbf{I}$, through R_2 is $(\mathbf{V}_2 - \mathbf{V}_{in})/R_2$, while the current through R_3 must also be $\mathbf{I}$ as no current goes into the op amp inputs, and is $(\mathbf{V}_{in} - \mathbf{V}_1)/R_3$. Thus

$$(\mathbf{V}_2 - \mathbf{V}_{in})/R_2 = (\mathbf{V}_{in} - \mathbf{V}_1)/R_3 \tag{11.29}$$

so that

$$\mathbf{V}_{in} - \mathbf{V}_2 = \frac{-R_2}{R_3}(\mathbf{V}_{in} - \mathbf{V}_1) = \frac{R_2}{R_3}\frac{\mathbf{V}_{in}}{j\omega CR_4} \tag{11.30}$$

We have used Equation 11.28 to substitute for $\mathbf{V}_{in} - \mathbf{V}_1$. But

$$\mathbf{I}_{in} = (\mathbf{V}_{in} - \mathbf{V}_2)/R_1 = R_2\mathbf{V}_{in}/j\omega CR_1R_3R_4 \tag{11.31}$$

Then $\qquad \mathbf{Z}_{in} = \mathbf{V}_{in}/\mathbf{I}_{in} = j\omega CR_1R_3R_4/R_2 \tag{11.32}$

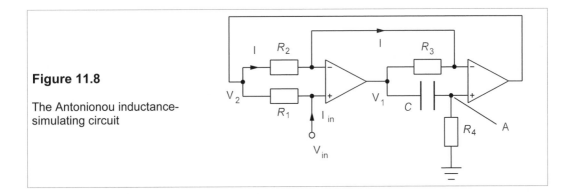

Figure 11.8

The Antonionou inductance-simulating circuit

By making all the resistances equal, the circuit has a reactance of $j\omega CR^2$, that is it behaves like an inductance, $L = CR^2$. Large inductances can be simulated easily, since if $C = 1\ \mu F$ and $R = 10\ k\Omega$, $L = 100\ H$! However, the capacitances needed for large inductances is still too high for ease of manufacture as an IC and they must be provided externally.

11.1.5 *The state-variable or KHN filter*

The name derives from the use of state-variables in solving the differential equations that are used to produce the filtering functions. The alternative name KHN comes from the initials of its originators, Kerwin, Huelsman and Newcomb. The circuit of Figure 11.9 shows a KHN circuit with three op amps that gives high, low and bandpass filters by means of feedback from two integrators: it is virtually an analogue computer.

By the usual method of analysis it can be shown that the highpass response is

$$V_1/V_{in} = \frac{K\Omega^2}{1 - \Omega^2 + j\Omega/Q_p}$$
(11.33)

where $K = -(R_6/R_4)(R_4 + R_5)/(R_1 + R_6)$, $\Omega = \omega/\omega_c$ and $Q_p = R_6/KR_1$ and $\omega_c = \sqrt{(R_5/C^2R_2R_3R_4)}$.
The bandpass response is

$$V_2/V_{in} = \frac{-(R_6/R_1)j\Omega/Q_p}{1 - \Omega^2 + j\Omega/Q_p}$$
(11.34)

And the lowpass response is

$$V_3/V_{in} = \frac{-KR_5/R_4}{1 - \Omega^2 + j\Omega/Q_p}$$
(11.35)

There are seven parameters under the designer's control (one capacitor and six resistor values), giving a good degree of flexibility.

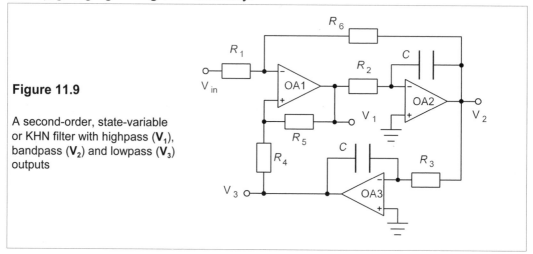

Figure 11.9

A second-order, state-variable or KHN filter with highpass (V_1), bandpass (V_2) and lowpass (V_3) outputs

State-variable filters can be obtained as ICs such as the Burr Brown UAF42AP, which is a second-order filter in 14-pin DIL form costing about £13 in single quantities. The 20-pin DIL MAX275ACPP made by Maxim, which can be configured as a fourth-order filter, costs about £6 and the 8th order MAX274ACNG costs £10. By suitably choosing external resistor values Butterworth, Chebyshev and Bessel filters can be made. The cut-off frequency range is typically from 100 Hz to 150 kHz and can be achieved within about ±1%. These are often called universal filters because non-inverted outputs can be obtained with variable gains, but they are only slightly-modified versions of the state-variable filter.

11.1.6 *Switched-capacitor filters*

Switched-capacitor filters are most ingenious devices that use only IC technology; the entire filter is made in a single silicon chip. The capacitance limitations of only a few picofarads with IC technology are cleverly overcome by switching a capacitor between a voltage source and another capacitor to simulate a *resistance*. Consider the circuit of Figure 11.10, which is a switched-capacitor integrator.

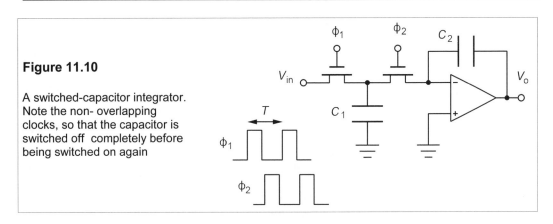

Figure 11.10

A switched-capacitor integrator. Note the non- overlapping clocks, so that the capacitor is switched off completely before being switched on again

MOSFET 1 is first switched on and MOSFET 2 off, so capacitor C_1 charges up instantaneously (in a few ns) to the voltage V_{in}. Then MOSFET 1 is switched off and after a short delay MOSFET 2 switches on, so that capacitor C_2 is charged up. The current flowing into the circuit, I_{in}, is on average Q/T, which is $C_1 V_{in}/T$. But the input resistance, R_{in}, is V_{in}/I_{in} = T/C_1, so the integrator's time constant, $R_{in}C_2$, is TC_2/C_1. Importantly, it depends only on the *ratio* of two capacitances, not their absolute values; and this ratio can be controlled to within 0.1%. The only problem is that large capacitance ratios are impractical. In addition the clock frequency, $1/T$, has to be much higher than the frequency of the input waveform. Filter circuits of great sophistication can now be made on a single chip. National Semiconductor's LMF100, for example, is a dual 1-pole, 20-pin DIL, CMOS IC which can be configured by external resistors as a 4th order filter of any type, operating at frequencies from 0.1 Hz to 100 kHz. The most rapidly growing market now for switched-capacitor filters is in ASICs (Application-Specific ICs).

11.2 Oscillators

Almost any piece of electronic equipment requires an oscillator somewhere in its construction or externally provided. Oscillators range from simple square-wave op amp RC types to quartz-crystal-controlled, high-stability, pure sinewave oscillators. All oscillators are based on the principles of a simple positive-feedback circuit as shown in the block diagram of Figure 11.11.

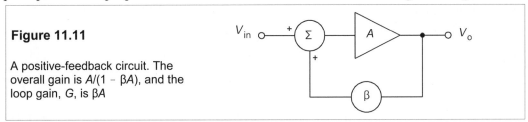

Figure 11.11

A positive-feedback circuit. The overall gain is $A/(1 - \beta A)$, and the loop gain, G, is βA

If the amplifier gain is A and the feedback ratio (the fraction of the output that is fed back) is β, then

$$V_o = A(V_{in} + \beta V_o) \qquad (11.36)$$

From which we find

$$V_o = \frac{AV_{in}}{1 - A\beta} = \frac{AV_{in}}{1 - G} \qquad (11.37)$$

where G is the loop gain. When $A\beta = G = 1$, the overall gain will be infinite, that is it will have an output *when there is no input*, which is the essence of an oscillator.

Since A, G and β are functions of ω, we can write the oscillation criterion (or *Barkhausen criterion*) as

$$G(j\omega_o) = \beta(j\omega_o)A(j\omega_o) = 1 + j0 \qquad (11.38)$$

that is at some frequency, ω_o, the phase of the loop gain is zero and its amplitude is unity. The problem with these conditions is they lead to an inherently unstable system: if the loop gain is just less than one, the oscillations cease, and if it is just greater than one they become infinite in amplitude (the op amp produces square waves, not sine waves). For this reason a non-linear element is necessary which will reduce the gain to unity when the output becomes large. Some examples will make this clearer.

11.2.1 RC oscillators

RC oscillators use op amps to provide feedback, and as a result are only suitable for frequencies below 1 MHz. The first one described here is very simple since it does not require stabilization because its output is a square wave.

The RC square-wave generator or relaxation oscillator

Consider the circuit of Figure 11.12. The op amp will detect an initial voltage between its input terminals and go into saturation at $+V_{CC}$ (or $-V_{CC}$ – it is immaterial) and start to charge the capacitor at a rate determined by the time constant, RC. When the capacitor charges to $\frac{1}{2}V_{CC}$ it will mean that the op amp's input terminals are at the same potential (the non-inverting input is held at a potential $\frac{1}{2}V_o$) so its output will drop to zero, by which time the capacitor voltage will have exceeded $\frac{1}{2}V_{CC}$ and the op amp will go into saturation, this time to $-V_{CC}$. The capacitor then discharges to $-\frac{1}{2}V_{CC}$ according to

$$v_C = 1.5 V_{CC} \exp(-t/RC) - V_{CC} \qquad (11.39)$$

and the cycle repeats. (To obtain this equation ask: What v_C will be at $t = 0$ and $t = \infty$?)

Setting v_C to $-V_{CC}$ in Equation 11.39 leads to

$$1.5 V_{CC} \exp(-t/RC) - V_{CC} = -0.5 V_{CC} \qquad (11.40)$$

So that $\exp(-t/RC) = 1/3$, or $t = 1.1RC$. This time is half the period of oscillation, T, which is $2.2RC$, or

$$f_o = 1/T = 1/2.2RC \qquad (11.41)$$

There is no need for gain limitation as the output is not sinusoidal and cannot exceed the supply voltage. The square wave can be filtered to make a sinewave if 1% or 2% harmonic distortion is permissible.

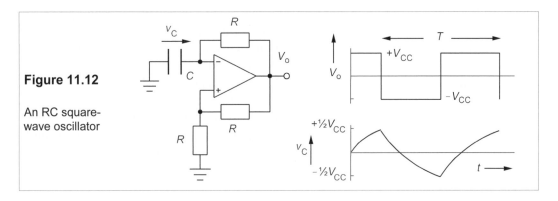

Figure 11.12

An RC square-wave oscillator

The Wien-bridge oscillator

The Wien-bridge oscillator provides a sine-wave output which is reasonably stable in frequency and relatively free from distortion. The Wien-bridge circuit shown in Figure 11.13a has no amplitude limitation and is therefore unstable.

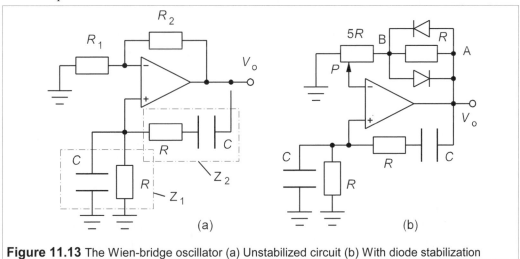

(a) (b)

Figure 11.13 The Wien-bridge oscillator (a) Unstabilized circuit (b) With diode stabilization

By the voltage-divider rule, the voltage at the inverting input, V_2, is $V_oR_1/(R_1 + R_2)$, and this must equal the voltage at the non-inverting input, V_1, by golden rule 1. Now $V_1 = V_o$ $Z_1/(Z_1 + Z_2)$, where $Z_1 = R/(1 + j\omega CR)$ and $Z_2 = R + 1/j\omega C$, so that

$$V_1 = \frac{V_o Z_1}{Z_1 + Z_2} = \frac{V_o}{3 + j(\omega CR - 1/\omega CR)} \tag{11.42}$$

Setting $V_1 = V_2$ gives

$$\frac{V_o R_1}{R_1 + R_2} = \frac{V_o}{3 + j(\omega CR - 1/\omega CR)} \tag{11.43}$$

which can only be so if $\omega = \omega_o = 1/CR$ and $R_2 = 2R_1$. The circuit is unstable, however, and will either not oscillate or will swing between the supply voltages, since the loop gain will be either just less than one or just greater than one, because of component tolerances, etc. This is remedied with the stabilized circuit of Figure 11.13b, in which the potentiometer has a total resistance of $5R$ and is set to P ohms above ground. Then $R_2 = R + 5R - P = 6R - P$ and $R_1 = P$, so that $P = 2R$ is the setting for oscillation. The diodes clamp the voltage across R to ± 0.7 V, limiting $\mathbf{V_o}$ to ± 3.5 V and $\mathbf{V_A}$ to ± 4.2 V. If $|\mathbf{V_A}|$ rises above 4.2 V the diodes shunt the extra current past R so the voltage at the inverting input rises and $|\mathbf{V_A}|$ falls. If $|\mathbf{V_A}|$ falls below 4.2 V, then the current through the resistive network falls, the voltage at the inverting input falls and $|\mathbf{V_A}|$ must rise. Although the output is smoother when taken from node B rather than node A, node B is a high-impedance node and therefore requires buffering.

11.2.2 LC oscillators

LC oscillators are especially useful for high frequencies as they use transistors to provide feedback. At low frequencies they require rather large inductances and therefore become fairly heavy. There are many types of LC oscillator, two of the most popular being

The Hartley and the Colpitts oscillators

Both of these widely-used oscillators rely on the parallel LC circuit shown in Figure 11.14a, which is the essential part of the Hartley oscillator (for the Colpitts oscillator, replace L by C and vice versa).

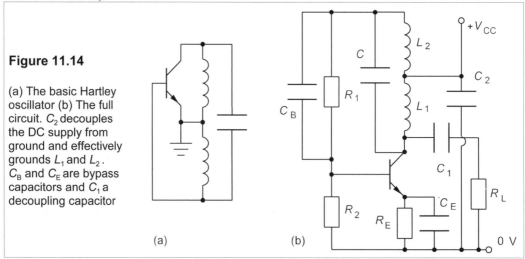

Figure 11.14

(a) The basic Hartley oscillator (b) The full circuit. C_2 decouples the DC supply from ground and effectively grounds L_1 and L_2. C_B and C_E are bypass capacitors and C_1 a decoupling capacitor

(a)   (b)

    Feedback from the tank circuit to the base of the transistor is via L_2. Let the transistor's voltage gain be A_v, while the feedback ratio (β in Equations 11.36 and 11.37) is $j\omega L_2 / j\omega L_1 = L_2/L_1$. Then for oscillations to start $A_v\beta$ must be greater than 1, or $A_v > L_1/L_2$. L_1 and L_2 are in series so their combined inductance is $L_1 + L_2$ and the resonant frequency, $\omega_0 = 1/\sqrt{(C[L_1 + L_2])}$. The circuit is a self starter and once oscillations have begun they build up until the transistor's gain is reduced to exactly L_1/L_2. Even though this may not occur before the output is distorted, the LC circuit has a high Q and an undistorted sinewave is produced. Figure 11.14b shows the Hartley oscillator with biasing; C_B is a bypass capacitor ensuring that AC feedback is only via L_2.

Example 11.2

A Hartley oscillator is to operate at 1 MHz with $C = 100$ pF, and a feedback ratio of 0.01. If the transistor in Figure 11.14b has $h_{oe} = h_{re} = 0$ and $R_L = 1$ kΩ, what is the least collector bias current for oscillations to start, and what are L_1 and L_2? (Ignore the inductor losses.)

    Since $L_2/L_1 = 0.01$, $\omega_0 = 1/\sqrt{[C(L_1 + L_2)]} = 1/\sqrt{(100 \times 10^{-12} \times 1.01 L_1)} = 2\pi \times 10^6$ rad/s, and $L_1 = 251$ µH, $L_2 = 2.51$ µH. The transistor must have a voltage gain greater than 100 for oscillations to start. As shown in Chapter 8, Section 8.6, A_v will be $h_{fe}R_L/h_{ie}$, with h_{ie}, the input resistance, given by $0.026/I_B \approx 0.026 h_{fe}/I_C$. Then $A_v = R_L I_C /0.026 = 3.85 \times 10^4 \times I_C > 100$, leading to $I_C > 2.6$ mA. Had R_L been bigger, the collector current could have been less. As we saw in Chapter 2, Section 2.5.2, the losses in L_1 can be represented as a resistance of $Q_c^2 r$ in parallel with R_L, where Q_c is the inductor's Q-factor and r is its resistance. If the coil Q is not fairly high (> 10) these losses must be taken into account.

11.2.3 *Quartz-crystal oscillators*

The frequency stability and distortion of an LC oscillator are reasonably good – about 0.1% – but for many purposes better frequency stability in particular is necessary. What is needed is a high-Q circuit whose component values do not change with temperature or time. A quartz crystal supplies these requirements. Quartz is a piezoelectric material that acquires a surface charge when it is strained, making it a useful material for electroacoustic transducers. It has modes of vibration which can be modelled electrically as a series RLC circuit in parallel with a capacitance, as in Figure 11.15a.

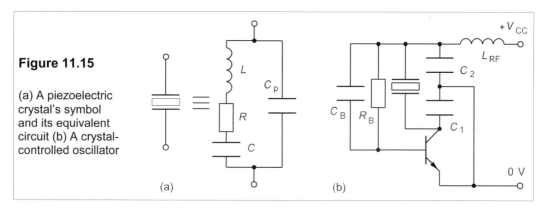

Figure 11.15

(a) A piezoelectric crystal's symbol and its equivalent circuit (b) A crystal-controlled oscillator

    Crystal Qs of 10^4 are usual and the series resistance, R, is about 1 kΩ giving a 1MHz crystal a series inductance of $L = RQ/\omega_0 = 1 \times 10^3 \times 10^4/2\pi \times 10^6 = 1.6$ H, and a series capacitance of $C = 1/\omega_0^2 L = 0.016$ pF. The parallel capacitance, C_p, is of the order of 1 pF for a small crystal: much greater than the series capacitance. When the crystal is placed in an LC oscillator circuit as in Figure 11.15b, it resonates at a frequency determined by L and the equivalent capacitance of C, C_p, C_1 and C_2, which is almost $1/\sqrt{(LC)}$ because $C_2 \gg C_1 \gg C_p \gg C$. Figure 11.15b is simply a Colpitts oscillator with a crystal in place of the inductor. Quartz-crystal oscillators cannot be 'tweaked' much in frequency, though some frequency 'pulling' with small adjustable capacitors is possible. They are extremely stable, being based on mechanical resonance that varies little with temperature; stabilities of 1 ppm/K are usual. Crystals mounted in two-terminal metal cans be bought for £1 to £2 off the shelf for numerous frequencies ranging from 32.768 kHz (2^{15} Hz) to 20 MHz.

11.3 **Phase-locked loops**

Phase-locked loops (PLLs), after a slow start (they originated in 1922), have lately become very widely used in IC form, cheap devices costing under £1, and their frequency ranges up to several GHz. They are a very good means of extracting clean sine or other waves from noisy input signals, and they can be used also for AM and FM demodulation, tone decoding, frequency-shift keying, waveform generation, tracking filters, voltage-to-frequency conversion and motor speed control: the uses are limited only by the designers' imaginations, it seems. Figure 11.16 shows a block diagram of a PLL.

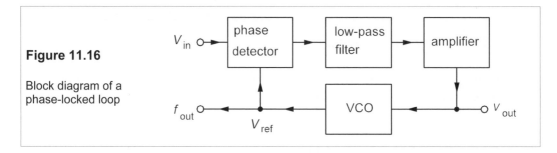

Figure 11.16

Block diagram of a phase-locked loop

The input signal's phase is compared to that generated by a voltage-controlled oscillator (VCO) and the phase detector's output is low-pass filtered and amplified. The phase detector's output depends on the difference in phase (and therefore frequency also) between the input and the signal from the VCO, V_{ref}. The input of the VCO is the filtered and amplified signal from the phase detector. When there is no input signal there is no error voltage and the VCO outputs its free-running frequency, f_O. If there is an input signal of frequency f_s which is nearly the same as f_O then the phase detector generates an error voltage which forces the VCO to synchronise with f_s. The PLL is then said to be *locked* onto the input frequency and $f_{out} = f_s$. Any noise in the input signal is removed and the VCO output is a cleaned-up version of V_{in}. If the input frequency varies slowly the PLL stays locked, but the error signal, V_{out}, will follow the changes in the input frequency. Thus an FM signal is readily demodulated.

One can visualise the operation of a phase-locked loop by imagining two heavy discs free to rotate about an axle, which is split in the middle and loosely coupled with a spring or a magnetic clutch, as in Figure 11.17.

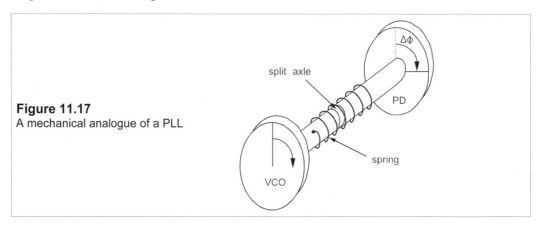

Figure 11.17
A mechanical analogue of a PLL

When one disc, the phase-detector (PD), is moved rapidly to a new position, the loose coupling means that initially the other disc, the VCO, remains stationary for a time before slowly accelerating to catch up with the first. When it does so it will overshoot and then stop and reverse direction. This will result in oscillations of the VCO disc about the mean position until it eventually comes to halt at a position slightly out of phase with the phase detector, giving rise to a characteristic lock-in signal. If the phase-detector disc moves slowly from its stationary position, the VCO disc will track it with a phase error which depends on the speed of the phase-detector, that is its rate of change of phase. A rapidly-rotating phase detector which varies its speed ($\equiv$ frequency) will cause the VCO to vary its speed too, but with a phase difference. This phase difference produces an output which is the demodulated FM signal.

The tracking VCO must still receive an error signal to keep it in lock, otherwise it will fall out of lock and into its free-running frequency, f_O, sometimes called the centre frequency. Thus the lock-in condition is essentially one of constant phase difference between the signal and its locally-generated, noise-free replica. The *lock range* is the range of frequencies over which the VCO, once locked on, can remain locked onto the input and runs from $f_O - f_L$ to $f_O + f_L$, that is the lock range is $2f_L$. The lock range must be greater than the *capture range*, $2f_C$, which is the range of frequencies over which the VCO can lock onto the input, starting out of lock. The capture range is also centred on f_O. The time taken to lock on is the *lock time*, t_L which is inversely proportional to bandwidth.

The loop gain, K_v, of the PLL is made up of three parts – the gains of the three component boxes in Figure 11.16 – that is the phase detector's conversion gain, K_D (in V/rad), the low-pass filter gain, K_{LP}, and the VCO conversion gain, K_O (in rad/s/V):

$$K_v = K_D K_{LP} K_O \qquad (11.44)$$

in units of s^{-1}, the unusual units arising because the VCO's gain, K_O, is given by

$$\Delta\omega_O = \frac{d\Delta\phi}{dt} = K_O V_{LP}$$

$$\Rightarrow \quad \Delta\phi = \int K_O V_{LP} dt$$

since angular frequency is the rate of change of phase. If the output of the low-pass filter, V_{LP}, is of the form $V_m \sin\omega t$ then the phase output of the VCO is

$$\Delta\phi = \frac{K_O V_m (-\cos\omega t)}{\omega} = \frac{K_O V_m \sin(\omega t - 90°)}{\omega} = \frac{K_O V_m \sin\omega t}{j\omega}$$

The $j\omega$ term is in the denominator because the VCO has introduced a phase shift of $-90°$.

In designing a PLL it is necessary to write down the exact expression for the loop gain and then select a cut-off frequency (or frequencies) which will give unity loop gain at or a little above the frequency of interest. The VCO can be non-linear in output without seriously affecting the operation of the PLL, it just means that K_O is also a function of ω and the effective loop gain Equation should take this into account. However, in most applications the variations in frequency are such that the VCO is very linear.

The damping factor, ζ, the stiffness of the spring in Figure 11.17, is given by $\zeta = (4\pi K_v)^{-1}$. If the loop gain is large, the damping factor is small and the PLL will respond in an oscillatory manner to transients and is hardly stable. On the other hand, if the damping factor is large the response is slow. Ideally $0.2 < \zeta < 2$. Thus we should make K_v only as large as necessary in order to prevent excessive variations in the output of the PLL. This is largely a matter of adjusting the component values in the low-pass filter as the other terms are outside of our control, unless we are building the whole thing from scratch.

Figure 11.18 shows National Semiconductor's 4046 PLL in more detail. The inhibit input when HIGH causes the PLL to go into standby mode to reduce power consumption. One of two phase detectors can be selected; PD1 is a phase detector which is just an XOR gate. Its output is a square wave ‑ shown in Figure 11.19a ‑ when in the locked-on condition, since the VCO and the phase detector will be out of phase by 90° because of the phase shift of the VCO (discussed later on). It is possible with this phase detector to get lock-in when the input is a harmonic of the VCO's centre frequency, f_0.

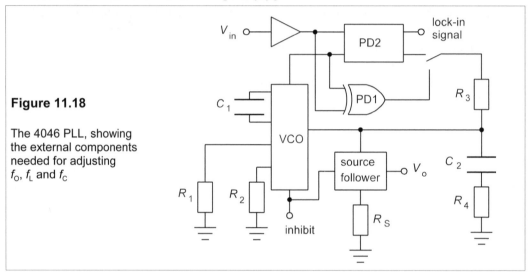

Figure 11.18

The 4046 PLL, showing the external components needed for adjusting f_0, f_L and f_C

Figure 11.19b shows the waveforms of PD2 when the VCO is locked-on. The slight residual phase difference between the output of the VCO and the input signal causes PD2 to output a small error voltage and a locked-in signal at the frequency of the input. Without an input signal PD1 causes the VCO to output its centre frequency, f_0, and if PD2 is used it outputs the minimum frequency, $f_0 - f_L$.

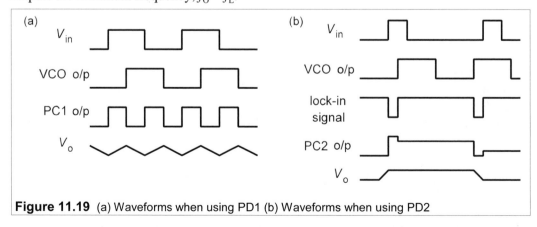

Figure 11.19 (a) Waveforms when using PD1 (b) Waveforms when using PD2

The adjustable components in Figure 11.18 are the resistors R_1, R_2, R_3, R_4 and R_S, and the capacitors C_1 and C_2. In practice all the resistors should be above 10 kΩ and then R_S has no effect on the PLL. R_4 prevents the high-frequency gain of the low-pass filter from falling off at high frequencies, and can be omitted. R_3 and C_2 determine the cut-off of the low-pass filter. R_1 and C_1 determine the centre frequency of the VCO, f_O; while R_2 and C_1 determine the lock range, $2f_L$. Unfortunately these two frequencies are markedly dependent on the supply voltage and there may be considerable variation from chip to chip, so some experimentation is usually needed. The centre frequency with $R_2 = \infty$ is approximately

$$f_O \approx \frac{0.15V_{DD}}{R_1 C_1}$$

Thus if $R_1 = 100$ kΩ, $C_1 = 1$ nF and $V_{DD} = 5$ V, $f_O = 7.5$ kHz. The practical limit for f_O is about 1 MHz, but may be only half of this. The lock range is given by

$$2f_L \approx \frac{0.5V_{DD}}{R_2 C_1}$$

so if $C_1 = 1$ nF, $V_{DD} = 5$ V and $R_2 = 500$ kΩ, then $2f_L = 5$ kHz and the VCO can vary its output frequency from 5 kHz to 10 kHz.

The capture range is the same as the lock range when PD2 is used, but with the XOR detector, PD1, the capture range depends on the components of the low-pass filter and if $R_4 = 0$ it is

$$2f_C = \sqrt{\frac{2f_L}{\pi R_3 C_2}}$$

If $R_4 \neq 0$ the capture range is a more complicated function of the component values. If $f_L = 2.5$ kHz, $C_2 = 1$ nF and $R_3 = 220$ kΩ, then $2f_C = 2.69$ kHz $< 2f_L$ as it must be. The low-pass cut off is at $\omega_{LP} = 1/R_3 C_2 = 4.55$ krad/s or $f_{LP} = 723$ Hz. Thus at 10 kHz the low-pass gain is about -23 dB (at -20 dB/decade), that is $K_{LP} = 0.072$, which gives us a condition for the phase detector's gain using Equation 11.44

$$K_D = \frac{K_v}{K_O K_{LP}}$$

taking K_O, the VCO gain, as 6 krad/V, we find (if the loop gain, K_v is to be unity at 10 kHz) that $K_D \approx 2.3$ mV/rad.

11.4 Waveform-generator ICs

Reasonably stable and cheap waveform generators can be purchased as integrated circuits chips. Two examples are the MAX038 from Maxim (up to 40 MHz, £15) and the ICL8038 from Harris (up to 100 kHz, £5). The MAX038CPP device is shown in Figure 11.20. They can produce square, sine and triangular waves simultaneously, as well as FM sweeps (see Section 23.4), by connecting suitable external resistors and a timing capacitor. Table 11.2 gives a few of the chip specifications for the MAX038. The logic inputs A0 and A1 control the type of output waveform. Applying a suitable voltage to the DADJ pin controls the duty cycle of square wave and the shape of triangular or sawtooth waveforms.

Table 11.2 *MAX038 Waveform generator specifications*

Parameter	min	typical	max
Supply voltage	±4.75	-	±5.25
Supply current, mA	-	+35/−45	+45/−55
Frequency, *f*	0.2 Hz	-	20–40 MHz
Stability of *f* with temperature	200 ppm/K	-	600 ppm/K
Δ*f* with supply voltage	-	0.4%/V	-
Output resistance	-	100 Ω	-
Square wave p-p	-	2 V	-
Square wave rise/fall time	-	12 ns	-
Duty cycle	5%	-	95%
Triangle/sawtooth/ramp p-p	-	2 V	-
Sinewave p-p	-	2 V	-
Distortion	-	0.75%	-

Figure 11.20

Pin configuration of the MAX038 20-pin plastic DIL-packaged waveform generator

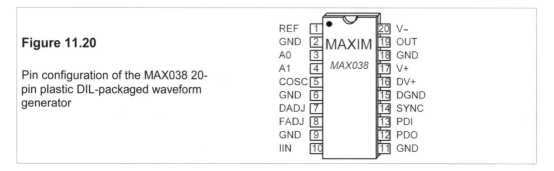

11.5 **Voltage regulators**

The next most widely used type of IC after op amps, which are overwhelmingly the most widely used, is the voltage regulator. The reasons are plain: with one poorly regulated, ripply power supply it is possible to feed all kinds of different circuits with ripple-free, well-regulated voltages. Table 11.3 lists a selection of voltage regulator ICs, together with a few performance parameters. Power supplies derived from the mains are always ripply and the average voltage regulator will reduce this by between 50 and 80 dB. If a 9V supply has a ripple of 1 V_{pp}, then it will be only 1.6 mV_{pp} (or about 0.02%) when reduced by 50 dB. Voltage regulators are tolerant devices, though wasteful of power.

Figure 11.21

A simple voltage regulator

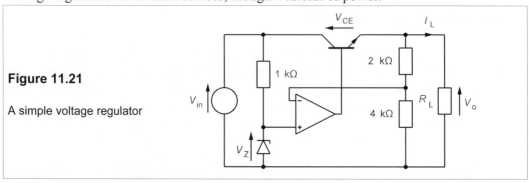

Table 11.3 *Voltage regulators*

Device	Package	V_{in} (V)[A]	V_o (V)	ΔV[B]	I_o (A)	Power[C]	$R\theta_j$[D]	Price[E]
LM2931AZ-5	TO92	+5.8/+26	+5	-	0.1	0.6	200	80p
UA78L05ACLP	TO92	+7/+20	+5	±0.2	0.1	0.6	200	34p
L78L08ACZ	TO92	+10/+30	+7	±0.8	0.1	0.6	200	22p
L78SO5CV	TO220	+8/+35	+5	±0.2	2	20	4	53p
MC7806CT	TO220	<+35	+6	±0.25	1	15	4	37p
BA178M12T	TO220	+15/+27	+12	±0.5	0.5	2	5	41p
LM78L15ACZ	TO92	+18/+30	+15	±0.6	0.1	0.6	200	20p
KA7908TU	TO220	-10/-35	-8	±0.3	1	8	5	45p
LM217T	TO220	+4.2/+40	+1.2/+37[F]		1.5	20	4	87p
VR350	TO3	+5/+35	+1.2/+38[F]		3	30	2	£18.43
TPS7201QP	DIP8	+3/+10	+1.2/+9.75[F]		0.25	0.8	100	£1.14
LM317LZ	T092	+4/+40	+1.2/+37[F]		0.1	0.6	200	29p
MAX663CSA	SOIC8[G]	+2/+16	+1.3/+16[F]		0.04	0.5	300	£3.71

Notes: All devices have short circuit protection and thermal cut-outs [A] Input voltage range [B] Output voltage variation [C] Max power dissipation in device in W [D] Thermal resistance from junction to case for T03 and TO220 , junction to ambient others, in °C/W [E] In single quantity [F] User adjustable [G] Surface-mount device

The principles on which they operate can be seen by referring to Figure 11.21, in which a DC source is regulated to $1.5V_Z$ by using the op amp to control the transistor output current. The output voltage divides across the 2kΩ and 4kΩ resistances to feed back 8 V to the inverting input, so the output voltage has to be 12 V. If it is less the transistor will drop its V_{CE} until the output *is* 12 V. If the output voltage rises above 12 V, the transistor increases V_{CE} to force it back to 12 V once more. Limitation of the load current, I_L, is required, since if $V_{in} = 35$ V, say, then V_{CE} will be 23 V and the power dissipation of the transistor is roughly $V_{CE}I_L$ or 23 I_L. Thermal protection is also advisable and most voltage regulators provide both. As usual, high-power devices are expensive, but the low-power devices are remarkably inexpensive. The 3-terminal devices give a fixed output voltage (V_o) on one pin with another pin for the variable input voltage, V_{in}. The third pin is ground.

11.6 **Analogue-to-digital (A/D) and digital-to-analogue (D/A) converters**

A/D and D/A converters – also known as ADCs and DACs – are the essential first step in digitising analogue waveforms and vice versa. They are based on the sample-and-hold circuit of Figure 11.22a, which samples a voltage waveform with a MOSFET switch at regular intervals and charges up a capacitor to 'hold' the voltage.

If the highest frequency of the sampled waveform is f, then the minimum sampling frequency needed to permit its reconstruction is $2f$. This sampling rate is often known as the *Nyquist frequency* (which is discussed in Chapter 25). Figure 11.22b shows the input and output of a sample-and-hold circuit, in which the sampling frequency is $f_s = 1/\tau$. The output voltage levels can be encoded in digital form as a *word* comprising, for example, 8, 12 or 16 binary digits or bits. If there are 8 bits in the word, then the number of discrete voltage levels will be $2^8 - 1$ or 255. Taking a signal with a minimum level of 0 V and a maximum of 5 V, then the least significant bit of its digital representation corresponds to $5/255 = 0.0196$ V, and the resolution of the 8-bit word is $\pm\frac{1}{2}$ bit, or 9.8 mV.

The output of an 8-bit ADC from each sample of the signal shown in Figure 11.22b will be a set of high (1) and low (0) outputs on its 8 output pins. For example the first sample at the end of the acquisition time is 3.9 V, or $3.9 \times 255/5 = 199$ in decimal or 11000111 in binary. The ADC output for this sample will be HHLLLHHH on its output pins. Notice how small is the error caused by the discrete levels of the digital representation (so-called *quantization error*) – far less in this case than the *sampling error* caused by the finite sampling rate. In fact the maximum error for an 8-bit ADC is only $1/[2 \times (2^8 - 1)]$, or 0.2%, of the full-scale voltage.

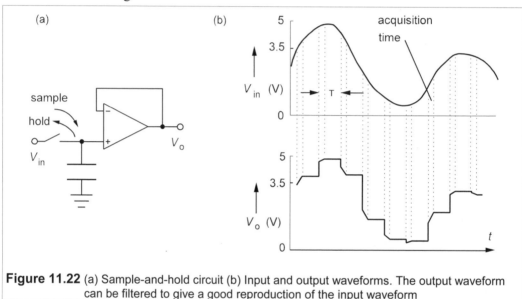

Figure 11.22 (a) Sample-and-hold circuit (b) Input and output waveforms. The output waveform can be filtered to give a good reproduction of the input waveform

To make a DAC one needs is a switching network that will scale each bit output according to its position in the word. Using a 4-bit DAC as an example, we want a number such as 1010 – representing say, 3.3 V, if 1111 is 5 V – to be output correctly as 3.3 V. A ladder resistance network can be used for this as in Figure 11.23, where the switches are controlled by the digital word input to the DAC.

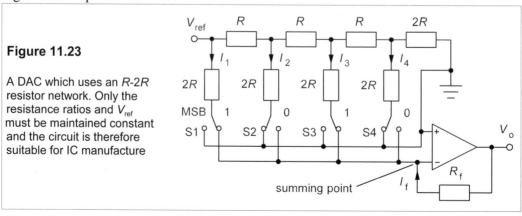

Figure 11.23

A DAC which uses an *R-2R* resistor network. Only the resistance ratios and V_{ref} must be maintained constant and the circuit is therefore suitable for IC manufacture

S1 is controlled by the most significant bit (MSB) and S4 by the least significant bit (LSB). A 1 causes a switch to be set to the summing position and a 0 causes it to be set to the ground position. In either case, since the summing amplifier's input is a virtual ground, the current flowing through the switch is the same. The network presents a resistance of $2R \mathbin{/\mkern-5mu/} 2R$ at each node, so the currents divide there into two equal halves: one through the switch and one to the rest of the network. Thus $I_1 = 2I_2 = 4I_3 = 8I_4 = V_{ref}/R$, and the currents are in correct proportions. In our example, I_1 and I_3 are sent through to the summing amplifier, which will output a voltage of $-(I_1 + I_3)R_f$, where R_f is the feedback resistance. But as $I_3 = \frac{1}{4}I_1$, $V_o = -1.25I_1R_f = -1.25V_{ref} R_f /R$. The accuracy of the result therefore depends chiefly on maintaining an accurate reference voltage; maintaining the resistance ratio R_f/R (and the $R : 2R$ ratio) constant to 0.1% is relatively easy.

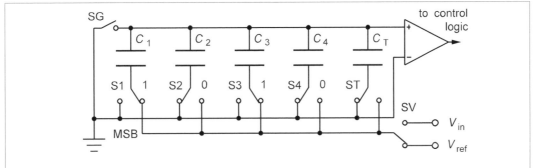

Figure 11.24 An ADC using switched capacitors. Only the capacitance ratio matters, not the actual values, so the circuit is well suited to IC manufacture

The implementation of an ADC is best achieved with a switched capacitor ladder network as in Figure 11.24. The capacitances are made so that $C_1 = 2C_2 = 4C_3 = 8C_4$ and $C_T = C_4$. The capacitance of the whole in parallel is then $2C_1$. At first SV is switched to v_{in} (the voltage to be sampled) and SG to ground allowing the capacitors to acquire a charge of $2C_1v_{in}$. SG is then opened and after that S1 to ST are all switched to ground. The capacitors cannot discharge, so the potential at the top plates is $-v_{in}$. SV is next set to V_{ref} and S1 to V_{ref} too, so that C_1 is in series with the rest of the parallel capacitors, which also total to C_1. The potential at the top plate of the capacitors is then $\frac{1}{2}V_{ref} - v_{in}$. If this is positive, then $v_{in} < \frac{1}{2}V_{ref}$, and S1 must be set to ground and S2 connected to V_{ref}. The detector senses the polarity of the top plates and controls the logic for the switches. Then C_2 is in series with all the rest of the parallel capacitors, that is $\frac{1}{2}C_1$ in series with $(C_1 + \frac{1}{4}C_1 + \frac{1}{8}C_1 + \frac{1}{8}C_1) = 1.5C_1$. The voltage on the top plates would then be increased by $\frac{1}{4}V_{ref}$. If the resulting voltage is positive S2 is switched back to ground and S3 is tried.

The voltage on the top plate is progressively reduced to zero, or within $\frac{1}{8}V_{ref}$ of it. In our example, where $v_{in} = 3.3$ V and $V_{ref} = 5$ V, the first switch will be set to V_{ref} and the voltage at the top plates will be $-3.3 + 2.5 = -0.8$. S2 will produce a voltage increment of $1\frac{1}{4}$ V, so as this is too big, S2 is set to ground and S3 switched in to increase the top plate voltage by $\frac{5}{8}$ V to -0.175 V. S3 is left connected to V_{ref} and S4 is switched so that it increments the top plate by $\frac{5}{16}$ V, which is too large. At the end of the switching process the switches set to V_{ref} are S1 and S3, so the digital representation in 4-bit form of 3.3 V is 1010.

Voltage-to-frequency (V/F) converters

These devices are special types of ADCs which also work in reverse as frequency-to-voltage (F/V) converters. The output is either a voltage which is directly proportional to the input frequency or vice versa. They can be found as 8, 14 or 16-pin plastic or ceramic DIL packages and can be comparatively expensive if high linearity and high frequencies are required (about £10, though devices with poorer performance can be found for £2). Cheaper models work at up to 100 kHz and more expensive ones at up to several MHz. Their uses are legion; almost any instrument which has a voltage output from a sensor that is proportional to the measured property (temperature, magnetic field, pressure, etc) can be made with a digital display in this way. Since the linearity even of poor V/F converters is better than 0.1%, they do not introduce large errors. There are also special-purpose tachometer ICs that convert input pulses to analogue voltages.

Suggestions for further reading

Analysis and design of analog integrated circuits by P R Gray *et al.* (Wiley, 4th ed., 2001)
Microelectronic circuits by A S Sedra and K C Smith (OUP USA, 4th ed., 1997) Excellent.
Analog integrated circuit design by D Johns and K W Martin (Wiley, 1996)
Design of analog integrated circuits and systems by K Laker and W Sansen (McGraw-Hill, 1994)

Problems

1. The circuit of Figure P11.1 is Widlar's current-source, used in the 741 op amp (originally Widlar's design). If $I_1 = 3$ mA, what must R be to make $I_2 = 5$ µA at 300 K? What will I_2 become then if the temperature goes up to 350 K? (Take V_{BE1} to be 0.7 V and calculate V_{BE2} from Equation 7.1) *[R = 33 kΩ, I_2 = 5.7 µA]*

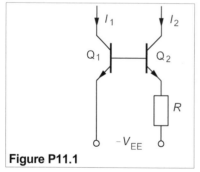

Figure P11.1

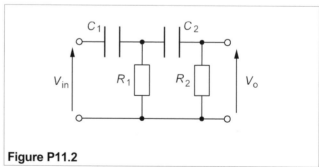

Figure P11.2

2. Show that the normalized frequency response of the filter of Figure P11.2 is given by

$$\mathbf{H}(j\Omega) = \frac{-\Omega^2}{1 - \Omega^2 + j\Omega/Q_p}$$

where $\Omega = \omega/\omega_c$, $\omega_c^2 = 1/R_1C_1R_2C_2$ and $Q_p = 1/\omega_c(R_1C_1 + R_2C_2 + R_1C_2)$. Hence show that Q_p(max) $= 1/2$, and then the normalized amplitude response is

$$H(j\Omega) = \Omega^2/(1 + \Omega^2)$$

and that at the cut-off frequency this must always be < -6 dB.

3. What must L and C be in an RLC filter with the same frequency response as the filter in the previous problem, if $f_c = 1$ kHz, $R = 1$ kΩ and $Q_p = 1/2$. What are the exact and approximate responses in dB at 300 Hz? *[318 mH, 79.6 nF. -21.7 dB, -20.9 dB]*

4. Show that the Sallen-and-Key filter of Figure P11.4 has a frequency response given by

$$\mathbf{H}(j\Omega) = \frac{-jK\Omega}{(1 - \Omega^2) + j\Omega/Q_p}$$

where $\Omega = \omega/\omega_0$, $\omega_0 = \sqrt{2}/RC$, $Q_p = \sqrt{2}/(4 - K)$, $K\sqrt{2} = (1 + R_1/R_2)$.

5. Using the results of the previous problem, suggest preferred component values to produce a Sallen-and-Key bandpass filter which has a Butterworth response and a centre frequency at 2 kHz. What is the mid-band gain? What will the filter's response (in dB) be at 20 kHz and at 50 Hz? If the mains pick-up at the filter input is 20 mV, what will it be at the output? *[+3 dB, -14 dB, -26 dB, 1 mV]*

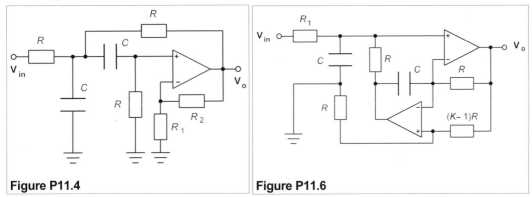

Figure P11.4   **Figure P11.6**

6. The circuit of Figure P11.6 is a Fliege filter. Show that its frequency response is

$$\mathbf{H}(j\Omega) = \frac{jK\Omega/Q_p}{(1 - \Omega^2) + j\Omega/Q_p}$$

where $\Omega = \omega/\omega_0$, $\omega_0^2 = (K - 1)/R^2C^2$ and $Q_p^2 = (K - 1)(R_1/R)^2$. Suggest preferred component values to achieve Butterworth response, a centre frequency, $f_0 = 500$ Hz, and a mid-band gain of 20 dB.

7. The circuit of Figure P11.7 is an Åkerberg-Mossberg filter. Show that its responses are given by

$$\frac{\mathbf{V}_1}{\mathbf{V}_{in}} = \frac{-jK\Omega/Q}{(1 - \Omega^2) + j\Omega/Q} \quad \text{and} \quad \frac{\mathbf{V}_2}{\mathbf{V}_{in}} = \frac{1}{(1 - \Omega^2) + j\Omega/Q}$$

where $\Omega = \omega/\omega_c$, $\omega_c = 1/\sqrt{(R_3R_4C_1C_2)}$, $K = R_2/R_1$ and $Q^2 = C_1R_2^2/C_2R_3R_4$. (Note that the output of OA3 is just $-\mathbf{V}_2$.) Choose suitable preferred component values to make a bandpass filter centred on 12 kHz with a midband gain of 10 dB, and a lowpass filter which has a 10 dB attenuation at 20 kHz.

8. A Colpitts oscillator is a Hartley oscillator with an inductor instead of a capacitor and capacitors in place of inductors (see example 11.2). It is to work at 100 kHz with a feedback ratio of 1/20 into a 50Ω load. What are C_1 and C_2 if $L = 100$ μH? If the transistor has $h_{ie} = 0.04/I_B$ and $h_{fe} = h_{FE}$, what is the minimum value for I_C? What effect will temper-ature changes have on stability and frequency? *[$C_1 = 26.6$ nF, $C_2 = 0.53$ μF. 16 mA]*

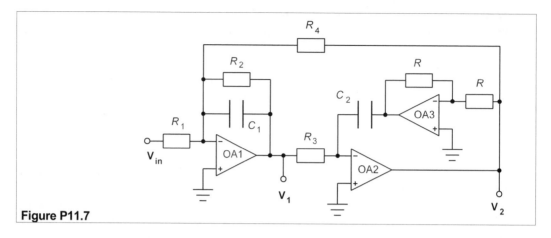

Figure P11.7

9. The circuit of Figure P11.9 is a Wien-bridge oscillator. Find two different expressions for V_o in terms of V_A. Hence prove that the conditions for the circuit to have an output are $K = 1 + R_2/R_1 + C_1/C_2$ and $\omega^2 = 1/R_1R_2C_1C_2$. These are the conditions for oscillation.

10. The circuit of Figure P11.10 is a phase-shift oscillator. Show that the circuit has an output only if $R_1 = 12R$ and $\omega = 1/(RC\sqrt{3})$.

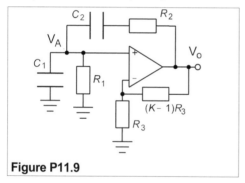

Figure P11.9

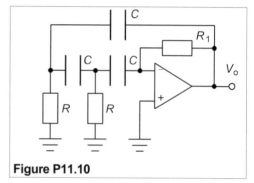

Figure P11.10

11. A PLL is to be built using the CD4046 device shown in Figure 11.18, and is to have a centre frequency of 500 Hz, a lock range of 500 Hz and a capture range of 300 Hz. Choose the components necessary for this, assuming $V_{DD} = 9$ V, making sure all the resistances are as small as possible but greater than 10 kΩ. If $K_v = 1$ at 750 Hz, $K_O = 6$ krad/V and $K_D = 1$ mV/rad, what must the low-pass filter's gain be at 750 Hz, taking $R_4 = 0$? What component values are required for this?

12   Power amplifiers, power supplies and batteries

A MPLIFIERS have already been discussed on several occasions in this work – FET amplifiers, BJT amplifiers and operational amplifiers. However, the loads driven by these were such that the power delivered was low, just as well in view of their low efficiencies. High efficiencies are vital for power amplifiers, not because of the cost of the wasted power itself, but because it produces heat that must be removed, causing inefficient amplifiers not only to be bulkier than need be but also more expensive. Power amplifiers are divided into classes according to the duty cycle of the output transistors, that is the proportion of the time they spend in the active parts of their output characteristics; the higher the duty cycle, the lower the efficiency. But the most rapidly developing power source is the battery, driven by the consumers demand for mains-free operation of more and more electrical equipment. A move towards greater power density than batteries can provide is leading to significant progress in fuel cells of very low power (a few W), as well as MW installations in the USA. Considerable work has also been done on fuel-cell-powered cars, which are non-polluting and need no hydrocarbons, but the sales of fuel sales in 2001 were only £150M.

12.1 Class-A amplifiers

Class-A amplifiers are those in which the transistor at the output stage is continually operating in its active region, never being switched off: it is said to have a duty cycle of 100%. Certain advantages accrue from this mode of operation, but they are more than offset, so far as all but low-power amplifiers are concerned, by low efficiency. The common-emitter and common-source amplifiers seen in Chapters 8 and 9 are both class-A. There are three ways of using the CE amplifier, each having different maximum efficiencies. The first considered is

12.1.1 The series-fed common-emitter amplifier

In this amplifier the load is considered to be the collector resistor, $R_C \equiv R_L$, as in Figure 12.1a. The biasing resistors are assumed to consume negligible power. The power from the collector supply, $P_S = V_{CC} i_C$, where $i_C = I_{CQ} + I_m \sin \omega t$ (see Figure 12.1b). Then $P_S = V_{CC} I_{CQ}$ as the average value of $V_{CC} I_m \sin \omega t$ is zero. R_L consumes instantaneous power $p_L = i^2 R_L$. The r.m.s. value of $I_m \sin \omega t$ is $I_m / \sqrt{2}$, so the average load power is

$$P_{\mathrm{L}} = (I_{\mathrm{CQ}}^2 + I_{\mathrm{m}}^2/2)R_{\mathrm{L}} \tag{12.1}$$

Now the efficiency of the amplifier is given by

$$\eta = \frac{\text{signal power in load}}{\text{total power supplied}} = \frac{I_{\mathrm{m}}^2 R_{\mathrm{L}}/2}{V_{\mathrm{CC}}I_{\mathrm{CQ}}} \tag{12.2}$$

The greatest possible value of I_{m} is I_{CQ}, making the efficiency

$$\eta = I_{\mathrm{m}}^2 R_{\mathrm{L}}/2V_{\mathrm{CC}}I_{\mathrm{CQ}} = I_{\mathrm{CQ}}R_{\mathrm{L}}/2V_{\mathrm{CC}} \tag{12.3}$$

while I_{CQ} is at most $V_{\mathrm{CC}}/2R_{\mathrm{L}}$. Substituting for I_{CQ} in Equation 12.3 leads to

$$\eta = \frac{(V_{\mathrm{CC}}/2R_{\mathrm{L}})R_{\mathrm{L}}}{2V_{\mathrm{CC}}} = \frac{1}{4} = 25\% \tag{12.4}$$

Only the amplified signal power is useful power in the load, the rest is DC quiescent power, so the useful power is just $\tfrac{1}{2}I_{\mathrm{m}}^2 R_{\mathrm{L}}$. Some distortion occurs under these conditions; the efficiency will be even lower for an undistorted signal.

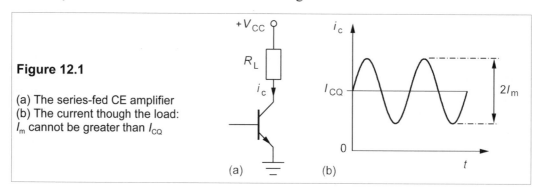

Figure 12.1

(a) The series-fed CE amplifier
(b) The current though the load: I_{m} cannot be greater than I_{CQ}

12.1.2 *The capacitively-coupled CE amplifier*

We looked earlier at this in the section on BJTs, but we did not consider its efficiency then. Figure 12.2a shows the output side (once more we ignore the small losses on the input side) and Figure 12.2b the DC and AC load lines.

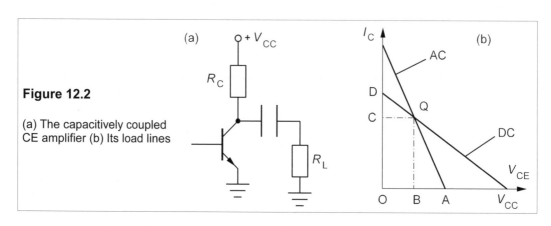

Figure 12.2

(a) The capacitively coupled CE amplifier (b) Its load lines

The greatest output voltage swing will be obtained when the Q-point is in the middle of the AC load line. The output resistance is $R_C // R_L$, or R_L', making the AC load line's slope $-1/R_L'$. Looking at ΔABQ in Figure 12.2b, we see that $BQ = I_{CQ}$, and because the slope is $BQ/AB = 1/R_L'$, then $AB = I_{CQ}R_L' = V_m$, the peak output voltage swing. The slope of the DC load line is $-1/R_C$, which means that in ΔCDQ, $CD/CQ = 1/R_C$. But $OC = I_{CQ}$ and $CD = OD - OC = V_{CC}/R_C - I_{CQ}$. Thus $CQ = CDR_C = V_{CC} - I_{CQ}R_C = OB = AB$, if Q is in the mid-point of the AC load line, and therefore

$$V_{CC} - I_{CQ}R_C = I_{CQ}R_L' \tag{12.5}$$

which leads to

$$I_{CQ} = V_{CC}/(R_C + R_L') \tag{12.6}$$

The power from the source is $P_S = V_{CC}I_{CQ}$ as before. All of the current through the load is signal (useful) current this time and is given by $V_o^2R_L$, where $V_o = V_m/\sqrt{2} = I_{CQ}R_L'/\sqrt{2}$. Thus the ratio of useful power to power input is

$$\eta = \frac{V_m^2/2R_L}{V_{CC}I_{CQ}} = \frac{I_{CQ}^2R_L'^2}{2R_LV_{CC}I_{CQ}} \tag{12.7}$$

But Equation 12.6 gives $I_{CQ}/V_{CC} = 1/(R_C + R_L')$, so putting this into Equation 12.7 produces

$$\eta = \frac{R_L'^2}{2R_L(R_C + R_L')} = \frac{R_LR_C}{2(R_C + R_L)(R_C + 2R_L)} \tag{12.8}$$

where R_L' has been replaced by $R_CR_L/(R_C + R_L)$. From this it can be shown (see Problem 12.1) that the maximum efficiency of the capacitively coupled CE amplifier is only 8.6%. It is left as an exercise (Problem 12.2) to show that the efficiency is 8.3% when R_C is adjusted to give maximum load power. When incorrectly biased, and with reduced gain to lessen distortion, the efficiency might be only 1–2%. The third method of feeding the load, discussed next, results in the highest efficiency.

12.1.3 The transformer-coupled CE amplifier

Connecting the load to the collector circuit via a transformer, as in Figure 12.3a, the DC load line is vertical (or nearly so, we ignore the small resistance of the primary). The AC load line then has a slope of $-1/R_L'$, where R_L', the transformed load resistance, is

$$R_L' = R_LN_1^2/N_2^2 = n^2R_L \tag{12.9}$$

N_1 is the number of turns on the primary (collector) side and N_2 the number on the secondary (load) side, while $n \,(= N_1/N_2)$ is the *turns ratio* of the transformer.

Transformers are discussed in Chapter 16, but here we shall just say that $V_1 = nV_2$ and consider the transformer to be 100% efficient: all the power going into its primary comes out of its secondary. Now the quiescent collector voltage must be V_{CC}, since at DC the collector is connected directly to V_{CC}, so the output voltage swing must be $\pm V_{CC}$ as shown in Figure 12.3b. The power into the load is $\frac{1}{2}V_2^2/R_L$ (V_2 is peak, not r.m.s. voltage), and this is $\frac{1}{2}V_1^2/n^2R_L = \frac{1}{2}V_{CC}^2/R_L'$ (as $V_2 = V_1/n = V_{CC}/n$ and $R_L' = n^2R_L$). The power supplied to the circuit is $V_{CC}I_{CQ}$, which means that the efficiency is

$$\eta = \frac{V_{CC}^2/2R_L{}'}{V_{CC}I_{CQ}} = \frac{V_{CC}}{2I_{CQ}R_L{}'} = \frac{1}{2} \qquad (12.10)$$

because from Figure 12.3b it can be seen that $I_{CQ} = V_{CC}/R_L{}'$, hence $I_{CQ}R_L{}' = V_{CC}$. The maximum efficiency of the transformer-coupled CE amplifier is a fairly respectable 50%. Transformers give flexibility in load resistance as the turns ratio can be adjusted for maximum efficiency.

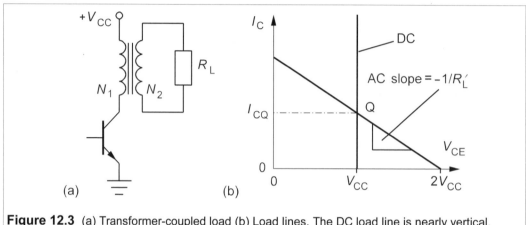

Figure 12.3 (a) Transformer-coupled load (b) Load lines. The DC load line is nearly vertical. $R_L{}' = n^2R_L$, where $n = N_1/N_2$

There are some caveats: the transformer primary carries the DC quiescent collector current, which must not be great enough to saturate the core lest the transformer ceases to function properly. Unfortunately the saturation current is proportional to the size of the transformer and large output transformers are costly. The maximum collector voltage is $2V_{CC}$; the transistor must be rated to take this. For a given power output the product of $I_{sat}V_{CE\,max}$ is constant; a low saturation current results in a high transistor voltage rating and vice versa. High voltage transistors used to be disproportionately expensive, but advances in technology have considerably reduced their cost: the 400V, 150W BUH150 npn transistor in a TO220 package costs only 88p in single quantities.

Example 12.1

The transistor in a transformer-coupled CE amplifier has V_{CE}(max) = 60 V and the transformer primary can take a maximum of 0.7 A DC. If the load is 8 Ω what must the turns ratio of the transformer be to give maximum efficiency?

    Taking a risk we could make $V_{CC} = \frac{1}{2}V_{CE}$(max) = 30 V, but it would be wiser to make this less. Suppose we take V_{CC} = 28 V, then the power supplied will be 28 × 0.7 = 19.6 W. At maximum efficiency the load power will be half this, 9.8 W. The load power is $\frac{1}{2}V_2{}^2/R_L$, where V_2 is the peak secondary voltage. Thus $V_2{}^2/16 = 9.8$ and $V_2 = 12.5$ V. Since the peak primary voltage, $V_1 = V_{CC} = 28$ V, $V_1/V_2 = 28/12.5 = 2.24 = n$. The turns ratio is 2.24:1, with the greater number on the primary. The load has been transformed into an effective load of $n^2R_L = 2.24^2 \times 8 = 40\ \Omega$ on the primary side, which is just $V_{CC}/I_{1\,max}$.

12.2 **Class-B amplifiers**

Class-B amplifiers are often called *push-pull* amplifiers because current is pushed into the load on the positive half-cycle and pulled out of it on the negative by a pair of matched, complementary npn and pnp transistors; each transistor conducts for just less than 50% of the time. The standard audio amplifier belongs to class B.

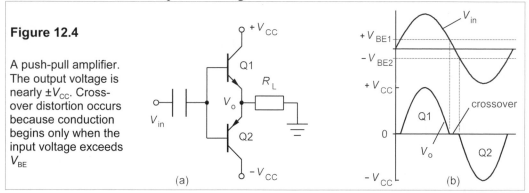

Figure 12.4

A push-pull amplifier. The output voltage is nearly $\pm V_{CC}$. Cross-over distortion occurs because conduction begins only when the input voltage exceeds V_{BE}

The prototype push-pull amplifier of Figure 12.4a uses two power supplies and is seen to comprise two emitter followers. It suffers from *crossover* distortion caused by the sharp reduction in current near the 0 V output level as shown in Figure 12.4b. By connecting the transistor bases with two diodes in series crossover distortion is much reduced as the base-emitter junctions are then in conduction at 0 V input, and since the transistors now have duty cycles of just over 50% this is sometimes called class-AB operation, but it differs little from class B. The diodes should match closely the base-emitter junctions' characteristics and be mounted on the transistors' cases to track their temperatures as closely as possible. A biassing network can be used to permit operation from a single DC power supply, giving an output voltage swing of $\pm\frac{1}{2}V_{CC}$. This is shown in Figure 12.5.

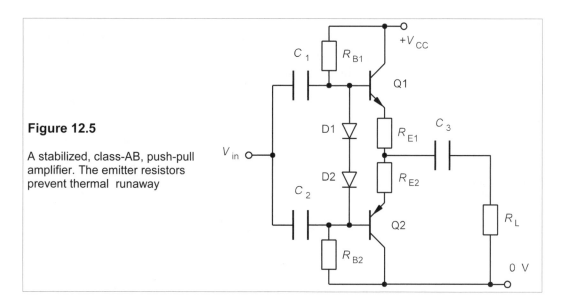

Figure 12.5

A stabilized, class-AB, push-pull amplifier. The emitter resistors prevent thermal runaway

The emitter resistors are necessary to counteract thermal instability. The single voltage source is only on when Q1 conducts and is cut off when Q2 conducts, the necessary energy during the latter half cycle coming from the coupling capacitor, C_2. The average power delivered from V_{CC} is $V_{CC}I_{Lm}/\pi$. (The time average of a sinewave is $2/\pi$, and V_{CC} is on half the time and I_{Lm} is the peak load current.) The average power in the load will be $I_L^2R_L = \frac{1}{2}I_{Lm}^2R_L$. Thus the maximum efficiency (if $R_L \gg R_E$) is

$$\eta = \frac{I_{Lm}^2R_L/2}{V_{CC}I_{Lm}/\pi} = \frac{\pi I_{Lm}R_L}{2V_{CC}} \tag{12.11}$$

Now the maximum value of $I_{Lm}R_L$ (the peak load voltage) is $\frac{1}{2}V_{CC}$ and substituting this into Equation 12.11 leads to a maximum efficiency of $\pi/4$ or 78.5%.

Example 12.2

A class-B push-pull amplifier as in Figure 12.5 has $R_L = 8\ \Omega$, $R_E = 1\ \Omega$ and $V_{CC} = 60$ V. What is the maximum average load power? What is the maximum efficiency? What must C_2 be if the lowest frequency to be amplified is 20 Hz?

The peak load current, I_{mL}, is given by $\frac{1}{2}V_{CC}/(R_E + R_L)$, which is 30/9 = 3.33 A. The maximum average load power is $\frac{1}{2}I_{Lm}^2R_L = 0.5 \times 3.33^2 \times 8 = 44.4$ W. The average power supplied is then $V_{CC}I_{Lm}/\pi = 63.7$ W, giving a maximum efficiency of 44.4/63.7 = 70%. Note that under standby conditions, the voltage drop across the emitter resistors will be almost zero, so that I_{CQ} is almost zero and the standby power consumption is almost zero. Since 20 Hz is the designed lower cut-off frequency, which is given by $f_c = 1/[2\pi C_2(R_E + R_L)]$, then

$$C_2(min) = 1/[2\pi f_c(R_E + R_L)] = 1/(2\pi \times 20 \times 9) = 884\ \mu F \tag{12.12}$$

This coupling capacitor is fairly large.

12.2.1 The Darlington transistor

Two drawbacks of the circuit of Figure 12.5 are the low input resistance of the transistors and the fairly low βs of power transistors, deficiencies which are overcome by using Darlingtons instead of single transistors. Though they are made up of two BJTs as in Figure 12.6a, they are sold in a single package with three terminals and can be used just like any BJT.

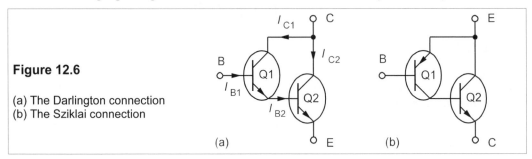

Figure 12.6

(a) The Darlington connection
(b) The Sziklai connection

In Figure 12.6a we see that $I_{E1} \approx I_{C1} = I_{B2}$, so that the overall beta is

$$\beta_{DP} \approx I_{C2}/I_{B1} = \beta_1 I_{C2}/I_{C1} \approx \beta_1 I_{C2}/I_{B2} = \beta_1\beta_2 \tag{12.13}$$

β values for single transistors can range from 20 to 300 or more, so that β_{DP} might be anything from 400 to 100 000. For example the BDX33 npn transistor (and its complementary pnp BDX54) has an h_{fe} ($\approx \beta$) of 750+. It can be operated from a 45 V supply at a maximum current of 10 A and has a power dissipation rating of 70 W, yet costs only about 60p.

The input resistance of Q1 is $h_{ie1} = V_{th}/I_{B1}$ and its output current is $I_{E1} = (\beta_1 + 1)I_{B1}$. Thus the input resistance of Q2 is amplified by a factor of $\beta_1 + 1$, so the overall input resistance is

$$h_{ieDP} = h_{ie1} + (\beta_1 + 1)h_{ie2} \tag{12.14}$$

But $h_{ie2} = V_{th}/I_{B2} = V_{th}/I_{E1} \approx V_{th}/\beta_1 I_{B1} = h_{ie1}/\beta_1$, so that Equation 12.14 becomes

$$h_{ieDP} \approx h_{ie1} + \beta_1 h_{ie2} \approx h_{ie1} + \beta_1 h_{ie1}/\beta_1 = 2h_{ie1} \tag{12.15}$$

The output resistance of a Darlington pair is very low, since it is an emitter follower with

$$r_{eDP} = h_{ieDP}/\beta_{DP} \approx 2h_{ie1}/\beta_1\beta_2 \tag{12.16}$$

Example 12.3

A Darlington transistor has $I_{CQ} = 1$ A, while its component transistors have $h_{fe1} = 300$ and $h_{fe2} = 25$. Find the input resistance, the forward current transfer ratio, β_{DP} and the output resistance. (Take $V_{th} = 30$ mV.)

Taking $h_{fe} = \beta$ and using Equation 12.13 we find $\beta_{DP} = 300 \times 25 = 7500$. The collector current of the pair is essentially just I_{C2} since $I_{C1} \ll I_{C2}$. Now $I_{B1} \approx I_{C2}/\beta_1\beta_2 \approx I_{CQ}/\beta_1\beta_2 = 1/(300 \times 25) = 0.133$ mA. Then the input resistance, $h_{ieDP} = 2h_{ie1} = 2V_{tn}/I_{B1} = 2 \times 30/0.133 = 450$ Ω, working in mV and mA. $\beta_{DP} \approx \beta_1\beta_2 \approx h_{fe1}h_{fe2} = 300 \times 25 = 7500$. The output resistance is $r_{eDP} \approx h_{ieDP}/\beta_{DP} = 450/7500 = 0.06$ Ω. Thus a source of moderate internal resistance can drive loads of very low resistance via this Darlington.

Instead of a complementary pnp Darlington, the connection of Figure 12.6b is often used. It is sometimes called the Sziklai connection and sometimes a pseudo-complementary Darlington. It behaves exactly like a pnp transistor with the terminals as labelled. Sziklai-connected pairs are usually listed as complementary pnps; they are easier to make than true complementary pnp Darlingtons, and have only one V_{BE} drop and not two.

12.3 Class-C amplifiers

Class-C amplifiers operate with their output transistors in conduction for less than half the time, so raising the efficiency to between 78.5% and almost 100%. The usual class-C amplifier has a parallel RC circuit (a tank circuit) connecting the transistor collector to the DC supply as shown in Figure 12.7a. The tank circuit supplies power to the load when the transistor is not conducting. V_{BB} is a negative base-biassing voltage which controls the conduction angle, θ_c. When the input voltage rises above $V_{BE} + |V_{BB}|$ the transistor conducts and refills the tank circuit. Because the Q-factor is high, the frequency is restricted to the tank circuit's resonant frequency, and input and bias voltages are adjusted to drive the transistor just into saturation when the input voltage is maximum.

The transistor conducts when the phase angle of the input is

$$\sin\phi = (V_{BE} + |V_{BB}|)/V_m \tag{12.17}$$

as shown in Figure 12.7b, and conduction continues until the input's phase angle is $\phi - \pi$ radians. The transistor conducts for a total of θ_c rads/cycle where $\theta_c = \pi - 2\phi$.

The tank circuit oscillates between $+V_{CC}$ and $-V_{CC}$ so that the peak load current is V_{CC}/R_{Ceq}, where R_{Ceq} is the total resistance in the collector circuit, that is $R_L \mathbin{/\mkern-5mu/} Q_c^2 r$, where Q_c is the Q-factor of the tank coil and r is its series resistance. The transistor waveform is a series of pulses the same as I_C in Figure 12.7b, but this is filtered by the tank circuit to give a sinewave output of angular frequency, $\omega_0 = 1/\sqrt{(LC)}$. Since the transistor waveform is not sinusoidal, all the Fourier components of it are lost in the filtering, which reduces efficiency.

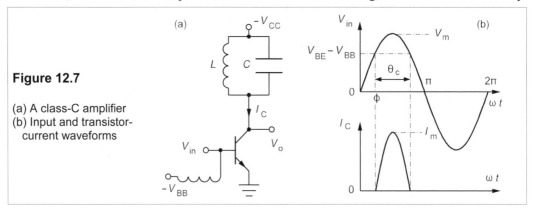

Figure 12.7

(a) A class-C amplifier
(b) Input and transistor-
    current waveforms

The output power depends on the ratio, γ, of the peak of the fundamental component, I_{fm} to the peak transistor current, I_m, which is given approximately by

$$\gamma = I_{fm}/I_m \approx 0.2\theta_c + 0.001\theta_c^2 - 0.00446\theta_c^3 \tag{12.18}$$

for $0 < \theta_c < \pi$. The load power is then $P_L = \tfrac{1}{2}I_{fm}V_{CC} = \tfrac{1}{2}\gamma I_m V_{CC}$. The power from the source is $P_S = I_{AV}V_{CC}$ and I_{AV} is given by

$$I_{AV} = (I_m/\pi)(\theta_c/\pi) = I_m\theta_c/\pi^2 \tag{12.19}$$

Thus the source power is

$$P_S = I_{AV}V_{CC} = \theta_c I_m V_{CC}/\pi^2 \tag{12.20}$$

and the maximum efficiency is (using Equation 12.18)

$$\eta = \frac{P_L}{P_S} = \frac{0.5\gamma I_m V_{CC}}{\theta_c I_m V_{CC}/\pi^2} = \frac{\pi^2\gamma}{2\theta_c} \approx 1 + 0.005\theta_c - 0.0234\theta_c^2 \tag{12.21}$$

which ranges from 1 when $\theta_c = 0$ to 0.785 when $\theta_c = \pi$ and the amplifier is then class B.

The class-C amplifier can be adapted for amplitude modulation (see Section 24.2) by means of the circuit shown in Figure 12.8a, and transformer-coupled loads can be driven by the circuit of Figure 12.8b. The latter circuit has no base bias, and the AC input is capacitively coupled and clamped by the base-emitter junction to V_{BE} above ground.

Example 12.4

A transformer-coupled class-C amplifier such as that in Figure 12.8b has $V_{CC} = 60$ V. It is drives an 8 Ω load via a 15:1 step-down transformer with a primary resistance of 6 Ω and $Q = 50$ at the tank circuit's resonant frequency of 150 kHz. V_{in} is sinusoidal of r.m.s. value 3 V. Calculate (a) C and L in the tank circuit (b) θ_c if $V_{BE} = 0.6$ V (c) the duty cycle (d) the efficiency from Equation 12.21 (e) the load power (f) the source power (g) I_C (peak) (h) the collector current at 150 kHz.

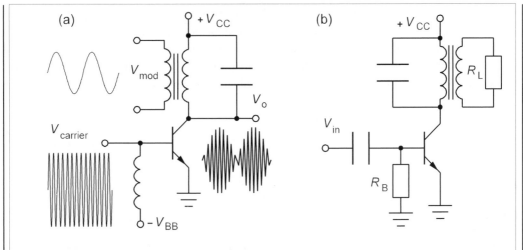

Figure 12.8 Class-C amplifiers (a) Amplitude modulation (b) A transformer-coupled load

(a) We can find the primary inductance from

$$Q = \omega_0 L/R \ \rightarrow \ L = QR/\omega_0 = 50 \times 6/(2\pi \times 150 \times 1000) = 318 \ \mu\text{H} \tag{12.22}$$

Then C is found from

$$\omega_0 = 1/\sqrt{LC} \ \Rightarrow \ C = 1/\omega_0^2 L = 3.54 \ \text{nF} \tag{12.23}$$

(b) The input has a peak value of $V_m = 3\sqrt{2} = 3.55$ V, so the transistor is turned on at a phase of $\phi = \sin^{-1}[(V_m - V_{BE})/V_m] = 56.2° = 0.98$ rads. Then $\theta_c = \pi - 2\phi = 1.18$ rads.

(c) The duty cycle of the transistor is $\theta_c/2\pi = 1.18/2\pi = 0.188 = 18.8\%$.

(d) Using Equation 12.21 the efficiency is 0.973 or 97.3%.

(e) The primary voltage will oscillate between $+2V_{CC}$ and 0 V like a transformer-coupled class-A amplifier, and thus the r.m.s. primary voltage is $V_1 = V_{CC}/\sqrt{2} = 60/\sqrt{2} = 42.4$ V. This makes the load power, $V_1^2/R_L' = 42.4^2/1800 = 1$ W.

(f) The source power should be $P_L/\eta = 1/0.973 = 1.03$ W.

(g) Rewriting Equation 12.20 as $I_m = \pi^2 P_S/\theta_c V_{CC}$ gives $I_m = 0.144$ A and this is the peak collector current.

(h) Using Equation 12.18 we find $I_{fm} = \gamma I_m = 0.23 \times 0.144 = 0.033$ A, or 23.4 mA r.m.s.
Equation 12.21 assumes ther are no losses in the transformer, which is not the case, so that the efficiency calculated here is substantially in excess of the true value of about 90%.

12.4 Class-D amplifiers

In class-D amplifiers the output transistors (usually MOSFETs) are either cut-off or in full saturation, resulting in efficiencies near 100%. With BJT transistors the power lost in the transistor is $I_C V_{CEsat}$, while the load power is $I_C^2 R_L$, so that the maximum efficiency is

$$\eta = \frac{P_L}{P_L + P_{BJT}} = \frac{I_C^2 R_L}{I_C^2 R_L + I_C V_{CEsat}} = \frac{I_C R_L}{I_C R_L + V_{CEsat}}$$

$$= (V_{CC} - V_{CEsat})/V_{CC} = 1 - V_{CEsat}/V_{CC} \tag{12.24}$$

Again, the only losses considered are those in the transistor(s). With $V_{CC} = 30$ V and $V_{CEsat} = 0.4$ V, the maximum efficiency is 98.7%. Most class-D amplifiers use MOSFETs instead of BJTs because they have high input resistances and can switch large currents more rapidly. The output of a class-D amplifier is a square wave, but by using pulse-width modulation (PWM, discussed in Section 24.6) and suitable filtering a fairly good sinusoid can be reconstructed.

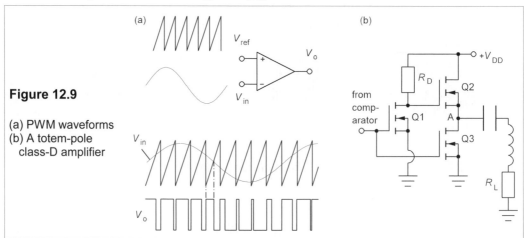

Figure 12.9

(a) PWM waveforms
(b) A totem-pole
class-D amplifier

Figure 12.9a shows how a comparator with a sawtooth wave reference produces a square-wave output whose mark-space ratio is proportional to the amplitude of the input. This output can be fed into the circuit of Figure 12.9b, known as a totem-pole, which employs n-channel enhancement MOSFETs. The load current is made sinusoidal by an LC filter. The output of the comparator is high when the input signal is greater than the sawtooth signal, so that when the input signal is a small voltage the sawtooth wave soon exceeds its amplitude and the output of the comparator goes low for the rest of the sawtooth ramp. Conversely, when the input voltage is large the sawtooth voltage takes longer to exceed it and the output of the comparator stays high for most of the sawtooth ramp. Thus the signal is encoded into a series of pulses whose widths are proportional to the signal voltage. The sawtooth frequency must be about ten times the largest frequency of interest in the input signal. The comparator then switches the output transistors on and off. When the comparator goes high, Q1 and Q3 are switched on (a positive gate voltage induces an n-type layer in the silicon beneath it and the device conducts), while Q2 is off as its gate is connected to the drain of Q1, which is low when Q1 is on: the output is low as a consequence. When the comparator goes low Q2 is switched on while Q1 and Q3 are off and the load is connected to V_{DD}.

12.5 **Power supplies**

It is frequently necessary to supply DC at various voltages to whole or part of electronic equipment from a mains electricity supply at 240 V AC, or similar. The principal elements of a small DC power supply are:

1. Protection devices such as fuses and transient suppressors at the input, and a snubber.
2. A transformer to produce the desired voltage from the 240 V supply.

3. A bridge rectifier and smoothing capacitor producing unregulated DC with some ripple.

4. A voltage regulator to reduce the voltage to the specified value without ripple.

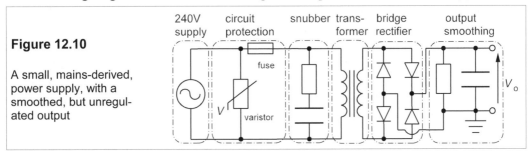

Figure 12.10

A small, mains-derived, power supply, with a smoothed, but unregulated output

Figure 12.10 shows a power supply without an output voltage regulator (discussed in Section 11.3). The fuse protects the user as well as the equipment, and must be rated for a current about 50% higher than the maximum continuously supplied in normal service in order to give an adequate margin. Fuses working near their ratings much of the time will eventually blow. The one shown here should be slow blowing to allow for current surge at switch-on. Transient suppression is needed to remove large voltage spikes of either polarity from the input as most mains supplies have spikes of a few kV from time to time. The usual device is the bidirectional zinc oxide varistor looking like a ceramic capacitor, that conducts in either direction when its terminal p.d. exceeds its rated voltage. Varistors (or transient voltage suppressors, TVS) are cheap and can dissipate safely large amounts of spike energy. For example, the P6KE400CA from Fairchild is rated at 342 V continuous peak voltage, will dissipate 500 W in a 1 ms transient and costs only 47p. Silicon transient suppressors (in effect Zener diodes) are available with ultra-fast response times and can protect MOS devices.

Line filtering is sometimes necessary to remove high-frequency components that are often large near thyristor-controlled power circuits. Because these are usually passive LC filters they attenuate only higher frequencies (typically by 35 to 75 dB from 150 kHz to 30 MHz) and are fairly dear. The price depends on current rating and attenuation specified, but around £10 is usual. A snubber is used to absorb the energy in the transformer when the supply is switching off. They are usually omitted in low-power supplies.

Transformers are discussed more fully in Chapter 16. They consist of an input coil (primary) and output coil (secondary) wound on a ferromagnetic core which links the flux in the two. 50Hz/240V transformers need a high primary inductance to draw as little current as possible on standby, so they are fairly bulky. The weight depends on the power rating, but it works out at about 1 kg/50 VA for small transformers. All things considered, they are reasonably priced: a 20VA transformer with a 240V primary costs only £4. Across the transformer secondary goes a diode bridge to give full-wave rectification, and on the DC side of the bridge is a smoothing capacitor and bleed resistor; the latter only being used in high-voltage DC supplies.

The final stage for low-voltage supplies is often a voltage regulator. These can cause trouble if the unregulated DC is not at least 2 V greater than the regulated DC, *including* ripple *and* at maximum rated current. If the transformer is small and cheap it may drop its voltage by 10% on full load and the supply voltage can fluctuate 5%, making a total of 15%. Allied with the ripple of 10% peak to peak, the voltage from the secondary can be 25% less than nominal. If +9 V is required from the regulator, its unregulated input must be at least 11 V, and the transformer secondary must provide 15 V. A low drop-out device can save a volt or two here.

12.6 Switch-mode power supplies

The transformers used in small mains-derived power supplies are fairly heavy – 1 kg per 50 VA – because the inductance required for 50Hz operation is reasonably large (2 H is a typical value). If the operating frequency can be increased then the transformer weight and volume will decrease roughly in inverse proportion (see Section 16.6). This is one reason – another is their greater efficiency – for using switch-mode power supplies (SMPS) wherein mains-frequency AC is rectified to DC which is switched on and off to give AC at from 100 kHz to 1 MHz. The AC can be input to a small ferrite-cored transformer to produce the voltage required before the final-stage rectification and smoothing.

   The transformer must be designed to take this current without saturating, which often means that the core's permeability must be reduced. In non-toroidal ferrite cores the permeability is reduced by increasing the gap between the core halves with plastic spacers. With toroidal cores a low-permeability material must be used. In Section 16.4, it is shown that the core volume, Λ, is given by

$$\Lambda = \gamma S/f \qquad\qquad (12.25)$$

where γ is a numerical constant, about 2×10^{-3} m^{3}Hz/VA for a small ferrite core, S is the power rating of the transformer and f is the switching frequency. With a ferrite core and $P = 1$ kVA and $f = 100$ kHz, we find $\Lambda = 2 \times 10^{-5}$ m^2 = 100 cm^3, a very small core for such a power.

   With a high-enough switching frequency, very small transformers will suffice for quite a large output power. In consequence switch-mode power supplies weigh much less than conventional ones of the same rating. Switch-mode power supplies are now frequently used for supplies of up to 3 kW, even though they are more complex than conventional power supplies, because while their overall cost is similar, their weight is considerably lessened.

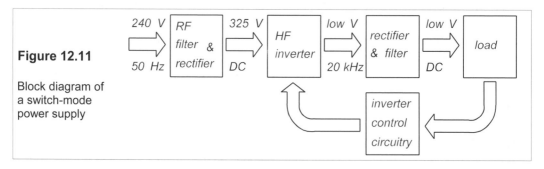

Figure 12.11

Block diagram of a switch-mode power supply

   A block diagram of a SMPS is shown in Figure 12.11, the key features being the high-frequency inverter, which turns DC to AC, and most importantly its associated control circuitry. There are various types of inverter, all of which use a transistor to switch the rectified, smoothed DC on and off and a diode to maintain current flow in the load when the transistor is off, using energy stored in the reactive elements. The inverter shown in Figure 12.12a is known as a *buck converter* or sometimes as a *forward converter*. When the transistor is on it transfers energy to both the inductor and the capacitor (and to the load) by means of a freewheeling diode (see Chapter 7). But when the transistor is off the inductor supplies energy to both the load and the capacitor, assuming that the current in the inductor is at all times greater than zero. The input voltage, V_{in}, from the rectifier and the load voltage,

V_o, are essentially constant if the supply is to be DC. The resulting current and voltage waveforms for the diode and the inductor are shown in Figure 12.12b.

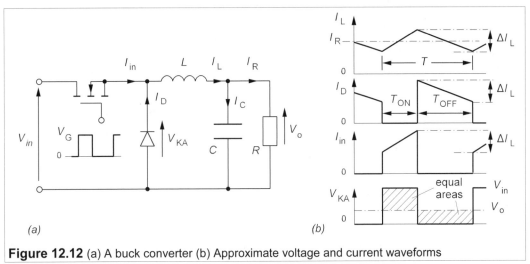

Figure 12.12 (a) A buck converter (b) Approximate voltage and current waveforms

When the transistor is ON, the voltage across the inductor is $V_{in} - V_o$ and so the change in the inductor current if the ON time is T_{ON} is

$$\Delta I_L = \frac{(V_{in} - V_o)T_{ON}}{L} \tag{12.26}$$

by rearranging $V = L\Delta I/\Delta t$. When the transistor is OFF the voltage across the inductor must be V_o (apart from a diode drop) and so if the OFF time is T_{OFF} the current change is

$$\Delta I_L = \frac{V_o T_{OFF}}{L} \tag{12.27}$$

Equating the two expressions for ΔI_L in Equations 12.26 and 12.27 leads to

$$V_o = \frac{V_{in}T_{ON}}{T} = V_{in}\delta \tag{12.28}$$

where the duty cycle, $\delta = T_{ON}/T$, and $T_{ON} + T_{OFF} = T$. Thus the output (load) voltage is proportional to the ON time of the transistor and must always be less than V_{in}

Example 12.5

A buck converter like that of Figure 12.12a is to operate at a switching frequency of 25 kHz with a DC input at 320 V and an output of 40 V across a 2Ω load. What is the ON time for the transistor? What is the value of inductance required if the current through it changes by 4.5 A during charge and discharge? Draw the capacitor current as a function of time. What must C be if the peak-to-peak load-voltage ripple is to be 1%? If the duty cycle can vary from 0–100%, when is the change in inductor current greatest? What is the load voltage and its ripple then?

The ON time for the transistor is, using a rearranged Equation 12.28,

$$T_{ON} = V_o T/V_{in} = V_o/fV_{in} = 40/(25 \times 10^3 \times 320) = 5 \ \mu s \tag{12.29}$$

The change in inductor current, by Equation 12.26, is

$$\Delta I_L = (V_{in} - V_o)T_{ON}/L = (320 - 40) \times 5 /L = 4.5 \text{ A} \tag{12.30}$$

working in μs and μH, we find L = 311 μH.

If the load voltage is constant, so is the load current, I_R, and then by KCL at the top right-hand node of Figure 12.12a, $I_C = I_R - I_L$. Then $I_{Cmax} = I_R + \tfrac{1}{2}\Delta I_L$ and $I_{Cmin} = I_R - \tfrac{1}{2}\Delta I_L$. Thus $I_{Cpp} = \Delta I_L$ and the capacitor current-time graph is as in Figure 12.13, a triangular waveform since the ripple is small and the exponential rise and decay is nearly linear.

Figure 12.13

The *I-t* graph for the capacitor in Example 12.5

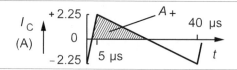

The load voltage is given by

$$V_C(t) = V_R(t) = \frac{1}{C}\int_0^t I_C \, dt \tag{12.31}$$

which is the area under the current-time graph divided by C. The positive part of the current-time graph, shown hatched in Figure 12.13, has an area of

$$A_+ = \tfrac{1}{2}I_{Cmax}T/2 = I_{Cmax}/4f = 0.5I_{Cpp}/4f = I_{Cpp}/8f \tag{12.32}$$

which gives A_+ = 22.5 μC. The peak-to-peak ripple voltage required is $0.01 \times 40 = 0.4$ V. And the increase in voltage during the charging of C is half the peak-to-peak ripple voltage, hence

$$0.5V_{rpp} = A_+/C \;\;\Rightarrow\;\; C = 2A_+/V_{rpp} = 2 \times 22.5/0.4 = 112.5 \text{ μF} \tag{12.33}$$

The inductor current as a function of T_{ON} is found by substituting Equation 12.28 into Equation 12.26 to obtain

$$\Delta I_L = V_{in}(1 - \delta)T_{ON}/L = V_{in}(1 - \delta)T\delta/L = V_{in}(1 - \delta)\delta/fL \tag{12.34}$$

This is greatest when $\delta = \tfrac{1}{2}$ and then ΔI_L = 10.3 A. The load voltage is $V_{in}\delta = 320 \times \tfrac{1}{2} = 160$ V and the average load current must be 160/2 = 80 A. Since the change in inductor current has increased by a factor of 10.3/4.5 = 2.29 and the capacitor current now goes from +5.15 A to −5.15 A, the ripple voltage has also increased by a factor of 2.29 to 0.92 V, which is $0.92 \times 100/160 = 0.575\%$. The percentage load-voltage ripple has decreased with increase in duty cycle (see Problem 12.9).

12.6.1 The switching regulator

The switching regulator shown in Figure 12.14 makes use of a PWM comparator to control the switching transistor. The reference voltage into the comparator is a sawtooth wave whose peak value is V_m. The comparator output goes high and switches on the transistor when $V_{ref}(\text{max}) > V_{R2}$ and so the OFF time is TV_{R2}/V_m and the ON time is thus $(1 - V_{R2}/V_m)T$. The duty cycle is

$$\delta = T_{ON}/T = 1 - V_{R2}/V_m \tag{12.35}$$

Then the output voltage is

$$V_o = V_{in}\delta = \left(1 - V_{R2}/V_m\right)V_{in} \tag{12.36}$$

wherein

$$V_{R2} = V_o R_2/(R_1 + R_2) = V_o/(1 + K) \tag{12.37}$$

and $K = 1 + R_1/R_2$. Substituting for V_{R2} in the expression for V_o leads to

$$V_o = V_{in}(1 - V_o/[(1+K)V_m])$$ (12.38)

This gives the output voltage as

$$V_o = \frac{V_{in}}{1 + V_{in}/[(1+K)V_m]} = \frac{(1+K)V_m}{(1+K)V_m/V_{in} + 1}$$ (12.39)

If $V_m = 10$ V and $V_{in} = 200$ V, then $V_o = 18$ V when $K = 1$ rising to 167 V when $K = 100$.

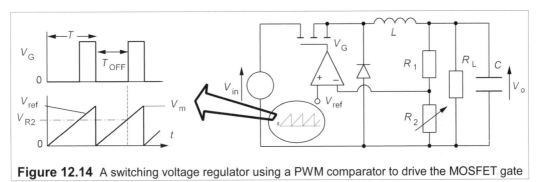

Figure 12.14 A switching voltage regulator using a PWM comparator to drive the MOSFET gate

12.7 **Batteries**

Batteries are sources of electrical power which rely on the reaction of chemical components within an electrochemical cell when the external circuit is completed. They are classified as *primary batteries* that are not rechargeable, and *secondary batteries* that are. The worldwide demand for batteries is enormous: sales were estimated at £26 *billion* in 2001 – larger than semiconductor sales, and expanding at 10% per annum. Much of this demand comes from the automotive industry and a significant amount is standby batteries, but the most important sector, amounting to about half of the battery market, is consumer batteries. These are used in all manner of electronic equipment, such as mobile telephones, TV remote controls, laptop computers, camcorders, timepieces, torches, heart pacemakers, radios, calculators, smoke alarms etc. Consumer batteries sales are about equally divided between throw-away primary batteries, with an increasing use of secondary batteries for all purposes, especially mobile telephones and equipment driven by electrical motors in which the power demand is high for relatively short periods. The expansion in battery use has been greatly assisted by the development of ICs and electronic equipment of all kinds. The discussion here is mainly restricted to consumer batteries[1].

12.7.1 *Types of battery*

Table 12.1 lists some of the most popular consumer batteries, which can be classified according to the chemical system which provides the electrical power. The negative electrode is invariably a metal such as zinc (Zn), cadmium (Cd) or lithium (Li), while the positive

---

[1] Strictly speaking a battery comprises two or more cells connected in series or parallel or both, while a cell has just two terminals and cannot be split up further; but in common usage this distinction has long gone.

electrode (the source of conventional current in the external circuit) is an oxidant such as manganese dioxide (MnO_2). The two electrodes are immersed in a conducting electrolyte which in the so-called 'dry' cell is actually a paste. The chemical system is then written with the negative electrode material first, then the electrolyte, then the positive electrode material as in $Zn/ZnCl_2/MnO_2$. Often the electrolyte is omitted as in the Zn/MnO_2 system, which is popularly called zinc-carbon. Most of the commonly-used chemical systems were discovered long ago; the Leclanché cell (Zn/MnO_2) was first described in 1868 and the Planté (or lead-acid, Pb/PbO_2) cell in 1859, but they have been considerably improved, especially in the last decade. Secondary consumer batteries are now mostly the nickel-metal-hydride type, which has rapidly replaced the very robust nickel-cadmium battery.

Table 12.1 *A selection of commonly-used consumer batteries*

Type[A]	Size[B]	Voltage[C]	Capacity[D]	Price[E]	p/Wh[F]
[G]Zn/MnO_2	D	1.5	3.8 (2 Ω)	50p	9
[H]$Zn/ZnCl_2/MnO_2$	D	1.5	3.8 (2 Ω)	67p	12
Alkaline Zn/MnO_2	D	1.5	7.5 (2 Ω)	£1.65	15
'Super' alkaline	D	1.5	10 (2 Ω)	£1.98	13
Alkaline Zn/MnO_2*	D	1.5	7.5 (2 Ω)	£3.00	27
[K]$Cd/NiO.OH$*	D	1.2	5	£6.89	115
mobile phone Ni/Cd	various	3.6	0.7	£6.90	274
[L]Ni/MH*	AA	1.2	1.2	£3.90	271
laptop comp. NiMH	Non-std pack	6	4.2	£47	224
camcorder NiMH	Non-std pack	6	1.8	£24	222
[M]Pb/PbO_2*	157 × 60 × 107H	6	12 (60 Ω)	£13	18
car battery[N]	200 × 165 × 170H	12	45 (5 Ω)	£30	5.6
'Li ion' batteries					
Camcorder Li	various	7.2	1.65	£24	202
Li/V_2O_5*	3.2H × 30Ø	3	0.1	£4.60	1533
$Li/(CF)_x$	3H × 23Ø	3	0.26	£1.54	197
Li/FeS_2	AA	1.5	2.5 (3.9 Ω)	£3	80
Li/MnO_2	PP3	9	1.2 (75 Ω)	£6.44	60
$Li/SOCl_2$	D	3.5	16.5 (25 Ω)	£15	26
Zn/air	5.4H × 7.9Ø	1.4	0.21 (250 Ω)	78p	265
	26 × 17 × 45H	8.4	1.5 (500 Ω)	£7	56
Zn/Ag_2O	5.4H × 11.6Ø	1.5	0.2 (6.5 kΩ)	£1.23	410

Notes: * = Rechargeable [A] –ve electrode active material/+ve electrode active material or –/electrolyte/+ [B] Sizes in mm. H = height, Ø = diameter of cylinder. The standard D (R20) cell is 57.2H × 31.8ɸ and the AA (R6) cell is 47.8H × 13.5Ø; the PP3 is 26.5 × 17.5 × 48.5H. [C] The working voltage may be considerably less [D] In Ah with load in brackets [E] Single quantities [F] Price ÷ (Ah capacity × nominal voltage) [G] Zinc-carbon or Leclanché cells [H] Chloride battery [K] Ni/Cd or Nicad [L] MH = metal hydride [M] Lead-acid [N] For comparison, not strictly a 'consumer' battery

Batteries are an expensive way to provide power unless they are very large and can be repeatedly recharged many times. The choice of battery depends on how and where it is used: the maximum power and current, the required voltage, the duty cycle, the service temperature and the shelf life. A heavy-duty battery should not be used where a light-duty one would suffice, since it will cost more and will offer little better performance if lightly discharged. For example alkaline 'heavy-duty' batteries will repay their extra cost when delivering 1–2 A cont-inuously, but not when delivering 50 mA intermittently. Rechargeable batteries are especially dear and should only be used when frequently-repeated recharging is required, as in portable power tools and shavers. Remember that the cost of the charger has to be recovered and that the rechargeable batteries themselves can only undergo a certain number of recharge cycles. As a rough guide, if you need to replace a battery every month and twenty times over is it worthwhile buying rechargeable batteries.

The development of the Leclanché cell has been continuous since its invention in 1868. In fact the D-sized zinc-carbon battery of today has about eight times the capacity it had in 1920 and the alkaline-manganese battery has tripled its capacity in the last 16 years. A great deal of money is now being spent developing rechargeable, high-power-density batteries for portable electrical equipment, especially mobile telephones, laptop computers, camcorders and power tools. Compared to integrated circuits the pace of improvement of batteries is slow but steady.

The most recent commercial development has been the used of nickel-metal hydride rechargeable batteries instead of nickel-cadmium. Not only does this replace cadmium – a potentially-harmful heavy metal pollutant – with a safe alternative, but it also offers a 50% increase in capacity for a given size of battery. Lead-acid batteries, though rechargeable, are less robust than Nicad batteries, and will not last long if deeply discharged, which is why they have been less popular as consumer batteries. This may change soon.

In many systems the battery e.m.f. declines continuously during discharge; for example the Leclanché cell's voltage can fall steadily to zero, but is reckoned to have reached its practical discharged state when its e.m.f. has fallen to 0.9 V. Where constancy of e.m.f. is necessary one must choose lithium, zinc-air or zinc-silver oxide batteries. Batteries employing lithium (Li) as the negative electrode have become popular since their introduction in about 1970. There are many different materials used for the positive electrode in lithium batteries and none seems to have achieved commercial dominance. Lithium batteries are generally only used if space is at a premium, since they are relatively expensive. Ultra-thin, flexible lithium/polymer batteries are now used in smartcards and can be made by thin-film technology at low cost. If the battery has to be held in reserve for an emergency it must have a good shelf life: zinc-air batteries can be stored indefinitely without loss of capacity since they only work when exposed to air. Some batteries have rather short shelf lives, especially in hot climates.

12.7.2 Battery characteristics

Among the many characteristics of batteries are the voltage-time discharge curves as a function of temperature and discharge rate, the charge retention as a function of time and temperature and the charging characteristics. All of these vary, even from battery to battery of the same type, though modern production methods have reduced this variation considerably. Some performance curves for the more popular batteries are shown in Figure 12.15, about which a word of caution: they are only a very small sample of huge amounts of manufacturers' data, which should always be consulted when designing battery-powered equipment of any kind. The Figures in Ah are the battery's capacity when discharged through the load indicated in ohms or at the rate indicated in amps.

Figures 12.15a and 12.15b show a big drawback of zinc-carbon or alkaline manganese batteries for powering electronic equipment: their terminal voltages drop continuously during discharge, and so they are usually reserved for driving small motors and lamps. Figure 12.15b also shows the advantage of the alkaline battery over the zinc-carbon battery when the current drain is heavy, an advantage that is almost entirely absent when the current drain is light.

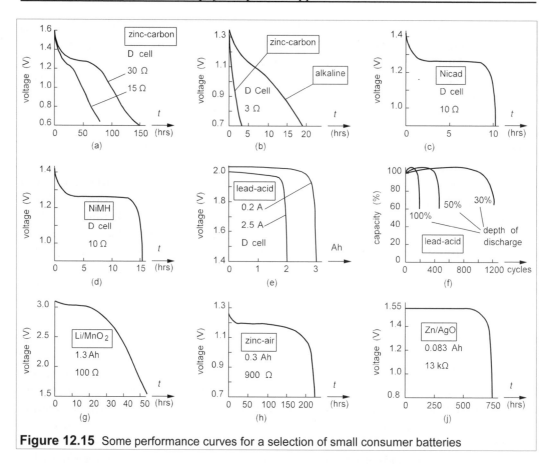

Figure 12.15 Some performance curves for a selection of small consumer batteries

Nicad batteries maintain their voltages much better on discharge than zinc-carbon batteries, as Figure 12.15c shows, but they have only about a quarter of the capacity of even ordinary zinc-carbon cells. The great advantage of Nicad batteries is their rechargeability. Figure 12.15d shows the superior capacity of rechargeable nickel/metal-hydride batteries (NiMH) compared to Nicads, probably an outdated Figure, given the rapid improvement in NiMH batteries.

The other common rechargeable battery is the lead-acid ($Pb/H_2SO_4/PbO_2$) battery, which is now available in sealed form owing to the development of jellied electrolyte. Lead-acid batteries have much greater capacity than Nicads and work at the much higher cell voltage of 2 V; they are also capable of very high discharge rates: a single D-cell can deliver 25 A for a minute or two. Unfortunately, Figure 12.15f shows that if the depth of discharge exceeds about 30%, then the battery life is much reduced: deep (50–100%) discharges must be avoided. Figure 12.15 does not show a further drawback of lead-acid batteries, which is their inability to remain discharged for long without suffering permanent reduction in capacity, in contrast to Nicads which suffer little loss on being left discharged. Lead-acid batteries should therefore always be connected to a charger; their characteristics are well suited to car ignition systems.

The development of MOS devices for performing logical functions has led to a proliferation of small battery-powered devices for all manner of purposes. The advantage of the CMOS family for example, is its very small current consumption and above all, its wide operating voltage range, typically from 3–18 V, which makes it ideal for battery operation.

The advent of low-voltage CMOS capable of operating for considerable periods from a single 1.5V cell has accelerated this trend. Well-designed CMOS circuits are so economical that the frequency of changing batteries is more dependent on their shelf life than their capacity.

Lithium batteries, exemplified by the Li/MnO_2 system in Figure 12.15g, have high terminal voltages and reasonably flat discharge characteristics. The figure does not show their superior capacity compared to all the other batteries. A lithium D-cell has a capacity of about 16 Ah, four times that of a zinc-carbon D cell, and at twice the voltage, so the capacity in Wh is actually eight times greater (see also Problem 12.8). Lithium batteries also have very long shelf lives (up to ten years) because their self-discharging rates are very low, but they suffer from an inability to supply large currents. Zinc-air and zinc-silver batteries are used where constancy of voltage or long shelf life are of paramount importance and are usually supplied as small button cells, but they have been made in very large sizes by specialist manufacturers for standby service. In terms of shelf life the zinc-air battery is unsurpassed since it can be kept indefinitely by simply blocking the air-holes. Service temperatures have not been mentioned since it is assumed that most electronic equipment will be used at room temperature, but when it must be used outside there are serious problems with reduced battery power at low temperatures. Table 12.2 summarises the chief properties of these battery systems.

Table 12.2 *Consumer batteries compared*

System	Shelf life[A]	Temperature[B]	Emf/cell[C]	Cost[D]	Typical use
Zn/MnO_2	2	$-7°C/+50°C$	1.5 (1.2)	low	clocks, alarms, TV remote
$Zn/ZnCl_2/MnO_2$	2	$-20°C/+70°C$	1.5 (1.2)	low	tape recorders, radios
$Zn/$alkaline MnO_2	4	$-30°C/+54°C$	1.5 (1.25)	low	toys, torches, motors
$Cd/NiO.OH$ ('Nicad')	recharge	$-40°C/+50°C$	1.35 (1.2)	high	tools, camcorders
NiMH	recharge	$-40°C/+50°C$	1.4 (1.25)	high	laptops, mobile phones
$Pb/H_2SO_4/PbO_2$ (lead-acid)	recharge	$-45°C/+60°C$	2.1 (2.0)	high	standby power, lighting
$Li/SOCl_2$	10	$-55°C/+85°C$	3.6 (3.3)	medium	remote data collection
Li/MnO_2	10	$-40°C/+70°C$	3.2 (2.7)	medium	sonobuoys, mines
Li/V_2O_5	recharge	$-55°C/+70°C$	3.4 (2.5)	high	backup power on PCBs
Li/CuO	10	$-40°C/+60°C$	2.3 (1.5)	high	oil-well loggers
Zn-air	indefinite	$-20°C/+60°C$	1.45 (1.25)	medium	hearing aids, watches
Zn/Ag_2O	2	$-40°C/+75°C$	1.6 (1.5)	med/high	instruments, calculators

Notes: [A] In years at 20°C; lower temperatures will increase and higher diminish this figure [B] Service temperature range. Capacity may be greatly reduced at the lower temperatures [C] Open-circuit e.m.f. with typical on-load voltage in parentheses [D] Initial cost for batteries of identical size

12.8 Cooling

The heat dissipated in electrical machines and circuits must ultimately be removed, or they will suffer breakdown through excessive temperature rise. This is particularly important for devices based on p-n junctions since all the heat is effectively produced at the junction, whose temperature can therefore rise to the point that either the device fails through excessive leakage current, or is destroyed by thermal runaway. In the case of silicon devices the junction temperature is limited to about 150°C. The heat produced at the junction is removed by conduction through the silicon to the case, which is usually made of a highly thermally-conductive material in power devices. When the heat reaches the out-side of the case it can be lost to the air by convection and to remote objects by thermal radiation. Unless the heat lost by these processes equals the heat produced in the junction, the junction temperature will rise until a balance is achieved.

The rate of loss of heat is proportional to the temperature difference between the junction and case, which can be found from the thermal resistance specified by the manufacturer. Thermal resistance, θ, is defined by

$$\theta = \Delta T / P \tag{12.40}$$

where ΔT is the rise in temperature produced by the power dissipated, P. Thus if the thermal resistance from junction to ambient, θ_{JA}, of a small transistor is given as 200°C/W, it means that a power dissipation of 100 mW will produce a rise in the junction temperature of

$$\Delta T_J = \theta_{JA} P = 200 \times 0.1 = 20°C \tag{12.41}$$

Then if we know the ambient air temperature to be 35°C, we can see that the junction temperature is

$$T_J = T_A + \Delta T_J = 35 + 20 = 55°C \tag{12.42}$$

The rate of heat removal from the case may be insufficient without the use of a heat sink, which is a piece of metal that is stuck onto the case to increase its surface area. This increases the convection and radiation in proportion to the increased surface area, provided the heat sink has effectively infinite thermal conductivity. Aluminium is the preferred material for heat sinks because its thermal conductivity is very high and it is much cheaper than other high-conductivity metals. The thermal resistance of heat sinks is always quoted by the suppliers and must be added to that of the device from its junction to its case, as in the next example.

Example 12.6

A silicon transistor has $\theta_{JC} = 10°C/W$ and has a thermal resistance from case to ambient, $\theta_{CA} = 30°C/W$. What is the maximum power dissipation permitted if the ambient is at 30°C and the junction limited to 150°C? When it is fitted with a heat sink with a thermal resistance, $\theta_{HA} = 7°C/W$, what is the junction temperature if the power dissipated is 5 W?

The maximum temperature difference from junction to ambient, $\Delta T_{JA} = 150 - 30 = 120°C$ and so the maximum power dissipation is

$$P_{max} = \Delta T_{JA} / \theta_{JA} = \Delta T_{JA} / (\theta_{JC} + \theta_{CA}) = 120/(10 + 30) = 3 \text{ W} \tag{12.43}$$

When the heat sink is fitted θ_{CA} is replaced by θ_{HA}, the thermal resistance from heat sink to ambient. The maximum power dissipation then becomes

$$P_{max} = \Delta T_{JA} / (\theta_{JC} + \theta_{HA}) = 120/(10 + 7) = 7 \text{ W} \tag{12.44}$$

When the power is 5 W, the temperature rise is

$$T_J = T_A + \Delta T_{JA} = T_A + (\theta_{JC} + \theta_{HA})P = 30 + 17 \times 5 = 115°C \tag{12.45}$$

The thermal contact between case and heat sink must be of low thermal resistance and a thermally

> conducting paste should be used to bond the heat sink to the device. Without this a small air gap of very high thermal resistance will exist between case and heat sink and the heat transfer will be poor. Take note that nothing can be done to reduce the thermal resistance from junction to case as this is determined by the way the device is made. In this example even if the heat sink had zero thermal resistance the maximum permitted dissipation would be restricted to $120/\theta_{JC} = 120/10 = 12$ W.

Heat sinks are made in many shapes and sizes and it is necessary to choose with care because an over-specified heat sink can waste a lot of money. For example a TO3 heat sink with $\theta_{HA} = 7°C/W$ costs only 40p, while one with $\theta_{HA} = 4°C/W$ costs £2. Large heat sinks with very low thermal resistances can be used when many devices require cooling. The one shown in cross-section in Figure 12.16 for example costs £14, has a thermal resistance of 0.5°C/W and can accommodate a dozen TO3 packages. Heat sinks are often painted matt black to aid radiative heat transfer. For individual devices one can buy clip-on heat sinks of relatively high thermal resistance. One such for a small TO5 transistor costs 20p and has $\theta_{HA} = 48°C/W$. Those made from aluminium sheet for TO3 packages are about 50p with $\theta_{HA} = 7°C/W$. Sometimes it is possible to mount a power device on the outside of the instrument case which provides fairly efficient heat sinking.

Figure 12.16

Cross-section of an extruded aluminium heat sink. With L = 150 mm, θ_{HA} = 0.5°C/W (L is the dimension normal to the page)

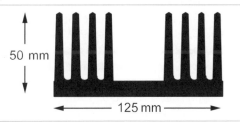

50 mm

125 mm

If the heat sink gets too hot it may be necessary to use forced convection, that is, to cool it by a fan. Fans are capable of improving the dissipation of a heat sink by a factor of two or three and may be worthwhile economically, since a smaller heat sink will be cheaper and will occupy less space perhaps than a small fan. A mains operated fan of 80mm × 80mm × 30mm dimensions costs only about £6, can blow 13 l/s of air and consumes 13 W. Axial fans are now available with thermistor sensors for speed control so that the air temperature remains almost constant. This cuts down both the noise and the power consumption. Peltier-effect heat pumps can also be used to cool or control the temperature of especially sensitive devices, but they are not cheap: about £1 per watt extracted. When optimally adjusted a temperature difference of 65°C can be maintained across the devices, which are only 3–5 mm thick. Again a heat sink, and possibly a fan too, is needed on the hotter face of the heat pump. Peltier pumps are best used for precise control of the temperature of critical devices.

The thermal resistance quoted for a heat sink does not allow for the thermal resistance between it and the device. When a heat sink is properly bonded to a device with a thermally-conductive paste, a small thermal resistance of around 0.1°C/W is introduced. Without this precaution the thermal resistance between device and heat sink may well be greater than that between the heat sink and ambient. Of course if θ_{JC} is for argument's sake, 10°C/W, no heat sink can bring the overall thermal resistance below that figure.

Suggestions for further reading

Active and non-linear electronics by T Schubert and E Kim (Wiley, 1996). Lengthy.
The art of electronics by P Horowitz and W Hill (CUP, 2nd ed., 1989). Getting on in years, but good.
The modern amplifier circuit encyclopedia by R F Graf (Tab books, 1992)
Battery reference book by T R Crompton (Newnes, 3rd ed., 2000). Comprehensive.

Problems

1. From Equation 12.8 show that the max. efficiency of a capacitively-coupled CE amplifier is 8.6%.
2. Show that the maximum efficiency of a capacitively-coupled CE amplifier in which R_C is adjusted to give maximum power in the load resistance, R_L, is 1/12 or 8.33%.
3. Show, by evaluating the time average of the product of v_{CE} and i_C, that 50% of the power supplied to a transformer-coupled CE amplifier is dissipated in the transistor when working at max. efficiency.
4. A transformer-coupled CE amplifier is operated from a 12V supply and must deliver maximum power to a 50 kΩ load. What must the turns ratio be if the transformer's maximum primary current is 2 A? *[n = N_1/N_2 = 0.011]*
5. A class-B amplifier is as in Figure 12.5, with $R_E = 2$ Ω and $R_L = 10$ Ω. What must the supply voltage be if the maximum average load power is 100 W? What is the efficiency? How much power is dissipated in each transistor? What is the standby power consumption? (The base-biassing resistors may be neglected, while the diode drops and V_{BE} are both 0.7 V.) *[107.3 V, 65.4%, 16.4 W, 18.8 W]*
6. A class-C amplifier works from a 200V supply, and delivers 250 W to a load. The transistor's max. current is 8 A when the input is 5 V. What is the conduction angle, θ_c, the power lost in the amplifier and the base-bias voltage, V_{BB}? (Take V_{BE} to be 0.7 V.) *[θ_c = 100.4°, 34 W, V_{BB} = -2.5 V]*
7. A class-C amplifier operates at 12 MHz with $V_{CC} = 60$ V, $V_{BB} = -10$ V. A peak input signal of 15 V gives a peak collector current of 3.2 A. Find the conduction angle, θ_c, the input power, P_S, the output power, P_L and the efficiency. If an LC tank circuit is used on the output with a capacitance of 50 pF, what is the inductance needed? (Take $V_{BE} = 0.7$ V.) *[89°, 30.2 W, 27 W, 89.4%. 3.52 µH]*
8. A series voltage regulator has an output voltage of 6 V and a dropout of 1 V. It feeds a load of 90 Ω and is supplied by 6 × 1.5 V carbon-zinc D-cells. These have the characteristics shown in Figure 12.15a. How long will they last? How many Wh will they have delivered to the load? If they cost 85p each, how much is the cost/Wh? If 3 × 3 V Li/MnO$_2$ D-cells are used instead with the characteristics of Figure 12.15g, but with the timescale multiplied by 6, how long will they last and at what cost/Wh, given that they are £12 each? Which is the better choice? *[35 h, 14 Wh, 36p/Wh; 250 h, 100 Wh, 36p/Wh]*
9. Show that the percentage peak-to-peak load voltage ripple for the forward converter of Figure 12.12 is given by

$$V_{rpp}(\%) = 25(1 - \delta)/LCf^2$$

where $f = 1/T$ and $\delta = T_{ON}/T$.
10. The output stage of a power amplifier dissipates 50 W. It comprises four identical transistors with $\theta_{JC} = 2.8°C/W$. They are all bolted onto one heat sink with a paste joint having a thermal resistance of 0.2°C/W. If the maximum permissible transistor junction temperature is 150°C and the maximum ambient temperature is 50°C, what must the thermal resistance of the heat sink, θ_{HA}, be? What is the maximum temperature of the heat sink and the transistor cases? *[1.25°C/W, 112.5°C, 115°C]*
11. The thermal resistance of the best available heat sink for the power amplifier of the previous problem is 2°C/W. Fan cooling is therefore used, rather than increasing the number of transistors, effectively reducing θ_{HA} to 0.7°C/W. What would be the junction temperature now? By what percentage can the power output be increased before the maximum permissible junction temperature is reached? *[122.5°C, 38%]*

13 Magnetism and electromagnetism

T HE STUDY of magnetism, with origins far back in history, was impelled by the urgent need for a compass to navigate with. Peter Peregrinus discovered in 1269 that lodestone had N and S poles like the earth and the first scientific book[1] on magnetism was published in 1600. A knowledge of elementary magnetism and electromagnetism is fundamental for studying electrical machines and transformers. We shall accordingly look at some basic magnetic phenomena and then some essential laws of electromagnetism.

Every material responds to a magnetic field in some way, but in the vast majority the effect is small. Only in ferromagnetic materials, such as iron, nickel and cobalt, is the effect large. If we take a bar of steel and place it in a strong magnetic field, we find that a N pole is induced near one end and a S pole near the other. The N pole is so called because it will point N if the magnet is suspended by a thread. Peregrinus found that unlike poles attract each other while like poles repel, so that the earth's N magnetic pole must be a S pole for it to attract the N pole of a magnet. If the magnetized bar of steel is cut in half two magnets are created each with its own N and S poles, and no matter how finely divided the bar, a N pole or a S pole cannot be isolated. Poles are always paired.

13.1 Magnetic units and quantities

Magnetic fields can be produced by electric currents. Consider the magnetic field and related quantities that are associated with a current-carrying coil wound on a ferromagnetic toroid as in Figure 13.1.

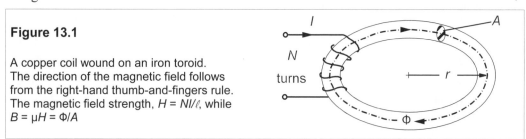

Figure 13.1

A copper coil wound on an iron toroid. The direction of the magnetic field follows from the right-hand thumb-and-fingers rule. The magnetic field strength, $H = NI/\ell$, while $B = \mu H = \Phi/A$

With the coil of N turns wound in the sense shown, a magnetic field, **H**, is generated which points in a direction given by the right-hand rule which states:

[1] *De Magnete* by William Gilbert, Queen Elizabeth I's physician. It was the first scientific book of any kind to be published in England and describes many important observations by Gilbert.

With the right hand, grasp the coil with fingers pointing in the direction of the current; the thumb then points in the direction of the magnetic field.

The magnitude of the magnetic field is given by

$$H = NI/l \qquad (13.1)$$

where I is the current through the coil and $l\ (= 2\pi r)$ is the path length of the magnetic flux, Φ, which in the case of a ferromagnetic toroid, or *core*, is its average circumference. If I is in amps and l is in metres, then H is in A/m.

The magnetic flux is defined by

$$\Phi = BA \qquad (13.2)$$

where B is the flux density (or magnetic induction) in tesla (T), A is the cross-sectional area of the toroidal core in m^2 and Φ is in weber (Wb). When a magnetic field is applied to a ferromagnetic material its magnetization, $\mathbf{M}$, lines up with the field and the resultant flux density is

$$\mathbf{B} = \mu_0(\mathbf{H} + \mathbf{M}) \qquad (13.3)$$

Dividing by H and using magnitudes for the vectors, $\mathbf{B}$, $\mathbf{H}$ and $\mathbf{M}$, we find

$$B/H = \mu = \mu_0(1 + M/H) = \mu_0\mu_r \qquad (13.4)$$

where μ, the *magnetic permeability* of the toroid, is $\mu_0\mu_r$. μ_0 is the magnetic constant (or *permeability of free space*) whose value is $4\pi \times 10^{-7}$ H/m. The *relative permeability*, μ_r, depends on the core material, ranging from 500 for ferrites to about 3000 for silicon steel[2].

Example 13.1

A toroid has $\mu_r = 750$ and is wound with 500 turns of copper wire carrying a current of 2 A. It has a cross-sectional area of 3×10^{-3} m^2 and a mean radius of 150 mm. Find the magnetic field, the magnetic flux, the magnetization and the flux density in the core.

The path length of the flux in the core is $2\pi r = 0.3\pi$, then using Equation 13.1 we find

$$H = NI/l = 500 \times 2/0.3\pi = 1061 \text{ A/m} \qquad (13.5)$$

Equation 13.4 yields

$$B = \mu_0\mu_r H = 4\pi \times 10^{-7} \times 750 \times 1061 = 1 \text{ T} \qquad (13.6)$$

which is a typical operating flux density in iron-silicon alloys. The magnetic flux, $\Phi = BA = 1 \times 3 \times 10^{-3}$ = 3 mWb. The magnetization can be found from Equation 13.3:

$$B = \mu_0(H + M) \quad \Rightarrow \quad M = B/\mu_0 - H = 1/4\pi \times 10^{-7} - 1061 = 795 \text{ kA/m}$$

This is virtually the same as B/μ_0 since $H << M$.

---

[2] In old texts, and many research papers, cgs-emu units are used with the magnetic field strength in oersteds (Oe) and the magnetic flux density in gausses (G). In a non-magnetic material (where $\mu_r = 1$) in the cgs-emu system, $B = H$, while in the SI system, $B = \mu_0 H$. The conversion factors are 1 T = 10,000 G and 1 A/m = 0.0126 Oe. Relative permeabilities are, of course, the same in both systems of units.

We might suppose that, by increasing the current through the coil in Example 13.1, any desired value of flux can be achieved, but in practice the core *saturates* because all its magnetic dipoles are aligned with the field. The magnetization, M, in Equations 13.3 and 13.4 has a maximum value for any material, known as the saturation magnetization, M_s. Pure iron, for example, has a saturation magnetization at 300 K of 1.717 MA/m (implying a maximum flux density of about $\mu_0 M_s = 4\pi \times 10^{-7} \times 1.717 \times 10^6 = 2.16$ T) and the silicon steel used in transformer cores has a saturation magnetization of about 1.4 MA/m. Thus there comes a point where any increase in the current through the coil merely increases H while M remains at its saturation value.

Example 13.2

The current in the coil of Example 13.1 is increased to 9 A and the core is saturated. If the saturation magnetization is 1.25 MA/m, what is the flux density in the core? What is the relative permeability of the core now?

Equation 13.1 gives the new magnetic field strength:

$$H = NI/l = 500 \times 9/0.3\pi = 4775 \text{ A/m} \tag{13.7}$$

Then the magnetic induction is found from Equation 12.3 to be

$$B = \mu_0(H + M) = 4\pi \times 10^{-7}(4775 + 1.25 \times 10^6) = 1.577 \text{ T} \tag{13.8}$$

where $M = M_s = 1.25$ MA/m.

Dividing Equation 13.8 by $\mu_0 H$ produces

$$\frac{B}{\mu_0 H} = \mu_r = \frac{1.579}{4\pi \times 10^{-7} \times 4775} = 263 \tag{13.9}$$

The relative permeability has been greatly reduced because the core has saturated. Provided machines and transformers are not driven into saturation, B and H can be taken to be proportional and the permeability as constant.

13.2 The magnetic circuit

When a small air gap is cut into the toroid of Figure 13.1, as in Figure 13.2, virtually all of the flux in the toroid passes through the gap and normal to its cross-sectional area, A. In this case we can calculate that flux from the equation for the magnetic circuit, which can be derived from Ampère's law in the form:

$$F = NI = \sum H_k l_k \tag{13.10}$$

where F is the *magnetomotive force* or m.m.f. We sum the field-length products in the differing regions round the magnetic circuit, in this case just the iron and the air gap. If the gap is short all the flux passes through it, that is Φ is constant, and we can write Equation 13.10 as

$$F = \sum B_k l_k/\mu_k = \sum B_k A_k\, l_k/A_k\mu_k = \sum \Phi_k l_k/A_k\mu_k = \Phi \times \sum l_k/A_k\mu_k = \Phi \times \mathbb{R} \tag{13.11}$$

using Equations 13.2 and 13.4 and writing $\mathbb{R}$, the *reluctance*[3] of the circuit for $\sum l_k/A_k\mu_k$.

---

[3] Another term – permittivity is one more – invented by Heaviside.

Equation 13.11 is of the same form as Ohm's law: m.m.f. replaces e.m.f., flux replaces current and reluctance replaces resistance. For a toroid with an air gap, the reluctance of the iron path can be calculated and added to the reluctance of the air gap as they are in series. Reluctances in series add like resistances in series.

Example 13.3

If the toroid of Example 13.1 has a gap 1 mm wide cut into it, as in Figure 13.2, and the coil still carries 2 A, what are the flux, the flux density, the magnetic field strength and the overall relative permeability of the magnetic circuit?

Figure 13.2

A toroid with an air gap

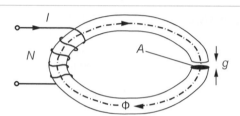

We first calculate the reluctance of the iron path:

$$\mathbb{R}_{Fe} = \frac{l}{\mu_{Fe}A} = \frac{0.3\pi}{\mu_0 \times 750 \times 3 \times 10^{-3}} = \frac{0.419}{\mu_0} \qquad (13.12)$$

Then the reluctance of the air gap is

$$\mathbb{R}_G = \frac{g}{\mu_g A} = \frac{1 \times 10^{-3}}{\mu_0 \times 1 \times 3 \times 10^{-3}} = \frac{0.333}{\mu_0} \qquad (13.13)$$

as the relative permeability of air is one. The total reluctance is therefore

$$\mathbb{R} = \mathbb{R}_{Fe} + \mathbb{R}_G = \frac{(0.419 + 0.333)}{\mu_0} = \frac{0.752}{\mu_0} \qquad (13.14)$$

Notice that the rather small air gap has a reluctance comparable to the whole of the iron toroid. Substituting this reluctance into Equation 13.11 leads to

$$NI = \Phi\mathbb{R} = \Phi \times 0.752/\mu_0 \qquad (13.15)$$

Thus

$$\Phi = \frac{NI\mu_0}{B} = \frac{2 \times 500\mu_0}{0.752} = 1.671 \text{ mWb}$$

And

$$B = \frac{\Phi}{A} = \frac{1.671 \times 10^{-3}}{3 \times 10^{-3}} = 0.577 \text{ T}$$

while $H = NI/l = 1061$ A/m as before, which means that $\mu = \mu_0\mu_r = 0.557/1061$, or $\mu_r = 418$. The flux density and the overall relative permeability have both been reduced to 56% of their previous magnitudes. Putting an air gap into a core increases the reluctance, reduces the flux density, and helps to avoid saturation.

13.3 Faraday's law of electromagnetic induction

Faraday's researches into the laws of electromagnetism are possibly the greatest of all experimental investigations. His mathematics were limited but he had a profound insight into the physics of electromagnetism, and his lucid explanations of hitherto baffling phenomena flowed from the mind of a genius. Before his discoveries the only voltage sources were primary cells – batteries of relatively low power and considerable cost. After Faraday[4] electricity was able to move out of the laboratory and into the street, the factory and the home.

Faraday's law of electromagnetic induction states that

The e.m.f. induced in a coil is directly proportional to the rate of change of magnetic flux through it

Expressed symbolically it is

$$e \propto -\,d\Phi/dt = -\,Nd\Phi/dt \qquad (13.16)$$

e, the e.m.f., is in volts if Φ is in Wb, t in s and N the number of turns on the coil. The minus sign indicates that the e.m.f. opposes the change in flux. If the flux is decreasing, $d\Phi/dt$ is negative and e is positive.

Figure 13.3

The current induced in the coil must flow in the direction which maintains the flux within it. By the right-hand rule the current flows upwards in the far side of the loop and downwards in the near

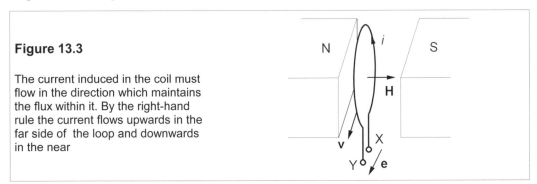

Figure 13.3 shows a loop of copper wire between the N and S poles of a magnet. The magnetic field is directed from N pole to S pole. The copper loop is then pulled towards the reader, thus decreasing the magnetic flux within the loop. The induced e.m.f. must be of a polarity to send a current round the loop that tries to keep the flux constant. By the right-hand rule the induced current must flow from X to Y as this will make **H** point from N pole to S. The polarity of the terminals is as marked. To see that this is correct, imagine an external resistance connected between X and Y with the same direction of current flowing, terminal Y must be positive.

---

[4] Michael Faraday (1791–1867) was born near London. On leaving school at thirteen, he became an apprentice bookbinder and gained his scientific knowledge from such books as passed through his hands. Faraday had been to public lectures given by Sir Humphrey Davy and the notes he wrote from those lectures so impressed Davy that he made Faraday his assistant. Faraday developed slowly: he was over forty when he made his greatest discoveries. J G Crowther in *British scientists of the 19th century* has written a short, engrossing biography.

Example 13.4 ─────────────────────────────────────

A rectangular loop of copper wire is located between the poles of a magnet as shown in Figure 13.4. The flux density, **B**, is uniform between the poles and normal to the plane of the loop. If the loop is removed from the poles at a steady velocity, **v**, what is the magnitude and direction of the induced e.m.f.? If the resistance of the loop is R, what power is dissipated in the wire?

Figure 13.4

A rectangular copper loop is pulled out of a uniform magnetic field at a constant velocity, **v**. The area of the loop initially within the field is ly

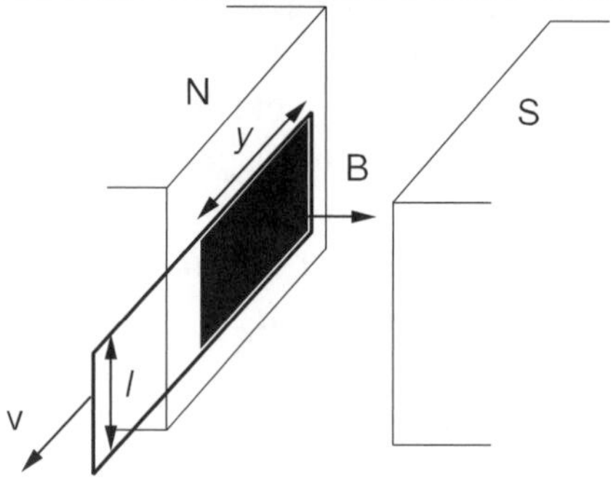

By Equation 13.16 the e.m.f. is $d\Phi/dt$, while $\Phi = BA = Bly$, because the area of the loop within the magnetic field is ly. Therefore the e.m.f. is

$$e = d\Phi/dt = d(Bly)/dt = Bl\,dy/dt \qquad (13.17)$$

Neither B nor l changes with time, but y does. However $dy/dt = v$, the speed at which the loop is moving, giving

$$e = Bl\,dy/dt = Blv \qquad (13.18)$$

The e.m.f. is the product of the length of the wire normal to the velocity, the flux density and the speed of the wire. The power dissipated within the loop is

$$p = ei = ilBv = (Blv)^2/R \qquad (13.19)$$

13.3.1 The Lorentz force

When the copper loop is pulled from the magnetic field in Figure 13.4, it experiences a force opposing the motion called the Lorentz[5] force. There must be a force opposing the motion for work to be done on the loop of wire. This work is converted to electrical energy. If the force opposing the motion is F, then the mechanical power put into the loop in Figure 13.4 is Fv. If all of this power is converted to electrical power, given by Equation 13.19, then

$$Fv = ei = Blvi \qquad (13.20)$$

That is $F = ilB$, which is the magnitude of the Lorentz force.

The Lorentz force, however, acts in a direction normal to the plane containing **B** and **l**, the length vector of the wire carrying the current, and it is best expressed in vector form. Consider the conductor in Figure 13.5 carrying a current, i, in a magnetic field whose flux density is B. The force, **F**, is given by the vector product

---

[5] H A Lorentz (1853–1928) was born in Leyden, Holland. He was awarded the Nobel prize for physics in 1902 for his researches in electromagnetism and was the author of the Lorentz contraction in the theory of relativity.

$$\mathbf{F} = i\mathbf{l} \times \mathbf{B} \tag{13.21}$$

B is the flux density and **l** is a vector whose magnitude is the length of the wire and whose direction is the same as the current's. The magnitude of **F** is $ilB\sin\theta$, where θ is the angle between **l** and **B**; in Example 13.4 $\theta = 90°$, so that $F = ilB$. The direction of **F** is found by the *right-hand corkscrew rule*: **l** is rotated towards **B** with a right-hand corkscrew. The direction that the corkscrew moves in is the direction of **F**, which is into the plane of the paper. Note that $\mathbf{B} \times i\mathbf{l}$ is a vector of the same magnitude as **F**, but pointing in the opposite direction. The force on a current-carrying conductor in a magnetic field can be harnessed in a motor.

Example 13.5

A straight copper wire of length 0.1 m carries a current of 10 A when wholly immersed in a field of uniform flux density 1.3 T, as in Figure 13.5. If the angle between the field and the conductor is 105° what is the force experienced and in which direction is it?

Since the angle between the field and the current, θ, is 105°, $F = ilB\sin\theta = 10 \times 0.1 \times 1.3 \times 0.966 = 1.26$ N. By the right-hand corkscrew rule, its direction must be normal to the page and into it as **l** is directed upwards and **B** from left to right.

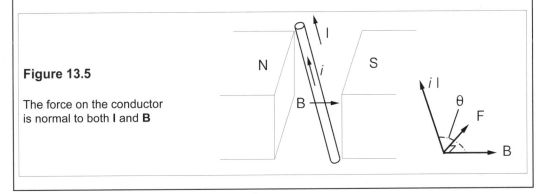

Figure 13.5

The force on the conductor is normal to both **l** and **B**

13.3.2 Fleming's rules

Fleming's left-hand (motor) rule gives the direction of the force (and therefore the motion) and can be used for this purpose instead of the vector form of the Lorentz force. The thumb and first two fingers of the left hand are held orthogonally (each digit at right angles to the other two) to represent the motion (thumb), the field (first finger) and the current (second finger), as in Figure 13.6.

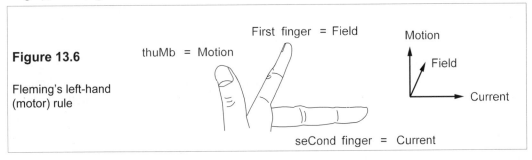

Figure 13.6

Fleming's left-hand (motor) rule

This rule is equivalent to the vector formulation of Equation 13.21, the motion being in the direction of the Lorentz force. Fleming's right-hand (generator) rule is identical to Fleming's left-hand (motor) rule except that the digits of the right hand are used to give the direction of current flow in a conductor moving in a magnetic field.

13.4 **Magnetic hysteresis**

When an unmagnetized long bar of iron is placed in a magnetic field whose strength is increased from zero up to a sufficiently large value, its magnetization eventually reaches a maximum value, the saturation magnetization, M_s. The saturation magnetization is constant for a given material and temperature - pure iron for example always has $M_s = 1.717$ MA/m, while $M_s = 490$ kA/m for pure nickel[6]. When the applied field is reduced from its saturation value the flux density falls, but not fast enough to retrace the magnetization curve. And when H is reduced to zero B has a positive value known as the *remanence*, B_r. The applied field must be reversed in order to make B zero and its magnitude is then called the *coercivity* of the material, H_c. Further increase of reverse field then causes the magnetization to saturate in the reverse direction, attaining the same saturation magnitude as previously, but in the opposite direction. Reducing the field to zero and then increasing it to its former positive maximum causes a complete loop to be traced which is called the *hysteresis* loop.

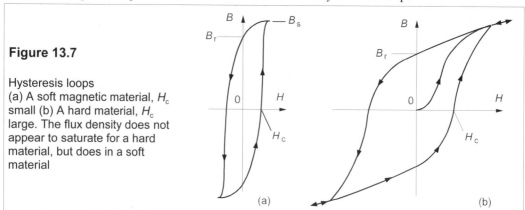

Figure 13.7

Hysteresis loops
(a) A soft magnetic material, H_c small (b) A hard material, H_c large. The flux density does not appear to saturate for a hard material, but does in a soft material

Figure 13.7a shows a hysteresis loop for a long rod of soft iron and Figure 13.7b the hysteresis loop for a permanent magnet, with different scales for the magnetic field, H. The significant difference between the two is the size of the coercivity, which is small for pure iron (50 A/m) and large for the permanent magnet (200 kA/m). Materials with low coercivities are called *soft* and those with large coercivities are called *hard*.

Soft magnetic materials have high relative permeabilities and hard materials low. For example soft iron may have $\mu_r = 3000$, while strontium hexaferrite ($SrFe_{12}O_{19}$, a permanent magnet) might have a relative permeability of only 2. As a result when the material is saturated the B-H curve slopes upwards (see Figure 13.3b) because H is not small compared

---

[6] Sometimes *magnetic polarization*, **J**, is used in place of **M**, where $\mathbf{J} = \mu_0 \mathbf{M}$. The saturation polarization of iron then becomes $1.717 \times 10^6 \times 4\pi \times 10^{-7} = 2.16$ T.

to M_s. When a soft magnetic material is saturated, the value of H required for saturation is so small compared to M_s that the B-H curve appears to saturate and the saturation flux density, B_s is very nearly equal to M_s.

One consequence of magnetic saturation is the generation of harmonics. For example if the current in the primary of a transformer is too large, the core may become saturated, with the consequence that its flux density is not directly proportional to the current. The secondary e.m.f. is then distorted, that is contains harmonics of the primary frequency. The distortion can be very considerable and causes much increased core losses and overheating, besides other undesirable effects. By restricting the maximum flux density to about 75% of its saturation value the generation of harmonics is reduced.

Example 13.6

A silicon-iron alloy used in transformer cores requires an applied field of 600 A/m to cause it to saturate at a magnetization of 1.5 MA/m. What is the value of the saturation flux density? What is the relative permeability at this point? A hard ferrite requires a field of 400 kA/m to cause it to saturate at a magnetization of 180 kA/m. What is the flux density at this point and the relative permeability?
Using Equation 13.3 we find

$$B_s = \mu_0(H + M_s) = 4\pi \times 10^{-7}(600 + 1.5 \times 10^6) = 1.886 \text{ T} \qquad (13.22)$$

And from Equation 13.4

$$\mu_r = B/\mu_0 H = 1.886/(4\pi \times 10^{-7} \times 600) = 2501 \qquad (13.23)$$

The same calculations for the hard material yield $B_s = 0.729$ T and $\mu_r = 1.45$.

13.4.1 Hysteresis losses

The area within a hysteresis loop has the dimensions of B times H or energy per unit volume. The area within the loop in fact represents the work done per cubic metre in taking the magnetic material once around the loop. This work appears as heat. If the frequency of the applied field is f, then the rate of expenditure of energy, or the power lost per unit volume, will be

$$P_h = f\oint B\,dH \qquad (13.24)$$

in W/m^3, as the sample will go round the hysteresis loop once per cycle.

Example 13.7

If a ferromagnetic transformer core operates at a maximum flux density of 1.3 T and has a coercivity of 60 A/m, estimate the hysteresis losses at 50 Hz in a core of volume 0.3 m^3. Assume a square loop as shown in Figure 13.8.
A 'square' loop is actually a rectangular one in which the area of the loop is $4B_rH_c$ as can be seen from Figure 13.4. If Λ is the core volume the hysteresis losses are

$$P_h = f\Lambda \oint B\,dH = 4f\Lambda B_r H_c \qquad (13.25)$$

$$= 5 \times 50 \times 1.3 \times 60 = 4680 \text{ W}$$

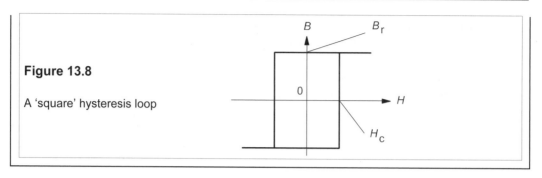

Figure 13.8

A 'square' hysteresis loop

13.5 Eddy currents

Besides hysteresis losses alternating magnetic fields induce *eddy currents* in the core. Eddy currents are caused because the conduction electrons in the core react so as to reduce the change in applied field by Lenz's law. Therefore if the coil current is increasing in the direction shown in Figure 13.9, eddy currents will circulate to oppose it, that is they will circulate in loops as shown.

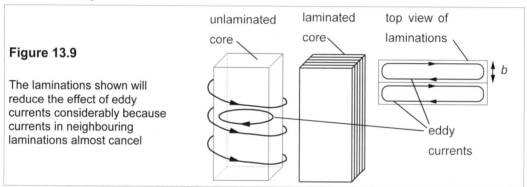

Figure 13.9

The laminations shown will reduce the effect of eddy currents considerably because currents in neighbouring laminations almost cancel

    The magnetic field of the eddy currents is given by the right-hand rule and is seen to oppose the increasing applied field. By splitting the core into thin laminations whose plane is normal to the eddy currents, the losses are reduced as the eddy currents in neighbouring laminations cancel out along most of the path. Eddy currents will cause Joule heating loss of magnitude

$$P_e = \frac{J^2}{\sigma} = \frac{(\pi b f B_m)^2 \sigma}{6} \tag{13.26}$$

in W/m^3. σ is the electrical conductivity of the core in S/m, J is the eddy-current density in A/m^2, b is the thickness of the laminations in m and B_m is the maximum flux density in the core in T. For example if $B_m = 1.3$ T, $f = 50$ Hz, $b = 1$ mm and $\sigma = 1$ MS/m, we find $P_e = 7$ kW/m^3. If the core material's density is 7500 kg/m^3 (a typical figure for silicon-steel transformer cores) then the eddy-current losses work out to about 1 W/kg. Obviously if the core were to be an electrical insulator (such as a ferrite), these losses would be nil. Silicon alloyed with iron causes a considerable decrease in electrical conductivity which assists in reducing eddy-current losses. Metal glasses also have a low conductivity, partly because they are disordered and partly because they are highly alloyed.

13.6 **Inductors**

Most inductors comprise a coil of insulated wire wound onto a ferromagnetic core, which is often a toroid. Suppose a coil of N turns is wound on the toroid of cross-sectional area, A and carries a current, i, which induces a magnetic flux, Φ, as in Figure 13.1. By Faraday's law of electromagnetic induction the e.m.f. induced in the coil is $-Nd\phi/dt$. But we also know that the e.m.f. induced in an inductance is $-Ldi/dt$. Equating the two gives

$$L di/dt = Nd\Phi/dt \tag{13.27}$$

which leads to

$$L = N\Phi/I = NBA/I = \mu HNA/I \tag{13.28}$$

since $B = \mu H$. But $H = NI/l$, where l is the magnetic path length, so that

$$L = \mu AN^2/l = \mu AlN^2/l^2 = \mu_0\mu_r\Lambda n^2 \tag{13.29}$$

where Λ $(= Al)$ is the volume of the core and n $(= N/l)$ is the number of turns per unit path length. The inductance is thus seen to be dependent on the size and permeability of the core and on the *square* of the number of turns/m. For example, if $A = 5 \times 10^{-4}$ m^2, $l = 0.3$ m, $N = 250$ and $\mu_r = 500$, then we find L to be 65.45 mH. Doubling the number of turns quadruples the inductance to 262 mH.

Example 13.8

An air-cored solenoid has an effective cross-sectional area of 8×10^{-3} m^2 and is 300 mm long. How many turns are required to produce an inductance of 20 H? What current in this solenoid will produce a magnetic field of 6 MA/m?

Equation 13.29 can be rearranged to give $n = \sqrt{(L/\mu\Lambda)}$, where the volume, $\Lambda = 8 \times 10^{-3} \times 0.3 = 0.0024$ m^3 and in air $\mu = \mu_0$ so that

$$n = [20/(4\pi \times 10^{-7} \times 0.0024)]^{1/2} = 81400 \text{ turns/m} \tag{13.30}$$

The number of turns is $N = nl = 81400 \times 0.3 = 24400$ turns. The current is found from Equation 13.28

$$I = N\Phi/L = \mu_0 NHA/L$$
$$= 4\pi \times 10^{-7} \times 24400 \times 6 \times 10^6 \times 0.008/20 = 73.6 \text{ A} \tag{13.31}$$

With this introductory electromagnetism we can now go on to study electrical machines in the chapters that follow.

Suggestions for further reading - see Chapter 18

Problems

1. Show that the moment of the force on a circular conductor, area, A, carrying a current, i, and whose plane is inclined at an angle, θ, to the magnetic flux density, **B**, is $M = BAi\cos\theta$.
2. The earth's magnetic field strength is about 40 A/m. What e.m.f. would be induced in a conducting rod 2.65 m long moving normal to it at 112 km/h? If the rod's conductivity is 50 MS/m and it has a uniform cross-sectional area of 10^{-5} m^2, what power is dissipated in it? What force is exerted on

the rod? (Assume the ends are connected by a wire of zero resistance.)
[4.14 mV, 3.23 mW, 104 μN]

3. The English Channel flows through the straits of Dover at a speed of 4 km/h. The sea is an electrical conductor, so if the vertical component of the earth's magnetic field strength is 25 A/m, what is the potential difference across the Channel? If the water is flowing into the N sea, is England or France at the higher potential? (The straits are 35 km wide.) *[1.22 V, England]*

4. A fluxmeter is made from a thin circular coil 25 mm in diameter with 10 turns of copper wire. This coil is connected to an op-amp integrator as shown in Figure P13.4. The coil is inserted between the poles of a soft-iron poled electromagnet and the voltage at the output of the fluxmeter is found to change by 0.46 V after insertion. The coil resistance is negligible. Show that the output voltage of the integrator is given by $V_o = N\Phi/RC$ where Φ is the magnetic flux through the coil. What is the field strength between the poles if $R = 10$ kΩ and $C = 1$ μF. *[746 kA/m]*

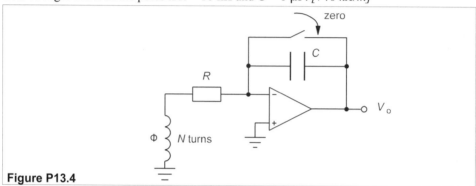

Figure P13.4

5. A rectangular ferromagnetic core is wound with 330 turns of wire carrying a current of 1.2 A. The core has a magnetic path length of 0.5 m and a cross-sectional area of 4×10^{-4} m². If the core is saturated with a saturation magnetization of 200 kA/m, what are the magnetic flux density and magnetic flux in the core, and the relative permeability of the core?
    [B = 0.252 T, Φ = 0.101 mWb, μ_r = 253]

6. The core in the previous problem is cut into two halves separated by an air gap 0.05 mm wide. If the core's relative permeability and the current remain unchanged, what are the overall relative permeability and the flux density? Repeat the calculation for the case where the two halves are separated by non-magnetic spacers 2 mm thick. *[μ_r = 241, B = 0.24 T. μ_r = 83.7, B = 0.0833 T]*

7. What are the hysteresis losses at 24 kHz in a transformer core which has a maximum flux density of 0.25 T and a square hysteresis loop with a coercivity of 5 A/m. The core volume is 3×10^{-3} m³
    [360 W]

8. A transformer core has laminations 0.3 mm thick and carries a maximum flux density of 0.68 T what are its eddy-current losses (a) at 50 Hz (b) at 2 kHz, given that its volume is 4×10^{-4} m³ and its conductivity is 0.7 MS/m? What are the eddy current densities?
    [(a) 0.048 W (b) 76.6 W; 9.16 kA/m², 366 kA/m²]

9. An inductor is to be made with a toroidal core as in Figure 13.1. The core has a relative permeability of 325, a cross-sectional area of 8×10^{-5} m² and a radius of 20 mm. If the inductance is to be 60 mH, how many turns are required? If the winding error is ±0.5 turn, what is the corresponding error in the inductance? *[480, ±0.125 mH]*

10. A transformer core has laminations 0.25 mm thick. It is made of material having a square hysteresis loop with $B_r = 0.8$ T and $H_c = 40$ A/m under operating conditions. The conductivity is 1.5 MS/m. At what frequency will the eddy-current losses equal the hysteresis losses? If this frequency is to be 5 kHz, what must the lamination thickness be, assuming the same operating flux density?
    [1.296 kHz, 0.127 mm]

14 DC machines

B Y 'DC MACHINES' we mean either generators which convert mechanical energy into electrical energy and deliver a unidirectional current, or motors which do the converse. Usually a DC machine can serve either purpose. AC machines are more widely used than DC machines, but in certain applications DC machines used to be preferred, especially in traction and whenever speed control over a wide range was desired. With the development of high-power semiconductor devices such as the thyristor and gate-turn-off thyristor (GTO) it has now become much easier to control the speed and torque of AC motors so that these are now preferred even for traction. However, the same devices have made possible the development of brushless DC motors of up to 50 kW. Regardless of the mode of excitation, what is indispensable in all electrical machines is relative motion of an electrical conductor and a magnetic field and we shall start by looking at a very simple example.

14.1 A prototype generator

Consider the single rectangular loop of conducting wire in Figure 14.1, which is of area A and is being rotated in a uniform magnetic field whose flux density is **B**. In order to supply current to an external load, the ends of the coil are connected to *slip rings* that are contacted by *brushes*, which in turn are connected to copper wires going to the generator's terminals and thence to the load. Brushes are actually spring-loaded blocks of graphite that form a low-resistance contact with the slip ring, while the slip rings are made from brass or hardened copper. Graphite is soft and produces relatively little wear in the slip rings, besides having other advantages discussed in due course.

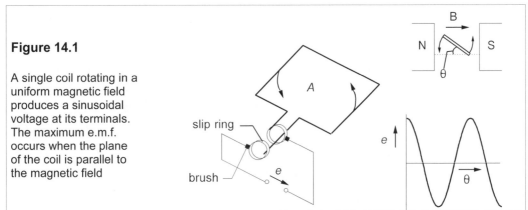

Figure 14.1

A single coil rotating in a uniform magnetic field produces a sinusoidal voltage at its terminals. The maximum e.m.f. occurs when the plane of the coil is parallel to the magnetic field

If the coil is at angle, θ, to the field, its area normal to **B** is $A\sin\theta$ and the e.m.f. is

$$e = d\Phi/dt = BA\,d(\sin\theta)/dt \qquad (14.1)$$

by Faraday's law. But we can write

$$\frac{d}{dt} = \frac{d}{d\theta}\frac{d\theta}{dt} = \omega\frac{d}{d\theta} \tag{14.2}$$

since $d\theta/dt = \omega$, whence

$$e = BA\,d(\sin\theta)/dt = BA\omega\,d(\sin\theta)/d\theta = BA\omega\cos\theta \tag{14.3}$$

The e.m.f. produced by the rotating coil is sinusoidal, reaching its maximum value of $BA\omega$ when $\theta = 0°$ (when **B** lies in the plane of the coil) and falling to zero when $\theta = 90°$ (when **B** is normal to the plane of the coil). Hence this simple machine is an AC generator.

14.1.1 Energy conversion

We have seen that the electrical power developed in a conductor in motion relative to a magnetic field is derived from the work done against the Lorentz force. Suppose the terminals of the coil in Figure 14.1 are connected to a resistance, R, while the coil itself is of negligible resistance. The instantaneous current flowing is i and the instantaneous electrical power generated, p_e, is ei, or e^2/R. We can calculate the mechanical power with the aid of Figure 14.2.

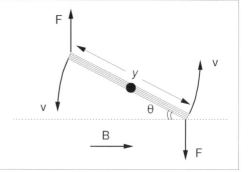

Figure 14.2

A rectangular coil rotating in a uniform magnetic field. The Lorentz force, **F**, has a component $F\cos\theta$ along the direction of motion

If the length of the (rectangular) coil parallel to the field is y, the work done to rotate the pair of conductors through an angle $\delta\theta$ is $T\delta\theta$. T is the torque, which is $yF\cos\theta$ as the component of **F** in the direction of motion is $y\cos\theta$. The rate of working is $p_m = T\delta\theta/\delta t = T\omega$ if it takes δt seconds to rotate $\delta\theta$ radians, thus

$$p_m = T\omega = \omega yF\cos\theta \tag{14.4}$$

Then substituting ilB for F (**l** is normal to **B** so $|i\mathbf{l}\times\mathbf{B}| = ilB$) gives

$$p_m = \omega yilB\cos\theta = i\omega BA\cos\theta = ei = p_e \tag{14.5}$$

since $yl = A$ and $e = BA\omega\cos\theta$ from Equation 14.3. Thus the mechanical power put in is equal to the electrical power out, ideally.

14.1.2 Commutation

In order to produce DC from the coil and magnet shown in Figure 14.1, it is necessary to provide *commutation* in the form of split slip-rings as in Figure 14.3.

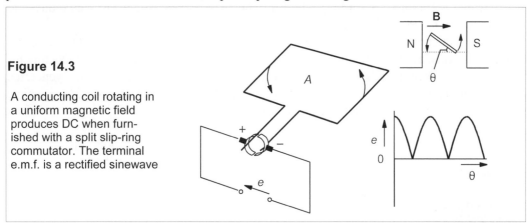

Figure 14.3

A conducting coil rotating in a uniform magnetic field produces DC when furnished with a split slip-ring commutator. The terminal e.m.f. is a rectified sinewave

At the point in the rotation of the coil when the current in brush A would reverse, the brush breaks contact with one split ring and makes contact with the other. Thus brush A is in contact only with the conductor moving down through the magnetic field on the left and must therefore be connected to the positive terminal of the generator. The coil e.m.f. is made up of a positive e.m.f. from the arm of the coil which passes in front of the N pole and an equal negative e.m.f. from the coil which passes a S pole. The total e.m.f. of the generator, though unidirectional, is sinusoidal and varies in magnitude from zero to a maximum of $BA\omega$.

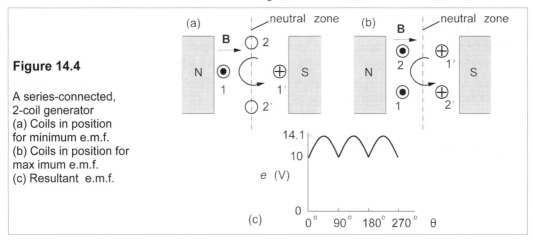

Figure 14.4

A series-connected, 2-coil generator
(a) Coils in position for minimum e.m.f.
(b) Coils in position for max imum e.m.f.
(c) Resultant e.m.f.

The large ripple of this rectified sinewave could be reduced by employing more coils connected in series, each with its own commutator segment. Consider, for example, a 2-coil, 2-pole generator as shown in Figure 14.4, which we shall suppose gives an e.m.f. of 10 V in each coil when it passes the centre of a pole face. When the coils are in the position shown in Figure 14.4a, the e.m.f. from coil 1 is 10 V while that from coil 2 is zero and the combined e.m.f. will be 10 V. Coil 2 gives no e.m.f. as it is in the magnetically neutral plane, or in the *neutral zone* of the generator.

On rotating a further $10°$, the e.m.f. of coil 1 will be $10\sin 80°$ V, while that of coil 2 will be $10\sin 10°$ V, for a total e.m.f. of $10(\sin 80° + \sin 10°)$ V or 11.6 V. Clearly, the maximum e.m.f. will be obtained from the pair of coils when they are in the position shown in Figure 14.4b, which is $10(\sin 45° + \sin 45°) = 14.1$ V. The peak-to-peak ripple has been reduced to 29% compared to 100% with one coil.

Further reduction in ripple can be obtained by using shaped pole pieces and a permeable core for the coils, as in Figure 14.5a. The flux is normal to the motion of the conductors for almost the whole time that the coil is inside the pole pieces, consequently the e.m.f. generated is nearly maximal for most of this time too, as in Figure 14.5b. We will now consider practical DC generators that incorporate these and other improvements.

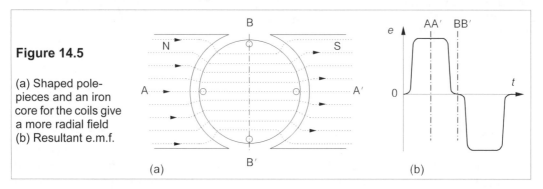

Figure 14.5

(a) Shaped pole-pieces and an iron core for the coils give a more radial field
(b) Resultant e.m.f.

(a)   (b)

14.2 **DC generators**

DC generators, or dynamos (a name invented by Lord Kelvin), are no longer so common as they once were; the advent of solid-state power-controlling devices like the SCR has allowed AC generators, which are generally lighter and cheaper for a given power output, to be used instead. A DC generator is a reversible machine and can be used both as generator and a motor; in fact many DC motors operate as generators part of the time. The principles of DC generators are therefore applicable to DC motors and provide an introduction to them. The speed of a DC motor may vary with load while the terminal voltage is constant; conversely, the speed of a DC generator is usually constant and the terminal voltage may vary under load.

14.2.1 *Practical DC generators*

The prototype DC generator described in Section 14.1 is not a very useful machine because it requires a great many ampere-turns to drive the magnetic flux across the large air-gap between the poles. The air gap can be reduced by placing the coils in slots cut into a core that consists of a cylinder of permeable steel, the whole assembly being known as the *armature*. The permeable armature offers a much smaller reluctance than air to the ing amp-turns of the field coils. The armature core is usually made from thin ($\approx$ 1 mm) laminations separated by even thinner insulating layers to reduce eddy-current losses, like a transformer core.

The flux distribution is made more uniform by shaping the pole pieces, as in Figure 14.6, and by increasing the numbers of coils and poles (always in pairs of N and S poles) the ripple can be reduced to 1 or 2%. The poles are held in place by a steel cylinder which also forms the outer case or *stator* of the dynamo.

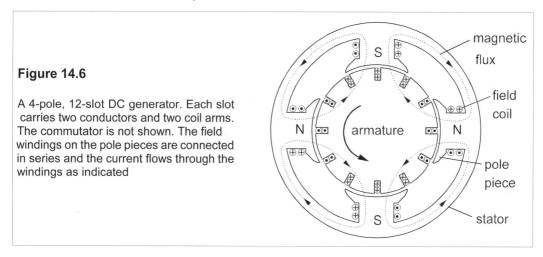

Figure 14.6

A 4-pole, 12-slot DC generator. Each slot carries two conductors and two coil arms. The commutator is not shown. The field windings on the pole pieces are connected in series and the current flows through the windings as indicated

Flux passes through the steel case and links neighbouring poles – see Figure 14.7, which shows the magnetic flux in a 4-pole, 12-slot machine. Notice that the poles alternate in polarity around the circumference of the stator. The magnetic flux is produced by passing current through the field coils wound round each pole piece. The direction of the magnetic field is given by the right-hand thumb-and-fingers rule. Fleming's right-hand (generator) rule enables one to check that the current is flowing in the correct direction in the armature conductors, that is out of the page when a conductor is next to a N pole and into the paper when it is adjacent to a S pole. A conductor equidistant from a N pole and a S pole it is in a neutral zone where ideally no e.m.f. is generated in it.

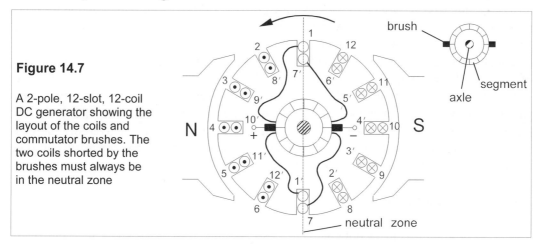

Figure 14.7

A 2-pole, 12-slot, 12-coil DC generator showing the layout of the coils and commutator brushes. The two coils shorted by the brushes must always be in the neutral zone

In the position of the armature shown in Figure 14.7, the sum of the e.m.fs. generated in the coils is a maximum; the armature must rotate 15° for the total e.m.f. to be a minimum. The ripple after commutation is greatly reduced by using many more armature coils than pole pairs.

For the moment we will consider the commutation in the 2-pole, 12-slot, 12-coil DC generator shown in Figure 14.7. Each coil has two arms which cut the field and so each slot contains two arms from different coils. Since the coils are identical they must be placed asymmetrically about the centre of the armature, with one arm in an outer position in the slot and the other in an inner position. This has little effect on the generator's operation. Only the connections of the neutral-zone coils to their commutator segments are shown, but the other coils are wired so that the ends of each are soldered to different, neighbouring segments of the commutator as shown in Figure 14.8a. Neighbouring segments are separated by thin layers of insulating material such as mica. The coil numbers are the same as the slot numbers and refer to the outermost coil in the slot. Neighbouring coils are connected to the same commutator segment so all the coils are joined in series by the commutator.

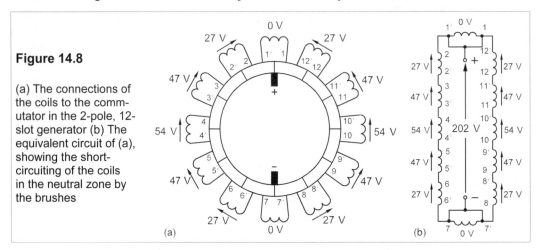

Figure 14.8

(a) The connections of the coils to the commutator in the 2-pole, 12-slot generator (b) The equivalent circuit of (a), showing the short-circuiting of the coils in the neutral zone by the brushes

As mentioned earlier, the directions of the currents in the coils can be found by Fleming's right-hand (generator) rule. With the commutator placed as shown, the e.m.fs. generated in the coils going past the N pole of the stator are positive and the e.m.fs. generated in the coils passing the S pole are negative, so that the sum of e.m.fs. round the whole of the armature coils is zero as can be seen in Figure 14.8. The brushes of the commutator short-circuit neighbouring segments of the commutator as they pass and in doing so each short-circuits one of the coils. By placing the brushes in the neutral zone of the generator, no loss of e.m.f. occurs due to this short-circuit. Because there are two brushes there are effectively two parallel sets of coils as shown in Figure 14.8b, which give a maximum e.m.f. of 202 V each. Each set of coils carries the same current, which is half the total current delivered. When the armature rotates through another 15° the brushes are in the middle of a segment, no coil is in the neutral zone nor is any shorted by a brush.

14.2.2 Calculating the e.m.f.

The average e.m.f. of this 2-pole generator can be calculated if we know the number of conductors (Z) on the armature, its speed of rotation and the flux per pole, since

$$e = \frac{Z}{2}\frac{d\Phi}{dt} = \frac{Z\Phi}{2T} = \frac{Z\Phi_{\mathrm{p}}n}{60} \tag{14.6}$$

Since there are two parallel paths in the armature winding the number of conductors per path is $Z/2$. The total flux, Φ, is twice the flux per pole (Φ_p) as there are two poles. This flux is cut by the Z coil conductors every T seconds where T is the time for one revolution. If the coil rotates at n r/min, then $1/T = n/60$. In machines with more than one pair of poles, the coils can be joined together by the commutator in different ways. In a *lap-wound* machine the coils are joined together so that the number of parallel paths is the same as the number of poles. Thus a 4-pole lap-wound machine will have four pairs of windings in parallel. Equation 14.6 therefore holds for *any* lap-wound generator.

It is possible, though less common, to arrange the coil connections to the commutator so that there are only two parallel paths no matter how many poles on the stator. Then the machine is said to be *wave wound*, and the e.m.f. it generates is greater by a factor of p than that given in Equation 14.6, where p is the number of *pole pairs (not* the number of poles). Wave-wound generators produce high voltages and relatively low currents, while lap-wound generators are for high currents and low voltages; however, the current in a conductor for a given power output will be the same whatever way it is wound.

Example 14.1

An 8-pole, lap-wound DC generator has an armature with 120 slots and has coils with 2 turns each. What will be the average e.m.f. if it rotates at 1000 r/min and the flux per pole is 28 mWb? If the current in each conductor is 25 A, what is the power output? What will be the voltage and current from the generator if it is wave connected instead? Roughly how big is such a machine?

There will be 60 coils in 60 slots, but each slot will contain two arms from different coils, for a total of 2 × 2 = 4 conductors/slot or 4 × 60 = 240 conductors altogether. Using Equation 14.6 we find

$$e = Z\Phi_p n/60 = (240 \times 0.028 \times 1000)/60 = 112 \text{ V} \qquad (14.7)$$

An 8-pole lap-wound generator will have 8 brushes (the same as the number of poles) arranged in 4 pairs, each pair comprising a positive and a negative brush. The 4 positive brushes are then wired in parallel as are the 4 negative brushes. There will be 8 parallel paths through the armature winding. The total current is therefore 8 × 25 = 200 A and the power output is given by $P = EI = 112 \times 200 = 22.4$ kW. Wave-connecting the generator will result in an e.m.f. which is larger by a factor of p, where p, the number of pole pairs is 4 in this case, so the generated e.m.f. will be 448 V, while the current will be from two parallel paths and not eight, so it will be reduced by a factor of 4 to 50 A. The power output stays the same whether the machine be wave wound or lap wound.

We estimate the physical size of the machine from the flux per pole, since

$$\Phi_p = BA_p = 0.028 \text{ Wb} \qquad (14.8)$$

where A_p is the area of a pole face and B is the flux density. Taking B to be 1 T will not be far wrong for an iron-cored machine, so Equation 14.8 amounts to $A_p = 0.028$ m^2, and as there are 8 poles the total pole area is 0.224 m^2. We can allow about 50% on the circumference for inter-pole spaces, and another 50% for the commutator and axle bearings for a surface area at the position of the pole faces (that is just outside the armature) of 1.5 × 1.5 × 0.224 = 0.5 m^2. Now this surface is cylindrical and so its area is given by

$$A = 2\pi RL = 2\pi R \times 4R = 8\pi R^2 \qquad (14.9)$$

assuming a reasonable length/diameter ratio of 2:1. Solving Equation 14.9 for R, we find $R = 0.14$ m and $L = 0.56$ m. The OD (outside diameter) will be about $4R$, or 0.56 m.

14.2.3 *Windings on multi-pole DC machines*

The coils, one of which is shown in Figure 14.9a, in a DC machine must be arranged so that when one arm of a coil in the centre of a N pole, the other arm is in the centre of a S pole. Thus if a machine has 12 poles and the armature has 72 slots, the coil *pitch* (as the number of slots spanned by a coil is called) has to be 72/12 = 6 slots.

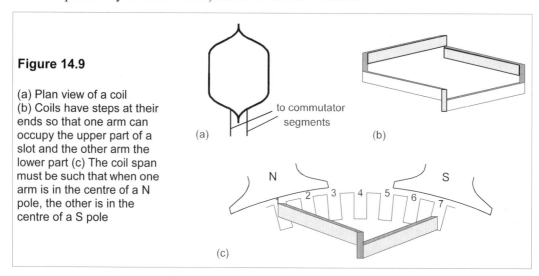

Figure 14.9

(a) Plan view of a coil
(b) Coils have steps at their ends so that one arm can occupy the upper part of a slot and the other arm the lower part (c) The coil span must be such that when one arm is in the centre of a N pole, the other is in the centre of a S pole

In double-layer armatures there are two coils per slot, as mentioned earlier, arranged so that one arm occupies the top part of one slot and the bottom part of the other. This necessitates putting a kink or step in the ends of the coils as shown in Figure 14.9b. Figure 14.9c shows how the coil fits into the upper half of slot 1 and the lower half of slot 7.

14.2.4 *The separately-excited generator*

In separately-excited machines the field coils that produce flux in the poles are supplied with current from a source that is independent of the generator. A special case of separate excitation is the permanent-magnet generator, but these are few and far between. The advantage of separate excitation is that the e.m.f. generated is under greater control than with self excitation. Figure 14.10 is the equivalent circuit of a separately-excited generator.

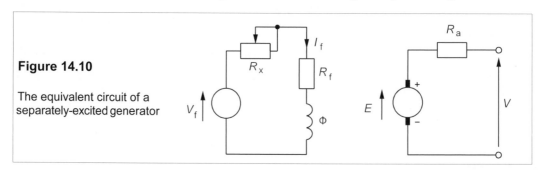

Figure 14.10

The equivalent circuit of a separately-excited generator

The resistance, R_a, in series with the brushes is the armature winding resistance. It should be placed inside the brushes strictly, but conventionally it is placed outside them. The field current is controlled by a variable resistor, or rheostat, placed in series with a constant-voltage supply. The field-winding resistance, R_f, is in series with the rheostat resistance, R_x, so that the total resistance in the field circuit is $R_t = R_f + R_x$.

The e.m.f. of the generator on open circuit is determined by its speed of rotation and the magnetic flux, so at constant speed, the flux determines the e.m.f. In turn the magnetic flux is determined by the exciting current, I_f. As I_f is increased from zero the magnetic flux rises linearly from its remanent value (the machine's magnetic circuit always has a remanent induction) until the magnetic circuit begins to saturate, and the generator e.m.f. does likewise as in Figure 14.11. When the speed is changed by a factor, α, the curve in Figure 14.11 is scaled by the same factor, which enables the em.f. to be calculated at any speed. (See Problem 14.4, for example.)

Figure 14.11

The e.m.f. of an open-circuit, separately-excited, constant-speed generator against field current. The remanent magnetization has been exaggerated. When I_f is reduced from 5 A there will be some hysteresis below point B, so that the e.m.fs. for decreasing values of I_f are greater than those for increasing values. This is shown by the dashed line. In practice, generators operate in the saturation region BC

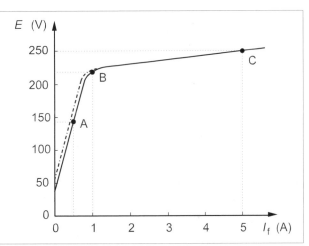

One chooses the operating point normally to be somewhere beyond the knee of the E-I_f curve in the saturation region BC so that changes in field current have less effect on the e.m.f. Suppose the generator in Figure 14.10 has the E-I_f curve of Figure 14.11, and $V_f = 100$ V and $R_f = 20\ \Omega$. Thus when $R_x = 180\ \Omega$, $R_t = 200\ \Omega$ and $I_f = 100/200 = 0.5$ A. The operating point, A, is in the initial linear region and the generator e.m.f. is 140 V. Reducing R_x to $80\ \Omega$ makes $R_t = 100\ \Omega$ and $I_f = 1$ A and the operating point moves up to the knee of the curve at point B, giving a no-load e.m.f. of 220 V. When R_x is reduced to zero, $I_f = 5$ A and the operating point, C, is in the saturation region, corresponding to an e.m.f. of 250 V. If R_x is now increased once more to $180\ \Omega$, the e.m.f. at A is found to be slightly greater than before due to hysteresis in the magnetic circuit. Below the knee of the curve the displacement is constant at about 20 V in this example, as shown by the dotted line. The remanent magnetization, which produces an e.m.f. of about 50 V, has been exaggerated. Assuming the field current and rotational speed are constant, when the generator is loaded its terminal voltage falls slightly because of the voltage drop across the armature winding, just as a battery's does. The terminal voltage is given by

$$V = E - I_a R_a \qquad (14.10)$$

where I_a is the armature (and load) current and R_a the armature resistance (see Figure 14.10). Equation 14.10 gives the V-I_a graph of Figure 14.12, however, because of saturation in the magnetic circuit, the terminal voltage falls rather faster than linearly, as shown by the lower, dashed line in Figure 14.12.

It is easy to see that the armature winding resistance must be kept down to very small values if the terminal e.m.f. is not too fall too fast. In the case of a 120V generator rated at 24 kW, the load current is 200 A and if a terminal drop of up to 10 V is permissible, the armature resistance must be 50 mΩ or less.

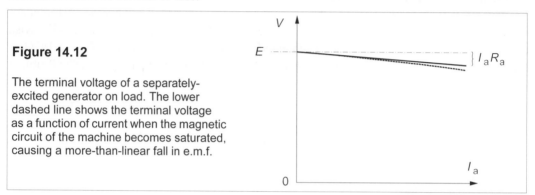

Figure 14.12

The terminal voltage of a separately-excited generator on load. The lower dashed line shows the terminal voltage as a function of current when the magnetic circuit of the machine becomes saturated, causing a more-than-linear fall in e.m.f.

14.2.5 *The shunt-wound generator*

It is inconvenient in many cases to supply separate excitation to the field coils and self excitation can be used instead, in which the generator's own e.m.f. supplies the field ampere turns. The field current is usually only 1–2% of the rated output current so the power lost in this way is negligible (the same amount of power is also lost in separately-excited generators), which is one reason why permanent magnet machines have not become more common. The equivalent circuit of a shunt-wound generator is shown in Figure 14.13.

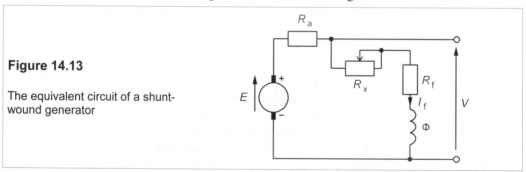

Figure 14.13

The equivalent circuit of a shunt-wound generator

The current through the shunt field coils is once more controlled by a rheostat. The no-load terminal e.m.f. is dependent on the total resistance of the field rheostat and the field winding resistance as in the separately-excited generator, but in the shunt-wound generator, the field current will now fall as the generator is loaded. The no-load e.m.f. of the generator will again depend on the magnetization curve of the magnetic circuit, and the E - I_f graph will look like Figure 14.14, that is the same shape as for a separately-excited generator.

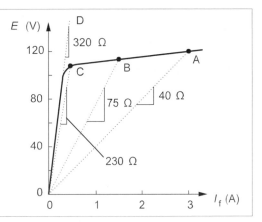

Figure 14.14

The no-load (open-circuit) characteristic of a shunt-wound DC generator. The shape is determined by the magnetization curve of the pole pieces. The critical total field resistance, R_t, is 320 Ω (line D)

The field winding resistance, $R_f = 40$ Ω and the field rheostat has a maximum resistance, $R_x(\text{max}) = 300$ Ω. What happens to the no-load e.m.f. as R_x is increased from zero to 300 Ω? When the field rheostat is set to zero, the total resistance, R_t, in the field circuit is 40 Ω, so we can draw a line of this slope on the E-I_f graph, which intercepts the generator characteristic at A, giving an open-circuit e.m.f. of 120 V and a field current of 3 A. When the rheostat is set to 35 Ω, $R_t = 75$ Ω and a line of this slope drawn from the origin will cut the characteristic at B, corresponding to a no-load e.m.f. of 113 V and $I_f = 113/75 = 1.5$ A. When R_x is set at 190 Ω, the field load line cuts the characteristic at C to give an e.m.f. of 100 V. This point is on the knee of the curve and increasing R_x still further will eventually cause the field load line to lie to the left of the characteristic, when the terminal e.m.f. will fall to zero, were it not for the residual magnetization in the magnetic circuit. In practice the terminal e.m.f. would fall to about 10 V.

The characteristic has an initial slope of 320 Ω, which is the critical resistance for R_t; correspondingly the critical value of R_x is 280 Ω. Once the field load line enters the region of the knee of the generator's characteristic, the e.m.f. will change rapidly with any change in field resistance, which is undesirable. On load, the terminal e.m.f. of the shunt-wound generator will fall by I_aR_a, and the field current will be less than its value at no load.

Example 14.2

Suppose a generator has a no-load characteristic as in Figure 14.14, a total field resistance, R_t, of 75 Ω and an armature resistance of 50 mΩ, while supplying a load of 0.6 Ω. What will the load current be?

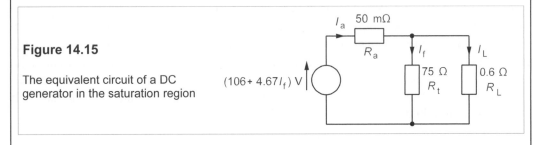

Figure 14.15

The equivalent circuit of a DC generator in the saturation region

The no-load e.m.f. with this value for R_t will be 113 V from Figure 14.14. The total resistance in the circuit is $R_L + R_a$ = 0.6 + 0.05 = 0.65 Ω, if we neglect the shunt field resistance of 75 Ω. Thus the armature and load currents are almost equal at 113/0.65 = 173.8 A. The terminal voltage will be nearly $ER_L/(R_L + R_a)$, or 104.3 V and I_f = 104.3/75 = 1.39 A. The exact solution can be made using the circuit of Figure 14.15, in which the generator's no-load e.m.f. is dependent on the field current.

In the saturation region of Figure 14.14 the no-load e.m.f. is given by

$$E = 106 + 4.67I_f \text{ V} \tag{14.11}$$

And by Kirchhoff's voltage law

$$E = I_aR_a + (I_a - I_f)R_L \tag{14.12}$$

Substituting known values into this equation yields

$$106 + 4.67I_f = 0.05I_a + 0.6(I_a - I_f) \quad \Rightarrow \quad 106 + 5.27I_f = 0.65I_a \tag{14.13}$$

But we can find I_f in terms of I_a by the current-divider rule:

$$I_f = \frac{I_aR_L}{(R_t + R_L)} = \frac{0.6I_a}{75.6} = 0.00794I_a \tag{14.14}$$

Then substituting for I_f into Equation 14.13 results in I_a = 174.3 A and I_f = 0.00794 × 174.3 = 1.384 A. The terminal voltage will be I_fR_t = 1.384 × 75 = 103.8 V. There is no practical difference between the approximate and the exact solutions when I_f is above the knee of the no-load characteristic.

Example 14.3

With the same generator as in Example 14.2, what will be the terminal voltage and load current in a 0.6Ω load if R_t = 275 Ω?

The operating point is now on the knee of the E-I_f curve, so that we cannot use the saturation E-I_f relation. A field resistance load line of slope 275 Ω intercepts the curve at E = 105 V. The circuit resistance is almost $R_a + R_L$ = 0.65 Ω, so I_a = 105/0.65 = 161.5 A and the armature's voltage drop is I_aR_a = 0.05 × 161.5 = 8 V. The terminal e.m.f. will then be 105 – 8 = 97 V. Thus the field current is 97/275 = 0.353 A. Though it is difficult to estimate from the graph, this field current corresponds to E = 103 V, and then the actual armature current is 103/0.65 = 158.5 A and the armature voltage drop becomes 158.5 × 0.05 = 8 V so the true terminal voltage is 102 – 8 = 94 V. The terminal voltage of a separately-excited generator on rated load will be some 10% less than its no-load e.m.f., compared to 13% less for a shunt-wound generator.

14.2.6 *The compound-wound generator*

The remedy for the terminal voltage drop of a generator on load is to add a field coil which is in series with the load, then as the load current increases so do the series field amp-turns. This is called *compound winding* or *compounding*. The additional amp-turns cancel the fall in shunt field amp-turns. If the compensation is insufficient to maintain the terminal voltage under load the machine is said to be *under compounded,* if it is just sufficient it is said to be *level compounded* and if it causes the terminal voltage to rise as more load current is drawn it is said to be *over compounded.* The characteristics of variously wound DC generators on load are shown in Figure 14.16.

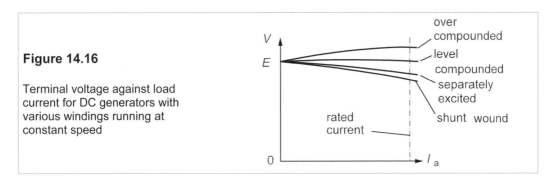

Figure 14.16

Terminal voltage against load
current for DC generators with
various windings running at
constant speed

Example 14.4

A DC generator has a no-load e.m.f. of 120 V when its field current is 2.5 A and each pole has 2000
turns on it. When drawing its rated current of 100 A the field current has to be increased to 3 A. How
many turns/pole in series with the load are needed to produce a level-compounded machine?

   The no-load amp-turns/pole are 2.5 × 2000 = 5000 and the amp-turns/pole are 3 × 2000 = 6000
when running at full load. Thus the series winding has to provide 6000 – 5000 = 1000 amp-turns/pole
when drawing 100 A, so the amp-turns are 1000/100 = 10. The series winding has very few turns
compared to the shunt winding.

14.3 DC motors

A DC generator can be run as a DC motor simply by applying a voltage to its terminals. DC
generators have almost entirely been superseded by AC generators and semiconductor rectifiers,
but DC motors are still valued for their wide range of speed and torque, accompanied by high
overall efficiency. Traction is often performed by DC motors and many special applications
such as driving conveyer belts and lifts in mines and quarries employ DC motors too. There are
essentially three types of DC motor according to the manner of field excitation: (1) shunt-
wound, including permanent-magnet and separately-excited motors, (2) series-wound and (3)
compound-wound. Each type has distinctive torque-speed characteristics.

14.3.1 The back e.m.f.

As the armature starts to rotate in the magnetic field of the pole pieces, the armature conductors
generate an e.m.f., E, which is in opposition to the applied voltage, V, and is therefore called the
back e.m.f. (counter e.m.f. in N America). The armature current, I_a, flows through the armature
winding of resistance R_a, as in Figure 14.17, and so by Kirchhoff's voltage law

$$V = E + I_a R_a \tag{14.15}$$

Therefore the current drawn by the motor is $I_a = (V - E)/R_a$. The electrical power that is
converted to mechanical work is EI_a.

   The back e.m.f. of a lap-wound DC motor is given by the same expression as the DC
generator's e.m.f., Equation 14.6:

$$E = Z\Phi_p n/60 \tag{14.16}$$

(Multiply by p, the number of pole pairs for a wave-wound motor.) From these equations it is possible to deduce the speed of a motor if the excitation is constant, as for example it is in a permanent-magnet DC motor.

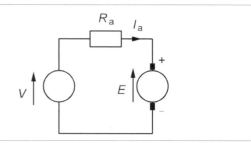

Figure 14.17

The equivalent circuit of a DC motor. The circuit is the same as that of a DC generator, only the direction of I_a is reversed as $V > E$ in a motor

14.3.2 Speed and torque in shunt-wound motors

Separately-excited and shunt-wound motors operate at speed only a few per cent less than their no-load speed, but as their speeds fall the torque they produce rises quickly. They produce maximum power at half the no-load speed, but at this point the Joule heat produced in the armature winding is as great as the mechanical work done by the motor. Thus sustained operation is only possible at speeds near the no-load speed, n_0.

Example 14.5

The armature resistance of a permanent-magnet DC motor is 0.5 Ω. With a supply voltage of 12 V its no-load speed is 1000 r/min. What is the starting current? What is the armature current at 500 r/min? What is the no-load speed if the supply voltage is 18 V?

On starting, the motor's back e.m.f. is zero, so $V = I_a R_a$ and the starting current is $I_a = V/R_a = 12/0.5$ = 24 A. At 1000 r/min the back e.m.f. of the unloaded motor will be equal to the supply voltage, 12 V. Thus at 500 r/min the back e.m.f. will be half this (by Equation 14.16), or 6 V and the voltage drop across the armature winding is 12 - 6 = 6 V, and I_a = 6/0.5 = 12 A. Increasing the supply voltage to 18 V means the no-load back e.m.f. must also be 18 V, so the speed will be 1000 × 18/12 = 1500 r/min. Running at speeds much less than the no-load speed produces large power losses in the armature winding.

The power delivered by a DC motor, P_m, is EI_a, but if it rotates at ω rad/s and supplies a torque of T N/m, that power in watts must also be $T\omega$:

$$P_m = EI_a = T\omega \tag{14.17}$$

Usually speeds are given in r/min where n r/min = $n/60$ r/sec = $2\pi n/60$ rad/s = ω; then Equation 14.17 becomes

$$P_m = EI_a = T\omega = 2\pi nT/60 \tag{14.18}$$

The torque developed at 500 r/min by the motor in Example 14.5 must be given by

$$T = 60P_m/2\pi n = 60 \times 72/(2\pi \times 500) = 1.375 \text{ Nm} \tag{14.19}$$

We can find an equation for the torque in terms of speed and supply voltage as follows:

$$E = Z\Phi_p n/60 = Kn \tag{14.20}$$

where $K = Z\Phi_P/60$ = constant. But the motor's mechanical power is given by

$$P_m = EI_a = E(V - E)/R_a = Kn(V - Kn)/R_a \tag{14.21}$$

And from Equation 14.18,

$$T = 60P_m/2\pi n = 60K(V - Kn)/2\pi R_a \tag{14.22}$$

For the motor of Example 14.5, we find K = 12/1000 and Equation 14.22 is then

$$T = 0.229(V - 0.012n) \tag{14.23}$$

The starting torque (when n = 0) from Equation 14.22 is

$$T_0 = 60KV/2\pi R_a \tag{14.24}$$

which is 2.75 Nm for V = 12 V and 4.125 Nm for V = 18 V.

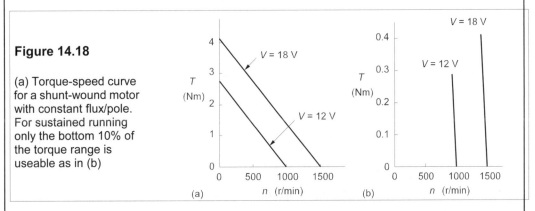

Figure 14.18

(a) Torque-speed curve for a shunt-wound motor with constant flux/pole. For sustained running only the bottom 10% of the torque range is useable as in (b)

(a) n (r/min)

(b) n (r/min)

Figure 14.18a shows a graph of Equation 14.23 for V = 12 V and V = 18 V. Other than for very small motors the range of sustainable speeds is small and near the zero-torque end of the graph. Since $T = P_m/\omega = EI_a/\omega$, substituting for E from Equation 14.16 and ω from Equation 14.18 yields the relation

$$T = Z\Phi_p I_a/2\pi \propto \Phi_p I_a \tag{14.25}$$

The flux/pole will be proportional to I_f in a machine with field windings, which is usually operated on the linear first part of the magnetization curve (unlike generators which operate on the saturated part of the curve to ensure a stable e.m.f.), so that $T \propto I_f I_a$.

14.3.3 Power

Equation 14.15 gives a DC motor's terminal voltage: $V = E + I_a R_a$; multiplying by I_a gives

$$V = E + I_a R_a \quad \Rightarrow \quad VI_a = EI_a + I_a^2 R_a \tag{14.26}$$

which can be written as

$$P_s = P_m + P_a \tag{14.27}$$

where, overlooking the power consumed by any shunt field, VI_a is the power supplied to the motor, P_s, while EI_a is the mechanical power delivered, P_m. The difference between these two, P_a, gives the copper losses caused by Joule heating in the armature winding ($= I_a^2 R_a$). Not all the mechanical power developed will be delivered to the load: there will be losses due to friction in the bearings and at the brushes, *windage* losses due to air resistance, and armature iron losses caused by hysteresis and eddy currents. There will also be copper losses in the field winding.

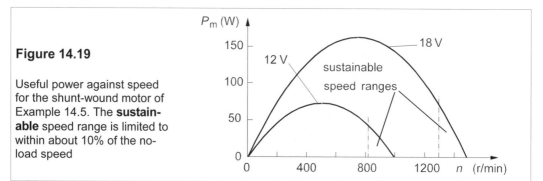

Figure 14.19

Useful power against speed for the shunt-wound motor of Example 14.5. The **sustainable** speed range is limited to within about 10% of the no-load speed

The mechanical power, P_m, is given by

$$P_m = EI_a = E(V - E)/R_a \qquad (14.28)$$

But as E is proportional to n, or $E = Kn$ (where K is a constant), for this motor with its constant field flux, and thus we have

$$P_m = Kn(V - Kn)/R_a \qquad (14.29)$$

Now $P_m = 0$ when $n = n_0$; then $V = Kn_0$ and Equation 14.29 can be written

$$P_m = Kn(Kn_0 - n)/R_a \propto nn_0 - n^2 \qquad (14.30)$$

P_m is a maximum when $n = \frac{1}{2}n_0$, as shown in Figure 14.19. The machine cannot run long at such a speed because the excessive armature current would soon burn it out.

Example 14.6

A DC motor runs off a 240V supply, has 4 poles with a flux/pole of 0.1 Wb. The armature is lap connected, carrying 288 conductors with $R_a = 0.04\ \Omega$. If $I_a = 200$ A, find the speed, the mechanical power, torque and copper losses in the armature under these conditions.

The back e.m.f., $E = V - I_aR_a = 240 - 0.04 \times 200 = 232$ V. But if $E = Z\Phi_p n/60$, then $n = 60E/Z\Phi_p = 60 \times 232/(288 \times 0.1) = 479$ r/min. The mechanical power is $EI_a = 232 \times 200 = 46.4$ kW. But this is also $2\pi nT/60$, so $T = 60P_m/2\pi n = 60 \times 46400/(2\pi \times 479) = 925$ Nm. The copper losses in the armature are $(V - E)I_a = (240 - 232) \times 200 = 1.6$ kW.

14.3.4 Speed control

The back e.m.f. of a DC motor running normally at rated load will be very nearly equal to the supply voltage (or else the efficiency will suffer); thus

$$V \approx E = Z\Phi_p n/60 \quad \Rightarrow \quad n \approx 60V/Z\Phi_p \propto V/\Phi_p \qquad (14.31)$$

The speed is nearly proportional to the supply voltage and inversely proportional to the flux/pole. We can therefore obtain speed control either by altering the supply voltage or the flux. It is because the speed of DC motors can be varied at will over large ranges whilst maintaining reasonable efficiency that they are still widely used despite the need for expensive commutation. By reducing the field current in a separately-excited or in a shunt-wound motor, the speed may be increased from its minimum value that occurs at maximum magnetic flux, a process sometimes called field weakening. Since reliable semiconductor thyristors became available, chopper circuits have become increasingly preferred for reducing the average voltage level of a DC supply and thus for controlling motor speeds. DC choppers are described in Chapter 19, Sections 19.2.2 and 19.3.2.

Example 14.7

A shunt-wound, DC motor operating from a 660V supply has a flux/pole of 60 mWb and rotates at 800 r/min. If the armature current is 110 A and its resistance is 0.2 Ω, what will the instantaneous armature current be if the flux/pole is suddenly changed to 55 mWb? What are the original torque, the final torque and the torque at the instant the flux is changed, if the mechanical power is constant? What will the speed change to eventually?

The back e.m.f. is $E = V - I_a R_a = 660 - (110 \times 0.2) = 638$ V at first, so when the flux changes from 60 to 55 mWb, this will fall proportionally to $638 \times 55/60 = 585$ V. The armature voltage drop then is $660 - 585 = 75$ V, and the armature current is $I_a = 75/0.2 = 375$ A. Note that an 8% fall in flux has produced a more-than-threefold increase in armature current.

The original torque, T_1, can be found from

$$P_m = EI_a = 2\pi Tn/60 \quad \Rightarrow \quad T = 60EI_a/2\pi n \qquad (14.32)$$

where $E = 638$ V, $I_a = 110$ A and $n = 800$ r/min, giving $T_1 = 838$ Nm. The instantaneous torque when the flux changes, T_2, is also found from Equation 14.32, with $E = 585$ V, $I_a = 375$ A and $n = 800$ r/min, to give $T_2 = 2619$ Nm. This sudden tripling of torque will rapidly accelerate the armature and load. However, as the speed rises, so will the back e.m.f., while the armature current and the torque will both fall until equilibrium is re-established. The eventual speed will be $800 \times 60/55 = 873$ r/min and the eventual torque will be $838 \times 55/60 = 768$ Nm.

14.3.5 Starting a DC motor

When the motor is stationary the back e.m.f. is zero, and should nothing be done to limit it, the starting current of the motor will be given by Equation 14.15 with $E = 0$, that is $I_a = V/R_a$, a current some ten or twenty times the rated current of the motor. For small motors where the moment of inertia of the armature is small too, acceleration is rapid and the duration of the overload is small. However, larger armatures accelerate more slowly and the overload current must be limited by a *starter*. A starter comprises several resistors connected in series with the armature winding as in Figure 14.20.

As speed builds up the resistors are switched out one by one by moving the starting handle from contact to contact. The current may thus be limited to 150% to 200% of normal full-load current. For example, if the armature resistance is 0.15 Ω, the excitation voltage is 200 V and the safe starting current is 100 A, the total resistance must be $200/100 = 2$ Ω and the starter's total resistance must be $2 - 0.15 = 1.85$ Ω. If there are four studs on the starter then it has three resistors of $1.85/3 = 0.62$ Ω each. Although the starter is in series with the shunt field winding when the handle is on stud 4 in Figure 14.20, the winding has such a large resistance compared to the starter that the reduction in field current is negligible.

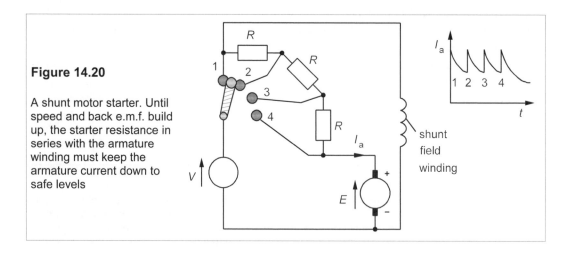

Figure 14.20

A shunt motor starter. Until speed and back e.m.f. build up, the starter resistance in series with the armature winding must keep the armature current down to safe levels

The starter always incorporates an overload trip which uses an iron-cored coil carrying a current proportional to that drawn by the motor. When the current in the coil rises above a preset limit, the coil short-circuits the 'no-volt' release. The no-volt release coil ensures that when the supply is cut off the starter arm is returned to the starting position; otherwise, when the supply is restored the starter resistance would be too low. It consists of a coil which, when energised, holds the starter arm against the pull of the return spring. When the no-volt coil is de-energised by either a supply failure or an overcurrent trip, the starter arm is released and the spring returns it to the start position.

14.3.6 *The series-wound DC motor*

The series-wound motor has distinct advantages over the shunt-wound motor for traction applications such as locomotives, where a large starting torque is highly desirable for rapid acceleration. The torque of a motor is given by $T = P_m/\omega = EI_a/\omega$. But $E \propto \omega\Phi_p$, so $E \propto \Phi_p I_a$, and as $\Phi_p \propto I_a$ in a series-wound motor, $T \propto I_a^2$. A starter, shown as R_x in Figure 14.21, is necessary because the combined resistance of the armature winding and field coils is still very small. The torque falls rapidly as speed builds up. The field coils comprise only a few turns of heavy-duty wire as they carry the full armature current.

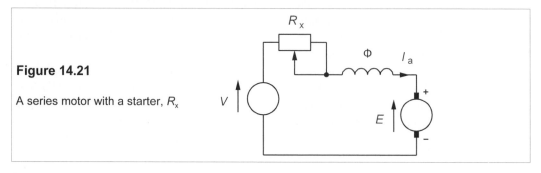

Figure 14.21

A series motor with a starter, R_x

The torque-speed relation for a series motor can be derived from the motor/generator power and e.m.f. equations: $P_m = EI_a = 2\pi Tn/60$ and $E = Z\Phi_p n/60$. Then if $\Phi_p \propto I_a$, so $E = KnI_a$, where K is a constant of proportionality. Thus

$$2\pi Tn/60 = EI_a = KnI_a^2 \quad \Rightarrow \quad T = K_1 I_a^2 \tag{14.33}$$

where K_1 is another constant. However, $V = E + I_a R_a$ (here, R_a is the *total* resistance in the armature circuit), giving

$$V = E + I_a R_a = K_1 n I_a + I_a R_a$$

$$\Rightarrow \quad I_a = V/(K_1 n + R_a) \tag{14.34}$$

Substituting for $I_a (= [V - E]/R_a)$ in Equation 14.34 yields the final torque-speed equation in terms of the starting torque, T_0 (when $n = 0$):

$$T = K_1 I_a^2 = \frac{K_1 V^2}{(Kn + R_a)^2} = \frac{T_0 R_a^2}{(Kn + R_a)^2} \tag{14.35}$$

Equation 14.35 shows that there is *no* limiting speed, n_0, as in the shunt motor: a series motor which is disconnected from its load while still supplied with power will run away. It may reach such a speed that the centrifugal force destroys the armature winding.

Figure 14.22 shows a graph of T and P_m against n. The power-speed graph is derived from $P_m = KnI_a^2$ and $I_a = V/(Kn + R_a)$, giving

$$P_m = \frac{KV^2 n}{(Kn + R_a)^2} \tag{14.36}$$

This expression is maximum when $n = R_a/K$ and $P_m(\text{max}) = (V/2R_a)^2$, as we found before with the shunt motor. Substitution of $n = R_a/K$ in Equation 14.36 leads to a torque of $\frac{1}{4}T_0$ at maximum power output.

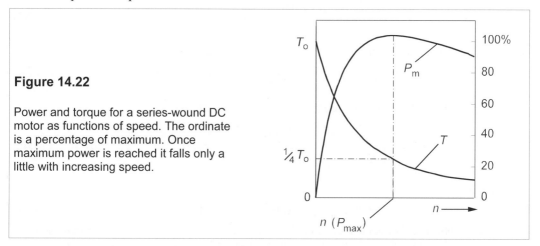

Figure 14.22

Power and torque for a series-wound DC motor as functions of speed. The ordinate is a percentage of maximum. Once maximum power is reached it falls only a little with increasing speed.

Compound-wound motors have characteristics between those of shunt and series motors, with the important advantage of high starting torque while not running away on no-load. Their precise characteristics depend on the ratio of shunt field turns to series field turns and may be more akin to one than the other. Figure 14.23 compares the torque speed characteristics of shunt, series and compound-wound DC motors.

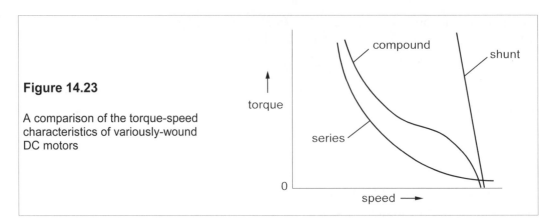

Figure 14.23

A comparison of the torque-speed characteristics of variously-wound DC motors

---

Example 14.8

A series-wound DC motor draws 100 A from a 200V supply when running at 1000 r/min. If the armature resistance is 0.1 Ω and the field winding resistance is 0.03 Ω, what additional series resistance must be placed in the armature circuit to reduce the speed to 800 r/min with a torque that is 80% of its former value?

The total armature circuit resistance is 0.13 Ω, so the voltage drop across it when drawing 100 A from the supply is 13 V, which implies that the back e.m.f. is 200 − 13 = 187 V. Taking the flux to be proportional to the armature current, then $E = KnI_a$, or

$$K = \frac{E}{nI_a} = \frac{187}{1000 \times 100} = 1.87 \times 10^{-3} \text{ V/A/(r/min)} \tag{14.37}$$

But Equation 14.32 yields for the original torque

$$T = \frac{60EI_a}{2\pi n} = \frac{60 \times 187 \times 100}{2\pi \times 1000} = 179 \text{ Nm} \tag{14.38}$$

Hence the required torque is 0.8 × 179 = 143 Nm. The mechanical power becomes

$$P_m = 2\pi Tn/60 = 2\pi \times 143 \times 800/60 = 12 \text{ kW} \tag{14.39}$$

But $P_m = EI_a = KnI_a^2$, K having been found in Equation 14.37. Therefore

$$I_a = \sqrt{\frac{P_m}{Kn}} = \sqrt{\frac{12000}{1.87 \times 10^{-3} \times 800}} = 89.6 \text{ A} \tag{14.40}$$

The back e.m.f. is

$$E = P_m/I_a = 12000/89.6 = 134 \text{ V} \tag{14.41}$$

and the armature circuit's voltage drop is $V - E$ = 200 − 134 = 66 V. Its total resistance must therefore be $(V - E)/I_a$ = 66/89.6 = 0.737 Ω. A series resistance of 0.737 − 0.13 = 0.607 Ω is needed.

14.3.7 *Commutation problems: the armature reaction*

Commutation is not as straightforward as we have made it seem. There are several problems that must be addressed to improve the commutation process, including the armature reaction, non-uniform current distribution in the brushes and the effects of coil inductance. Poor commutation causes arcing at the trailing edges of the brushes, leading to erosion and higher contact resistance. The brushes are made from graphite to reduce commutator wear and to increase the contact resistance, which leads to a more uniform current distribution. Coil inductance must be kept small since the current in the coils is reversed frequently. As a result a large e.m.f. is produced: $e = L\mathrm{d}i/\mathrm{d}t$. As the inductance goes as the square of the number of turns, the coils are usually restricted to just one turn.

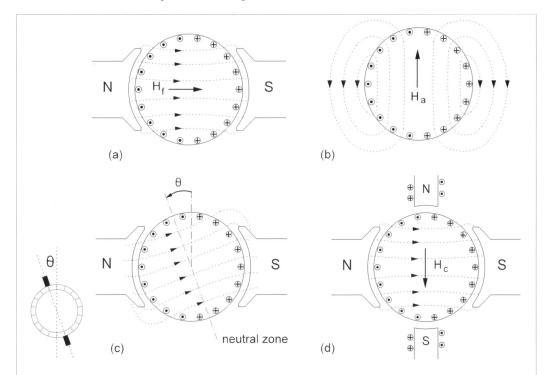

Figure 14.24 (a) The field produced by the poles alone (b) The field produced by the armature amp-turns: the armature reaction (c) Armature reaction rotates the field by θ. The brushes should be rotated by the same angle (d) Compoles provide a field, $\mathbf{H}_c$, which is equal and opposite to $\mathbf{H}_a$

Thus far we have not considered the effect on the flux distribution of the current flowing in the armature of a DC machine. This *armature reaction* causes the magnetic flux in the armature to rotate *against* the direction of rotation for a *motor* and *with* the direction of rotation for a *generator*. The field produced by the normal field winding, $\mathbf{H}_f$, is shown in Figure 14.24a and that produced by the armature current, $\mathbf{H}_a$, is shown in Figure 14.24b. The resultant flux is rotated by the armature reaction through an angle, θ, and depends on the relative magnitudes of the currents in the armature and the field windings (see Figure 14.24c). The effect of this flux rotation is to rotate in turn the neutral zone where commutation should take place, and so the brushes must be rotated by θ also. Selecting the best value for θ requires an accurate knowledge of the operational parameters of the machine. In small machines this method may be used to improve the commutation and poor commutation in some circumstances is tolerated, but in large machines it is usual to place smaller poles between the main field poles to counteract the armature reaction. These are called *compoles* or *interpoles* and are shown in Figure 14.24d. The amp-turns of the armature reaction, $\mathbf{H}_a$, are cancelled by the equal and opposite amp-turns, $\mathbf{H}_c$, of the compoles. The compole winding is in series with the armature winding ensuring cancellation at all armature currents.

14.4 **Efficiency and losses**

The efficiency of any power conversion system is defined by

$$\eta = \text{Useful Power} \div \text{Input Power} \tag{14.42}$$

which can be written also in the form

$$\eta = \frac{\text{Input Power} - \text{Losses}}{\text{Input Power}} = 1 - \frac{\text{Losses}}{\text{Input Power}} \tag{14.43}$$

Besides emphasising the role of the losses in reducing efficiency, this equation is in the right form for calculating it. The sources of power loss in DC motors and generators are identical and we shall discuss each in a separate subsection, starting with the copper losses.

14.4.1 *Copper losses*

Conductors are nearly always made of copper (though aluminium is sometimes used) and losses are caused by Joule heating in the armature winding ($= I_a^2 R_a$), the series and compole field windings (when present) and the shunt field winding (when present). In a series machine, R_a includes all the windings in series with the armature winding, since they all carry the full armature current. The copper losses will vary with temperature even when the armature and shunt-field currents are constant because of the change in resistivity of copper with temperature:

$$\rho_T = \rho_0(1 + \alpha T) \tag{14.44}$$

ρ_T is the resistivity (in Ωm) of copper at temperature T (in °C), ρ_0 is its resistivity at 0°C and α is its temperature coefficient of resistivity, which is 0.00427/°C, for annealed copper wire. Though α seems small, a rise of 40°C or 50°C makes a significant difference to the resistance of copper wire. Equation 14.44 means that the armature resistance is a convenient way to estimate the running temperature of a machine.

┌─ **Example 14.9** ───

A series DC motor is run from a constant 200V supply and takes 150 A at 20°C, when its combined armature and series field resistance is 60 mΩ. After running for an hour its temperature rises to 70°C. If the mechanical power, P_m, ($= EI_a$) is constant, what is the current at 70°C? What are the copper losses at each temperature? What fraction of the mechanical power are they?

We must use Equation 14.44, first to find the resistance at 0°C and then at 70°C. The resistance of a piece of wire is found from its length, *l*, its cross-section area, *A* and its resistivity, ρ:

$$R = \rho l / A \quad \Rightarrow \quad \rho = R A / l \tag{14.45}$$

If we ignore tiny dimensional changes with temperature, then Equation 14.45 becomes

$$R_T = R_0(1 + \alpha T) \tag{14.46}$$

At 20°C, $R_{20} = 0.06 = R_0(1 + 0.00427 \times 20) = 1.0854 R_0$, giving $R_0 = 0.06/1.0854 = 0.0553$ Ω. Then

$$R_{70} = 0.0553(1 + 0.00427 \times 70) = 0.072 \text{ Ω} \tag{14.47}$$

At 20°C the copper losses are $150^2 \times 0.06 = 1.35$ kW. The mechanical power, P_m, is

$$P_m = E I_a = (V - I_a R_a) I_a \tag{14.48}$$

At 20°C, $R_a = 0.06$ Ω, $I_a = 150$ A and $P_m = (200 - 150 \times 0.06) \times 150 = 28.65$ kW. The copper losses are $1.35 \times 100/28.65 = 4.7\%$ of the mechanical power. At 70°C, $R_a = 0.072$ Ω and the power is constant at 28.65 kW, so that Equation 14.48 yields

$$28650 = (200 - 0.072 I_a) I_a \tag{14.49}$$

Solving, $I_a = 151.5$ A (the other solution is $I_a = 2626$ A, which is physically impossible). The armature losses are then $151.5^2 \times 0.072 = 1.65$ kW, which is 5.8% of the mechanical power. The armature losses have gone up by 22% because of the increased resistance.

14.4.2 Iron losses

After copper losses, iron losses are probably the most important. The current in the armature conductors alternates as the armature rotates and consequently there must be eddy current and hysteresis losses in the armature core. It is therefore made from laminations of low-loss transformer steel which are insulated from each other, usually by surface oxide. Iron losses will depend on the speed of rotation and the armature current. There are also losses in the poles because the field they experience alters with the passage of conductors. Iron losses cause an additional drag on the armature which is much the same as frictional drag, so part of the mechanical power developed by a motor will be spent overcoming this.

14.4.3 Friction and windage losses

The frictional losses are those of the bearings and the brushes and will be proportional to the speed of the machine. Windage losses are causes by air resistance and vary as the cube of the speed. Most motors of any but the smallest power rating have an integral fan attached to the shaft to force cooling air over the windings. The power required to drive the fan is a significant source of mechanical power loss. Generally speaking the frictional losses are small and so are windage losses when fan cooling is absent.

14.4.4 Voltage drop at the brushes

There is a small voltage drop, V_b (≈ 1 V), at each of the positive and negative brushes, for a total power loss of $2V_b I_a$. In a machine where $I_a = 100$ A, this amounts to about 200 W and may be 10% of the total losses. Contact-voltage drop at the brushes is essential if uniform current distribution is to be established.

14.4.5 Maximum efficiency

Machines operating at a fairly constant output are designed so that they are then operating at maximum efficiency. When the output is variable, some idea of the duration and size of the load is needed before deciding where the point of maximum efficiency should be. However, the efficiency is not sharply peaked at any output, as we shall see. In shunt-wound, separately-excited and permanent-magnet machines the iron, frictional and windage losses will be nearly independent of the armature current, as will the field-winding losses. The remaining losses that depend on the armature current are the armature's copper losses and the brushes' electrical contact losses. Hence we can express the efficiency in the form of Equation 14.43, taking the input power as VI_a approximately:

$$\eta = 1 - \frac{I_a^2 R_a + 2V_b I_a + W_c}{VI_a} = 1 - \frac{I_a R_a}{V} - \frac{2V_b}{V} - \frac{W_c}{VI_a} \tag{14.50}$$

where $W_c = W_{Fe} + W_{fw} =$ constant, because W_{Fe}, the iron losses, and W_{fw}, the friction and windage losses, are constant. Differentiating with respect to I_a and setting to zero to find a maximum for η gives

$$d\eta/dI_a = -R_a/V + W_c/VI_a^2 = 0 \tag{14.51}$$

As a check that this is the condition for a maximum we can differentiate again:

$$d^2\eta/dI_a^2 = -2W_c/VI_a^3 \tag{14.52}$$

The second differential coefficient is negative and therefore Equation 14.51 *is* the condition for maximum efficiency. It leads to $I_a^2 = W_c/R_a$ or $I_a = \sqrt{(W_c/R_a)}$, and then $I_a^2 R_a = W_c$: the constant losses must equal copper losses in the armature for maximum efficiency.

Example 14.10

A shunt-wound DC motor operating at full load draws a current of 80 A from a 400V supply. The field current is negligible, the armature resistance is 0.3 Ω and the fixed losses (friction, windage, etc.) are 1.3 kW, while the voltage drop across each brush is 1.1 V. What will be the efficiency at full load and at half load? What is the armature current when the efficiency is maximal? What will the maximum efficiency be?

The armature copper losses at full load, $I_a^2 R_a$, are $80^2 \times 0.3 = 1.92$ kW. The brush electrical contact loss is $2 \times 1.1 \times 80 = 176$ W, making the total losses $1920 + 1300 + 176 = 3.396$ kW. The power input is $400 \times 80 = 32$ kW giving an efficiency of $1 - 3.396/32 = 89.4\%$. At half load, the constant losses, W_c, will still be 1.3 kW, but the armature current will be halved at 40 A. The brush contact losses will be halved also to 88 W, while the armature copper losses will be one quarter of their previous value (assuming the temperature is unchanged), or 480 W. Thus the total losses now become $480 + 1300 + 88 = 1.868$ kW, while the input power is halved to 16 kW. The efficiency at half load is thus $1 - 1.868/16 = 88.3\%$.

For maximum efficiency the armature copper losses must equal the fixed losses, that is $I_a^2 R_a = 1300$, or $I_a = \sqrt{(1300/0.3)} = 65.83$ A. The brush contact losses are $2 \times 1.1 \times 65.83 = 145$ W, yielding losses to a total of $1.3 + 1.3 + 0.145 = 2.745$ kW. The input power is $400 \times 65.83 = 26.33$ kW and the maximum efficiency is $1 - 2.745/26.33 = 89.6\%$. Figure 14.25 is a graph of efficiency for this motor plotted as a function of rated load. The curve is pretty flat around the point of maximum efficiency and the motor will be efficient between about 25% and 200% of rated output.

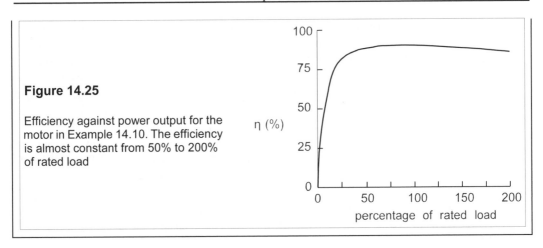

Figure 14.25

Efficiency against power output for the motor in Example 14.10. The efficiency is almost constant from 50% to 200% of rated load

η (%)

percentage of rated load

Suggestions for further reading – see Chapter 18

Problems

1. A separately-excited DC generator has six poles each providing a magnetic flux of 30 mWb. The open-circuit e.m.f. of the generator at a certain excitation field is to be 450 V when the speed is 1000 r/min. How many conductors are required if it is (a) lap wound (b) wave wound? If the armature resistance is 0.05 Ω what is the terminal voltage when it gives 200 A at 825 r/min with constant excitation field? *[(a) 900 (b) 300, 361 V]*

2. A lap-wound DC generator has six pole pairs and produces an e.m.f. of 660 V. If the armature has 300 slots and two conductors/slot what is the flux/pole at 720 r/min? What would it be if the generator were wave wound? *[91.7 mWb, 15.3 mWb]*

3. A shunt-excited DC generator has a terminal voltage of 660 V when its output is 200 kW. The field-circuit resistance is 240 Ω and the generated e.m.f. is 690 V. Find the armature resistance. What is the terminal voltage when the output is 100 kW if the e.m.f. generated is constant? *[0.0981 Ω, 675.2 V]*

4. The open-circuit characteristic of a shunt-wound DC generator running at 900 r/min is:

Terminal voltage (V)	38	75	123	154	170	183
Field current (A)	0.25	0.5	1.0	1.5	2.0	3.0

Draw a graph of these values and use it to answer these questions. What is the critical resistance of the field-coil circuit? If the generator is run at 900 r/min and an open-circuit voltage of 160 V, what is the value of resistance needed in the field-coil circuit? What would be the terminal voltage and generated e.m.f. if, with the same field-coil resistance, the generator were running at 900 r/min and supplying 20 A to a load? The generator's armature resistance is 0.46 Ω. Suppose the speed is increased to 1200 r/min, while the resistance in the field-coil circuit is changed to 110 Ω, what would the open-circuit e.m.f. be? *[152 Ω, 94 Ω, 145 V, 155 V, 228 V]*

5. A compound generator has to maintain a constant voltage of 120 V between its full load of 3 kW and no load. Without a series winding the shunt current has to be 0.278 A on full load and 0.25 A unloaded to maintain the terminal voltage at 120 V. If the number of turns/pole on the shunt winding is 1800, how many turns/pole are needed on a series winding to keep the voltage constant? Suppose the number of turns/pole were 1400 and the series winding had 2 turns/pole and a

resistance of $0.02\,\Omega$, what resistance would be needed in parallel with the series winding to maintain the terminal voltage? *[2, 0.0726 Ω]* **6** A shunt-wound DC motor runs a 600 r/min and draws an armature current of 95 A from a 440V supply when its field resistance is set to $480\,\Omega$. Its armature resistance is $0.32\,\Omega$. What is the field resistance when the speed increases to 675 r/min for the same armature current? What will the speed be if the field resistance is $480\,\Omega$ and the armature current falls to 50 A? What is the power output in each case? (Assume a linear field current relationship with magnetic flux.) *[540 Ω, 621 r/min, 38.9 kW, 21.2 kW]*

7. If the motor in the previous problem outputs 32 kW when the supply voltage is 400 V, what is the armature current? Then what is the motor speed if the field resistance is $480\,\Omega$ and the field resistance if the motor speed is 660 r/min? *[85.9 A, 600 r/min, 528 Ω]*

8. A DC shunt-wound motor runs at 1320 r/min when the armature current drawn from a 200V supply is 18 A and the field resistance is $400\,\Omega$. The armature resistance is $0.73\,\Omega$. What field resistance is required for a no-load speed of 1320 r/min? What is the no-load speed when the field resistance is $510\,\Omega$? (Assume the flux is proportional to the field current.) *[374 Ω, 1801 r/min]*

9. The pole flux in a shunt-wound DC motor follows the relation $\Phi = 60I_f - 50I_f^2$ mWb, for $0 < I_f <$ 0.5 A, where I_f is the field current. If the supply voltage is 240 V and the field resistance is $900\,\Omega$, the motor runs at 1222 r/min and takes an armature current of 20.5 A. The armature resistance is $0.96\,\Omega$. What is the no-load speed for this field resistance? What field resistance is required to produce a no-load speed of 1100 r/min? What would be the motor speed if the field resistance were 1 kΩ and the armature current 20.5 A? What is the armature current when the motor runs at 900 r/min and the field resistance is $500\,\Omega$? *[1331 r/min, 671 Ω, 1320 r/min, 15.3 A]*

10. A shunt-wound DC motor runs on a 600V supply with an armature current of 200 A when running at 740 r/min, with an armature resistance of $0.125\,\Omega$. What is the torque and the power output of the motor and the efficiency if the copper and fixed losses are equal? What will the efficiency be if the armature current is (a) 90 A and (b) 250 A, if the speed is unchanged? (Neglect all other losses.) *[1.484 kNm, 115 kW, 91.7%, 88.9%, 91.5%]*

11. A shunt-wound DC motor armature's resistance is given by $R_T = R_0(1 + 0.005T)$, where R_0 is its resistance at 0°C and T is the temperature in °C. It is designed to deliver its rated power when operated from a 440V supply with a current of 14 A. Under these conditions at steady state the armature temperature rises from ambient 20°C to 70°C where its resistance is $0.816\,\Omega$. If the field excitation and motor speed are constant, what will the armature resistance and temperature be when the power output drops to half at an ambient temperature of 20°C? What will the armature resistance and temperature be if the motor delivers its rated power output and the ambient temperature is 40°C? (Assume that the temperature rise is proportional to the copper losses.) *[0.697 Ω, 30.7°C. 0.89 Ω, 94.5°]*

12. A series-wound DC motor runs from a 650V supply and has a total armature and field coil resistance of $0.05\,\Omega$, giving a speed of 550 r/min and an armature current of 466 A. What extra resistance in series with the armature will reduce the speed to 400 r/min (a) if the torque is constant (b) if the torque is reduced by 25%? What are the corresponding power outputs? (The magnetic circuit is unsaturated.) *[(a) 0.37 Ω (b) 0.58 Ω; 212, 159 kW]*

13. If I_a(max) = 1 kA in the motor of the previous problem, what is the maximum torque at a speed of 200 r/min? What is the total armature-circuit resistance? Find the efficiency considering only the armature-circuit losses. *[23.35 kNm, 0.161 Ω, 75.2%]*

15 Three-phase systems

S AN FRANCISCO saw the opening of the first central power station in 1879, supplying DC from two Brush dynamos to arc lamps for lighting the streets. Similar systems were soon installed all over the world. AC generators began to be used for the same purpose and the two varieties of electricity supply coexisted for a time, with the advantage perhaps lying with DC generation. The invention by Tesla of the induction motor in 1887 altered the balance decisively in favour of AC generation. However, the induction motor would not start with single-phase AC unless an auxiliary coil was used, and a polyphase supply was needed. Presently three-phase generators were installed when it became apparent that considerable savings in copper were to be had, compared to single or two-phase AC systems[1]. By 1896 the Westinghouse company in the USA had settled exclusively for the three-phase system, partly because it gave smoother motor operation, and very gradually DC systems lapsed into obsolescence.

15.1 The generation of three-phase electricity

In principle, three-phase generation is achieved simply by rotating three coils or sets of coils between the poles of a pair of electromagnets as in Figure 15.1a.

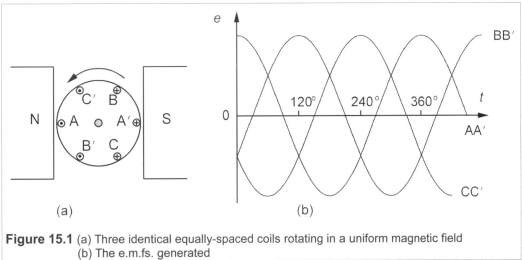

(a)   (b)

Figure 15.1 (a) Three identical equally-spaced coils rotating in a uniform magnetic field
(b) The e.m.fs. generated

---

[1] But not compared to DC; DC transmission lines, installed in several huge N American hydro-electric projects, are considerably more economical in copper.

If the coils are equally spaced at 120° on the armature, then the e.m.fs. generated will be like those shown in Figure 15.1b. Each coil is responsible for generating one of the phases and the phase sequence will be ABC, that is the e.m.f. generated in coil AA′ will be at its positive maximum 120° before that in coil BB′, while the e.m.f. of coil BB′ will be at a (positive) maximum 120° before the e.m.f. of coil CC′. Making $\mathbf{E_{AA'}}$, the e.m.f. of coil AA′, the reference phasor we can therefore write

$$\mathbf{E_{AA'}} = E_p\angle 0°; \quad \mathbf{E_{BB'}} = E_p\angle -120°; \quad \mathbf{E_{CC'}} = E_p\angle -240° \tag{15.1}$$

This is known as the *positive phase sequence*, ABC, in N America. In the UK the three phases are called, after the colour of the wire insulation, red (R), yellow (Y) and blue (B) and the positive phase sequence is RYB, or YBR or BRY, taking the letters always in the same cyclic order. If any pair of phases is interchanged then a *negative* phase sequence results, such as RBY, BYR or RBY. We shall in future always use the letters R, Y and B to designate the phases of a three-phase supply.

The sum of the three e.m.fs. is zero:

$$\begin{aligned} \mathbf{E_{AA'}} + \mathbf{E_{BB'}} + \mathbf{E_{CC'}} &= E_p\angle 0° + E_p\angle -120° + E_p\angle -240° \\ &= E_p[1 + (-1/2 - j\sqrt{3}/2) + (-1/2 + j\sqrt{3}/2)] \\ &= 0 \end{aligned} \tag{15.2}$$

15.1.1 The star-connected generator

We could connect a load to each phase with a return wire making six wires in all, but some economy is possible by using a single return (or neutral) wire, so that in effect the negative ends of the three phases are connected. This is the star (or Y) form of generator connection and is shown in Figure 15.2.

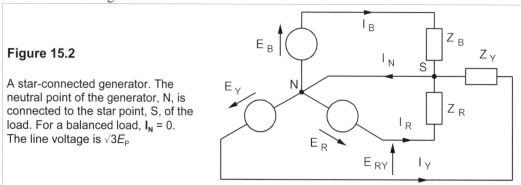

Figure 15.2

A star-connected generator. The neutral point of the generator, N, is connected to the star point, S, of the load. For a balanced load, $I_N = 0$. The line voltage is $\sqrt{3}E_P$

The phase voltages, $\mathbf{E_{RN}}$, $\mathbf{E_{YN}}$ and $\mathbf{E_{BN}}$, are all equal in magnitude and are the same as the line-neutral voltages. The line currents, $\mathbf{I_R}$, $\mathbf{I_Y}$ and $\mathbf{I_B}$, pass through the load and return through the neutral so that by Kirchhoff's current law, $\mathbf{I_N} = \mathbf{I_R} + \mathbf{I_Y} + \mathbf{I_B}$. Now if the load is *balanced*, its impedances are all equal, that is $\mathbf{Z_R} = \mathbf{Z_Y} = \mathbf{Z_B} = \mathbf{Z}$, and the line currents are $\mathbf{E_{RN}}/\mathbf{Z}$, $\mathbf{E_{YN}}/\mathbf{Z}$ and $\mathbf{E_{BN}}/\mathbf{Z}$. The neutral current is then

$$\mathbf{I_N} = \frac{\mathbf{E_{RN}} + \mathbf{E_{YN}} + \mathbf{E_{BN}}}{\mathbf{Z}} = 0 \tag{15.3}$$

That is,

$$\textit{for a balanced load, } \mathbf{I_N} = 0.$$

Thus the neutral wire could just as easily be left out to give a three-wire system, and providing the load currents are balanced, the load line-to-neutral voltages will be the same as at the generator.

Although the line-to-neutral voltage is the same as the phase voltage, the line-to-line voltage (usually known simply as the *line voltage*, $\mathbf{E_L}$) is different. From Figure 15.2 we can see that the voltage between the red and yellow phases, for example, is given by

$$\begin{aligned}
\mathbf{E_{RY}} = \mathbf{E_{RN}} - \mathbf{E_{YN}} &= E_p\angle 0° - E_p\angle -120° \\
&= E_p[1 - (-0.5 - j\sqrt{3}/2)] = E_p(3/2 + j\sqrt{3}/2) \\
&= \sqrt{3}E_p\angle 30°
\end{aligned}$$

$$(15.4)$$

where $\mathbf{E_{RN}}$ has been taken as the reference phasor. Therefore

$$\textit{for a star-connected generator, } E_L = \sqrt{3}E_p.$$

The line voltages lead the phase voltages by 30°. Like the phase voltages, the line voltages are 120° apart in phase from each other and their sum is also zero. The phasor diagram of all the voltages may be seen in Figure 15.3, which indicates how the line voltages are derived from the phase voltages.. The standard convention for describing a three-phase supply specifies the line-to-line voltages; thus a three-phase, 415V, 50Hz supply has line-to-line voltages of 415V and line-to-neutral voltages of $415/\sqrt{3} = 240$ V.

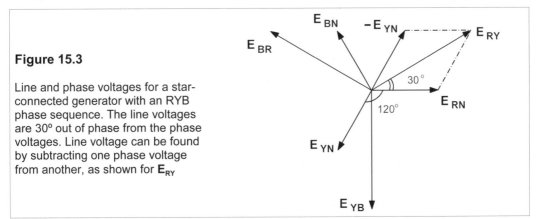

Figure 15.3

Line and phase voltages for a star-connected generator with an RYB phase sequence. The line voltages are 30° out of phase from the phase voltages. Line voltage can be found by subtracting one phase voltage from another, as shown for $\mathbf{E_{RY}}$

15.1.2 *The delta or mesh-connected generator*

Instead of connecting all the generating coils together at one end, they could be joined in series, that is to say, in mesh or delta connection as in Figure 15.4. There is no way of connecting a neutral wire to this form of generator so it must always be a three-wire system. The voltages generated are equal and 120° apart in phase as before:

$$\mathbf{E_R} = E_p \angle 0°; \quad \mathbf{E_Y} = E_p \angle -120°; \quad \mathbf{E_B} = E_p \angle -240° \qquad (15.5)$$

when the phase sequence is RYB. Clearly the phase voltage is the same as the line voltage:

$$\textit{for a delta-connected generator } E_L = E_P$$

Generators are almost invariably star-connected to reduce the phase voltage (and hence the insulation required) for a given line voltage. They also prevent additional loss due to the 3rd and 'triple n' (9th, 15th, etc.) harmonics circulating around the delta loop.

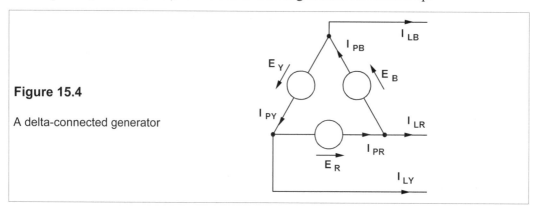

Figure 15.4

A delta-connected generator

15.2 **Balanced loads**

A balanced load is a load in which all three phases have identical impedances, which may be connected together in star or delta form. Because the loads are balanced the currents and voltages are the same for each phase and we need only do calculations for one phase.

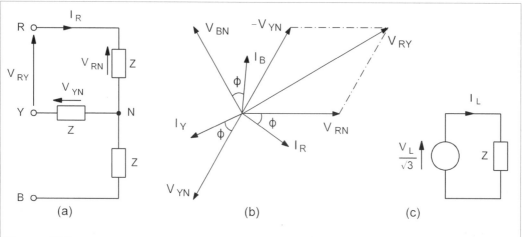

Figure 15.5 (a) A star-connected, balanced load (b) Its phasor diagram (c) The equivalent circuit per phase

15.2.1 The star-connected balanced load

Each phase of a star-connected load is connected between one of the lines and neutral as in Figure 15.5a. The line voltage is equal to the difference between two load phase voltages as shown in Figure 15.5b. Thus the line voltage, $V_L = \sqrt{3}V_P$, or $V_P = V_L/\sqrt{3}$, just like the star-connected generator. We also see that the line and phase current are identical and the equivalent circuit per phase is as in Figure 15.5c. Thus for a *star-connected, balanced load*,

$$I_P = I_L \qquad \text{and} \qquad V_P = V_L/\sqrt{3}$$

Example 15.1

Each phase of a balanced, star-connected load has an impedance of $4\angle 35°\ \Omega$ and is fed by a generator whose line voltage is 400 V. What are the line currents and the currents and voltages in each phase of the load if it is star connected? What power is consumed by the load? What is the reactive power and the apparent power in the load?

   We need only find the current and voltage for one phase of a balanced load; the other phases will have the same voltages and currents except for a phase shift of ±120°. If the line voltage is 400 V, then the voltage across each phase of the load (see Figure 15.5b) will be $400/\sqrt{3}$ = 231 V. The line and phase currents are the same, so taking the red phase as reference gives

$$\mathbf{I_R} = \mathbf{V_R}/\mathbf{Z_R} = \frac{231\angle 0°}{4\angle 35°} = 57.75\angle -35°\ \text{A} \qquad (15.6)$$

The other phase currents will be 120° from this, that is $57.75\angle -155°$ and $57.75\angle 85°$ A..

   The power consumed by each phase of the load the load is I^2R, where I is the load phase current and R is the load resistance, given by

$$R = Z\cos\phi = 4\cos 35° = 3.277\ \Omega$$

Then the power per phase is

$$P = I^2R = 57.75^2 \times 3.277 = 10.93\ \text{kW} \qquad (15.7)$$

The total power consumed will be three times the power/phase, or 32.8 kW. The reactive power per phase is

$$Q = I^2X = I^2Z\sin\phi = 57.75^2 \times 4\sin 35° = 7.652\ \text{kvar}$$

The total reactive power is 3 × 7.652 = 22.96 kvar. The apparent power per phase of the load is

$$S = V_PI_P = 231 \times 57.75 = 13.34\ \text{kVA}$$

The total apparent power in the load is therefore 40 kVA. (Check that $S^2 = P^2 + Q^2$.)

15.2.2 The balanced, delta-connected load

Each phase of a delta-connected load is placed between a pair of lines, as in Figure 15.6a. It is clear from the figure that the phase voltage of the load is identical to the line voltage, but the line current is the difference between two phase currents, as shown in the phasor diagram of Figure 15.6b. Suppose the line voltage is 415 V with a load of $4\angle 35°\ \Omega$ per phase. The red-yellow phase current is

$$\mathbf{I_{RY}} = \frac{\mathbf{V_{RY}}}{\mathbf{Z}} = \frac{400\angle 0°}{4\angle 35°} = 100\angle -35°\ \text{A} \qquad (15.8)$$

taking the red-yellow line voltage as reference. The blue-red phase current is

$$\mathbf{I_{BR}} = \frac{\mathbf{V_{BR}}}{\mathbf{Z}} = \frac{400\angle120°}{4\angle30°} = 100\angle90° \text{ A} \qquad (15.9)$$

because $\mathbf{V_{BR}}$ is +120° from $\mathbf{V_{RY}}$ (see Figure 15.6b). Thus the line current is

$$\mathbf{I_R} = \mathbf{I_{RY}} - \mathbf{I_{BR}} = 100\angle-35° - 100\angle85° = 73.2 - j157 = 173.2\angle-65° \text{ A}$$

Therefore $I_R/I_{RY} = I_L/I_P = 173.2/100 = 1.732 = \sqrt{3}$.

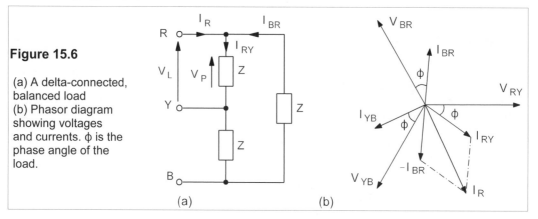

Figure 15.6

(a) A delta-connected, balanced load
(b) Phasor diagram showing voltages and currents. ϕ is the phase angle of the load.

(a)   (b)

There are two important points to note in a *balanced, delta-connected load*:

$$V_P = V_L \qquad \text{and} \qquad I_P = I_L/\sqrt{3}$$

(Compare this with the balanced, star-connected load.)

Example 15.2

What power is consumed by a balanced, delta-connected load of $4\angle35°$ Ω per phase if the line voltage is 400 V? What is the reactive power of the load? What is the apparent power in the load? What is the ratio of power consumed when the load is delta-connected and when star-connected?

The power in each phase of the load is $I_P^2R = I_P^2Z\cos\phi = 100 \times 100 \times 4\cos35° = 32.77$ kW, making the total power consumed by the load $3 \times 32.77 = 98.3$ kW. The reactive power in each phase is $I_P^2Z\sin\phi = 100 \times 100 \times 4\sin35° = 22.94$ kvar, so the total reactive power is three times this, 68.8 kvar. The apparent power/phase is $S = V_PI_P = 400 \times 100 = 40$ kVA, and the total apparent power is $3 \times 40 = 120$ kVA. Again, it is better to check this result using $S^2 = P^2 + Q^2$. The ratio of power consumed is $98.3/32.8 = 3:1$, that is the delta-connected load consumes as much power/phase as the entire star-connected load.

This result is to be expected from the star-delta transformation: if three equal impedances, $\mathbf{Z}$, in the star form are transformed to the delta form the delta impedances must each be $3\mathbf{Z}$ for the circuits to be equivalent and dissipate the same power, as in Figure 15.7. By keeping the impedances the same in the star and in the delta-connected load, the power developed in the latter is tripled.

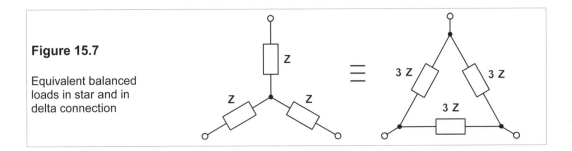

Figure 15.7

Equivalent balanced
loads in star and in
delta connection

Whether the connection is star or delta the *total* power consumed by a balanced three-phase load is *always*

$$P_T = \sqrt{3} V_L I_L \cos\phi \qquad (15.10)$$

since for a star connection

$$P_T = 3 V_P I_L \cos\phi = 3(V_L/\sqrt{3}) I_L \cos\phi = \sqrt{3} V_L I_L \cos\phi \qquad (15.11)$$

and for a delta connection

$$P_T = 3 V_L I_P \cos\phi = 3 V_L (I_P/\sqrt{3}) \cos\phi = \sqrt{3} V_L I_L \cos\phi \qquad (15.12)$$

15.3 Unbalanced loads

When the loads are unbalanced, calculating the load voltages and currents with Kirchhoff's laws becomes slightly tedious. We have not the space to discuss here the use of symmetrical component analysis by which the system is broken down into three balanced loads, but we can show how unbalanced three-phase loads may be tackled by standard circuit analysis.

15.3.1 The unbalanced, star-connected load

The star-connected load may be a three-wire or a four-wire configuration. If the neutral wire is used to connect the star point of the load with the neutral of the generator, then we cannot assume they are at the same potential because of the impedance of the neutral line. Even if this is small, in an unbalanced load the neutral current may be large and the potential difference, V_{SN}, consequently may also be substantial.

Figure 15.8 shows an unbalanced, four-wire, star-connected load. The easiest way to analyse the circuit is to find the star-neutral voltage, V_{SN}, by transforming the circuit using Norton's theorem. The circuit of Figure 15.8 then becomes that of Figure 15.9, in which the three Norton-equivalent current sources are given by $I_R = E_{RN} Y_R$, $I_Y = E_{YN} Y_Y$ and $I_B = E_{BN} Y_B$, where $Y_R = 1/Z_R$, etc. The three current sources in parallel can be added to form a single current source of magnitude

$$I_N = I_R + I_Y + I_B \qquad (15.13)$$

And the four parallel admittances can be added

$$Y = Y_R + Y_Y + Y_B + Y_N \qquad (15.14)$$

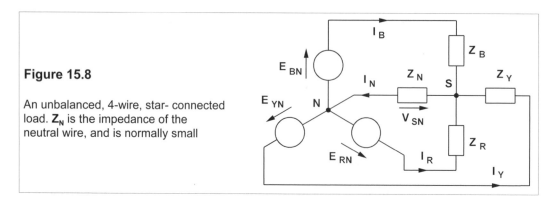

Figure 15.8

An unbalanced, 4-wire, star- connected load. Z_N is the impedance of the neutral wire, and is normally small

Hence the potential difference between the star point of the load and the neutral point of the generator is

$$V_{SN} = \frac{I_N}{Y} = \frac{I_R + I_Y + I_B}{Y_R + Y_Y + Y_B + Y_N} \tag{15.15}$$

Or, substituting for the Norton-equivalent currents,

$$V_{SN} = \frac{E_{RN}Y_R + E_{YN}Y_Y + E_{BN}Y_B}{Y_R + Y_Y + Y_B + Y_N} \tag{15.16}$$

Once V_{SN} has been found all the currents and voltages are soon calculated.

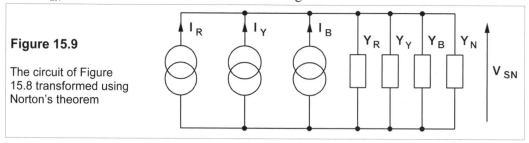

Figure 15.9

The circuit of Figure 15.8 transformed using Norton's theorem

The unbalanced, three-wire, star-connected load can be solved by using Equation 15.16, but setting $Y_N = 0$, since the neutral line is not connected (see Problems 15.5 and 15.8).

15.3.2 *An unbalanced, delta-connected load*

This is tackled in the same way as the unbalanced, three-wire, star-connected load. In Figure 15.10 the line currents entering the load sum to zero by Kirchhoff's current law:

$$I_R + I_Y + I_B = 0 \tag{15.17}$$

The phase currents are

$$I_{RY} = \frac{V_{RY}}{Z_{RY}}; \quad I_{YB} = \frac{V_{YB}}{Z_{YB}}; \quad I_{BR} = \frac{V_{BR}}{Z_{BR}} \tag{15.18}$$

where $\mathbf{V_{RY}}$ etc. are the line voltages, while the line currents from Kirchhoff's current law are

$$\mathbf{I_R} = \mathbf{I_{RY}} - \mathbf{I_{BR}}; \quad \mathbf{I_Y} = \mathbf{I_{YB}} - \mathbf{I_{RY}}; \quad \mathbf{I_B} = \mathbf{I_{BR}} - \mathbf{I_{YB}} \qquad (15.19)$$

One of these last three equations is redundant and Equations 15.17, 15.18 and 15.19 yield six independent equations to be solved for six unknown currents.

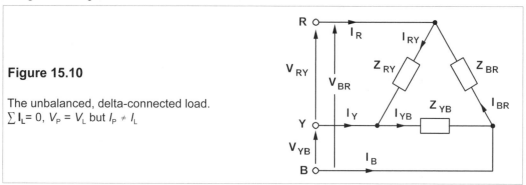

Figure 15.10

The unbalanced, delta-connected load.
$\Sigma I_L = 0$, $V_P = V_L$ but $I_P \neq I_L$

15.4 **Power measurement in three-phase circuits**

Power can be measured in any circuit by using a wattmeter. This instrument has two pairs of terminals: one pair for measuring voltage and the other for current. If the load is a balanced with four wires, only one wattmeter reading is required, since it may be connected across any phase and its reading multiplied by three to give the total power in the load. However, in three-wire loads two wattmeters are required to measure the power as shown in Figure 15.11.

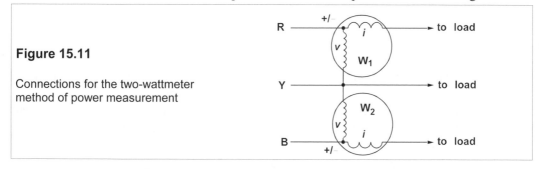

Figure 15.11

Connections for the two-wattmeter method of power measurement

The terminals designated +/− are connected to the supply side of, say, the red phase and the other current terminal connects to the load side of the same phase while the other voltage terminal connects to, say, the yellow phase. The second wattmeter is connected similarly between the blue phase and the yellow. The wattmeter readings are

$$P_1 = V_{RY}I_R\cos\phi_1 \quad \text{and} \quad P_2 = V_{BY}I_B\cos\phi_2 \qquad (15.20)$$

which for sinusoidal waveforms will result in average values of

$$p_1 = v_{RY}i_R \quad \text{and} \quad p_2 = v_{BY}i_B \qquad (15.21)$$

ϕ_1 and ϕ_2 are the phase differences between $\mathbf{V_{RY}}$ and $\mathbf{I_R}$ and between $\mathbf{V_{BY}}$ and $\mathbf{I_B}$ respectively. The wattmeters will give an average power reading for *any* three-phase load whether the waveform is sinusoidal or not, but when the load is balanced and the waveform is sinusoidal, the two wattmeter readings can be used also to measure the p.f., the reactive power and the phase angle of the load.

Consider the phasor diagram of Figure 15.12:

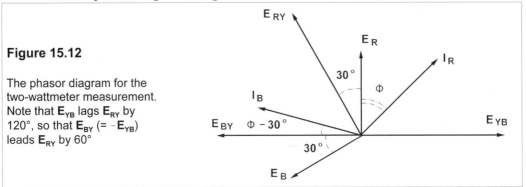

Figure 15.12

The phasor diagram for the two-wattmeter measurement. Note that E_{YB} lags E_{RY} by 120°, so that E_{BY} (= $-E_{YB}$) leads E_{RY} by 60°

The phase angle of the load is ϕ and lagging, implying that I_B lags E_B and I_R lags E_R by ϕ. The line voltages are displaced 30° from the phase voltages, as we have seen before in Figure 15.3. Thus if the first wattmeter is connected between red and yellow phases as in Figure 15.11, the angle between E_{RY} and I_R will be $\phi + 30°$ and the power it reads is

$$P_1 = E_{RY}I_R\cos(\phi + 30°) \tag{15.22}$$

Now $E_{BY} = -E_{YB}$, and we can deduce the angle between E_{BY} and I_B to be $(30° - \phi)$, making the power measured by the second wattmeter

$$P_2 = E_{BY}I_B\cos(30° - \phi) \tag{15.23}$$

The total power is the sum of P_1 and P_2, the two readings. Examining Equations 15.22 and 15.23, we can see that, even when the load is balanced, the readings of the wattmeters will differ unless $\cos(\phi + 30°) = \cos(30° - \phi)$, or $\phi = 0°$, that is when the p.f. is unity. And when $\cos(\phi + 30°) < 0$, or $\phi > 60°$, that is the p.f. is less than a half, the reading of wattmeter, W_1, is negative and the voltage coil connections must be reversed to give a reading. The load power is then the difference between the wattmeter readings. Three wattmeter readings are required to give the power in an unbalanced four-wire load. The total power is

$$P = P_1 + P_2 = E_LI_L[\cos(\phi + 30°) + \cos(30° - \phi)]$$
$$= \sqrt{3}E_LI_L\cos\phi = 3E_PI_P\cos\phi \tag{15.24}$$

Subtraction of the first wattmeter's reading from the second's produces

$$P_2 - P_1 = E_LI_L[\cos(\phi - 30°) - \cos(\phi + 30°)]$$
$$= E_LI_L\sin\phi = \sqrt{3}E_PI_P\sin\phi \tag{15.25}$$

Now the reactive power in the load, Q, is $\sqrt{3}V_LI_L\sin\phi$, which means that

$$Q = \sqrt{3}(P_2 - P_1) \tag{15.26}$$

The phase angle of the load is $\tan^{-1}(Q/P)$, or

$$\phi = \tan^{-1}(Q/P) = \tan^{-1}[\sqrt{3}(P_2 - P_1)/(P_1 + P_2)] \qquad (15.27)$$

Thus two wattmeters in a *balanced* 3-phase circuit suffice to find all the power parameters.

Problems

1. Find the power consumed by a balanced, three-phase load, each phase of which is 6 Ω in series with an inductance of 9 mH, connected in star to a 2.4kV, 60Hz supply. What is the apparent power in the load? What is the p.f. of the load? What delta-connected capacitance/phase is required to correct the p.f. to 0.95? *[727 kW, 836 kVA, 0.87, 27 μF]*
2. If the power consumed by a balanced, three-phase, inductive load that is connected to an 11kV, 50Hz supply is 1.2 MW, and the line current is 105 A, what are the components of each phase of the load (a) if it is star connected? and (b) if it is delta connected? *[36.3 Ω, 154 mH; 109 Ω, 462 mH]*
3. Two wattmeters are used to measure the power consumed in a balanced three-phase load. The reading of one wattmeter is 30.9 kW. What is the reading of the second wattmeter if the load is known to have a lagging phase angle of 35°? *[72.8 kW]*
4. A three-phase, four-wire, star-connected load comprises an impedance of 10∠30° Ω attached to the red wire, an impedance of 15∠–60° Ω attached to the yellow and a resistance of 12 Ω attached to the blue wire. If the neutral wire has a resistance of 0.1 Ω, find the voltage between the star point of the load and the generator, and hence all the currents in the load, if the phase voltage is 240 V. Check that the three load currents add up to the correct neutral-line current. What is the power consumed in (a) each load and (b) the neutral wire? *[V_{SN} = 2.022∠–24.5° V, I_R = 23.8∠–29.8° A, I_Y = 16∠–60.5° A, I_B = 20.14∠120.3° A, (a) P_R = 4.914 kW, P_Y = 1.922 kW, P_B = 4.867 kW (b) 40.9 W]*
5. Repeat the previous problem, but without the neutral wire. Check that the three load currents sum to zero. Note that if the rated load of the blue phase is 5 kW and the neutral connection is omitted, a big overload results. *[V_{SN} = 101.45∠–26.63° V, I_R = 15.61∠–13.1° A, I_Y = 17.73∠–82.4° A, I_B = 27.46∠129.75° A. P_R = 2.11 kW, P_Y = 2.358 kW, P_B = 9.049 kW]*
6. Two wattmeters are connected to a balanced, three-phase, delta-connected load, in which each of the three load impedances is 30∠36° Ω. What do the wattmeters read if the line voltage is 415 V? If the load reactance alone can be varied at what inductive reactance will one wattmeter read zero? What does the other one then read? If the reactance is kept at its first value and the resistance alone is varied until one wattmeter reads zero, what is the resistance then? What does the other wattmeter read? *[4.044 kW, 9.889 kW, X_L = 42.04 Ω, 5.323 kW, R = 10.18 Ω, 12.69 kW]*
7. A 415V, three-phase system supplies a balanced, delta-connected motor which takes 10 kW at a lagging phase angle of 30°. If resistive loads, each of 3 kW, are connected between the R and B and the B and Y lines, what will wattmeters read if they have their current coils in the R and B lines? The phase sequence is BYR. *[P_R = 8.165 kW, P_B = 7.835 kW]*
8. Figure P15.8 shows a phase sequencer. Find the currents in each phase of the sequencer and show that the phase sequence (RYB) is indicated by the connections of the bright lamp, then the dim lamp followed by the capacitor. (The supply frequency is 50 Hz, and the lamps have resistances of 1 kΩ.)

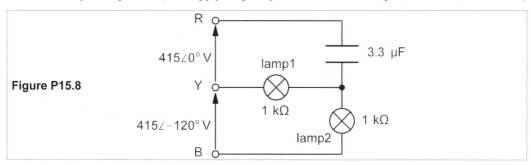

Figure P15.8

16 Transformers

P ROPERLY-DESIGNED transformers are highly efficient (up 99.5% for some multi-MVA transformers) devices that are chiefly used to step down (or step up) alternating voltages and also for matching loads. They range in size from multi-tonne, multi-MVA transformers in electricity distribution systems to 1VA, PCB-mounted transformers for portable instruments. Besides their high efficiencies, they are also notable for their reliability and freedom from the need for maintenance – virtues stemming from their lack of moving parts. AC became standard partly because transformers have these attractive properties, relegating DC to special purposes.

16.1 The ideal transformer

A transformer comprises two coils wrapped around a magnetic (more correctly ferromagnetic) core which channels nearly all the magnetic flux through them. Figure 16.1 shows a transformer, one coil of which – the *primary* – is connected to the supply, while the other coil – the *secondary* – is connected to the load. The coils are wound onto the *limbs* of the core, which are connected by a *yoke*. Transformers are reversible because primary can become secondary and vice versa: the primary is the coil connected to the primary supply, that is all.

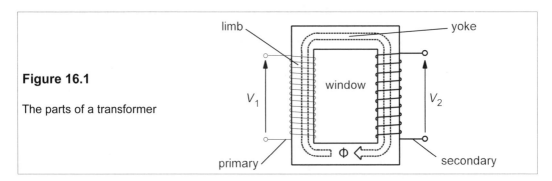

Figure 16.1

The parts of a transformer

Current passing through the primary coil produces magnetic flux in the core; if all this flux passes through the secondary coil then the e.m.f. induced in it can be found from Faraday's law of electromagnetic induction:

$$V_1 = N_1 \mathrm{d}\Phi/\mathrm{d}t \; ; \qquad V_2 = N_2 \mathrm{d}\Phi/\mathrm{d}t$$

$$\Rightarrow \qquad V_1/V_2 = N_1/N_2 \tag{16.1}$$

The voltages are proportional to the numbers of turns in the coils: the coil with the *most* turns has the *higher voltage.*

An *ideal* transformer is 100% efficient and the power put into the primary coil is equal to the power put out by the secondary: $P_1 = P_2$. That is

$$V_1 I_1 = V_2 I_2 \tag{16.2}$$

$$\text{hence} \quad V_1/V_2 = I_2/I_1 \tag{16.3}$$

Substituting for V_1/V_2 from Equation 16.1 gives

$$I_2/I_1 = N_1/N_2 = n \tag{16.4}$$

where *n* is the *turns ratio* of the transformer. From Equation 16.4 we find

$$N_1 I_1 = N_2 I_2 \tag{16.5}$$

That is

$$PRIMARY\ AMP\text{-}TURNS = SECONDARY\ AMP\text{-}TURNS$$

The current in a coil is inversely proportional to the number of turns: the *lower-current* side is the *higher-voltage* side and has the *most* turns.

Example 16.1

An ideal transformer is to be used to step down a primary voltage of 14.4 kV to a secondary voltage of 120 V, what is the required turns ratio? If the secondary current is 500 A, what is the primary current?
  Since $V_1/V_2 = N_1/N_2 = n$, then n = 14 400/120 = 120 and the required turns ratio is 120:1. The current ratio, $I_2/I_1 = n$ = 120, so that $I_1 = I_2/120$ = 500/120 = 4.17 A.

16.1.1 The dot convention

The conventional symbol for a transformer wound on a solid core is shown in Figure 16.2a.

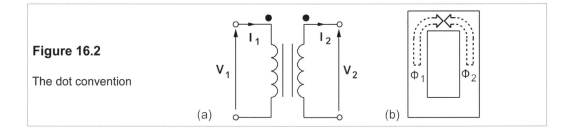

Figure 16.2

The dot convention

(a)   (b)

The coils are represented by four equal, semicircular arcs and the core by two vertical lines between them, which may optionally be omitted. If the primary terminal marked with a dot in Figure 16.2a is positive instantaneously, the current flow is *into* the dotted terminal. The instantaneous flux due to the primary current (see Figure 16.2b) circulates clockwise and the instantaneous voltage induced on the dotted secondary terminal is also positive. This voltage will cause the current to flow *out of* the dotted terminal and induce a flux which circulates anticlockwise in opposition to the flux caused by the primary current. If the bottom secondary is positive when the dotted primary terminal is positive, then *it* must be dotted. The dots are usually omitted in circuit diagrams and it is then understood that both dots are on the upper terminals.

16.1.2 Impedance transformation: load matching

Transformers enable loads to be matched with sources because they change the impedance seen by the source. Consider the voltage source of Figure 16.3a, which has an internal resistance of 50 Ω, and suppose we wish to maximise the power into an 800Ω load.

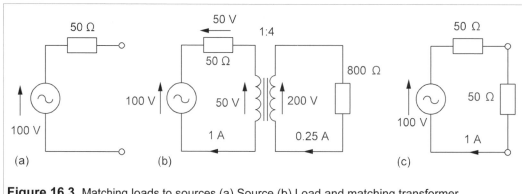

Figure 16.3 Matching loads to sources (a) Source (b) Load and matching transformer
(c) The equivalent circuit of (b)

The maximum power transfer theorem says the load required is 50 Ω, in which case the current through it would be 1 A and the power consumed 50 W. If load matching is accomplished with an ideal transformer, the power developed in the 800Ω load must be 50 W also, while the source would supply 1 A to the transformer primary. The load current is then found from

$$P_L = I_2^2 R_L = 50 \text{ W} \tag{16.6}$$

with $R_L = 800$ Ω, yielding $I_2^2 = 50/800$ and $I_2 = 0.25$ A. Since the primary current is 1 A, by Equation 16.4 the turns ratio must be 1:4, as in Figure 16.3b.

The load power, $I_2^2 R_L$, is $(nI_1)^2 R_L$, also from Equation 16.4, or $I_1^2 \times n^2 R_L$, so that

$$I_1^2 \times n^2 R_L = I_1^2 R_{eq} \tag{16.7}$$

where R_{eq} is the load resistance *referred to the primary*:

$$R_{eq} = n^2 R_L \tag{16.8}$$

The transformer has transformed the load, as far as the source is concerned, into one of 50 Ω, as in Figure 16.3c. The voltage on the secondary side is four times that on the primary side, but it is not 400 V, but 200 V, because 50 V is dropped across the source resistance in the primary circuit, leaving a 50 V drop across the transformer primary – besides, if the current through the 800Ω load is to be 0.25 A, the voltage across it must be 200 V. For an impedance the analysis identical, but R_L must be replaced by Z_L.

16.1.3 *Flux, e.m.f. and applied voltage*

We have said before, a transformer obeys Faraday's law of electromagnetic induction:

$$e = -N \mathrm{d}\Phi/\mathrm{d}t \tag{16.9}$$

If the flux through the coil is given by $\Phi = \Phi_m \cos \omega t$, then

$$e = -N \mathrm{d}\Phi/\mathrm{d}t = \omega N \Phi_m \sin \omega t \tag{16.10}$$

Taking the r.m.s. value of e gives

$$E = \omega N \Phi_m/\sqrt{2} = 2\pi f N \Phi_m/\sqrt{2} = 4.44 f N \Phi_m \tag{16.11}$$

Writing $B_m A$ for Φ_m, where B_m is the core's peak flux density in Tesla (T) and A is its cross-sectional area, the r.m.s. or effective e.m.f., E, is given by

$$E = 4.44 B_m N A f \tag{16.12}$$

Example 16.2

A transformer core has a 50 mm × 50 mm square cross-section with a maximum flux density of 1.2 T. If the e.m.f. in the primary is 230 V at 50 Hz, when the secondary is open circuit, find the number of turns on primary and secondary if the turns ratio is 2:1.

The flux, $\Phi_m = B_m A$, where B_m is the maximum flux density and A the cross-sectional area of the core. Therefore, using Equation 16.12,

$$E_1 = 230 = 4.44 \times 1.2 \times N_1 \times (50 \times 10^{-3})^2 \times 50 \tag{16.13}$$

leading to $N_1 = 345$ turns and then $N_2 = 0.5 N_1 = 173$ (there must be whole numbers of turns). Notice that if the frequency were to be much higher, say 2 kHz, the required number of turns for the secondary would decrease by 2000/50, or 40 times, to 9! A transformer of a given rating working at a higher frequency needs a smaller core than one of the same rating operating at a lower frequency.

When a transformer is on no load the secondary is open circuit, and application of an alternating voltage, V_1, to the primary results in a flow of AC which produces an alternating flux in the core. This alternating flux induces an e.m.f., E_1, in the coil almost exactly equal and opposite to the applied voltage. However, a small magnetizing current ($\approx$ 1–2% of the full load primary current) flows, making E_1 slightly less than V_1. When the secondary is on full load the secondary current produces a flux in opposition to the primary flux, so reducing both it and the primary e.m.f.. The difference between primary and secondary amp-turns is

$$N_1 I_1 - N_2 I_2 = \Phi \mathcal{R} = \frac{\Phi l}{\mu A} \tag{16.14}$$

$\mathcal{R}$ is the reluctance of the magnetic circuit, l is its length, A its cross-sectional area and μ its permeability. The reduction in the induced e.m.f. in the primary means that there is now a much larger difference between V_1 and E_1, say 2%, in which case the primary current is 20 times the magnetizing, or no-load current. But because the difference between V_1 and E_1 is still small the ideal transformer equations are very nearly exact.

16.2 **Transformer testing**

Though all but small and special-purpose transformers are nearly ideal under normal operating conditions they still exhibit small but significant departures from ideality. These are seen in the consumption of current and power when on standby (that is with no load) and in the change in secondary voltage when loaded. Whether loaded or not the transformer always consumes power and heats up to a greater or lesser degree. An adequate working model of the trans-former can be constructed from the results of two standard tests: the *open-circuit test* and the *short-circuit test*. This working model can be used to predict the transformer's performance with an accuracy sufficient for all practical purposes.

16.2.1 The open-circuit test

The test is made using the circuit of Figure 16.4a, in which the transformer's secondary is open circuit while the primary is connected to its rated supply. The power consumed is measured by the wattmeter, W, and the current, I_0, drawn from the supply is measured by the ammeter, A. When the transformer is not connected to a load, the secondary coil does nothing; the primary coil nevertheless takes a small current, I_0, made up of two components: the magnetizing current which provides the useful flux linking the coils, I_M, and the current supplying the power lost in the core, I_R. I_R remains constant whether the transformer is loaded or not. I_M lags I_R by 90°, so the transformer's equivalent circuit must contain a core-loss equivalent resistance, R_M, in parallel with an inductive reactance, X_M, the *magnetizing reactance*. These are placed across the primary supply in parallel with the primary winding of an ideal transformer, which draws no current.

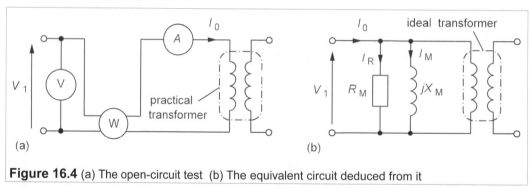

Figure 16.4 (a) The open-circuit test (b) The equivalent circuit deduced from it

The primary voltage, V_1, is measured by the voltmeter and then R_M is found from

$$P = V_1^2/R_M \qquad (16.15)$$

P being the power reading of the wattmeter. The apparent power drawn from the supply is $S = V_1 I_0$, the product of the voltmeter and ammeter readings. The reactive power is then

$$Q = \sqrt{S^2 - P^2} = V_1^2/X_M \qquad (16.16)$$

from which X_M is found. An example will help clarify the method.

Example 16.3

The results of an open-circuit test on a 240V/120V, 1 kVA transformer were:

Voltmeter reading: 240 V. Wattmeter reading: 23 W. Ammeter reading: 196 mA

Find the transformer's equivalent core-loss resistance, R_M, and magnetizing reactance, X_M.
From Equation 16.15 we find $R_M = V_1^2/P = 240^2/23 = 2.5$ kΩ. The apparent power drawn is $S = V_1 I_0$ = 240 × 0.196 = 47 VA, so that $Q = \surd(S^2 - P^2) = \surd(47^2 - 23^2) = 41$ var. Then from Equation 16.16 we find $X_M = V_1^2/Q = 240^2/41 = 1.4$ kΩ.

16.2.2 The short-circuit test

In this test the secondary is short-circuited and then the primary voltage is carefully increased until the secondary current attains its rated value. The primary voltage and power consumed are both measured; Figure 16.5a shows the experimental arrangement. Because the primary voltage is only about 10% of its rated value, the current drawn by the parallel equivalent components, R_M and X_M, is negligible. The power measured by the wattmeter is the power consumed by the resistances of the primary and secondary windings, which can be combined into an equivalent series resistance, R_e, referred to the primary, as shown in Figure 16.5b. The apparent power is greater than the power measured by the wattmeter because there is an equivalent leakage reactance, X_e, which draws vars from the supply. This is in series with R_e as shown in Figure 16.5b. (Leakage reactance is discussed further in Section 16.3.2) Thus the equivalent circuit for the short-circuit test comprises R_e and X_e in series with the primary of an ideal transformer taking rated primary current, I_1.

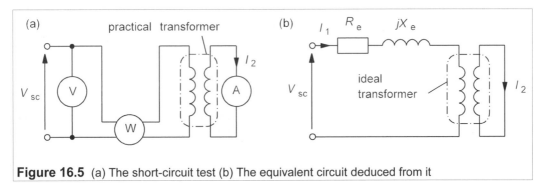

Figure 16.5 (a) The short-circuit test (b) The equivalent circuit deduced from it

The primary current, I_1, in Figure 16.5b is the rated primary current which can be found from the nameplate data of the transformer. The equivalent series resistance, R_e, in the primary is found from

$$P = I_1^2 R_e \qquad (16.17)$$

And the leakage reactance, X_e, is found from the reactive power, Q, since

$$Q = \sqrt{S^2 - P^2} = I_1^2 X_e \qquad (16.18)$$

An example will make the method of calculation clearer.

Example 16.4

The results of a short-circuit test on a 240V/120V, 1 kVA transformer were:

Voltmeter reading: 24 V. Wattmeter reading: 42 W. Ammeter reading: 8.33 A

Find the equivalent winding resistance and leakage reactance, referred to the primary.
   The transformer turns ratio is $V_1/V_2 = 240/120 = 2:1$, so the rated primary current is $I_1 = I_2/n = 8.33/2 = 4.17$ A. Then Equation 16.17 gives $R_e = P/I_1^2 = 42/4.17^2 = 2.4\ \Omega$. The apparent power is $S = V_{sc}I_1 = 24 \times 4.17 = 100$ VA, so that $Q = \sqrt{(S^2 - P^2)} = \sqrt{(100^2 - 42^2)} = 91$ var. And Equation 16.18 gives $X_e = Q/I_1^2 = 91/4.17^2 = 5.2\ \Omega$.

Per-unit values

These are often used to simplify calculations in transformers and machines. They are based on rated values of power, current and voltage and measured Z_e, sometimes called the short-circuit impedance, referred to the primary. The rated or base impedance is $Z_n = V_n/I_n = V_n^2/S$, where $n = 1$ or 2 for primary or secondary respectively. The per-unit impedance is $z = Z_e/Z_1$. If a current or voltage or power is half the rated value it is said to be 0.5 per unit. Some per-unit values are given below for well-designed transformers with iron-silicon cores:

$$r_e = 0.03 - 0.004 \quad x_e = 0.05 - 0.1 \quad r_M = 40 - 160 \quad x_M = 50 - 500$$

The first Figure is for kVA and the second for MVA transformers and we multiply by Z_1 to find the values referred to the primary. Thus a 15kVA, 240V/60V transformer will have $Z_1 = 240^2/15000 = 3.84\ \Omega$ and then $R_e = 0.03 \times 3.84 = 115$ mΩ, $X_e = 192$ mΩ, $R_M = 154\ \Omega$, $X_M = 192\ \Omega$, and $Z_e = \sqrt{(R_e^2 + X_e^2)} = 224$ mΩ, so that $z = Z_e/Z_1 = 0.224/3.84 = 0.058$, or 5.8%. Transformer impedances are often given as percentages in data sheets.

16.2.3 The transformer on load: voltage regulation

One consequence of the winding resistance and leakage reactance is that the secondary voltage varies slightly according to the current drawn by the load. This is called voltage regulation and is defined by

$$reg = (V_2 - V_{2L})/V_2 \tag{16.19}$$

where V_2 is the magnitude of the rated secondary voltage and V_{2L} is the magnitude of the secondary voltage on load. Normally the voltage regulation is a few percent, except for special types or very small transformers.

   We can see how voltage regulation arises from the equivalent circuit by examining Figure 16.6a which shows a load attached to the secondary of the ideal transformer of Figure 16.5b. Since the transformer is ideal, $V_{2L} = E_2 = E_1/n$, and for any transformer rated $V_2 = $ rated V_1/n, so that Equation 16.19, expressed as a percentage, becomes

$$\%reg = \frac{100(V_1 - E_1)}{V_1} \tag{16.20}$$

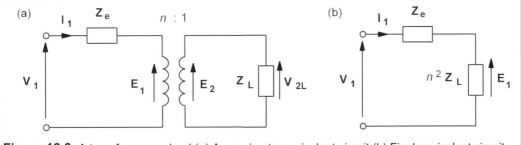

Figure 16.6 A transformer on load (a) Approximate equivalent circuit (b) Final equivalent circuit

But the load impedance, $\mathbf{Z}_L$ $(= \mathbf{V}_2/\mathbf{I}_2)$, can be referred to the primary of the ideal transformer as an impedance of $n^2\mathbf{Z}_L$ $(\approx \mathbf{V}_1/\mathbf{I}_1)$, giving the equivalent circuit of Figure 16.6b. By the voltage-divider rule we see that

$$\mathbf{E}_1 = \frac{n^2\mathbf{Z}_L\mathbf{V}_1}{n^2\mathbf{Z}_L + \mathbf{Z}_e} = \frac{\mathbf{V}_1}{1 + \mathbf{z}} \tag{16.21}$$

where $\mathbf{z}$ is the relative impedance of the transformer, or the per-unit impedance if the load is rated, that is

$$\mathbf{z} = \frac{\mathbf{Z}_e}{n^2\mathbf{Z}_L} = \frac{Z_e\angle\phi_e}{n^2Z_L\angle\phi_L} = z\angle(\phi_e - \phi_L) = z\angle\theta = z\cos\theta + jz\sin\theta \tag{16.22}$$

We need only the magnitude of $\mathbf{E}_1$ to substitute into Equation 16.20 for the percentage voltage regulation:

$$E_1 = \frac{V_1}{|1 + \mathbf{z}|} = \frac{V_1}{|1 + z\cos\theta + jz\sin\theta|}$$

$$= \frac{V_1}{\sqrt{(1 + z\cos\theta)^2 + z^2\sin^2\theta}} \approx \frac{V_1}{1 + z\cos\theta} \tag{16.23}$$

Then Equation 16.20 becomes

$$reg = \frac{V_1 - E_1}{V_1} = 1 - \frac{V_1}{E_1} \approx 1 - \frac{1}{1 + z\cos\theta} \approx z\cos\theta \tag{16.24}$$

The regulation is therefore worst for a given load when $\cos\theta = 0$, that is when $\phi_e = \phi_L$.

Example 16.5

A 1kVA, 240V/120V transformer has a short-circuit impedance, referred to the primary, $\mathbf{Z}_e = 2.4 + j5.2\ \Omega$. Find the secondary voltage and the percentage regulation when connected to (a) rated load at unity p.f. (b) rated load at a lagging p.f. of 0.8 (c) rated load at a leading p.f. of 0.8 and (d) an overload of 30% at unity p.f.

(a) Rated load implies a load of $Z_L = V_2/I_2$ (both rated) in the secondary, or a load of $n^2Z_L = V_1/I_1 = V_1^2/V_1I_1 = V_1^2/S$ (all rated values) referred to the primary. This is $240^2/1000 = 58\ \Omega$. In polar form $\mathbf{Z}_e = 5.7\angle65°$ so that $z = Z_e/(n^2Z_L) = 5.7/58 = 0.098$. Then $\theta = \phi_e - \phi_L = 65°$, since $\phi_L = 0°$ for unity p.f.;

thus $z = 0.098\angle 65°$. Equation 16.24 gives the voltage regulation as $0.098\cos 65° = 0.041 = 4.1\%$, so $V_{2L} = (1 - 0.041)V_2 = 0.959 × 120 = 115$ V.

(b) A lagging p.f. of 0.8 means that $\cos\phi_L = 0.8$ and $\phi_L = 37°$. Then $\theta = \phi_e - \phi_L = 65° - 37° = 28°$ and $z\cos\theta = 0.098\cos 28° = 0.087 = 8.7\%$. Thus $V_{2L} = (1 - 0.087)V_2 = 0.913 × 120 = 110$V. The regulation with rated load and a lagging p.f. is quite large.

(c) A leading p.f. of 0.8 implies that $\phi_L = -\cos^{-1}0.8 = -37°$ and $\theta = 65° + 37° = 102°$. Thus $z\cos\theta = -0.02 = -2\%$, and $V_{2L} = (1 + 0.02)V_2 = 122$ V. The regulation is negative because the secondary voltage is *higher* than rated.

(d) An overload of 30% means that the primary is taking 30% more than rated current which is 1.3 × 4.17 = 5.42 A. The load impedance referred to the primary is therefore $V_1/I_1 = 240/5.42 = 44\ \Omega$ and $z = 5.7/44 = 0.13$. Since $\phi_L = 0°$, $\theta = \phi_e = 65°$ and the voltage regulation is $0.13\cos 65° = 0.055 = 5.5\%$. Then $V_{2L} = (1 - 0.055) × 120 = 113$ V.

16.2.4 Transformer efficiency

The results of the open-circuit and short-circuit tests can be combined to give an approximate equivalent circuit for a practical transformer, as shown in Figure 16.7.

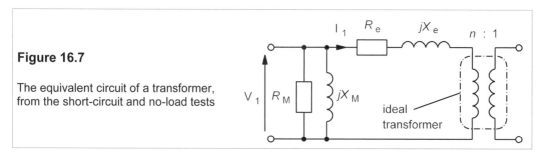

Figure 16.7

The equivalent circuit of a transformer, from the short-circuit and no-load tests

This takes into account:

(a) Core losses, $R_M = 2.5\ k\Omega$, for the example chosen
(b) Magnetizing reactance, $X_M = 1.4\ k\Omega$
(c) Winding resistance, $R_e = 2.4\ \Omega$
(d) Leakage reactance, $X_e = 5.2\ \Omega$

From this equivalent circuit we can calculate the efficiency of the transformer under various loading conditions, which is given by

$$\eta = P_{out}/P_{in} \qquad (16.25)$$

power being real power in W. But the power input is V_1I_1 (at unity p.f.), and the power out is given by

$$P_{out} = P_{in} - \text{losses} \qquad (16.26)$$

The losses have two components: the core losses, modelled by R_M, which are constant no matter what the load, and the winding losses, which are modelled by R_e and which are load dependent. The core losses, P_{Fe}, are also called *iron losses* since most transformer cores are made from iron alloys. The winding losses, P_{Cu}, are also known as *copper losses* because windings were made of copper (aluminium is often used in power transformers). Looking at Figure 16.7 we see that

$$P_{\text{Fe}} = V_1^2/R_M \tag{16.27}$$

and that

$$P_{\text{Cu}} = I_1^2 R_e \tag{16.28}$$

Then Equation 16.26 can be expressed as

$$P_{\text{out}} = V_1 I_1 - P_{\text{Fe}} - P_{\text{Cu}} = V_1 I_1 - V_1^2/R_M - I_1^2 R_e \tag{16.29}$$

And this can be substituted into Equation 16.25 to produce

$$\eta = \frac{V_1 I_1 - V_1^2/R_M - I_1^2 R_e}{V_1 I_1} = 1 - \frac{V_1}{I_1 R_M} - \frac{I_1 R_e}{V_1} \tag{16.30}$$

The maximum efficiency is found by differentiating with respect to I_1 and equating to zero:

$$\frac{d\eta}{dI_1} = \frac{V_1}{I_1^2 R_M} - \frac{R_e}{V_1} = 0$$

which gives

$$V_1^2/R_M = I_1^2 R_e \tag{16.31}$$

The term on the left we recognise as the core losses and that on the right as the copper losses. The efficiency is therefore maximum when

Iron losses = Copper losses

Example 16.6

(a) At what load is the transformer of the previous examples most efficient and what is its maximum efficiency? What is its efficiency on (b) full load at unity p.f (c) half load at unity p.f. and (d) 30% overload at a p.f. of 0.8?
    The iron (core) losses are independent of the load and are given by

$$P_{\text{Fe}} = V_1^2/R_M = 240 \times 240/2500 = 23 \text{ W}$$

These are of course the same as the wattmeter reading in the open-circuit test.
    (a) For maximum efficiency the copper losses must also be 23 W, but these are given by

$$P_{\text{Cu}} = I_1^2 R_e = 2.4 I_1^2 = 23 \implies I_1 = \sqrt{(23/2.4)} = 3.1 \text{ A}$$

The rated primary current is 4.17 A; that for maximum efficiency is 74% of that, and so the load for maximum efficiency is 74% of rated. At this load the power in is $V_1 I_1 = 240 \times 3.1 = 744$ W, and the losses are 23 + 23 = 46 W so that the power out is 744 − 46 = 698 W and the maximum efficiency is

$$\eta_{\text{max}} = 698/744 = 93.8\%$$

    (b) On full load the power in is $V_1 I_1 = 240 \times 4.17 = 1$ kW. The iron losses are still 23 W and the copper losses are
$$P_{\text{Cu}} = I_1^2 R_e = 4.17^2 \times 2.4 = 42 \text{ W}$$

which is the same as the wattmeter reading in the short-circuit test. The total losses are 23 + 42 = 65 W, so the power out is 1000 − 65 = 935 W and the efficiency is 93.5%, virtually the same as the maximum efficiency.

(c) At half load the input power is 500 W. The iron losses are constant, 23 W, and the copper losses are $P_{Cu} = I_1^2 R_e = (4.17/2)^2 \times 2.4 = 10.4$ W, which is a quarter of the full-load figure because the copper losses are proportional to I_1^2 while the power input is proportional to I_1. The total losses are 23 + 10.4 = 33.4 W, so the power out is 500 − 33.4 = 466.6 W and the efficiency is 466.6/500 = 93.3%.

(d) At a 30% overload, $I_1 = 1.3I_1$(rated), and so the copper losses are $1.3^2 \times 42 = 71$ W, while the iron losses stay at 23 W, giving total losses of 94 W. The apparent power in is $S = 1.3$ kVA, so the real power in is $S\cos\phi = 1300 \times 0.8 = 1040$ W, and the real power out is 1040 − 94 = 946 W, giving an efficiency of 946/1040 = 91%.

Note that the efficiency is pretty close to maximal in all these examples.

All-day efficiency is the efficiency of a transformer which is permanently connected to its supply, and thus consuming power even when on standby. It is found by considering the energy output over a 24h period and dividing this by the energy input over the same period. The all-day efficiency is clearly going to be less than the maximum efficiency because the transformer load varies. (See Problem 16.13.)

16.3 Practical transformers

16.3.1 Losses

Though small, transformer losses are of great economic and practical importance: all the electricity from power stations is transformed up from 25 kV to 400 kV for long-distance transmission, down to 132 kV for local distribution and then to 11 kV at substation level, finally to 415 V or 240 V for light industrial and domestic use. If the losses at each transformer were 1%, it would result in 4 or 5% of all our electricity production being wasted as heat in transformers[1]. Coil, winding or copper losses are caused by Joule heating in the primary ($I_1^2 R_1$) and secondary ($I_2^2 R_2$) coils, and are minimised by making the windings from highly conducting material, usually copper. The coil losses are dependent on the coil resistances (a function of the number of turns and the wire diameter) and should be arranged to be about the same as the core losses. The resistance and leakage reactance of the secondary winding are made so that after referring to the primary (that is multiplying by n^2) they are equal to those of the primary winding.

Core or iron losses, being independent of the load on the transformer, occur all the time the supply is connected; minimizing these losses is therefore especially important. They comprise hysteresis and eddy-current losses, both of which must be kept to a minimum. Eddy currents induced in the core circulate in planes normal to the core flux, so may be reduced by making the core of thin laminations, insulated from each other, as indicated in Figure 16.8. When the laminations are made from iron-silicon alloys they are about 0.3 mm thick in a 50 Hz transformer. Of late very thin metal-glass (based on iron-boron alloys) tape has been produced with very low eddy current and hysteresis losses, and this is not only used in high-

---

[1] It has been estimated by Allied Signal, a US transformer maker, that core losses in distribution transformers in AD 2000 were about 60 TWh in the USA, costing about $4.5 billion

frequency transformers but is finding its way into medium-power transformers (up to 100kVA). Since the eddy-current losses are proportional to f^2, high-frequency cores are usually made from non-conducting materials such as ferrites.

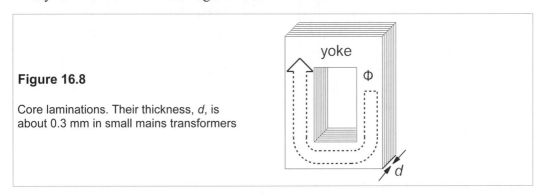

Figure 16.8

Core laminations. Their thickness, d, is about 0.3 mm in small mains transformers

Hysteresis losses are proportional to the area of the hysteresis loop of the core material. Making the loop smaller (a) by reducing the coercivity of the core material and (b) by reducing B_m below the saturation flux density, B_s, (typically $B_m = 0.8B_s$) minimises hysteresis losses.

16.3.2 Leakage reactance

Some of the flux generated by the primary magnetizing current does not link with the secondary and is called leakage flux. We can imagine it as something like that shown in Figure 16.9.

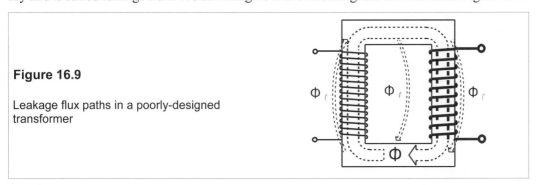

Figure 16.9

Leakage flux paths in a poorly-designed transformer

Leakage flux and hence leakage reactance plays a large part in voltage regulation, so it is essential to minimise it. In practice no transformer would be wound like that in Figure 16.9, with primary wound on one limb and secondary on the other, as its excessive leakage flux would lead to very poor regulation. There are several ways, shown in Figure 16.10, to bring about a reduction in leakage flux; they are often used in combination, such as (a), (c) and (d).

    (a) Wind primary and secondary over each other – interleaving the layers is best
    (b) Sandwich the coils
    (c) Use long, thin coils by making the limbs longer and the yoke shorter
    (d) Use a core which completely encloses the windings

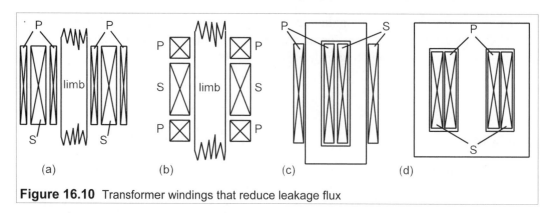

Figure 16.10 Transformer windings that reduce leakage flux

To facilitate manufacture the transformer core is usually made in two halves enabling the coils to be wound on formers and dropped onto the limbs. The yoke can then be clamped into position by an external framework. To avoid anything but the most minute air gap between the two halves of the core the mating surfaces are usually ground as flat as possible, then polished. An air gap of 0.1 mm increases the reluctance significantly, since it is equivalent to a path length of about 2–300 mm in the magnetic core material.

16.4 Transformer design

The most important consideration is the coil voltage, given by Equation 16.12: $E = 4.44B_m NAf$. Unless this equation can be satisfied without B_m approaching saturation the transformer will not work properly. This happens if the ampère-turns product, NA, is too small: to reduce saturation one must increase the turns. A saturated core has a highly non-linear I-Φ relationship leading to a non-sinusoidal secondary voltage. The core losses are also increased as they go roughly as B^2. In addition, as the induced e.m.f. is too small, the primary current becomes excessive and increases the copper losses; efficiency is reduced and the transformer overheats.

The magnetizing field is given by

$$H = NI_M/\ell = B_m/(\sqrt{2}\mu)$$

where $\mu = \mu_r\mu_0 \approx 0.004$ for iron-silicon and $B_m = 1.2$ T, so that $H \approx 200$ A/m. The magnetizing current, I_M, is about $0.02I_1$, where I_1 is rated primary current.

The core volume, Λ, and hence the overall size of a transformer is proportional to the power rating, S, and inversely proportional to the excitation frequency, f,

$$\Lambda = \gamma S/f \qquad (16.32)$$

where γ ranges from about 4×10^{-5} m³Hz/VA for iron-silicon cores used in transformers up to 100 kVA, and about half this for multi-MVA transformers, to about 2×10^{-3} m³Hz/VA for ferrite cores below 100 kHz. The difference is mainly due to the much lower flux density in ferrites: core volume is proportional to the square of the flux density.

The only variable left to the judgement of the designer is what to make the flux path length or the cross-sectional area of the core. Having fixed A and ℓ the designer has to adjust the window size to accommodate the winding. In practice most ab initio transformer design is for small special-purpose devices and relies on existing core shapes and sizes. An important consideration in power transformers is cooling, and the larger the transformer the harder it is to keep cool. A balance has to be struck between conflicting requirements. A larger transformer core and coils for a given rating would not heat up so much and would be more efficient, but would cost more. The allowable temperature rise depends on the class of insulation used and its maximum operating temperature. Above about 200 kVA, transformers are oil cooled and above a few MVA they have forced oil circulation as well (a 10MVA transformer will be dissipating 100 kW in its core and coils). The power rating of a transformer is considerably increased by oil cooling – roughly by a third for unforced oil coiling and by a further third if forced oil cooling is used. Again, the larger the transformer, the more economic it becomes to use forced cooling as the saving in copper and iron is great.

Example 16.7

What core size is required for a 240V/60V, 15 kVA, 60 kHz transformer? What must the number of turns on primary and secondary be if the maximum core flux density is 1.2 T and the mean flux path length is 1.2 m? If the core losses are 3 W/kg and the density of the core is 7700 kg/m^3, what are the core losses? If the short-circuit test gave a wattmeter reading of 210 W and V_{1sc} = 9.6 V, what is the maximum efficiency of the transformer? At what percentage of rated load is the efficiency a maximum? What is the voltage regulation on full load at unity p.f.? Draw the full equivalent circuit showing primary and secondary resistances and leakage reactances, assuming the copper losses and leakage flux are equally distributed between them.

The core size comes from Equation 16.32 with $\gamma = 4 \times 10^{-5}$ m^{3}Hz/VA, and is found to be 0.01 m^3. The core volume, $\Lambda = A\ell$, so that the core cross-sectional area, $A = \Lambda/\ell = 0.01 \div 1.2 = 8.3 \times 10^{-3}$ m^2. We can then find the number of turns on the primary by using Equation 16.12

$$E_1 = 240 = 4.44B_m AfN_1 = 4.44 \times 1.2 \times 8.3 \times 10^{-3} \times 60N_1$$

whence N_1 = 5400 (and N_2 = 1350 turns, though not asked for).

The core weight is 7700Λ = 77 kg, so its losses are 77 × 3 = 231 W. Maximum efficiency occurs when these are matched by the copper losses, so the total is then 462 W. But at rated secondary current the power consumed was 210 W, not 231 W, which means that maximum efficiency must occur when I_2 is 231/210 (= 110%) times rated and the output power is likewise 110% of rated, that is 16.5 kW. Then the total losses are 462 W and the input power is 16.96 kW and the maximum efficiency is 16.5/16.96 = 97.3%.

Rated primary current is $I_1 = S/V_1$ = 15000/240 = 62.5 A. In the short-circuit test the wattmeter reading was 210 W, so that $R_e = P_{sc}/I_1^2 = 210/62.5^2 = 54$ mΩ. The apparent power in this test is 9.6 × 62.5 = 600 VA, so that $Q = \sqrt{(S^2 - P^2)} = \sqrt{(600^2 - 210^2)} = 562$ var. Then $X_e = Q/I_1^2 = 562/62.5^2 = 144$ mΩ. Thus $\mathbf{Z_e}$ = 154∠69.5° mΩ. The primary impedance on full load is $Z_1 = V_1^2/S = 240^2/15000 = 3.84$ Ω, so that $z = Z_e/Z_1 = 0.154/3.84 = 0.0401$ per unit. At unity p.f. $\phi_L = 0°$ and then $\theta = \phi_e - \phi_L = 69.5°$ and the regulation is $z\cos\theta = 0.014$ or 1.4%. The secondary voltage is 59.2 V on full load.

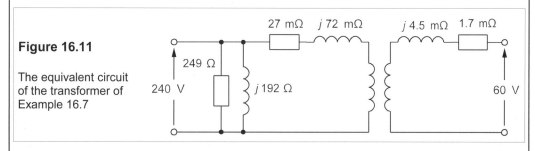

Figure 16.11

The equivalent circuit of the transformer of Example 16.7

We can find R_M from the iron (core) losses which are 231 W, so that $R_M = V_1^2/P_{Fe} = 240^2/231 = 249$ Ω. The next parameter to find is X_M which is V_1/I_M. We can estimate this by taking $I_M = 0.02I_1 = 0.02 \times 62.5 = 1.25$ A, then $X_M = 240/1.25 = 192$ Ω.

The primary winding contributes half of the copper losses to R_e, ideally, so that the primary winding resistance, $R_1 = R_e/2 = 27$ mΩ. The secondary winding, referred to the primary, contributes the other half, or 27 mΩ, too. In the secondary this is divided by n^2 where $n = V_1/V_2 = 4$, making $R_2 = R_1/4^2 = 1.7$ mΩ. The same with X_1 and X_2; $X_1 = 0.5X_e = 72$ mΩ, and $X_2 = 72/4^2 = 4.5$ mΩ. The full equivalent circuit is shown in Figure 16.11.

16.5 Special types of transformer

We have dealt with 'ordinary' single-phase transformers so far, but there are several other types of transformers designed for specific purposes that we shall look at briefly.

16.5.1 Three-phase transformers

Three-phase transformers consist of an E-core and yoke with windings on each of the three limbs as in Figure 16.12. These transformers can have primaries and secondaries connected in star or delta, so there are four possible arrangements for the winding connections. Three-phase transformers are the same in practice as three separate single-phase transformers, but considerable weight and cost reduction is achieved by using a single core. When the windings are star-star or delta-delta there is no phase difference between the primary phases and their corresponding secondary phase, but when the connection is star-delta or delta-star, the secondaries and primaries of each of the phases differ by 30°.

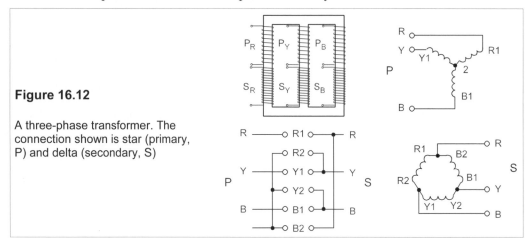

Figure 16.12

A three-phase transformer. The connection shown is star (primary, P) and delta (secondary, S)

16.5.2 Auto-transformers

Auto-transformers have just one winding, part of which is common to both primary and secondary, as shown in Figure 16.13. Any transformer with a separate primary and secondary winding can be connected as an auto transformer by connecting the two windings in series, thereby giving a secondary voltage of either $V_1 + V_2$ or $V_1 - V_2$, depending on the polarity of the connection, V_1 and V_2 being the original, rated primary and secondary voltages. In auto-transformers where the tapping of the secondary is variable these transformers are popularly called 'Variacs'. Variable auto-transformers are usually operated from the 240 V mains supply and are equipped with a scale showing secondary voltage as 0–115% of the mains voltage. They are handy for control of power for heating or motor speed. Variable trans-formers require brushes for contacting bare conductors, so they are dearer and less trouble-free than fixed-ratio transformers. They also require a heavier winding and are less efficient than conventional transformers, and cannot provide isolation from the supply, which is sometimes desirable.

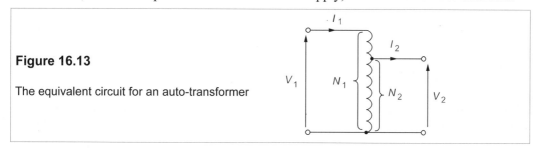

Figure 16.13

The equivalent circuit for an auto-transformer

16.5.3 Current transformers

These are used for measuring alternating currents without connecting an ammeter into the circuit in which the current is flowing. They are especially useful for measuring currents in circuits carrying high voltages or high currents, or where connection of a current meter is impractical. A current transformer often consists of a ferromagnetic toroidal core wound around with a sensing coil that connects with a meter as in Figure 16.14.

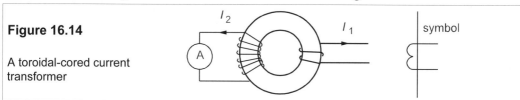

Figure 16.14

A toroidal-cored current transformer

The wire carrying the current to be measured is wound round the toroid, or the toroid is made up from two halves which can be opened to allow a wire to pass through. The primary consists of just one turn in the latter case and if the secondary carries N_2 turns the current flowing through it is $1/N_2$ times that in the circuit to be measured. The meter is scaled accordingly. With suitable cores, current transformers will work up to at least 300 kHz. The ammeter forms the load, which is customarily called the *burden*. It is important never to open circuit the secondary when primary current is flowing because the core saturates. The core then switches rapidly from saturation in one direction to saturation in the other as the primary current switches from plus to minus, thereby inducing very high secondary voltages.

┌─ **Example 16.8** ───

A 40:1 current transformer is connected to a 400 Hz line carrying 25 A and the burden resistance is 0.6 Ω. What are the primary and secondary voltages? It has a ferrite core is 10^{-4} m² in cross-sectional area and has a saturation flux density of 0.4 T. If the secondary is open circuited and the core is then saturated for 98% of the time, what is the maximum secondary voltage?

The secondary current is 25/40 = 0.625 A, so the secondary voltage is I_2R_2 = 0.625 × 0.6 =

0.375 V. The primary voltage is $V_2/n = 0.375/40 = 0.0094$ V or 9.4 mV, negligible. The core flux when saturated is $\Phi_s = B_sA = 0.4 \times 10^{-4} = 40$ µWb. Assuming that the primary has only a single turn, the secondary will have 50 turns and we can find the e.m.f. from $E = Nd\Phi/dt$. The flux change is from +40 to –40 µWb, or 80 µWb in one half cycle and then from –40 to + 40 µWb in the next. Half a cycle at 400 Hz is 1.25 ms, and the flux is saturated for 98% of this time, so the flux changes in 2% of 1.25 ms or 25 µs. Therefore

$$d\Phi/dt \approx \Delta\Phi/\Delta t = 80 \times 10^{-6}/25 \times 10^{-6} = 3.2 \ \text{V/turn}$$

There are 40 turns so the secondary voltage is 40 × 3.2 = 128 V.

16.5.4 Voltage transformers

These are also known as potential transformers and are normally used to measure high voltages in transmission lines. Figure 16.15 shows a voltage transformer with secondary grounded to avoid potentially-lethal high voltages between secondary and ground that could be produced by the inter-winding capacitance.

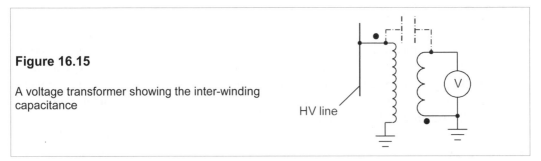

Figure 16.15

A voltage transformer showing the inter-winding capacitance

HV line

16.5.5 High-frequency transformers

Iron-cored transformers are generally used only up to a few kHz because of eddy-current losses, given by

$$W_v \approx 1.6\sigma(B_m fd)^2 \tag{16.33}$$

W_v is the eddy current loss/m³, σ is the conductivity of the core and d is the thickness of the core laminations. The importance of keeping the laminations thin is clear, 0.3 mm is typical. Lately, magnetic glasses have been produced in thin tapes ($d \approx 0.1$ mm or less) that are more resistive than iron-silicon mains transformers. These cores can be used at up to 500 kHz. Ferrites have effectively zero conductivity and hence no eddy-current losses; some types can be used up to 30 MHz. The core's size is limited by the heating effect of hysteresis losses which in practice means that B_m is reduced as the frequency increases: a rule of thumb is B_m = 15/√f. However, the biggest problem at high frequencies is the capacitance associated with the winding which leads to self resonance of the coil. And the larger the inductance the lower the self-resonant frequency: high-frequency transformers tend to be small and have low power ratings.

16.5.6 *High leakage-reactance transformers*

These are designed to have poor coupling between primary and secondary and therefore poor regulation. The high leakage reactance means the short-circuit impedance, Z_{SC}, is large and so the short-circuit current is much smaller than in a conventional design. This is necessary with arc-welding transformers because the arc must be struck with a relatively high secondary voltage, but when struck the secondary load has low resistance and the voltage must be considerably reduced. The large leakage reactance effectively stabilises the arc once it is struck. Fluorescent-lamp ballasts and low-voltage transformers used in toys require similar properties.

Suggestions for further reading – see Chapter 18

Problems

1. A transformer operating from a 500V supply has a core of effective cross-section 15 mm × 15 mm and is to operate at a maximum flux density of 0.2 T. If the secondary voltage is to be 2.2 kV and the frequency 20 kHz, how many turns are required on the primary and secondary? *[125, 550]*

2. What primary and secondary turns are required for a 240V/1kV, 50 Hz transformer if its core cross-sectional area is 0.008 m² and its maximum flux density is 1.2 T? *[113, 471]*

3. What turns ratio is required in a transformer used to match an 8Ω load to a 1 kΩ source? What are the primary and secondary voltages when the load power is 5 W? What is the maximum core flux density if the operating frequency is 3 kHz, the secondary winding has 100 turns and the effective core cross-section is 6 mm × 6 mm? *[11.2:1; 70.7 V, 6.3 V; 0.13 T]*

4. An open-circuit test on a 75kVA, 240V/14.4 kV transformer gave an ammeter reading of 12 A and wattmeter reading of 975 W. Find the magnetizing reactance, the equivalent core-loss resistance and the maximum efficiency of the transformer. *[21 Ω, 59 Ω; 97.4%]*

5. A short-circuit test on a 75 kVA, 14.4kV/240V transformer gave a primary voltage of 580 V and a wattmeter reading of 1.1 kW. Find $\mathbf{Z}_e$ and the primary and secondary series impedances assuming they are equal when referred to the primary. What is the voltage regulation for rated load at p.f.s of (a) unity (b) 0.9, lagging (c) 0.75, leading?
    [111.4∠68.6° Ω, 55.7∠68.6° Ω, 15.5∠68.6° mΩ; (a) 1.5% (b) 2.9% (c) –2.6%]

6. A 300MVA, 132kV/33 kV transformer has $\mathbf{Z}_e = 4.5∠86°$ Ω referred to the primary. What is its voltage regulation at p.f.s of (a) unity (b) 0.9, lagging (c) 0.7, lagging (d) 0.98, leading (e) 0.8, leading? *[(a) 1.8% (b) 5% (c) 6.7% (d) 0.34% (e) –3%]*

7. A no-load test on a 10kVA, 440V/120V, 60Hz transformer gives an ammeter reading of 0.8 A and a wattmeter reading of 150 W. If the primary has 148 turns, what is the maximum flux in the core? If $B_m = 1.3$T what is the effective area of the core? What are the magnetizing current, the equivalent core-loss resistance, R_M, and magnetizing reactance, X_M? Maximum efficiency occurs at a 20% overload, what is it? What is the equivalent series resistance referred to the primary, R_e? What is the voltage regulation at full load and unity p.f.? *[11.1 mWb. 0.0093 m². 0.545 A, 1.29 kΩ, 608 Ω. 97.5%. 0.2 Ω. 1%]*

8. A 69kV/4.16kV, 3MVA transformer has a primary leakage reactance of 50 Ω and a primary coil resistance of 6 Ω, while its secondary leakage reactance is 0.22 Ω with a coil resistance of 0.029 Ω. Calculate the percent regulation when it supplies (a) a resistive load of 2 MW (b) rated load at a p.f. of 0.9 lagging and (c) rated load at a p.f. of 0.95 leading. *[(a) 0.48% (b) 3.65% (c) –1.57%]*

9. A 100MVA, 25kV/275kV transformer has per-unit leakage reactances of 0.05 in its primary and

secondary and coil resistances of 0.002 per unit. What is the percentage regulation when it supplies (a) rated MVA at p.f. = 1 (b) half rated load a lagging p.f. of 0.8? (c) rated load at a leading p.f. of 0.99? (d) What is the load if the regulation is 1% at a lagging p.f. of 0.9?
[(a) 0.4% (b) 3.2% (c) −1% (d) 21% of rated]

10. The per-unit magnetizing reactance of a 1MVA transformer is 50 and its per-unit core loss resistance is 100. If the primary voltage rating is 33 kV, what are the no-load current and power loss? *[0.678∠−63. 4° A, 10 kW]*

11. A 60MVA, 225kV/26.4kV transformer has an impedance of 8%, almost all of which is reactive. What are the leakage reactances of primary and secondary if they have the same per-unit values? What are the full-load copper losses if the per-unit resistances of primary and secondary are 0.002? What is the efficiency if it is maximal on full load? *[33.75 Ω, 0.464 Ω, 240 kW, 99.2%]*

12. A 60Hz, 330kV/22kV, 10 MVA transformer has copper losses that equal the core losses when it delivers rated current. If the no-load primary current is 0.55 A at a p.f. of 0.3, what is the efficiency at unity p.f. on full load? If the core losses are 2 W/kg, what is the volume of the core if its density is 7.7 tonnes/m^3? What is the magnetizing reactance? What is the secondary coil's resistance if the copper losses are evenly divided between primary and secondary?
[98.9%, 3.54 m^3, 629 kΩ, 0.132 Ω]

13. A transformer is designed to work at a maximum efficiency of 99% on full load. It operates fully loaded 4 hours a day and half loaded 8 hours a day, both at a p.f. of unity. If it is unloaded the remainder of the day, what is its all-day efficiency? *[98.16%]*

14. A 5kVA, 240V/24V transformer takes a primary current of 1.2 A and consumes 65 W when its secondary is open-circuited and it is connected to a 240V supply. When its secondary is short-circuited and carries rated current, the power consumption is 77 W and the primary voltage is 9.8 V. What is the efficiency of the transformer when supplying full load at a p.f. of 0.9? What is the secondary voltage here? What are the components of the approximate equivalent circuit, referred to the primary? *[96.94%, 23.3 V, Z_e = 0.47∠67.8° Ω, R_M = 886 Ω, X_M = 205 Ω]*

17 Induction motors

THREE-PHASE induction motors are the workhorses of industry. About 90% of electrical motors used in industry are of this type. Single-phase induction motors running from the 240V supply are among the commonest in domestic use. All induction motors are cheap, reliable, rugged, lightweight and reasonably efficient; in fact large (> 1 MW) induction motors have efficiencies of up to 98%. Their biggest drawback hitherto has been their small range of operating speeds, though this is changing with the introduction of solid-state, variable-frequency drives. Table 17.1 shows typical characteristics for a small and a large squirrel-cage motor.

Table 17.1 *Three-phase induction motors*

Output power, kW	4	4000
Supply voltage, V	415	6900
Power factor at rated load	0.85	0.90
Speed, r/min	2840	596
Poles per phase	2	12
Torque, Nm	13.4	64000
Dimensions, m	0.3L 0.2D	2.6L 1.8D
(L = length, D = diameter)		
Weight, kg	30	21000
Efficiency, %	80	97

The smaller motor in Table 17.1 has a power output of about 133 W/kg, while the larger one has an output of 190 W/kg. The larger machine is significantly more efficient than the smaller. The difference between the input power and the output appears as heat in the motor and it must somehow be removed. Its rate of removal depends on the surface area of the motor and this determines the size of machines as much as anything.

In DC machines it is customary to move current-carrying conductors relative to a stationary magnetic field. However, with induction motors (and synchronous machines, discussed in the next chapter) it is more convenient to move a magnetic field relative to current-carrying conductors, although both are in motion. The magnetic field of an induction motor is rotated by distributing the windings of two or more phases around a stator that is essentially a steel cylinder. When an alternating current flows through the stator winding a rotating magnetic field is produced. The rotor holds conductors in which currents are induced by this rotating magnetic field. The interaction of the field set up by these induced currents and the stator field causes a torque to be exerted on the rotor which then turns and endeavours to keep up with the rotating stator field.

There are two types of conventional three-phase motors depending on the arrangement of conductors in the rotor:

   1. Squirrel-cage motors
   2. Wound-rotor motors with slip rings

Squirrel-cage motors are simpler and cheaper, but harder to control than wound-rotor types.

17.1 The construction of a squirrel-cage motor

Figure 17.1a shows the stator and rotor of a small squirrel-cage motor. The stator and rotor conductors are wound on laminated silicon-iron to reduce eddy-current losses. The copper conductors are placed in slots parallel to the axis of rotation. The rotor carries a squirrel cage winding consisting of a number of heavy-duty conductors whose ends are short circuited by a copper or aluminium ring (Figure 17.1a). In small (< 10 kW) motors the conductor assembly and integral cooling fan blades are usually die-cast in one piece by forcing the iron laminations of the rotor core into a bath of molten aluminium. This simplicity of construction means the squirrel-cage motor is cheap, but the current flowing in the rotor cannot be controlled as it can in a wound rotor. Speed control is therefore only achievable by varying the frequency of the supply.

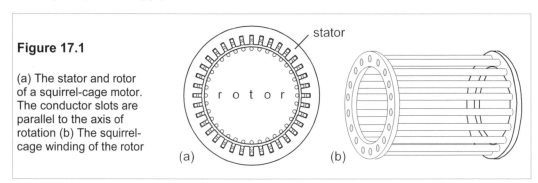

Figure 17.1

(a) The stator and rotor of a squirrel-cage motor. The conductor slots are parallel to the axis of rotation (b) The squirrel-cage winding of the rotor

(a)

(b)

17.2 Rotation of the stator field

Consider the six-pole stator in Figure 17.2 (the poles are shown as salient for convenience, though in practice they would not be) in which each of the three pairs of opposite poles is connected to one of the phases of the supply. The ends of the windings, R', Y' and B', are connected to the star point and the phase sequence is RYB. The winding is such that the polarity is as shown, that is opposite poles have opposite polarity. Suppose that in the red phase the current is $i_R = I_m \cos \omega t$ and the corresponding flux density is $\mathbf{B_R} = B_m \cos \omega t$. The flux densities due to the other phases will be $\mathbf{B_Y} = B_m \cos (\omega t - 120°)$ and $\mathbf{B_B} = B_m \cos (\omega t - 240°)$, assuming that the currents and poles are identical and the iron is not saturated. However, because the poles excited by the different phases are at 120° to each other, the resultant flux density rotates at the same frequency as the supply. With the phase sequence RYB the rotation is clockwise and if any two phases are swapped the rotation becomes anticlockwise. The magnitude of the resultant field remains constant throughout.

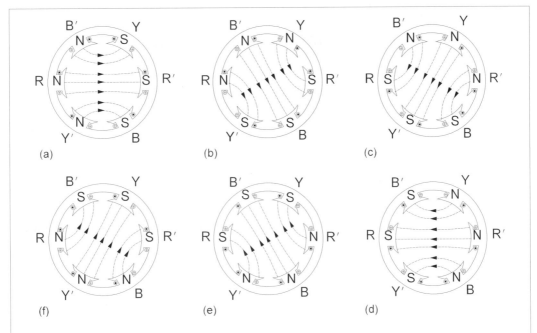

Figure 17.2 Flux rotation by three-phase field excitation (a) $\omega t = 0°$ (b) $\omega t = 60°$ (c) $\omega t = 120°$ (d) $\omega t = 180°$ (e) $\omega t = 240°$ (f) $\omega t = 300°$

We can show how the field rotates mathematically by splitting the field into rectangular components and summing. If the x-axis lies along the line joining the centres of poles RR′, then the x-components of the fields due to the R, Y and B phases are:

$$\begin{aligned} B_{Rx} &= |\mathbf{B_R}| \cos 0° = |\mathbf{B_R}| \\ B_{Yx} &= |\mathbf{B_Y}| \cos(-120°) = -0.5|\mathbf{B_Y}| \\ B_{Bx} &= |\mathbf{B_B}| \cos(-240°) = -0.5|\mathbf{B_B}| \end{aligned} \tag{17.1}$$

The sum of these is the x-component of the resultant field, B_x:

$$B_x = |\mathbf{B_R}| - 0.5(|\mathbf{B_Y}| + |\mathbf{B_B}|) \tag{17.2}$$

The y-components are

$$\begin{aligned} B_y &= |\mathbf{B_R}|\sin 0° + |\mathbf{B_Y}|\sin(-120°) + |\mathbf{B_B}|\sin(-240°) \\ &= 0.5\sqrt{3}(|\mathbf{B_B}| - |\mathbf{B_Y}|) \end{aligned} \tag{17.3}$$

Looking at the field at an instant in time when $\omega t = \theta$ (see Figure 17.3), so that

$$|\mathbf{B_R}| = B_m \cos\theta; \quad |\mathbf{B_Y}| = B_m \cos(\theta - 120°); \quad |\mathbf{B_B}| = B_m \cos(\theta - 240°) \tag{17.4}$$

Then the x-component of the resultant field, from Equation 17.2, is

$$B_x = B_m \cos\theta - 0.5B_m[\cos(\theta - 120°) + \cos(\theta - 240°)]$$

$$= B_m \cos\theta - 0.5B_m[-0.5\cos\theta - 0.5\cos\theta]$$

$$= 1.5B_m \cos\theta \tag{17.5}$$

And the *y*-component is

$$B_y = 0.5\sqrt{3}[-B_m\cos(\theta - 120°) + B_m\cos(\theta - 240°)]$$

$$= -1.5B_m\sin\theta \tag{17.6}$$

The magnitude of **B** is

$$|\mathbf{B}| = \sqrt{B_x^2 + B_y^2} = 1.5B_m \tag{17.7}$$

which is independent of θ and hence of *t* also. The direction of **B** is given by

$$\phi = \tan^{-1}(B_y/B_x) = \tan^{-1}[-\tan\theta] = -\theta \tag{17.8}$$

That is to say, **B** is rotated *clockwise* (positive angles are measured anticlockwise) by θ radians. Figure 17.3 shows the time-varying magnetic fields and the instantaneous values of the field vectors at $\omega t = \theta$.

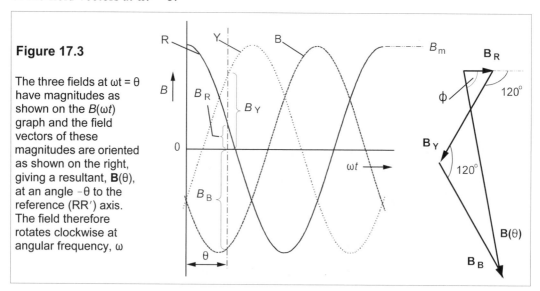

Figure 17.3

The three fields at $\omega t = \theta$ have magnitudes as shown on the $B(\omega t)$ graph and the field vectors of these magnitudes are oriented as shown on the right, giving a resultant, **B**(θ), at an angle $-\theta$ to the reference (RR′) axis. The field therefore rotates clockwise at angular frequency, ω

17.3 **Slip**

A six-pole, three-phase induction motor has a stator field which rotates at the supply frequency, known as the *synchronous* frequency, f_s, with a corresponding synchronous speed, n_s ($= 60f_s$ r/min). The rotor will be dragged round by this field and, if not loaded in any way, will eventually rotate at very nearly the same speed, that is at 50 revolutions per second or 3000 r/min if the supply frequency is 50 Hz – quite a high speed even in a moderate-sized motor. However, if twice as many pole pairs are provided for each phase – 12 in all – the synchronous frequency will be halved, since the pole-pole angle becomes 30° and the field will rotate 30° for a 60° increase in the phase angle of the supply. The synchronous speed is

$$n_s = 60f_s/p \tag{17.9}$$

where f_s is the supply frequency and *p* is the number of pole pairs per phase. The synchronous speed is still fixed by the supply frequency and the number of poles, but it can be made 3000, 1500, 1000 or 750 r/min by having 1, 2, 3 or 4 pole pairs per phase and a 50Hz supply. However, the synchronous speed cannot be made an arbitrary value like 1800 r/min if we want to operate it from a 50Hz supply.

The frequency of the e.m.f. induced in the stationary rotor will be the same as the synchronous frequency, and when the rotor turns at synchronous speed the induced e.m.f. must be zero since the rotor conductors do not move relative to the stator field. The frequency of the rotor e.m.f. is therefore given by

$$f_r \propto (n_s - n_r)f_s = (n_s - n_r)f_s/n_s = sf_s \qquad (17.10)$$

where n_r is the rotor speed. The constant of proportionality must be $1/n_s$ to make the frequency of the rotor e.m.f. equal to f_s when $n_r = 0$. The parameter, s, is called the (fractional) *slip*. Often the slip is expressed as a percentage:

$$s = 100(n_s - n_r)/n_s \ \% \qquad (17.11)$$

When a large induction motor is running at its rated load and speed the slip will be only about ½%. Smaller machines may run at 3% slip and very small machines at 20%. Induction motors are thus nearly constant-speed machines, though because there *is* slip, they are sometimes called *asynchronous* machines.

The e.m.f. induced in the rotor conductors per phase will be proportional to the slip also, so that if the rotor e.m.f. at standstill is E_0 per phase, at fractional slip, s, it will be

$$E_2 = sE_0 \qquad (17.12)$$

In squirrel-cage machines each conductor is effectively a phase, but in wound rotors there will be usually three phases.

The leakage reactance per phase of the rotor is proportional to the leakage inductance times the rotor e.m.f.'s frequency, that is

$$X_r = 2\pi f_r L = 2\pi s f_s L = sX_{r0} \qquad (17.13)$$

where X_{r0}, the leakage reactance at standstill, is $2\pi f_s L$.

Example 17.1

Slip can be positive or negative and can take, in theory, any value depending which way the rotor moves relative to the stator field and at what relative speed. For example if a 12-pole induction machine works from a 50Hz, three-phase supply, what will be the slip at a rotor speed of 1800 r/min, 1479 r/min and −300 r/min? (A negative rotor speed implies that the rotor is moving in the opposite direction to the stator field and the machine operates as a brake). What are the corresponding frequencies of the rotor e.m.fs.?

The number of pole pairs is 12/2 = 6, and the number of pole pairs per phase, p, is then 6/3 = 2. Equation 17.9 gives the synchronous speed

$$n_s = 60f_s/p = 60 \times 50/2 = 1500 \ \text{r/min} \qquad (17.14)$$

If n_r = 1800 r/min the slip will be

$$s = (n_s - n_r)/n_s = (1500 - 1800)/1500 = -0.2 \qquad (17.15)$$

from Equation 17.11. A negative slip implies that the machine is acting as a generator. When n_r = 1470 r/min, s = 100(1500 − 1470)/1500 = 2%, which indicates the machine is running as a motor, possibly at rated load.

Finally, when $n_r = -300$ r/min, $s = 100(1500 + 300)/1500 = 120\%$. A value for slip greater than 100% must indicate the machine is acting as a brake.

The frequencies of the rotor e.m.fs. are found from Equation 17.10 and they are, firstly, $f_r = -0.2 \times 60 = -12$ Hz. The negative sign merely shows that the e.m.f. induced is in the opposite direction to that induced in the stationary rotor. Secondly, $f_r = 0.02 \times 60 = 1.2$ Hz and thirdly, $f_r = 1.2 \times 60 = 72$ Hz.

17.4 The equivalent circuit of an induction machine

The steady-state performance of an induction machine can be derived quite readily from the equivalent circuit of its rotor, Figure 17.4.

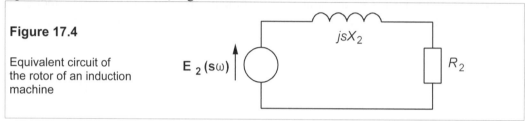

Figure 17.4

Equivalent circuit of the rotor of an induction machine

The induction machine acts like a special type of transformer, with a turns ratio of one. The leakage reactances are rather large because of the air gap between stator and rotor which makes the coupling much less tight than in a conventional power transformer. Let the resistance of the rotor be R_2 and let its inductance be L_2, so that its leakage reactance is $\omega L_2 = X_2$ when the supply frequency is ω and the rotor is stationary. Then, when the slip is s, the secondary frequency becomes $s\omega$ and the secondary leakage reactance becomes sX_2, while its resistance is unchanged. This is coupled with the stator circuit via a 1:1 transformer, making the overall equivalent circuit that of Figure 17.5. However, the primary and the secondary have different frequencies because of slip, the primary frequency being ω and the secondary $s\omega$.

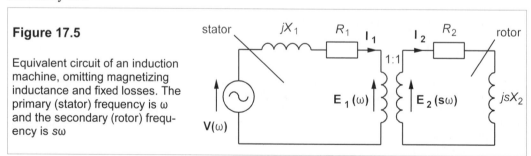

Figure 17.5

Equivalent circuit of an induction machine, omitting magnetizing inductance and fixed losses. The primary (stator) frequency is ω and the secondary (rotor) frequency is $s\omega$

If the secondary impedance is $\mathbf{Z}_2$, then $\mathbf{I}_2 = \mathbf{E}_2/\mathbf{Z}_2$. When $\mathbf{Z}_2$ is transformed into the primary, it becomes $\mathbf{Z}_2'$, and carries the primary current, $\mathbf{I}_1 = \mathbf{E}_1/\mathbf{Z}_2'$. Now in a 1:1 transformer, $\mathbf{I}_1 = \mathbf{I}_2$, that is

$$\mathbf{E}_2/\mathbf{Z}_2 = \mathbf{E}_1/\mathbf{Z}_2' \tag{17.16}$$

But $\mathbf{Z}_1 = R_1 + jX_1$ and $\mathbf{E}_2 \approx s\mathbf{E}_1$ (the e.m.f. induced in the rotor is almost proportional to the slip), which means that Equation 17.16 is

$$\frac{\mathbf{E_1}}{\mathbf{Z_2}'} = \frac{\mathbf{E_2}}{\mathbf{Z_2}} = \frac{s\mathbf{E_1}}{R_2 + jsX_2} = \frac{\mathbf{E_1}}{R_2/s + jX_2} \qquad (17.17)$$

Thus $\mathbf{Z_2}' = R_2/s + jX_2$ and the equivalent circuit of Figure 17.5 becomes that of Figure 17.6.

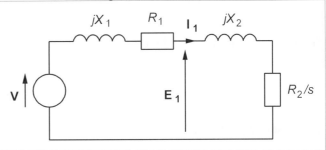

Figure 17.6

When the rotor equivalent circuit is transformed into the primary, the rotor leakage reactance becomes X_2 and the rotor resistance becomes R_2/s

The power developed in R_2/s is $I_1^2 R_2/s = P_{rj}/s = P_r$, which is the total power transferred to the rotor, while $P_{rj} = I_1^2 R_2$, the copper losses in the rotor. The mechanical power in the rotor is

$$P_m = P_r - P_{rj} = P_{rj}/s - P_{rj} = I_1^2 R_2 (1 - s)/s \qquad (17.18)$$

Thus the resistance R_2/s can be split into two series resistances: R_2 and $R_2(1-s)/s$, the former representing the rotor copper losses, P_{rj}, and the latter the mechanical power in the rotor, P_m.

Now the leakage reactances, X_1 and X_2, can be combined as a single reactance, X_0 so that the final equivalent circuit is that of Figure 17.7. The magnetising reactance, X_M, and the no-load losses (friction, windage and iron losses), R_0, have been included as parallel components across the supply, since the latter will be almost independent of the current drawn from the supply. Since these components are relatively large, we can normally make the stator current, $\mathbf{I_s} = \mathbf{I_1}$. From the equivalent circuit of Figure 17.7 we can calculate torque-speed curves as a function of slip, as we shall see in the next section.

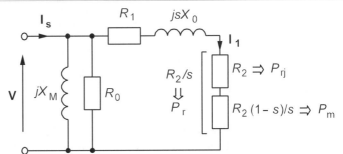

Figure 17.7

The equivalent circuit of an induction machine, including magnetizing reactance and no-load losses

17.5 Torque and slip in an induction machine

The important relationship between torque and slip in an induction motor may be derived from the equivalent circuit of Figure 17.7. If the stator resistance, R_1, can be neglected (the slip is small in normal operation so that $R_2/s \gg R_1$), the electromagnetic power transferred to the rotor is given by

$$P_r = \omega_s T = I_1^2 R_2 / s \qquad (17.19)$$

But from Figure 17.7 we can see that

$$\mathbf{I_1} = \frac{\mathbf{V}}{R_2/s + jX_0} \quad \rightarrow \quad I_1^2 = \frac{V^2}{(R_2/s)^2 + X_0^2} \qquad (17.20)$$

Then from Equation 17.19 the torque is

$$T = \frac{I_1^2 R_2}{\omega_s s} = \frac{V^2 R_2}{\omega_s s[(R_2/s)^2 + X_0^2]} = \frac{sV^2 R_2/\omega_s}{R_2^2 + (sX_0)^2} \qquad (17.21)$$

which can be put in the form

$$T = \frac{Ks\alpha}{s^2 + \alpha^2} \qquad (17.22)$$

where $K = V^2/\omega_s X_0$ and $\alpha = R_2/X_0$. Equation 17.22 is shown as torque-speed and torque-slip curves for various values of α in Figure 17.8 – note that α is normally $\ll 1$. Low values for α are found in large induction machines and high values can be obtained by increasing the resistance in series with the rotor in wound-rotor machines, or by increasing the resistance of the cage in squirrel-cage motors.

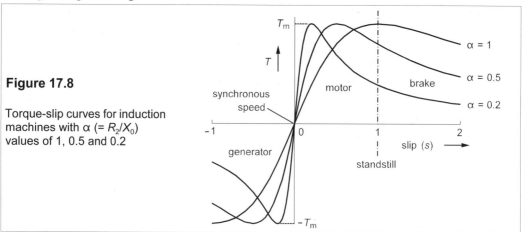

Figure 17.8

Torque-slip curves for induction machines with α (= R_2/X_0) values of 1, 0.5 and 0.2

By differentiating Equation 17.22 with respect to s, the turning points can be found (see Problem 17.4). One at $s = +\alpha$ is a maximum and the other at $s = -\alpha$ is a minimum. Putting $s = \alpha$ in Equation 17.22 we find the maximum torque or *breakdown* torque is

$$T_m = \frac{K\alpha^2}{\alpha^2 + \alpha^2} = \frac{K}{2} = \frac{V^2}{2\omega_s X_0} \qquad (17.23)$$

Note that the breakdown torque, T_m, does not depend on R_2, the rotor resistance, only on the leakage reactance, X_0, for a given supply voltage and frequency. This is true even when R_1 is included (see Problem 17.14). The smaller the value of X_0, the larger the breakdown torque. If the load produces a torque of this value or greater, the motor stalls.

When $s = 1$, the rotor is at standstill (*locked-rotor* condition) and the starting torque is $T_0 = K\alpha/(1+\alpha^2)$, which is small for small values of α. The ratio of standstill torque to breakdown torque is

$$T_0/T_{\mathrm{m}} = \frac{K\alpha/(1+\alpha^2)}{K/2} = \frac{2\alpha}{1+\alpha^2} \qquad (17.24)$$

which is roughly 2α for small α (or large leakage reactance). For example a 10 MW induction motor may have $\alpha = 0.025$ ($X_0 = 40R_2$) and $T_0/T_{\mathrm{m}} = 0.05$. The slip at maximum torque is 2.5% and the slip at normal operating torque will be only about 1%.

There are three operating regions in Figure 17.8:

1. $s > 1$ ($n < 0$)   The machine is a *brake*.
2. $0 < s < 1$ ($0 < n < n_{\mathrm{s}}$)   The machine is a *motor*.
3. $s < 0$ ($n > n_{\mathrm{s}}$)   The machine is a *generator*.

17.5.1 Braking

In this mode the rotor is turning in the opposite direction to the field and $s > 1$. Because $R_2/s < R_2$, the joule dissipation in the rotor is larger than the power transferred from the stator. The excess power is drawn from the mechanical energy of the rotor plus load. The motor quickly comes to rest. By reversing two of the phases on a motor running under load, the stator field reverses direction and the motor is then braked because $s > 1$, which is called *plugging*. The stator supply is switched off when the rotor stops to prevent the rotor motoring in reverse.

An alternative method of braking is to supply DC to the stator, causing the rotor conductors to pass through a stationary magnetic field. The advantage of this method is that the rotor dissipates only the kinetic energy of the load whereas during plugging it also absorbs power from the stator. Plugging can therefore bring about an excessive rise in rotor temperature. In addition the machine operates at less slip (because for DC in the stator, $s = 0$ at standstill), resulting in a higher braking torque according to Equation 17.22.

17.5.2 Motoring

Though the slip lies between zero and one in this region, the normal operating point lies around $0.01 < s < 0.05$, where the torque is given by

$$T = \frac{K\alpha s}{s^2 + \alpha^2} \approx \frac{Ks}{\alpha} \qquad (17.25)$$

The torque therefore varies linearly with slip near $s = 0$.

17.5.3 Generating

When the speed exceeds synchronous speed, the slip becomes negative, R_2/s is negative and the rotor supplies electrical energy to the stator. Provided the generator is connected to a normal three-phase supply (so that the stator has magnetising current) it will deliver power. Alternatively, reactive (magnetising) volt-amperes can be provided by connecting three capacitors across the stator phases.

17.6 Evaluating the components of the equivalent circuit

The components of the equivalent circuit of Figure 17.9 can be found from two tests analogous to the open and short-circuit tests on a transformer, namely the no-load and locked-rotor tests. In the no-load test the machine runs at very near synchronous speed and so $s \rightarrow 0$ and $R_2/s \rightarrow \infty$. The current supplied goes into X_M and R_0. Normally one measures the power consumed with a wattmeter, the line-line voltage, the line current and the resistance between any two terminals of the stator. With a star-connected stator the stator phase resistance is then half the measured terminal resistance, and with a delta-connected stator, the phase resistance is 1.5 times the measured terminal resistance.

Example 17.2

A no-load test on a three-phase induction motor with star-connected stator gave the following results:

Resistance between two stator terminals = 0.8 Ω   Power consumed = 1 kW
Line-line voltage = 415 V   Line current = 9 A

Find R_0, X_M and the series resistance of the stator, R_1. (All per phase.)
   The power consumed by the stator winding is negligible under no-load conditions and so is the stator current. The phase voltage $415/\sqrt{3} = 240$ V. This is impressed across $R_0 \parallel X_M$. Now the power consumed per phase, $P = 1000/3 = 333$ W. But $P = V_p^2/R_0 = 240^2/R_0$. Thus $R_0 = 240^2/333 = 173$ Ω. The apparent power per phase is $240 \times 9 = 2.16$ kVA. The power consumed per phase is $1000/3 = 333$ W, so the reactive power per phase is

$$Q = \sqrt{S^2 - P^2} = \sqrt{2.16^2 - 0.333^2} = 2.13 \ \text{kvar} \qquad (17.26)$$

But $Q = V_p^2/X_M$, so that $X_M = V_p^2/Q = 240^2/2\,130 = 27$ Ω. The series resistance/phase of the stator is half the resistance between two of its terminals, so $R_1 = 0.4$ Ω.

In the locked-rotor test the motor is connected to a variable-voltage supply which is adjusted to give the rated current through the stator. The voltage, current and power consumed are measured. In this test the voltage is reduced well below rated level and so the current through the parallel branch of the circuit in Figure 17.9 is negligible. However, in finding the resistance of the series branch, we cannot ignore R_1, the stator's per-phase resistance since $s = 1$ and R_2/s can no longer be said to be much greater than R_1.

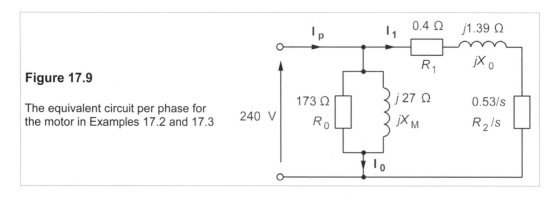

Figure 17.9

The equivalent circuit per phase for
the motor in Examples 17.2 and 17.3

Example 17.3

The motor in Example 17.2 is found to draw its rated current of 60 A at standstill (locked rotor) when the line voltage is 173 V. The power consumption is then found to be 10 kW. What are the series components of its equivalent circuit? If the motor has a synchronous speed of 1500 r/min, what will its breakdown and starting torques be approximately?

The phase voltage is $173/\sqrt{3} = 100$ V, so the apparent power per phase is $100 \times 60 = 6$ kVA. The power consumed per phase is $10/3 = 3.33$ kW. Then $Q = \sqrt{(S^2 - P^2)} = \sqrt{(6^2 - 3.33^2)} = 5$ kvar. But $Q = I_p^2 X_0 = 60^2 X_0$, giving $X_0 = 5000/60^2 = 1.39$ Ω. And $P = I_p^2 R$, giving $R = R_1 + R_2 = 3333/3600 = 0.93$ Ω. Therefore $R_2 = 0.93 - 0.4 = 0.53$ Ω. The equivalent circuit of each phase looks like Figure 17.9.

The approximate breakdown torque, T_m, from Equation 17.23 is $V^2/2\omega_s X_0$, where $V = V_p = 240$ V and $\omega_s = 2\pi n_s/60 = 157$ rad/s, yielding $T_m = 132$ Nm *per phase*, so the total torque is three times this or about 500 Nm. Equation 17.24 shows that $T_0/T_m = 2\alpha/(1 + \alpha^2)$, where $\alpha = R_2/X_0 = 0.53/1.39 = 0.38$; the ratio of torques is then 0.66 and $T_0 = 0.66 \times 132 = 88$ Nm, again per phase, for a total starting torque of 264 Nm.

17.7 Power and efficiency

The power input to the stator, P_{in}, is less than the power transferred to the rotor, P_r, because of the stator copper (and iron) losses, P_{sl}. The power transferred to the rotor must be given by the product of rotor torque, T, and the angular velocity of the stator field, ω_s:

$$P_r = \omega_s T = 2\pi n_s T/60 \tag{17.27}$$

But the mechanical power developed by the rotor must be given by the product of rotor torque and its angular velocity:

$$P_m = \omega T = 2\pi n T/60 \tag{17.28}$$

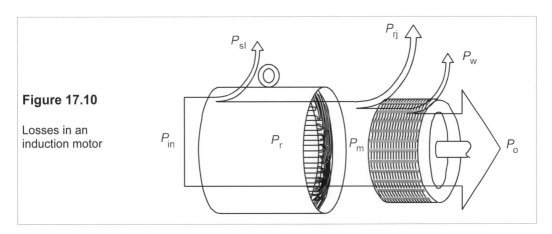

Figure 17.10

Losses in an induction motor

The difference between P_r and P_m is the joule loss of the rotor, P_{rj}, since the iron losses will be small when the slip is small, thus

$$P_{rj} = P_r - P_m = 2\pi(n_s - n)T/60 = 2\pi s n_s T/60 = sP_r \tag{17.29}$$

Another way of writing this is

$$P_\mathrm{m} = (1 - s)P_\mathrm{r} \qquad (17.30)$$

The useful power out, P_o, is less than P_m because of friction and windage losses, P_w. Then

$$P_\mathrm{in} = P_\mathrm{sl} + P_\mathrm{r} = P_\mathrm{sl} + P_\mathrm{rj} + P_\mathrm{m} = P_\mathrm{sl} + P_\mathrm{rj} + P_\mathrm{w} + P_\mathrm{o} \qquad (17.31)$$

Figure 17.10 shows these losses.

Example 17.4

The line current drawn by a three-phase induction motor is 14.7 A, while the line voltage is 415 V. The stator is delta wound with a phase resistance of 1.2 Ω. The total friction and windage losses are 210 W. Assuming the iron losses in the stator are half the copper losses in the stator, calculate the overall efficiency if the motor operates at a power factor 0.87, a synchronous speed of 750 r/min and a running speed of 734 r/min.

Let us look at Figure 17.10 to see where the losses are and then proceed to find the efficiency. The phase current, I_p, is 14.7/√3 = 8.5 A. The input power per phase is

$$P_\mathrm{in} = V_\mathrm{p}I_\mathrm{p}\cos\phi = 415 \times 8.5 \times 0.87 = 3069 \text{ W/phase} \qquad (17.32)$$

The stator copper losses are

$$P_\mathrm{sl} = I_\mathrm{p}^2R_\mathrm{s} = 8.5^2 \times 1.2 = 86.7 \text{ W/phase} \qquad (17.33)$$

so the iron losses are half this, 43.4 W/phase, for total stator losses of 130 W/phase. Thus the power transferred from stator to rotor is (3069 – 130) = 2939 W/phase. The slip is

$$s = (750 - 734)/750 = 16/750 = 0.0213 \qquad (17.34)$$

Therefore the mechanical power developed by the rotor is

$$P_\mathrm{m} = (1 - s)P_\mathrm{r} = 0.9787 \times 2939 = 2876 \text{ W/phase} \qquad (17.35)$$

Of this mechanical power, 70 W (210 W in total = 70 W/phase) are lost by windage and friction, resulting in a useful output power of 2876 – 70 = 2806 W/phase. The efficiency is then

$$\eta = \text{useful output power} \div \text{input power} = 2806/3069 = 91.4\%$$

17.7.1 *Torque and stator current versus slip*

We can use the equivalent circuit of Figure 17.9 to calculate the torque more accurately than in Section 17.5 (though qualitatively the results are much the same). When the rotor is stationary, $s = 1$ and the series-branch resistance is 0.93 Ω. $\mathbf{I}_1$, the series-branch current

$$\mathbf{I}_1 = \frac{240}{0.93 + j1.39} = \frac{240\angle 0°}{1.67\angle 56°} = 144\angle -56° \text{ A} \qquad (17.36)$$

The start-up power in the rotor, P_r0, is $I_1^2R_2 = 144^2 \times 0.53 = 11$ kW. But $P_\mathrm{r0} = \omega_\mathrm{s}T_0$, giving $T_0 = 11000/157 = 70$ Nm per phase, 210 Nm in total. The difference from the approximate value – 20% less – is significant because X_0 is not much greater than $R_1 + R_2$. The breakdown torque will also be somewhat less than the approximate value. Figure 17.11 shows a graph of I_1 and T against s for the motor, calculated from the equivalent circuit of Figure 17.9. The torque comes from the Equation $P_\mathrm{r} = \omega_\mathrm{s}T = I_1^2R_2/s$.

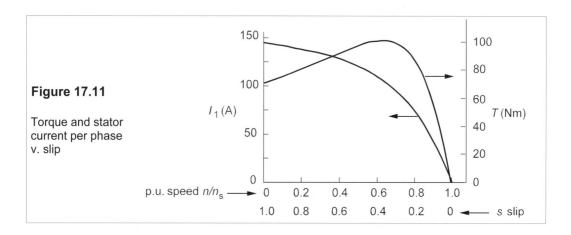

Figure 17.11

Torque and stator
current per phase
v. slip

17.7.2 *Efficiency and power factor versus slip*

The efficiency, power factor and line current can also be calculated as functions of slip and speed using the equivalent circuit per phase of Figure 17.9. To find the efficiency we must calculate the power developed in each resistance. The fixed losses are represented by $R_0 =$ 173 Ω. Inevitably these losses include friction and windage losses, which form part of the power transferred to the rotor, as well as iron losses in the stator. However, we shall overlook this complication as its effect on the calculation of efficiency is not great. The fixed losses must then be $F = V^2/R_0 = 240^2/173 = 333$ W per phase.

The joule losses in the stator winding, P_s, are $I_1^2 R_1$. The mechanical power developed by the rotor, which is nearly equal to the useful power, will be

$$P_m = (1 - s)I_1^2 R_2/s \tag{17.37}$$

The joule losses in the rotor are $P_{rj} = I_1^2 R_2$ and the power transferred to the rotor is $P_r = I_1^2 R_2/s$. The total power is must be the sum of the three resistive terms: $P_{in} = F + P_s + P_r$. The efficiency is then

$$\eta \approx \frac{P_m}{F + P_s + P_r} \tag{17.38}$$

Now I_1 is given by

$$\mathbf{I_1} = \frac{240}{0.4 + 0.53/s + j1.39} \Rightarrow I_1^2 = \frac{240^2}{(0.4 + 0.53/s)^2 + 1.39^2} \tag{17.39}$$

so both stator and rotor joule loss are dependent on the slip. For example when $s = 0.2$

$$I_1 = \frac{240}{\sqrt{(0.4 + 0.53/0.2)^2 + 1.39^2}} = \frac{240}{\sqrt{3.05^2 + 1.39^2}} = 71.6 \text{ A} \tag{17.40}$$

leading to $P_s = 71.6^2 \times 0.4 = 2.05$ kW, $P_r = 71.6^2 \times 0.53/0.2 = 13.6$ kW and $P_m = (1 - s)P_r$ = 10.9 kW. The total power input per phase is

$$P_{in} = F + P_s + P_r = 0.33 + 2.05 + 13.6 = 16.0 \text{ kW} \qquad (17.41)$$

The efficiency is

$$\eta \approx P_m/P_{in} = 10.9/16 = 0.68 \qquad (17.42)$$

Similarly we can find that when the slip is 0.02, $I_1 = 8.91$ A, $P_s = 32$ W, $P_r = 2.104$ kW and $P_m = 2.062$ kW, $P_{in} = 0.333 + 0.032 + 2.104 = 2.469$ kW and $\eta = 2.062/2.469 = 0.84$. Maximum efficiency occurs at $s = 0.04$ where it reaches 86.3%. Figure 17.12 shows a graph of efficiency against slip for this motor.

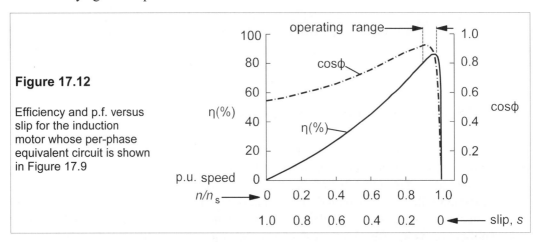

Figure 17.12

Efficiency and p.f. versus slip for the induction motor whose per-phase equivalent circuit is shown in Figure 17.9

The phase angle, ϕ, of the total current drawn by the motor, I_p, is found by calculating $\mathbf{I_0}$ ($= 240/173 + 240/j27$) and adding it to $\mathbf{I_1}$. As can be seen in Figure 17.11, I_1 is large at large slip, but falls quite rapidly as the slip approaches zero. Thus at low speeds and high slip $I_1 \gg I_0$ and ϕ is almost entirely determined by the stator current. However, as the slip approaches zero, the contribution to ϕ of I_0 becomes more important and eventually predominant. A graph of power factor against slip is shown in Figure 17.12. Though the power factor is always lagging for an induction motor, under normal operating conditions it is quite high: from about 0.8 to 0.97.

17.8 Practical induction machines

The constructional characteristics of induction motors are now considered in more detail.

17.8.1 Stator windings

Though Tesla's original induction motor had salient poles, the stator of today's machines is a smooth hollow cylinder with slots on the inside for the conductors. Each slot carries two halves of two different coils, making one coil/slot. Poles are formed by connecting coils in series to form a *phase group* or *group* for short. The poles are thus formed by coils which are not on top of each other, but next to each other: this is called a *distributed winding*. For example, a two-pole, three-phase machine will have two groups for each of the supply phases to form a N pole and a S pole, making 6 groups altogether. Thus

$$G = P\phi \qquad (17.43)$$

where G is the number of groups, ϕ is the number of supply phases and P is the number of poles. If the red phase has groups at mechanical angles 0° and 180°, the yellow phase groups must be at +60° from these and the blue phase groups at +120° as in Figure 17.13 More poles can be formed by making more groups. Thus a 120-slot stator can have 2 poles by joining 20 coils together to form a group, or 4 poles by connecting 10 coils in a group. The mechanical angle between the red, yellow and blue groups is 60° in the latter case.

The number of slots/pole is the *pole pitch*. In the case of a 120-slot, 4-pole machine the pole pitch is 30. One could arrange each coil to span the same number of slots as the pole pitch, that is to have coil pitch the same as the pole pitch, but in practice it is found that the flux distribution is more nearly sinusoidal and the machine smoother running if the coil pitch is about 10–20% less than the pole pitch. Suppose we decide to make the coil pitch 26 slots, or 87% of the pole pitch. Then if one side of a red coil of group 1 goes in slot 1, the other side must go into the top of slot 27. The red group 1 coils fill the bottoms of slots 1–10 (or 1–10B) and the tops of slots 27–36 (27–36T). Supposing this group to form a N pole (let us call the group R1N) when the current is positive, the next red group must be connected in reverse to form a S pole (let us call this group R2S) and must fill slots 90° physically from the R1N group, that is slots 31–40B and 57–66T. Then the rest of the red phase groups follow. The yellow and blue phase groups must be equally spaced about the stator and at 60° either side of the red phase groups. Figure 17.13 indicates the arrangement.

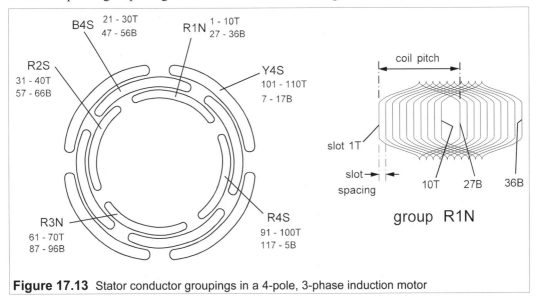

Figure 17.13 Stator conductor groupings in a 4-pole, 3-phase induction motor

17.8.2 Rotor design

The rotor resistance determines the torque-slip characteristic of the motor. A low rotor resistance results in a low starting torque and a large starting current. With a wound rotor, the windings of the rotor are connected to three slip rings and so external resistance can be inserted into the rotor circuit. Not only does this control the starting torque, but also the running speed may be varied since the slip depends on rotor joule losses, while the torque depends on the total power transferred to the rotor. However, increasing the slip in this way leads to considerable power wastage in the external resistors.

Squirrel-cage rotors are less flexible. Nevertheless, some compromise between high starting torque and good running efficiency is possible. By using brass instead of copper or aluminium in the cage conductors the rotor resistance is increased, and hence the starting torque is raised, at some cost in efficiency at running speed; the running speed may also vary without changing the stator current much. The dual-cage rotor is an attempt to get the best of both worlds. The rotor conductors are given a cross section as in Figure 17.14.

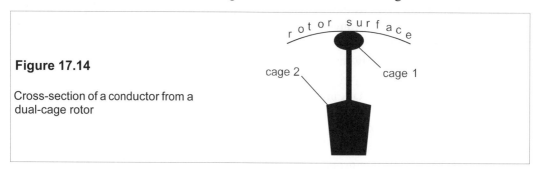

Figure 17.14

Cross-section of a conductor from a dual-cage rotor

On starting, when the rotor e.m.f.'s frequency is relatively high, the current is confined to the outer part of the conductor, because the penetration depth (see Chapter 28, Section 28.4.1) of the current is proportional to $1/\sqrt{f}$; hence the effective rotor resistance is high. When the rotor attains running speed the frequency of the rotor e.m.f. is much smaller, so the penetration depth of the current is much greater. It is carried by the whole of the rotor conductor whose effective resistance is now low.

17.8.3 *Speed control*

Speed control is inherently more difficult with induction motors than with DC motors. The most fundamental way of controlling the speed is to control the supply frequency, a method of much greater importance today when solid-state frequency converters are available. Two-speed motors can be made by reversing the connections on all the poles of like polarity. Thus if a two-pole motor runs at a speed of 3000 r/min with normal N-S pole connections, it can be made to run at 1500 r/min – half the speed – if the polarity is made N-N by reversing the connections to one pole. S poles are formed in the stator iron between the wound N poles. To make room for these *consequent* poles the stator winding should cover only half the circumference.

In the nearly-linear operating region near synchronous speed the slip is given by

$$s \propto TR_2V^2 \tag{17.44}$$

Speed control is therefore obtainable by varying the supply voltage as well as the rotor resistance. However, if the torque is constant, any increase in slip (no matter how achieved) must be accompanied by greater joule losses in the rotor as $P_{rj} = sP_r = s\omega_s T$.

17.9 **Domestic-supply induction motors**

There are many places where a three-phase supply is unavailable: homes, shops, offices and light industrial premises, for example. In these places induction motors have to be run from a single-phase supply, though the rating is usually under 1 kW and the efficiency is not high. There are several ways of producing the necessary field rotation in the stator each giving rise to a different type of motor. Single-phase and split-phase motors vibrate because power is supplied to them in pulses, and not continuously as in a three-phase motor, while their loads remain constant.

17.9.1 The single-phase induction motor

The construction of these motors is very like that of three-phase induction motors: the stator and rotor look exactly the same and the stator winding is distributed in many slots. We can wind a motor with one or more pole pairs all supplied from a single phase and the synchronous speed will still be given by Equation 17.9. The polarity of the poles will alternate with the supply, but it is not immediately obvious that there should be rotation of the rotor in consequence and, in fact, the starting torque of such a motor is zero. However, once the rotor is set in motion, its conductors will cut the stator field and currents will be induced that produce a flux, Φ_r, normal to the stator flux, Φ_s; and because the rotor conductors are inductive, the rotor flux will lag 90° behind the stator flux. A rotating field is thereby set up, as in Figure 17.15, and exerts a torque on the rotor which increases with the rotor speed until synchronous speed is almost attained.

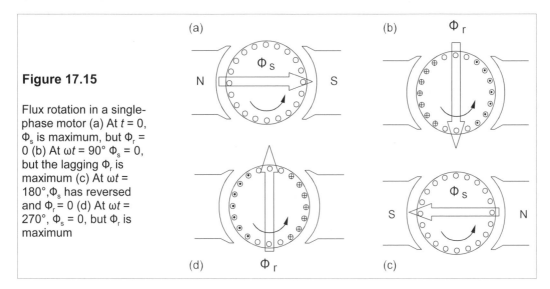

Figure 17.15

Flux rotation in a single-phase motor (a) At $t = 0$, Φ_s is maximum, but $\Phi_r = 0$ (b) At $\omega t = 90°$ $\Phi_s = 0$, but the lagging Φ_r is maximum (c) At $\omega t = 180°$, Φ_s has reversed and $\Phi_r = 0$ (d) At $\omega t = 270°$, $\Phi_s = 0$, but Φ_r is maximum

Another way of looking at the single phase induction motor is to consider the field to be the resultant of two equal counter-rotating fields which produce identical, but opposing, torque-speed curves as shown in Figure 17.16. The torques produced will cancel only when the speed is zero.

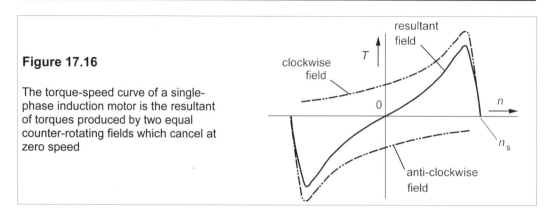

Figure 17.16

The torque-speed curve of a single-phase induction motor is the resultant of torques produced by two equal counter-rotating fields which cancel at zero speed

17.9.2 Resistance split-phase motors

These are often just called split-phase motors. Starting torque can be provided by making the stator field rotate. The stator field will rotate if there is a set of alternate poles excited by a current which is out of phase with the main stator field. Auxiliary poles are inserted between the main poles and in the resistance split-phase motor a phase difference is produced between main and auxiliary fields by making the auxiliary winding of higher resistance and similar inductance. The auxiliary coils are wound of much finer wire than the main winding to increase their resistance. Once the motor has started the auxiliary winding is disconnected by a time-delay, or a centrifugal switch. If δ is the phase difference between the current in the main field, I_s, and the current in the auxiliary field, I_f, the starting torque, T_0, is then

$$T_0 \propto I_s I_f \sin \delta \qquad (17.45)$$

17.9.3 Capacitor motors

Obviously, from Equation 17.45, the starting torque would be greater if phase difference between main and auxiliary currents were greater in a split-phase motor. In capacitor motors the auxiliary winding is placed in series with an electrolytic capacitor. In *capacitor-start* motors the auxiliary is disconnected by either a time-delay switch or a centrifugal switch, but in *capacitor-run* motors both windings remain connected to the supply at all times. In some cases extra capacitance is employed for starting. Capacitor motors have tended to replace resistance split-phase motors as better capacitors have become available. They are relatively cheap, costing from £60 for a 200W to £120 for a 1kW motor. Because the phase angle between the currents in the windings is near 90°, the capacitor-run motor is really a two-phase motor and therefore suffers much less from the vibration inevitable in single-phase motors.

17.9.4 Shaded-pole motors

Shaded-pole motors are simple, cheap (as little as £6) and inefficient – typically around 10% for smaller and 30% for larger motors. Because of their inefficiency they are seldom found in ratings above 50 W, and usually much less. The design is shown in Figure 17.17.

They are single-phase, two-pole motors, part of each pole having a single ring of copper round. The changing flux of the poles causes e.m.fs. to be set up in the rings which induce currents that produce a counteracting field, so 'shading' part of the pole. The result is a weak, rotating component of the field which drags the rotor from the unshaded part towards the shaded part of each pole. The direction of rotation is thus fixed by the shading and is invariable: you have to buy a clockwise or an anticlockwise motor as needed.

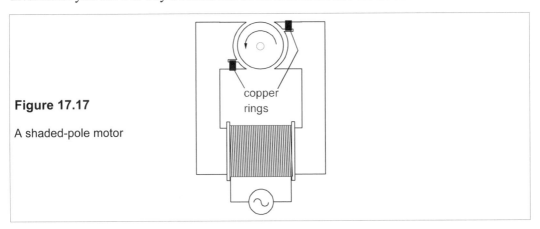

Figure 17.17

A shaded-pole motor

Example 17.5

A shaded-pole motor runs from a 240V, 50Hz supply and draws 200 mA when working at its rated torque of 12.5 mNm and running at 2400 r/min If the input power is 20 W, what are the power factor, the slip and the efficiency of the motor?

   The apparent power, $S = 240 \times 0.2 = 48$ VA, making the p.f., $\cos\phi = P/S = 20/48 = 0.42$. The low p.f. is typical. The synchronous speed is given by $n_s = 60f_s/p$, where $f_s = 50$ Hz and p, the number of pole pairs is one, so $n_s = 3000$ r/min The slip is $s = 1 - n/n_s = 1 - 2400/3000 = 0.2$, or 20%. The high slip is also typical. The mechanical power, $P_m = 2\pi n T/60 = 3.14$ W, leading to the efficiency, $\eta = 3.14/20 = 16\%$, very low, but of little consequence when the power input is only 20 W.

Suggestions for further reading – see Chapter 18

Problems

1. What is the synchronous speed of a three-phase induction machine that runs from a 60Hz supply, if it has 18 poles? What is the slip if the rotor speed is (a) 1150 r/min (b) 1500 r/min (c) –16 radians/s? What is the machine operating as in each case?
    [1200 r/min (a) 0.0417, motor (b) –0.25, generator (c) 1.127, brake]
2. The synchronous speed of an induction machine is 2000 r/min. If the machine is at standstill, what is the slip? If it is to run with –33% slip, what is its speed? If the slip is 110%, what is its speed? What is the machine acting as in each case? *[1; 2660 r/min, generator. –200 r/min, brake]*
3. What are the frequencies of the rotor e.m.fs. in Problem *17.1? [(a) 2.5 Hz (b) –15 Hz (c) 67.6 Hz]*
4. Find the turning points of Equation 17.22, and show that the minimum value of the torque is $-V^2/2\omega_s X_0$.
5. If the standstill torque of an induction motor is 300 mNm and its breakdown torque is 800 mNm, what is the approximate torque developed at a slip of (a) 0.03 (b) –50% (c) 1.2? At what fractional

slip will the torque be the same as at standstill? *[(a) 241 mNm (b) –541 mNm (c) 253 mNm; 0.0379]*

6. The total power consumed by a three-phase induction motor in a no-load test was 106 W, with a line current of 0.9 A and a line voltage of 415 V. If the resistance between two of the stator terminals was 0.69 Ω and the stator was star-connected, what are the magnetising reactance, the parallel loss resistance and the stator series resistance, all per phase?
    [X_M = 270 Ω, R_0 = 1.625 kΩ, R_1 = 0.345 Ω]

7. The motor in the previous problem draws its rated line current of 15 A when its rotor is locked and the line voltage is 258.5 V. The total power consumption is 1.2 kW. If the stator is star-connected, what are the series components of the equivalent circuit per phase? (Neglect the no-load losses.)
    [R_2 = 1.433 Ω, X_0 = 9.79 Ω]

8. What is the maximum torque that the motor in the previous problems can exert if the synchronous speed is 3000 r/min? Neglect the stator resistance, R_1, and assume it runs at a line voltage of 415 V. What is the slip at this torque? What is the maximum power in the rotor? What is then the current? What are the copper losses in the rotor? *[28 Nm, 14.6%, 8.8 kW, 17.3 A, 1.29 kW]*

9. Calculate the maximum torque that the motor of the previous problem can exert, including the effect of the stator resistance. At what slip does this occur? What are the stator's copper losses at this torque? What is the current? *[27 Nm, 14.6%, 299 W, 17 A]*

10. Maximum torque is exerted by a three-phase induction motor, whose synchronous speed is 600 r/min, when the slip is 3%. The motor has a star-connected stator and operates from a line voltage of 6.9 kV. If the stator resistance of 100 mΩ per phase can be neglected and the rotor equivalent resistance is 80 mΩ per phase, what is the maximum rotor power and torque? What is the approximate efficiency at this torque? What is the maximum rotor power if R_1 is included? (Neglect the no-load losses.) *[8.93 MW, 142 kNm, 97%, 8.6 MW]*

11. In the motor of the previous problem the no-load losses are represented by a parallel resistance, R_0, of 590 Ω. What is the maximum efficiency of the motor and at what slip does it occur? If the motor is rated to operate at maximum efficiency, what is its power rating? (Maximum efficiency is obtained when the no-load losses are equal to the copper losses in rotor and stator.)
    [η_{max} = 96.4%, s = 0.812%, 4.6 MVA]

12. A three-phase, 415V, 1500r/min, star-connected, induction motor has per-phase equivalent resistances of R_1 = 1.4 Ω and R_2 = 1.6 Ω and a leakage reactance, X_0, of 6.6 Ω. What are the starting and breakdown torques? What are these if a resistance of 5 Ω is placed in series with R_1, R_2 and X_0? What are the torques if a inductive reactance of 5 Ω is placed in series instead of a resistance? (The no-load losses can be ignored.) *[33.4 Nm, 67.3 Nm; 16.3 Nm, 35.2 Nm; 12.2 Nm, 41.9 Nm]*

13. A 415V, three-phase, induction motor has a synchronous speed of 1800 r/min. If the maximum rotor power is 12 kW at a slip of 18%, what are the leakage reactance, X_0, and the resistance, R_2, per phase? What is the maximum torque? What series resistance is required to operate at maximum torque and a slip of 10%? What are the maximum power and torque then? Assume a delta-connected stator and ignore the stator and no-load losses.
    [X_0 = 21.5 Ω, R_2 = 3.875 Ω, 63.7 Nm, 32.2 Ω, 3.64 kW, 19.3 Nm]

14. Show that maximum torque is obtained in a three-phase induction motor when the slip is R_2/β, where $\beta = \sqrt{(R_1^2 + X_0^2)}$, R_2 is the equivalent resistance per phase of the rotor, X_0 is the equivalent leakage reactance per phase and R_1 is the series resistance per phase in the stator circuit. Hence show that the maximum rotor power is $1.5V_p^2/(\beta + R_1)$, which is independent of R_2, where V_p is the phase voltage.

18 Synchronous machines

IN SYNCHRONOUS machines the rotor turns in synchronism with the e.m.f. developed or the alternating voltage applied. A synchronous machine can function either as a motor or as a generator and sometimes does both, as in the 1.8GW pumped-storage scheme at Dinorwig in N Wales, which uses six, 300MVA, reversible pump-generators. Virtually all the electrical power we use is generated by synchronous generators, some of which are among the largest of all electrical machines. They are large, not only because we need vast quantities of electricity, but also because large generators are cheaper per MW to construct and to operate, and are more efficient than small generators. A synchronous generator rated at 1000 MVA should turn more than 98% of the mechanical power input into electrical power.

18.1 Synchronous generators

Synchronous generators, also known as alternators, consist of a rotor and a stator. We can make the stator a magnet, or set of magnets, and the rotor an armature, or we can make the rotor a magnet and the stator an armature. The magnets can be wound or permanent. Generally speaking, permanent magnets and rotating armatures are found only in small machines. Because a rotating armature requires slip rings and brushes, it is inconvenient for high-power generators where voltages and currents are very large indeed. All the generators in power stations are synchronous machines with stator armatures and DC-excited rotors.

The frequency of the e.m.f. generated can be calculated if we recognise that the motion of a N pole past a conductor will generate a maximum e.m.f. once per cycle and the accompanying S pole will develop a minimum e.m.f. also one per cycle. The frequency of the generated e.m.f. is the same as the frequency of rotation. If the speed of the rotor is n_s r/min, then $f_s = n_s/60$. If there are p pole pairs, the frequency will be

$$f_s = pn_s/60 \qquad (18.1)$$

Thus to generate at 50 Hz with only one pair of poles, the speed of rotation required is $50 \times 60 = 3000$ r/min, the greatest possible speed at which 50 Hz can be generated in a synchronous machine. Steam-driven turbo-alternators usually turn at high speeds, 1500 or 3000 r/min, while hydro-electric alternators turn at relatively low speeds, from 100 to 600 r/min. Turbo-alternator rotors therefore have two or four poles whereas hydro-electric rotors need from 10 to 60. Because the rotor is very large in generators used by public utilities, high-speed rotors must be a long, thin cylinder to reduce the radius and hence the centrifugal forces which are very great. Low-speed rotors of the same volume can be short, fat cylinders with a greater radius.

18.2 **Synchronous torque**

If the rotor is magnetised with a direct current and rotates within the stator, it induces e.m.fs. and currents in the stator conductors which give rise to a stator magnetic field and stator poles. At synchronous speed both rotor and stator poles will rotate together. If no torque is applied to the rotor the two sets of poles will line up exactly and there will be no electromagnetic torque. If now a torque is applied to the rotor it will accelerate momentarily and then slow down to synchronous speed again. The poles of the rotor and stator are now displaced by an angle, θ, and the mechanical torque on the rotor is balanced by the electromagnetic torque of the displaced poles. If there are $2p$ poles the *torque angle*, δ, is $p\theta$ (see Figure 18.1). The torque is proportional to $\sin\delta$ and will have a maximum value, called the *pull-out torque*, when $\delta = 90°$. Synchronism is lost when the torque angle reaches 90° and the mechanical and electromagnetic torques cannot be balanced.

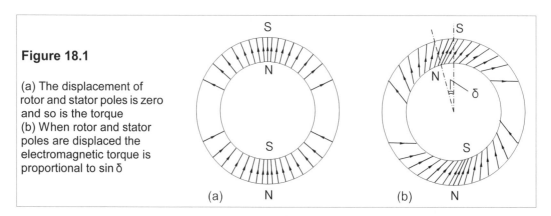

Figure 18.1

(a) The displacement of rotor and stator poles is zero and so is the torque
(b) When rotor and stator poles are displaced the electromagnetic torque is proportional to $\sin\delta$

(a)   (b)

18.3 **The equivalent circuit of a synchronous generator**

The stator of a synchronous generator is identical to that of a three-phase induction motor. Normally the stator generates a three-phase e.m.f. and may be connected in star or delta. A star connection reduces the conductor voltage by $\sqrt{3}$ and therefore needs less insulation between the conductors and the grounded stator core than by a delta connection. The reduced insulation permits larger conductors to be used for a given size of core. Even very large alternators employ relatively low voltages in the stator conductors: line voltages of 22 kV (line-neutral = 12.7 kV) are typical. Star connection also helps to reduce the third harmonic content (a component of the induced e.m.f. which has three times the fundamental frequency) of the generated waveform. In a star connection the third harmonics tend to cancel, whereas in a delta connection they tend to add. For these reasons most generators are star-connected.

Each phase in the stator is identical and generates an identical e.m.f. to the others except for a phase difference of 120°. The rotor is excited by a direct current, I_f, and the magnitude of the e.m.f. generated in each phase, E, is proportional to I_f, provided the magnetic circuit is not saturated. The armature winding will have a reactance, X_s, per phase which is known as the *synchronous reactance*. There will also be a phase resistance, R_s, but since $R_s \ll X_s$ (by a factor of 50 or more in large machines), it may be neglected and the equivalent circuit per phase takes the simple form of Figure 18.2. The magnetising inductance and the core losses have been omitted as they too are small. The voltage regulation is given by

$$\% \text{ reg} = 100(E - V)/V \tag{18.2}$$

where E is the open-circuit voltage and V the rated terminal voltage. Regulation is usually poor because of the relatively large synchronous reactance.

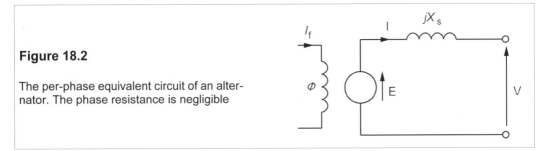

Figure 18.2

The per-phase equivalent circuit of an alternator. The phase resistance is negligible

18.4 Per-unit values and the short-circuit ratio

It is customary to express generator parameters in terms of the line-neutral voltage, V, and the apparent power per phase, S. These are the *base* values, and from them we can calculate the base line-neutral impedance, Z_B

$$Z_B = V^2/S \tag{18.3}$$

The synchronous reactance can then be expressed in terms of Z_B, that is per unit.

Example 18.1

A three-phase, 23.5kV generator is rated at 660 MW with a power factor of 0.85 and has a synchronous reactance of 1.2 per unit. What is its synchronous reactance in ohms? If the resistance per phase is 0.01 p.u., what will the copper losses be at rated output?

The voltage rating given will be the line voltage and so the line-neutral base voltage, $V_B = 23.5/\sqrt{3}$ = 13.57 kV. The power rating is 660/3 = 220 MW per phase, so the base apparent power per phase, $S_B = P_B/\cos\phi$ = 220/0.85 = 259 MVA, and the base impedance is

$$Z_B = E_B^2/S_B = \frac{(13.57 \times 10^3)^2}{259 \times 10^6} = 0.71 \ \Omega \tag{18.4}$$

The synchronous reactance is therefore $X_s = x_s Z_B$ = 1.2 × 0.71 = 0.85 Ω. (x_s is the per-unit value of X_s.) The rated current will be 1 p.u. and the copper losses will be I^2r (i, r being p.u.) or 1^2 × 0.01 = 0.01 p.u. = 0.01 × 259 = 2.59 MW per phase = 7.8 MW in total.

The *short-circuit ratio* (SCR) is often used instead of the per-unit value of the synchronous reactance. It is the ratio of the field current, I_{fo}, required to produce the rated open-circuit voltage, E, to the field current, I_{fs}, required to produce the rated current, I, when the terminals are shorted. If the generated e.m.f. is E_s when the terminals are shorted and carrying rated current, then $I_B = E_s/X_s$ and $E_s = I_B X_s$. Since the generated e.m.f. is proportional to the field current, $E_s = kI_{fs}$ and $E_B = kI_{fo}$ and

$$SCR = \frac{I_{fo}}{I_{fs}} = \frac{E_B}{E_s} = \frac{E_B}{I_B X_s} = \frac{Z_B}{X_s} = \frac{1}{x_s} \qquad (18.5)$$

The SCR is merely the reciprocal of the per-unit value of the synchronous reactance.

18.5 The generator under load

The load is connected across the terminals of Figure 18.2, as in Figure 18.3. If the power factor of the load is cosϕ, lagging, and it takes a current, **I**, the phasor diagram is that shown in Figure 18.3. Note that the generator e.m.f., **E**, leads the terminal voltage, **V**, by the torque angle, δ. The terminal voltage, **V**, and current, **I**, can be measured and by Kirchhoff's voltage law

$$\mathbf{E} = \mathbf{V} + jX_s\mathbf{I} \qquad (18.6)$$

hence all the components of the phasor diagram may be determined.

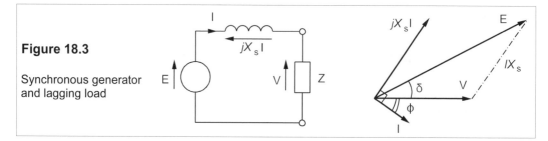

Figure 18.3

Synchronous generator
and lagging load

Example 18.2

The generator of Example 18.1 supplies a load at its rated power and a p.f. of 0.85. Determine the phasor diagram and find the voltage regulation. What is the torque angle? If the generator operates at 50 Hz and 1500 r/min, what is the angle between the rotor and stator poles? What is the rotor torque? What is the pull-out torque? What is the maximum possible output of the generator if the excitation and terminal voltage are unchanged?

　　Rated power implies that $v = 1$ and $i = 1$, and then the p.u. voltage across the synchronous reactance is $ix_s = 1 \times 1.2 = 1.2$ p.u. This voltage must lead **i** by 90° and **i** lags **v** by $\cos^{-1}0.85 = 31.8°$. The phasor diagram is as shown in Figure 18.4, which is in p.u. values. The torque angle can be measured from this diagram and so can e, the generated per-unit e.m.f. However, we can calculate them from the diagram too, since

$$\tan\delta = BC/OB = BC/(OA + AB)$$

Then substituting for BC, OA and AB gives

$$\tan\delta = \frac{ix_s\cos\phi}{v + ix_s\sin\phi} = \frac{1.2 \times 0.85}{1 + 1.2 \times 0.527} = 0.625 \qquad (18.7)$$

leading to $\delta = 32°$. As $\sin\delta = BC/OC = ix_s\cos\phi/e$, we find $e = ix_s\cos\phi/\sin\delta = 1.2 \times 0.85/\sin32° = 1.925$ p.u. or $1.925 \times 13.57 = 26.1$ kV. The regulation is $(e - v)/v = e - 1 = 0.925 = 92.5\%$, very poor. Because the regulation is poor with alternators, some form of automatic voltage regulation is usually required, especially when they are feeding isolated loads, and even more so when the speed varies as it does, for example, in a car alternator.

The torque angle can also be calculated with phasor arithmetic, as $\mathbf{i} = 1\angle-\cos^{-1}0.85 = 1\angle-31.8°$ and $\mathbf{x_s} = 1.2\angle90°$, then $\mathbf{ix_s} = 1.2\angle58.2°$. Since $\mathbf{v} = 1\angle0°$, then $\mathbf{e} = \mathbf{v} + \mathbf{ix_s} = 1\angle0° + 1.2\angle58.2° = 1.925\angle32°$.

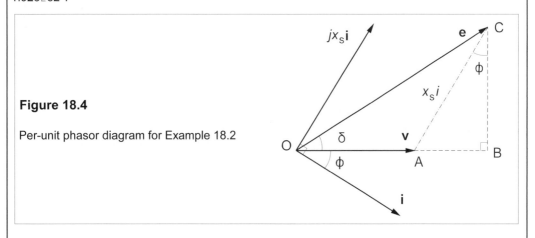

Figure 18.4

Per-unit phasor diagram for Example 18.2

For the generator to operate at 50 Hz and a speed of 1500 r/min it must have four poles and the angle between stator and rotor poles is then $\delta/2 = 16°$. The rotor torque is found from $P = \omega_s T$, or $T = P/\omega_s = 60P/2\pi n_s = 60 \times 220/2\pi \times 1500 = 1.4$ MNm/phase or 4.2 MNm in total. With constant excitation, the torque is proportional to $\sin\delta$, which means the ratio of the pull-out torque to the rated torque is $\sin90°/\sin32° = 1.89$, giving a pull-out torque of 7.94 MNm in total. The maximum power output will be 1.89 times the rated power or 1.25 GW. In practice an overload of this magnitude would not be tolerable, but overloads of 20% are.

18.6 **The generator on an infinite bus**

Electrical supply systems are so enormous that a single generator cannot influence the voltage and frequency; the generator is then said to be connected to an *infinite bus* and the options open to the operator are just two. He cannot vary the speed of the generator as it must rotate at the speed determined by the bus frequency: nor can he vary the terminal voltage which is also determined by the bus. He is left with the excitation current, I_f (and therefore the generator e.m.f., $\mathbf{E}$) and the torque on the rotor as his only controllable parameters. To connect his generator to the bus he must synchronise its speed exactly and make its open-circuit terminal voltage ($\mathbf{E}$) exactly equal to the system voltage ($\mathbf{V}$) in both magnitude and phase. When the generator is so connected it is not delivering power as $\mathbf{E} = \mathbf{V}$ and $\mathbf{I} = 0$; it is said to be *floating* or *spinning*. The torque angle, $\delta = 0$.

If the operator now alters the exciting current he will change $\mathbf{E}$ in magnitude, but not in phase, let us say to $\beta\mathbf{V}$, where β is a constant. Making $\mathbf{V}$ the reference phasor, the current flowing through the synchronous reactance is given by

$$\mathbf{I} = \frac{\mathbf{E} - \mathbf{V}}{jX_s} = \frac{(\beta - 1)\mathbf{V}}{jX_s} = \frac{(\beta - 1)V}{X_s}\angle-90° \tag{18.8}$$

But this current is exactly 90° out of phase with $\mathbf{V}$ and so the power factor is $\cos90° = 0$ and the generator delivers no power. It can be seen that if $\beta > 1$, the generator supplies vars to the bus and if $\beta < 1$, it receives vars.

When the torque angle is zero, no power can be supplied to (or received from) the bus. The only parameter left for the operator to vary is the rotor torque. If he causes the torque to increase (by supplying more steam) the rotor will accelerate and move ahead of the stator poles, thus introducing a phase angle, δ, between **E** and **V** and so supplying the bus with power since **V** and **I** no longer differ by 90° in phase. If the operator cuts off the steam completely, the rotor will lag behind the stator poles and the machine will absorb power from the bus; it will become a motor. When **E** and **V** are no longer in phase ($\delta \neq 0$) changing the excitation *will* change the power.

Figure 18.5 (which is identical in form to Figure 18.4) shows the phasor diagram for a synchronous generator supplying current at a lagging phase angle, ϕ, with a torque angle of δ. The power delivered by the generator is therefore $P = VI\cos\phi$. Now in the diagram we see that BC is $E\sin\delta$, while AC is parallel to, and the same length as, $jX_s\mathbf{I}$. But BC is also $AC\cos\phi = IX_s\cos\phi$ and thus

$$E\sin\delta = IX_s\cos\phi \quad \Rightarrow \quad \frac{EV\sin\delta}{X_s} = VI\cos\phi = P \tag{18.9}$$

The power from the generator is $(EV\sin\delta)/X_s$.

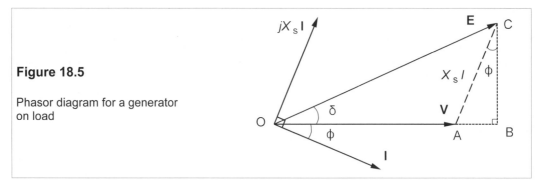

Figure 18.5

Phasor diagram for a generator on load

Example 18.3

A three-phase, 750MVA generator is to supply its rated power at a lagging p.f. of 0.9 to an infinite bus whose voltage is 22 kV. If the SCR of the generator is 0.8, what must the torque angle and e.m.f. be? What is the greatest possible leading phase angle that can be achieved when supplying rated power?

Working in per-unit values is simplest. The per-unit synchronous reactance is the reciprocal of the SCR; $x_s = 1/0.8 = 1.25$, $\mathbf{x_s} = 1.25\angle90°$ p.u. A lagging p.f. of 0.9, taking **v** as the reference phasor, means that the current's phase angle is $-25.84°$, and $\mathbf{i} = 1\angle-25.84°$. Thus $\mathbf{ix_s} = 1.25\angle64.16°$. Now from Figure 18.3 we see that

$$\mathbf{e} = \mathbf{v} + \mathbf{ix_s} = 1\angle0° + 1.25\angle64.16° = 1.911\angle36.06° \text{ p.u.} \tag{18.10}$$

The base voltage is the line-neutral voltage, $22/\sqrt{3} = 12.7$ kV, making $E = 1.911 \times 12.7 = 24.3$ kV and **E** $= 24.3\angle36.06°$ kV. To check this, we can calculate the power according to Equation 18.9:

$$p = (ev/x_s)\sin\delta = (1.911/1.25)\sin36.06° = 0.9 \text{ p.u.} \tag{18.11}$$

The base power is 250 MW/phase and the power delivered is $0.9 \times 250 = 225$ MW/phase.

To make ϕ leading and as large as possible we must rotate the $jx_s\mathbf{i}$ phasor anticlockwise as far as possible. Supplying rated power means that $i = 1$ p.u, $v = 1$ p.u. and $ix_s = 1.25$ p.u. as before. But as

$jx_s\mathbf{i}$ rotates it moves **e** with it and increases δ to 90°, its maximum possible value, shown in Figure 18.6. Thus φ is maximum when δ = 90°. From Figure 18.6, sinφ = v/ix_s = 1/1.25 and φ = 53°. We can find the generated e.m.f. in this case by Pythagoras's theorem:

$$e = \sqrt{(ix_s)^2 - v^2} = \sqrt{1.25^2 - 1^2} = 0.75 \quad \text{p.u.} \tag{18.12}$$

which means E = 0.75 × 12.7 = 9.5 kV. The generated e.m.f. is always less than the terminal voltage when supplying capacitive loads.

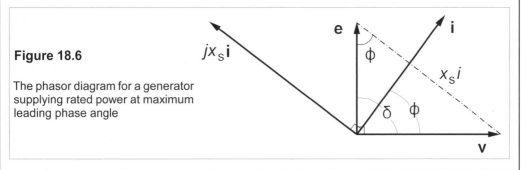

Figure 18.6

The phasor diagram for a generator supplying rated power at maximum leading phase angle

18.7 The construction of synchronous machines

We shall consider only those in which the stator is the armature and the rotor is DC excited. The stator winding is the same as that of an induction motor. Large turbo-alternators have copper tubes for stator conductors which carry de-ionised, high-resistivity water as an internal coolant. Hydrogen under pressure (about 4 bar) is force-circulated as an external coolant for both stator and rotor. Hydrogen is chosen as the coolant gas for two reasons: it has a high thermal conductivity (seven times that of air) and it has low density (7% of air's), so that not only is cooling more efficient than with air, but also windage losses are greatly reduced.

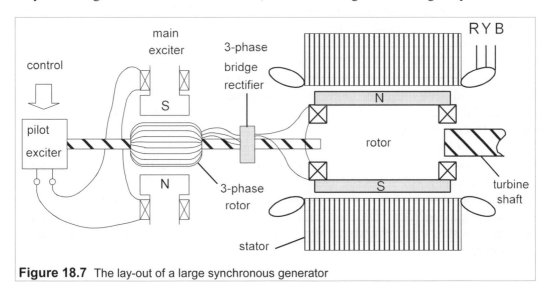

Figure 18.7 The lay-out of a large synchronous generator

On large machines the rotor excitation is two-stage. An AC, three-phase, pilot exciter supplies (after commutation) the stator field of the main exciter. The main exciter is a DC generator which is connected to the rotor windings via a commutator and brushes. Brushes and slip rings have lately given way to rectifiers mounted on the shaft of the rotor and turning with it, as shown in Figure 18.7. Two-stage excitation is used because the pilot exciter carries much smaller currents and is thereby more readily controlled than the main exciter. For example, the main exciter of a 776MVA generator provides a maximum current of 5.2 kA at 750 V while its pilot exciter runs at a maximum of 260 A and 350 V. Clearly, 90 kVA is much easier to control than 4 MVA and the rotor excitation can be rapidly adjusted (from zero to maximum in under a second) in response to changes in demand.

There are two types of rotor: salient pole and cylindrical. Cylindrical rotors have to be used on large high-speed turbo-generators to withstand the enormous centrifugal forces. Figure 18.8 shows a salient-pole rotor with 12 poles.

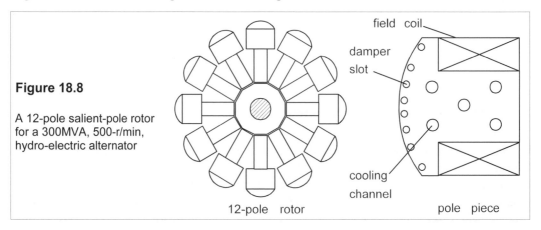

Figure 18.8

A 12-pole salient-pole rotor for a 300MVA, 500-r/min, hydro-electric alternator

12-pole rotor

pole piece

The pole pieces and coils are bolted onto a metal framework called a spider and the whole assembly rotates within the stator. The rotor poles are provided near their outer surfaces with a squirrel-cage winding which runs parallel to the rotational axis. This damper winding is not excited and comes into play only if the rotor speed differs from the synchronous speed. It damps oscillations caused by sudden changes in the load. The rotor of a synchronous machine acts as if it were connected to the stator by springs (the magnetic field) and any imbalance in mechanical and electromagnetic torques sets up oscillations like a weight on a coiled spring. The natural damping in a synchronous machine is quite small and the oscillations would persist unless damped by the currents induced in the cage winding.

The rotors of large generators cannot have a large radius because the centrifugal forces will exceed safe limits. At 3000 r/min the largest rotor radius is about 0.7 m and a salient-pole construction is not feasible. Instead the rotor is machined from a single billet of silicon steel (laminations are not needed as there are no eddy-currents). Since the power generated is proportional (among other things) to the rotor surface area, long rotors are necessary in high-speed machines. In large cylindrical rotors the coils are held in place by end rings as well as slot geometry.

18.8 **Synchronous motors**

A synchronous motor is just a synchronous generator driven the opposite way by the line voltage. Synchronous motors must run at synchronous speed, no matter what the load or excitation. Their great advantage over induction motors is that their power factor may be controlled rather easily, and they are on occasion used not as motors but as *synchronous capacitors* to improve the power factor of a large load or supply system. Synchronous motors are *salient-pole* machines because they are then self starting.

18.8.1 *Salient-pole motors*

Motors usually have salient poles. With a cylindrical rotor the gap between stator and rotor is essentially constant all round, but this is not the case for salient-pole machines where there is a large gap between poles. The magnetic flux then has easy paths (low reluctance, see Figure 18.9a, position A) when the poles of stator and rotor are lined up and difficult paths (high reluctance, position B in Figure 18.9) when they are not. In consequence as well as the synchronous reactance a further reactance term is required. The high-reluctance reactance is X_h and in the low-reluctance state it is X_e, where $X_e > X_h$. The *reluctance torque* developed is

$$T_r = \frac{V^2 \sin 2\delta}{2X_r} \tag{18.13}$$

where X_r, the reluctance reactance, is given by $1/X_r = 1/X_h - 1/X_e$.

The reluctance torque's maximum value is about 20% of the synchronous torque's maximum. The total torque developed is the sum of synchronous and reluctance torques: $T = T_s + T_r$. Since the expression for T_r does not contain E, the excitation e.m.f., it follows that a rotor winding can be omitted on a salient-pole machine; reluctance motors make use of this property. The $\sin 2\delta$ in the reluctance torque arises because the maximum and minimum reluctance are at a separation of 90° whereas the synchronous torque has maxima and minima separated by 180° as seen in Figure 18.9b.

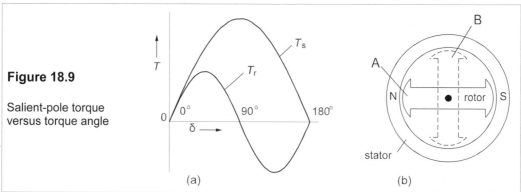

Figure 18.9

Salient-pole torque versus torque angle

(a)   (b)

To start a synchronous motor, a damper winding of the squirrel-cage variety is incorporated in the poles and, on start-up, torque is provided by the induced currents flowing in this winding. The rotor excitation is switched off during starting, however, as the rotor accelerates its winding cuts the stator flux and a current flows as in the induction motor. The starting torque can be increased by connecting external resistors in the rotor winding circuit just as in an induction motor. Near synchronous speed the DC rotor excitation must be switched on at precisely the right time so that the rotor is smoothly pulled into synchronism. The damper winding contributes nothing to the torque once synchronous speed is attained. Large motors are often started with an auxiliary motor called a *pony engine*.

18.8.2 V-curves: reactive power as a function of excitation

By varying the excitation of a synchronous motor we can absorb or deliver reactive volt-amperes to the supply. This is shown in the so-called V-curves named after their characteristic shape. The reactive power as a function of excitation may be found from the equivalent circuit of the motor and its accompanying phasor diagram, Figure 18.10.

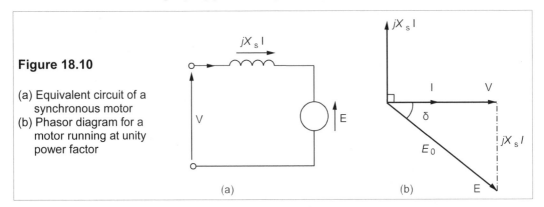

Figure 18.10

(a) Equivalent circuit of a synchronous motor
(b) Phasor diagram for a motor running at unity power factor

These are the same as those of the synchronous generator except that the direction of **I** is reversed and the governing equation becomes

$$\mathbf{V} = \mathbf{E} + jX_s\mathbf{I} \tag{18.14}$$

where **V** is the voltage applied to the stator winding and **E** is the excitation e.m.f. produced in the stator by the current in the rotor field winding. The power supplied to the motor is given by the same expression as the power delivered by the generator. The phasor diagram is also identical except that **E** lags **V** by δ rather than leading by δ.

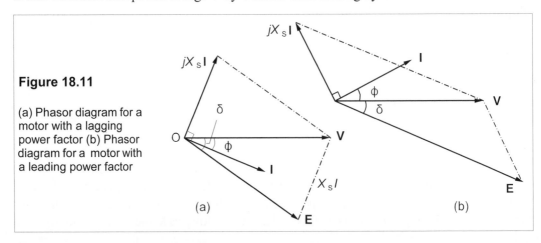

Figure 18.11

(a) Phasor diagram for a motor with a lagging power factor (b) Phasor diagram for a motor with a leading power factor

The motor's power factor is unity when the angle between **V** and **I** is zero (Figure 18.10a), and $P = VI = S$ while $Q = 0$. When the excitation current is reduced and E falls below E_0, the phase diagram looks as in Figure 18.11a: **I** must now lag **V** and the motor looks inductive to the supply and takes reactive volt-amperes from it. And when the excitation is increased as in Figure 18.11b, **I** must lead **V** and the motor delivers vars to the supply. The motor is now acting as a capacitive load.

The V-curve can be plotted by finding δ from

$$P = \frac{VE\sin\delta}{X_s} \quad \Rightarrow \quad \sin\delta = \frac{PX_s}{VE} \tag{18.15}$$

which is identical to the generator's power equation. We can then write **E** as $(E\cos\delta - jE\sin\delta)$ and find **I**

$$\mathbf{E} + jX_s\mathbf{I} = \mathbf{V} \quad \Rightarrow \quad \mathbf{I} = \frac{\mathbf{V} - \mathbf{E}}{jX_s} = \frac{V - E\cos\delta + jE\sin\delta}{jX_s} \tag{18.16}$$

Rationalising this expression, we have

$$\mathbf{I} = \frac{E\sin\delta - j(V - E\cos\delta)}{X_s} \tag{18.17}$$

From this we can find I^2, and then $S^2 = (VI)^2$, which leads to

$$S^2 = (VE\sin\delta/X_s)^2 + (V[V - E\cos\delta]/X_s)^2 = P^2 + Q^2 \tag{18.18}$$

When $E\cos\delta < V$, the p.f. is lagging or inductive, and when $E\cos\delta > V$, it is capacitive.

Figure 18.12 shows V-curves for a motor whose supply voltage is constant at 100 V. When $P = 0$, Equation 18.15 shows that $\delta = 0$, then $\cos\delta = 1$ and Equation 18.18 shows that $S = \pm V(V - E)$, which is the equation of a vee. As P increases, the minimum value of E must be increased and the V-curve is more distorted and shifted up and along the E-axis. The curves in Figure 18.12 are also I-E curves scaled by a factor, V, since $S = VI$ and V is constant.

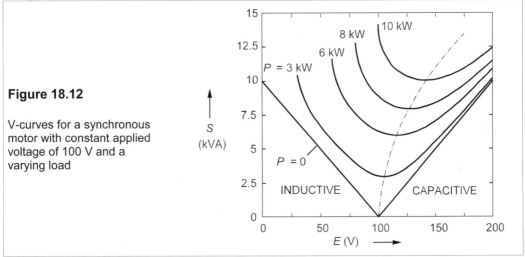

Figure 18.12

V-curves for a synchronous motor with constant applied voltage of 100 V and a varying load

Synchronous motors are used for power-factor correction as it is easy to deliver vars to the supply by increasing the excitation, but the saving in electricity costs produced by the improvement in power factor must be set against the cost of the machine. Synchronous capacitors (which could as well be called synchronous inductors) are synchronous motors run with no load that are used in electricity supply systems for power-factor improvement; they can have ratings of 200 Mvar or more. Synchronous motors are as efficient, or more efficient than induction motors when running at their rated loads.

Example 18.4

A three-phase synchronous motor runs at 250 r/min from a 23kV line at unity p.f. and takes 20 MW, its rated power. If its stator reactance is 0.5 p.u., what will be the line current when the excitation e.m.f. is 1.5 p.u. and the power consumed is unchanged? What will then be the reactive power supplied by the motor? What will the pull-out torque be?

Using per-unit values, we can find δ from

$$p = \frac{ve\sin\delta}{x_s} \ ; \quad \Rightarrow \quad \sin\delta = \frac{px_s}{ve} = \frac{1 \times 0.5}{1 \times 1.5} = 0.333$$

From which we find $\delta = 19.5°$. Then using **v** as reference phasor, **e** is

$$\mathbf{e} = e\cos\delta - je\sin\delta = 1.5\cos 19.5° - j1.5\sin 19.5°$$

$$= 1.414 - j0.5 \text{ p.u.}$$

And

$$\mathbf{v} - \mathbf{e} = 1 - (1.414 - j0.5) = -0.414 + j0.5 = 0.65\angle 129.6° \text{ p.u.}$$

Now

$$\mathbf{i} = \frac{\mathbf{v} - \mathbf{e}}{jx_s} = \frac{0.67\angle 127°}{0.5\angle 90°} = 1.3\angle 37° = 1 + j0.828 \text{ p.u.}$$

The complex p.u. power is

$$\mathbf{vi}^* = p + jq = 1 - j0.828 \text{ p.u.}$$

Thus $q = -0.828$ p.u., the minus sign indicating that q is capacitive (leading). The rated power is 20 MW at unity p.f. so that $S = 20$ MVA, and from this we find the reactive power:

$$Q = qS = 0.828 \times 20 = 16.56 \text{ Mvar}$$

The rated line current is $P/3V_p = 20 \times 10^6/(3 \times 23000/\sqrt 3) = 502$ A $= 1$ p.u., so the actual current of 1.3 p.u. is 653 A.

The pull-out torque occurs when $\delta = 90°$ and $p = ve/x_s = 1.5/0.5 = 3$ p.u. But the per-unit torque is given by $t = p$ as $\omega_s = 1$ p.u., and therefore the pull-out torque must be 3 p.u. If $n_s = 250$ r/min, $\omega_s = 2\pi \times 250/60 = 26.2$ rad/s. The rated torque is $P/\omega_s = (20 \times 10^6)/26.2 = 764$ kNm and the pull-out torque is $3 \times 764 = 2.29$ MNm.

Suggestions for further reading – the following texts cover all the material in Chapters 13–18

Electrical machines, drives and power systems by T Wildi (Prentice Hall, 5th ed., 2002) An excellent introductory text.
Electric machines: theory, operating applications and controls by C I Hubert (Prentice Hall 2nd ed., 2002)

Principles of electrical machines and power electronics by P C Sen (J Wiley, 2nd ed. 1997)
Electrical machines and drives by G R Slemon (Addison-Wesley, 1992)

Problems

1. A three-phase, star-connected generator gives an open-circuit line voltage of 6.9 kV at a certain excitation level. It produces a short-circuit current per phase of 1.2 kA at the same level of excitation. What is the synchronous reactance per phase, neglecting the phase resistance? With the same excitation a resistive, delta-connected, load of 24 Ω per phase is attached. If the phase resistance of the generator is 0.05 Ω, what are the line currents, the power in the load, the power dissipated in the generator and the regulation? *[3.32 Ω, 457.5 A, 5.02 MW, 31.4 kW, 8.85%]*

2. A three-phase, synchronous generator produces its rated, open-circuit, line voltage of 11 kV when the DC excitation current is 188 A. When it is short-circuited, it produces its rated line current of 1.575 kA at an excitation current of 95 A. What is the short-circuit ratio? What are the per-unit synchronous reactance, the base impedance and the synchronous reactance (all per phase)? *[1.98, 0.505, 4.03 Ω, 2.03 Ω]*

3. The per-unit synchronous reactance of a three-phase synchronous generator is 0.33 per phase. What is its voltage regulation when it supplies rated load at (a) unity p.f. (b) a lagging p.f. of 0.8 (c) a leading p.f. of 0.8? Repeat the calculation if the per-unit phase resistance of the generator is 0.05. *[5.3%, 22.7%, –15.6%; 10.1%, 26%, –10.8%]*

4. What are the torque angles in the previous problem? *[18.3°, 12.4°, 18.2°; 17.45°, 10.7°, 19.25°]*

5. Show that a synchronous generator which has a per-unit synchronous reactance of magnitude x_s will have zero regulation when delivering rated power at rated voltage, provided the p.f. of the load is $\sqrt{(1 - x_s^2/4)}$, leading. Hence show that the regulation of the generator in Problem 18.3 is zero when the rated load has a leading phase angle of 9.5°. (Take the generator's phase resistance to be zero.)

6. A star-connected synchronous generator is connected to an infinite bus whose line voltage is 33 kV. If the SCR of the generator is 2.5, what are the e.m.f. and torque angle when it supplies rated power at a leading p.f. of 0.7? What is the per-unit pull-out torque here? *[14.61 kV, 21.4°, 1.92 p.u.]*

7. If the generator in the previous problem is operating at its maximum possible torque angle when supplying rated power at leading p.f., what is the p.f.? What are the torque angle and the e.m.f.? *[0.916, 23.6°, 17.5 kV]*

8. A star-connected, three-phase, synchronous motor with a synchronous reactance of 0.7 Ω per phase is connected to a 415V supply and consumes 50 kW, its rated power. If an exciter current of 10 A produces an e.m.f. of 250 V in the stator, what are the exciter current and torque angle at a p.f. of (a) unity (b) lagging 0.9 (c) leading 0.8? (Assume the machine operates on a linear portion of the magnetisation curve.) *[(a) 9.78 A, 11.5° (b) 8.86 A, 12.7° (c) 11.2 A, 10°]*

9. A synchronous motor rated at 20 kVA has a synchronous reactance of 0.8 p.u. What is the torque angle if the power consumption is 16 kW and the stator e.m.f. is 1 p.u? What reactive power is supplied to the motor? What is the pull-out torque p.u.? If the power consumption is 20 kW and the e.m.f. is raised to 1.85 p.u., what reactive power does the motor supply? If the motor is to deliver 20 kvar and consume 20 kW, what must the per-unit stator e.m.f. be?
[39.8°, 5.82 kvar, 1.25 p.u., 16.7 kvar, 1.97 p.u.]

10. A star-connected, 100MVA, synchronous capacitor has a per-unit synchronous reactance of 0.8 and is on a 22kV line. If it is to operate at rated current and a p.f. of 0.1, what is the reactive power supplied by the capacitor? What are the stator e.m.f. and the torque angle? If the stator e.m.f. is reduced to 1 p.u. at the same power consumption, what are the torque angle and the reactive power supplied by the capacitor? *[99.5 Mvar, 22.84 kV, 2.55°; 4.6°, –0.4 Mvar]*

19 Power electronics

POWER ELECTRONICS is the name given nowadays to the study of circuits and systems using solid-state devices to control electrical power. Though power devices such as SCRs became common in high-voltage traction motors as long as thirty years ago, it is only in the last ten years that the marriage of microprocessors and power devices has brought about a revolution in the control of electrical power. This is especially noteworthy in electric motor drives. We have seen that speed control of AC motors was not easy from a supply of fixed frequency, and that DC motors were preferred when a large range of speeds was required. Power electronics has removed that restriction by making it possible to operate AC and DC machines over a large range of speeds by varying the frequency or the voltage, or both, of its supply. In essence the methods are

1. $AC(f_1, V_1) \rightarrow$ converter $\rightarrow AC(f_2, V_2)$
2. $AC(f_1, V_1) \rightarrow$ rectifier $\rightarrow DC \rightarrow$ converter $\rightarrow AC(f_2, V_2)$

Besides machine control, power electronics is used in lighting and heating controllers, in high-voltage DC transmission systems – in fact almost anywhere there that more than a few watts of power is consumed. Here we have no space to go into the details of control circuitry, but some of the essentials of power electronics can be discussed, starting with the three-phase bridge rectifier.

19.1 The three-phase bridge rectifier

A three-phase bridge rectifier attached to a star winding is shown in Figure 19.1.

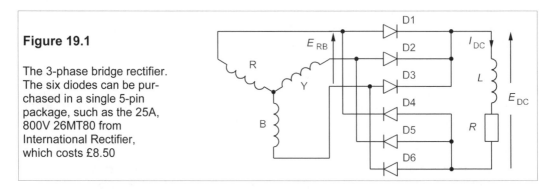

Figure 19.1

The 3-phase bridge rectifier. The six diodes can be purchased in a single 5-pin package, such as the 25A, 800V 26MT80 from International Rectifier, which costs £8.50

Figure 19.2 shows the line voltages with the negative excursions omitted. The numbers of the conducting diodes are given under the corresponding line voltage maxima. Thus the peak of the line voltage, e_{RB}, causes diodes D1 and D6 to conduct, followed by diodes D2 and D6 and so on. The maximum voltage, E_m, is $\sqrt{2}$ times the line voltage. The ripple voltage even without the smoothing inductor is fairly small, as can be seen in Figure 19.2. If E_m is the peak voltage at 90°, then the minimum at 60° will be $E_m\sin 60° = 0.866E_m$. The peak-to-peak ripple is only $0.134E_m$ with no smoothing at all.

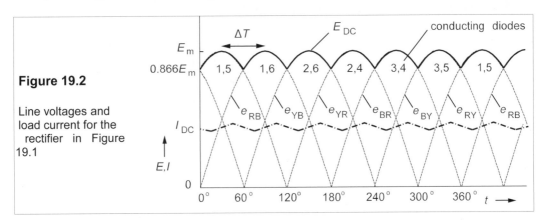

Figure 19.2

Line voltages and load current for the rectifier in Figure 19.1

The average DC voltage is

$$E_{DC} = \frac{\int E_m \sin \omega t\, dt}{\int dt} = \frac{E_m \int_{\pi/3}^{\pi/2} \sin\theta\, d\theta}{\int_{\pi/3}^{\pi/2} d\theta} = \frac{3E_m}{\pi} = \frac{3\sqrt{2}E}{\pi} = 1.35E \qquad (19.1)$$

where $\theta \equiv \omega t$ and E is the line voltage. The inductor is chosen to smooth the waveform to the desired degree by supplying energy during half the ripple cycle, or one twelfth of the supply cycle time. The energy supplied is

$$W = 0.5L(I_{max}^2 - I_{min}^2) \approx LI_{DC}\Delta I \qquad (19.2)$$

This energy 'tops up' the resistor's current from $0.866I_{max}$ (what it would have been without the inductor) to I_{min} during half a ripple cycle, and so it is given by

$$W = [I_{min}{}^2 R - (0.866I_{max})^2 R]\Delta T/2$$

$\Delta T/2$ being half a ripple cycle, or one sixth of a full mains cycle, that is $1/(6f)$. Using Equation 19.2,

$$LI_{DC}\Delta I = LI_{DC}^2\Delta I/I_{DC} = LI_{DC}^2 i = [I_{min}^2 - (0.866I_{max})^2]R\Delta T/2 \qquad (19.3)$$

Where $i = \Delta I/I_{DC}$. Equation 19.3 can be rearranged to give

$$L = \frac{[(I_{min}/I_{DC})^2 - (0.866I_{max}/I_{DC})^2]R}{12fi} \qquad (19.4)$$

But $I_{min} = I_{DC} - \Delta I/2$ and $I_{max} = I_{DC} + \Delta I/2$ so that $I_{min}/I_{DC} = 1 - i/2$ and $I_{max}/I_{DC} = 1 + i/2$. Substituting these into Equation 19.4 yields

$$L = \frac{[(1 - i)^2 - 0.75(1 + i)^2]R}{12fi} \approx \frac{(0.25 - 3.5i)R}{12fi} \qquad (19.5)$$

If the ripple is very small the inductance is about 4% of that required for the same ripple in a single-phase bridge rectifier, but if the ripple is 5%, the value drops to only 1%! The inherent smoothness of three-phase supplies is apparent.

19.1.1 Harmonic distortion and power factor

Each diode conducts for a third of the time in the unsmoothed three-phase bridge rectifier, so that the line currents flow for two thirds of the time (1/3 forward, 1/3 reverse). The line currents are virtually square pulses whose r.m.s. value is

$$I_{rms}^2 = 2I_{DC}^2/3 \quad \Rightarrow \quad I_{rms} = 0.816I_{DC} \tag{19.6}$$

Though these currents are in phase with the phase voltages, $\mathbf{E_{RN}}$ etc, the power factor is not unity because of the harmonic content of the square-wave currents. If the fundamental component of the current is I_F, the power (in watts) delivered to the rectifier is

$$P = 3E_p I_F = \sqrt{3}EI_F = E_{DC}I_{DC} \tag{19.7}$$

since the phase voltage, $E_p = E/\sqrt{3}$ (E is the line voltage) and assuming all the power is delivered to the load. But Equation 19.1 gives $E_{DC} = 3\sqrt{2}E/\pi$, so that substituting into Equation 19.7 yields

$$\sqrt{3}EI_F = 3\sqrt{2}EI_{DC}/\pi \tag{19.8}$$

which leads to

$$I_F = \sqrt{6}I_{DC}/\pi = 3I_{rms}/\pi = 0.955I_{rms} \tag{19.9}$$

Now the power factor is defined as

$$\text{p.f.} = P/S = \frac{\sqrt{3}EI_F}{\sqrt{3}EI_{rms}} = \frac{I_F}{I_{rms}} = 0.955 \tag{19.10}$$

The non-sinusoidal current has caused the p. f. to be less than unity. For this reason it is sometimes known as the *distortion power factor*.

19.2 **Power semiconductor devices**

In the last thirty years there has been a great change in the power field as reliable high-voltage and high-current semiconductor devices have been manufactured. Chief among these devices is the silicon-controlled rectifier (SCR) and more recently the gate-turn-off thyristor (GTO). Thyristors – SCRs and GTOs – are devices that conduct when forward biased on receipt of a suitable control signal. We shall discuss thyristors here and not power diodes or power transistors, since the latter work the same whether in high-current or low-current applications. The power capability of switching devices is roughly in the order MOSFET < IGBT[1] < BJT < GTO; the switching speeds are in reverse order.

---

[1] Insulated-Gate Bipolar Transistor; a cross between a MOSFET and a BJT and having a high input resistance combined with a low on-state resistance and high blocking voltage (typically > 600 V).

19.2.1 Thyristors (SCRs)

The terms thyristor and SCR are now used interchangeably, though at one time 'thyristor' denoted any member of a class of multi-layer devices, including SCRs, diacs, triacs, silicon-controlled switches (SCS) and Shockley diodes, that are used especially for switching currents to control lights, motors, arc welders and a variety of other equipment. The reason they are used instead of switches and relays are firstly, they do not arc when switched off – thus saving power and wear and tear on switches – and secondly, they can be electronically controlled to conduct at a specific phase angle of the supply waveform. Thus a specialised field of electronics emerged: power electronics.

Figure 19.3 shows the circuit symbol and semiconductor structure for an SCR, which is essentially four alternating layers – pnpn – of silicon. It can be thought of as two interconnected transistors, Q1 and Q2, sharing the same material for bases and collectors, that is the base of Q1 is the collector of Q2 and vice versa. Q1 is a pnp transistor while Q2 is an npn transistor. Figure 19.3 also shows the currents flowing when the anode is positive with respect to the cathode ($V_{AK} > 0$) and the gate terminal is open circuit.

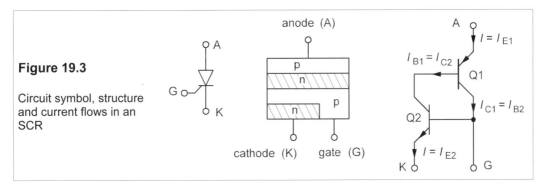

Figure 19.3

Circuit symbol, structure and current flows in an SCR

Suppose the anode current is I and the transistors have current gains α_1 and α_2, making $I_{C1} = \alpha_1 I_{E1} + I_1$ and $I_{C2} = \alpha_2 I_{E2} + I_2$, where I_1 and I_2 are the respective collector-base leakage currents. But $I_{E1} = I_{E2} = I$, so $I_{C2} = \alpha_2 I + I_2$. Also

$$I_{C2} = I_{B1} = I_{E1} - I_{C1} = (1 - \alpha_1)I - I_1 \tag{19.11}$$

and then

$$I_{C2} = \alpha_2 I + I_2 = (1 - \alpha_1)I + I_1 \tag{19.12}$$

Therefore
$$I = \frac{I_1 + I_2}{1 - \alpha_1 - \alpha_2} \tag{19.13}$$

When V_{AK} is small α_1 and α_2 are $\ll 1$, and $I \approx I_1 + I_2$, the sum of the leakage currents. As V_{AK} is increased, though, I increases and so do both α_1 and α_2, further increasing I and so on until I is as large as Ohm's law will permit. Turn on is very rapid, peak current being reached in 1 µs or less. After turning on, Q1 and Q2 are saturated, causing V_{AK} to fall to about 1 V. The minimum value of V_{AK} that will keep the device in conduction is called the *holding voltage*, V_h, while I_h is the holding current. As long as $V_{AK} > V_h$ the SCR is said to be *latched up*. The resulting characteristic for the SCR is shown in Figure 19.4.

The *forward breakover* voltage, V_{BF}, is quite large and the reverse breakdown voltage, V_{BR}, is very large: from 400 V to 1200 V in relatively cheap devices. When a small pulse of current, I_G, is applied to the gate, the device conducts at much lower values of V_{AK}, depending on I_G. Usually I_G is pulsed to a specified maximum value which results in very rapid switch-on at a value of V_{AK} little more than V_h, shown by the dashed line in Figure 19.4. When the gate current returns to zero the device will stay on until the current drops below I_h, the holding current when $I_G = 0$. Typically, I_h is about 1% of the maximum average on-state current, $I_T(AV)$.

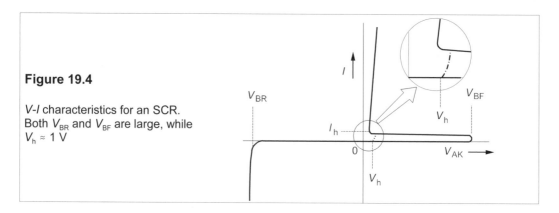

Figure 19.4

V-I characteristics for an SCR. Both V_{BR} and V_{BF} are large, while $V_h \approx 1$ V

Example 19.1

In the circuit of Figure 19.5a, in which a sinusoidal voltage source is in series with a resistive load and an SCR, what is the firing angle that gives a power consumption of 2 kW in the load, if $R_L = 10\ \Omega$ and the source voltage is 240 V r.m.s? What then is R_G if I_{GT} is 100 mA?

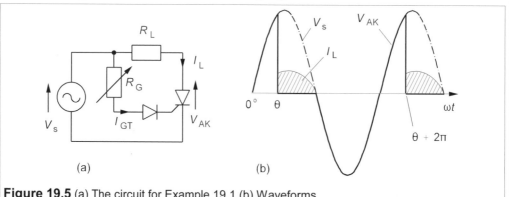

(a)   (b)

Figure 19.5 (a) The circuit for Example 19.1 (b) Waveforms

The SCR's gate is connected to a variable resistor via a diode (to prevent negative bias being applied to the gate). The point on the voltage waveform (usually expressed in degrees as a *firing angle*, θ) at which the SCR will fire, or turn on, depends on the resistance, R_G, in series with the gate, the peak source voltage, V_m, and the gate firing (or triggering) current, I_{GT}.

The thyristor will only fire during the positive voltage cycle and will unlatch when the source voltage is near zero. Control is therefore only achievable for firing angles between 0° and 180°. At a firing angle, θ, the voltage across the thyristor, V_{AK}, is equal to the source voltage, $V_m \sin\theta$. The SCR then fires and conducts, so that V_{AK} drops to near zero. The current waveform will be like that in Figure 19.5b.

The power developed in the 10Ω resistance is $I_L^2 R$, and the average power is

$$P_R = \frac{1}{T}\int_0^T I_L^2 R\,dt = \frac{1}{T}\int_{\theta/\omega}^{\pi/\omega} (V_m^2/R)\sin^2\omega t\,dt = \frac{V_m^2}{2RT}\int_{\theta/\omega}^{\pi/\omega} (1-\cos 2\omega t)\,dt \qquad (19.14)$$

The limits on the integral correspond to phase angles of θ and 180° (when $t = \theta/\omega$ and $t = \pi/\omega$). Integration gives

$$P_R = \frac{V_m^2}{2RT}\left[t - \frac{\sin 2\omega t}{2\omega}\right]_{\theta/\omega}^{\pi/\omega} = \frac{V_m^2}{2RT}\left[\frac{\pi}{\omega} - \frac{\theta}{\omega} - \frac{\sin 2\pi - \sin 2\theta}{2\omega}\right] \qquad (19.15)$$

As $\omega T = 2\pi$ and $V_m^2 = 2V^2$ (V is the r.m.s. value), this becomes

$$P_R = V^2(2\pi - 2\theta + \sin 2\theta)/4\pi R \qquad (19.16)$$

Substituting 2000 W for P_R, 240 V for V_{rms} and 10 Ω for R leads to

$$2\theta - \sin 2\theta = 2\pi - 8000\pi R/V^2 = 1.92 \qquad (19.17)$$

As an analytical solution is impossible, we must guess and use the Newton-Raphson method of approximation. Let us take $\theta_1 = 1$ radian as a first guess. The error term is $2\theta_1 - \sin 2\theta_1 - 1.92$, so the first error is $\epsilon_1 = 2 - \sin 2 - 1.92 = -0.83$. Then the next approximation to θ is

$$\theta_2 = \theta_1 - \frac{\epsilon_1}{f'(\theta_1)} = \theta_1 - \frac{\epsilon_1}{2 - 2\cos 2\theta_1} = 1 + \frac{0.83}{2 - 2\cos 2} = 1.29 \qquad (19.18)$$

where f' stands for differentiation with respect to θ and $f(\theta) = 2\theta - \sin 2\theta$. The third approximation is sufficiently close: $\theta_3 = 1.255$ or 72°. This angle corresponds to a voltage of $240\sqrt{2}\sin 72°$ or 323 V. Neglecting the voltage drops across the thyristor and the diode, we have $I_{GT}R_G = 323$ V, or $R_G = 323/0.1 = 3.23$ kΩ.

The r.m.s. voltage across the load is found from

$$V_L(\text{r.m.s.}) = \sqrt{R_L P_R} = \sqrt{10 \times 2000} = 141.4 \text{ V} \qquad (19.19)$$

and the power factor is

$$P/S = V_L(\text{r.m.s.})I_L/VI_L = V_L(\text{r.m.s.})/V \qquad (19.20)$$

since the current from the supply is (apart from the small current through R_G) the same as that through the load. Thus in this case the p.f. is 141.4/240 = 0.589.

If firing when the supply goes negative is wanted, a *triac* could be used, which is effectively two side-by-side SCRs of opposite polarity. The reversed SCR then fires at the same angle as its companion, but on the negative half-cycle. In practice I_{GT} is not a reliable parameter and varies from one device to another and also with temperature changes, accordingly R_G would also vary.

One way around this problem is to use a *diac*, which is just two back-to-back diodes that will conduct when the potential across them reaches a certain, constant value. With a capacitor in series with R_G, as in Figure 19.6, variable, controlled, phase-angle firing is achievable.

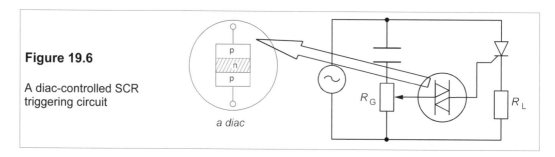

Figure 19.6

A diac-controlled SCR
triggering circuit

a diac

Crowbar protection

A very common use for thyristors is the *crowbar*, an overvoltage protection circuit (Figure 19.7) in which the output terminals are effectively short-circuited by a thyristor.

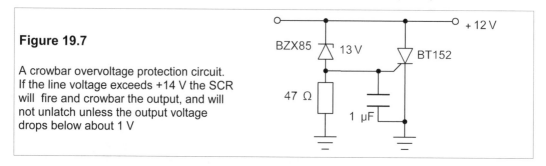

Figure 19.7

A crowbar overvoltage protection circuit. If the line voltage exceeds +14 V the SCR will fire and crowbar the output, and will not unlatch unless the output voltage drops below about 1 V

Once brought into conduction to crowbar the output, the thyristor will not unlatch until the equipment is disconnected from the supply (often because a fuse has blown). In Figure 19.7, the firing voltage level is set by the Zener which is in series with a resistor that pulls up the thyristor's gate voltage. The capacitor guards against accidental crowbarring by transient spikes. When fired, the BT152 device will take up to 13 A continuously.

19.2.2 Gate turn-off thyristors: GTOs

In GTOs full control of conduction is given to the gate electrode, which is pulsed positively with respect to the cathode to turn it on, and negatively to turn it off. Unlike the SCR the GTO passes current in both directions when turned on: blocking diodes must be used if this is undesirable. GTOs can be used in much the same applications as SCRs. We have only the space here to illustrate a single use – as a DC chopper.

A DC chopper

By chopping DC with a GTO it is possible to control a great range of power in loads which would otherwise be ill-matched to the power source. Rheostatic (variable resistance) control is inefficient for most high-power loads. Examples of such loads are battery chargers, DC motors and resistive loads such as lights and heaters. Often one wants to maintain a fairly constant current through the load even when the source has been cut off by the thyristor, and then a series inductor may provide energy at such times.

Figure 19.8 shows an inductor in series with a voltage source (such as a battery or a DC motor), both shunted by a freewheeling diode, D1. When the thyristor is fired it pumps up the current through the inductor, and when it is cut off the freewheeling diode allows the current to flow unbroken through the inductor and load. The inductor must have sufficient inductance to give adequate smoothing.

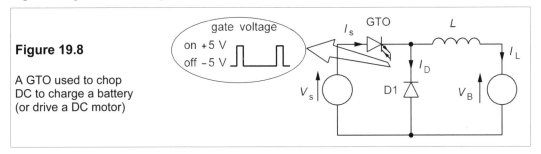

Figure 19.8

A GTO used to chop
DC to charge a battery
(or drive a DC motor)

Example 19.2

A 340V DC supply is to be used to charge a 48V battery using the circuit of Figure 19.8. If the average charging current is 20 A and the peak-to-peak ripple is 10%, what must the inductance be if the thyristor is on for 2 ms in every firing cycle? What is the chopping frequency? Draw the current waveform.

After the thyristor fires, the voltage across the inductor is $V_s - V_B = L\,di/dt$, giving:

$$L\,di/dt = L\Delta I/T_{on} = V_S - V_B \tag{19.21}$$

where ΔI is the peak-to-peak ripple, 10% of 20 A, or 2 A. Rearranging this gives

$$L = (V_s - V_B)T_{on}/\Delta I = (340 - 48) \times 2 \times 10^{-3}/2 = 0.292\ \text{H} \tag{19.22}$$

The 340V source supplies energy equal to $V_s I T_{on}$ per cycle and this must equal the energy absorbed by the battery, $V_B I T = V_B I/f$, T being the cycle period and f the chopping frequency. Therefore $V_S I T_{on} = V_B I/f$, that is

$$f = V_B/V_S T_{on} = 48/340 \times 2 \times 10^{-3} = 70.6\ \text{Hz} \tag{19.23}$$

The current waveforms are shown in Figure 19.9.

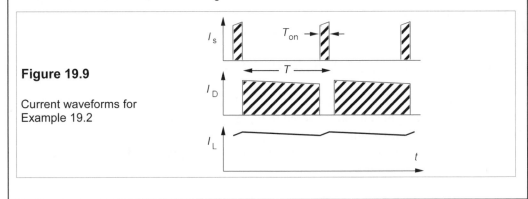

Figure 19.9

Current waveforms for
Example 19.2

19.3 **Power-control circuits using thyristors**

There are a number of basic thyristor power circuits used to drive AC and DC motors and other loads. Normally most of these would be three-phase circuits, but the principle will be explained here with just a single phase.

19.3.1 DC matching of a resistive load

The power dissipated in a resistive load may be controlled by chopping the current with the circuit of Figure 19.10.

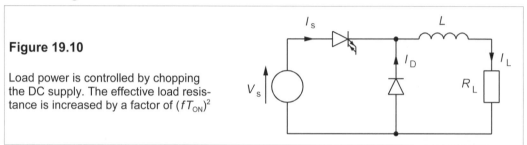

Figure 19.10

Load power is controlled by chopping the DC supply. The effective load resistance is increased by a factor of $(fT_{ON})^2$

Such load matching avoids the inefficiencies caused by merely adding series resistance to reduce the load current. We assume that the ripple is small so that the load current may be assumed constant. Let T_{on} be the time during each cycle that the thyristor is fired and let the chopping frequency be f. The energy supplied per cycle by the source is $V_s IT_{on}$, while the energy consumed by the load is $I^2 R_L/f$ per cycle. Equating these energies gives

$$V_s IT_{on} = I^2 R_L/f \quad \Rightarrow \quad I = V_s T_{on} f/R_L \tag{19.24}$$

But the load power is

$$P_L = I^2 R_L = (V_s T_{on} f/R_L)^2 R_L = V_s^2/R_{LE} \tag{19.25}$$

where R_{LE} is the effective load resistance. Comparing both sides of the last equality, $R_{LE} = R_L /(fT_{on})^2$.

For example suppose a 2Ω load is to dissipate 100 W when supplied with 120 V with a chopping frequency of 50 Hz. We must find the time for which the thyristor fires during each cycle. The effective resistance has to be $120^2/100 = 144\ \Omega$. Thus $(fT_{on})^2 = R_L/R_{LE} = 1/72$ and $fT_{on} = 0.118$. If f is 50 Hz, then $T_{on} = 1/50\sqrt{72} = 2.4$ ms.

19.3.2 AC source supplying DC to an active load

A thyristor can be used to provide DC to an active load, such as a motor or battery, from an AC source. This rectifying circuit is shown in Figure 19.11a. An inductor is used to provide current during the negative half cycle.

Suppose the thyristor fires at θ_{on} as in Figure 19.11b, then by Kirchhoff's voltage law we can write

$$V_{AC} = V_L + V_{DC} \tag{19.26}$$

Thus as $V_{AC} > V_{DC}$, the inductor voltage, V_L, is positive and the current builds up (since $V_L = Ldi/dt$) until $V_{AC} < V_{DC}$. At this time, Equation 19.26 shows that V_L is negative and the current decays until it is zero. When this happens the anode (A) of the thyristor is below the cathode (K) potential and the thyristor turns off. Figure 19.11b shows that the current lags the voltage which means that the source is supplying reactive power (Q) to the load as well as active power (P). The maximum current can be calculated from

$$V_L = V_m \sin \omega t - V_{DC} = L \, di/dt \qquad (19.27)$$

which gives

$$i(\theta) = \frac{1}{L} \int (V_m \sin \omega t - V_{DC}) dt = \frac{1}{\omega L} \int_{\theta_{on}}^{\theta} (V_m \sin x - V_{DC}) dx \qquad (19.28)$$

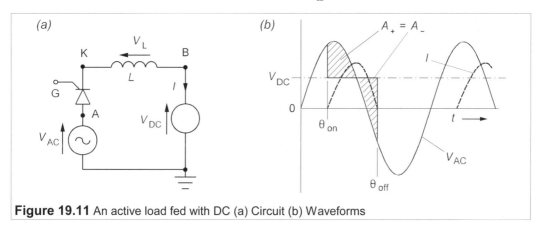

Figure 19.11 An active load fed with DC (a) Circuit (b) Waveforms

The current can be limited either by choosing the inductance value (which may not be possible with a motor) or by controlling the firing angle, θ_{on}. The commutation in this circuit is said to be natural. By commutation we mean turning off the thyristor and natural commutation means that the thyristor turns off because the voltage between its anode and cathode has dropped to zero during the normal working of the circuit. The three-phase bridge rectifier is another example of natural commutation, as is the next circuit.

19.3.3 DC source supplying power to an AC source

The circuit – Figure 19.12a – is the same as the previous one with the thyristor reversed. When the thyristor conducts the DC source delivers power to the AC source. Now any device which turns DC into AC is an *inverter* and so must this circuit be. During conduction the inductor is supplied with energy which it must release before the thyristor can turn off. If a decreasing current is flowing in the inductor the voltage across the inductor is in the same direction as V_{DC} and keeps $V_K < V_A$. The cessation of current flow makes $V_L = 0$ and then $V_K > V_A$ and the thyristor is turned off. If too much current flows in the inductor and it cannot discharge, then the thyristor cannot commutate and current will build up more and more every cycle. Current flow will cease if the positive and negative areas under the $V_A(t)$ graph in Figure 19.12b are equal. If, during one cycle, the thyristor is fired before θ_{min} it cannot be turned off, and it will not fire after θ_0.

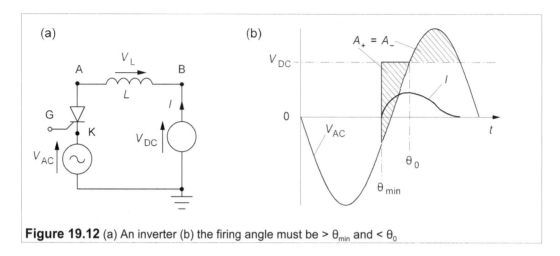

Figure 19.12 (a) An inverter (b) the firing angle must be > θ_{min} and < θ_0

19.3.4 The cycloconverter

A cycloconverter is a device which converts three-phase AC of the input frequency, f_{in}, to single-phase AC of a lower frequency, f_o. The circuit of Figure 19.13a is a cycloconverter when the thyristors are fired in the right sequence. Let gate pulses be applied to Q1, Q2 and Q3 at 0°, 150° and 270°, and then to Q6, Q4 and Q5 420°, 570° and 690°, the cycle being repeated from 840° on. The waveform resulting is as in Figure 19.13b, that is a waveform of frequency 180/420 = 3/7 of (about 40%) the input frequency. This is the maximum frequency for three-phase operation.

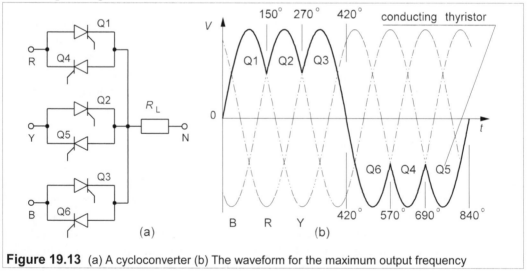

Figure 19.13 (a) A cycloconverter (b) The waveform for the maximum output frequency

Lower frequencies are achieved by increasing the number of successive positive (and negative) firings in a group. Thus the firing sequences Q1, Q2, Q3, Q1 (positive pulses) and Q4, Q5, Q6, Q4 (negative pulses) would produce a half cycle of width 150° + (2 × 120°) + 150° = 540°, resulting in $f_o = f_{in}/3$. The output frequencies are given by the formula

$$f_o = \frac{3 f_{in}}{(2n + 5)} \qquad (n = 1, 2, 3,...) \tag{19.29}$$

The waveform can be made more sinusoidal by suitable filtering.

19.3.5 The H-bridge

The H-bridge circuit of Figure 19.14a is very often used to invert the power from a DC source into an AC load. The principle of operation is simple. Thyristors Q1 and Q4 are fired together so that current flows through the load from A to B. Then Q1 and Q4 are switched off and Q2 and Q3 are fired causing the current flow to reverse and go from B to A. The diodes allow freewheeling of inductive loads after the thyristors are switched off. The load voltage is not sinusoidal, however, but almost rectangular. The maximum frequency of operation is limited by the device-switching times, which can be as high as 100 kHz if low-power MOSFETS are used, or as low as a few hundred hertz if large GTOs are used. The H-bridge can also be used as a DC chopper and can be adapted for three-phase operation by adding another two pairs of diodes and thyristors.

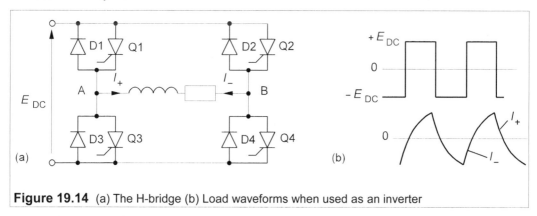

Figure 19.14 (a) The H-bridge (b) Load waveforms when used as an inverter

19.3.6 Problems caused by phase-angle firing

Generally the use of phase-angle firing and thyristors results in two serious problems:

1. Waveform distortion, resulting in harmonics, power loss and electromagnetic interference (see Chapter 28).
2. A lagging p. f. taking vars from the supply, causing higher wiring and electricity costs.

These problems cannot always be avoided, but when, for example, resistive loads are being controlled for heating purposes, phase-angle firing should not be used, but any method of control that uses complete half-cycles can be adopted. Some interference is still caused when large loads are switched on and off like this, but the two problems above are solved.

19.4 **Motor control with power electronics**

Here we can provide only the briefest outline of this huge and rapidly-expanding field. Much of the development has centred on an area about which nothing has been said: the programming of the gate-firing sequences by which control is achieved. The controller is programmed to receive inputs from sensors in the motor such as speed, power output, temperature and so forth and, comparing these with the desired values, to send appropriate trigger pulses to the gate electrodes of the thyristors. Safety limits such as maximum current or maximum speed can be programmed in so that an inexperienced operator cannot harm the motor by starting too quickly or running too fast. This desirable feature has contributed to the successful application of power electronics to machine operation.

19.4.1 The three-phase, six-pulse converter

This thyristor converter looks just like the three-phase bridge rectifier (Figure 19.1) with thyristors instead of diodes. If the thyristors are fired at the times when the diodes would conduct, then the output voltage waveform is identical to that of the diode bridge. If, however, the thyristor firings are each delayed until an angle α after these times, then the voltage waveform is more ragged as in Figure 19.15, where α, the *delay angle*, is 17° so that firing occurs at 17°, 77° etc. Because the current flows every 60° there are six current pulses in every cycle and the converter is called a six-pulse converter.

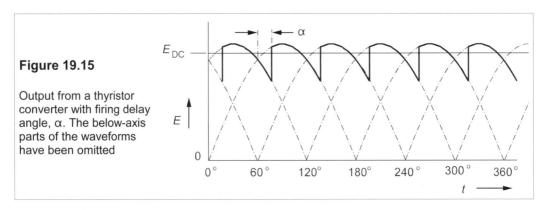

Figure 19.15

Output from a thyristor converter with firing delay angle, α. The below-axis parts of the waveforms have been omitted

In consequence of delayed firing the average output voltage is less than that of the three-phase bridge rectifier and may be shown to be given by

$$E_{DC} = 1.35\,E\cos\alpha \tag{19.30}$$

where E is the line voltage and the numerical constant is the same as in Equation 19.2, $3\sqrt{2}/\pi$. By delaying α by between 90° and 0° the voltage can be continuously varied from 0 to $1.35E$. At the same time the output current is delayed so that the power factor is lagging and the reactive power is

$$Q = P\tan\alpha \tag{19.31}$$

These vars are drawn from the supply, which is not normally desirable. In practice it is difficult to fire at small delay angles and the normal firing range is from 10° to 160°. Note that when the firing angle is greater than 90° the converter becomes an inverter but the thyristors will not conduct without forward DC bias as they can only pass forward current

---

Example 19.3

A three-phase, six-pulse converter fed from a 415V supply is used to control the input power to a DC source whose e.m.f. is constant, as in Figure 19.11. Suppose the source has an e.m.f. of 480 V and an internal resistance of 1.5 Ω. The inductor may be ignored. If the minimum delay angle is 10°, what is the maximum converter output power and maximum input power to the DC source? What will these be if the delay angle is 21°? What is the power loss in the internal resistance in each case? At what delay angle is the converter power output zero?

Both power maxima occur when α = 10° and then

$$E_{DC}(\text{max}) = 1.35 \times 415\cos 10° = 552 \text{ V} \tag{19.32}$$

The voltage drop across the 1.5Ω internal resistance is 552 – 480 = 72 V, so the current through it is I_{DC} = 72/1.5 = 48 A. The power supplied is 48 × 560 = 26.9 kW and the power into the source is $E_s I_{DC}$ = 480 × 48 = 23 kW. The difference is the power lost in the internal resistance, 3.9 kW, or 14.5% of the converter output power.

When α = 21°

$$E_{DC} = 1.35 \times 415\cos 21° = 523 \text{ V} \tag{19.33}$$

The voltage drop across the internal resistance is 523 – 480 = 43 V, so the current is 43/1.5 = 28.7 A and the output power is 523 × 28.7 = 15 kW. The power into the load is 480 × 28.7 = 13.8 kW. The power lost is 1.2 kW or 8% of the converter output power.

The converter will put out zero power when E_{DC} = 480 V, that is when

$$1.35 E\cos \alpha = 480 \text{ V} \quad \Rightarrow \quad \cos\alpha = 480/560 = 0.857 \tag{19.34}$$

which gives α = 31°. In practice the firing angle will have to be a little less as some voltage is dropped across the thyristors. If the firing angle is increased further no power will be transferred from the converter as the thyristors will stay switched off because their cathodes will be at higher potentials than their anodes.

19.4.2 *The three-phase, 12-thyristor converter*

The three-phase, six-pulse converter cannot act as an inverter without DC bias, but, by providing another set of six thyristors to conduct during the negative half cycles, this difficulty may be overcome. The resulting converter (a cycloconverter since it can reduce the frequency of the supply) has the useful ability to act as an inverter and can therefore return power to the supply, for example when a motor is braking. This makes not only for more efficient operation, but also produces faster braking. Figure 19.16 shows a three-phase, 12-thyristor cycloconverter.

The group of six thyristors indicated as '+ group' in Figure 19.16 conduct on positive half cycles and the six marked '– group' conduct during negative half cycles of the supply. Only one group can operate at any moment and the other group must be blocked. By adjusting the firing angles it is possible to control the voltage of the output as well as its frequency. If a three-phase motor is to be operated with this type of converter each motor phase is separately connected to its own 12-thyristor cycloconverter, making 36 thyristors in all. The motor phases cannot be connected together. The output frequency ranges from 3/7 of the supply frequency down to zero.

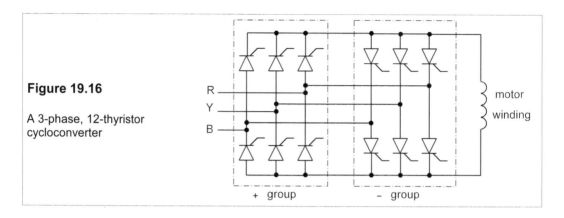

Figure 19.16

A 3-phase, 12-thyristor
cycloconverter

R
Y
B

motor
winding

+ group   − group

19.4.3 A three-phase converter with DC link

This also uses 12 thyristors as in Figure 19.17. The first group of six act as a controlled rectifier to supply DC to the second group of six via a smoothing inductor (the DC link). The DC current is determined by the firing angle of the rectifier group while the second converter's firing sequence controls the frequency of the AC output. The advantage of this arrangement is that frequencies higher than the supply frequency can be generated.

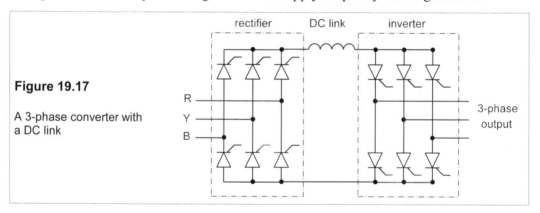

rectifier   DC link   inverter

Figure 19.17

A 3-phase converter with
a DC link

R
Y
B

3-phase
output

19.4.4 The brushless DC motor

Small brushless DC motors, Figure 19.18, are used for example in hard-disc drives and in laser printers; large ones are used for various specialised tasks. It is clearly possible to apply power electronics (such as the DC chopper circuit described above) to a 'conventional' DC motor and achieve more efficient operation than with 'conventional' control. But it is also possible to go a step further and get rid of the troublesome commutator. Not only does this make the motor more efficient, it also makes it more reliable and cheaper to manufacture. If the commutator segments in a DC motor are replaced by slip rings the commutation which reverses the current flow in each armature coil could be achieved by firing thyristors at the appropriate times. However, if the poles in the stator were to rotate, the slip rings could also be eliminated. The stator then becomes the armature while the rotor can be a magnet (permanent or DC coil excited), as in Figure 19.18.

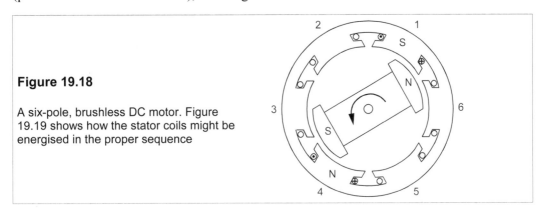

Figure 19.18

A six-pole, brushless DC motor. Figure 19.19 shows how the stator coils might be energised in the proper sequence

A brushless DC motor is a conventional DC motor turned inside out. The field coils can be sequentially switched by a circuit such as that in Figure 19.19. Sensors are required to detect the position of the rotor so that the stator coils can be energised at the right time. The coils are energised here by a push-pull amplifier. The transistor bases are connected to photodetector diodes which are turned on when the cut-away sector of the shaft-mounted disc (shown on the left of Figure 19.19) passes them. The correct firing sequence is therefore 'hard wired' into the electronics. Speed control would only be obtained by varying the current drawn by the stator coils. This type of motor is sometimes called 'bipolar' because the coil currents reverse, but the electronics can be made even simpler by having only three stator coils in a unipolar arrangement. The rotational sense is fixed by the wiring and it would be necessary to install more switches to reverse it. Employing microprocessor-controlled thyristors would allow much more flexibility in operation.

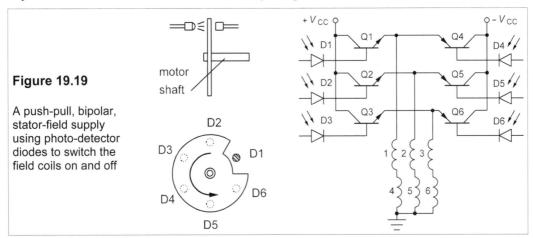

Figure 19.19

A push-pull, bipolar, stator-field supply using photo-detector diodes to switch the field coils on and off

Hall-effect sensors can be used to detect the position of the rotor and control the base current to the stator-field transistors. The back e.m.f. generated in the coils by the rotor is proportional to the speed of rotation and can be detected and fed into a comparator together with the desired speed setting. The comparator output is used to control the current through the Hall sensors, hence the base and collector (= stator) currents.

Large brushless DC motors employ thyristors to control the stator field. For example, a three-phase AC supply can be used with a cycloconverter to vary the frequency and hence the speed of rotation. These large machines are of exactly the same physical form as a synchronous motor. The essential difference between a synchronous motor and a brushless DC motor lies in the control of the stator field. In a synchronous motor the speed is determined by the frequency of the stator-field supply, whereas in a brushless DC machine it is controlled by feedback from speed and rotor-position sensors. Wound rotors permit control of the rotor field in both types.

19.4.5 Electronic drives for AC motors

AC motors, especially induction motors, are simpler to construct, are more rugged, cheaper and more reliable than DC motors. Power electronics has enabled the use of these in applications such as electric locomotives in which they were previously excluded. Three principal means for electronic control of AC motors exist:

> 1. Varying the voltage
> 2. Varying the frequency
> 3. Varying both

The first method is simplest and is used for squirrel-cage motors. Two thyristors are required per phase – one for the positive half cycle and one for the negative – as in Figure 19.20.

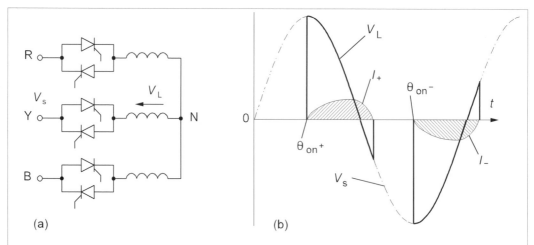

Figure 19.20 (a) Simple voltage control with two thyristors per phase (b) Waveforms for operating at roughly half rated voltage. Switching off is delayed by the motor inductance

The firing angle determines the average voltage. Figure 19.20b shows the voltage and current waveforms for one of the phases when operating at about half the maximum voltage ($\theta_{on} \approx 90°$). The current is lagging greatly, so this method of control produces a low power factor at low voltages and large losses. Consequently it is used with small motors or medium-power motors with intermittent use.

The second method uses cycloconverters and if the 12-thyristor converter (Figure 19.16)

is used the efficiency can be high as power can be transferred both to and from the motor. Squirrel-cage and synchronous motors can readily be operated this way. A 2-pole motor running from a 50Hz supply could be run at a maximum of $3/7 \times (50 \times 60) = 1286$ r/min.

The third method has greater flexibility at the expense of greater cost and complexity in the electronics. Each phase of the supply can be fed into a rectifier with a DC link to a converter as in Figure 19.17. The output frequency is now limited by the switching times of the thyristors – not the supply frequency – and can be up to a few kHz with medium-power devices. The cycloconverters produce a lot of harmonics at low speeds which can cause difficulties with vibration and torque fluctuations. By using pulse-width modulation (PWM) techniques it is possible to construct relatively smooth sinewaves with low harmonic content. (PWM is described more fully in Chapter 24.) PWM is now used with induction motors in electric traction where precise speed control is needed when starting and stopping.

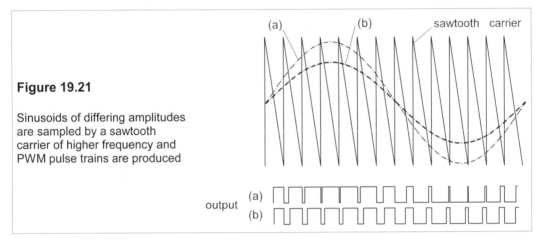

Figure 19.21

Sinusoids of differing amplitudes are sampled by a sawtooth carrier of higher frequency and PWM pulse trains are produced

Figure 19.21 shows two sinewaves which are used to modulate a high-frequency, sawtooth, carrier wave. The resultant PWM waves, when smoothed by the motor's inductance, will produce voltage waveforms fairly similar to the original sinusoids. PWM is a sophisticated and flexible technique for power control. It has other advantages too, particularly in the reduction of unwanted harmonics. Harmonics in a PWM converter tend to be harmonics of the sampling (or carrier) frequency rather than the fundamental and are therefore pushed much further up the spectrum. It is now possible with microprocessors to use PWM to produce optimal harmonic content, a process known as *power conditioning*.

Suggestions for further reading

Introduction to modern power electronics by A M Trznadlowski (Wiley, 1998)
Introduction to power electronics by D W Hart (Prentice Hall, 1997)
Principles of electrical machines and power electronics by P C Sen (Wiley, 2nd ed., 1996)
Power electronics by N Mohan, W Robbins and T Undeland (Wiley, 2nd ed., 1995)
Power electronics: principles and applications by J Vithayathil (McGraw-Hill, 1995)

Problems

1. A purely resistive load of 9.5 Ω is fed with DC from a 415V three-phase supply by means of a three-phase, diode-bridge rectifier. Neglecting forward-voltage drops across the diodes, what is the power consumption? If the diode drop is 0.7 V, what fraction of the total power is lost in the bridge?
    [33 kW, 0.25%]

2. A load of 9.5 Ω in series with 30 mH is connected to a three-phase, diode-bridge rectifier fed by a 415V, 50 Hz supply. Neglecting the forward-voltage drops across the diodes, what is the average direct voltage across the load? What is the average DC load current and the peak-to-peak ripple current in the load? *[560 V, 59 A, 2.7 A]*

3. In the previous problem, what are the r.m.s. line current and the power factor? What reactive power is taken? *[48.1 A, 0.955, 10.3 kvar]*

4. A 6-pulse thyristor converter, fed from a three-phase, 415V, 50 Hz supply is connected to a resistive load of 9.5 Ω. If the thyristor delay angle is 30°, what is the DC voltage across the load, the power consumed by the load and the reactive power supplied? What capacitance must be delta-connected across the 3-phase lines to give a lagging power factor of 0.95?*[485 V, 24.8 kW, 14.3 kvar, 38 μF]*

5. What delay angle would be needed to reduce the power consumption of the load in the previous problem to 10 kW? What would the reactive power taken then be? *[57°, 15.2 kvar]*

6. A 12V battery is to be charged using the circuit of Figure 19.8, a charging current of 15 A and a DC supply voltage of 40 V. If the chopping frequency is 500 Hz and the inductance is 45 mH, what is the peak-to-peak ripple in percent and for how long does the thyristor conduct each firing cycle? *[2.5%, 0.6 ms]*

7. The firing angle for the thyristor in Example 19.1 is 45°. What must R_G be? What is the power consumption if R_L is 200 Ω and $V_m = 240\sqrt{2}$ V? What is the r.m.s. voltage across R_L? What is the p.f. of the circuit? What vars are drawn from the supply? *[2.4 kΩ, 131 W, 162 V, 0.674, 143 var]*

8. In the circuit of Figure 19.11a, what is the current when $\theta = 120°$, $\theta_{on} = 45°$, $L = 100$ mH, $\omega = 100\pi$ rad/s, $V_m = 339$ V and $V_{DC} = 100$ V? When is the current a maximum? What would L have to be to make the current 15 A at $\theta = 135°$? *[8.9 A, $\theta = 162.8°$, 11.4 A, 68 mH]*

9. In the circuit of Figure P19.9 the delay angle of the thyristor firing is α radians. Show that the r.m.s. voltage across the load is

$$V_{rms} = V_m \sqrt{\frac{(2\pi - \alpha)}{4\pi} + \frac{\sin 2\alpha}{8\pi}}$$

Does this give the right answer if the thyristor is (a) replaced by a diode or (b) not fired?

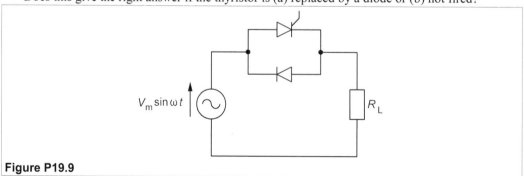

Figure P19.9

10. Using the result of the previous problem, show that the r.m.s. load voltage when $V_m = 339$ V and $\alpha = 60°$ is 228 V. What is the power factor? What reactive power is drawn from the supply if $R_L = 8 \Omega$? *[0.95, 2.14 kvar]*

20 Combinational logic

COMBINATIONAL LOGIC in electronic engineering enables the processing of information that is effectively in the form of binary numbers, since the inputs and output(s) of a logical circuit can be in only one of two states. The output state is a function of the input states, and can be used to control some other device or machine or circuit, or can just be used as information. Figure 20.1 shows schematically what combinational logic does. Two-state inputs can be expressed as two-state variables, or Boolean[1] variables, which can be combined according to the rules of Boolean algebra. In formal terms we can write

$$Q = f(A,B,C,D...)$$

where Q is the output and A, B, C, D etc. are the inputs.

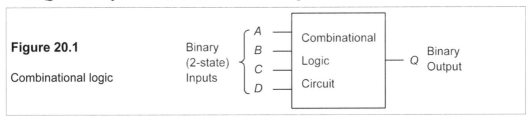

Figure 20.1

Combinational logic

Logic circuits are made up of component circuits, called *logic gates* or *gates*, which perform the operations of Boolean algebra, such as AND, OR and NOT. Besides logical functions, combinational logical circuits are used for binary arithmetic, and extremely fast logic gates are used in digital computers to add, subtract, multiply and divide binary numbers. We therefore start with an account of both these and hexadecimal numbers.

20.1 Binary and hexadecimal numbers

Normally one counts and performs arithmetic in decimal numbers made up of the set of digits {0,1,2,3,4,5,6,7,8,9}, that is numbers are represented as powers of 10. Decimal numbers are said to be to *base* 10. For example the decimal number 11.25d represents[2]

$$10^1 + 10^0 + 2 \times 10^{-1} + 5 \times 10^{-2}$$

---

[1] George Boole (1815–64) was born in Lincoln and taught himself mathematics. He was a schoolmaster in that city when he devised what he called *The Laws of Thought* in the years 1847–54. He became Professor of Mathematics at Queen's College, Cork, where he died. His school in Lincoln can still be seen today close by the cathedral.

[2] We shall use the suffix 'd' for decimal (base 10), 'b' for binary (base 2) and 'h' for hexadecimal (base 16) numbers.

Computers and calculators perform arithmetic with the set {0,1} of *binary* numbers, to base 2. The advantage of using binary numbers is that a digit can be either a 1 or a 0 and nothing else so that two-state devices can be used to store and operate on them. A binary number such as 1011.01b represents

$$1 \times 2^3 + 0 \times 2^2 + 1 \times 2^1 + 1 \times 2^0 + 0 \times 2^{-1} + 1 \times 2^{-2}$$
$$= 8 + 0 + 2 + 1 + 0 + 0.25$$
$$= 11.25d$$

Whatever the base, the point indicates where the power of the base changes from 0 to −1.

20.1.1 Conversion from decimal to binary and vice versa

Conversion from integer decimal numbers to binary is done by dividing successively by 2 and storing the remainders; for example 23d can be converted as follows:

$$
\begin{aligned}
23 \div 2 &= 11, &\text{remainder } 1 &\ (= \text{LSB}) \\
11 \div 2 &= 5, &\text{remainder } 1 \\
5 \div 2 &= 2, &\text{remainder } 1 \\
2 \div 2 &= 1, &\text{remainder } 0 \\
1 \div 2 &= 0, &\text{remainder } 1 &\ (= \text{MSB})
\end{aligned}
$$

The remainders written down from the bottom (MSB, most significant bit) up to the top (LSB, least significant bit) give the binary form as 10111b.

Fractional decimal numbers such as 11.25d must be split into the integral part, 11d, which is converted to binary as above, and the fractional part 0.25d, which is converted by multiplying successive fractional parts by 2 until no fractional part remains:

$$
\begin{aligned}
0.25 \times 2 &= 0.5, &\text{integral part } 0 &\ (= \text{MSB}) \\
0.5 \times 2 &= 1.0, &\text{integral part } 1 &\ (= \text{LSB})
\end{aligned}
$$

Thus 0.25d is 0.01b and 11.25d is 1011.01b; however, most decimal fractional numbers give recurring binary fractional numbers. For example, 0.2d gives

$$
\begin{aligned}
0.2 \times 2 &= 0.4, &\text{integral part } 0 &\ (= \text{MSB}) \\
0.4 \times 2 &= 0.8, &\text{integral part } 0 \\
0.8 \times 2 &= 1.6, &\text{integral part } 1 \\
0.6 \times 2 &= 1.2, &\text{integral part } 1 \\
0.2 \times 2 &= 0.4, &\text{integral part } 0
\end{aligned}
$$

The digits 0011 are recurring so that 0.2d is 0.0011b

20.1.2 Binary arithmetic

Addition is performed on two binary numbers at a time and if three or more binary numbers are to be added the third is added to the sum of the first two and so on, to avoid carries of more than 1. For example 1001b is added to 1101b as follows:

$$
\begin{array}{r}
1001 \\
+\ 1101 \\
\hline
10110
\end{array}
$$

since $1 + 1 = 0$, carry 1. Subtraction is similar, with a borrow if needed. Multiplication and division are the same as for decimal numbers, but in computers these operations are done by shifting the binary point.

Subtraction is performed in computers by addition of the complement of the *subtrahend* (the number that is subtracted). Thus instead of performing the subtraction $X * Y$, the computer adds: $X + (*Y)$. Two types of complement can be used, the 1's complement and the 2's complement. The 1's complement of an n-digit binary number, N, is defined as $(2^n * 1) * N$, and the 2's complement is defined as $2^n * N$, which is the 1's complement plus 1. For example the 1's complement of 101100b is

$$(2^6 - 1) - 101100b = 111111b - 101100b = 010011b$$

We can see that forming a 1's complement is easy, just change all the 0s to 1s and vice versa. The 2's complement of 101100b is

$$2^6 - 101100b = 1000000b - 101100b = 010100b$$

But it is easier to add one to the 1's complement:

$$010011b + 1 = 010100b$$

To subtract 101100b from 1000101b directly we have

$$\begin{array}{r} 1000101 \\ - \quad 101100 \\ \hline 11001 \end{array}$$

The borrowing is complicated here! Using 2's complement addition is much easier:

$$\begin{array}{r} 1000101 \\ + \quad 010100 \\ \hline 1011001 \end{array}$$

The leading 1 is then dropped to give the correct result, 11001b. What has been done is like saying $100d * 89d = 100d + 11d * 100d$.

Figure 20.2 shows the process of machine multiplication of binary numbers. It is accomplished by shifting the multiplicand (the number being multiplied by another) to the left and adding the shifted parts together.

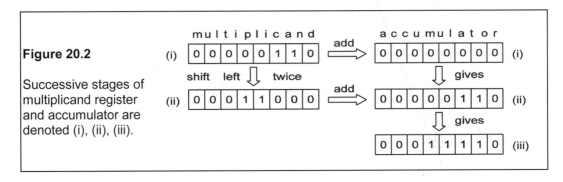

Figure 20.2

Successive stages of multiplicand register and accumulator are denoted (i), (ii), (iii).

For example 110b × 101b would be interpreted by a machine firstly as placing 110b in a multiplicand *register* (a place to store a number) and then adding it to the contents of another register called an *accumulator*, which is initially set to zero. This is the action 110b × 1b, that is multiplication by the LSB of the multiplier. Secondly, since the next multiplier digit is zero, nothing needs doing. Thirdly, the final multiplier digit is 1, so the contents of the multiplicand register are shifted two places left (equivalent to multiplying by 100b) and added to the accumulator. Division is akin, but requires shifting right and subtraction.

20.1.3 Hexadecimal numbers

Hexadecimal numbers have 16d as a base and the set {0,1,2,3,4,5,6,7,8,9,A,B,C,D,E,F} as digits, where the hexadecimal numbers A, B, C, D, E and F represent the decimal numbers 10, 11, 12, 13, 14 and 15. They can be formed from binary numbers by grouping them in fours:

$$10110011b = 1011b'0011b = 11d'3d = B3h$$

20.1.4 Binary-coded decimal (BCD) numbers

When numbers are to be output to a display or other device as decimal numbers, it is convenient to represent each decimal digit as a 4-bit binary number. This is said to be *binary-coded decimal* or BCD form; for example the number 395d is coded in BCD as

$$0011b'1001b'0101b$$
$$3 \qquad 9 \qquad 5$$

The binary numbers 1010, 1011, 1100, 1101, 1110 and 1111 are meaningless in BCD.

20.2 Logic functions and Boolean algebra

Boolean algebra was first applied to logical switching circuits by Shannon of Bell Labs[3], who also made great contributions to information theory. Consider the variables A and B each of which can be in one of two states, and suppose we wish to combine them according to just two rules symbolised by '+' and '•', which bear a resemblance to 'ordinary' addition and multiplication, but are called OR and AND in Boolean algebra. What does '$A + B$' (A OR B) mean? We can only tell by looking at all possible outcomes of that operation; fortunately there are only four combinations, shown in Table 20.1.

Table 20.1 *Truth tables for the OR function*

A	B	$A + B$	A	B	$A + B$	A	B	$A + B$
T	T	T	ON	ON	ON	1	1	1
T	F	T	ON	OFF	ON	1	0	1
F	T	T	OFF	ON	ON	0	1	1
F	F	F	OF	OF	OF	0	0	0

---

[3] *A symbolic analysis of relay and switching circuits* Trans. Am. IEE **57**, 713–23 (1938)

There are three ways we can fill in the truth table depending on how we wish to represent the states of the variables. In the first the two states are called 'true' (T) and 'false' (F) as in symbolic logic, in the second they are denoted ON and OFF corresponding to the switching circuit of Figure 20.3b, and in the third they are given binary values 1 and 0^4. It is not important what we call the two states: if the operation obeys Table 20.1, it must be the OR function.

These tables may be taken as a *definition* of the operation $A + B$ (A OR B). When the logic is implemented with logic ICs (or 'gates'), the ON or TRUE state is associated with a HIGH voltage (often +5 V), and the OFF or FALSE state with a LOW voltage (usually ground or 0 V) in the positive logic convention normally used. The logic circuit symbol for an OR gate is shown in Figure 20.3a, together with a parallel switching arrangement which will turn the lamp ON in agreement with Table 20.1.

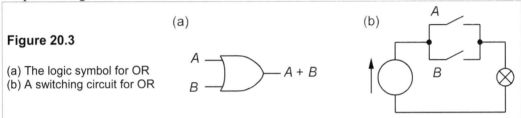

Figure 20.3

(a) The logic symbol for OR
(b) A switching circuit for OR

The second operation we wish to make use of is AND: $A{\bullet}B$, defined in Table 20.2, with its logic circuit symbol in Figure 20.4a and a switching circuit in Figure 20.4b. By examining the binary forms of Tables 20.1 and 20.2, it can be seen that the AND operation looks very much the same as ordinary multiplication, while OR is like ordinary addition except that 1 OR 1 $(1 + 1)$ is defined as 1, since only the states 0 and 1 exist. We shall see a little later that the truth table where this last outcome is defined as 0 also exists and is called the *exclusive OR* or XOR, denoted $A{\oplus}B$.

Table 20.2 *Truth tables for the AND function*

A	B	$A{\bullet}B$		A	B	$A{\bullet}B$		A	B	$A{\bullet}B$
T	T	T		ON	ON	ON		1	1	1
T	F	F		ON	OFF	OFF		1	0	0
F	T	F		OFF	ON	OFF		0	1	0
F	F	F		OFF	OFF	OFF		0	0	0

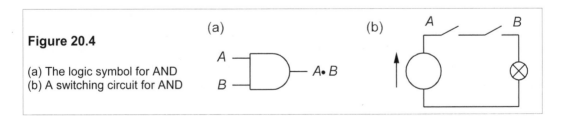

Figure 20.4

(a) The logic symbol for AND
(b) A switching circuit for AND

---

[4] In digital circuits 1 is generally taken to mean a HIGH voltage and 0 a LOW (usually near 0 V or ground). This is the positive logic convention. In negative logic the HIGH voltage state is taken as 0 and the LOW as 1. It is possible to use one convention for inputs and the other for outputs, which is called *mixed logic*. Though sometimes convenient this can be confusing.

As well as AND and OR it is convenient to use the NOT function, also called *inversion* or *complementation*, whereby A becomes NOT-A or $\bar{A}$ or $A\,'$ (the inverse or complement of A), AND becomes NAND and OR becomes NOR. Table 20.3 is the very brief truth table for NOT and the full logic circuit symbol is shown in Figure 20.5, though whenever it is possible the inversion operation will be shown by a circle placed on the appropriate input or output of the logic gate.

Table 20.3 *Truth tables for NOT*

A	A′	A	A′
T	F	1	0
F	T	0	1

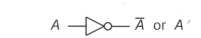

Figure 20.5 The circuit symbol for NOT

For example, Table 20.4 shows the inversion of OR and AND, denoted $\overline{A+B}$, or $(A+B)'$ or A NOR B, and $\overline{A \cdot B}$ or $(A\cdot B)'$ or A NAND B, respectively. So the circuit symbols for these devices, shown in Figure 20.6, have circles on their outputs to indicate inversion. In most of what follows we will use only the binary form of the truth tables.

Table 20.4 *Truth tables for NOR and NAND*

A	B	A + B	(A + B)′	A	B	A•B	(A•B)′
1	1	1	0	1	1	1	0
1	0	1	0	1	0	0	1
0	1	1	0	0	1	0	1
0	0	0	1	0	0	0	1

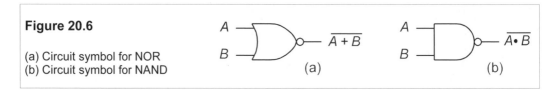

Figure 20.6

(a) Circuit symbol for NOR
(b) Circuit symbol for NAND

These five operations – OR, AND, NOT, NOR and NAND – turn out to be almost all that are needed in combinational logic. However, fewer than five functions in the form of logical devices are required because the NOT function, for example, can be realised by joining both inputs of either a NAND or a NOR gate as shown in Figure 20.7. Later on we shall see that by using de Morgan's theorems (sometimes called de Morgan's law) any logical expression can be realised with just *one* type of gate.

Figure 20.7

Inversion using NOR and NAND gates

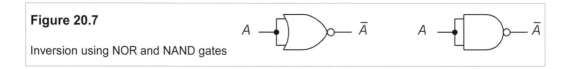

20.2.1 The laws of Boolean algebra

Having looked at some operations in Boolean algebra, its laws can be formally stated:

$$A + 0 = A$$
$$A \bullet 0 = 0$$
$$A + 1 = 1$$
$$A \bullet 1 = A$$

These define an element, 0, much like zero in 'normal' algebra, and another element, 1, that behaves in some ways like unity does normally. The rules for the complement of A are:

$$A + A' = 1$$
$$A \bullet A' = 0$$

The *idempotency* (from the Latin *idem* = the same and *potens* = power) laws are:

$$A + A = A$$
$$A \bullet A = A$$

The laws of *association*:
$$A + (B + C) = (A + B) + C = (A + C) + B$$
and
$$A \bullet (B \bullet C) = (A \bullet B) \bullet C = (A \bullet C) \bullet B$$

And finally the *distributive* laws:
$$A \bullet (B + C) = A \bullet C + A \bullet C$$
and
$$(A + B) \bullet (A + C) = A + B \bullet C$$

The last seems a little strange until it is realised that

$$(A + B)(A + C) = AA + AB + AC + BC = A(1 + B + C) + BC = A + BC$$

since $1 + B + C = 1$. Note that the dot is omitted by custom between ANDed variables. In the absence of parentheses AND takes precedence over OR. All of the above can be proved using truth tables and the definitions of AND, OR and NOT.

Example 20.1

Prove that $A + A'B = A + B$.
   Since $A = A(1 + B) = A + AB$, the equation above can be written as

$$A + AB + A'B = A + B(A + A') = A + B$$

which is the desired result. The student should verify this result by drawing up a truth table for $A + A'B$ and $A + B$.

20.3 **Logic ICs**

Devices that implement binary logical operations are made from transistors, diodes and resistors forming an integrated circuit (IC) and are called logic gates or just gates. These operate at voltage levels which are either HIGH (anything from about +3 V to +18 V) or LOW (usually ground or 0 V, but possibly as high as +2 V). In the positive logic convention, HIGH corresponds to 'true' or 1 and LOW to 'false' or 0. Usually several gates are put on a single chip which is then packaged in plastic with externally available tinned copper leads. An example is the quad, dual-input NAND, which has four 2-input gates in a 14-pin DIL (Dual In Line) package.

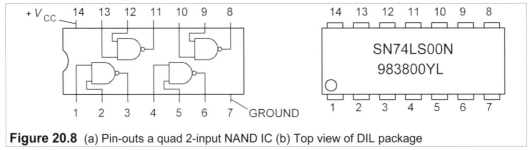

Figure 20.8 (a) Pin-outs a quad 2-input NAND IC (b) Top view of DIL package

The 14-pin DIL packages contain four, two-input, single-output gates (taking 12 pins) together with a ground pin and a positive supply pin. The external appearance of a 74LS (this is the logic family identifier, discussed in Section 20.11) quad 2-input NAND is shown in Figure 20.8b, while the pin arrangements of the gates are shown in Figure 20.8a. Logic devices are oriented with pin 1 at the bottom left of a DIL package, marked by a circle or dot on the top of the device (as in Figure 20.8b), or with an indentation on the left-hand side viewed from the top (as in Figure 20.8a). The DIL package must be the right way round to read the numbers stamped on it. Then the pin numbering runs from 1 at the bottom left to 7 (nearly always ground) at the bottom right, then 8 at the top right to 14 (nearly always $+V_{CC}$) at the top left. Inverters are single-input devices, so six of them can be arranged in one 14-pin DIL package. It is also possible to buy multi-input devices, such as the triple, 3-input, NAND from Harris Semiconductor, CD74HCT10E, and the MC74LS21N dual, 4-input AND from Motorola Semiconductors. 74HCT and 74LS are the logic families to which the devices belong, 10 and 21 are the device code numbers, and E and N denote package variations, such as plastic, ceramic, metal etc. Letters preceding the logic family are the manufacturer's identification code, and letters following the device number are a code for the type of package or encapsulation process. Any alphanumerics following on subsequent lines are the batch and date codes stamped on by the manufacturer, so that subsequent failures can be identified.

20.3.1 *The exclusive OR (XOR)*

As an example of the use of these logic gates in combination, let us look at the truth table given in Table 20.5, defining the XOR function, $A \oplus B$. Calling the output function Q (that is $Q \equiv A \oplus B$), it is apparent that $Q = 1$ when $A \neq B$, that is when $A = 1$ and $B = 0$, or when $A = 0$ and $B = 1$. The logical expression for Q is therefore $Q = AB' + A'B$, which is confirmed by Table 20.6. Since $A \oplus B = AB' + A'B$, the logic circuit for $A \oplus B$ will be that of Figure 20.9a.

Table 20.5 *The truth table for A ⊕ B*

A	B	A ⊕ B
0	0	0
1	0	1
0	1	1
1	1	0

Table 20.6 The truth table for *A'B + AB'*

A	B	A'B	AB'	Q
0	0	0	0	0
1	0	0	1	1
0	1	1	0	1
1	1	0	0	0

In the next section we shall see that this circuit can be changed to one using different gates by means of de Morgan's theorems. The XOR function can be used as a controlled inverter (see Section 21.3.4), since when $A = 1$ the output, $Q = B^*$, and when $A = 0$, $Q = B$, as can be seen from the truth table. A is the controlling variable and B the controlled. Thus the XOR function is frequently required and has been given its own circuit symbol (Figure 20.9b). It is available as one quarter of the 7486, a quad XOR.

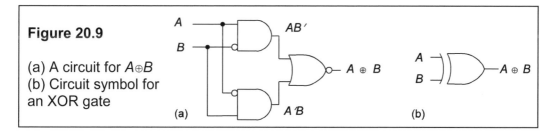

Figure 20.9

(a) A circuit for A⊕B
(b) Circuit symbol for an XOR gate

(a)

(b)

20.4 De Morgan's theorems[5]

De Morgan's theorems state:

$$\overline{A + B} = \overline{A} \cdot \overline{B}$$
$$\overline{A \cdot B} = \overline{A} + \overline{B}$$

Sometimes these equations are called de Morgan's law, best remembered as '*change the sign and break the line*'. In the left-hand side of the first of the equations, if the + sign is changed to a * and the line above is broken, so that half appears over the A and half over the B, then the right-hand side results. In the left-hand side of the second, the * is replaced by a + and the line broken, then the right-hand side follows. De Morgan's law enables us to manip-ulate logical expressions into more convenient forms. The first of the theorems shows that NORing A and B is identical to -ANDing their complements A^* and B^*, while the second shows that NANDing is identical to ORing the complements. When the dot is omitted from ANDed variables it must be (mentally) inserted when using de Morgan's law.

Remember that the line to be broken is the one *immediately* above the sign that is altered, and that lines crossing a bracket, or an implied bracket, must first be broken outside the

---

[5] Augustus de Morgan (1806–1871) was Professor of Mathematics at the University of London. He was a great logician and champion of George Boole.

bracket. For example, in the expression

$$Q = \overline{AB + A'B'}$$

AB and $A*B*$ have implied brackets round them, so the line must be broken at the $+$ to give

$$Q = \overline{(AB)}\,\overline{(A'B')}$$

Here the brackets are optional. The lines above AB and $A*B*$ can then be broken further, but now we *must* insert brackets:

$$Q = (A' + B')\overline{(\overline{A'} + \overline{B'})} = (A' + B')(A + B)$$

Complementing a variable twice leaves it unchanged: $(A*)* = A$.

Example 20.2

The circuit of Figure 20.10a uses NOR gates to form the logic func-tion W, given by

$$W = \overline{\overline{A + B} + \overline{A' + C'}}$$

Use de Morgan's law to turn it into a form suitable for implementing in NAND gates. By de Morgan's law

$$\overline{A + B} + \overline{A' + C'} = A' \cdot B' + \overline{A'} \cdot \overline{C'} = A'B' + AC$$

Then we can invert the result twice, leaving its value unchanged and apply de Morgan's law to the result:

$$W = \overline{\overline{W'}} = \overline{\overline{A'B' + AC}} = \overline{\overline{(A'B')} \cdot \overline{(AC)}}$$

This expression is in all-NAND form, and leads directly to the circuit of Figure 20.10b.

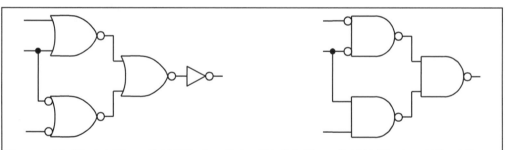

Figure 20.10 (a) The all-NOR circuit for W (b) The all-NAND circuit for W

20.4.1 *Redrawing circuits with de Morgan's theorems*

To restate, de Morgan's theorems are

$$\overline{A + B} = A' \cdot B' \qquad \text{and} \qquad \overline{A \cdot B} = A' + B'$$

In the first, a NOR operation becomes an AND with inverted inputs and in the second a NAND operation becomes an OR with inverted inputs. Thus the gates in Figure 20.11 are equivalent, and circuits containing one may be directly redrawn using its equivalent.

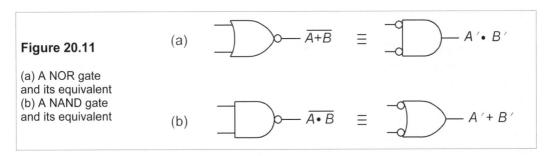

Figure 20.11

(a) A NOR gate
and its equivalent
(b) A NAND gate
and its equivalent

For example consider the circuit of Figure 20.12a, which is the circuit for the XOR function previously seen in Figure 20.9a. The ANDs can be replaced by their equivalent NOR forms, and the final OR can be replaced by a NOR and an inverter to give the all-NOR circuit of Figure 20.12b. Alternatively the final OR of Figure 20.12a can be replaced by its NAND equivalent (OR ≡ negative-input NAND) to give the circuit of Figure 20.12c. Then the circles on the input of the output NAND in this figure can be moved back to the outputs of the AND gates to give the all-NAND form of Figure 20.12d.

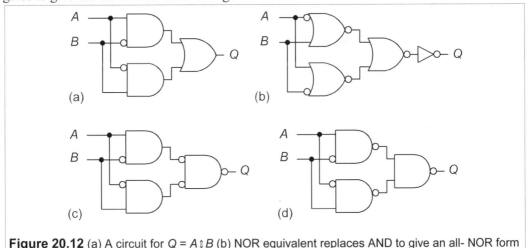

Figure 20.12 (a) A circuit for $Q = A \updownarrow B$ (b) NOR equivalent replaces AND to give an all- NOR form (c) The OR in Figure 20.12a is replaced by its NAND equivalent to give (d) The all-NAND form

20.5 **Minterms and maxterms**

An expression for the output of a logic circuit can be deduced from its truth table. Here we shall explain how this may be done, but first we must introduce two new words: *minterm* and *maxterm*. Suppose we have a function of three variables: $W = f(A,B,C)$ (we could have had as many variables as we liked, but three will suffice); then the minterms are the result of ANDing each of the three variables or its complement, while the maxterms are the result of ORing the same. Thus a minterm is an expression such as ABC, or $A'B'C$, and a maxterm is an expression such as $A' + B + C$, or $A' + B' + C$. There are 2^3 or eight possible minterms and eight possible maxterms for W. The minterm expression is often known as the *Sum Of Products* (SOP) form and the maxterm expression the *Product Of Sums* (POS) form.

20.5.1 The sum of products (SOP)

Consider the truth table (Table 20.7), for W in Example 20.2; we note that there are four entries in the output column where $W = 1$. The first of these has for inputs $A = 0$, $B = 0$ and $C = 0$, so that the min-term $A*B*C* = 1$. If any other combination of values for A, B or C is put into this minte-rm, such as $A = 0$, $B = 1$, $C = 1$, it will be zero. The next corresponds to $A = 0$, $B = 0$ and $C = 1$, so that the minterm $A*B*C = 1$; again, any other combination of values for A, B or C will give zero for this minterm. The other two minterms corre-sponding to $W = 1$ are $AB*C = 1$ and $ABC = 1$. If we now form the sum of these four minterms, we obtain

$$f(A,B,C) = ABC + AB'C + A'B'C + A'B'C' = W$$

We can see that $f(A,B-,C)$ is the required function, since each term will be zero except for the sole combina-tion of values of A, B and C that is required for W to be 1. For example, if $A = 1$, $B = 0$ and $C = 0$, each of the four terms is zero (and $W = 0$); while if $A = 0$, $B = 0$ and $C = 1$, the first two terms and the last term of $f(A,B,C)$ are 0, while the third term is 1 * as it must be since $W = 1$ for this combination. W has thus been expressed as a *sum of minterms-* or SOP. The number of minterms is the same as the number of times $W = 1$ in the truth table.

Table 20.7 *The truth table for W*

A	B	C	W
0	0	0	1
0	0	1	1
0	1	0	0
0	1	1	0
1	0	0	0
1	0	1	1
1	1	0	0
1	1	1	1

The SOP can be simplified with Boolean algebra:

$$f(A,B,C) = ABC + AB'C + A'B'C + A'B'C'$$

$$= AC(B + B') + A'B'(C + C') = AC + A'B'$$

By using de Morgan's law this can be put in all-NAND form. Complementing it twice produces:

$$f(A,B,C) = W = \overline{W'} = \overline{\overline{AC} + \overline{A'B'}} = \overline{(A\,C) \cdot (A'B')}$$

This W is identical to that of Example 20.2.

20.5.2 The product of sums (POS)

W can also be expressed as a *product of maxterms* or POS. To do this we look for the com-binations producing a zero for W. The first of these is $A = 0$, $B = 1$ and $C = 0$, so the OR function $(A + B* + C)$ will be 0 for this combination, but 1 for *all of the rest*. The next time

$W = 0$ occurs when $A = 0$, $B = 1$ and $C = 1$, then $A + B' + C' = 0$ for this combination only. The other two maxterms formed in this way are $(A' + B + C)$ and $(A' + B' + C)$. The product of these four maxterms is:

$$F(A,B,C) = (A+B'+C)(A+B'+C')(A'+B+C)(A'+B'+C) = W$$

It can be seen that $F(A,B,C)$ is a valid expression for W, since each maxterm is 0 only at one combination of values for A, B and C, and that is the combination for which $W = 0$. The number of maxterms must therefore be the same as the number of zero entries for W in the truth table.

The POS expression above for $F(A,B,C)$ could also have been obtained from the SOP or minterm expression for W', which is

$$W' = A'BC' + A'BC + AB'C' + ABC'$$

Then by de Morgan's law

$$F(A,B,C) = \overline{W'} = \overline{A'BC' + A'BC + AB'C' + ABC'}$$

$$= \overline{A'BC'} + \overline{A'BC} + \overline{AB'C'} + \overline{ABC'}$$

$$= (A + B' + C)(A + B' + C')(A' + B + C)(A' + B' + C)$$

$F(A,B,C)$ can also be simplified by Boolean algebra:

$$(A + B' + C)(A + B' + C') = A + B'$$

and $$(A' + B + C)(A' + B' + C) = A' + C$$

so that $$F(A,B,C) = (A + B')(A' + C) = A'B' + AC + B'C$$

Perhaps surprisingly this expression for W differs from the one found from the sum of minterms, yet it is undoubtedly an expression which obeys the truth table for W, as can quickly be checked. The extra term $B'C$ is redundant and can logically be omitted, that is $A'B' + AC = A'B' + AC + B'C$. (As an exercise prove this by Boolean algebra.) In Section 20.9.6 we see that the redundant term may be needed for the circuit to work properly. The algebra required to produce the simplest expression for W is fairly lengthy and can be greatly reduced or eliminated all together by Karnaugh mapping.

20.6 Karnaugh mapping and circuit minimization

Karnaugh mapping is a means by which examination of a truth table can produce an efficient, or minimised, logical expression for a function of several variables. First let us examine a very simple specimen, the Karnaugh map (or KM) for the function $Z = XY + XY' + X'Y'$, in Figure 20.13. Along the top row are the values for X (just two, of course) and down the first column the two values for Y. The four cells forming the KM contain the values of Z corresponding to all four combinations of X and Y. We can circle a horizontal group of cells containing ones (1-cells) and a vertical group of 1-cells also. These groups are known as *subcubes*. Since each subcube corresponds to a minterm such as XY, the circled horizontal group corresponds to XY'

$+ XY'$, which is just Y'. The circled vertical subcube corresponds to $XY + XY'$, which is just X. Thus the logical expression for Z must be $Z = X + Y'$. In the horizontal subcube, X changes value, and therefore must give rise to a term in the logical expression for Z which does *not* contain X. Because this occurs when $Y = 0$, that term can only be Y'.

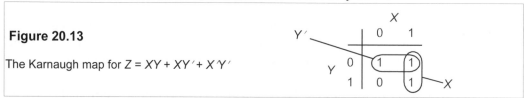

Figure 20.13

The Karnaugh map for $Z = XY + XY' + X'Y'$

Similarly for the vertical subcube: when $X = 1$, Y is both 0 and 1, so this term must be independent of Y, that is just X. There are then only two terms in the expression for Z: $Z = X + Y'$. The same result could have been found by manipulating the original expression for Z, but we might have overlooked that $X + X'Y'$ was the same as $X + Y'$.

To be sure, in such a simple Karnaugh map we can see immediately that only one entry is zero, when $X = 0$ and $Y = 1$, so that $Z = X + Y'$ is the POS form, having just one maxterm.

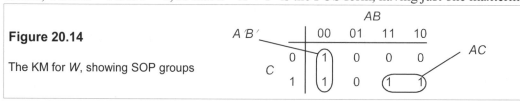

Figure 20.14

The KM for W, showing SOP groups

Look now at the Karnaugh map for W (from Table 20.7) in Figure 20.14: note that because there are three variables, we have had to group two together (A with B) along the top row. In this row only one variable is different in adjacent cells. Then we proceed to circle the groups of 1-cells. The bottom horizontal subcube where $C = 1$, $A = 1$ and B varies, must correspond to the term AC. The left-hand vertical subcube, wherein C varies, must correspond to the term $A'B'$. We grouped all the 1-cells and so the minimized expression for W is $W = AC + A'B'$. The SOP expression is well-suited to form all-NAND logic circuit implementations of a function, as we saw in Example 20.2.

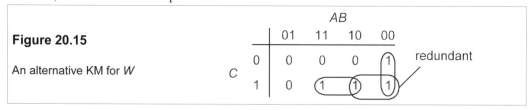

Figure 20.15

An alternative KM for W

We could have drawn up a different KM for W, obeying the rule that the variables change one at a time, as in Figure 20.15. Here the leftmost column of Figure 20.14 has been moved to the right end. This makes it clearer that there is a grouping of 1-cells which is redundant, namely that where $C = 1$, $B = 0$ and A varies, corresponding to the term $B'C$. We see that a KM *has no edges*, and we can imagine it as drawn on a cylinder in this case so that the columns at the edge are adjacent.

We can group the 0-cells instead of 1-cells, to find this time a minimized POS expression for W, as in Figure 20.16. Note here that the top two corner cells are adjacent (no edges to a KM). This group has B varying and $A = 1$, $C = 0$, so the maxterm is $(A' + C)$. The other group has C varying, $A = 0$ and $B = 1$, corresponding to $(A + B')$, and as all the 0-cells are accounted for in these two groups, the minimized POS expression for W is

$$W = (A' + C)(A + C')$$

This expression is well suited for implementing an all-NOR form for W by complementing twice, producing the circuit of Figure 20.10a.

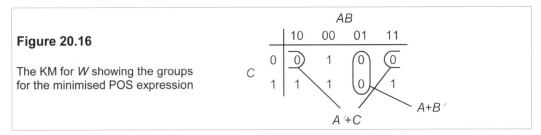

Figure 20.16

The KM for W showing the groups
for the minimised POS expression

Example 20.3

Suppose we have a lamp operated by three switches as shown in Figure 20.17. What is the function which describes the operation of the circuit?

First we make up a truth table: when switch A is closed we shall set $A = 1$ and when it is open we shall put $A = 0$ and the same for the other switches. When the light is on the output function $S = 1$. The result is the truth table on the right of Figure 20.17.

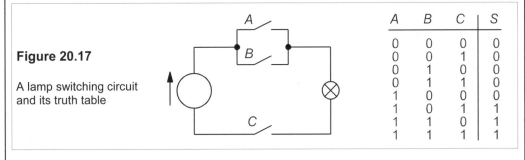

Figure 20.17

A lamp switching circuit
and its truth table

A	B	C	S
0	0	0	0
0	0	1	0
0	1	0	0
0	1	1	0
1	0	0	0
1	0	1	1
1	1	0	1
1	1	1	1

The truth table is not particularly helpful as it stands, so the KM for S, Figure 20.18, must be examined. At once it is clear that the combinations which switch the light on are those of the subcubes where $C = 1$, $AB = 01, 11$ and $C = 1$, $AB = 11, 10$. The first of these corresponds to $BC = 1$ (as A can be 0 or 1) and the other corresponds to $AC = 1$ (since B can be 0 or 1). Thus the light switching function is $S = BC + AC$. It is hardly surprising that C is part of each term in S, since it is the switch controlling the only return path.

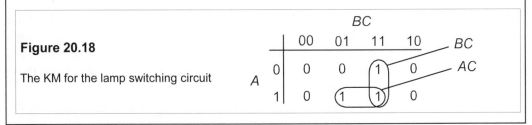

Figure 20.18

The KM for the lamp switching circuit

20.6.1 Karnaugh maps with four input variables

The principles are the same as for three variables, except that the variables are grouped in pairs. The like cells are grouped together in the largest-possible blocks. Consider for example the Boolean function

$$Q = (A'B + C)(BC + D)\overline{AD'}$$

whose Karnaugh map is shown in Figure 20.19.

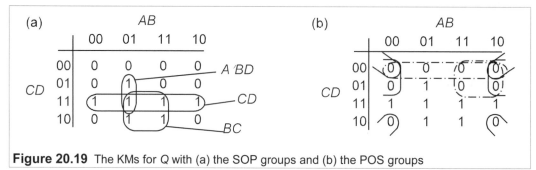

Figure 20.19 The KMs for Q with (a) the SOP groups and (b) the POS groups

In Figure 20.19a the four 1s in a square block have been grouped, corresponding to the term minterm BC, and a group of two 1s vertically which correspond to the minterm $A'BD$. Finally we can make a group of the four 1-cells running right across the table, which corresponds to the minterm CD, as it is independent of the variables A and B. Thus the minimal minterm expression for Q is

$$Q = A'BD + BC + CD \tag{20.1}$$

This form is suitable for implementing with NAND gates, since by complementing twice we find

$$Q = \overline{A'BD + BC + CD} = \overline{(A'BD) \bullet (BC) \bullet (CD)}$$

The maxterm expression for Q is found by grouping the zeros in the KM, Figure 20.19b, and is much more difficult to find, but it is

$$Q = (A' + C)(C + D)(B + C)(B + D) \tag{20.2}$$

and can be readily implemented in all-NOR form by complementing twice. The four corner zeroes, corresponding to the term $(B + D)$, are easy to miss!

20.6.2 Don't cares

There are cases when the output or switching function does not exist for some combination of the input variables, or the input variables cannot be set to a particular combination (such as 1100 in BCD coding), so we don't care whether it is 0 or 1. In these cases the *don't care* combinations are marked with a cross on the Karnaugh map and we are at liberty to set these cells to 0 or 1 as convenient. For example examine the KM of Figure 20.20.

The simplest solution is to call all the don't care cells 0 except the 0101 cell, giving the subcube of 1s shown. In this subcube $C = 0$ and $B = 1$, while A and B change; so the switching function is $S = BC*$. When Karnaugh maps contain more 1s than 0s the switching function is simply expressed as a *POS* or in terms of its inverse and the *SOP*.

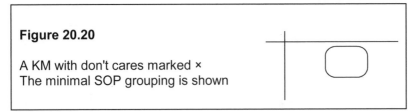

Figure 20.20

A KM with don't cares marked ×
The minimal SOP grouping is shown

20.7 **Practical examples**

We shall give just a few examples of circuits using combinational logic, starting with one of the basic building blocks in the arithmetic unit of digital computers.

20.7.1 *The half adder*

The half adder circuit's function is to add two binary digits, A and B, which gives the truth table of Table 20.8. There are two outputs: S, which is the least significant digit in the sum of the two added digits and C, the carry which must be added into the sum of the next, more significant, digit. S is seen to be just the XOR function while C is the AND function.

Table 20.8 *The truth table for a half adder*

A	B	S	C
0	0	0	0
1	0	1	0
0	1	1	0
1	1	0	1

The Karnaugh map for S leads to no simplification, so the *SOP* expression is, as we have seen before, $S = AB* + A*B$. Loo-king for the *POS* expression we find

$$S = (A + B)(A' + B')$$

so that

$$S' = \overline{(A + B)(A' + B')} = \overline{A + B} + \overline{A' + B'}$$

by de Morgan's law. Inverting this produces

$$\overline{S'} = S = \overline{\overline{A + B} + \overline{A' + B'}}$$

which is in all-NOR form. The advan-tage of this representation is that it gives the carry, C, which is AB. By de Morgan's law once more

$$C = \overline{C'} = \overline{\overline{AB}} = \overline{A' + B'}$$

and this is one of the terms in the all-NOR form of S. Hence the circuit for the half adder with carry is as in Figure 20.21a, and requires three dual-input gates. If an XOR gate is used to form S and an AND gate to form C, the circuit of Figure 20.21b results.

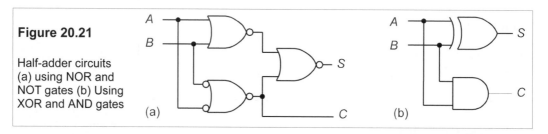

Figure 20.21

Half-adder circuits
(a) using NOR and
NOT gates (b) Using
XOR and AND gates

20.7.2 The full adder

The half adder is not much use on its own as it cannot cope with a carry in from a previous stage, C_i. Thus three inputs are required for a full adder, as shown in the truth table of Table 20.9. The outputs from the full adder are a sum, S, and a carry out, C_o. From the truth table we can form the SOP for S:

$$S = A'BC_i' + AB'C_i' + A'B'C_i + ABC_i$$

which rearranges to

$$S = C_i'(A'B + AB') + C_i(A'B' + AB)$$

$$= C_i'(A \oplus B) + C_i(\overline{A \oplus B}) = (A \oplus B) \oplus C_i$$

The circuit for S therefore requires two XOR gates, as in Figure 20.22.

Table 20.9

Truth table for a full adder

A	B	C_i	S	C_o
0	0	0	0	0
0	0	1	1	0
0	1	0	1	0
0	1	1	0	1
1	0	0	1	0
1	0	1	0	1
1	1	0	0	1
1	1	1	1	1

The carry out is

$$C_o = ABC_i' + A'BC_i + AB'C_i + ABC_i$$

$$= C_i(A'B + AB') + AB(C_i' + C_i)$$

$$= C_i(A \oplus B) + AB$$

This expression for C_o can be obtained by ANDing C_i and the output of one XOR, $A \oplus B$, and ORing this with A AND B. The circuit of Figure 20.22 performs the logic for S and C_o and comprises two half adders with an OR gate to form C_O from the carry outs.

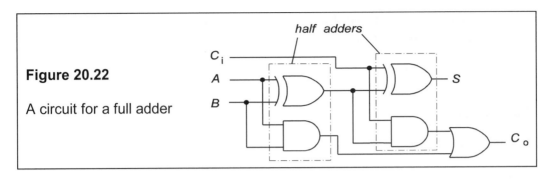

Figure 20.22

A circuit for a full adder

20.7.3 Encoders and decoders

It is often necessary to encode and decode data: for example one might wish to code keyboard alphanumericals into ASCII binary code prior to transmission, or to encode decimal digits into BCD form. One can then decode transmitted data back into the form in which it is required. The block diagram for the coder/decoder is shown in Figure 20.23.

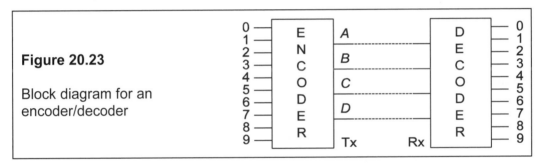

Figure 20.23

Block diagram for an encoder/decoder

The 2-line-to-4-line decoder

An n-bit binary number can convey up to 2^n different messages. A decoder takes an n-bit binary number at its input and decodes it into one of up to 2^n outputs. Thus a 3-bit decoder can have 8 outputs (or fewer) and a 2-bit decoder four. The truth table for the 2-line-to-4-line decoder is shown in Table 20.10.

Table 20.10

The truth table for a 2-line to 4-line decoder

A	B	W	X	Y	Z
0	0	1	0	0	0
0	1	0	1	0	0
1	0	0	0	1	0
1	1	0	0	0	1

From the truth table we see that

$$W = A'B' = \overline{A+B} \; ; \quad X = A'B = \overline{A+B'}$$

$$Y = AB* = \overline{A'+B} \; ; \quad Z = AB = \overline{A*+B'}$$

Hence the logic for W, X, Y and Z requires only the four two-input NOR gates and two inverters shown in Figure 20.24

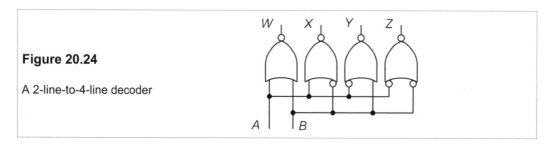

Figure 20.24

A 2-line-to-4-line decoder

20.7.4 BCD to 7-segment display

In Section 20.1.4 we saw how decimal numbers could be conveniently represented in binary form by encoding them digit by digit in BCD form. In devices like calculators decimal numbers are displayed by switching on segments from a 7-segment array, as in Figure 20.25a.

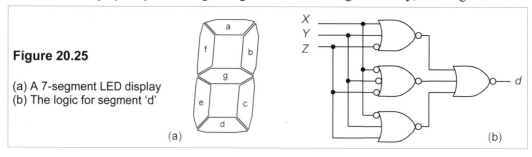

Figure 20.25

(a) A 7-segment LED display
(b) The logic for segment 'd'

(a)   (b)

If we take the decimal number, 6 (0110 in binary), as an example, we can see that segments c, d, e, f and g are to be lit, while a and b are not. Considering all the numbers for which a particular segment, such as d, must be lit, a truth table can be drawn up as in Table 20.11, where W is the MSB and Z is the LSB.

Table 20.11

The truth table for segment 'd'

		B i n a r y				
		W	X	Y	Z	d
	0	0	0	0	0	1
d	1	0	0	0	1	0
e	2	0	0	1	0	1
c	3	0	0	1	1	1
i	4	0	1	0	0	0
m	5	0	1	0	1	1
a	6	0	1	1	0	1
l	7	0	1	1	1	0
	8	1	0	0	0	1
	9	1	0	0	1	0

		WX			
		01	11	10	00
	01	1	×	0	0
YZ	11	0	×	×	1
	10	1	×	×	1
	00	0	×	1	1

Figure 20.26 The KM for segment 'd'

The Karnaugh map for d in Figure 20.26 shows that we can make three POS terms by lumping some of the don't cares in with the 0s, none of which includes W. The POS for d is then

$$d = (X+Y+Z')(X'+Y'+Z')(X'+Y+Z)$$

We could AND three 3-input ORs with a 3-input AND gate, or we could turn it into all-NOR form by de Morgan's law:

$$d = \overline{d'} = \overline{\overline{X+Y+Z'} + \overline{X'+Y'+Z'} + \overline{X'+Y+Z}}$$

This can be achieved with four 3-input NORs as shown in Figure 20.25b, and does not require an input from W, the most significant BCD bit. We could use, for example, the 74ALS27 device, a triple 3-input NOR, to implement the circuit, or a pair of dual 4-input NORs such as the 74HC4002. Each IC package costs about 50p in single quantities. In the 74HC4002 the spare inputs to the 4-input NORs should be tied to logical 0, in case they drift to 1. Likewise, the spare inputs of AND or NAND gates should be tied to logical 1 in case they become grounded. The LEDs themselves will sink quite a lot of current and it will be necessary to use the logic output to turn on a transistor to drive the display. Then all that remains is the logic design for the other six segments. We can best accomplish this kind of task with a programmable logic array (see Section 20.11), though in practice a BCD 7-segment LED decoder/driver chip would be used.

20.7.5 Gray-to-binary decoder

Gray code is used to convert decimal numbers into a form of binary number in which only neighbouring digits change in counting up or down. For example the three-bit binary counting sequence is 001, 010, 011, 100, 101 etc. and the count from 3_{10} to 4_{10} involves the change of all three binary digits. In Gray code the same decimal sequence would become 001, 011, 010, 110, 111 etc. so that no more than one digit changes at a time. The advantage of the Gray code is that errors in one bit of an encoded decimal number can only affect the decimal count by one, which would not be the case with unencoded binary representation. The conversion from Gray to binary is made by a combinational logic circuit. Consider the 3-digit Gray code and its binary equivalent in Table 20.12, from which we can draw up three Karnaugh maps – one for each binary bit – as in Figures 20.27a–c.

Table 20.12 *Gray code and binary equivalent*

		G_1	G_2	G_3	B_1	B_2	B_3
	0	0	0	0	0	0	0
d	1	0	0	1	0	0	1
e	2	0	1	1	0	1	0
c	3	0	1	0	0	1	1
i	4	1	1	0	1	0	0
m	5	1	1	1	1	0	1
a	6	1	0	1	1	1	0
l	7	1	0	0	1	1	1

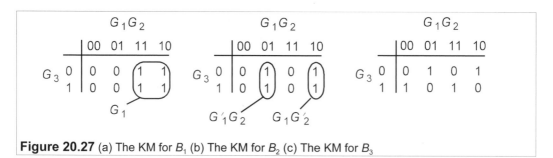

Figure 20.27 (a) The KM for B_1 (b) The KM for B_2 (c) The KM for B_3

The Karnaugh map in Figure 20.27a shows that $B_1 = G_1$ and the KM of Figure 20.27b that

$$B_2 = G'_1 G_2 + G_1 G'_2 = G_1 \oplus G_2$$

The KM of Figure 20.27c has no groupings of 1s or 0s, but the pattern suggests (see problem 20.18) an XOR function:

$$B_3 = G_1 \oplus G_2 \oplus G_3$$

Therefore the logic circuit for the 3-bit Gray-to-binary decoder is that of Figure 20.28 below.

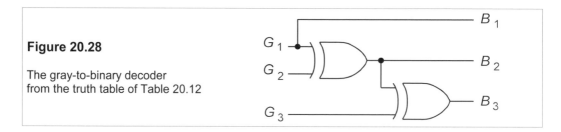

Figure 20.28

The gray-to-binary decoder
from the truth table of Table 20.12

20.8 **Logic families**

Integrated circuits or ICs can be made in a variety of ways some of which are indicated in Figure 20.29. This shows that ICs are made not only from the semiconductors silicon and gallium arsenide, but also from thick and thin films on insulating substrates. The latter are not considered here as they are not used for logic ICs. Gallium arsenide (GaAs) ICs are restricted to a few specialised uses, mainly at very high frequencies: all the ICs supplied by popular stockists are made from silicon.

The oldest of the present-day digital logic families is bipolar TTL (transistor-transistor logic, a successor to resistor-transistor logic, RTL, and diode-transistor logic, DTL). ECL (emitter-coupled logic) gates are very fast, but consume much more power than other types of logic and are largely restricted to special purposes such as high-speed computing. I^{2}L (integrated injection logic) was an attempt to increase the packing density and speed of TTL which has faded like the NMOS and PMOS precursors of CMOS (complementary metal oxide semiconductor). Virtually all readily-available logic ICs now are 74 series TTL, CMOS or BiCMOS, as listed in Table 20.13.

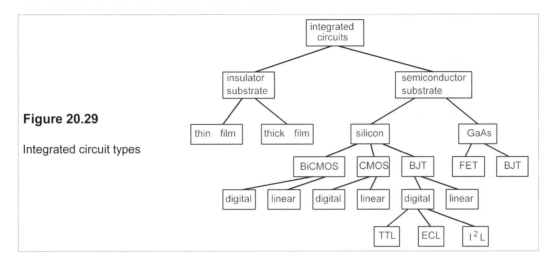

Figure 20.29

Integrated circuit types

20.8.1 TTL

TTL is still widely used because it is cheap, operates at well-defined and well-known voltage levels and is robust. It suffers from a larger power consumption than CMOS but it is normally faster. Both of these aspects of performance have been enhanced by the advent of Schottky TTL, but probably not by enough to ensure TTL's ultimate survival. Drop-in CMOS replacements for advanced low-power Schottky TTL have been available for some time, offering greatly reduced power consumption at comparable speeds. TTL has standard operating voltages which means that all TTL devices, no matter where made, can be used interchangeably. The supply voltage is normally in the range from 4.5–5.5 V (though low-voltage TTL operates with 2.7–3.6 V), and the HIGH-LOW transition (V_{IH}) is at 2.0 V (minimum) while the LOW-HIGH transition (V_{IL}) is at 0.8 V (maximum).

Table 20.13 *Some logic families*

Family	Type	Gate Delay[a]	QP[b]	OP[c]	Current[d]	Price[e]
74LS	5V Low-power Schottky TTL	9	5	5	8	6p
74MG	3-18 V CMOS	65	0.0006	0.04	1	5p
74ALS	5V Advanced 74LS TTL	7	1.2	1.2	8	7p
74ABT	5V Advanced BiCMOS TTL	3	0.005	1	32	10p
74FAST	5V High-current TTL	3	12.5	12.5	20	8p
74HC	2-6V CMOS	8	0.003	0.6	4	5p
74FACT	1.2-3.6V Fast Advanced CMOS	5	0.0001	0.6	24	8p
74LVC	2-3.6V Low-voltage CMOS	3	0.003	0.8	24	8p
74LCX	2-6V TTL-compatible CMOS	3.5	0.0001	0.3	24	10p
10H	−5V ECL	1	25	25	50 Ω	25p
100K	−5V ECL	0.75	50	50	50 Ω	£5
ECLinPS	−5V ECL	0.33	25	25	50 Ω	£10
Elite	−5V ECL	0.22	73	73	50 Ω	£12

Notes: [a] Propagation delay/gate in ns [b] quiescent power in mW/gate [c] Operating power at 1 MHz in mW/gate [d] drive current in mA, load for ECL [e] Approximate cost/gate in small quantities

20.8.2 Schottky TTL

In Schottky TTL (74S, 74LS, 74AS and 74ALS) the transistors are replaced by Schottky transistors, whose symbol is shown in Figure 20.30a. These have Schottky diodes (see Section 7.2) placed across their base-collector junctions to reduce the base current flowing when the device is in saturation, as shown in Figure 20.30b.

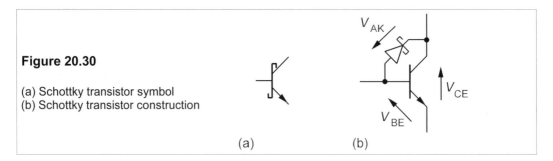

Figure 20.30

(a) Schottky transistor symbol
(b) Schottky transistor construction

(a)   (b)

The Schottky diode has a low forward voltage drop of about 0.3 V and ensures that $V_{CE}(\text{min}) = V_{BE} - V_{AK} \approx 0.4$ V. The transistor cannot be taken fully into saturation, thereby reducing the diffusion capacitance of the transistor and hence the time taken to charge when switched. The result is a considerable improvement in speed.

20.8.3 CMOS

Besides the TTL-compatible CMOS families 74HCT, 74ACT and 74LVT, there are a considerable number of other CMOS logic families. Switching voltages depend on the supply voltage, and the present standard for CMOS is that $V_{IL} = 0.2V_{DD}$ and $V_{IH} = 0.7V_{DD}$ with the output HIGH and LOW voltages being 0.1 V from V_{DD} and 0 V respectively.

20.8.4 Low-voltage logic

The overall power dissipation in CMOS goes as $(V_{DD})^n$, where $3 > n > 2$, so there is a great incentive to reduce V_{DD}, the supply voltage. There has been a great proliferation of low-voltage logic families in the last few years, such as the 74LV, 74LVX, 74LCX, 74LVC and 74LVCT series, most of which require supply voltage in the range from 1.2 V to 3.6 V.

Low-voltage ICs are intended for use with batteries and both regulated and unregulated supplies. Several of these families will operate at voltages down to 1.2 V, though at a lower speed, and these can be powered by a single cell. Recently, specialised devices for battery operation have been produced which run from a supply of less than 1 V. Several low-voltage families now operate on TTL logic levels, though with a reduced supply voltage. The standard for full interfacing and compatibility with TTL logic is still under discussion. Operation at lower supply voltages mean that the switching power consumption is reduced, but the need for tight parameter control and hence tight process control is accentuated. Another advantage of low-voltage operation is reduced EMI (electromagnetic interference) because of much smaller switching transients. Low supply voltages will cause an increase in propagation delay unless device sizes can be reduced and so the MOS channel length on low-cost standard ICs is now only about 1 μm and the BJT basewidth of BiCMOS devices is about 0.1 μm, compared to values three times as great five years ago. There is no doubt that low-voltage CMOS will become standardized and further developments will be in the direction of lower voltages yet, though these are already under 1V.

The need for small propagation delays has meant that the package size must be reduced, so that only the slowest of the principal low-voltage logic families, the LV series, has logic gates in standard DIL packages; all the others use the smaller SO (Small Outline) SSOP (Shrink Small Outline Package) or the TSSOP (Thin SSOP) packages. Figure 20.31 shows the 14-pin plastic DIL versions of these packages: the SOT27 is the standard package, the SOT108 is the plastic small outline package, the SOT337 is the plastic shrink small outline package and the SO402 is the plastic thin shrink small outline package. The latter three are surface-mount devices and are drawn to twice the scale of the standard DIL package.

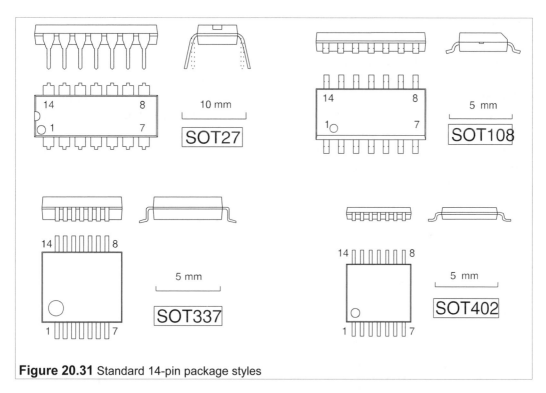

Figure 20.31 Standard 14-pin package styles

20.9 **Practical aspects of logic integrated circuits**

We shall briefly consider next a few important practical points that are family-related.

20.9.1 *Logic voltage levels: noise immunity*

Logic devices need a certain minimum input voltage before they will register a logical 1, and a certain maximum input voltage below which they will register a 0. These are specified by the manufacturers, but a standard +5V TTL or TTL-compatible device will normally operate on an input 1-level (one-level) of +2.0 V and an input 0-level (zero-level) of about +0.8 V. The *noise immunity* specified by the manufacturers is 0.4 V, which means that 0.4 V can be added to, or subtracted from, the output levels before they produce malfunctions in subsequent logic. Thus the 0-level output voltage must then be +0.4 V or less and the 1-level output voltage 2.4 V or more (0.8 – 0.4 V and 2.0 + 0.4 V). The 1-level is usually determined by the supply voltage, which must therefore be not less than that specified by the data sheets. CMOS has a greater noise immunity than TTL and ECL has a lower noise immunity.

Spare inputs to logical devices should either be connected to one of the inputs that *is* connected, or they should be connected to logical zero (usually ground, for OR and NOR gates) or logical 1 (usually +V_{CC}, for AND and NAND gates). If this is not done there is a strong possibility that the floating input will either respond to noise, or will just drift and go to a voltage level which will negate the valid inputs.

20.9.2 Power and speed

The power consumed by logic gates depends on whether their inputs and outputs are HIGH or LOW, but more importantly depends on the device family being used. If the IC is a fast ECL type, it will consume a lot more power than a slower TTL type. CMOS has been slowest and least power consuming, but the position has changed since the introduction of advanced low-power Schottky (ALS) CMOS, which has an average power consumption of 1 mW per gate and is faster than TTL. The AC (advanced low-power CMOS) series is even faster with a power consumption of 1 µW/gate.

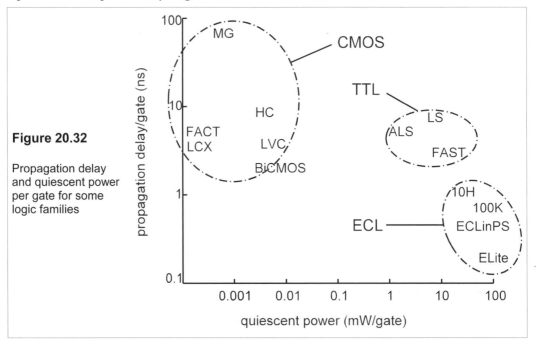

Figure 20.32

Propagation delay and quiescent power per gate for some logic families

The speed of logic devices is expressed as *propagation delay*, the time taken for the signal to pass from input to output at some percentage (usually 50%) of the maximum signal. Figure 20.32 shows propagation delays and power consumptions for some logic families, mostly 74-

series. See also Table 20.13.

The speed-power product is often used as a figure of merit (the lower the better) and this is simply the product of average power/gate and propagation delay. The advent of CMOS and Schottky transistors has resulted in a considerable reduction in speed-power product – from around 100 pJ for standard TTL to around 5–10 pJ for advanced, low-power, Schottky (ALS) devices.

20.9.3 Fan-out

Logic devices require current to operate. The current required depends on the device type and the logic level. The amount of current they can supply from their outputs is also a function of device type and logic level. If a logic gate can supply sufficient current under worst operating conditions to a maximum of ten inputs to gates of the same logic family (logic families should not be mixed), then it is said to have a fan-out of ten. Fan-outs are typically ten or more.

If a logic circuit is interfaced with analogue circuitry the current drawn by, or supplied by, that circuit from the digital part must be less than the maximum specified. When the logic devices are sourcing current, they can supply typically from 10 mA to 65 mA, and when sinking current from –10 mA to –25 mA, depending on the logic family.

20.9.4 Propagation delay

For each logic gate a finite time, known as the propagation delay, elapses before the output changes in response to a change of input. This delay can have serious consequences if the designer of a circuit assumes that the output is at 1 (or 0) when it may for a short time be at 0 (or 1) instead. In the case of the now-obsolescent, slow, 4000B MOS family, the typical propagation delay for a NAND gate transition from LOW to HIGH, t_{PLH}, is 40 ns at a supply voltage of +15 V, rising to 125 ns with a +5V supply. These times should be doubled to allow for worst-case devices. The 74 series has a typical propagation delay of 11 ns, while the 74HC and 74HCT series are a little faster. Fastest at the moment is the 74ABTC (advanced BiCMOS) family with $t_{PLH} = 3.6$ ns, maximum.

20.9.5 Three-state logic

The word 'three-state' is slightly misleading for there are still only two logic states, 1 and 0, as in normal logic, but three output states: HIGH, LOW and open circuit. The reason for this is that some types of TTL gate cannot be connected together, because when one is HIGH and the other LOW they try to pull each other into their state: the outputs are said to be in contention. Three-state gates connected to a common bus are enabled one at a time, the others being disabled by an enabling input so that they present open circuits to the bus. Three-state logic allows the use of the same devices to transmit or receive, or multiplexers to multiplex or demultiplex according to the state of the enabling input.

Figure 20.33 shows some three-state logic gates which are enabled when $E = 0$ and disabled when $E = 1$, and a three-state bus with data transmitting (Tx) and receiving (Rx) lines controlled by three-state inverters.

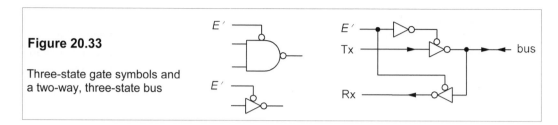

Figure 20.33

Three-state gate symbols and
a two-way, three-state bus

20.9.6 Hazards

Hazards are states caused by propagation delays which arise at the output of a logical circuit which should not be present according to the truth table. Consider the circuit of Figure 20.34a which has a logical output of $Q = AB' + BC$ and whose Karnaugh map is shown in Figure 20.35.

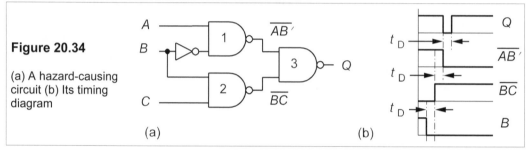

Figure 20.34

(a) A hazard-causing
circuit (b) Its timing
diagram

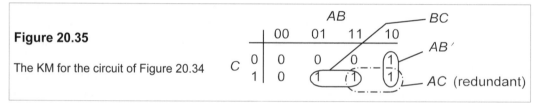

Figure 20.35

The KM for the circuit of Figure 20.34

There is a propagation delay caused by the inverter which produces B' at the input of NAND1. If the inputs are initially $A = 1$, $C = 1$ and $B = 1$, then when B switches from 1 to 0 the output at NAND2 switches from 0 to 1 after one gate propagation delay, t_D. But the output of NAND1 at $t = t_D$ is 1, and so at $t = 2t_D$ the output of NAND 3 is 0, not 1 as suggested by the KM. After $3t_D$ the output at NAND3 will be 1 as required. The timing diagram for this is shown in Figure 20.34b. While the output is 'wrong' a problem can arise such as lift doors closed, alarm bells rung, sprinklers turned on etc.

The changing variable which causes the hazard is that in the redundant loop corresponding to the term AC, and it can be overcome quite simply by including the redundant term in the circuit, that is we form

$$Q = AB' + BC + AC$$

as in Figure 20.36. We can see from this figure that the output must remain at 1 when B switches from 0 to 1 while A and C are 1. The term AC, which is 1, ensures that the output of NAND3 is 0 during the time B switches and so the output of NAND5 stays at 1. Hazards arise whenever adjacent 1s in a Karnaugh map are not looped together. Logically-redundant terms might not be redundant practically.

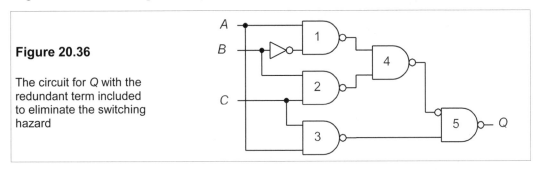

Figure 20.36

The circuit for Q with the redundant term included to eliminate the switching hazard

20.10 **Multiplexers and demultiplexers**

A multiplexer is a device with a number of data inputs which can be selected one at a time as output. The selection is performed by address (or data selection) inputs in accordance with the desired output sequence. The output can then be transmitted and demultiplexed into the original inputs.

Figure 20.37a shows a multiplexer with four inputs, W, X, Y and Z, an output, B, and two address inputs A_1 and A_2. If W is selected by $A'_1A'_2$, X by $A_1A'_2$, Y by A'_1A_2 and Z by A_1A_2 then the logic for the output, B, is

$$B = WA_1'A_2' + XA_1'A_2 + YA_1A_2' + ZA_1A_2$$

and this can be realised with the circuit of Figure 20.37b.

It is clear that if there are 2^N data lines then N address (data-selection) lines are needed. If the address inputs are switched so that the data on the data inputs stays constant during one complete address cycle, then the data is presented to the output in multiplexed bit form, and parallel-to-serial conversion has been effected. If the address cycle time is a multiple of the cycle time for bit multiplexing then the multiplexed data stream comprises several bits of each data input. Thus the data can also be multiplexed in bytes (normally of eight bits) or nibbles (half a byte) or half nibbles

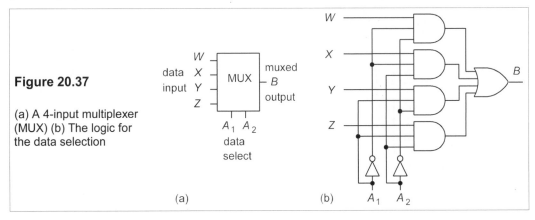

Figure 20.37

(a) A 4-input multiplexer (MUX) (b) The logic for the data selection

The demultiplexer (which is usually the same IC as the multiplexer) takes the multiplexed serial data stream and converts it to the component data streams once more as shown in Figure 20.38a. The truth table to demultiplex the data from the circuit of Figure 20.37b is like that of the 2-line-to-4-line decoder and leads to the logic circuit shown in Figure 20.38b.

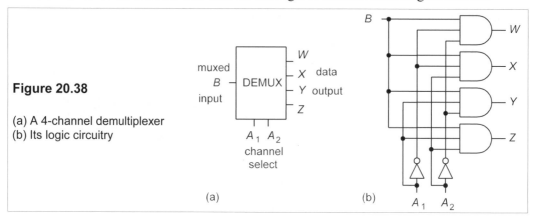

Figure 20.38

(a) A 4-channel demultiplexer
(b) Its logic circuitry

(a)

(b)

20.10.1 Boolean functions using multiplexers

By using the data input and data select lines it is possible to form Boolean functions at the output of a multiplexer. Almost any combination of three variables is possible using a 4-channel multiplexer. For example consider the truth table of Table 20.14, which can be expressed in SOP form as

$$Q = A'B'C' + A'B'C + A'BC' + AB'C + ABC$$

If we make A and B the address lines, A_1, A_2, of a 4-to-1 MUX, the first two terms are

$$A'B'C' + A'B'C = A'B'1 = A_1'A_2'W$$

Thus the W-input must be connected to logical 1. The third term is $A'BC' \equiv A_1'A_2X$ so that $X = C'$. The fourth and fifth terms give $Y = Z = C$. The multiplexer must then be configured as shown in Figure 20.39 to give Q.

Table 20.14 *The truth table for Q*

A	B	C	Q
0	0	0	1
0	0	1	1
0	1	0	1
0	1	1	0
1	0	0	0
1	0	1	1
1	1	0	0
1	1	1	1

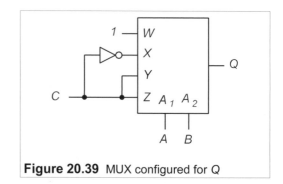

Figure 20.39 MUX configured for Q

20.11 **Programmable logic arrays**

Programmable logic arrays (PLAs), programmable logic devices (PLDs), or gate-array logic devices (GALs) can be purchased in IC form. They are programmed by the user to perform a variety of logic functions that might have used many logic ICs of the NAND, NOR and NOT type. Figure 20.40 shows the arrangement of the connections in a PLA which has programmable input connections to AND gates and programmable OR outputs.

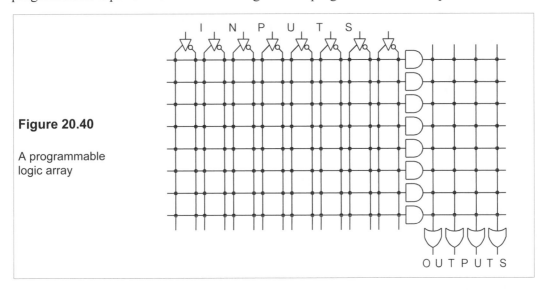

Figure 20.40

A programmable logic array

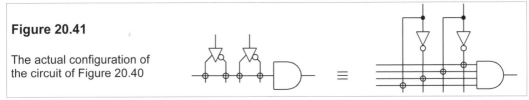

Figure 20.41

The actual configuration of the circuit of Figure 20.40

The inputs can be connected to all the AND gates in both complemented and uncomplemented form. Figure 20.41 indicates the actual connection symbolised in Figure 20.40. The cross-links are connected where desired by a suitable PLA programming device, which can be instructed by a microcomputer. Some PLAs are one-time programmable (OTP) and some may be reprogrammed when desired, like an erasable programmable read-only memory (EPROM), by exposing them to ultra-violet light, and some are electrically erasable as in EEPROMs. There are programs which will convert a logical expression into instructions to make the required connections, but otherwise one has to tell the programming device exactly which links to make. In Figure 20.42 the connections are shown for forming the logical expression $Q = ABC' + A'BC$, for example.

Commonly-available PLAs have anything from 8 to 12 inputs and from 4 to 10 outputs. Some have outputs which can be reconfigured as inputs. The cheaper PALs are in 20, 24 or 28-pin packages. Some have programmable OR outputs as in Figure 20.30, while others are hard wired. Prices range from about £1 for the ATF16V8B-15PC device (8 inputs, 8 programmable outputs, 15 ns propagation delay in a 20-pin DIL package) up to £10 for a PALCE22V10H-7PC/4 device (12-input, 10-programmable output, 7 ns propagation delay, 28-pin package). Increasingly complex logic arrays are being used, known as complex programmable logic devices (CPLDs) in the £10–£50 range. And a large and rapidly-evolving group of devices are the powerful and still more complex field-programmable gate arrays (FPGAs), costing from £10–£100, which we have no space to discuss.

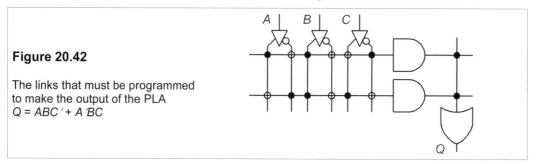

Figure 20.42

The links that must be programmed to make the output of the PLA
$Q = ABC' + A'BC$

Suggestions for further reading – See Chapter 22

Problems

1. Convert the hexadecimal numbers A0C1, B2E7 and FD10 to binary and thence to decimal form. *[1010000011000001b, 41153d, 1011001011100111b, 45799d, 1111110100010000b, 64784d]*
2. Use the 2's complement to subtract 11010110.111b from 11100101.100b. *[1110.101b]*
3. Add in binary arithmetic 1000111_2, 110011.11b and 11101.101b. *[1011000.011b]*
4. A BCD number is 100101001.01110101, what is its decimal form? *[129.76]*
5. Convert 38.7d to binary form and then to hexadecimal. *[100110.101100110011..., 26.B33...]*
6. Convert 245.6d to binary, octal (base 8) and hexadecimal form *[11110101.1001...,365.46314631.. F5.9999...]*
7. Simplify $F' = A'BC' + ABC' + A'BC + ABC$, $G = (A + B' + C)(A + B + C)(A + B' + C')(A + B + C')$ and $H = (A' + B + C)(A' + B + C')(A' + B' + C)$ using Boolean algebra.
   [F = B', G = A, H = A' + BC]
8. Find the simplest Boolean expression for Y if $Y' = ABC' + AB'C' + A'B'C + ABC + AB'C$, firstly by Boolean algebra and then by using a Karnaugh map. Use de Morgan's law to derive an all-NAND expression, and hence a logic circuit, for Y that uses only 2-input NAND gates. Repeat the last part using only NOR gates.
9. Show by Boolean algebra and truth tables that Equations 20.1 and 20.2 in Section 20.6.1 are the same.
10. Draw up a truth table for the expression

$$f = \overline{BD'}(AB' + C)(A + B'C)$$

then draw a Karnaugh map to find the minimal expression for f. Draw a logic circuit for f which uses only NOR gates or only NAND gates, whichever is simplest.
11. The truth table for a Boolean function, Q, is given in Table P20.11. Write down a SOP expression for Q and simplify by using Boolean algebra. Use a Karnaugh map to find the minimal logical expression for Q. From this draw the logic diagram for Q. Then use de Morgan's law to find an all-NOR expression and a logic circuit for Q that uses only 2-input NOR gates.
12. Draw up a truth table for the logic circuit of Figure P20.12 and then make a Karnaugh map for Q. Hence draw logic circuits which use solely NAND or solely NOR gates to perform the same fun-

ction as the circuit of Figure P20.12.

13. Show that the XOR function obeys the law of association, that is $(A \oplus B) \oplus C = A \oplus (B \oplus C) = (A \oplus C) \oplus B$.

14. Prove that $A \oplus B \oplus B = A$, (a) with a truth table and (b) by Boolean algebra. (This property was used by the German Lorenz teleprinter cypher in WW II, making decypherment by the recipient easy.)

15. Show that the XNOR function (an XOR followed by inversion) obeys the law

$$\overline{(A \oplus B)} \oplus (C \oplus D) = (A \oplus B) \oplus \overline{(C \oplus D)} = \overline{(A \oplus B) \oplus (C \oplus D)}$$

Table P20.11

A	B	C	Q
0	0	0	0
0	0	1	1
0	1	0	0
0	1	1	0
1	0	0	1
1	0	1	1
1	1	0	1
1	1	1	1

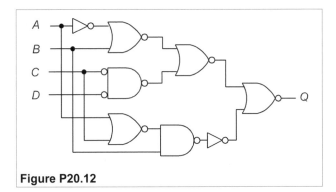

Figure P20.12

16. Show by a truth table and then by Boolean algebra that

$$AB + AC + BC = (A \oplus B)C + AB$$

17. The Boolean burglary squad of a certain police force notices that in one neighbourhood houses are burgled ($B = 1$) at any time when a door is left open ($D = 1$), but only at night ($N = 1$) if a window is left open ($W = 1$), and never if a dog is in the house ($H = 1$). Draw up a truth table and a Karnaugh map for B. Derive a Boolean expression for the burgling function in terms of D, H, N and W, then draw a logic circuit for B using only 2-input NOR gates. $[B = \overline{H}(D + NW)]$

18. The inputs, A and B, in the circuit of Figure P20.18 are either 0 V or +5 V. Make out a truth table for the output, C. What sort of logic gate is it?

19. For the circuit of Figure P20.19, the inputs, A and B, are 0 V or +5 V. Draw up a truth table for the output, C. What sort of logic gate is it?

20. Show that the KM in Figure 20.27c leads to $B_1 = G_1 \oplus G_2 \oplus G_3$.

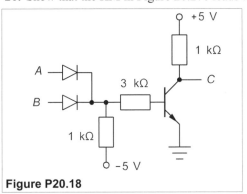

Figure P20.18

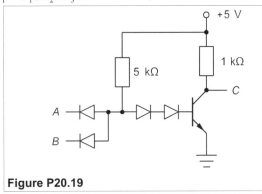

Figure P20.19

21. Deduce a logical function for Q from its truth table, Table P20.21, and implement it with XOR gates only.

22. Draw up the truth table for lighting segment g in the BCD LED of Figure 20.25a and from it deduce the simplest Boolean expression for lighting segment g, and then draw the logic diagram for this. Draw a circuit for lighting segment g which uses only 2-input NAND gates.

Table P20.21

Q	0	1	1	0	1	0	0	1	1	0	0	1	0	1	1	0
A	0	0	0	0	0	0	0	0	1	1	1	1	1	1	1	1
B	0	0	0	0	1	1	1	1	0	0	0	0	1	1	1	1
C	0	0	1	1	0	0	1	1	0	0	1	1	0	0	1	1
D	0	1	0	1	0	1	0	1	0	1	0	1	0	1	0	1

23. Deduce which linkages need to be made in a PLA such as that of Figure 20.40 so that all seven segments of a BCD LED can be selected, assuming there are seven output ORs and sufficient ANDs.

24. Draw up the Karnaugh map for the function, Q, whose truth table is given in Table P20.24. Draw a logic circuit for Q which eliminates the switching hazard and uses (a) fewest 2-input gates and (b) fewest 2-input gates of the same type.

25. The 4-bit XS3 code for the decimal numbers 0–9 in sequence is 0011, 0100, 0101, 0110, 0111, 1000, 1001, 1010, 1011, 1100. Design a converter to turn this into binary.

26. Draw up a Karnaugh map for the function W whose truth table is that of Table P20.26, and hence find a logic circuit for W which uses (a) the smallest number of gates and (b) the smallest number of 2-input gates.

Table P20.24

A	B	C	Q
0	0	0	1
0	0	1	1
0	1	0	1
0	1	1	0
1	0	0	0
1	0	1	1
1	1	0	0
1	1	1	0

Table P20.26

W	1	0	1	1	1	1	1	0	0	1	1	0	1	0	1	1
A	0	0	0	0	0	0	0	0	1	1	1	1	1	1	1	1
B	0	0	0	0	1	1	1	1	0	0	0	0	1	1	1	1
C	0	0	1	1	0	0	1	1	0	0	1	1	0	0	1	1
D	0	1	0	1	0	1	0	1	0	1	0	1	0	1	0	1

27. Design a demultiplexer for the 4-bit multiplexer of Figure 20.37, that is a device which has a single input, B, two control inputs, C_1 and C_2 and four outputs, A_1, A_2, A_3 and A_4.

21 Sequential logic

IN COMBINATIONAL logic the output of a circuit is determined solely by the state of its inputs at any moment, but in sequential logic the output of a circuit depends on previous inputs as well as present ones; a sequential circuit has *memory*. The simplest devices with this property which can be used to build much more complicated circuits are called *flip-flops*. Sequential logic is used to generate from a set of initial input states a sequence of output states. These can be in response to a succession of inputs (asynchronous mode) or clock pulses (synchronous mode). The sequences can be used to store or read information, or to count, and possibly to act as a result of these.

21.1 Unclocked flip-flops

A flip-flop in its simplest form has two outputs, one the complement of the other, which can adopt only a HIGH (or 1) state or a LOW (or 0) state, that is they are *bistable* devices. They are building blocks for many counting and dividing circuits. The basic types are exceedingly cheap – as little as 15p for a dual, plastic, DIL-packaged IC. Unclocked flip-flops change their output states as soon as their input states change appropriately.

21.1.1 The SR (set-reset) flip-flop or latch

Consider the circuit of Figure 21.1 and suppose that the R and S inputs are both LOW or logical 0. Then if $Q = 0$, $Q' = 1$ and vice versa, but we cannot decide from the circuit diagram which state the output will be in because it depends on its former state.

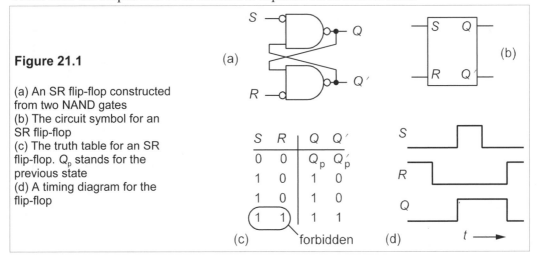

Figure 21.1

(a) An SR flip-flop constructed from two NAND gates
(b) The circuit symbol for an SR flip-flop
(c) The truth table for an SR flip-flop. Q_p stands for the previous state
(d) A timing diagram for the flip-flop

S	R	Q	Q'
0	0	Q_p	Q_p'
1	0	1	0
1	0	1	0
1	1	1	1

forbidden

Now if S is made 1 while R is still 0, whatever the state of Q, it will be set to 1 and Q' to 0. When S is returned to 0, the outputs remain unchanged. And likewise, if R is 1 while S is 0, Q' will be set to 0 and Q to 1. Returning R to 0 leaves the outputs unchanged. The state of the outputs indicates which of R or S was last set to zero and in this sense the circuit has memory. If $R = S = 1$, then $Q = Q' = 1$, not a permissible state since Q can never equal Q'. Instead of NAND gates we could have used NOR gates to build the SR flip-flop, in which case the truth table is slightly different (see Problem 21.1). The SR latch is found in the 74 logic series in 16-pin DIL quad form, numbered 74279.

We can draw up a Karnaugh map (Figure 21.2) for the SR flip-flop, with the forbidden combination $SR = 11$ represented by don't cares ($\times$), and then we see that the output is

$$Q = S + R'Q_\mathrm{p}$$

The output is a function of the previous output, Q_p, as it must be in a sequential circuit.

Figure 21.2

The Karnaugh map (KM) for an SR flip-flop

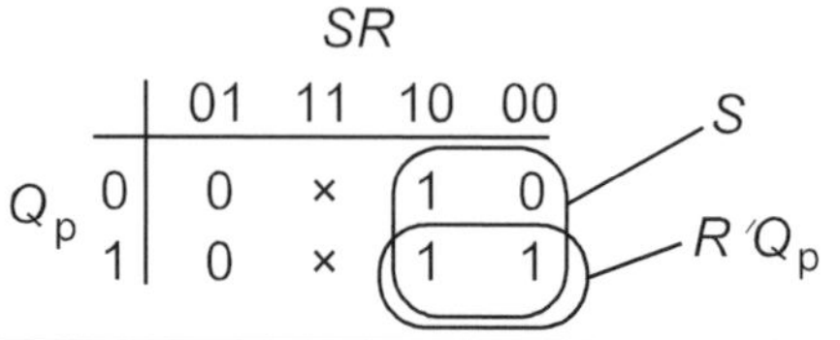

$$
\begin{array}{c|cccc}
 & \multicolumn{4}{c}{SR} \\
 & 01 & 11 & 10 & 00 \\
\hline
Q_\mathrm{p}\ 0 & 0 & \times & 1 & 0 \\
\phantom{Q_\mathrm{p}}\ 1 & 0 & \times & 1 & 1 \\
\end{array}
$$

The SR flip-flop can be used to build other, more complicated, flip-flops; but it can be used on its own, for example, to 'de-bounce' switches. Most switches do not make a single clean transition but many (perhaps ten or more) make-and-break sequences instead as the contact bounces on and off. In many logic circuits the devices used will follow all the make-and-break sequences and the transition from state to state is not clean. If an SR flip-flop is connected as in Figure 21.3, the transitions of Q are clean because once contact is made the flip-flop changes state and stays there when the switch bounces open and closed.

Figure 21.3

Debouncing a switch with an SR flip-flop

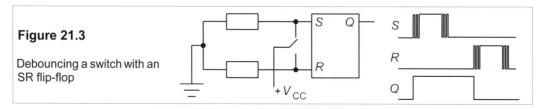

21.1.2 The delay and the data latch

An inverter can be used to prevent the inputs to the SR flip-flop from being in the same state, as in Figure 21.4.

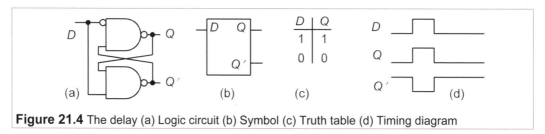

Figure 21.4 The delay (a) Logic circuit (b) Symbol (c) Truth table (d) Timing diagram

This type of flip-flop is called a *delay* or D flip-flop. It has only a single input and its truth table is particularly simple because the input is identical to the output with a small time delay. The delay flip-flop does not latch the output, which merely follows the input. In order to latch the D input, an enabling input, E, must be provided which latches the D input when E is low, as in the circuit of Figure 21.5a, which can be made from four NAND gates and an inverter. Its conventional circuit symbol is shown in Figure 21.5b. The truth table for this latch is shown in Figure 21.5c and a timing diagram in Figure 21.5d. The timing diagram shows that the output is latched when E is low and the flip-flop ignores all further transitions at the D input. Because the output can respond to ('sees') the input when $E = 1$, the device is called a *transparent data latch*.

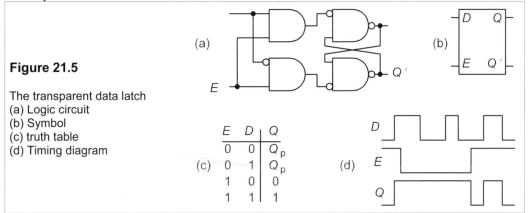

Figure 21.5

The transparent data latch
(a) Logic circuit
(b) Symbol
(c) truth table
(d) Timing diagram

E	D	Q
0	0	Q_p
0	1	Q_p
1	0	0
1	1	1

21.2 Clocked flip-flops

Most flip-flops are *clocked*, that is they change state only when a signal is received from a timing device or clock. This is because it is usually necessary to know precisely when a change of state comes about. The enabled data latch can be clocked by connecting the clock to the enable input, but the result is not altogether satisfactory since the output follows the input when the clock is HIGH. What is required is a device that can change state only on receipt of the correct clock pulse and at no other time. There are two principal types of clocked flip-flop: the D-type and the JK flip-flop.

21.2.1 The D-type master-slave flip-flop

One of the commonest ways of achieving output transitions only on receipt of the correct clock pulses is the master-slave method, because it is tolerant of slowly rising or slowly falling signals (or slow 'edges'). Most common flip-flops are master-slave types and available as cheap ICs. The principles on which they work can be shown with a logic diagram as in Figure 21.6a, which is triggered by negative-going clock edges.

The 'master' and the 'slave' are both SR flip-flops whose inputs are controlled by AND gates like the data latch. When the clock is HIGH the master's inputs are enabled, allowing the data on the D input to be transferred to the master's outputs. However, the slave's inputs are disabled by the inverted clock signal and so its output is latched. When the clock goes LOW the reverse happens: the master's inputs are disabled while the slave's inputs are enabled. Thus the master remains latched in its previous state while the slave's outputs become the same as the master's. The timing diagram of Figure 21.6c shows this (*M* is the master's output). Because data is transferred to the output only on receipt of a negative-going clock input, the device is negative-edge triggered, as indicated by the arrow on the clock waveform. Note that the overall delay in transferring data from input to output can be as much as 1.5 clock cycles. The circuit symbol for this device is shown in Figure 21.6b, the circle on the clock input indicating an active negative-going transition and the wedge inside the clock input indicating edge triggering.

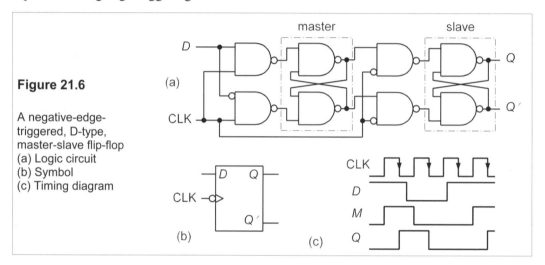

Figure 21.6

A negative-edge-triggered, D-type, master-slave flip-flop
(a) Logic circuit
(b) Symbol
(c) Timing diagram

The D-type flip-flop is available in many forms, such as the positive-edge-triggered 7474 with preset (PR) and clear (CR) inputs, shown in Figure 21.7. This advanced, low-power, Schottky TTL IC contains two D-types in a 14-pin DIL package and costs just 40p. Many data books are now using the American IEEE Standard 91-1984 as outlined in IEC Publication 617-12, which indicates negative-going and active-low transitions with a flag, as shown in Figure 21.7b. The circles (or flags in IEEE symbols) on the preset and clear inputs indicate that they operate when LOW and would normally be held HIGH at $+V_{CC}$. In Figure 21.7c active-low pins are primed: CR′.

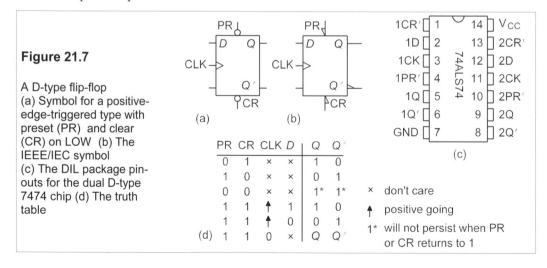

Figure 21.7

A D-type flip-flop
(a) Symbol for a positive-edge-triggered type with preset (PR) and clear (CR) on LOW (b) The IEEE/IEC symbol
(c) The DIL package pin-outs for the dual D-type 7474 chip (d) The truth table

PR	CR	CLK	D	Q	Q′
0	1	×	×	1	0
1	0	×	×	0	1
0	0	×	×	1*	1*
1	1	↑	1	1	0
1	1	↑	0	0	1
1	1	0	×	Q	Q′

(d)

×   don't care

↑   positive going

1* will not persist when PR or CR returns to 1

21.2.2 The JK flip-flop

Though the JK flip-flop is also of the master-slave type, it differs from the D-type by having two data inputs (apart from preset, clear and clock). Its circuit symbol and truth table are shown in Figure 21.8.

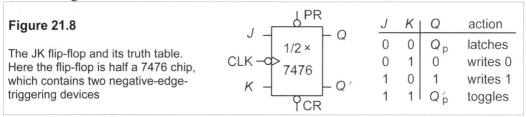

Figure 21.8

The JK flip-flop and its truth table. Here the flip-flop is half a 7476 chip, which contains two negative-edge-triggering devices

J	K	Q	action
0	0	Q_p	latches
0	1	0	writes 0
1	0	1	writes 1
1	1	Q'_p	toggles

The JK flip-flop comes in various IC packages. The one shown in Figure 21.8 is the 7476, a dual device (two flip-flops in the package) with negative-edge triggering and active-low preset (PR) and clear (CR). Thus if PR = 0 and CR = 1, $Q = 1$; and if PR = 1 and CR = 0, $Q = 0$, regardless of the clock or what J or K are; preset and clear over-ride all other inputs. In normal operation with CR and PR tied HIGH, the output is latched when $J = K = 0$. When $J = K'$ then on receipt of a negative-going clock pulse $Q = J$ and $Q' = K$. The device *toggles* when $J = K = 1$, that is the output inverts on receipt of a negative-going clock pulse. T-type flip-flops (T for toggle) have just one data input and can be made from a JK flip-flop by connecting the J and K inputs to $+V_{CC}$. A D-type flip-flop can be made from a JK flip-flop by connecting J' to K. Thus the JK flip-flop is more versatile than either the D or the T flip-flop.

21.3 Counters and shift registers

Among other things, flip-flops can be arranged to count, to divide frequencies and to shift binary numbers. Many of the circuits below are available cheaply as IC packages in the 74 series.

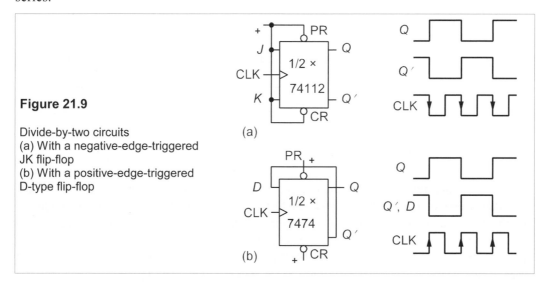

Figure 21.9

Divide-by-two circuits
(a) With a negative-edge-triggered JK flip-flop
(b) With a positive-edge-triggered D-type flip-flop

21.3.1 Divide-by-two circuits

The JK flip-flop can be used in the divide-by-two circuit of Figure 21.9a. Since the flip-flop is triggered by only negative clock edges, the output will toggle at the end of each clock pulse if $J = K = 1$ and preset and clear are held HIGH (indicated by the $+$ sign on these inputs). A D-type flip-flop will do the same if the Q' output is connected to the D input as in Figure 21.9b. On receipt of the first clock pulse Q goes HIGH, while D and Q' go LOW. The next clock pulse transfers D to Q, so Q goes LOW and the frequency of Q is half that of the clock.

21.3.2 Ripple counters

A binary up-counter counts up on receipt of a pulse and stores each bit of the count in a single flip-flop. The first stage of the counter registers the least significant bit (LSB) of the count and successive stages must count at half the rate of the previous stage. By connecting n divide-by-two circuits in series, n-bit ripple counters can be constructed. These count up from 0 to $2^n - 1$, so a 4-bit counter counts from 0 to 15d or from 0000 to 1111b. Though the clock inputs are used in ripple counters they are nevertheless *unclocked* or *asynchronous* counters, because the count is not changed on receipt of a single clock pulse, but ripples through all the flip-flops.

  Figure 21.10 shows a 3-bit ripple counter made from JK flip-flops such as contained in the 74HCT112 dual, high-speed, CMOS, TTL-compatible, 16-pin, plastic DIL-packaged IC. These are negative edge triggered devices that are arranged with $J = K = 1$ so they toggle at the end of a positive-going pulse. The timing diagrams show the counting sequence. When the output is taken from Q' instead of Q the counter counts down instead of up.

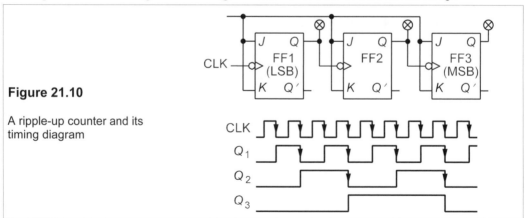

Figure 21.10

A ripple-up counter and its timing diagram

21.3.3 The synchronous binary counter

The ripple counter responds to an input at the first stage which is then passed down through each stage of the counter with a delay at each stage. In a clocked or synchronous counter all the flip-flops change simultaneously on receipt of a clock pulse and so not only do delays not build up in multi-stage counters, but the exact timing of the change is known. A synchronous binary counter is best constructed with JK flip-flops that are arranged to toggle at the right times. J and K inputs of each flip-flop are connected together and toggling occurs when $J =$

$K = 1$. Control logic must be designed for each stage, so the state of the counter after each clock pulse, Table 21.1, must be examined. The clock count is given on the top and the corresponding states of the outputs underneath. Toggling on the *next* clock pulse is indicated with a T.

Table 21.1 *The counting sequence for a 4-bit, binary up-counter*

	0	1	2	3	4	5	6	7	8	9	10	11	12	13	14	15
Q_1	0	1	0	1	0	1	0	1	0	1	0	1	0	1	0	1
Q_2	0	0T	1	1T	0	0T	1	1T	0	0T	1	1T	0	0T	1	1T
Q_3	0	0	0	0T	1	1	1	1T	0	0	0	0T	1	1	1	1T
Q_4	0	0	0	0	0	0	0	0T	1	1	1	1	1	1	1	1T

Looking at the Ts of Q_2 and comparing them with the state of Q_1 we can see that if Q_1 is connected to J_2 (and K_2) correct toggling will result. The Ts in column 3 indicate that Q_3 must toggle only when $Q_1 = Q_2 = 1$ and not at any other time, that is an AND function: $J_3 = Q_1 Q_2$. Likewise Q_4 must toggle when $Q_1 = Q_2 = Q_3$ and at no other time, that is $J_4 = Q_1 Q_2 Q_3$. The control logic is two AND gates between stages 2 and 3, and 3 and 4, as in Figure 21.11. The 4-bit synchronous (or clocked) binary counter is available as an IC, such as the 74LVC161 (low-voltage CMOS) which costs only 29p.

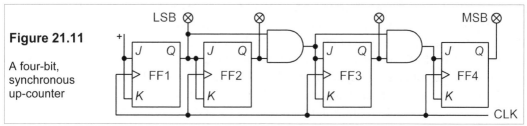

Figure 21.11

A four-bit, synchronous up-counter

21.3.4 A bi-directional (up/down) counter

The same counter that was used for counting up can be used to count down; all that is needed is some additional logic circuitry. A 4-bit binary up/down counter can be derived from the synchronous 4-bit binary counter of Figure 21.11 by adding extra gates to control the direction of the count. Looking at the counting sequence of Table 21.1, we see that the second stage will toggle correctly whether counting up or down. But for counting down the third and fourth stages must now toggle when $Q_1 = Q_2 = 0$ and $Q_1 = Q_2 = Q_3 = 0$ and not at any other time, that is we require

$$J_3 = Q_1' Q_2' \quad \text{and} \quad J_4 = Q_1' Q_2' Q_3'$$

Table 21.2

Truth table for a controlled inverter, in which F controls A to give B. It is identical to that of an XOR

F	A	B
0	0	0
0	1	1
1	0	1
1	1	0

An XOR gate can be used as a controlled inverter as can be seen from Table 21.2. Then F is connected to LOW, the XOR output is the same as the other input ($B = A$). But when F is connected to HIGH, the output is the complement of the other input ($B = A$ ′).To make an up/down counter XOR gates must be inserted after Q_1, Q_2 and Q_3 to invert them for counting down. The outputs of the XOR gates are ANDed and the AND outputs connected to the JK inputs as in Figure 21.12. The up/down counter is available as an IC, for example the 74LS191 (Schottky TTL) costs about 50p.

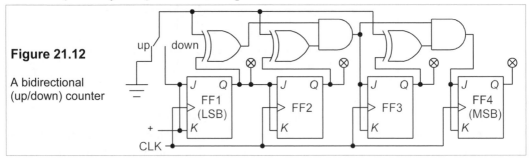

Figure 21.12

A bidirectional (up/down) counter

21.3.5 A decade counter

A counter that counts from 0 to 9 in binary and then resets is sometimes called a decade counter and sometimes a modulo-10 counter. It must count through the binary numbers 0000, 0001, ... 1001 and then reset to 0000. One might think the easiest way of doing this would be to detect 1010 (10d) on the counter by ANDing stages two and four, then resetting the counters almost instantaneously with the AND's output. This is bad practice, as the count of 1010 appears only momentarily and may not last long enough to ensure resetting. If it does last long enough to reset all the counters it may be passed on to the next decade counter or some other device and produce a wrong count or other incorrect action. Therefore we need some logic circuitry to produce the correct counting sequence on each clock pulse without using preset or clear.

Table 21.3 gives the counting sequence, showing where toggling on the next clock pulse must take place.

Table 21.3

The counting sequence for a decade counter.

	Q_1	Q_2	Q_3	Q_4	J_2	J_3	J_4
1	0	0	0	0	0	0	0
2	1	0T	0	0	1	0	0
3	0	1	0	0	0	0	0
4	1	1T	0T	0	1	1	0
5	0	0	1	0	0	0	0
6	1	0T	1	0	1	0	0
7	0	1	1	0	0	0	0
8	1	1T	1T	0T	1	1	1
9	0	0	0	1	0	0	0
10	1	0	0	1T	0	0	1

The T's are omitted in the first column since the output of FF1, Q_1, must toggle every clock pulse, and we connect J_1 (and K_1) to logical 1 to achieve this. The states required for

J_2, J_3 and J_4 are also shown in Table 21.3, with ones written at all the toggling places, that is wherever there are Ts on the corresponding outputs, and zeroes elsewhere. We next draw up Karnaugh maps for J_2 etc. as shown in Figure 21.13, remembering to put $\times$ for the unused (don't care) states (0101 to 1111).

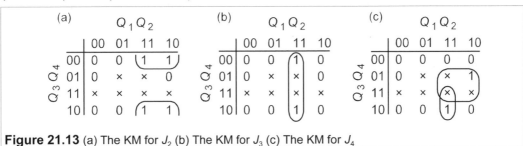

Figure 21.13 (a) The KM for J_2 (b) The KM for J_3 (c) The KM for J_4

From Figure 21.13a we deduce $J_2 = Q_1 Q'_4$, from Figure 21.13b that $J_3 = Q_1 Q_2$ and from Figure 21.13c that $J_4 = Q_1 Q_4 + Q_1 Q_2 Q_3$. Thus the circuit for the complete decade counter is that of Figure 21.14, which requires five 2-input gates.

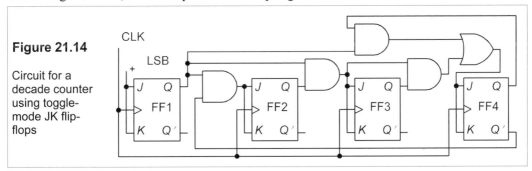

Figure 21.14

Circuit for a decade counter using toggle-mode JK flip-flops

21.3.6 Designing counters

With the help of some combinational logic it is possible to build counters to count through any arbitrary sequence until the pattern repeats. The principles are explained in the following examples.

Example 21.1

Design a synchronous (clocked) counter with D-type flip-flops that will go through the sequence 00, 11, 10, 01, 00 etc.

(a)   (b)

Table 21.4

(a) Successive states of the flip-flops: the outputs are one clock pulse behind the inputs
(b) The KM for D_2

	D_1	Q_1	D_2	Q_2
1	1	0	1	0
2	0	1	1	1
3	1	0	0	1
4	0	1	0	0
5		0		0

		Q_1	
		0	1
Q_2	0	1	0
	1	0	1

We first write down the output sequence and the D inputs that will produce the *next* state on clocking as indicated in Table 21.4a, which also shows that $D_1 = Q_1'$. We cannot see easily what function of Q_1 and Q_2 is needed to give D_2, so we must draw a Karnaugh map as in Table 21.4b. We cannot group any of the entries in this and $D_2 = Q_1'Q_2' + Q_1Q_2$, This is the inverse XOR function (or XNOR), which can be obtained by inverting either of the inputs of an XOR (or its output, see Problem 20.14), that is $D_2 = Q_1 \oplus Q_2'$ (or $D_2 = Q_1' \oplus Q_2$). The required circuit is that of Figure 21.15.

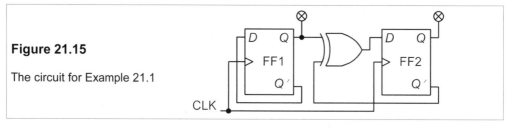

Figure 21.15

The circuit for Example 21.1

The next example uses JK flip-flops and is a little more complicated because more flip-flops are required and because the JK flip-flop has two inputs to be controlled, which while making it more versatile, also makes it more difficult to design with.

Example 21.2

Design a counter with JK flip-flops which will generate the sequence 000, 010, 100, 110, 011, 001, ... etc.

Table 21.5

(a) JK values for stated transitions (b) Output sequence (c) JK values required for (b)

(a)

Transition	JK
0 → 0	0×
0 → 1	1×
1 → 1	×0
1 → 0	×1

(b)

	Q_1	Q_2	Q_3
1	0	0	1
2	0	1	0
3	1	0	0
4	1	1	0
5	1	1	1
1	0	0	1	etc.

(c)

	Q_1	JK_1	Q_2	JK_2	Q_3	JK_3
1	0	0×	0	1×	1	×1
2	0	1×	1	×1	0	0×
3	1	×0	0	1×	0	0×
4	1	×0	1	×0	0	1×
5	1	×1	1	×1	1	×0
1	0		0		1	

The JK flip-flop obeys the rules given in the Table in Figure 21.8, which are summarized in Table 21.5a. The × entries are don't cares as usual. The transition from 0 to 0 can be achieved by latching, that is $JK = 00$, or by writing 0, that is $JK = 01$, hence $JK = 0\times$ will serve. In the same way one can derive the requirements for the other three transitions. Table 21.5b shows the output sequence for clock pulses 1–5, at which point the sequence repeats. Table 21.5c shows the JK_n $(= J_nK_n)$ inputs needed to give the next state of Q_n according to Table 21.5a.

Figure 21.16

KMs for Example 21.2
(a) For J_2 (b) For K_2

(a)

$$Q_1 Q_2$$

Q_3	00	01	11	10
0	×	×	×	1
1	1	×	×	×

(b)

$$Q_1 Q_2$$

Q_1	00	01	11	10
0	×	1	0	×
1	×	×	1	×

Now we must draw up the Karnaugh maps for the six inputs. Figure 21.16a shows that for J_2 and Figure 21.16b that for K_2, giving $J_2 = 1$ and $K_2 = Q_1' + Q_3$. In these KMs the unused sequence states 000, 011, 110 are also entered as don't cares. The reader may care to confirm that $J_1 = Q_3'$, $K_1 = Q_3$, $J_3 = Q_1Q_2$ and $K_3 = Q_1'$ and that the circuit required for the sequence generator is that of Figure 21.17.

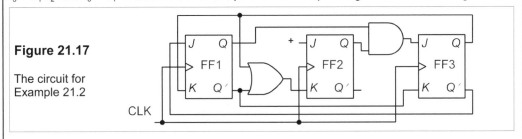

Figure 21.17

The circuit for
Example 21.2

A word of caution here! The unused sequence states 000, 011, 110 may accidentally appear on power up, and it is possible that they will cycle round and the desired sequence then never appears. The designer must check that this does not happen, and if it does he should put in additional circuitry to avoid it. It is sensible to ensure the circuit starts in one of the desired states by presetting the flip-flops. The reader should verify that the circuit of Figure 21.17 will always revert to the desired sequence no matter what the starting point.

21.3.7 Shift registers

We have noted in the section on binary arithmetic that multiplication and division by two can be achieved by shifting a binary number to the left or right respectively. Shift registers can also be used to convert parallel data to serial data and vice versa. A series of D-type flip-flops with outputs connected to inputs as in Figure 21.18 will shift data along one gate at each clock pulse.

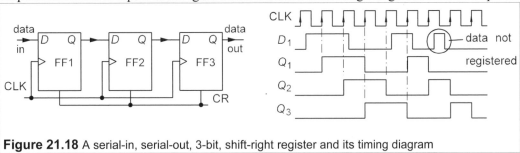

Figure 21.18 A serial-in, serial-out, 3-bit, shift-right register and its timing diagram

The data is read in serially on receipt of a positive-going clock pulse, at which point Q_1 goes high if D_1 is high and stays high until D_1 is low when a positive-going clock pulse is received. If D_1 is high for a time in which no positive-going clock pulses occur, the data is not read into the register. Data on Q_1 is shifted into Q_2 on the next clock pulse and eventually appears as serial data at the output as shown in the timing diagram.

A 3-bit, parallel-input/serial-output (PISO) shift register is shown in Figure 21.19, which can also be used as a serial-in/serial-out (SISO) shift register. The data is loaded in parallel on receipt of a clock pulse when the SH/LD′ is LOW and is shifted right on receipt of a clock pulse when the SH/LD′ is HIGH. Serial data can be read into stage one and shifted too. The serial output comes from the output of stage 3. The CR signal will override all others and set all the flip-flops to zero. Shift registers are available in IC form, for example the 74166 is an

8-bit parallel/serial-to-serial shift register and costs only 30p in 100+ quantities in a 16-pin DIL plastic package.

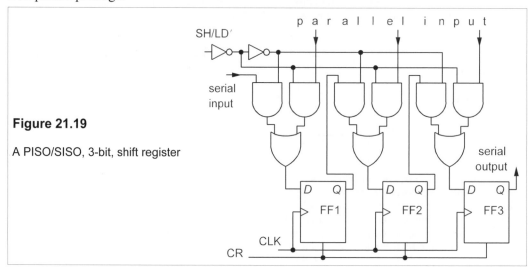

Figure 21.19

A PISO/SISO, 3-bit, shift register

21.4 **The monostable multivibrator**

Whereas the flip-flop is a bistable multivibrator with two stable output states, the monostable multivibrator (or just *monostable*) has only one: HIGH or LOW. This might seem to restrict the usefulness of the device, but in fact the monostable is widely used to produce a single pulse on triggering, hence its other nickname: *one shot*. The output pulse width can be altered by varying the time constant of an external RC network.

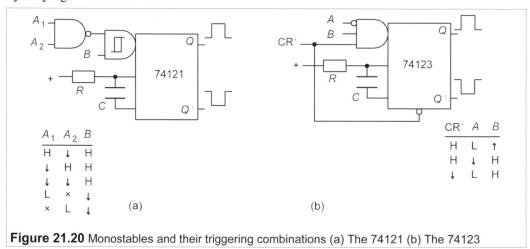

Figure 21.20 Monostables and their triggering combinations (a) The 74121 (b) The 74123

Although a monostable can be constructed from an op amp nobody does so because they are readily available as cheap ICs with all kinds of trigger inputs. Examples are the 74121 and 74123 (retriggerable with clear), though there are plenty more, varying in price from about 20p upwards. Figure 21.20 shows the input (triggering) logic for these devices and the combinations which result in pulse outputs. The non-retriggerable nature of the 74121 means that it will ignore further valid trigger inputs while it is outputting a pulse. The 74123 will accept these, with consequent pulse lengthening. Sometimes one requires retriggerability and sometimes one does not. The pulse width is given by $T = RC \ln 2 \approx 0.7RC$, where $2 \text{ k}\Omega \leq R \leq 40 \text{ k}\Omega$ and $10 \text{ pF} \leq C \leq 10 \text{ \textmu F}$, implying that it can be fairly accurately controlled over the range 14 ns to 0.4 s, which is more than 10 decades.

21.5 **Timers**

Timers such as the cheap and popular 555 (an IC introduced as long ago as 1972) use a pair of comparators, a flip-flop and a transistor to generate a pulse (monostable mode) or pulse train (astable mode) whose length and duty cycle are controlled by an external RC network. The circuit diagram for the 555 bipolar IC is shown in Figure 21.21a and the external RC network in Figure 21.21b; the pin outs of the 8-pin DIL package are shown in Figure 21.21c.

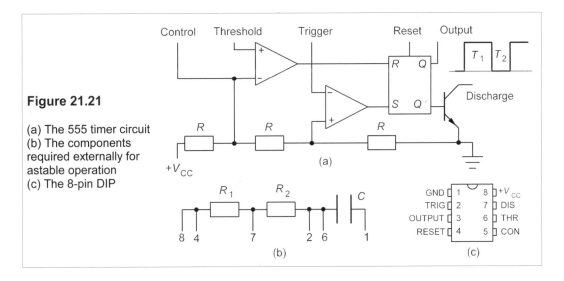

Figure 21.21

(a) The 555 timer circuit
(b) The components required externally for astable operation
(c) The 8-pin DIP

The three resistances, R, provide the threshold voltage of $\frac{2}{3}V_{CC}$ and the trigger voltage of $\frac{1}{3}V_{CC}$. When any voltage $< \frac{1}{3}V_{CC}$ is applied to the trigger terminal, the trigger comparator goes high ($\approx V_{CC}$) and sets the flip-flop (that is Q goes high and Q' low). The output transistor then turns off and the external capacitor, C, charges up to $\frac{2}{3}V_{CC}$ at which point the threshold comparator resets the flip-flop (that is Q goes low and Q' high). This turns on the output transistor and the capacitor discharges through R_2. The capacitor charges through R_1 and R_2 and so the voltage across C is

$$v_C(t) = V_{CC} - \frac{2}{3}V_{CC}\exp\left(\frac{-t}{C(R_1 + R_2)}\right)$$

because at $t = 0$ (the commencement of charging), $v_C = \frac{1}{3}V_{CC}$. If the output is high until v_C reaches $\frac{2}{3}V_{CC}$, which is at time $t = T_1$, then

$$\tfrac{2}{3}V_{CC} = V_{CC} - \tfrac{2}{3}V_{CC}\exp\left(\frac{-T_1}{C(R_1+R_2)}\right) \quad \Rightarrow \quad T_1 = C(R_1+R_2)\ln 2$$

During discharge through R_2, the capacitor voltage drops from $\tfrac{2}{3}V_{CC}$ to $\tfrac{1}{3}V_{CC}$, during which time, T_2, the output is low (that is ≈ 0 V). The capacitor's voltage during discharge is given by

$$v_C(t) = \tfrac{2}{3}V_{CC}\exp\left(-t/CR_2\right)$$

And so $T_2 = CR_2\ln 2$.

Thus the frequency of the output pulse train is

$$f = \frac{1}{T_1 + T_2} = \frac{1}{C(R_1 + 2R_2)\ln 2} \approx \frac{1.4}{C(R_1 + 2R_2)}$$

When R_1 and R_2 lie between 1 kΩ and 50 kΩ and $C = 0.1$ μF the timing error is about 2.25%. The frequencies of best accuracy lie between 100 Hz and 5 kHz, but the extreme range is from 0.2 Hz to 500 kHz.

21.5.1 The mark-space ratio

The mark-space ratio, $T_1/T_2 = 1 + R_1/R_2 > 1$, during operation as described above. However, R_2 may be short-circuited with a diode as in Figure 21.22, and then the mark-space ratio can be found as follows.

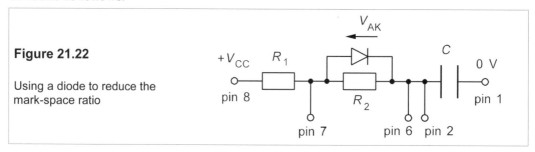

Figure 21.22

Using a diode to reduce the mark-space ratio

Looking at Figure 21.22 we see that during discharge R_2 operates as normal and the discharge time is unchanged at $T_2 = CR_2 \ln 2$. But during charging the capacitor charges according to

$$v_C(t) = A\exp\left(-t/CR_1\right) + B$$

since the charging current now passes only through R_1. Once more A and B are constants to find. When $t = \infty$ the voltage across C would be $V_{CC} - V_{AK}$, so that $B = V_{CC} - V_{AK}$. And when $t = 0$, $v_C = \tfrac{1}{3}V_{CC}$, making $A = -\tfrac{2}{3}V_{CC} + V_{AK}$. Hence

$$v_C(t) = (V_{AK} - \tfrac{2}{3}V_{CC})\exp\left(-t/CR_1\right) + V_{CC} - V_{AK}$$

and this must be equal to $\tfrac{2}{3}V_{CC}$ at $t = T_1$. Substituting $V_C = \tfrac{2}{3}V_{CC}$ and $t = T_1$ into the equation above gives

$$T_1 = CR_1 \ln\left(\frac{2V_{CC} - 3V_{AK}}{V_{CC} - 3V_{AK}}\right)$$

Problem 21.5 indicates how the mark-space ratio can be changed in this way.

21.5.2 The control pin 5

When a voltage, $V_{CON} < V_{CC}$, is applied to the control pin the thresholds of the comparators become V_{CON} and $\frac{1}{2}V_{CON}$ instead of $\frac{2}{3}V_{CC}$ and $\frac{1}{3}V_{CC}$. Then the capacitor charges from $\frac{1}{2}V_{CON}$ to V_{CON} and discharges from V_{CON} to $\frac{1}{2}V_{CON}$. The charging time is found from

$$v_C(t) = A\exp\left[-t/C(R_1 + R_2)\right] + B$$

Now $v_C(t) = V_{CC}$ when $t = \infty$ and $v_C(t) = \frac{1}{2}V_{CON}$ when $t = 0$, so that $B = V_{CC}$ and $A = \frac{1}{2}V_{CON} - V_{CC}$. Then when $t = T_{CON1}$, $V_C = V_{CON}$, so that

$$T_{CON1} = C(R_1 + R_2)\ln\left(\frac{V_{CC} - \frac{1}{2}V_{CON}}{V_{CC} - V_{CON}}\right)$$

And the discharge time remains at $T_2 = CR_2\ln 2$. We can therefore also vary the mark-space ratio using the control pin. However, the minimum value permitted for V_{CON} is $0.35V_{CC}$, making the minimum value of $T_{CON1} = 0.24C(R_1 + R_2)$. It is best to derive V_{CON} from V_{CC} so that variations in supply voltage will not affect the timing.

Though cheap in itself (about 25p), the 555 timer requires a good low-leakage capacitor for reasonably accurate performance and these are relatively expensive when values are above 1–2 µF. It is not well-suited to long timing intervals.

Suggestions for further reading (Chapters 20 and 21)

Introduction to digital electronics by J Crowe and B Hayes-Gill (Butterworth-Heinemann, 1998)
Digital electronics by J Carr (Prompt, 1999) Succinct
Digital electronics by D Green (Longman, 5th ed. 1998)
Digital electronics by R L Tokheim (McGraw-Hill, 5th ed., 1999)
Digital fundamentals by T L Floyd (Prentice Hall, 7th ed., 1999) Clear, but expensive and long.
Basic digital electronics by A J Evans (Prompt, 1997). Inexpensive.
The essence of digital design by B Wilkinson (Prentice-Hall, 1997). Not too long. Inexpensive.

Problems

1. The D-type flip-flops in the synchronous circuit of Figure P21.1 are initially set to 1. What will the subsequent output states Q_1 and Q_2 be at successive clocked transitions? Draw the timing diagram. Will the repetitive sequence eventually be any different if the flip-flops are initially reset to $Q_1 = 0$ and $Q_2 = 0$?
2. Design a synchronous counter with D-type flip flops that cycles through the states 111, 011, 101, 010, 110, 000, 111 etc. Will the circuit hang up if the initial state is 100 or 001? (Neither is in the sequence and can be written as don't cares in the Karnaugh map.)
3. The JK flip-flops in Figure P21.3 are initially all set by making $S = 1$. What will be the successive

states of each of the flip-flops as clock pulses are subsequently received?

4. The D-type flip-flops of Figure P21.4 are initially set to 1, then the clock is started. Sketch the output at Q_2.

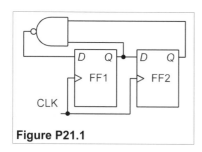

Figure P21.1

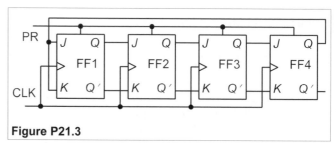

Figure P21.3

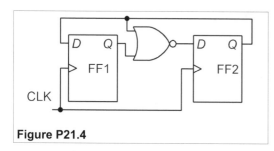

Figure P21.4

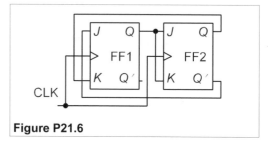

Figure P21.6

5. Show that the output of the 555 timer in astable mode is HIGH for a time $T_1 \approx 0.7(R_A + R_B)$ and that it is LOW for a time $T_2 \approx 0.7R_BC$.

6. The outputs of the JK flip-flops in Figure P21.6 are initially set to 1, then they are clocked. Draw the timing diagram for Q_1 and Q_2. What does the circuit do? Will the circuit get stuck in one state if the initial conditions are altered? Design another circuit using JK flip-flops and no logic gates which will have the same output at Q_2. Will it work no matter what the initial values of Q_1 and Q_2 are?

7. The counter of Figure 21.14 uses XOR gates to count up or down as required. Design an up/down counter using the Q and Q' outputs of the flip-flops and no XORs.

8. Design a synchronous, 3-bit, binary up-counter using D-type flip-flops.

9. Design a divide-by-five counter using D-type flip-flops. Repeat using JK flip-flops.

10. The shift register of Figure P21.10 feeds back the output of flip-flops FF3 and FF4 via an XOR gate and thus generates a pseudo-random-number sequence. What is the sequence if the starting state is 1111? If there are 10 flip-flops, what is the longest possible sequence before repetition must occur?

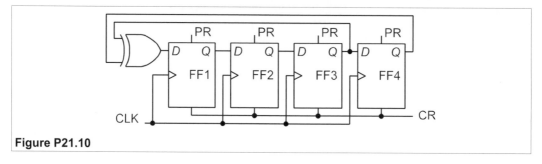

Figure P21.10

22 Computers

CALCULATING machines are as old as history, and punched-card control of machines dates back to the early eighteenth century when they were used in France to control looms used for weaving elaborate patterns. In England, Charles Babbage[1] produced designs for a mechanical computer that were too far in advance of technology to produce a working machine. Babbage's proposed 'analytical engine' of c. 1840 had an arithmetic unit and a memory and used punched cards, not only for input and output, but also to store programs capable of performing iterative calculations with conditional branching. It was only when the Mark I computer at Manchester University began working in 1948 that Babbage's vision became reality. There is not time enough nor the space to go into the origins of the computer; those wishing to can try *The Origins of Digital Computers*, edited by B Randell, 3rd ed., Springer-Verlag (1982).

Hollerith used punched cards to tabulate the data from the US census of 1890 and a tabulator company was formed in 1896, which eventually became the International Business Machine (IBM) corporation in 1924. By about 1900 an adding facility was incorporated into the Hollerith tabulator to produce a punched-card-controlled calculator. Telephone relays were used in various calculating machines during the 1930s, but were eventually superseded by much faster thermionic valves (or tubes). However, the unreliability of vacuum tubes hampered any machine incorporating more than a few.

The beginning of the computer age can be dated to 1937 when a very remarkable paper[2] was published by a young English mathematician, A M Turing[3]. The paper presented a solution to a difficult problem in pure mathematics, though as fate decreed it had been solved by several others independently at about the same time. But what stayed forever in the minds of those that read it, was the technique by which Turing reached his conclusions – the conception of a universal computer, known as the Turing machine. Turing stated: 'It is possible to invent a single machine which can be used to compute any computable sequence'. Moreover any other machine could be emulated by his universal machine. It appears that Turing saw no essential difference between the calculating processes of a human computer and those of his hypothetical machine and that it would be soon be possible to build machines that were, by all objective criteria, *intelligent*. To Turing the manipulation of numbers and the manipulation of non-numerical data were similar: a computing machine would play chess and write music as well as perform calculations accurately at lightning speed. This is the crucial distinction between a computer and a mere calculating machine.

---

[1] Charles Babbage (1792–1871). Lucasian Professor of Mathematics at Cambridge, 1828–39. M V Wilkes claims that Babbage invented microprogramming well over a hundred years before Wilkes himself reinvented it.

[2] 'On computable numbers, with an application to the Entscheidungsproblem', *Proc. Lond. Math. Soc.* **42**(2), 230–265 (1936–7). The paper was received in May, 1936. Turing was 23 when he wrote it.

[3] Alan Mathison Turing (1912–1954). BA in Mathematics 1934. Fellow of King's College, Cambridge 1935. PhD Princeton 1938. From 1939–1945 played an important part in breaking German cyphers, among them Enigma. Turing became Reader at Manchester University in 1949 where he worked with Tom Kilburn, dying in controversial circumstances. *Alan Turing: The Enigma*, by Hodges (Burnett Books ,1983) is a good biography.

Turing's conception of a universal machine led a Bletchley Park ('Station X') mathematician, Max Newman, to design a Boolean computer to help crack the most difficult German Lorenz cypher. The first machine was made of electromechanical telephone switching components and only worked for short periods. T H Flowers (1905-98), a Post Office engineer, took the design and transformed it into a working programmable computer within ten months, delivering Colossus to Bletchley Park in December 1943, where it was an instant success. Colossus contained 1500 electronic valves and used an optical, punched-paper-tape reader to read 5000 characters (not bits) per second, and is now considered to be the first programmable electronic computer. Flower destroyed the designs for the Colossi immediately after the war, on the orders of the Government, and most of the computers were destroyed. Flowers stayed in the PO Research Centre at Dollis Hill until he retired in 1965. He received £1000 and the MBE for his war work.

The first general-purpose computer to perform arithmetic and logical operations was ENIAC, built by Eckert and Mauchly at the University of Pennsylvania between 1943 and 1946. The number of electronic valves (19000) was such that under normal conditions a failure would have occurred long before any useful computation could have been completed. By careful selection of the valves and by using them at minimal current the mean time between failures was increased enormously and computation became practical.

The first stored-program-controlled computer was designed and built at the University of Manchester by F C Williams and T Kilburn[4]. It had a random-access store using cathode-ray tubes and ran its first program on 21st June 1948. The design was manufactured commercially by Ferranti (with considerable help from Kilburn) as the Mark I computer. The first of these was delivered to a customer site in February 1952, beating UNIVAC in the USA by about five months. Turing, who had moved to the University of Manchester in September 1948, was one of the first users of the computer and also wrote the first programming manual. In parallel with the work at Manchester, a group at Cambridge under M V Wilkes[5], designed the EDSAC I computer, using mercury delay lines as the storage medium. The system was completed in 1949 and was reliable enough to be used by non-specialists, who could be taught programming techniques. The first commercial manifestation of EDSAC was the Leo ('Lyons Electronic Office') computer system of December 1951. EDSAC I was retired in July, 1958 (and unfortunately destroyed), though by then EDSAC II had been in service some time. EDSAC II ceased working in 1965. From the mid 1950s the USA assumed a position of overwhelming dominance in computers, particularly in the business and commercial field, where IBM had a near monopoly in spite of anti-trust legislation and protracted litigation. Computer development was greatly assisted by the invention of the transistor and later the integrated circuit, leading to decreased size and increased reliability and speed.

---

[4] Kilburn died aged 79 in 2000. He founded Manchester University's computer science department and was its first professor. He never owned a computer.

[5] Wilkes took a degree in mathematics at Cambridge the same year that Turing did. His very readable autobiography, *Memoirs of a Computer Pioneer* (MIT Press), came out in 1985.

22.1 **Computer architecture**

The way in which the component parts of a computer (the hardware) interact with each other is called computer *architecture*. Computers require input and output ports, a *central processor unit* (CPU) and a memory, as Figure 22.1 shows. The CPU carries out all the arithmetic, logic and control operations and is the fastest and most complex part of the hardware. The memory in Figure 22.1 is shown in two parts, but often the secondary memory is itself split into several levels. The main or primary memory gives read and write access to random addresses. It is the fastest level, but also the most expensive and most limited in capacity. The secondary memory is generally divided into blocks of several thousand *bytes* (nearly always a group of 8 bits) with transfer of blocks between primary and secondary memories being managed by the CPU. The first level of secondary memory is normally magnetic discs and the second level magnetic tape. Secondary memory is slower, cheaper and of much higher capacity than primary memory. One of the most important tasks of a computer designer is to arrange for the fast memory to be used most effectively as the program is executed.

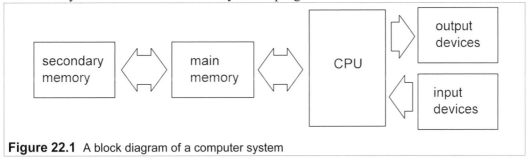

Figure 22.1 A block diagram of a computer system

Input and output devices, or *peripherals*, are attached to the input and output ports. In some cases, such as a disc drive, the input and output device is the same, whereas others can only be used for input (such as a keyboard) or output (a printer, for example). Magnetic tape and disc can serve as memory (though of very slow access time) as well as input and output devices. The CPU must monitor the input port continually for *interrupt* or *wait* commands and from time to time send data to the output port as instructed by the program or the input port.

The memory stores two types of information: data and instructions. The computer carries out a sequence of instructions that forms a program. The data are manipulated in some way by the program to produce the output data required. The program may be to carry out a complex mathematical operation or may be simply for sorting a list into alphabetical order. The CPU must be able to select locations, or *addresses*, in memory and write to them or read and decode the information in them.

22.2 **The CPU**

The architecture of a CPU might look like that of Figure 22.2. The CPU comprises four main parts each with associated *registers*, which are special-purpose stores containing one or two bytes:

    1. The control unit (CU) + registers (PC, IR)

2. The arithmetic and logic unit (ALU) + registers (Fn, In1, In2, Out, SR)
3. The memory + registers (MAR, MBR)
4. A set of fast registers (say 16) generally called R registers + registers (RAR, RBR)

It is now normal practice to make all register instruction and *operand* (what the instruction operates on) lengths, and all other dimensions of the system, as powers of two. In addition to the connections shown in Figure 22.2, all the registers are attached to a highway and data transfers between registers can be initiated by the CU.

The program counter (PC) contains the address of the next instruction to be executed. When an instruction is fetched into the instruction register, the PC is incremented by one so that it is always pointing to the next instruction. However, it is possible to take instructions out of order by issuing a command to *jump*, that is to move the pointer unconditionally up or down the memory by a certain number of places.

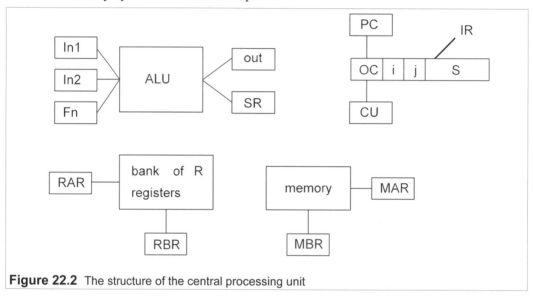

Figure 22.2 The structure of the central processing unit

The instruction register is connected to the control unit, which decodes the instruction. Decoding derives the information and addresses needed to generate the sequence of register-to-register transfers required for executing the instruction.

The R registers are loaded from or stored to the memory by instructions of the type F,i,j,S. F is the operation code, often one byte of 8 bits, i and j (4 bits, a *nibble*, each) identify the registers involved and S is a 16-bit integer, giving a 32-bit instruction. In one mode of operation the j-field is not used and S is sent to the memory-address register (MAR) to select the appropriate memory location. In the second mode of operation the contents of register R_j are added to S and the result is sent to the MAR. This process is called address modification. The memory-buffer register (MBR), sometimes known as the memory-data register, temporarily holds data fetched from memory or data which is written to memory.

The CPU has a set of short, 16-bit instructions to perform arithmetic and logical operations using the ALU and two of the R registers, such that $R_i = F(R_i,R_j)$. Instructions to perform these operations have the format F,i,j. The operations are carried out in the ALU, which has a function register (Fn), two input registers (In1 and In2), one output register (Out) and a status register (SR). The SR has *flags*, which are bits that change according to whether the result of an operation is positive, negative or caused an *overflow*. Overflow means that during an arithmetic operation a number was produced that was too large to be stored in a register. The jump command referred to above can be arranged to jump conditionally, depending on the status of individual flags in the SR.

Most modern computers also have a floating-point arithmetic unit which interprets the contents of the R registers as being in two parts, *a* and *b*, representing the number $a.2^b$. The longer part, *x*, represents a binary fraction and the remaining digits, *y*, a binary integer. The range of numbers that can expressed is thereby greatly expanded. In many computers the R registers are of 32 bits and are used for integers; in this case a set of separate 64-bit X registers is used for floating-point numbers.

22.2.1 A typical instruction sequence

The actions carried out by the CPU, in executing a single program instruction with the registers listed above, can be summarised with the help of a symbolic code called *register-transfer language*. In this language, the contents of a register, say the MAR, are written [MAR] and the contents of the memory location whose address is [MAR] is written as [[MAR]]. Transfers are indicated by a backwards arrow, so that [MAR] ← [PC] means 'the contents of the program counter are transferred to the MAR' or 'the contents of the PC become the contents of the MAR'. The following might be a typical sequence to add two numbers, one in register, R_2, the other in register, R_3, and put the result in register, R_2, using an instruction of the type *F,i,j*:

1. [MAR] ← [PC]   *the contents of the PC are moved to the MAR*
2. [PC] ← [PC] + 1   *the program counter is incremented by one*
3. [MBR] ← [[MAR]]   *the contents of the address in memory specified*
                         by the PC are moved to the MBR

The time from loading the MAR in step 1 to the loading of the MBR in step 3 is very long compared to one cycle of the CPU clock, accordingly it is necessary for step one also to make the MBR busy, for step 3 to conclude by making the MBR free and for step 4 to wait for the MBR to be free. All that has happened until now is that a single instruction has been read to the IR. After step 3 the contents of the IR will be 'add,2,3' and the following sequence is required to complete the process:

4. [RAR] ← [IR(R_i)]   *the first address of IR is transferred to RAR*
5. [RBR] ← [[RAR]]   *the contents of R_2 are transferred to RBR*
6. [In1] ← [RBR]   *the contents of RBR are transferred to ALU In1*
7. [RAR] ← [IR(R_j)]   *the second address of IR is transferred to RAR*
8. [RBR] ← [[RAR]]   *the contents of R_3 are transferred to RBR*
9. [In2] ← [RBR]   *the contents of RBR are transferred to ALU In2*
10. [Fn] ← [IR(OC)]   *the operation code, add, is transferred to ALU Fn*
11. [RBR] ← [Out]   *when ALU has placed the result in the Out register, it is*
                        then transferred to RBR
12. [RAR] ←[IR(R_i)]   *the first address of IR is transferred to RAR*
13. [[RAR]] ← [RBR]   *the contents of RBR are transferred to R_2*

Each of the thirteen steps above takes at least one cycle of the CPU clock. The operation codes are given mnemonics such as 'INC-PC', which means INCrement the contents of the PC by one, but the *machine code* is a binary word. If an *assembly-language* program of operation codes is written in mnemonics, it must be translated into machine code by an *assembler*. The complete set of operation codes is known as the *instruction set* of the computer. For an operation code that is an eight-bit word there are $2^8 = 256$ possible codes, a number normally sufficient for microprocessors and microcomputers.

Machine code is said to be the *lowest-level* language, assembly language is a *low-level* language and FORTRAN, C, BASIC, COBOL, etc. are said to be *high-level* languages. A high-level language may be translated into assembly language at first by a *compiler*, and the assembly language must then be translated into machine code. Some high-level languages such as BASIC, LISP and PROLOG are not usually compiled but are *interpreted* step by step by an interpreter which generates the machine code for each step. High-level languages are intended to be easier to use than lower-level languages, but there are wide variations in their level of difficulty. We shall return later to programming.

22.3 Memory

In the UK the word 'storage' was used at first, but the American term 'memory' has replaced it in popular usage. Almost anything that can be changed from one stable state to other stable states can be used to store data, and there is no need to store numbers in binary if more than two states are available. However, it was soon realised that two-state stores were more readily constructed and easier to use, so they became standard. At first a motley assortment of stores was pressed into service, but soon magnetic storage of one kind or another was used by all commercial computers[6]. Magnetic stores were of three kinds: ferrite toroids (known as ferrite cores) were fastest, rapidly spinning magnetic drums were of intermediate speed (though orders of magnitude slower than ferrite cores), while magnetic tape was the slowest and mainly used for back-up and file storage. Ferrite cores started off as toroids about the size of necklace beads and the read and write wires were threaded through by hand just like stringing a necklace. As a consequence they were expensive and ferrite-core memory capacity was very small, even on the most advanced computers. Programmers had to know where their variables were stored so that those frequently used were placed in the ferrite-core memory. It could happen that a small program change would cause one or more variables to be moved from the fast core store to the slow drum store and the program's execution time would increase enormously.

Ferrite cores lasted a long time (about 25 years) because they developed steadily to beat off competition: their size fell from 10 mm down to 1 mm in diameter and they were eventually stamped out of ferrite-impregnated plastic sheets; machines did the wire threading and Mbit core stores could be made. However, their demise was inevitable; everyone in the industry knew that semiconductor memories were ordained to take over the task of fast-access storage; they had done so by about 1980.

Memory is organised into cells which are of a size in bits determined by the computer designer. Almost all modern machines use cells of 8 bits, called bytes, a term introduced by

---

[6] In his autobiography, Wilkes says ferrite cores were mainly responsible for the computer's commercial success.

IBM in the 1960s. Each cell has associated with it a number, which is its unique address. As we have seen before, the address is quite distinct from its contents. While the contents of a cell may change often during the execution of a program, its address is the same from program to program. *Read-only memories* (ROMs) have invariant contents as well as invariant addresses.

22.3.1 Memory hierarchy

The memory can be divided into two types according to the method of access:

> (1) Random-access memory (RAM)
> (2) Serial-access memory (SAM)

In random-access memory the time taken to locate any address is constant, no matter if it is the next cell to the one just accessed or one furthest away in terms of address. RAM is the fastest (shortest access time) memory; semiconductor memories are usually RAM. In serial-access memory the next address to be accessed is always next to the one just accessed. To access a cell M addresses from the one being accessed takes M times as long as accessing the next cell. Magnetic tapes are an example. In direct-access memory the accessing is a combination of random and serial and the access time for a particular address is not predictable (though the average access time might be). Magnetic discs and drums are examples. Drums and discs are divided up into circular tracks produced by *formatting*. Within the tracks, access is serial, but if each track has a read/write head associated with it, then the track access is random.

Memory is organised into a hierarchy, as in Figure 22.3, because fast-access memory is much more expensive than slow. Registers are fewest and work fastest. Next in speed are *cache* memories[7], which are small blocks of very fast memory used to store frequently-accessed instructions and variables.

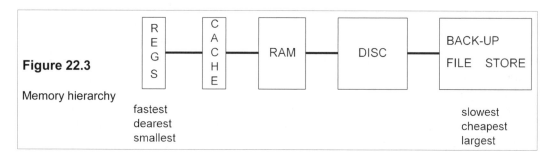

Figure 22.3

Memory hierarchy

Cache memories contain a small, continuously varying subset of the total addresses, together with the contents of those addresses; they are fast enough to be accessed in one or two CPU clock cycles. They are essential if the maximum computational speed of the CPU is to be realised. Management of cache memory is generally by hardware, as management by the operating system would be too slow. When an address leaves the CPU it goes first to the cache; if the address in the cache then the data access is very fast. If a match is not found in cache (a 'cache miss'), the address goes to the RAM and the contents of the required address

---

[7] The use of cache memories was first proposed by M V Wilkes in 1965.

and those of a few surrounding addresses, which form a block, are transferred to cache together with the block address. The reason for this is that neighbouring addresses are frequently required in succession. This process is slow, but subsequent accesses to addresses within the transferred block will of course be fast. When the cache is full, a block has to be removed before a new block can be accommodated, the choice being made by hardware, based on the activity of all the blocks in the cache. Cache misses cause a significant downgrading from peak performance attained when all data accesses are within the cache. If only 1% of accesses are not in the cache, performance is degraded by about 10%.

Most up-to-date machines operate with a virtual-address space which is larger than the real address space required to cover the whole of RAM and disc. Both of these memories are divided into pages of a fixed size of a few thousand bytes. The virtual address generated by the CPU is first presented to the cache as described above then, if not found in the cache, it goes to the RAM. If the real page containing the required virtual page is in RAM it is accessed, the matching of virtual and real pages being achieved by fast hardware. If the address is not in the RAM, the operating system finds the location on the disc of the real page containing the required virtual address and moves this page into an empty page of the RAM. The operating system maintains at least one empty page in the RAM by moving pages back to the disc when they are no longer actively used. The integration of RAM and disc into a 'one-level store' was first implemented at the University of Manchester in 1967[8].

22.3.2 Semiconductor memories

Semiconductor RAM and ROM are made from MOS devices, especially low-power CMOS. RAM used to be *static* or *dynamic*: SRAM or DRAM. DRAM is now almost obsolete, because the information stored in it gradually leaks away and must be refreshed every now and again, adding to the complexity of the controlling software. SRAM does not lose its information so long as power is supplied, but an SRAM bit requires four components instead of the one needed by DRAM, increasing cost and size. Both SRAM and DRAM are usually *volatile*, that is they lose their contents when power is lost, though non-volatile RAM (NVRAM) is available at extra cost. All memory ICs have increased in size very rapidly and 8Mbit chips are available for the same price as an 8kbit chip a few years ago.

Table 22.1 shows a selection of the many hundreds of memory ICs now available. The prices are for single quantities and are many times the price paid by manufactures of large quantities of equipment. One-time programmable (OTP) EPROM is a little as £2/Mbit, but SRAM and E^2PROM (electrically-erasable PROM) are slightly more expensive.

The capacity of a '1 Mbit' memory is actually $1\,048\,576\ (= 2^{20})$ bits, usually arranged as $131\,072 \times 8$-bit or sometimes $65\,536 \times 16$-bit words. Addressing as words rather than bits has the effect of speeding up the reading time by a factor comparable to the word length. Access times are around 100 ns, which is slow compared to fast CPU clocks. A chip requiring 1 mA of standby current at $V_{DD} = 5$ V can be backed up by a 5F electrolytic capacitor in emergencies for two or three hours.

---

[8] 'One-level storage system' by T Kilburn, D B G Edwards, M J Lanigan and F H Sumner, *IRE Trans. on Electronic Computers*, **EC-11**, No.2, 230–65 (1962). The first proposal for a virtual memory.

Table 22.1 *Semiconductor memory ICs*

Device	Type	Size^A	t_{acc} ^{B}	I_0 ^{C}	Price^D	Price/Mbit
AM27C512-120JC	OTP EPROM	64k × 8	120	1	£4.62	£9
AT27C040-15JC	OTP EPROM	512k × 8	150	1	£13.60	£3.40
M27C160-100M1	OTP EPROM	1M × 16*	100	1	£36.85	£2.30
M27C1001-12F1	UV EPROM	128k × 8	120	1	£5.22	£5
M27C801-100F1	UV EPROM	1M × 8	100	1	£32	£4
M24C01-MN6	E^{2}PROM	128 × 8	SA	0.001	£0.46	£450
NM24C04M8	E^{2}PROM	512 × 8	SA	0.05	£1.17	£286
AT24C64-10PC	E^{2}PROM	8k × 8	SA	0.035	£3.71	£58
AT28C256-15PC	E^{2}PROM	32k × 8	150	0.2	£11.59	£45
AM29F010A-70JC	Flash EPROM	128k × 8	70	30	£5.50	£5
AM29LV400BT-90SC	Flash EPROM	256k ×16*	90	60	£9.98	£2.44
IDT79321SA55J	SRAM	2k × 8	55	155	£8.22	£514
KM64258CJ-15	SRAM	64k × 4	15	140	£2.90	£11.32
KM684000ALP-7L	SRAM	512k × 8	70	90	£15.13	£3.69
M41T56M6	NVRAM	64 × 8	SA	0.3	£1.76	£3440
DS1330YP-100	NVRAM	32k × 8	100	85	£16.55	£65
DS1245Y-120	NVRAM	512k × 8	120	85	£31	£7.57
µPD42426OLE-80	DRAM	256k × 16	80	2	£36	£8.79

Notes: ^{A} Words × bits * can be configured with 8-bit words ^{B} Access time in ns, SA = serial access ^{C} Standby current in mA ^{D} For one device

Non-volatile ROM is available in various forms of PROM – programmable ROM. EPROMs have a window on the back which can be exposed to ultra-violet light to erase the data stored, but this can only be done a relatively few times (10–100). E^{2}PROMS can be erased electrically and in this way are like SRAMs, but can be write-protected. They can be put through many more cycles than EPROMs: up to 10 million. For some applications ferroelectric RAM has been developed which has even more cycling capability. Large computers now use much faster chips (1-2 ns access time) than those shown in Table 22.1, but at greater expense and with larger standby currents. The power consumption of these is quite high, necessitating special cooling methods such as Peltier-effect heat pumps.

Flash memory can be used as both RAM and ROM with restrictions on the blocks which can be deleted or written to. They are used in very large numbers in mobile telephones (or cellphones) where 256Mbit chips are now being used, especially for the third generation handsets, and 1Gbit chips are planned.

Figure 22.4 shows the block diagram for a 4Mbit SRAM arranged as a matrix of 1024 rows by 512 columns of 8-bit words. This chip can hold roughly 100,000 words of English text. The control logic truth table is also shown in the figure, in which CS = chip select, OE = output enable, WE = write enable.

The memory chips in a computer are connected to two *buses*: the data bus and the address bus, which are simply lines (usually several in each bus) to and from which messages and data are sent. The memory in Figure 22.4 can only be addressed when the chip-select control is low so that the CPU can select one chip at a time for connection to the data or address buses. The connections from chip to bus are via three-state gates, explained in Section 20.9.5. In so-called *von Neumann* architecture the data and instructions are stored in one block of RAM, so that a particular address may contain data or instructions. In *Harvard* architecture the RAM is divided into data RAM and instruction RAM so that a single address can refer to both data and instructions. The CPU has to decide from the context which is required when an address is specified.

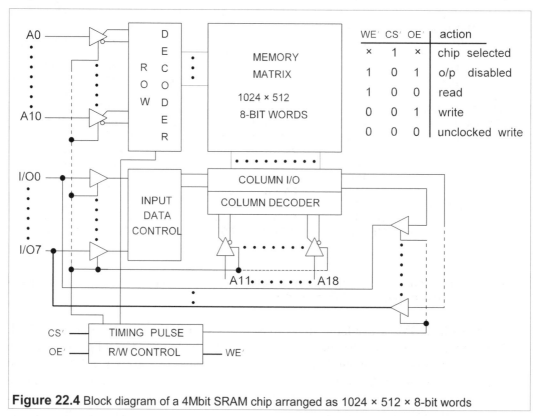

Figure 22.4 Block diagram of a 4Mbit SRAM chip arranged as 1024 × 512 × 8-bit words

22.4 **Input and output devices**

One reason computers have so many uses is the steady increase in types of input and output device. Though the devices differ greatly in their methods of operation, the techniques by which they are connected (or *interfaced*) to computers are more-or-less standard. Table 22.2 lists some of the more common input and output devices.

The problem with peripheral devices is that they are not synchronised with the computer, and the computer may have no idea of how long a peripheral will take to perform a task. For this reason they are linked to the computer via interface units, as Figure 22.5 shows. The organisation of the CPU, memory, interface units and buses can be arranged differently to that shown. But while the interface units may all differ when viewed from the peripheral device, they are all alike when viewed from the computer. Looking from the peripheral the interfaces must all be dissimilar, since the requirements of each type of device are quite distinct: the control commands for a laser printer are totally unlike those of a cathode-ray tube. Not all the interface registers in Figure 22.5 may be needed for a particular peripheral.

Table 22.2 *Some input and output devices*

Device	Purpose[A]	Speed[B]	Comments
keyboard	i/p	5	echoed to CRT
'paper' tape	i/p	500	fragile, obsolete
punched card	i/p	1000	easily damaged, obsolete
optical bar code	i/p	1000	very widely used in businesses
MCR[C]	i/p	300	robust
OCR[D]	i/p	1000	easily misread
scanners	i/p	100+	expensive when fast
mouse	i/p	slow	for menus
CD	i/o	250k+	fast, cheap, 0.8 - 4 Gbyte capacity
flexible disc	i/o	35k+	cheap, obsolescent
hard disc	i/o	350k+	fast, not dear, capacity has gone from 20 M to 60 Gbyte in 15 years
magnetic tape	i/o	10M - 1G	serial access, vast capacity, backing stores
ink-jet printer	o/p	120	fairly quiet
laser printer	o/p	0.25-25k	price $\propto$ speed
cathode-ray tube	o/p	1M+[E]	fastest, not dear

Notes: [A] i/p = input, i/o = input and output, o/p = output [B] very approximately, in characters per sec (cps), 100k = 100,000 cps [C] magnetic-character recognition [D] optical-character recognition [E] the limitation is the rate at which characters are sent to the CRT

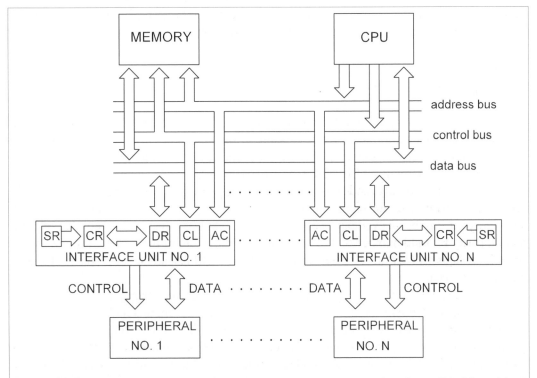

Figure 22.5 Computer with a common bus for memory and peripheral interface units. AC = address comparator, CL = control logic unit, CR = control register, DR = data register, SR = status register

22.4.1 Interrupts

We have mentioned that the basic problem of input and output is that it is device-dependent and the time taken is beyond the computer's control. In programmed input/output all transfers of data to and from the CPU to peripherals are under the control of the program. The programmer then has to set the control flags (for example if the printer is busy) at the interface unit and test them. The CPU has to be programmed to look at the status flags of all the peripherals to find out which is ready (*software polling*). Since the computer must be managed in such a way that it can perform other tasks while waiting for input and output, the programmer's task is complicated. Ideally the details of input and output should not be the concern of the programmer and interrupts or an interrupt structure can do the job. An interrupt signals to the CPU that it must carry out the instruction currently being acted on and then stop working on that program and store the contents of all the registers in memory. Each interrupt has a unique set of instructions stored in its own area of memory, and servicing is instituted by hardware to maximise speed of execution. The program counter is set to the start of the interrupt routine and the CPU then carries out the servicing of the interrupt. When it has finished, the registers are restored from memory and the CPU continues with the program that was interrupted.

   The system design must set priorities for interrupts, the priority being determined by how long the cause of interruption can wait before being serviced. The shorter the wait, known as the *crisis time*, the higher the interrupt's priority. If it is necessary to interrupt a low-priority interrupt because one of a higher priority has occurred, then the interrupt's register state must be stored, as well as the original register state before the first interrupt. There are usually no more than three, and sometimes fewer, levels of interrupt.

22.4.2 Character codes

The input and output of a computer comprise numbers and letters (alphanumerics); special characters like punctuation marks, brackets and symbols; and control codes. The two codes most commonly used are the seven-bit ASCII (American Standard Code for Information Interchange) and the eight-bit EBCDIC (Extended Binary Coded Decimal Interchange Code). The ASCII code is usually made up to eight bits by appending a most significant bit (MSB); either a zero in all cases and signifying nothing, or a *parity* bit to act as a simple error check. The parity bit may be an even parity bit (that is the eight bits have an even number of 1s) or an odd parity bit (odd number of 1s), depending on the whim of the system designer. When the 8-bit code is sent serially it is transmitted with the MSB last so that b_0 is sent first and the parity bit, b_7, last.

   Table 22.3 shows the ASCII code (the bits are b_{0-6}). Thus the character '6' is encoded as x0110110 in 8-bit binary (x is a zero, or parity bit) and is character 54 in decimal. For example in the word-processing program WordPerfect, the number 6 can be typed in by pressing the 6 on the keyboard or entered as character 54 from character set 0 (the ASCII set) by typing <Ctrl>v0,54<Enter>. The codes numbered 0–32 are control codes (including SP, the 'space' character) and the rest are called printing characters.

Table 22.3 *The ASCII code*

$b_3 b_2 b_1 b_0$		0	16	32	48	64	80	96	112
		000	001	010	011	100	101	110	111 ← $b_6 b_5 b_4$
0	0000	NUL	DLE	SP	0	@	P	'	p
1	0001	SOH	DC1	!	1	A	Q	a	q
2	0010	STX	DC2	"	2	B	R	b	r
3	0011	ETX	DC3	#	3	C	S	c	s
4	0100	EOT	DC4	$	4	D	T	d	t
5	0101	ENQ	NAK	%	5	E	U	e	u
6	0110	ACK	SYN	&	6	F	V	f	v
7	0111	BEL	ETB	'	7	G	W	g	w
8	1000	BS	CAN	(	8	H	X	h	x
9	1001	HT	EM	)	9	I	Y	i	y
10	1010	LT	SUB	*	:	J	Z	j	z
11	1011	VT	ESC	+	;	K	[	k	{
12	1100	FF	FS	'	<	L	\	l		
13	1101	CR	GS	-	=	M	]	m	}
14	1110	SO	RS	.	>	N	^	n	~
15	1111	SI	US	/	?	O	_	o	DEL

22.4.3 Interfacing methods

The computer is connected to each interface unit by a port and signals are transmitted along wires connecting the port to the interface unit. The transmission can be two-way with two separate channels (known as *full duplex* or FDX), two-way using one channel (*half-duplex* or HDX) and one way (*simplex*). The transmission can be synchronous or asynchronous. In asynchronous serial transmission each character is sent with its own start and stop codes, while in synchronous serial transmission a stream of characters is sent, which is broken up into blocks by synchronisation code at the start and an end-of-message character at the end of the block. A cheap asynchronous interface IC called the UART (Universal Asynchronous Receiver and Transmitter) can be obtained very cheaply (from £2 to £5). The receiver section converts input serial start, data, parity and stop bits to parallel data out and verifies correct transmission. The transmitter section converts parallel data to serial form and adds on the start, parity and stop bits automatically. The word length is selectable from 5 to 8 bits.

There are two commonly-used serial interface standards, the RS-232C (USA) and the V24 (CCITT, Europe), for connecting computers to peripherals and to *modems* (modulator-demodulator) used for telephonic data transmission. Both use 25-way connectors, but whereas the RS-232C standard defines connector-pin usage, the V24 standard refers to circuits. Although the cables have 25 wires in them, the data are transmitted and received serially on just two of them. The others are used for control and synchronisation signals.

Data loggers and other instruments are often interfaced to computers with the IEEE standard 488 bus, which uses a 24-way connector and one bus can support up to fifteen instruments. The cable length must be no more than 2 m per instrument, with a maximum length of 20 m.

22.4.4 Keyboards

Pressing a key on the standard QWERTY keyboard produces a binary code which is transmitted to the computer, which sends it to the CRT screen as a 'printed' character (known as *echoing*) and then interprets within the context of the operating program. Control codes are sent by pressing the <Ctrl> key plus another or <Ctrl> + <Shift> + key. The keyboard/CRT combination is a VDU (Visual Display Unit). Commonly keys operate switches of the contacting kind with a spring return, and these gradually wear out. The most popular non-contacting switch is the Hall-effect switch in which the depression of a key causes a small permanent magnet to be brought near to a piece of semiconductor. The semiconductor carries a current in a direction normal to the magnet's motion and a voltage is produced at right angles to this to indicate depression of the key. The absence of contacts improves switch life, and there is less possibility of contact bounce.

The switches are wired to a matrix (see Figure 22.6) which must be scanned to detect which switch has been closed. Suppose we use a 12 × 12 matrix and the switch on row 3, column 5 has been closed, thus connecting the row with the column. The rows are scanned in turn by putting 0 V onto the row in time with a clock. When row 3 is scanned, column 5 will go to 0 V and be detected by the column decoder. Thus the ROM array will output the binary code at row 3, column 5. The scanning must be sufficiently rapid that fast keying is detected. If a maximum of ten keys can be depressed in a second, then an adequate scanning time would be 20 ms, which with 144 keys would imply a clock frequency of $144/0.02 = 7.2$ kHz. In practice much faster clocks would be used. The keyboard's output is serial and is connected to the computer by a five-pin DIN (*Deutsche Industrie Normal* – German industrial standard) connector.

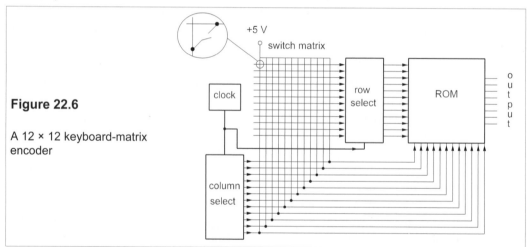

Figure 22.6

A 12 × 12 keyboard-matrix encoder

22.4.5 Printers

Printers have evolved very quickly in the last few years. The 9-pin dot-matrix printer was followed by the 24-pin printer, which in turn was superseded by the ink-jet, bubble-jet and laser printers. The 24-pin printer gave much better quality of lettering and could also be programmed easily, but was rapidly overtaken by ink-jet and laser printers offering very high print quality. Ink-jet printers work by projecting a spherical blob of ink towards a specified point on the paper. If the blob is small enough and of uniform size, the print quality will be good if the right type of paper is used. Unlike dot-matrix printers, ink-jet printers do not use ribbons to hold the ink but an ink reservoir, so the density of the printing remains constant. Unless ribbons are changed often, or the pressure of the needles is increased as the ink in the ribbon dries out or is used up, the printing will gradually fade, like that of a typewriter.

Laser printers contain a cylindrical drum coated with a photoconductive material. The photoconductive surface will hold an electric charge until exposed to light, so that if a laser diode is aimed at selected parts of the drum, the charge leaks away. The drum is rotated past a developer containing the ink powder, which is given the same polarity charge as the drum. The charged ink is repelled by the similarly charged areas of the drum and attracted to the uncharged areas that the laser has defined. The drum then passes over the paper which has an opposite charge to the ink, so that the ink powder is stuck to the paper. The paper then passes through a fuser which applies heat and pressure to the plastic ink and sticks it to the paper. The printing density therefore is reasonably constant with time. The printing resolution of a laser printer depends on the size of the light spot coming from the laser, but is normally about 40 μm, compared to 250 μm for 9-pin, and 90 μm for 24-pin, dot-matrix printers.

Dot-matrix colour printers use a four-colour ribbon and a program to 'mix' the colour specified by the computer. The printer can only print dots of the colours on the ribbon and these must be placed on the page individually, so that the eye averages out the coloured dots to give the illusion of a uniform shade. The problem is that there are not enough dots per unit area to prevent sudden jumps in colour shades that lead to visible bands on the printed page. Ink-jet printers 'mix' up to six colours to achieve the required shade (they still print separate dots of each colour like dot-matrix printers) and because the ink-blob is small, there is less of a jump from one shade to the next and better uniformity is achieved. Because there are so many dots to print, the printing times are very slow compared to black-and-white graphics (by a factor of 5–10). However, the colour printer market is in a state of rapid evolution, and there is little doubt that high-quality, low-cost, colour printers will soon be available, using other means than dot-matrix and ink-jet printing.

22.4.6 *Magnetic discs*

Magnetic discs have become one of the standard means of transferring information, data and programs from one computer to another. Rigid discs (or hard discs, or Winchesters) started off at a diameter of 14 inches (356 mm), but now quite small discs are being used as the bit density has gone up and diameters down to 33 mm have become common. Maximum bit densities in hard discs are now approaching 1 Mbit/mm^2. The magnetic medium of the disc is a thin layer of ferrite powder which is stuck to plastic for floppies or aluminium for hard discs. The magnetic surface layer can be magnetised in a certain direction by a write head to form a 1 bit and in the opposite direction for a 0 bit. These magnetised bits can then be read by a either a special read head or the same head that was used to write. In flexible disc drives the disc is actually pushed into physical contact with the head via a slot cut into the plastic envelope, and so the discs (and heads) gradually wear out.

Figure 22.7 indicates how the head is positioned over the disc like the needle arm of an old record player. The arm can be driven by a stepper motor to give accurate positioning. Since the tracks are 0.3 mm wide or less on a flexible disc and perhaps 5 μm wide on a hard disc, positioning is very critical. The head and arm assembly is made as light and as rigid as possible for fast acceleration and motion. The coil in the head which detects the magnetic field of the bits during reading may be only a few microns across. Most discs are now double sided and require at least two heads, one for the top and one for the bottom. Hard discs usually have multiple heads for each side.

Figure 22.7

A magnetic disc and its read/write arm. The head size is exaggerated: it is only a few microns in diameter

Figure 22.8 shows a 3½-inch flexible disc ('floppy') in its sealed plastic envelope. The hole in the middle accommodates the drive shaft, which rotates the disc at about 300 r/min. A small hole next to the centre hole is used as an index mark, which is can be detected by a photocell and tells the disc controller where the start of the track is. The tracks are rings on the disc which are produced by the disc formatting software.

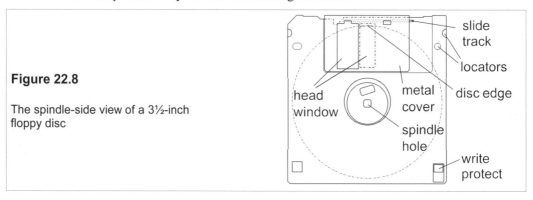

Figure 22.8

The spindle-side view of a 3½-inch floppy disc

In Figure 22.9 the original IBM 3740 disc format is shown, which illustrates the principles. In this format one side of the disc is given 77 tracks, three of which store information about the disc or can be used as alternatives to bad sectors. The remaining 74 tracks are each formatted so that each starts with a 46-byte preamble (a gap to give the read/write mechanism a breathing space), a 1-byte address and another 33-byte gap and then 26 sectors in which the data is written, followed by a 241-byte postamble gap. Figure 22.9 shows also the arrangement within a sector, which starts with a 1-byte address, followed by a 6-byte track-and-sector identifier (ID), another gap of 17 bytes and a data synchronisation byte. The data is then written or read from the next 128 bytes, which terminate with a 2-byte checksum error detector and a final gap of 33 bytes. Thus in this format the data bytes total only $128 \times 26 = 3.33$ kbyte/track, or about 250 kbyte a side and 500 kbyte for a two-sided disc. Higher density formatting, with 512 bytes in a data cell instead of 128, is now the norm and gives 1.2 Mbyte for a two-sided disc. The 3½-inch discs are kept in hard plastic containers. Disc technology could easily enable many Mbyte to be stored on a floppy, but these would no longer be compatible with 1.2-Mbyte disc technology. Floppies are rapidly giving way to CDs.

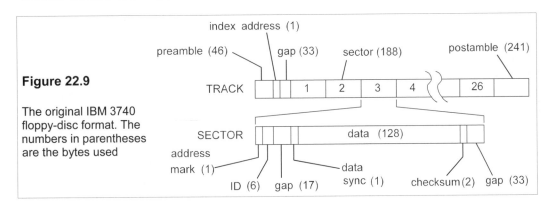

Figure 22.9

The original IBM 3740 floppy-disc format. The numbers in parentheses are the bytes used

Hard discs are sealed in containers filled with an inert gas to avoid dust contamination. The head does not need to contact the disc since the rigidity of the disc prevents disc wobble and possible damage by head contact. The gap between disc surface and head is critical for legible writing and reading, and must be maintained accurately, especially as the track widths on hard discs are smaller than on floppies and the speed of rotation is ten times as fast. Access times on discs are limited by the speed of rotation. On a flexible disc turning at 300 r/min, the average access time, when the head is in position over the right track, is $0.5 \times 60/300 = 0.1$ s, rather a long time. In addition, if the head has to be moved, there is a further delay of further delay of about 0.2 s. The data can be transferred at a relatively fast rate, since a single track holds about 7 kbyte in a 1.2-Mbyte disc and this can be read in one revolution, or 0.2 s, giving a data rate of 35 kbyte/s. Hard discs have data-transfer rates at least ten times as large. The average access time for hard discs is much less than for flexible discs, 10–15 ms being typical.

22.4.7 Backing stores and archives

The cheapest storage medium is magnetic tape, where the cost per bit stored has always been low, but is now being dramatically reduced by recent developments in helically-scanned tapes. In these the recording is not along the length of the tape, but at an angle to it, almost vertically. The recording head is rotated rapidly as the tape moves and achieves a bit density about a hundred times that of in single longitudinal track. These tapes are used in video cameras and in digital-audio tape (DAT) recorders. They are much cheaper than the special computer tape decks which have hitherto been used for mass storage. The use of automatic retrieval and tape changeover (robot-line storage) means that tapes can be changed in a few seconds. The 8mm helical tape cartridge holds 10–100 Gbyte and robotic mass storage systems have capacities of from 1 to 250 Tbyte. To have some idea of that size, consider that this book uses about 6 Mbyte of disc storage (including both text and graphics). The British Library contains 13 million volumes, or about 7 Tbytes of text. Thus all the text in the books of the largest library in Britain could easily be held in one 'small' mass storage unit. Each of these costs a few tens of thousands of pounds and if books could be scanned in automatically at a page a second, a hundred scanners could transfer 13 million books in about a year. Transmission over an optical-fibre network at 1 Gbit/s would enable remote stations to read a book in a few milliseconds, though access time would be ten seconds or so.

22.4.8 CRT displays

The construction of the CRT is described elsewhere in this book and will not be considered here, rather we shall say a little about the display of information on the CRT monitor. Colour displays have a screen coated with red, green and blue phosphors arranged in groups of three (a colour dot) – one of each colour. Three separate electron beams can be used to excite these into emitting light, or one beam which is colour modulated. The colour dots are separated by about 0.3 mm, which thereby becomes the smallest feature that it is possible to show on the screen. In fact the screen resolution may be worse than this because of the way in which it is addressed and the associated display memory of the computer, but the graphics cards now used with desk-top computers can now give approximately this resolution. Take for example a screen which is 250 mm by 200 mm. If the colour-dot spacing is 0.3 mm, that means the screen contains a total of about 500 kdots. Let us further suppose that we modulate the beam to mix the colour required at each dot and that this requires 8 bits (256 colours). We thus require a total of $8 \times 500 \text{ k} = 4$ Mbits, or 500 kbytes of video RAM to store the information necessary to drive the display. Only recently has this amount of video RAM become cheap enough for this use.

The first colour displays were driven by a colour graphics adaptor (CGA), which had the relatively poor resolution of 320×200 pixels (a pixel is the smallest part of the screen that could be written to) and only four colours (that is 2-bit colour), but required only 16 kbytes of video RAM. Soon the enhanced graphics adaptor (EGA) came on the market offering 640×350 pixels and 16 (4 bits) colours, using 112 kbytes of RAM. The video graphics adaptor (VGA) of IBM had slightly better resolution with 640×480 pixels but the same 16 colours. The latest graphics adaptors are the super VGA (SVGA) with 800×600 pixels and the enhanced SVGA (ESVGA) with 1024×768. This is about the maximum number of pixels available with a small, cheap monitor.

Another problem with CRT displays is flicker. The eye retains an image for about 40 ms, the image persistence time, so the display must be refreshed at a faster rate than this to avoid perceived intensity changes. The screen is written to one horizontal line at a time, the whole screen requiring 600 lines to make up one frame, or complete screen in the case of the SVGA. In some cases the screen is split into one group of odd-numbered lines and another of even-numbered and these are refreshed alternately (known as interlacing). This has the effect of slowing down the frame-refreshing frequency ('frame rate') by a factor of two, so that a 70Hz vertical-scanning frequency results in a 35Hz frame rate. Though interlacing at 70 Hz is better than not interlacing at 35 Hz, it is not as good as a non-interlaced 70Hz frame rate. Flicker-free displays require a frame rate of at least 50 Hz.

22.5 Computer networks

Linking computers facilitates rapid transfer of data, but it also enables smaller, less-powerful computers to make use of the greater power of others, and provides access potentially to much greater storage. The link may be via the telephone system using a modem, in which case the distance between computers can be 20 000 km, or they may be linked by cables installed for the purpose. In most of the latter cases the link is no more than a few hundred metres long and the set of linked computers is called a local-area network (LAN), as opposed to the wide-area network (WAN) that uses public telephone lines or perhaps a microwave link. LANs soon gained widespread acceptance because they made better use of expensive resources – for example a laser colour printer can be connected to all the computers at a company site. LANs can of course be connected to WANs to give global interconnection possibilities.

Figure 22.10 shows a variety of network types. The circles (called nodes or stations) represent physical devices – printers, computers and so on, while the lines represent the transmission path, which is usually a cable in a LAN. The bus network of Figure 22.10a has each station linked to a bus, which can be a single coaxial cable, down which the data must be serially transmitted. The Ethernet™ is an example of a bus network. Figure 22.10b shows a tree network, such as some parts of a telecommunications network might resemble, but which is not used in LANs. The star network of Figure 22.10d is the configuration of a PABX telephone exchange (private automatic branch exchange). The ring network of Figure 22.10c is used in the Cambridge ring LAN, one of the first put into use by M V Wilkes to link computers in Cambridge. The network of Figure 22.10e resembles parts of the telephone system.

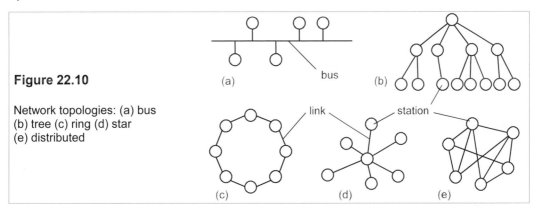

Figure 22.10

Network topologies: (a) bus
(b) tree (c) ring (d) star
(e) distributed

The critical points in a network depend on its topology. In a bus network, the bus is the critical point and when severed (for example, by unwittingly unplugging it) the network goes down. In the star network, the central node is vital, and in the tree network the nodes increase in importance the higher up the tree they are. Ring networks look most vulnerable, since all the stations receive and transmit all the data, and if any one breaks down, the network goes down too. This problem can be overcome by using bypass relays as Figure 22.11 shows.

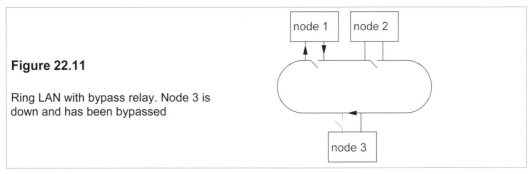

Figure 22.11

Ring LAN with bypass relay. Node 3 is
down and has been bypassed

22.5.1 Access to the network

Access to the network must be controlled by a signalling procedure known as a *protocol*. Protocols are arranged in a hierarchy called the open-systems interconnect (OSI) model. In this the protocols are arranged in a hierarchy of levels or layers (see Figure 22.12). The lowest level is the physical and the protocol for this layer would concern itself with connector layout, voltages, etc. The RS-232C serial interface specification is a physical-layer protocol. The next highest layer is the data-link layer in which the protocol must define the way data is moved from one station to another in the network. Examples are the high-level data-link control (HLDLC) and the binary synchronous communications (BISYNC) protocol, which define the format of message blocks. The third level is the network layer whose protocol defines how messages are broken into packets and routed from station to station (packet switching). The transport layer protocol defines the method of error detection and correction and the multiplexing methods used. The highest three layers are not concerned with the network itself, but with logging-in and out (session layer), character codes (presentation layer) and applications programs (applications layer).

Figure 22.12

The OSI protocol hierarchy. The bottom layer is the physical interface and the top three have nothing to do with the network proper

7	APPLICATION
6	PRESENTATION
5	SESSION
4	TRANSPORT
3	NETWORK
2	DATA LINK
1	PHYSICAL

When there is no control node, a station wishing to send a message first listens to the bus to see that it is not carrying traffic and then it places a packet of data on the bus containing the address to which it is sent and an error-checking code. The packet is received by all the other stations, but only the one with the right address opens it and reads its contents, checking that it is correct. The receiving station then sends an acknowledgment if all is well, or a request for retransmission if it is incorrectly sent. This is known as *handshaking*. Messages can be lost by noise or attenuation or when by mischance another station sends a message at the same time and there is a collision in which both messages are lost. The Ethernet LAN uses single-core, 5mm diameter, 50Ω coaxial cable with a maximum data rate of 10 Mbit/s. The cable attenuation is such that repeaters are necessary for station separations of more than 700 m. But for campus and local industrial use this is perfectly acceptable.

In some cases a control computer will be designated which is called the *host* and access to the network is only with its permission. The host computer looks at each of the other stations in some specified order to see if they have data for the network, a process called *polling*. The host then routes the data to its destination. In this way only one node can transmit data at any time and the host can determine the priority of the message or its source.

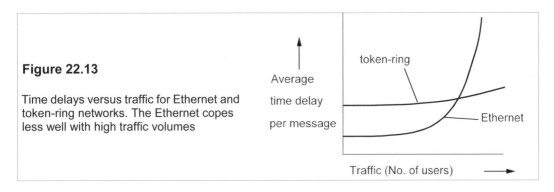

Figure 22.13

Time delays versus traffic for Ethernet and token-ring networks. The Ethernet copes less well with high traffic volumes

Average time delay per message

token-ring

Ethernet

Traffic (No. of users) →

Ring LANs often use a system of channel sharing called *token passing*. A token is a distinctive pattern of bits that circulates round the ring when no traffic is carried. A node wishing to transmit must wait for the token before it can do so. When the token arrives, the node seizes it and puts its message packet onto the ring. The message is delivered and its receipt is confirmed, the token then being freed. The token cannot be used more than once, nor can it be held indefinitely, so that no node can appropriate the channel for too long. Token-ring networks offer a better service than bus networks when the level of traffic become high and collisions become more frequent, as Figure 22.13 shows qualitatively.

22.6 Programming languages

There have been, in the main, three areas of application for programming languages that have quite different characteristics:

1. Scientific: complex algorithms needed
2. Office and business: masses of data, simple algorithms
3. Real-time control: for missiles, communications. Speed paramount

It seems unlikely that a single language could ever serve all three purposes equally well. Any language that does is likely to be unwieldy, possessing many facilities of little use most of the time, but which may lead into traps for the ignorant or unwary. Programming languages developed haphazardly at first, being poor relations to the hardware they ran on, but eventually they came to play a leading part in computing. Table 22.4 is a list of some of the more popular programming languages.

Originally computers were programmed by physically connecting various parts together. When the first stored-program computers came into existence, so did the art of programming on paper, but writing machine code in 1s and 0s was an exceedingly tiresome, time-consuming and error-prone activity. Wilkes wrote (of his first experience at programming the EDSAC I in 1949) 'the realization came over me with full force that a good part of the remainder of my life was going to be spent in finding errors in my own programs' – hardly a profitable use of time. The solution proposed by Wilkes in the first programming text[9] was to provide a 'pseudo code' which was relatively easy to learn and could be used with

---

[9] *The Preparation of Programs for an Electronic Digital Computer* by Wilkes, Wheeler and Gill, Addison-Wesley (1953)

considerably fewer errors than were produced by writing in machine code. Each instruction of the pseudo-code was translated into machine code by an interpreter, which was itself a machine-code program. Computer users at Manchester had stayed with machine code using an alphabet of 32 symbols, based on the teleprinter code, to represent groups of five binary digits. By March 1954, Tony Brooker had produced a working version of the Mark I autocode system, which was a 'high-level' language that had many similarities to FORTRAN I.

Table 22.4 *Programming languages*

Language	Year	Application	Comment
FORTRAN	1957	Scientific	Numerous updated versions
ALGOL	1958	Scientific	No input/output handling
COBOL	1960	Data processing	Still very widely used
LISP	1960	List processing	Used in artificial intelligence
BASIC	1964	Scientific	Easy to learn. Not 'serious'
PL/1	1965	All purpose	Too big for convenience
Pascal	1970	Scientific, teaching	Developed for teaching
C	1972	Control, scientific	From Bell Labs for UNIX
Ada	1979	Scientific, control	Named after Byron's daughter
C++	1985	Control, scientific	For object-oriented programming
Perl	1987	Text processing	Based partly on C, easy to use, not 'elegant'
HTML	1990	Web pages	Hypertext Markup Language
F	1995	Scientific	A subset of Fortran90

Statements in high-level languages are no different in principle from pseudo-code and can be executed by means of an interpreter; it is, however, usually more efficient to translate all the program statements into machine code by means of a program called a compiler. Once a program is working, the compiled code can be stored and used on subsequent occasions without any need for recompiling, while interpreted programs must be re-interpreted each time they are run. Most modern high-level languages are compiled, but there are still some occasions when interpreted programs might be desired as stated in Section 22.2.1

In the early days the time penalty of interpreted programs was hardly noticeable since the programs spent virtually all their time in floating-point subroutines – there were no hardware implementations of floating-point arithmetic. However, as early as 1953 floating-point arithmetic circuitry became available (on the IBM 704), when it was realised that interpreting pseudo code was slower by far than running a machine-code program. The need for a compiled language then became urgent.

The first widely-used high-level language was FORTRAN I (FORmula TRANslation) developed by J W Backus at IBM in a surprisingly casual way ('we made [it] up ... as we went along'). The reason for this was that designing the language was felt to be the easy part of the job, while the hard part was designing the compiler. To some extent this was true, if you overlooked the consequences of bad language design, and in the event the FORTRAN I compiler was very efficient. FORTRAN I came out in 1957, and was superseded by FORTRAN II in 1958, FORTRAN IV in 1962, FORTRAN77 in 1978, FORTRAN90 in 1991 and FORTRAN95 in 1995 (and it could be argued that PL/1 – Programming Language 1 – of 1965 was just a bloated version of FORTRAN). One might suppose that the later versions were simply fine tuning of the highly-successful original, but they are in fact almost different languages, made necessary by the competition of 'better' programming languages.

The initial expectations of Backus and IBM for FORTRAN were probably exceeded and the language became the industry's standard for scientific computation. The latest versions, FORTRAN 90 and FORTRAN 95, are still by far the most widely used for scientific computation using today's supercomputers. For business use, where there was much greater need for file and text-handling, COBOL was developed and it proved at least as successful, and if anything more durable, than FORTRAN. However, in the mid-1950s, FORTRAN was seen as a limited language, somewhat tied to IBM machines, which did not encourage the writing of well-structured programs. The perceived faults of FORTRAN led to the development of ALGOL (ALGOrithmic Language) as a second-generation language to replace it. The first specification, ALGOL-58, was described in a new formal syntax by Backus, but his description did not tally with that of P Naur, thereby indicating difficulties of describing what programming languages could do. Naur modified Backus's notation and the Backus-Naur Form (BNF) of notation came to light. BNF is now used to describe all programming languages. The complete description of ALGOL in BNF occupied only 15 pages and was issued as ALGOL-60 in 1960. Table 22.5 shows the definition of various types of number in BNF. The symbol ::= stands for 'is defined as' and | stands for 'or', so an 'unsigned integer' is defined as a 'digit or an unsigned integer and a digit'. As one progresses down the list the definitions make use of preceding definitions until 'number' is defined.

Table 22.5 *BNF definitions for numerical types*

```
<unsigned integer>   ::=   <digit>
                      |    <unsigned integer><digit>
         <integer>   ::=   +<unsigned integer>
                      |    -<unsigned integer>
                      |     <unsigned integer>
<decimal fraction>   ::=   .<unsigned integer>
   <exponent part>   ::=   10<integer>
 <decimal number>    ::=   <unsigned integer>
                      |    <decimal fraction>
                      |    <unsigned integer><decimal fraction>
<unsigned number>    ::=   <decimal number>
                      |    <exponent part>
                      |    <decimal number><exponent part>
         <number>    ::=   +<unsigned number>
                      |    <unsigned number>
                      |    <unsigned number>
```

In spite of all its advocates and its greatly-admired elegance, ALGOL failed to catch on, partly because IBM would not back it, and partly because it completely omitted the important field of input/output. Though not widely used, ALGOL soon became the standard by which other languages were judged ('a significant advance on most of its successors').

Of other languages there were legions, certainly several hundred and possibly over a thousand, of which Pascal, Ada and C have been prominent. Pascal originated as a language for teaching good programming techniques and became widespread in universities with some uptake by industry. Ada was the result of a US Department of Defense initiative when it came to the conclusion that no existing language would serve its need for embedded-computer control of missiles and missile systems. It was designed to promote the writing of programs in which most of the programmers' errors would be caught and pinpointed at compilation and not while running. However, it now appears that C (including C++) has become the most widely-used programming language in industry. C is a high-level language which has some of the ability of assembly language. C came out of the development of the UNIX operating system by Bell Labs; it is a small language that is, above all, fast. Much commercial software is written in C, as is the C compiler itself.

C has been criticised on the grounds that it permits sloppy programming and is not suited to proper software design. Modula2 was developed in an attempt to make a language which would not permit non-rigorous programming, and it succeeds, perhaps at the expense of flexibility. Nevertheless Modula2 has made some headway in teaching establishments, though it has so far failed to make a great impact on industry.

The development today of *object-oriented* programming resembles the development of structured programming twenty years ago in being fast evolving and fashionable. It is designed to enable programs readily to make use of other data classes defined by other programs. The end result should be the construction of programs from well-tested parts. The most widely used object-oriented programming language is C++ at present. More recently, Sun Microsystems have produced Java for use on the Internet. Java is a development of C++ which enables its users to create dynamic pages as well as download pages faster.

It is generally true that the software run on a computer is more expensive than the hardware; certainly it is true of the machine on which this is written. Software also makes more people wealthy than hardware. Yet software started out as a poor relation – the important thing was making bigger, faster computers. What happened was that the hardware evolved at a different rate to software and the problems of hardware manufacture proved easier to solve than those of software. It must be the aim for programmers to produce software which is

- *portable*, that is can be run efficiently on any machine without any or much modification
- *maintainable*, that is it can readily be altered to take care of new circumstances by someone without a detailed knowledge of the program
- *well-documented*, so that other programmers can easily understand the way it works

and this can only be done within the constraints of the language used.

Suggestions for further reading

Computers: from logic to architecture by R D Dowsing, F W D Woodhams and I Marshall (McGraw-Hill, 2nd ed., 2000). A good introduction
The architecture of computer hardware and systems software: an IT approach by I Englander (Wiley, 2nd ed., 2000)
Computer organisation and architecture by R S Chalk (Palgrave, 1996). Short
Computer architecture and organisation by J P Hayes (McGraw-Hill, 3rd ed., 1998)
Programming languages by T W Pratt and M V Zelkowitz (Prentice-Hall, 4th ed., 2001)

23 Microprocessors and microcontrollers

THE FIRST microprocessor to be made commercially was the 4-bit Intel 4004 of 1971. By 1980 a variety of cheap 8-bit microprocessors, such as the Rockwell 6502 and the Intel 8080/8085 series had come on the market. Faster 8-bit microprocessors and microcontrollers followed with more and more facilities added to fill what were seen as gaps in a highly-competitive market. 16-bit, 32-bit and even 64-bit devices have become available and are used mostly for specialised purposes such as PWM motor (mostly 16-bit microcontrollers) and robotic control. The robotics field in particular has urgent need of fast microcontrollers with large word sizes. Table 23.1 lists some of the devices, which is only a small sample of those available now and in the past.

The microprocessor is not a stand-alone computer, since it lacks memory and input/output control. These are the missing parts that the microcontroller supplies, making it more nearly a complete computer on a chip. Figure 23.1 shows a block diagram of the component parts of a microcontroller in which the component parts of a microprocessor are enclosed in a dashed rectangle. It can be seen that the microcontroller is a microprocessor with some additional on-chip features:

    1. A block of ROM to store program and data
    2. A small block of RAM
    3. Several ports for input and output

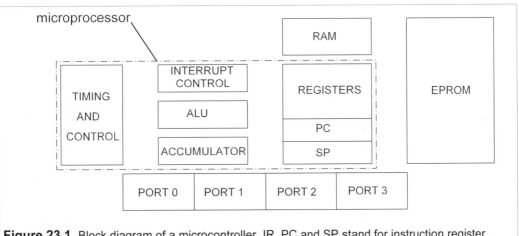

Figure 23.1 Block diagram of a microcontroller. IR, PC and SP stand for instruction register, program counter and stack pointer respectively

Table 23.1 *Some readily-available microprocessors and microcontrollers*

Device No.	Clock freq. (MHz)	Data size (bit)	Prog. Mem. (byte)	Prog. Mem. Type[A]	Internal RAM (byte)	No. of Ports	Serial I/O[B]	No. of Timers[C]	Spec. Feature[D]	Package Type[E]	Price (1+)
80C31/51 types											
P80C31SBPN	16	8	0	–	128	32	UART	2 × 16	–	DIP40	£1.76
TSC80C31-121A	12	8	0	–	128	32	UART	2 × 16	–	DIP40	£2
P87C52SBAA	16	8	8k	OTP	256	32	UART	3 × 16	-40°C	PLCC44	£3.24
AT89C51-24PC	24	8	4k	Flash	128	32	UART	2 × 16	–	DIP40	£8
AT89C2051-24SC	24	8	2k	Flash	128	15	UART	2 × 16	Analog	SOIC20	£5.60
P87C51RB+4N	16	8	16k	OTP	256	32	UART	3 × 16	PWM	DIP40	£5.10
P87LPC769FD	20	8	4k	OTP	128	15	UART/I²C	2 × 16	4 × 8 A/D	SOIC20	£4.87
PIC types											
PIC12C508-04/P	4	12	512	OTP	25	1/5	–	1 × 8 WD	25 mA i/o	DIP8	£1.60
PIC12CE519/JW	4	12	1k	UV	41+16E²	1/5	–	1 × 8 WD	25 mA i/o	CDIP8	£20
PIC12C74-JW	4	14	2k	OTP	128	1/5	–	1 × 8 WD	-40°C/25 mA	SOIC8	£3.50
PIC16C54/JW	20	12	512	UV	25	12	–	1 × 8 WD	20 mA o/p	CDIP18	£11.20
PIC16C505-04/P	4	12	1k	OTP	72	1/11	–	1 × 8 WD	20 mA o/p	DIP14	£2.19
PIC16C62-04/SP	4	14	2k	OTP	128	22	I²C/SPI	2 × 8 WD	PWM	DIP28	£7.55
PIC16C710-04/P	4	14	512	OTP	36	13	–	1 × 8 WD	4 × 8 A/D	DIP18	£4.11
Other 8-bit types											
ACE1101N	4	8	1k	E²	64+64E²	6	–	1 × 16 WD	Arithmetic	DIP8	£2.60
ST62T00CB6	8	8	1036	OTP	64	9	–	1 × 8 WD	4 × 8 A/D	DIP16	£1.98
ST62E28CF1	8	8	8k	UV	192	20	UART/SPI	2 × 8 WD	12 × 8 A/D	CDIP28	£23

Notes: [A] OTP = One-Time Programmable, UV = Ultra-violet erasable (=EPROM), E² = Electrically-Erasable (=E²PROM) [B] UART = Asynchronous Universal Receiver/Transmitter, I²C = Inter-Integrated Circuit bus interface, SPI = Serial/Parallel Interface, SCI = Serial Comms. Interface [C] WD = watchdog [D] Analog = Analogue computing, PWM = Pulse-Width Modulation (for motor control etc.), A/D = analogue-to-digital converter [E] DIP = dual-in-line plastic, CDIP = ceramic/plastic DIL (UV window), SOIC = surface mount, PLCC = Plastic Lead Chip Carrier.

Microprocessors are for moving and processing as bytes relatively large amounts of data internally and to and from external sources. A microprocessor is really the CPU of a computer. A microcontroller, however, is as likely to be required to perform bit manipulations as much as byte, and must be able to accept and output control and timing signals. The memory stack in a microcontroller is usually much smaller than that of a microprocessor, while a micro-processor needs quite a lot of additional memory to perform its functions.

Both microcontrollers and microprocessors are programmed by writing code into a microcomputer which is then translated by a compiler into machine code that is stored in ROM. The programming, ROMing and debugging require a 'development system' for the particular device or device family. Microcontroller development systems can be obtained which support high-level languages, but the code produced is less efficient. Software simulators are available that run on most microcomputers to help with program debugging. Emulators are also used, which carry out all the operations that the microcontroller can perform on external hardware. An emulator is the same as the microcontroller so far as external devices are concerned, while a simulator can only show what the contents of the RAM and ROM are at each stage in the program. In order to understand a microcontroller well enough to make it perform useful tasks there is no substitute for learning about all its features, which is why this chapter is full of detail.

23.1 The 8051 microcontrollers

The very-popular 8051 series of microcontrollers has dozens of variants, some of which are listed in Table 23.1. The 8031 has no on-board ROM and is therefore the cheapest, the 8051 includes one-time programmable ROM (OTPROM) and the 8751 is for prototype production, since it has EPROM (erasable PROM) on board. For most general purposes the 8-bit microcontroller is the best choice and we shall be looking in some detail at one of the most popular, the Intel 8051. The 8-bit word gives a maximum resolution of 0.4% (1 in 256), which is enough for all but high-precision tasks and those requiring great speed. The 8051/8751 comes in a 40-pin DIP shown in Figure 23.2, but the effective number of pins is increased to 64 by giving 24 of them dual functions.

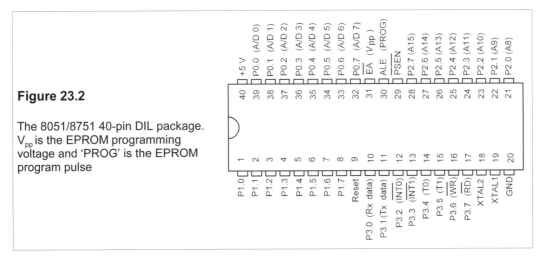

Figure 23.2

The 8051/8751 40-pin DIL package. V_{pp} is the EPROM programming voltage and 'PROG' is the EPROM program pulse

Figure 23.3 shows the architecture of the various parts of the 8051. Unlike computers, which can be programmed in high-level languages and require very little knowledge of the computer's organisation, microcontrollers require one to have considerable knowledge of their architecture. The most important parts of the microcontroller are the RAM, the special-function registers (SFRs) and the ports.

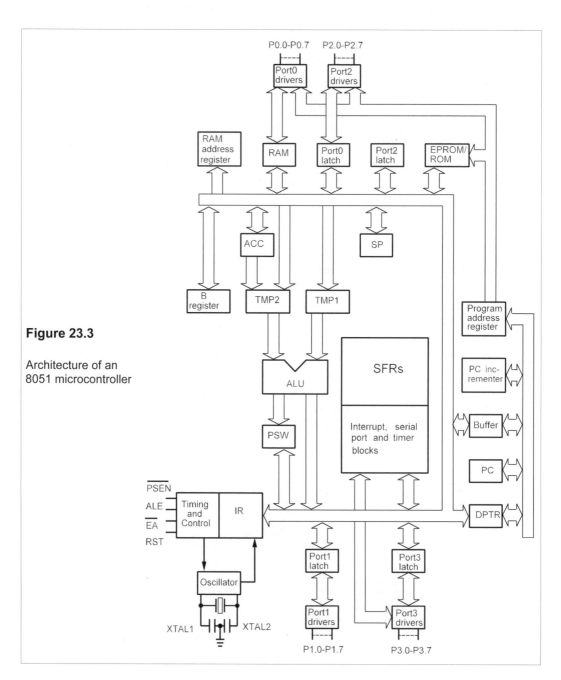

Figure 23.3

Architecture of an 8051 microcontroller

Figure 23.4 shows the layout of the RAM and the SFRs. To make the device do what we want requires us to know in more or less detail what is stored in this RAM and the SFRs. The RAM consists of 128 bytes (of 8 bits) arranged as four register banks, each containing 8 registers given the labels R0 to R7. Each RAM byte has a number associated with it which is its address, so one can either use a register's label, or its absolute RAM address. The register banks must be selected, so that the CPU knows in which bank R3, for example, is. On power up or reset, bank0 is selected; to select any other banks the appropriate settings must be made in the Program Status Word (PSW) SFR.

Above the register banks is an area of RAM (20h to 2Fh, inclusive[1]) which is bit addressable. This saves space when we want to store bits, as we do not have to use a whole byte to do so. In RAM locations from 30h to 7Fh is a general purpose area which can be used as a scratch-pad. The whole of the RAM area above 07h (register bank0) can be used as a memory stack to store variables, data, etc. It is best to put the stack above register bank3 and the bit-addressable bytes, that is above 2Fh, to avoid accidentally overwriting data.

RAM locations above 7Fh are reserved for the defined SFRs and can *only* be used when defined. Attempts to used undefined RAM above 7Fh will cause errors. We shall now look at these SFRs. ROM is used for storing code, RAM for data that can be altered as the program runs. The 8751 uses Harvard architecture where code and data are stored in different areas with the same address: the CPU decides which is which. If off-chip ROM is needed it will automatically be accessed by calls to addresses above 0FFF. If you wish all ROM calls can be made externally by connecting $\overline{EA}$ (pin 31) to GND.

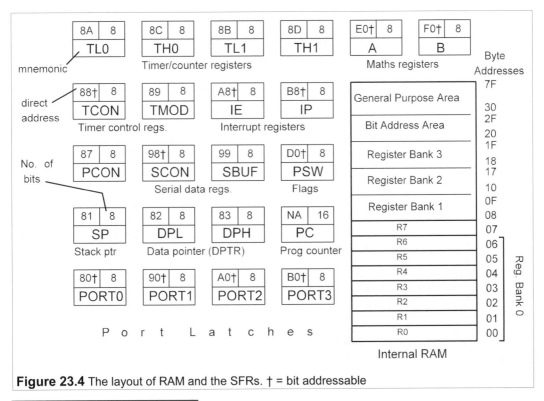

Figure 23.4 The layout of RAM and the SFRs. † = bit addressable

---

[1] Hexadecimal (base 16) numbers will be written with an 'h' appended, decimal numbers with a 'd' appended and binary with a 'b' appended, thus 52h = 82d = 01010010b

23.2 **The 8051 clock and the machine cycle time**

The master clock can be set to any frequency between about 1 MHz and 16 MHz (or 12 MHz for some types). The user selects the master clock frequency by connecting a quartz crystal between pins 18 and 19 on the DIP, together with some auxiliary capacitors. This oscillator frequency determines the speed at which any operation is carried out. However, the machine cycle is actually 12 oscillator cycles, not one, as each instruction involves a number of smaller steps.

The machine cycle is the smallest time in which the microcontroller can execute the simplest instruction. If the instruction takes C machine cycles to execute and the oscillator frequency is f Hz, then the time taken is

$$t_{op} = 12C/f \qquad\qquad (23.1)$$

Thus if the oscillator frequency is 12 MHz, the time of execution is C μs. The 8051 can perform at most 1 million instructions per second (1 Mips). The operation mnemonics (op codes) are all listed under the appropriate headings in this chapter, together with the number of bytes taken in memory and the time for execution in machine cycles. With this information the space used in memory and the time taken to perform subroutines can be calculated. It is often essential that the precise time taken to do something is known, as well as whether a location in memory is close enough to be called by its label. This is called relative addressing and is discussed later.

Table 23.2 *The 8051 SFRs*

Name	Function	RAM address
A	Accumulator	0E0h
B	Arithmetic	0F0h
DPL	Data Pointer Low byte; addressing external memory	82h
DPH	Data Pointer High byte; addressing external memory	83h
IE	Interrupt Enable control	0A8h
IP	Interrupt Priority	0B8h
P0	Port0 latch	80h
P1	Port1 latch	90h
P2	Port2 latch	0A0h
P3	Port3 latch	0B0h
PCON	Power down CONtrol, user flags	87h
PSW	Flags	0D0h
SBUF	Serial port data BUFfer	99h
SCON	Serial port CONtrol	98h
SP	Stack Pointer	81h
TCON	Timer/counter CONtrol	88h
TMOD	Timer/counter MODe	89h
TL0	Timer0 Low byte	8Ah
TL1	Timer1 Low byte	8Bh
TH0	Timer0 High byte	8Ch
TH1	Timer1 High byte	8Dh

23.3 **The special-function registers (SFRs)**

The defined special-function registers are given in Table 23.2, above. SFRs can be used in some operation codes by their names and in others by their addresses. Since any address in a program *must* start with a number, the addresses above, such as that of port3, B0, must be written as 0B0, with a leading zero. Assembly errors result if this is not done.

23.3.1 *The program counter (PC) and the data pointer (DPTR)*

These are the only 16-bit SFRs, all the others are 8-bit. Program bytes are fetched from memory address which is in the PC. These addresses can be in ROM with 16-bit addresses from 0000h to 0FFFh (on chip) or from 0000h to FFFFh off-chip or a mixture of the two. Any address above 0FFFh will result in an off-chip call. As soon as an instruction address is fetched from the PC it is incremented, so that it always points to the *next* instruction. Unlike all the other SFRs, the PC has no address. The DTPR register is made up of two 8-bit registers, DPH and DPL, which have individual addresses. They contain the addresses for both internal and external code and external data. The DPTR is under program control and can be specified by its name (for all 16 bits) or by the names of the high and low bytes (DPH and DPL). Its hexadecimal address is the two 8-bit bytes of DPH and DPL, 83h and 82h.

23.3.2 *The A and B arithmetic registers*

A is the accumulator. It is used to perform addition, subtraction, multiplication and division (of integers) and also Boolean algebra on bits. B is used as an auxiliary register for multiplication and division only.

23.3.3 *The program status word (PSW)*

The PSW byte is set out below.

The program status word SFR, PSW	7	6	5	4	3	2	1	0
	C	AC	F0	RS1	RS0	OV		P

The functions of the bits are as follows:

Bit address	Name	Function
PSW.7	C	Carry flag, used in arithmetic, in jumps and in Boolean operations
PSW.6	AC	Auxiliary carry flag, used only in BCD arithmetic
PSW.5	F0	User flag 0
PSW.4	RS1	Register bank select bit 1
PSW.3	RS2	Register bank select bit 2

RS1	RS2	Selects
0	0	bank0
0	1	bank1
1	0	bank2
1	1	bank3

PSW.2	OV	Overflow, used in arithmetic
PSW.1		Not used
PSW.0	P	Parity flag for the accumulator. P = 1 indicates that the sum of the bits in A is odd

23.3.4 The stack pointer (SP)

The stack is a location in internal RAM which is used for fast transfer of data. The SP points to the location of the item on the top of the stack. When an item is put on the stack the SP is incremented *before* it is placed, and *after* it is removed the SP is decremented. The stack starts at RAM location 07h on powering up the chip or on reset, and can run up to top of RAM (7Fh). If the stack is expanded beyond 7Fh, disaster can result, since other memory gets overwritten including program data. The stack pointer can be set by the programmer and normally is put as high as possible in RAM, above 2Fh.

23.3.5 Counters and timers

Counters and timers are the distinguishing feature of microcontrollers. The 8051 has two counter/timers which are set up and controlled by the TMOD and TCON SFRs. The timer control register, TCON, is shown below.

	7	6	5	4	3	2	1	0
The timer-control SFR, TCON	TF1	TR1	TF0	TR0	IE1	IT1	IE0	IT0

Bit address	Name	Purpose
TCON.7	TF1	Timer1 overflow flag. Set when timer1 overflows. Cleared when CPU vectors to the interrupt service routine.
TCON.6	TR1	Timer1 runs/stops when software sets this to 1/0
TCON.5	TF0	Timer0 overflow flag.
TCON.4	TR0	Timer0 run control bit.
TCON.3	IE1	External Interrupt1 flag. Set to 1 by external interrupt and cleared to 0 when interrupt is serviced.
TCON.2	IT1	Interrupt1 control bit. Set/cleared by software to 1/0 to specify falling edge/low level interrupt triggering.
TCON.1	IE0	As IE1, but for External Interrupt0.
TCON.0	IT0	AS IT1.

Counting and timing can be done with software, but this may take too much memory, so two 16-bit up counters, T0 and T1, are provided. When counting internal clock pulses they are called timers and when counting external pulses they are called counters, but the action is the same. They are each divided into two 8-bit registers called TL0, TL1, TH0, TH1 for low and high bytes. The timers count machine cycles, that is the oscillator frequency divided by 12d.

Timers have four modes of operation that can be set by bits 1 and 0 in the TMOD register, which is shown below.

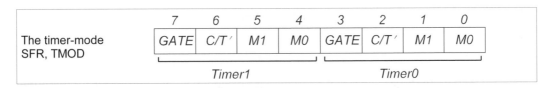

Notes:

GATE   When TRX in TCON is set and GATE = 1, timerX will run only while INTX' pin (port pin P3.2 or P3.3) is high (hardware control). When GATE = 0, timerX will only run while TRX = 1 (program control).

C/T'   Timer or Counter select. Cleared to 0 for timer and input from machine clock, set to 1 for counter and input from TX input pin.

M1   Mode selector bit 1.

M0   Mode selector bit 0.

M1	M0	Mode	Action
0	0	0	13-bit timer
0	1	1	16-bit timer/counter
1	0	2	8-bit autoreload timer/counter
1	1	3	Timer0: TL0 is an 8-bit timer/counter controlled by the standard timer0 control bits, TH0 is an 8-bit timer controlled by timer1 control bits. Timer1: stopped

Each timer can operate independently of the other in modes 0,1 and 2 but not in mode 3.

Mode 0

The 13-bit counter mode. It uses THX as an 8-bit counter and TLX as a 5-bit counter. The pulse input (machine cycle) is divided by 32d in TLX, so that THX counts the oscillator frequency divided by 384d (12d × 32d). Thus a 12MHz oscillator frequency would become a timer frequency of 31.25 kHz. The timer flag is set whenever THX goes from FFh to 00h (256 counts), or in 8.2 ms for this case.

Mode 1

This works like mode 0, but uses 8 bits in TLX, so dividing the machine frequency by 256d. A 12MHz oscillator frequency divides down to $12 \times 10^6/12/256 = 3906$ Hz, so 256 counts of this would be 65.5 ms.

Mode 2

The TLX counter is used as a stand-alone 8-bit counter, while THX is used to hold a value that is loaded into TLX every time it overflows from FFh to 00h. Thus the counter can be set up to count from anything from 00h to FEh. For example, to count exactly 200 machine cycles requires 56d (= 256d − 200d = 38h) to be loaded into THX and then the time elapsed is 200

$\times$ 12/12 $\times$ 10^6 = 0.2 ms if a 12MHz oscillator is used.

Mode 3

Choosing mode 3 for timer1 causes it to stop counting and place its control bit TR1 and flag TF1 in the service of timer0. Choosing mode3 for timer0 causes it to split into two independent 8-bit counters. TL0 sets TF0 and TH0 sets TF1. When timer0 is mode 3 timer1 can be used, but it cannot set its flag, but can still generate a clock or become a baud-rate generator.

Baud-rate generation

The standard baud rates are 19200, 9600, 4800, 2400, 1200, 600 and 300 bits/s, and the oscillator frequency must be capable of being divided down to these frequencies. The usual crystal frequency chosen is 11.0592 MHz since 11059200/12 gives a machine cycle frequency of 921.6 kHz = 48 $\times$ 19200. Normally the serial-port mode1 is selected in the serial port control register, SCON (described later), then timer mode2 is selected for timer1 in the TMOD register, and the baud rate is

$$\text{Baud rate} = B = \frac{2^{\text{SMOD}} f_{\text{osc}}}{32 \times 12 \times (256 - \text{TH1})} \tag{23.2}$$

SMOD is bit 7 of the PCON register that can be set to 1 or cleared to 0. TH1 is the reload value of the high byte of timer1, which can be loaded by programming the TH1 register (RAM address 8Dh). One usually knows what baud rate is desired and really wants to know the value of the number to load into TH1, which is given by

$$\text{TH1} = 256 - 2^{\text{SMOD}} f_{\text{osc}} / 384 B \tag{23.3}$$

For example if the baud rate is to be 600, with SMOD = 1 and f_{osc} = 11.0592 MHz, Equation 23.3 gives 160d, or A0h, as the number to be put into TH1. However, mode0 in SCON can be used in which case no timers need to be set up and the baud rate is just f_{osc}/12. For standard baud rates, this requires a very slow clock rate.

23.3.6 Interrupts

Interrupts force the program to call a subroutine located at some specified place in memory. Some interrupts are hardware generated to save time and are the only way in which real time control can be obtained. The interrupt enable (IE) SFR is shown below.

The interrupt enable SFR, IE	7	6	5	4	3	2	1	0
	EA		ET2	ES	ET1	EX1	ET0	EX0

Bit address	Name	Function
IE.7	EA	Enables/disables all interrupts when EA = 1/0.
IE.6	–	Not used
IE.5	–	Not used
IE.4	ES	Enables/disables serial port interrupt when ES = 1/0.
IE.3	ET1	Enables/disables timer1 when ET1 = 1/0.
IE.2	EX1	Enables/disables external interrupt1 when EX1 = 1/0.
IE.1	ET0	As ET1 but for timer0.
IE.0	EX0	As EX1 but for external interrupt0.

There are five interrupts, three of which are generated automatically by internal operations: timer0 flag, timer1 flag and the serial port interrupt (RI or TI). Two interrupts are triggered by external signals with circuitry connected to $\overline{INT0}$ and $\overline{INT1}$ (that is the P3.2 and P3.3 pins). Of course the program can interrupt itself simply by setting the appropriate flag.

The programmer can assign priority to interrupts with the interrupt priority (IP) SFR and can enable them with the interrupt enable (IE) SFR. All interrupts can be blocked by these controls. When the interrupt is serviced by the routine placed by the programmer at the location specified by the chip manufacturers, the interrupt is cleared and the program resumes where it left off. Program resumption is achieved by storing the interrupted PC address on top of RAM stack before changing PC to the interrupt address in ROM. The PC address is restored at the end of the interrupt subroutine with the RETI instruction.

Reset

Reset is the overriding interrupt: it cannot be blocked. In this case the PC is not saved and the reset causes a jump to ROM address 0000h and run from there.

SFR	Reset to	SFR	Reset to
A	00h	PSW	00h
B	00h	SP	07h
DPTR	0000h	SBUF	××h
IE	0××00000b	SCON	00h
IP	×××00000b	TCON	00h
P0	FFh	TMOD	00h
P1	FFh	TH0	00h
P2	FFh	TL0	00h
P3	FFh	TH1	00h
PCON	0×××××××b (HMOS)	TL1	00h
	0×××0000b (CHMOS)		

Note: × = undefined

When HIGH is applied to RST pin the reset is begun. When LOW is then applied to RST pin, the internal registers have the values listed in the table above. Internal RAM is NOT affected by reset, but in power up it is in a random state. Register bank0 is selected after reset as the PSW contains 00h.

Interrupt priority

Each source of an interrupt (except reset) can be controlled by the IE SFR, but the order of servicing interrupts is set by the IP SFR. However, the IP SFR can only give priority when the appropriate bit is set to 1. Thus if several interrupts are priority 1, then the order of servicing is IE0 > TF0 > IE1 > TF1 > serial port (RI or TI). Priority can only be given to TI (or RI) if it is set to 1 in the IP SFR and all the others to 0. The IP SFR layout is:

		7	6	5	4	3	2	1	0
The interrupt-priority SFR, IP					PS	PT1	PX1	PT0	PX0

The interrupts must be serviced by instructions located in ROM at the addresses given in the table under the 'Vector address' heading. Interrupts that are not serviced for any reason must persist until they are, or they are lost.

Bit address	Name	Priority to	Vector address
IP.7	–	Not used	–
IP.6	–	Not used	–
IP.5	–	Not used	–
IP.4	PS	Serial port	0003h
IP.3	PT1	Timer1	000Bh
IP.2	PX1	External interrupt1	0013h
IP.1	PT0	Timer2	001Bh
IP.0	PX0	External interrupt0	0023h

Input/output pins and ports

The 8051 DIP has 40 pins, which is not enough. 32 of the pins are used by 4 ports, and 24 of these can be used for dual functions, effectively expanding the pin count to 64.

Pin	Alternative use	Pin	Alternative use
P0.0	Address/data bit 0	P3.0	Receive data
...	...	P3.1	Transmit data
...	...	P3.2	External interrupt0
P0.7	Address/data bit 7	P3.3	External interrupt1
P2.0	Address bit 8	P3.4	Timer0 output
...	...	P3.5	Timer1 output
...	...	P3.6	Write strobe
P2.7	Address bit 15	P3.7	Read strobe

Strobes are activated by 0s, not 1s. Other pins on the DIP are 9 (reset), 18 (XTAl1), 19 (XTAL2), 20 (GND), 29 (program store enable, $\overline{PS\,EN}$) 30 (address latch enable, ALE, or the EPROM program pulse), 31 (external enable, $\overline{EN}$, or the EPROM programming voltage), 40 (V_{CC} or V_{DD}, +5 V).

The ports cannot drive big loads and are limited to a little over 1 mA. All the port pins have latches and the associated latch circuitry depends on the alternative function (if any). Port1 has no dual function. Ports P0 and P2 are used for accessing 16-bit addresses of external memory, P0 the low bytes and P2 the high bytes. P3 has multiple alternative uses such as serial data input/output, external interrupts and timers. Serial data is input or output in conjunction with the serial data buffer, SBUF. External interrupt0 is controlled via bit TCON.1, external interrupt1 via bit TCON.3 and the external counters via TMOD. All the ports can be used as input/output ports instead of these alternatives.

Input and output of serial data

Serial data must use port P3. The serial port control register, SCON, controls data communications, while the PCON register controls the data rate. The data is held in SBUF, which comprises two separate registers, one for write and one for read. Both have the same address: 99h. The SCON register is shown below:

The serial port control SFR, SCON	7	6	5	4	3	2	1	0
	SM0	SM1	SM2	REN	TB8	RB8	TI	RI

Bit address	Name	Function
SCON.7	SM0	Serial port mode specifier. Set/cleared by program to 1/0 to select mode
SCON.6	SM1	Serial port mode specifier. As SM0

SM0	SM1	Mode	Action
0	0	0	Shift register, baud rate = $f_{machine}$
0	1	1	Variable baud rate set by timer1
1	0	2	Fixed baud rate. If SMOD = 0 it is 1/64d of f_{osc} and if SMOD = 1 it is 1/32d of f_{osc}
1	1	3	Variable baud rate set by timer1

Bit address	Name	Function
SCON.5	SM2	Multiprocessor communication mode
SCON.4	REN	Set/cleared by program to enable/disable reception
SCON.3	TB8	9th bit sent in modes 2 and 3. Set/cleared by program
SCON.2	RB8	9th bit received in modes 2 and 3. In mode 1, if SM = 2, it is the received stop bit. Not used in mode 0
SCON.1	TI	Tx interrupt flag. Set by hardware, cleared by program
SCON.0	RI	Rx interrupt flag. Set by hardware, cleared by program

Serial data communications are slow, so serial data flags in SCON are required to aid data flow. Input data cannot be under program control, of course, and must generate an interrupt when received. The TI and RI flags are ORed to generate an interrupt. The program must respond and clear the flag, unlike timer flags that are cleared automatically.

Data transmission and reception

Any data written to SBUF is transmitted. TI (SCON.1) is set to 1 after transmission to signal that SBUF is empty so that further data can be sent to it. Data will be received if the receive enable bit, REN (SCON.4) is set to 1 for all modes. For mode 1 only, the receiver interrupt flag, RI (SCON.0), must be cleared to 0 also. RI is set after data is received in all modes. Setting REN by program is the only way of limiting the receipt of unexpected data. Reception can begin in modes 1, 2 and 3 if RI is set when the serial stream of bits begins. RI must have been reset by the program before the last bit is received or the incoming data will be lost. Incoming data is not transferred to SBUF until the last data bit has been received. Then the previous transmission can be read from SBUF while new data is being received.

Mode bits can be set with bits M0 (SCON.7) and M1 (SCON.6). Virtually any submultiple of the machine frequency can be set for the baud rate. The SMOD bit in PCON must be set with timer 1 in modes 1, 2 or 3 for variable baud rates. Serial data modes 2 and 3 are used for communicating with second processors.

23.4 **Moving data**

An important part of effective use of the microcontroller is moving data around the chip efficiently. There are several different ways of achieving this, depending on what and where the data is and the manner in which is addressed.

23.4.1 *Immediate addressing*

In this mode the data itself is entered in the program operation code. Numbers that are to be read literally (that is as pure numbers) are entered by using the symbol # followed by the number subscripted by h for hexadecimal, or d for decimal numbers. # is an *immediate data* identifier. Data is moved by the MOV mnemonic followed by the destination address and then the source address, separated by a comma:

```
MOV A,#12h
MOV R1,#0F2h
```

Note that a number must begin with a digit (0 to 9), so if a hexadecimal number such as F2h is to be used, it must be prefixed by a zero. Unfortunately, omitting the # causes the compiler to assemble the number as an opcode: there is no way of trapping this error other than by using a simulator. This lack of safety is a recurring problem for microprocessor programmers.

23.4.2 Register addressing

Some SFRs have mnemonics and can be called by them and they have addresses and can be called by them too. For example, here are several ways of MOVing data by immediate and register addressing:

Mnemonic	Comment
MOV A,#2Fh	;Copy the immediate data 2Fh into A. *(2B, 1C)*
MOV A,R1	;Copy the contents of register R1 into A. *(1B, 1C)*
MOV R1,A	;Copy the contents of A into register R1. *(1B, 1C)*
MOV R1,#0AFh	;Copy the immediate data AFh into register R1. *(1B, 1C)*
MOV DPTR,#0A2F1h	;Copy the immediate data A2F1h into the DPTR. *(3B, 2C)*

The numbers in parentheses after the comments respectively indicate the number of bytes of memory taken by the instruction and the number of machine cycles taken in execution. Comments must be preceded by a semicolon to be ignored by the assembler. The register bank that R1 is in must be specified by the appropriate bit settings in the PSW SFR (start up default setting is bank0).

23.4.3 Direct addressing

All 128 bytes of internal RAM and the SFRs can be addressed directly using the single-byte address assigned (see Figure 23.4). The 128 RAM addresses go from 00h to 7Fh. But RAM address 00h to 1Fh are also the locations of the 32 working registers in the banks:

bank	register	address
0	R0	00h
...	...	...
0	R7	07h
1	R0	08h
...	...	...
1	R7	0Fh
2	R0	10h
...	...	...
2	R7	17h
3	R0	18h
...	...	...
3	R7	1Fh

On reset RS0 and RS1 in the PSW SFR are set to 0 so bank0 is selected and the SP is set to 07h. Thus as the stack is used, register space in banks 1, 2 and 3 gets used up in turn. The programmer must ensure that the SP is set high enough not to corrupt data put in the register banks during the program.

Data can be copied from one address to another by direct addressing:

```
MOV addr1,addr2    ;Copy the contents of addr1 to addr2. In register-transfer notation,
                   ;[addr1] ← [addr2]. (3B, 2C)
MOV A,addr1        ;[A] ← [addr1]. (2B, 1C)
MOV addr1,A        ;[addr1] ← [A]. (2B, 1C)
MOV addr1,#0F1h    ;[addr1] ← #0F1h. (3B, 2C)
MOV addr1,R2       ;[addr1] ← [R2]. (2B, 2C)
MOV R2,addr1       ;[R2] ← [addr1]. (2B, 2C)
```

Moves involving the accumulator, A, are always the fastest.

23.4.4 Indirect addressing

With indirect addressing the register holds the address that is used to store the data. The MOV command only indirectly addresses registers 0 and 1, the so-called *pointing registers*. Indirect addresses are denoted by a preceding @, for instance

```
MOV @Rp,#0D3h      ;Copy the immediate data D3h into the address contained in Rp.
                   ;Or, [[Rp]] ← 0D3h. (2B, 1C)
MOV @Rp,addr1      ;[[Rp]] ← [addr1]. (2B, 2C)
MOV addr2,@Rp      ;[addr2] ← [[Rp]]. (2B, 2C)
MOV @Rp,A          ;[[Rp]] ← [A]. (1B, 1C)
MOV A,@Rp          ;[A] ← [[Rp]]. (1B, 1C)
```

The address in Rp must be a RAM or a SFR address. Indirect addressing is always used whenever external RAM is used, but with the special mnemonic MOVX to remind us that a call on external RAM is taking place. Since external RAM can go to 64 kbytes, the DPTR is most often used to store the address:

```
MOVX A,@DPTR    ;[A] ←[[DPTR]]. (1B, 2C)
MOVX @DPTR,A    ;[[DPTR]] ← [A]. (1B, 2C)
```

23.4.5 Moving data from ROM

Data in ROM can be accessed by indirect addressing using the accumulator and the PC or DPTR. The mnemonic is MOVC:

```
MOVC A,@A+DPTR     ;[A] ←[[A + DPTR]]. (1B, 2C)
MOVC A,@A+PC       ;[A] ← [[A + PC + 1]]. 1 is added because the PC increments
                   ;before the instruction executes. (1B, 2C)
```

The data pointer has separately-addressable bytes:

```
MOV DPTR,#0AF1Eh    ;Put AF1Eh into the DPTR
MOV DPH,#0AFh       ;Put AFh into the DPTR high byte
MOV DPL,#1Eh        ;Put 1Eh into the DPTR low byte
MOV 83h,#0AFh       ;Put AFh into the DPTR high byte
MOV 82h,#1Eh        ;Put 1Eh into the DPTR low byte
```

```
MOVC A,@A+DPTR          ;Copy the contents of ROM address AFC0h into A
```

Lines 2 & 3 and 4 & 5 above accomplish the same thing as line 1. The last line adds AF1Eh to A2h to form the address AFC0h. Suppose the PC contains B22Dh, then the code

```
MOV A,#0DEh       ;[A] ← DEh
MOVC A,@A+PC      ;[A] ← [[A+PC+1]]
```

would result in 3B0Ch being placed in A.

23.4.6 PUSH and POP

These are stack commands that PUSH data onto the top of the stack or POP (remove) it from the top of the stack. The stack pointer is incremented by one *before* a PUSH and decremented by one *after* a POP. For example:

```
PUSH addr1  ;Increment the PC then copy the data in addr1 into the stack address
            ;specified by the PC. (2B, 2C)
POP addr2   ;Copy the data at the internal RAM address specified by the stack pointer into
            ;addr2 then decrement PC. (2B, 2C)
```

The stack pointer resets to 07h, or is at 07h on start up. The first PUSH onto the stack therefore puts data in 08h, which is register R0 of bank1.

Examples:

```
MOV SP,#30h       ;[SP] ← 30h. The SP is set to the lowest address in the general
                  ;purpose area in RAM
MOV R0,#2Bh       ;[R0] ← 2Bh
PUSH R0           ;[SP] ← 31h. RAM address 31h now contains 2Bh
PUSH 00h          ;[SP] ← 32h. 00h is the same as R0 if register bank0 is selected.
                  ;Address 32h now contains 2Bh
POP 01h           ;[SP] ← 31h. R1 contains 2Bh
POP 0A0h          ;[SP] ← 30h. Port2 latch now contains 2Bh
```

'PUSH R0' is the same as 'PUSH 00h' only if register bank0 is selected. Direct addresses, not register names, must be used for most registers as the stack mnemonics have no means of telling which bank is in use. When the stack pointer reaches FFh it rolls over to 00h. RAM ends at 7Fh, so pushes above here give errors. The stack pointer is usually set to 30h or higher. The stack pointer can be PUSHed to and POPped from the stack.

23.4.7 Exchanging data: XCH

The mnemonic XCH can be used to swap data between the accumulator and RAM addresses or registers, for instance

```
XCH A,R3          ;[A] ← [R3], [R3] ← [A]. (1B, 1C)
XCH A,addr        ;[A] ← [addr], [addr] ← [A]. (2B, 1C)
XCH A,@Rp         ;[A] ← [[Rp]], [[Rp]] ← [A]. (1B, 1C)
XCHD A,@Rp        ;Exchange the lower nibbles between A and the address in Rp.
                  ;The higher order nibbles are unaffected. (1B, 1C)
```

Exchanges between A and any port location copy the data in port *pins* to A while data in A is copied to the *latch*. XCH is a very convenient way to save A without PUSHing or POPping. When using XCHD, the upper nibbles of A and the exchanged address are unchanged.

23.5 Logical operations

Boolean algebra can be performed on both bits and bytes. Bit operations are very useful in control applications. Only the bit-addressable registers can be specified in bit operations. The following Boolean byte-level operations can be performed:

mnemonic	Boolean
ANL (ANd Logical)	AND
ORL (OR Logical)	OR
XRL (eXclusive oR Logical)	XOR

The operations are written with two operands separated by a comma. The destination of the operation is the first operand, which must be an address in RAM or the accumulator. Indirect addresses and immediate data cannot be destinations. For example:

```
ANL A,#10h      ;AND each bit in A with the corresponding bit of the immediate number 10h, put
                ;the result in A. (2B, 1C)
ORL A,R1        ;OR each bit in A with the corresponding bits in R1 and put the result in A.
                ;(1B, 1C)
XRL A,addr      ;Exclusive OR the bits in A and the bits in the RAM address and store the result
                ;in A. (2B, 1C)
ANL addr,#2Bh   ;AND each bit in the RAM address with the corresponding bit of the immediate
                ;data 2Bh and put the result in the RAM address. (3B, 2C)
ORL A,@Rp       ;OR each bit in A with the corresponding bit in the RAM address stored in Rp.
                ;Store the result in A. (1B, 1C)
```

The following are unary (they have only one operand) operations that can only be used on the accumulator:

```
CPL A           ;ComPLement. NOT A. If A = B2h, CPL A makes A = 4Dh. (1B, 1C)
CLR A           ;CLeaR. Sets A to zero. (1B, 1C)
```

There are also byte manipulations that can only be carried out on the accumulator:

RL A	;Rotate the byte in A one bit to the left. The MSB becomes the LSB. *(1B, 1C)*
RLC A	;Rotate the byte in A and the carry bit to the left by one bit. The carry bit becomes ;the LSB and the MSB the carry. *(1B, 1C)*
RR A	;Rotate the byte in A one bit to the right. *(1B, 1C)*
RRC A	;Rotate the byte in A and the carry one bit to the right. The carry becomes the ;MSB and the LSB the carry. *(1B, 1C)*
SWAP A	;Exchange low and high nibbles in A. *(1B, 1C)*

For example:

CLR C	;C = 0
MOV A,#E6h	;[A] ← 11100110b
RL A	;[A] ← 11001101b = CDh, C = 0
RLC A	;[A] ← 10011010b = 9Ah, C = 1
RR A	;[A] ← 01001101b = 8Dh, C = 1
RRC A	;[A] ← 10100110b = A6h, C = 1
SWAP A	;[A] ← 01101010b = 6Ah, C = 1

When the *destination* of a logical operation is the port address, the *latch* register not the pins is used both as the source of the data and as the destination of the result. Logical operations that use the port as a source but not a destination use the pins as source. For example, suppose port0 has its pins grounded (zero) and its latch contains FFh, then

ANL P0,#0F0h	;Reads FFh from the latch, ANDs it with F0h to give F0h and sends this to the ;latch, which sets the lower latch bits of port0 to zero, so turning off the lower ;output transistors and pulling the lower pins up to 1
ANL A,P0	;Reads the port0 *pins* and not the port latch

23.5.1 Bit-level operations

These are useful for control. They can only be done with bit-addressable locations in RAM. The bytes of RAM that are bit addressable are from 20h to 2Fh, with bit address running from 00h to 7Fh. Thus the address of bit 4 of RAM address 28h is 44h.

byte address	bit address
20h	00h-07h
...	...
2Eh	70h-77h
2Fh	78h-7Fh

The SFRs do not all have bit addresses, those that do are

SFR	byte address	bit address
A	0E0h	0E0h-0E7h
B	0F0h	0F0h-0F7h
IE	0A8h	0A8h-0AFh
IP	0B8h	0B8h-0BFh

```
P0      80h         80h-87h
P1      90h         90h-97h
P2      0A0h        0A0h-0A7h
P3      0B0h        0B0h-0B7h
PSW     0D0h        0D0h-0D7h
TCON    88h         88h-8Fh
SCON    98h         98h-9Fh
```

When bit operations are used in the mnemonics, the assembler recognises hexadecimal addresses such as 9Fh as bit 7 of the timer control register, TCON. One can also write TCON.7 in the mnemonic and the assembler recognises this too. Using hexadecimal bit addresses makes a program more difficult for people to follow.

Bit-level Boolean operations are often done on the carry flag of the PSW, mnemonic C, which is a useful way of controlling the program according to the status of C.

```
ANL C,b     ;AND C with addressed bit, b, put result in C. (2B, 2C)
ANL C,/b    ;AND C and the complement of the addressed bit and put the result in C. b is
            ;unaffected. (2B, 2C)
ORL C,b     ;OR the carry flag with bit b, store in C. (2B, 2C)
ORL C,/b    ;OR C with the complement of b, store in C. (2B, 2C)
CPL C       ;Invert C. (1B, 1C)
CPL b       ;Invert b. (2B, 1C)
CLR C       ;Set C to zero. (1B, 1C)
CLR b       ;Set b to zero. (2B, 1C)
MOV C,b     ;Copy b into the carry. (2B, 1C)
MOV b,C     ;Copy the carry into b. (2B, 2C)
SETB C      ;Set the carry flag to 1. (1B, 1C)
SETB b      ;Set b to 1. (2B, 1C)
```

Other flags must be set using their bit addresses. As with byte operations the port bits used as destination are latches not pins. A port bit used only as a source is the pin. Latch bit logical operations are CLR, CPL, MOV, SETB only.

As an example, suppose we wish to double the contents of R0 and store the result in R1 (low byte) and R2 (high byte), the assembler instruction code for this is

```
MOV A,R0    ;[A] ← [R0]
CLR C       ;[C] ← 0, we must be sure C is 0
RLC A       ;Doubles [R0], with carry
MOV R1,A    ;Low byte in R1
CLR A       ;[A] ← 00h, clears A to receive carry
RLC A       ;The carry bit was bit 9, now bit 0 of A
MOV R2,A    ;[R2] ← [A], stores the carry (high byte) in R2
```

23.6 **Arithmetic operations**

There are 24 arithmetic operation codes of the following types:

```
INC A          ;Increment [A] by 1. (1B, 1C)
INC Rn         ;Increment [Rn] by 1. (1B, 1C)
INC addr       ;Increment the contents of RAM address by 1. (2B, 1C)
INC @Rp        ;Increment [[Rp]] by 1. (1B, 1C)
DEC A          ;Decrement [A] by 1. (1B, 1C)
...            ;etc. DEC is just like INC
ADD A,Rn       ;[A] ← [A] + [Rn]. C = 1 when [A] > FFh. (1B, 1C)
ADD A,addr     ;[A] ← [A] + [addr]. C = 1 when [A] > FFh. (2B, 1C)
ADD A,#n       ;[A] ← [A] + #n. C = 1 when [A] > FFh. (2B, 1C)
ADD A,@Rp      ;[A] ← [A] + [[Rp]]. C = 1 when [A] > FFh. (1B, 1C)
ADDC A,Rn      ;[A] ← [A] + [Rn] + [C]. (1B, 1C)
...            ;etc. ADDC is like ADD, but with carry
SUBB A,Rn      ;[A] ← [A] – [Rn], with carry. (1B, 1C)
...            ;etc. SUBB is just like ADDC, except – for +
MUL AB         ;Multiply [A] by [B] and store in A. Low order byte is left in A and the high
               ;order in B. If the product > FFh, the overflow flag is set, otherwise it is
               ;cleared. The carry is always cleared. (1B, 4C)
DIV AB         ;Divide [A] by [B], store in A. The integer part of the quotient is stored in A,
               ;and the integer remainder in B. The carry and overflow flags are both
               ;cleared. If [B] = 0 before division, [A] and [B] are both undefined, C = 0 and
               ;OV = 1. (1B, 4C)
DA A           ;Decimal adjust A. Can only be used after ADD or ADDC, not on its own.
               ;(1B, 1C)
```

There are several difficulties with arithmetic on the 8751. The first is that the numbers are of only 8 bits, so that accuracy is easily lost (and is at best ±0.4%). To maintain accuracy one must use two-byte arithmetic. In any case it is up to the programmer to take care of any carries and borrows. The flags affected are C, AC (auxiliary carry), OV (overflow), P (parity) all in the PSW. Another problem is that any arithmetic routine using division has to be protected by the programmer from inadvertent division by zero. The following is a list of the flag-affecting operations:

Mnemonic	Flags affected	Mnemonic	Flags affected
ADD	C, AC, OV	MOV C,addr	C
ADDC	C, AC, OV	MUL	C = 0, OV
ANL C,addr	C	ORL C,addr	C
CJNE	C	RLC	C
CLR C	C = 0	RRC	C
CPL C	C = C'	SETB C	C = 1
DA A	C	SUBB	C, AC, OV
DIV	C = 0, OV		

All flags are in the PSW. Flag C is set during borrowing. The parity flag is set at any time the 1s in A add up to odd, no matter what the operation.

23.6.1 Incrementing and decrementing

These instructions can be the best way of getting a desired result – simply keep INCing or DECing. No flags are affected.

```
INC A        ;[A] ← [A] + 1. (1B, 1C)
INC Rn       ;[Rn] ← [Rn] + 1. (1B, 1C)
INC @Rp      ;[[Rp]] ← [[Rp]] + 1. (1B, 1C)
INC addr     ;[addr] ← [addr] + 1. (2B, 1C)
INC DPTR     ;[DPTR] ← [DPTR] + 1. (1B, 2C)
DEC A        ;[A] ← [A] – 1. (1B, 1C)
...          ;etc. DEC is like INC, except that the DPTR cannot be DECed
```

The contents overflow from FFh to 00h. The DPTR cannot be DECremented, but there is INC DPTR, which violates the principle of regularity – rules should be made without exceptions.

23.6.2 Addition

The first operand must be A, which is also used to store the result. C is set to 1 if there is a carry out of bit 7, the MSB, otherwise it is cleared to 0. AC set to 1 if there is a carry out of bit 3 (lower nibble), otherwise it too is cleared. OV is set to 1 if there is a carry out of bit 7, but not bit 6; or a carry out of bit 6 but not bit 7: in Boolean terms, OV = C7 XOR C6.

23.6.3 Unsigned or signed addition

The programmer can decide if the numbers used in the program are unsigned or signed. Signed numbers use bit 7 as a sign bit in the most significant byte of the group of bytes chosen by the programmer to represent the largest number needed. Bits 0–6 of the most significant byte and 0–7 of any other bytes represent the magnitude and bit 7 of the MSB the sign. If bit 7 is 1, the number is negative and if bit 7 is 0, it is positive. All negative numbers are in 2's complement form. A single-byte number can therefore range from – 128d to +127d, and as 000d is positive, there are 128 negative and 128 positive numbers. The OV flag is not used for unsigned addition or unsigned subtraction. Unsigned addition uses the carry flag to show numbers > FFh, for example

```
MOV A,#0D5h
ADD A,#0A7h
```

The accumulator will contain 7Ch while C contains 1. The carry must be saved in a more significant byte.

Signed addition may be carried out in two ways: by adding like and unlike signed numbers. There is no way that addition of unlike signed numbers can exceed – 128d or +127d. For example

$$-002d = 11111110b$$
$$\underline{+045d = 00101101b}$$
$$+043d = 00101011b$$

The carry, C = 1, and OV = 0 as there was a carry from bit 6, the next most significant. The carry is ignored as OV = 0. When positive numbers are added you can get overflow:

$$+046d = 00101110b$$
$$\underline{+099d = 01100011b}$$
$$+145d \quad 10010001b = -121d, \text{ wrong!}$$

But C = 0 and OV = 1, so corrective action can be taken, which is to complement the sign bit. If two positive numbers are added without exceeding +127d, then C = 0 and OV = 0. Adding two negative numbers that do not exceed –128d gives:

$$-033d = 11011111b$$
$$\underline{-044d = 11010100b}$$
$$-077d = 10110011b$$

C = 1 and OV = 0, so the result is correct. Adding two negative numbers that give too big a result:

$$-096d = 10100000b$$
$$\underline{-064d = 11000000b}$$
$$-160d \quad 01100000b = +096d$$

C = 1 and OV = 1, so the result can be corrected by complementing the sign bit. To summarise:

C	OV	Action
0	0	None
1	0	None
0	1	Complement the sign
1	1	Complement the sign

The only action required is to complement the sign when OV = 1.

Multi-byte signed numbers work similarly but a chain of carries forms. Only the most significant byte requires a sign bit and then all other bytes are treated like unsigned numbers. To use multi-byte numbers we must employ a carry:

```
ADDC A,#0D2h
```

Note that the least significant bytes are added with ADD and only higher bytes are added with ADDC.

23.6.4 Subtraction

There are two ways in which subtraction can be achieved: (1) by 2's complement addition and (2) by direct subtraction using SUBB. With unsigned numbers the carry flag, C, must be cleared to zero before subtracting or else it is included. For multi-byte subtractions C is cleared for the first byte but is included for subsequent bytes, as in

$$
\begin{array}{rl}
                & 022d = 00010110b \\
\text{subtract} & \underline{101d = 01100101b} \\
                & -79d \quad 10110001b = 177d
\end{array}
$$

The carry, C = 1 and OV = 0. The 2's complement of the result is 079d. Subtracting 22d from 101d gives C = 0, OV = 0 and the correct result. Signed subtraction is like signed addition and if OV = 1 the sign bit of the result must be complemented.

23.6.5 Multiplication and division

These operations use registers A and B and treat *all* numbers as *unsigned*. In multiplication the high order byte of the result is placed in B and the low order in A. If the result is > FFh, OV = 1 indicating that you must look for a high order byte in B. There cannot be overflow from the B register as FFh × FFh is the largest product whose value is FE01h

```
MOV A,#0A1h
MOV B,#0C2h
MUL AB          ;A = 02h, B = 7Ah, OV = 1
```

In division DIV AB means divide [A] by [B] and put the integer part of the quotient into A and the integer remainder in B. For example, if A = 0FEh (254d), B = 23h (35d), then DIV AB gives [A] = 07h (07d), [B] = 09h (09d). Before division OV = 0 unless B contains zero, when OV is set to 1. The carry is always reset.

23.6.6 Decimal arithmetic

Decimal numbers can be used if they are put in BCD form. BCD numbers occupy 4 bits, so two can be put in a register. Unfortunately the CPU can only add in binary and the result needs to be adjusted to give the right answer. The CPU can adjust if given the DA A command (decimal adjust A). Only addition can be performed on BCD numbers. For example:

```
MOV A,#38h
ADD A,#24h      ;[A] ← 5Ch
DA A            ;[A] ← 62d
ADDC A,#19h     ;[A] ← 7Bh, C = 0
DA A            ;[A] ← 81d
ADDC A,#23h     ;[A] ← A4h, C = 0
```

DA A   ;[A] ← 04d, C = 1

The result of adding 38h to 24h is 5Ch, which is 62d. Adding 23h to 81d gives 104d, so there is a carry of 1. BCD numbers can be used only with ADD and ADDC.

23.7 **Jumps**

Jumps give the microcontroller its real power: that of decision making. Instead of executing the program instructions in sequence, the microcontroller can be made to go unconditionally or conditionally to a different point in the program by means of *jumps* or calls on subroutines. There are jumps made according to the condition of certain specified bits, or bytes, and unconditional jumps. Jumps and calls are also classified according to their ranges, that is, how much the program counter has to be changed. There is a *relative* address range of +127d to −127d bytes from the instruction following the jump or call; an *absolute* range of anywhere on the same 2kbyte page as the instruction following; and a *long* range address of 0000h to FFFFh − anywhere in the memory. Relative addresses are identified by *labels* written to the left of the instruction mnemonic.

23.7.1 Bit jumps

The bit jumps are

JC rad	;Jump to relative address if carry = 1. *(2B, 2C)*
JNC rad	;Jump to rad if C = 0. *(2B, 2C)*
JB b,rad	;Jump to rad if the addressable bit, b = 1. *(3B, 2B)*
JNB b,rad	;Jump to rad if b = 0. *(3B, 2C)*
JBC b,rad	;Jump to rad if the b = 1, then clear b to 0. *(3B, 2C)*

JBC is used for clearing flags. If a JBC mnemonic refers to a port bit, the state of the port latch SFR is changed, not cleared. The following example shows the function of JNC, JNB and JBC:

Label	Mnemonic	Comment
NEXT:	MOV R2,#32h	;[R2] ← 32h
	MOV A,R2	;[A] ← 32h, this initialises A
LOOP:	ADD A,R2	;[A] ← [A] + [R2], C = 0
	JNC LOOP	;Do LOOP until C = 1, then go on to next ;instruction
	MOV A,R2	;[A] ← 32h, initialising A
FLAG:	ADD A,R2	;[A] ← [A] + [R2]
	JNB 0D7h,FLAG	;The carry flag's bit address is D7h. The loop ;is repeated until C = 1
	JBC 0D7h,NEXT	;When C = 1, clear C to 0, go to NEXT and ;repeat the program

Using the JBC instruction instead of JB makes a CLR C instruction unnecessary. Labels cannot be words in the instruction set (*reserved* words). The label must be at an address within +127d and −128d of the next instruction to the jump. Jump ranges can be calculated by looking up the bytes used for each mnemonic. Normally the assembler should trap out-of-range addressing.

Example

In this example JNC is used to select different routines according to which byte is larger in 16-bit subtraction. The object is to subtract the smaller of two 16-bit numbers from the larger. We start off with one 16-bit number in R0 (high byte) and R1 (low byte) and the other in R2 (high byte and R3 (low byte) and the result is to be stored in R4 (high byte) and R5 (low byte). The whole subroutine is called by LCALL from the main program (LCALL is discussed in the next section).

```
SUBTR:      MOV A,R1        ;The low byte in R1 is copied to A
            MOV B,R3        ;The other low byte in R3 is copied to B
            CLR C           ;Make sure C = 0
            SUBB A,B        ;[A] ← [R1] − [R3], low byte subtracted
            MOV R4,A        ;Low byte of result is stored in R4
            MOV A,R0        ;High byte is copied into A
            MOV B,R2        ;Other high byte copied to B
            JNC SUBT1       ;Go to SUBT1 if C = 0, that is if [R1] > [R3]
            CLR C
            INC B           ;C = 1 if [R1] < [R3], so a carry of 1 is added to high
                            ;byte that was in R2
SUBT1:      SUBB A,B        ;[A] ← [R0] − [R2], high bytes subtracted
            JNC HI_SAV      ;Go to HI_SAV if C = 0, that is if [R0] > [R2]
            MOV A,R3        ;If C = 1, the order of subtraction must be reversed
            MOV B,R1
            CLR C
            SUBB A,B        ;[A] ← [R3] − [R1], low byte subtracted but numbers ;are
                            reversed
            MOV R5,A        ;Store low byte in R5
            MOV A,R2        ;High byte
            MOV B,R0        ;Other high byte
            JNC SUBT2       ;Go to SUBT2 if C = 0
            CLR C
            INC B           ;If C = 1, must increment high byte
SUBT2:      SUBB A,B        ;[A] ← [R2] − [R0], high byte subtracted but ;numbers are
                            reversed
HI_SAV:     MOV R3,A        ;high byte stored in R3
            RET             ;Return to main program
```

Note that the assembler treats upper case and lower case letters alike; CLR, CLr, cLR, ClR, Clr, cLr, clR and clr are all read as CLR. Clarity of presentation should determine which is used.

23.7.2 Byte jumps

The byte jumps are:

```
CJNE A,addr,rad    ;If [A] ≠ [addr], jump to relative address. If [A] < [addr], then C = 1, else C =
                   ;0. (3B, 2C)
CJNE A,#n,rad      ;If [A] ≠ n, jump to relative address. C = 1 if [A] < n, else C = 0. (3B, 2C)
CJNE Rn,#n,rad     ;If [Rn] ≠ n, jump to relative address. C = 1 if [Rn] < n, else C = 0. (3B, 2C)
CJNE @Rp,#n,rad    ;If [[Rp]] ≠ n, jump to relative address. C = 1 if [[Rp]] < n, else C = 0. (3B, 2C)
DJNZ Rn,rad        ;[Rn] ← [Rn] – 1, then jump to rad if [Rn] ≠ 0. (2B, 2C)
DJNZ addr,rad      ;[addr] ← [addr] – 1, jump to rad if [addr] ≠ 0. (3B, 2C)
JZ rad             ;Jump to rad if [A] = 0, flags and A unaffected. (2B, 2C)
JNZ rad            ;Jump to rad if [A] ≠ 0, flags and A unaffected. (2B, 2C)
```

For example:

```
                MOV R0,#FFh
WAIT:           DJNZ R0,WAIT                ;Does loop 254d times = 508d machine cycles
                MOV A,#00h                  ;initialises A
LONG_WAIT:      INC A
                CJNE A,#FFh,LONG_WAIT       ;Does loop 255d times = 765d machine cycles
```

Byte jumps are economical and powerful instructions.

23.7.3 Unconditional jumps

And finally the unconditional jumps:

```
JMP @A+DPTR     ;Jump to the address formed by adding [A] to [DPTR]. (1B, 2C)
AJMP sad        ;Jump to the absolute short-range address. (2B, 2C)
LJMP lad        ;Jump to the absolute long-range address. (3B, 2C)
SJMP rad        ;Jump to relative address. (2B, 2C)
NOP             ;Do nothing. This wastes time and can also be used to save places in the
                ;program for later use. (1B, 1C)
```

An example of the use of NOP is

```
CLR P1.0        ;Pin0 of port1 is driven low
NOP             ;1 (machine) cycle elapsed
NOP             ;2 cycles elapsed
NOP             ;3 cycles elapsed
SETB P1.0       ;Pin0 of port1 is driven high. 4 cycles elapsed.
```

This routine puts a low pulse on pin P1.0 for 4 machine cycles (48 oscillator cycles).

23.8 **Calls and subroutines**

Calls are made to subroutines which are located in short-range addresses (those on the same page as the instruction) or in long-range addresses (those anywhere in ROM). Interrupts require subroutines to be placed at the interrupt servicing address. For example, a program could be written which tested each port pin in turn to see what data had been received, a technique called polling. Polling takes time (a few ms) and the program could be made to respond immediately to the receipt of data at a port (within a few μs), by using an interrupt. Software interrupts can be written using *calls* to execute a subroutine. A call results in a jump to the address of the subroutine. When the subroutine has been executed the program resumes at the instruction immediately following the call. The PC is stored at the top of the RAM stack when a call is executed, requiring 2 bytes of RAM as the PC contains 16 bits. This is the location of the *next* instruction to be executed. When a *return* is encountered at the end of the subroutine the stack is automatically POPped twice to restore the PC. The subroutine return mnemonic is RET and the interrupt return instruction is RETI. The mnemonics for calls are:

```
ACALL sad    ;Call the subroutine located at the short-range address which is located on
             ;the same page as the instruction after ACALL. Push the PC contents onto
             ;the stack. (2B, 2C)
LCALL lad    ;Call the subroutine from anywhere in memory. Push the PC contents onto
             ;the stack. (3B, 2C)
RET          ;Pop two bytes from the stack to the PC, returning the program to the next
             ;instruction after ACALL or LCALL. (1B, 2C)
```

In the following example AJMP is used to skip the interrupt subroutine and start timer1.

```
      ORG 0000h      ;ORG is an assembler directive, meaning ORiGinate the code on the next
                     ;line at the address in ROM given. Here, it puts the first instruction of the
                     ;program at ROM address 0000h
      AJMP NEXT      ;Jump over interrupt subroutine
      ORG 001Bh      ;Put timer1 interrupt subroutine here
      CLR 8Eh        ;Stop timer1, reset TR1 = 0. (TR1 is timer1's control bit whose bit address
                     ;is 8Eh)
      RETI           ;Return and enable interrupt structure
            ...            ...
            ...            ...
NEXT: MOV 0A8h,#88h  ;Enable timer1 interrupt in the IE register. A8h is the IE SFR address. 88h
                     ;is 10001000 so that bit7 and bit3 are set to 1. Bit7 is the master interrupt
                     ;enable and bit3 is the timer1 interrupt enable
      MOV 89h,#00h   ;Set timer operation mode0. 89 is the TMOD SFR address
      MOV 8Bh,00h    ;Clear TL1 counter
      MOV 8Dh,00h    ;Clear TH1 counter
      SETB 8Eh       ;Timer1 control bit is bit5 in TCON, whose bit address is 8Eh. Start timer1,
                     ;set TR1 = 1
```

23.9 **Look-up tables**

Because arithmetic is very limited on an 8-bit microcontroller, it is better to compute the values of functions such as $\sin x$, $\ln x$, x^2 or $\sqrt{x}$, and put them into a table which can be looked up for the appropriate values. The table takes up some memory, but can be read much faster than the computation can be completed. The look-up table can be placed anywhere in ROM using the ORG directive and the bytes can be read in using the DB (define byte) directive. For example a table for $\sqrt{x}$ might begin

```
ORG 2000h       ;Start table at ROM location 2000h
DB 00h          ;ROM address 2000h contains 00d
DB 01h          ;2001h contains 01d
DB 01h          ;2002h contains 01d
DB 02h          ;2003h contains 02d, √3d = 2d
DB 02h          ;2004h contains 02d
...             ...
DB 10h          ;20F0h contains 16d, √F0h = 16d
...             ...
DB 10h          ;2100h contains 16d
```

If the numbers are restricted to one byte, or 8 bits, looking up is straightforward. For example, if the square root of 09h is wanted, the DPTR is set to 2000h, which is the *base address* (the table begins at 2000h). The correct number is fetched with an indirect-address instruction such as MOVX A,@A+DPTR, if 09h (the *offset*) is first placed in A.

```
MOV DPTR,#2000h     ;2000h is the base address in external ROM
MOV A,#09h          ;09h is the number whose square root is wanted
MOVX A,@A+DPTR      ;A contains the desired square root
```

If greater precision is needed, we must increase the number of bytes in the square root to two and look them up separately as high and low bytes. Though more tedious, it does not differ much from the example given. The same is true for other 16-bit tables.

Programs are easier to write and to understand if names are used for constants. The assembler directive EQU can be used as in

```
EQU base_addr,#10F0h
MOV DPTR,#base_addr
EQU hibyte,#1000h
EQU lobyte,#1200h
```

This has been a very detailed look at the 8051, but there is no middle way between the superficial discussion of microprocessors in general and the close examination of a particular model. Only by carefully studying the instruction set and facilities of a specific device can one begin to put it to work usefully. The 8051 series has much in common with other microcontrollers, so that a working knowledge of it will assist in using all of them.

23.10 **Interfacing**

Microcontrollers must send and receive signals from the external environment: that is their purpose. But most data comes in a form which is incompatible with the format required by the microcontroller, and many devices to be controlled either require data in a different form or have power or current sourcing or sinking requirements that are beyond a micro-controller's limited capacity. Interfacing devices are required to overcome these problems and so enable the microcontroller to perform its functions. Many ICs have been specifically designed for interfacing with particular microcontroller families, while other more simple devices have general applicability.

23.10.1 *Transferring data*

Data is transferred between a microcontroller and its peripherals via a collection of wires called a bus, which is divided into three subsections: the address bus, the data bus and the control bus. Figure 23.5 shows the set up with the number of lines in each bus written over them with an oblique line through the bus.

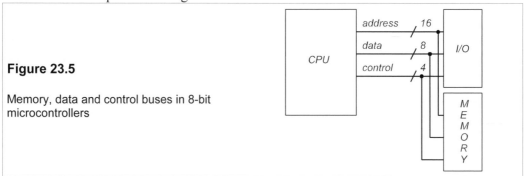

Figure 23.5

Memory, data and control buses in 8-bit microcontrollers

The address bus conveys the identity or number of the device to or from which data is transferred, and the control bus signals the type of data to be transferred and also provides synchronisation. The address bus has as many lines as there are bits in the memory address, usually 16 for 8-bit microcontrollers. The number of locations that can be addressed is then $2^{16} = 65536$. The data bus is the defining bus for microcontrollers: an *n*-bit microcontroller has *n* lines in its data bus, so the 8051 microcontrollers have eight data lines. The control bus will be smaller, normally with four lines for an 8-bit micro-controller, and a total of only 16 control codes. Different microcontrollers have different ways of signalling and each has to be learnt as the need arises.

Data may be written to or read from an external device according to the state of the write and strobe (an enabling pulse) signals. When the write signal is HIGH (logic 1) data is *read* and when it is LOW (logic 0) data is written, so that write is active low. This is indicated by writing it: write'. The timing diagram for a write instruction might look like Figure 23.6

which shows the address lines, data lines and write′ and strobe′ (which is also active low) signals.

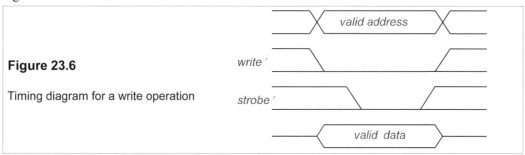

Figure 23.6

Timing diagram for a write operation

In Figure 23.6 neither the horizontal time axis nor the vertical voltage axis is indicated, which is the custom. The three kinds of transition shown in Figure 23.6 mean: firstly (Figure 23.7a) that the address data changes from one defined state to another at the time indicated by the crossover; secondly (Figure 23.7b) that the write′ and strobe′ lines change from whatever state they were in to a logical 0 at the instant indicated; thirdly (Figure 23.7c) that the state changes from whatever it was to a logical 1; and fourthly (Figure 23.7d) that the data lines change from an undefined state to a defined state at the times indicated.

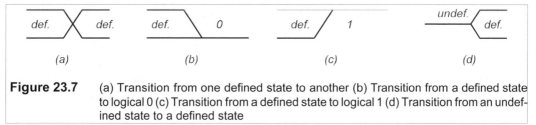

Figure 23.7 (a) Transition from one defined state to another (b) Transition from a defined state to logical 0 (c) Transition from a defined state to logical 1 (d) Transition from an undefined state to a defined state

A similar timing diagram to that of Figure 23.6 can be constructed for the process of reading data but the write′ signal is then set to logical 1. For a write cycle the master[2] microcontroller puts the address to be written to on the address bus and the data to be transferred on the data bus. It then sets write′ to logical zero and when all the signals have settled down it sets strobe′ to zero (it is said to *assert* strobe′). After a fixed length of time strobe′ is released to 1 and the master stops driving the data bus.

But the process of transferring data from the master CPU to the slave peripheral and vice versa takes time and that time will vary from one slave device to another, so how long must strobe′ be asserted? Obviously, as long as it takes the slowest slave to accept the data. This means that all data transfers must run at the pace of the slowest slave, which is clearly not a good thing as it will slow down the operation considerably: there is no point in spending 1 ms on an operation that need only take 10 ns. The solution is to communicate from the slave to

---

[2] 'Master' here means the device which initiates an action. 'Slave' means a device which is read from or written to by the master device. Devices which are written to are sometimes called receivers and devices which are read from are sometimes called transmitters. Masters and slaves can do either or both.

the master so that the master knows that the data has been received. This form of acknowledgement is called handshaking, and the data transfer is called asynchronous, in contrast to that just described which is termed synchronous. Thus an additional signal, acknowledge', is required and the timing diagram for the write operation is that of Figure 23.8.

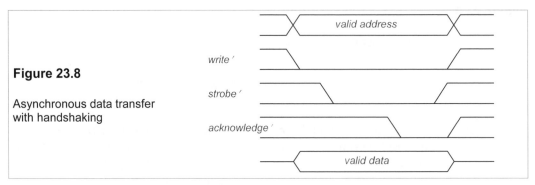

Figure 23.8

Asynchronous data transfer with handshaking

23.10.2 Data buffering

Devices are connected to the data bus via buffers which prevent the device from interfering with the bus while it carries out unrelated data transfers, that is it isolates the device from the bus except when data is to be transferred. Three-state buffers can be used for the purpose of isolation as they present a high impedance to the bus when not in use. Figure 23.9 shows an octal three-state CMOS buffer in the 74 series of logic circuits, the 74HCT240N, in 20pin DIP form, costing 38p for 1+. The truth table for the 74241 buffer is given in Table 23.3.

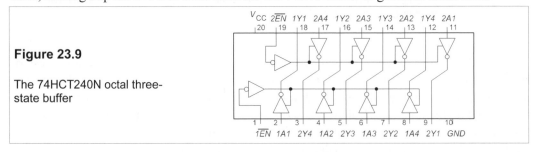

Figure 23.9

The 74HCT240N octal three-state buffer

Table 23.3

The truth table for the 74241 three-state buffer

1EN	1A	1Y	2EN	2A	2Y
1	1	1	1	1	1
1	0	0	1	0	0
0	1	Z	0	1	Z
0	0	Z	0	0	Z

Notes: Z = high-impedance state. The 74240 buffer inverts the data; the 74241 does not.

When EN is HIGH the data is transferred. When EN is LOW the outputs are high impedance regardless of the states of the inputs. For example, if we wish to connect a hexadecimal keypad to the bus we must connect the four data lines to one side of the octal buffer and also the keypad status line which tells whether a key has been pressed. This leaves three input lines spare which can be connected to ground (it is always unwise to leave unconnected inputs floating). Figure 23.10 shows these connections. Since the grounded inputs will read as zeros they are made the MSBs (pins 2A2–4) and the status line is the LSB of the upper nibble (2A1). The two enable pins are connected together.

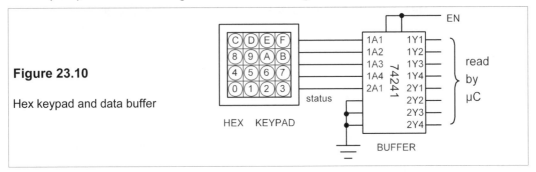

Figure 23.10

Hex keypad and data buffer

23.10.3 Latches

Ports 1, 2 and 3 of the 8051 microcontroller have output buffers which can drive up to four TTL loads each and port 0 can drive eight. But microcontrollers cannot supply much current to drive loads and one way around this problem is to use a latch to source or sink current. Generally speaking latches can sink more current than they can source. The 74 series of ICs contains a number of latches such as the 74HCT373 which is a three-state octal D-type latch, capable of supplying ±35 mA at each pin. The data pins of the latch can be written to and the data transferred to the output pins, thus either driving them HIGH (sourcing current) or LOW (sinking current). Two control pins are used on the 74373 latch, the latch enable (EN) and the output control (OC′). Its truth table is that of Table 12.7, which shows that the OC′ is normally LOW and when HIGH it causes the latch to present a high impedance to the bus. Maintaining both control pins at LOW keeps the data on the outputs. With OC′ LOW and EN HIGH the data on the input pins is transferred to the output pins.

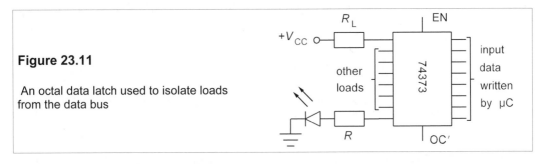

Figure 23.11

An octal data latch used to isolate loads from the data bus

Figure 23.11 shows the latch connected to one load which takes current (the LED) and one which supplies current. The latch will dissipate less power when sinking current and this is to be preferred if several loads are to be activated simultaneously.

Resistor R should be about 150 Ω if $V_{CC} = +5$ V, to restrict the LED current to about 0 mA. If the current into the latch pins is to be no more than 35 mA, then R_L should be not less than 130 Ω if V_{CC} is +5 V. This current is enough to drive many relays and so large loads can be accommodated.

Table 23.4 *Truth table for the 74373 three-state octal latch*

OC′	EN	Input	Output
0	1	1	1
0	1	0	0
0	0	×	Q_p
1	×	×	Z

Notes: × = don't care Q_p = previous value Z = high-impedance state.

23.10.4 Interfacing the asynchronous bus

When several devices are connected to the microcontroller each is given an identifying number which is its address so that it can be written to or read from as required. An asynchronous bus requires two signals to be generated by external logic: a valid′ and an acknowledge′ signal. The valid′ signal is generated when the address is the correct one and strobe′ is asserted (that is LOW). The acknowledge′ signal is generated when the data has been successfully read from or written to the device whose address is valid. If the address of the device is entered via a hexadecimal keypad then an 8-bit comparator (sometimes called an equality detector) will serve to generate the valid′ signal

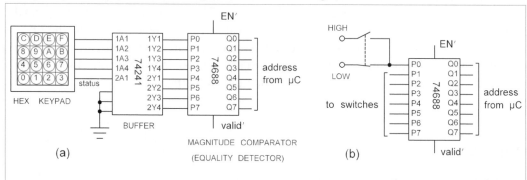

Figure 23.12 Using an equality detector to generate a valid′ signal (a) Address entered by keypad (b) Address entered by switches

The 74HCT688 is an 8-bit device which will serve the purpose. Figure 23.12a shows the circuit for this. Normally each device on a bus will be attached to an equality detector which has the address entered, perhaps by switches as in Figure 23.12b.

The timing of acknowledge' depends on the operational details of the devices being addressed. Figure 23.13 shows a timing diagram with delayed acknowledge'. In some cases it is necessary to delay assertion of acknowledge' and for this purpose an 8-bit shift register could be used to produce a delay varying from 1–8 clock cycles.

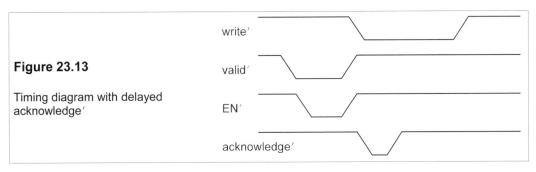

Figure 23.13

Timing diagram with delayed acknowledge'

Inverters can also be used (as well as other logic gates) to produce reasonably accurate unclocked delay times. The disadvantage of using gates, as in Figure 23.14a, is that the delay is variable depending on the device family, the supply voltage and manufacturing tolerances. Thus a double inversion using a standard 74HCT04 CMOS IC can have a delaying effect of from 50 ns up to 120 ns. If a reasonably constant and definite delay is wanted, the circuit of Figure 23.14b will work well. There are two inputs on this shift-register chip, either can be used. The delay at each output, Q_n, is $n/f_c + \Delta T$, where f_c is the clock frequency and ΔT is the propagation delay from clock input to gate output.

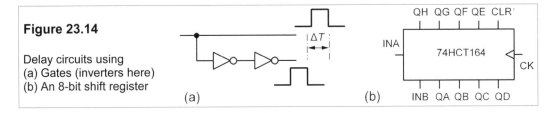

Figure 23.14

Delay circuits using
(a) Gates (inverters here)
(b) An 8-bit shift register

Suggestions for further reading

The best information on microprocessors and microcomputers is to be found in the manufacturers' data books, which are indispensable for the use of any of their products. These must be studied before any sensible choice from the numerous variants can be made. Some useful additional help can be found in the following:

8051 microcontrollers by D Calcutt, F Cowan and G. Parchizadeh (Butterworth-Heinemann, 1998)

24 Analogue communications

B Y 'COMMUNICATIONS' we really mean 'electronic communications', whose era began with the opening of a commercial telegraph by Cooke and Wheatstone on the Great Western Railway in London in 1839. In the USA Morse's telegraph opened between Baltimore and Washington, a distance of 65 km, in 1844. Telegraphy grew so rapidly that by about 1860 Dickens could remark in a novel on the wires festooned about London. The practical and theoretical demands of the electric telegraph produced a remarkable number of inventions and discoveries, particularly when submarine cables of great length were laid[1]. William Thomson, later Lord Kelvin, was knighted in 1866 for his substantial contributions to undersea telegraphy.

Telegraphs sent coded signals, but the telephone invented by Bell[2] could transmit speech itself. His US patent No. 174,465 of 1876, the most profitable ever granted, says the instrument worked by 'causing electrical vibrations similar in form to the vibrations of the air accompanying the said vocal sounds'. From then on telephonic communications grew exponentially and with them the need for better and better switching systems.

Radio telegraphy came on the scene in 1894 with Marconi's spark transmitter and made use of a new medium for electromagnetic communications. Improvements were so rapid that the Atlantic was spanned as early as 1901. But radio (and television) as we know it had to wait for the invention of the diode (Fleming, England, 1904) and the triode (de Forest, US, 1907). Microwaves were first used for communication in 1934 in a link across the English Channel operating at 1.7 GHz. After about 1960 microwaves became the standard carrier for long-distance telephone communications, particularly in N America. Since 1980 optical fibres have taken over more of this role.

24.1 The elements of a communications system

The three essential elements to a communications system are a transmitter, a channel and a receiver as shown in Figure 24.1. By *transmitter* we mean all the hardware and software necessary to encode the information that is to be sent down the channel, and by *receiver* we mean all the hardware and software necessary to decode the transmitted information to make

---

[1] The first transatlantic cable was completed on Aug. 6th, 1858; it failed in October the same year. The first transatlantic *telephone* cable was not laid until 1956 and ten years later the use of transatlantic cables exclusively for telegrams ended.

[2] Alexander Graham Bell (1847–1922) was born and educated in Edinburgh, where his scholastic attainment was not great, and moved to Canada with his parents in 1870. He became Professor of vocal physiology and elocution at Boston University in 1873, while teaching deaf mutes privately with outstanding success. He invented the telephone in his private laboratory in Boston in 1875/6.

it intelligible at the other end. Each of the three stages in the system presents an opportunity for information to be lost or corrupted by one of several processes: signal distortion, attenuation, noise and equipment malfunction. It is the business of the engineer to understand these processes and design or adapt systems to minimise their effects.

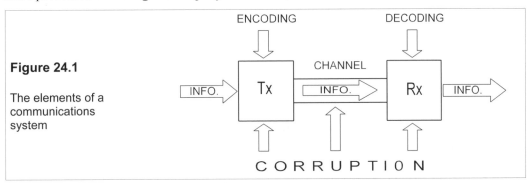

Figure 24.1

The elements of a communications system

Communications systems based on electronics employ a *carrier* wave of a chosen frequency which is *modulated* before transmission, that is the information to be sent is encoded by modifying the carrier wave in some way. For example, we may modulate the carrier's amplitude, frequency or phase; or we may send bursts of carrier as in Morse code. Modulation can be either analogue, where some property of the carrier wave is modulated in a continuous way, or digital, where the property is modified in discrete steps. Once modulated, the carrier is transmitted down the *channel*, defined as that part of the frequency spectrum being used in the medium through which communication occurs. A single physical link may contain numerous channels, but often 'channel' is used to refer to a particular medium, link or even a complete communications system: the context should indicate which.

The medium can be almost anything: a vacuum, air, water, earth, wire, waveguide, optical fibre – anything which can support wave motion of some kind, though electromagnetic waves and acoustic waves are the only ones used in practice and in electronic systems, electromagnetic waves are overwhelmingly predominant. Because electromagnetic waves are the carrier for nearly all commercial communications systems (a very minor exception is underwater telephony) the electromagnetic spectrum will first be examined.

24.2 The electromagnetic spectrum

Radio waves, infra-red waves, visible light, X-rays and γ-rays are all electromagnetic waves which differ only in frequency or wavelength since

$$f = c/\lambda \tag{24.1}$$

where f is the frequency, c is the speed of light and λ is the wavelength. Because the speed of light is so high (3×10^8 m/s) electronic communications are effectively instantaneous on earth. If light travelled only as fast as sound (300 m/s in air, 1500 m/s in water), distant communications would resemble a postal service.

The electromagnetic spectrum is shown in Figure 24.2, which shows how small a part can be readily apprehended by the human senses: only about an octave.

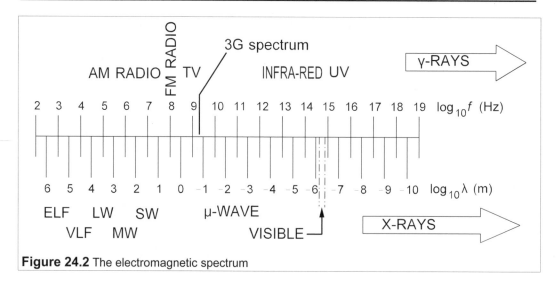

Figure 24.2 The electromagnetic spectrum

For communications purposes all of the spectrum below the ultra-violet region is used. Because the energy associated with high frequencies and short wavelengths is high, those parts of the spectrum beyond the visible are the domain of ionising radiation – potentially harmful and difficult to use for communication. Figure 24.2 is very misleading because it is drawn with a logarithmic scale to fit all the frequencies in. The capacity of a communications system – the amount of information that can be sent in a given time – is proportional to its bandwidth or the range of frequencies used. Thus if a system uses the slice of the spectrum from 100 kHz to 200 kHz it has only 0.1% of the capacity of a system using the frequencies between 100 MHz and 200 MHz. The capacity of a system operating in the ELF (extra-low frequency – below 3 kHz) region is extremely limited and low frequencies are of little practical importance.. Communications have been driven to use ever higher frequencies to expand capacity: in fact the history of electronic communications is a record of its progress from left to right of Figure 24.2. The logarithmic scale of the figure grossly exaggerates the importance of lower frequencies and diminishes that of higher frequencies.

Bandwidth is extraordinarily dear: in the 3G (third generation) spectrum auction Vodaphone paid £6 *billion* to the UK treasury for 30 MHz of bandwidth in two slices between 1.955–1.96 GHz and 2.135–2.15 GHz, or £200/Hz. At that price the whole wireless/microwave spectrum in the UK only would be worth £200 000 *billion*, which is rather more than 200 times the GNP. The entire 3G bandwidth of 140 MHz registers as a very thin line on Figure 24.2.

24.3 Amplitude modulation (AM)

Anyone entering a crowded room, where all and sundry are engaged in animated conversation, will know that communication is difficult when only a limited band of frequencies – in this case a few kHz – is available. It would be quite impractical to transmit these frequencies of electromagnetic wave directly through the air as all the signals would interfere with each other. Modulation enables us to use much higher frequencies and thereby establish numerous separate channels each operating on or near its own wavelength.

Consider the carrier sinewave

$$e_c = E_c \sin(\omega_c t + \phi_c) \qquad (24.2)$$

Three parameters can be altered in proportion to the instantaneous amplitude of the modulation: E_c, ω_c and ϕ_c. All three methods are forms of analogue modulation. If E is chosen, the amplitude of the carrier is changed and we have *amplitude modulation* or AM. If ω_c (or f_c) is chosen we have *frequency modulation*, FM, while choosing ϕ_c would be to use *phase modulation*, PM. In practice modulating ω_c amounts to much the same as modulating ϕ_c and so PM will not be discussed here as it is of less importance.

AM radio operates on wavelengths roughly between 10 m and 2 km and consequently the signal can pass around fairly large obstacles; it is not a *line-of-sight* system. This confers the advantage, especially with long-wave radio, of wide-area coverage from a single, powerful transmitter. AM radio stations use class-C amplifiers (or class-B amplifiers for very low power) as discussed in Chapter 12. With transistorised class-C amplifiers the carrier signal is applied to the base and the collector current is modulated by a suitably conditioned audio signal. Demodulation can be accomplished with a simple envelope-detector circuit as described in Section 7.6.8. Figure 24.3 shows a block diagram of the transmitter system.

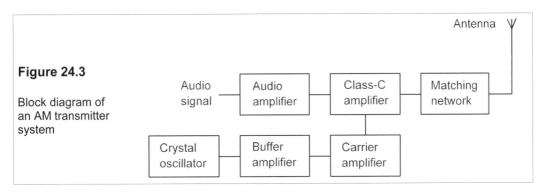

Figure 24.3

Block diagram of an AM transmitter system

In amplitude modulation the amplitude of the carrier is changed proportionally to the instantaneous amplitude of the modulating signal. For example, if the signal is the pure sinewave, $e_m = E_m \sin\omega_m t$, and the carrier is the pure sinusoid $E_c \sin\omega_c t$, the resultant AM wave will be

$$e = (E_c + E_m \sin\omega_m t)\sin\omega_c t = (E_c + e_m)\sin\omega_c t = e_n \sin\omega_c t \qquad (24.3)$$

where

$$e_n = E_c + e_m = E_c + E_m \sin\omega_m t \qquad (24.4)$$

e_n is the envelope of the carrier, as in Figure 24.4.

Dividing Equation 24.4 by E_c produces

$$e_{nn} = e_n/E_c = 1 + (E_m/E_c)\sin\omega_m t = 1 + m\sin\omega_m t \qquad (5)(24.5)$$

where e_{nn} is the normalised envelope, that is the same as taking the carrier amplitude to be 1 V, and m is the *modulation index*, defined by

$$m = E_m/E_c \qquad (6)(24.6)$$

The modulation index is often expressed as a percentage.

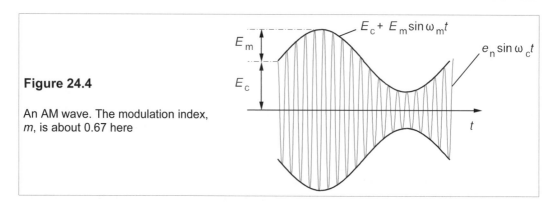

Figure 24.4

An AM wave. The modulation index, *m*, is about 0.67 here

When the modulation index exceeds 100%, the modulated carrier wave is suppressed during part of the transmission and the signal is distorted as shown in Figure 24.5. It is therefore important to ensure that $0 \leq m \leq 1$.

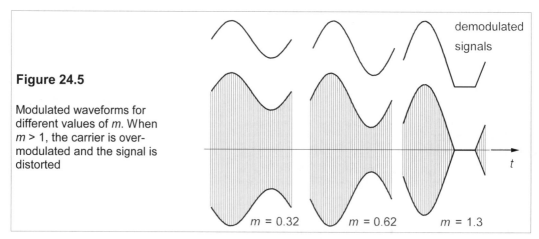

Figure 24.5

Modulated waveforms for different values of *m*. When *m* > 1, the carrier is over-modulated and the signal is distorted

24.3.1 Frequency components of an AM sinewave: sidebands

The normalised voltage of a carrier wave modulated by a single sinusoid becomes

$$e = e_{nn} \sin \omega_c t = (1 + m \sin \omega_m t) \sin \omega_c t \qquad (24.7)$$

Recalling that

$$\cos(A - B) - \cos(A + B) = 2 \sin A \sin B \qquad (24.8)$$

Equation 24.7 can be written in the form

$$e = \sin \omega_c t + 0.5m[\cos(\omega_c - \omega_m)t - \cos(\omega_c + \omega_m)t] \qquad (24.9)$$

An AM wave therefore has three component frequencies: the carrier frequency, ω_c, a difference frequency, $\omega_c - \omega_m$ and a sum frequency, $\omega_c + \omega_m$. The modulating signal will ordinarily contain a range of frequencies and the resultant sum and difference frequencies will occupy bands above and below the carrier frequency called the *upper sideband* (USB) and the *lower sideband* (LSB) as in Figure 24.6.

Figure 24.6

An AM sinewave has three
frequency components.
Modulation creates sidebands

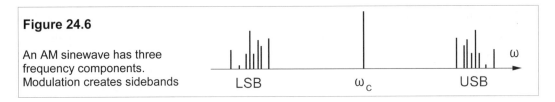

LSB ω_c USB

24.3.2 Power in an AM sinewave

The three component frequencies produced by modulating a sinewave with another sinewave can be represented by three voltage sources in series. By the superposition principle each may be taken separately in a circuit and the voltages and currents found by addition of the three separate components. However, the average power developed by the AM sinewave will be proportional to the sum of the squares of all the component amplitudes. Then when the voltage as given by Equation 24.9 is impressed across a 1Ω resistance, the average power developed will be

$$P_m = 0.5(1 + 0.25m^2 + 0.25m^2) = 0.5(1 + 0.5m^2) \qquad (24.10)$$

The half comes from the amplitude of the carrier being unity with an r.m.s. value of $1/\sqrt{2}$; and the amplitude of the side bands is $0.5m$, an r.m.s. voltage of $0.5m/\sqrt{2}$. Now the power of the unmodulated carrier is 0.5, so the ratio of the power of a modulated to that of an unmodulated sinewave is

$$\frac{P_m}{P_c} = \frac{0.5(1 + 0.5m^2)}{0.5} = 1 + 0.5m^2 \qquad (24.11)$$

As the maximum value of m is 1, this ratio has a maximum value of 1.5: an AM sinewave develops up to 50% more power than the unmodulated carrier.

Antenna currents are easily measured and from them the modulation index can be calculated with a modification of Equation 24.11:

$$\frac{I_m}{I_c} = r = \sqrt{1 + 0.5m^2} \quad \Rightarrow \quad m = \sqrt{2(r^2 - 1)} \qquad (24.12)$$

For example, if an unmodulated carrier produces a current of 600 mA in an antenna, which increases to 700 mA when the carrier is modulated by a single sinewave, the modulation index from Equation 24.12 is

$$m = \sqrt{2[(700/600)^2 - 1]} = 0.85 \qquad (24.13)$$

Audio signals contains many frequency components and the total power is then

$$\frac{P_m}{P_c} = 1 + 0.5m_1^2 + 0.5m_2^2 + \dots = 1 + 0.5m^2 \qquad (24.14)$$

where $m^2 = m_1^2 + m_2^2 + \dots$, is the square of the effective modulation index and m_1, m_2 etc. are the modulation indices of the component frequencies. The effective modulation index must still not exceed 100%, or signal distortion will result.

24.3.3 Single-sideband operation

Examination of Equation 24.9 indicates that the essence of the information in the AM sinewave lies in the sidebands, the carrier itself conveys no information and it may be suppressed without loss. There are twin benefits of doing this: less power is needed in the transmitted signal and less bandwidth is required. In practice the techniques for modulation and demodulation of full-carrier, double-sideband (FCDSB) transmissions are so easily accomplished that this form is standard for radio broadcasting. Suppression only of the carrier is achieved with a balanced modulator and the transmitted signal is known as a suppressed-carrier, double-sideband signal (SCDSB). The power saving is

$$1 - P_{\text{SCDSB}}/P_{\text{m}} = 1 - \frac{m^2/2}{1 + m^2/2} \geq 67\% \tag{24.15}$$

Only one sideband is actually required to transmit all the information in the modulating signal and so single-sideband (SSB) transmission is commonly used where bandwidth and power are at a premium, such as portable radio transmitters. The power savings appear to be at least 83%. Carrier suppression is achieved by using a balanced modulator, which puts out the modulating frequency and the upper and lower sidebands. The unwanted sideband may be removed by filtering (of limited use at high frequencies), by a phase-shift method and by the so-called Weaver's third method.

24.4 Frequency modulation

Frequency modulation (FM) came into radio transmission some time after AM because it was technically more difficult. FM radio uses wavelengths around 3 m and is therefore much more susceptible to the intervention of objects between transmitter and receiver; it is more akin to a line-of-sight system. A single, powerful transmitter is not sufficient to obtain wide-area radio coverage and geographical features may require subsidiary transmitters to be built to ensure adequate local reception. Even when the coverage in this way is adequate, the loss of signal caused by bridges and other obstacles makes in-car FM inferior to AM, though if the radio set is well located and not moved, FM reception is usually superior to AM.

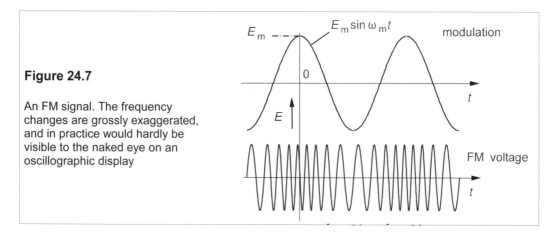

Figure 24.7

An FM signal. The frequency changes are grossly exaggerated, and in practice would hardly be visible to the naked eye on an oscillographic display

In FM the instantaneous frequency of the carrier is proportional to the instantaneous amplitude of the modulating signal, as shown in Figure 24.7. The amplitude of the FM signal is constant. If the modulating voltage is $E_m \cos \omega_m t$, and the unmodulated carrier frequency is ω_c, then the instantaneous angular frequency of the FM signal is

$$\omega = \omega_c(1 + kE_m \cos \omega_m t) \tag{24.16}$$

k is a constant and a cosine function has been used for convenience. The frequency of the FM wave deviates by a maximum of $\pm kE_m\omega_c (= \pm 2\pi\delta f)$ from the carrier angular frequency, where δf is the maximum deviation in Hz. The instantaneous voltage of the FM signal is

$$e = E_c \sin \theta(t) \tag{24.17}$$

where $\theta(t)$ is a function to be determined. Now $\omega = d\theta/dt$, so from Equation 24.16 we find

$$\theta = \int \omega \, dt = \int (\omega_c + 2\pi\delta f \cos \omega_m t) dt = \omega_c t + \frac{2\pi\delta f \sin \omega_m t}{\omega_m} + \phi \tag{24.18}$$

ϕ is a constant of integration and may be neglected, in which case the expression for the instantaneous voltage of an FM signal is

$$e = E_c \sin(\omega_c t + m_f \sin \omega_m t) \tag{24.19}$$

where $m_f (\equiv 2\pi\delta f/\omega_m = \delta f/f_m)$ is the *modulation index* for FM. This can be compared with the phase-modulated signal, $e = E_c \sin(\omega_c t + E_m \sin \omega_m t)$, and the near-equivalence is seen.

Example 24.1

In an audio-FM signal the modulating (audio) signal is a sinewave of frequency 1.5 kHz. The maximum deviation is 12 kHz, while the unmodulated carrier frequency is 90 MHz and its r.m.s. voltage is 10 V. Write down the expression for the instantaneous voltage of the FM signal. If the audio signal's amplitude is halved and its frequency is doubled, what then is the modulation index?

Equation 24.19 gives the instantaneous FM voltage. The maximum amplitude, $E_c = \sqrt{2}E_{rms} = \sqrt{2} \times 10 = 14.14$ V. The modulation index, $m_f = \delta f/f_m = 12/1.5 = 8$. The instantaneous voltage is

$$e = 14.14 \sin(2\pi \times 90 \times 10^6 t + 8 \sin 2\pi \times 10^3 t) \text{ V} \tag{24.20}$$

If the audio signal halves in amplitude, so must the deviation, that is $\delta f = 0.5 \times 12 = 6$ kHz, while f_m doubles to 3 kHz, making $m_f = 6/3 = 2$.

24.4.1 *Sidebands*

Equation 24.19 may be expressed as a sum of single-frequency sinewaves, but because it involves the sine of a sine, the number of terms is infinite and the amplitudes of the components must be evaluated from Bessel functions. The expression is

$$\begin{aligned} e = E_c\{&J_0(m_f)\sin \omega_c t + J_1(m_f)[\sin(\omega_c + \omega_m)t - \sin(\omega_c - \omega_m)t] \\ &+ J_2(m_f)[\sin(\omega_c + 2\omega_m)t + \sin(\omega_c - 2\omega_m)t] \\ &+ J_3(m_f)[\sin(\omega_c + 3\omega_m)t - \sin(\omega_c - 3\omega_m)t]...\} \end{aligned} \tag{24.21}$$

where $J_n(m_f)$ is a Bessel function of the first kind, of order n and argument m_f, found from the defining formula

$$J_n(2x) = x^n \sum_{m=0}^{\infty} \frac{(-1)^m x^{2m}}{m!(n+m)!}$$

(24.22)

These may be looked up in tables, such as Table 24.1. The sidebands are equally spaced about the carrier frequency at $\pm f_m$, $\pm 2f_m$ etc.

Table 24.1 *Bessel functions of the first kind*

$m_f \downarrow$	0	1	2	3	4	5	6	7	8	9	10	11	12	13	14	15	16	17
0	1.00	0.00	0.00	0.00	0.00	0.00	0.00	0.00	0.00	0.00	0.00	0.00	0.00	0.00	0.00	0.00	0.00	0.00
0.5	0.94	0.24	0.03	0.00	0.00	0.00	0.00	0.00	0.00	0.00	0.00	0.00	0.00	0.00	0.00	0.00	0.00	0.00
1	0.77	0.44	0.11	0.02	0.00	0.00	0.00	0.00	0.00	0.00	0.00	0.00	0.00	0.00	0.00	0.00	0.00	0.00
2	0.22	0.58	0.35	0.13	0.03	0.00	0.00	0.00	0.00	0.00	0.00	0.00	0.00	0.00	0.00	0.00	0.00	0.00
3	-0.26	0.34	0.49	0.31	0.13	0.04	0.01	0.00	0.00	0.00	0.00	0.00	0.00	0.00	0.00	0.00	0.00	0.00
4	-0.40	-0.07	0.36	0.43	0.28	0.13	0.05	0.02	0.00	0.00	0.00	0.00	0.00	0.00	0.00	0.00	0.00	0.00
5	-0.18	-0.33	0.05	0.36	0.39	0.26	0.13	0.05	0.02	0.01	0.00	0.00	0.00	0.00	0.00	0.00	0.00	0.00
6	0.15	-0.28	-0.24	0.11	0.36	0.36	0.25	0.13	0.06	0.02	0.01	0.00	0.00	0.00	0.00	0.00	0.00	0.00
7	0.30	0.00	-0.30	-0.17	0.16	0.35	0.34	0.23	0.13	0.06	0.02	0.01	0.00	0.00	0.00	0.00	0.00	0.00
8	0.17	0.23	-0.11	-0.29	-0.10	0.19	0.34	0.32	0.22	0.13	0.06	0.03	0.01	0.00	0.00	0.00	0.00	0.00
9	-0.09	0.25	0.14	-0.18	-0.27	-0.06	0.20	0.33	0.30	0.21	0.12	0.06	0.03	0.01	0.00	0.00	0.00	0.00
10	-0.25	0.04	0.25	0.06	-0.22	-0.23	-0.01	0.22	0.31	0.29	0.20	0.12	0.06	0.03	0.01	0.00	0.00	0.00
11	-0.17	-0.18	0.14	0.23	-0.02	-0.24	-0.20	0.02	0.22	0.31	0.28	0.20	0.12	0.06	0.03	0.01	0.00	0.00
12	0.05	-0.22	-0.08	0.20	0.18	-0.07	-0.24	-0.17	0.05	0.23	0.30	0.27	0.20	0.12	0.07	0.03	0.01	0.00
13	0.21	-0.07	-0.22	0.00	0.22	0.13	-0.12	-0.24	-0.14	0.07	0.23	0.29	0.26	0.19	0.12	0.07	0.03	0.01

Note: A negative coefficient indicates an inversion and can be ignored. The coefficients are relative *amplitudes* and the power in a component is proportional to the amplitude *squared*

In an FM signal the modulated carrier conveys the same power as the unmodulated carrier; modulation transfers power from the carrier to the sidebands. For some values of m_f, $J_0 = 0$, and *all* the power resides in the sidebands. These are the *eigenvalues* of m_f. (They are $m_f = 2.4, 5.5, 8.7, 8.7 + \pi, 8.7 + 2\pi, \dots$).

The relative power in a sideband is proportional to the square of its coefficient, and hence, though the number of sidebands is infinite, in practice only those in the table whose coefficients are greater than 0.1 need be considered (that is a fractional power of 1%). The effective bandwidth of an FM signal is given by Carson's rule:

$$B = 2(f_m + \delta f)$$

(24.23)

In normal (as opposed to narrow-band) FM, δf is substantially greater than f_m causing the bandwidth to be essentially independent of the modulating frequency. Wide-band FM is virtually a *constant bandwidth, constant amplitude* system, unlike AM.

Example 24.2

What are the bandwidths of the FM signals of Example 24.1 (a) from Table 24.1 and (b) from Carson's rule? What is the sideband power as a fraction of the total power in each case? Draw the spectra.

In the first case, $m_f = 8$ and Table 24.1 shows that the sidebands are significant (a fractional amplitude > 0.1) up to $n = 9$, giving a bandwidth of $2nf_m = 2 \times 9 \times 1.5 = 27$ kHz. In the second case, $m_f = 2$ and the sidebands are significant up to $n = 3$, giving a bandwidth of $2nf_m = 2 \times 3 \times 3 = 18$ kHz.

When δf = 12 kHz and f_m = 1.5 kHz, Carson's rule gives the bandwidth as 2(1.5 + 12) = 27 kHz. When δf = 6 kHz and f_m = 3 kHz, the bandwidth by Carson's rule is 2(3 + 6) = 18 kHz. The two sets of answers are identical.

The power in the sidebands is proportional to the square of the coefficients in Table 24.1, so in the first case the sum of the squares is

$$\sum_{n=1}^{9} [J_n(8)]^2 = 0.23^2 + (-0.11)^2 + (-0.29)^2 + (-0.11)^2 \qquad (24.24)$$
$$+ 0.19^2 + 0.34^2 + 0.32^2 + 0.22^2 + 0.13^2 = 0.48$$

There are upper and lower sidebands giving a total of 2 × 0.48 = 0.96. The carrier power is found by squaring the coefficient of $J_0(8)$, which is 0.17^2 = 0.03. To find the total power the sideband power is added to the carrier power, which gives 0.99: it should be 1, but we have neglected the more remote sidebands. Thus to find the ratio of sideband power to total power, all we need do is subtract $[J_0(m_f)]^2$ from unity:

$$P_{SB}/P_T = 1 - [J_0(m_f)]^2 \qquad (24.25)$$

In the second case $J_0(2)$ = 0.22 and Equation 24.25 yields P_{SB}/P_T = 1 - 0.22^2 = 0.95. The spectra of the two signals are shown in Figure 24.8.

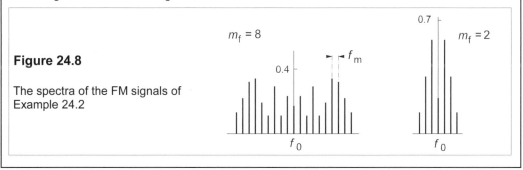

Figure 24.8

The spectra of the FM signals of Example 24.2

24.4.2 Narrow-band FM

Narrow-band FM is used for low-power mobile communications such as police and taxis use. By keeping the modulation index, m_f, below about 0.5 the number of sidebands is limited to just the first two at $\pm f_m$. The transmission bandwidth is therefore $2f_m$ just as in AM and the characteristics of narrow-band FM are much the same as full-carrier, double-sideband AM, especially regarding noise immunity.

24.4.3 FM transmitters

One of the simplest ways of achieving frequency modulation is to use the signal voltage to alter the capacitance in an oscillator circuit, thereby making a voltage-controlled oscillator or VCO. This is achieved by a varactor diode in the Colpitts oscillator of Figure 24.9.

The capacitance of a varactor diode is a function of the reverse voltage across it and may be used for the purpose. Varactor diodes came into use in the 1950s and gradually improved in their frequency range until now they are common in microwave equipment. If the Colpitts oscillator circuit is used (see Section 11.2.2), whose resonant circuit is shown in Figure 24.9, the frequency is

$$\omega = 1/\sqrt{LC_T} = L^{-1/2}(1/C_D + 1/C)^{1/2} \qquad (24.26)$$

where $1/C_T = 1/C + 1/C_D$ and C_D is the varactor diode's capacitance.

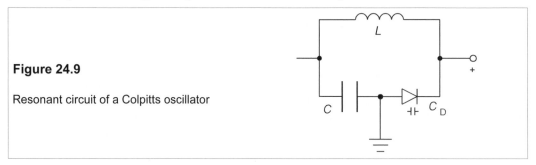

Figure 24.9

Resonant circuit of a Colpitts oscillator

We may calculate the rate of change of this frequency with diode capacitance by taking logs

$$\ln\omega = -0.5\ln L + 0.5\ln(1/C_D + 1/C) \qquad (24.27)$$

and then differentiating with respect to C_D:

$$\frac{1}{\omega}\frac{d\omega}{dC_D} = \frac{0.5}{1/C_D + 1/C} \times \frac{-1}{C_D^2} = \frac{-1}{2(C_D(1+c)} \qquad (24.28)$$

so that

$$\frac{d\omega}{dC_D} = \frac{-\omega}{2C_D(1+c)} \qquad (24.29)$$

where $c\ (= C_D/C)$ is the capacitance ratio *at the operating point*. If the diode's capacitance as a function of reverse bias is given by

$$C_D = KV^n \quad \Rightarrow \quad dC_D/dV = nKV^{n-1} = nC_D/V \qquad (24.30)$$

where K is a constant and n is a constant (usually in the range -0.5 to -0.3). The reverse bias, V, is taken as positive. The rate of change of frequency with voltage is

$$\frac{d\omega}{dV} = \frac{d\omega}{dC_D}\frac{dC_D}{dV} \qquad (24.31)$$

We can substitute $d\omega/dC_D$ from Equation 24.29 and dC_D/dV from 24.30 into this to obtain

$$\frac{d\omega}{dV} = \frac{-\omega}{2C_D(1+c)}\frac{nC_D}{V} = \frac{-n\omega}{2V(1+c)} \qquad (24.32)$$

The next example illustrates the extent of frequency changes from a Colpitts oscillator.

Example 24.3

The varactor diode used in an FM LC oscillator follows Equation 24.30 with $n = -1/3$ and has a capacitance of 50 pF at a reverse bias of 5 V. The unmodulated resonant frequency is 95 MHz and the diode's reverse bias at the Q point is 10 V. If the fixed capacitance, C, in the circuit is 500 pF and the modulation voltage changes by ±50 mV, what is the modulation index if the modulating frequency is 12 kHz?

We first must find the capacitance of the diode at the Q point by substituting $C_D = 50$ pF, $n = -1/3$ and $V = 5$ V into Equation 24.30 to find K, which is $50 \times \sqrt[3]{5} = 85.5$ pF.
Then when $V = 10$ V, $C_D = 85.5/\sqrt[3]{10} = 39.7$ pF. The capacitance ratio, $c = C_D/C = 39.7/500 = 0.0794$.
Equation 24.32 gives

$$\frac{d\omega}{dV} = \frac{-(-1/3)2\pi \times 95}{2 \times 10 \times (1 + 0.0794)} = 9.2 \text{ Mrad/s/V} \tag{24.33}$$

The modulating voltage is ±0.05 V, making the frequency change ±9.2 × 0.05 = ±0.46 Mrad/s or ±73 kHz, which is the deviation, Δf. Then the modulation index, $m_f = \Delta f/f_m = 73 \times 10^3/12 \times 10^3 = 6$.

24.4.4 FM receivers

The demodulation of FM signals used to be more difficult than with AM signals, and the methods used conveyed some distortion to the demodulated signal. The amplitude of the demodulator (usually called a *discriminator* in FM receivers) output voltage must be proportional to the frequency of the signal received, and this can be accomplished in several ways of which we shall briefly mention three: the tuned-circuit discriminator, the zero-crossing detector and the phase-locked loop (PLL), which is now preferred.

Tuned-circuit demodulation

Consider the parallel-resonant circuit of Figure 24.10(a) which is excited by a current source. The voltage across the circuit will follow the curve of Figure 24.10(b). If the FM carrier frequency lies above or below the resonant frequency on the linear part of the curve then, provided the circuit's bandwidth is much greater than the maximum frequency deviation of the signal, the output should have a component proportional to the instantaneous frequency deviation. There is inevitably some non-linearity, however, which distorts the demodulated signal.

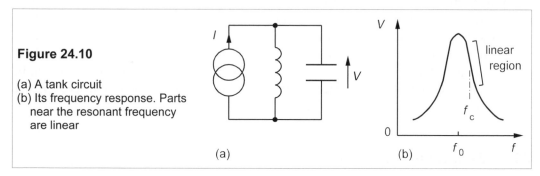

Figure 24.10

(a) A tank circuit
(b) Its frequency response. Parts near the resonant frequency are linear

In the *balanced demodulator* of Figure 24.11 two tuned circuits are used, one tuned above the carrier frequency and one an equal amount below. The two outputs are subtracted and better discrimination is achieved. Normally the incoming signal is shifted down from around 100 MHz to about 10 MHz before it is demodulated, thus easing the requirements on the tuned-circuit components. There is, however, still some distortion because of the tank circuit's inherent non-linearity.

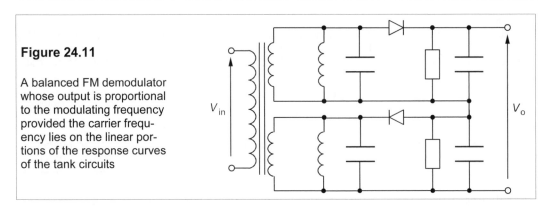

Figure 24.11

A balanced FM demodulator whose output is proportional to the modulating frequency provided the carrier frequency lies on the linear portions of the response curves of the tank circuits

The zero-crossing detector

The FM signal (exaggerated) of Figure 24.12 has consecutive positive-slope zero-crossings at times t_1 and t_2. In a normal FM signal the change in frequency will be negligible between these times, so that the change in phase of the signal is 2π radians. If the instantaneous angular frequency is ω_i, then $\omega_i(t_2 - t_1) = 2\pi$, or $f_i = 1/(t_2 - t_1)$.

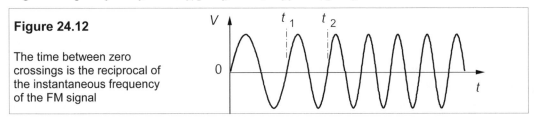

Figure 24.12

The time between zero crossings is the reciprocal of the instantaneous frequency of the FM signal

The Phase-locked loop detector

We have discussed PLLs in Section 11.3: they are well-suited to FM demodulation. Suppose a signal (a time-varying voltage of some sort) is input into a voltage-controlled oscillator (VCO) so that the broadcast output is $f_c + \delta f$, where f_c is the carrier frequency and δf the deviation produced by the signal. This signal is received and input to a PLL whose free-running frequency is f_c. The output error signal driving the VCO of the PLL will then be proportional to δf so that the error signal is the demodulated signal. The PLL detector is free from the inherent distortion of tuned-circuit demodulators.

24.5   AM and FM compared

The advantages of FM compared to AM are:

1. FM is a constant-amplitude system which means that amplitude limiters can be used in the receiver to remove noise-induced amplitude variations. FM is less subject to noise and fading.
2. All the transmitted power in an FM signal is useful, whereas in full-carrier AM most of the power is in the carrier which conveys no information.

3. By using more bandwidth further noise immunity can be gained, which is impossible in an AM system as the modulation depth cannot exceed 100%. Only by using more power can the effects of noise be reduced in an AM system.

There are disadvantages, however:

1. FM requires more transmitters for local coverage as it is closer to a line-of-sight system. Moving receivers are especially subject to interference by obstacles in the line of sight, though this is a consequence of the high frequencies used rather than an intrinsic property of FM.
2. A larger bandwidth is required by FM.
3. The transmitters and receivers are more complex for FM, though with the advent of cheap VCO and PLL ICs, they are no longer more expensive.

24.6 **Pulse modulation**

Instead of modulating the amplitude or frequency of a continuously transmitted carrier wave, one could sample the message signal and send bursts of the carrier frequency to represent its instantaneous amplitude at the moment of sampling. The 'missing' parts of the signal are reconstructed at the receiver by the smoothing processes of filtering.

24.6.1 *Sampling*

Sampling has advantages. Because the pulses from a single sampling circuit occupy only a fraction of the time, several pulse-modulated signals can be multiplexed before sending, thereby making more efficient use of a costly communication link. There is, however, clearly going to be a minimum sampling rate below which some of the information in the sampled message is lost. Suppose we are sampling a pure sinusoid of frequency, f, at a sampling rate of once every T_s seconds, a sampling frequency of $1/T_s = f_s$. Unless the sampling frequency is greater than $2f$, the sinusoid cannot be reconstructed from the samples. This is known as the *Nyquist rate*. If the signal to be sent contains frequency components up to f_{max}, the minimum sampling rate is $2f_{max}$:

$$f_s \geq 2f_{max} \tag{24.34}$$

Telephone channels use frequencies up to 3.4 kHz, giving a Nyquist rate of 6.8 kHz, though in practice the sampling rate is a little greater at 8 kHz.

Pulse modulation can be analogue or digital; analogue pulse modulation will be discussed here and digital pulse modulation in the next chapter.

24.6.2 *Analogue pulse modulation*

In analogue pulse modulation, some property of the pulse – duration, amplitude or position – is altered in proportion to the modulating signal's instantaneous amplitude at the moment of sampling. If the amplitude of the carrier pulses is made proportional to the amplitude of the signal sample we have what is called *pulse-amplitude modulation* or PAM. However, the pulse amplitude could be kept constant while its duration varies with the sample amplitude, which is *pulse-width modulation*, PWM. Thirdly the pulse's width and amplitude could be held constant, while its position is varied within a fixed time slot, which is *pulse-position modulation*, PPM. Each of these methods varies a pulse parameter continuously: the amplitude can take any value within certain limits and the duration and position likewise are continuously variable. In this sense these methods of pulse modulation are analogue rather than digital.

Figure 24.13 shows these three forms of pulse modulation. They are readily achieved in principle; PAM by means of a sample-and-hold circuit, PWM by integrating PAM and PPM by differentiating PWM and filtering out the leading-edge component, since the trailing edges of the PWM pulses are in fact position modulated. PAM is never sent through a channel, but is a necessary preliminary stage in the production of PWM or PPM. The PWM pulses can be demodulated with a detector diode and low-pass filter just like AM. PPM can be demodulated by regenerating PWM with a flip-flop controlled by sync pulses from the transmitter. These turn the flip flop on until the PPM signal switches it off, so generating a pulse whose width depends on the time difference between the sync pulse and the PPM pulse.

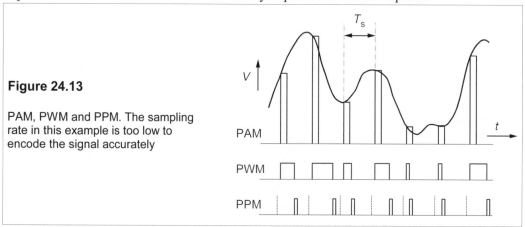

Figure 24.13

PAM, PWM and PPM. The sampling rate in this example is too low to encode the signal accurately

24.7 **Noise**

Noise is always present in any communications system; its sources are both internal and external, and while it may be reduced it can never be eliminated. External noise may be caused by atmospheric conditions (such as thunderstorms and solar e.m. radiation) or by interference from domestic and industrial equipment. Internal noise is more fundamental and is an intrinsic property of electrons. The effects of noise are to obscure the information being sent to a greater or lesser extent, though in systems employing analogue modulation these effects are not readily quantified. In systems using digital modulation, however, the effects of noise can be expressed in terms of the probability that a bit is detected at the receiver. Partly for this reason digital transmission of information has rapidly gained ground over analogue transmission, since the quality of the information received can be predicted.

24.7.1 Thermal noise

The most important internal-noise source is *thermal noise*, also known as *Johnson noise* and *white noise* (because it has a uniform spectral distribution), whose origin is random thermal energy. It occurs in any resistive device, not just in resistors, it also occurs for example in diodes and especially transistors with appreciable base resistance. The electrons in a resistor of resistance, R, have random thermal energy, kT, where k is Boltzmann's constant, 1.38×10^{-23} J/K, and T is the temperature in K. Thus there will be a fluctuating thermal voltage across the resistor whose r.m.s. value is proportional to the bandwidth of the system. The Johnson noise power is

$$P_n = kTB \qquad (24.35)$$

where B is the bandwidth. We can consider the resistor to be an ideal resistance in series with a noise-voltage source, r.m.s. value E_n, as in Figure 24.14.

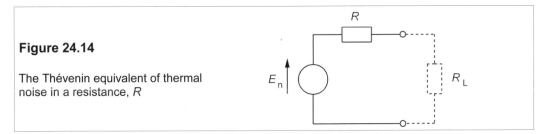

Figure 24.14

The Thévenin equivalent of thermal noise in a resistance, R

If a noiseless resistance, R_L, is connected across R, the power developed in R_L will be a maximum when $R = R_L$, and in that case it is

$$P_n = E_n^2/4R = kTB \qquad (24.36)$$

leading to the final expression for the Johnson noise:

$$E_n = \sqrt{4RkTB} \qquad (24.37)$$

Often the bandwidth is omitted and the noise voltage given in V/√Hz, in which case Equation 24.37 becomes

$$E_n = \sqrt{4kTR} \qquad (\mathrm{V}/\sqrt{\mathrm{Hz}}) \qquad (24.38)$$

The instantaneous value of the noise voltage has a gaussian probability distribution – the well-known bell-shaped curve – of mean zero and standard deviation E_n, as shown in Figure 24.15.

The gaussian distribution implies that the probability of the noise voltage lying between E and $E + \mathrm{d}E$ is

$$P(E, E + \mathrm{d}E) = f(E)\mathrm{d}E = \frac{1}{\sqrt{2\pi}E_n} \exp[-0.5(E/E_n)^2]\mathrm{d}E \qquad (24.39)$$

The probability that the noise voltage will exceed E is found by integrating Equation 24.39:

$$P(\geq E) = \frac{1}{\sqrt{2\pi}E_n} \int_E^{\infty} \exp[-0.5(E/E_n)^2]\mathrm{d}E = 0.5\,\mathrm{erfc}(E/\sqrt{2}E_n) \qquad (24.40)$$

erfc is the *complementary error function*. It is the area under the tail of the gaussian. The area under the rest of the gaussian is the *error function*, erf. The total area under the normalised gaussian curve is 1 (the probability of *some* voltage must be 1) and therefore

$$\text{erf}(x) + \text{erfc}(x) = 1 \qquad (24.41)$$

Neither erf nor erfc can be found by direct integration of Equation 24.40, but numerical integration is straightforward and they are tabulated as in Table 25.1, in the next chapter.

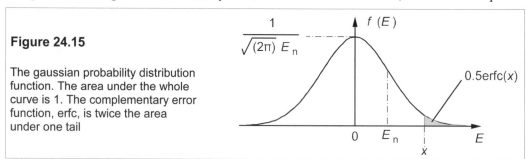

Figure 24.15

The gaussian probability distribution function. The area under the whole curve is 1. The complementary error function, erfc, is twice the area under one tail

Figure 24.16 shows a series of voltages measured across a resistor at equal intervals of time. The r.m.s. voltage is E_n and its time-averaged value zero. The thermal-noise voltage is significant, even though it is not very large. For example, if $R = 1\ \text{M}\Omega$, $B = 10\ \text{kHz}$, $T = 330$ K then Equation 24.37 yields $E_n = 13.5\ \mu\text{V}$. Signal voltages are often no greater than this.

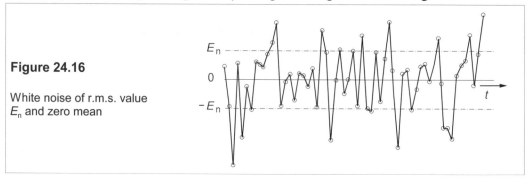

Figure 24.16

White noise of r.m.s. value E_n and zero mean

24.7.2 *Shot noise*

Shot noise originates from the random arrival of electrons at the output terminals of a device such as an amplifier. It received its name from the sound it produces when amplified in a loudspeaker – the sound of lead shot falling on a tin roof. Shot noise depends on the current flowing in the device. In diodes and transistors the shot-noise current is

$$I_n = \sqrt{2qI_D B} \qquad (24.42)$$

where q is the electronic charge, 1.6×10^{-19} C, and I_D the direct current. The relative noise current is $I_n/I_D = \sqrt{(2qB/I_D)}$, and accordingly shot noise is relatively more important when the direct current is small. To express the shot noise in volts one must know the appropriate resistance of the device.

24.7.3 Flicker (1/f) noise

Random fluctuations occur in resistances which are found to depend on the method of construction. These fluctuations give rise to a further noise voltage which is proportional to $1/f$ and is sometimes called *pink noise*, as it is concentrated at the lower frequencies and longer wavelengths, that is towards the 'red' end of the spectrum. The flicker noise for metal-film resistors is about 0.1 μV/V per frequency decade and 0.2 μV/V per decade for carbon-film resistors. Thus if a metal-film resistor has 8 V dropped across it and is used in an amplifier between 100 Hz and 100 kHz (3 decades), the flicker noise will be

$$E_n = 0.1 \times 8 \times 3 = 2.4 \ \mu V \qquad (24.43)$$

24.7.4 Transistor noise

The noise in bipolar transistors and integrated circuits is produced by Johnson noise in the spreading resistance[3] of the transistor's base, r_b, shot noise due to the current in the base-emitter junction resistance, r_e (see Chapter 8, Section 8.2.4) and $1/f$ noise also plays a part. The thermal-noise voltage is $\sqrt{(4kTr_bB)}$ and the shot-noise voltage is $I_n r_e = \sqrt{(2qI_E B)} r_e$, using Equation 24.43 with $I_E = I_D$. But $r_e = kT/qI_C$, making the shot-noise voltage per $\sqrt{Hz}$

$$r_e\sqrt{2qI_E} = \frac{kT}{qI_E}\sqrt{2qI_E} = \sqrt{\frac{2k^2T^2}{qI_E}} \qquad (24.44)$$

The total noise voltage per $\sqrt{Hz}$ is found by adding squares:

$$E_n^2 = 4kTr_b + 2k^2T^2/qI_E \qquad (24.45)$$

In FETs the noise is largely due to Johnson noise in the channel. If the channel transconductance is g_m, the noise voltage per $\sqrt{Hz}$ is

$$E_n = \sqrt{4kT/g_m} \qquad (24.46)$$

The channel transconductance is proportional to $\sqrt{I_D}$ (see Section 9.1.3), causing the noise slowly to increase with drain current.

Example 24.4

A BJT operating at 400 K with an emitter current of 1 mA has a base spreading resistance of 1 kΩ and is used in an amplifier with a bandwidth of 300 kHz. Calculate the noise voltage of the transistor.
    The noise voltage is made up of Johnson noise:

$$E_{thermal}^2 = 4 \times 1.38 \times 10^{-23} \times 400 \times 1000 = 2.21 \times 10^{-17} \ V^2/Hz \qquad (24.47)$$

and shot noise:

$$E_{shot}^2 = \frac{2k^2T^2}{qI_E} = \frac{2 \times (1.38 \times 10^{-23} \times 400)^2}{1.6 \times 10^{-19} \times 10^{-3}} = 3.81 \times 10^{-19} \ V^2/Hz \qquad (24.48)$$

---

[3] The internal resistance of the base rather than the external base resistor

The total noise voltage squared per Hz is

$$E_{tot}^2 = E_{thermal}^2 + E_{shot}^2 = 2.21 \times 10^{-17} + 3.81 \times 10^{-19}$$

$$= 2.25 \times 10^{-17} \ V^2/Hz$$

(24.49)

The bandwidth is 300 kHz, making $E_{tot}^2 = 2.25 \times 10^{-17} \times 3 \times 10^5 = 6.75 \times 10^{-12} \ V^2$ and $E_{tot} = 2.6 \ \mu V$. The shot-noise contribution is usually small.

24.7.5 Equivalent noise resistance

Consider a signal voltage source, E_s, of resistance, R_s, connected to the terminals of an amplifier of input resistance, R_{in}, and whose equivalent noise resistance is R_n, what is the input noise voltage? The equivalent circuit is shown in Figure 24.17a, and the noisy amplifier is replaced in 24.17b by a noise-free amplifier and an equivalent noise voltage source, E_n, in series with the input. The equivalent noise resistance is used to calculate the noise voltage, but *not* the signal voltage at the receiver terminals, and therefore does not appear in Figure 24.17b.

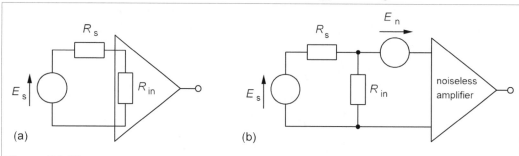

Figure 24.17 (a) A noisy amplifier and (b) Its equivalent circuit

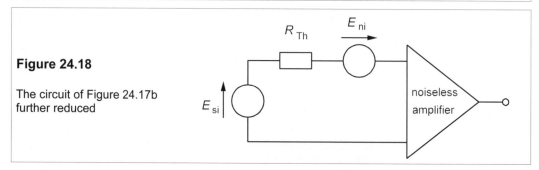

Figure 24.18

The circuit of Figure 24.17b further reduced

The signal voltage at the amplifier's terminals is

$$E_{si} = \frac{R_{in}E_s}{R_{in} + R_s}$$

(24.50)

While the noise voltage at the same terminal is

$$E_{ni} = \sqrt{4kTB(R_n + R_{Th})}$$

(24.51)

where

$$R_{Th} = R_s \; // \; R_{in} = R_s R_{in}/(R_s + R_{in}) \qquad (24.52)$$

The noise voltage at the amplifier's terminal is that produced by R_n and $R_s \; // \; R_{in}$. That is to say, we can replace Figure 24.17b with the Thévenin equivalent circuit of Figure 24.18.

24.7.6 *Signal-to-noise ratio*

The signal-to-noise ratio (SNR) is defined by

$$SNR = P_s/P_n \quad \rightarrow \quad SNR_{dB} = 20\log_{10}(E_s/E_n) \qquad (24.53)$$

The SNR largely determines the quality of the information received from a communications channel. If the SNRs in the components of a communications system are known then it is possible to calculate the likelihood of error in messages sent through it. As a consequence the SNR is important in communications theory and practice.

Example 24.5

The SNR in each of four cascaded amplifiers is 46 dB. What is the overall SNR? If a fifth amplifier of SNR 30 dB is added, what will the overall SNR be then?

If we take the signal power to be unity, the SNR gives us the noise power of each amplifier as −46 dB, that is

$$P_n = 10^{-4.6} = 2.5 \times 10^{-5} \quad \rightarrow \quad P_n(\text{tot}) = 4P_n = 10^{-4} \qquad (24.54)$$

The overall SNR is

$$SNR(\text{tot}) = -10\log_{10} 4P_n = 34 \quad \text{dB} \qquad (24.55)$$

Adding a fifth, much noisier amplifier brings the total noise power to

$$P_n(\text{tot}) = 10^{-4} + 10^{-3} = 1.1 \times 10^{-3} \qquad (24.56)$$

And the overall SNR becomes $-10\log_{10} P(\text{tot}) = -29.6 \approx -30$ dB. The SNR of a communications system is determined largely by the part with the lowest SNR.

Example 24.6

A signal source of resistance 600 Ω is connected to an amplifier of input resistance 2 kΩ, equivalent noise resistance 1.5 kΩ and bandwidth 20 kHz. If the r.m.s. signal voltage is 10 μV and the amplifier's temperature is 300 K, what is the equivalent noise voltage and the signal voltage at the amplifier's terminals. What is the input SNR?

We use the circuit of Figure 24.18, with $R_{Th} = 600 \times 2000/2600 = 462\ \Omega$ and $E_{si} = 2000 \times 10/2600 = 7.7\ \mu V$. The total noise resistance is $R_n + R_{Th} = 1.5 + 0.462 = 1.962\ k\Omega$, so that the noise voltage at the (noise-free) amplifier's input is

$$E_{ni}' = \sqrt{4kTB(R_n + R_{Th})}$$
$$= \sqrt{4 \times 1.38 \times 10^{-23} \times 300 \times 20000 \times 1962} = 0.806\ \mu V \qquad (24.57)$$

The overall SNR becomes

$$SNR = 20\log_{10}(E_{si}/E_{ni}') = 20\log_{10}(7.7/0.806) = 19.6\ \text{dB} \qquad (24.58)$$

24.7.7 *Noise figure and noise temperature*

Amplifiers and mixers – all circuit elements in fact – contribute noise to a communications system. The noisiness of an individual component or building block of a system is expressed as a noise factor or noise figure. The *noise figure* is $10\log_{10}F$, where F, the *noise factor*, is given by

$$F = \frac{\text{input SNR}}{\text{output SNR}} \tag{24.59}$$

which is

$$F = \frac{P_{si}/P_{ni}}{P_{so}/P_{no}} = \frac{P_{si}P_{no}}{P_{so}P_{ni}} \tag{24.60}$$

And hence

$$P_{no} = F(P_{so}/P_{si})P_{ni} = FGP_{ni} \tag{24.61}$$

where $G\ (\equiv P_{so}/P_{si})$ is the signal-power gain of the amplifier. If the amplifier had been noise free, $P_{no} = GP_{ni}$. The output noise is increased by the noise factor over what it would have been had the amplifier been noiseless, that is if R_n were zero.

Noise temperature is sometimes used to indicate how much noise there is in a component or system. Noise temperatures are directly proportional to noise powers and are therefore conveniently additive:

$$\begin{aligned} P &= P_1 + P_2 + P_3 + \ldots = kT_1B + kT_2B + kT_3B + \ldots \\ &= k(T_1 + T_2 + T_3 + \ldots)B = kTB \end{aligned} \tag{24.62}$$

$T = T_1 + T_2 + T_3 + \ldots$ is the overall noise temperature.

Example 24.7

Calculate the noise figure for the amplifier in Example 24.6.

The input SNR for the amplifier is the ratio of E_{si}^2 to the square of the noise voltage *without R_n*, since R_n comes after the input terminals of the real amplifier. The input noise voltage without R_n, E_{ni}, is $\sqrt{(4kTBR_{Th})}$. Now $E_{so} = A_vE_{si}$ and $E_{no} = A_vE_{ni}'$, where A_v is the voltage gain of the amplifier and E_{ni}' is the input noise voltage to the (fictitious) noiseless amplifier. Both noise and signal voltages must be subject to the same amplification. Here E_{ni}' is $\sqrt{[4kTB(R_n + R_{Th})]}$. Thus the noise factor of the amplifier becomes

$$F = \frac{P_{si}P_{no}}{P_{ni}P_{so}} = \frac{E_{si}^2(A_vE_{ni}')^2}{E_{ni}^2(A_vE_{si})^2} = \left(\frac{E_{ni}'}{E_{ni}}\right)^2 = \frac{R_{Th} + R_n}{R_{Th}} \tag{24.63}$$

which is unity if $R_n = 0$. The voltage gain does not enter in F, only the input side's Thévenin-equivalent resistance and the equivalent noise resistance of the amplifier do. Substituting for $R_{Th} = 462\ \Omega$ and $R_n = 1.5\ \mathrm{k\Omega}$ into Equation 24.63 leads to $F = 1962/462 = 4.25$, or $F_{dB} = 6.3$ dB.

The total noise referred to the input of an amplifier, P_{nri} is $P_{no}/G = FP_{ni}$, from Equation 24.61. By definition $P_{ni} = kT_0B$, where T_0 is the ambient temperature and so

$$P_{nri} = FkT_0B \tag{24.64}$$

Of this noise, the source's contribution, P_{ni}, is still kT_0B, which makes the amplifier's contribution

$$P_{na} = P_{nri} - P_{ni} = FkT_0B - kT_0B$$

$$= (F - 1)kT_0B = kT_{eq}B \tag{24.65}$$

The *equivalent noise temperature* of the amplifier, T_{eq}, is $(F - 1)T_0$. In Example 24.6, if the ambient temperature were 25°C, or 298 K, $T_{eq} = (4.25 - 1) \times 298 = 969$ K.

Example 24.8

When two white-noise sources of equivalent noise temperature, 500 K and 1000 K and are connected one at a time to an amplifier, it is found that the ratio of the total noise power outputs is 0.7. Find the equivalent noise temperature, T_A, of the amplifier.

The amplifier contributes a noise power of $P_A = kT_AB$, while the first noise source contributes a power of kT_1B, where P_1 is the noise power of the first white-noise source. The combined noise power is

$$P_1 = kB(T_A + 500) \tag{24.66}$$

When the first noise source is replaced by the second the output noise power is

$$P_2 = kB(T_A + 1000) \tag{24.67}$$

where P_2 is the noise power of the second white-noise source. Dividing Equation 24.66 by 24.67 gives

$$\frac{P_1}{P_2} = 0.7 = \frac{T_A + 500}{T_A + 1000} \tag{24.68}$$

which can be solved to yield $T_A = 667$ K.

Amplifiers in cascade

If a series of amplifiers is connected in cascade the overall noise figure may be found as follows: The noise power from the first amplifier is $G_1F_1kT_0B$, using Equations 24.61 and 24.65. This noise is present at the input of the second amplifier along with its own noise, $(F_2 - 1)kT_0B$, for a total input noise at stage two of

$$P_{2ni} = F_1G_1kT_0B + (F_2 - 1)kT_0B \tag{24.69}$$

which is then amplified by G_2 to give an output noise after stage 2 of

$$P_{2no} = G_2G_1F_1kT_0B + G_2(F_2 - 1)kT_0B \tag{24.70}$$

The overall gain, $G = G_1G_2$, and the overall noise factor is

$$F = P_{2no}/GkT_0B = F_1 + (F_2 - 1)/G_1 \tag{24.71}$$

Further amplifiers lead to an overall noise factor after k stages given by

$$F = F_1 + \frac{F_2 - 1}{G_1} + \dots + \frac{F_k - 1}{G_{k-1}..G_2G_1} \tag{24.72}$$

This is known as Friiss's formula and it shows that in general the earliest stages should be those of lowest noise and highest amplification. In terms of equivalent noise temperatures it is

$$T_{eq} = T_{1eq} + T_{2eq}/G_1 + ... + T_{keq}/G_1 G_2 .. G_{k-1} \qquad (24.73)$$

provided the amplifiers are matched and the noise figures have been measured under matched conditions at the same ambient temperature.

Example 24.9

A signal is amplified 30 dB by a preamplifier with a noise factor of 1.5 and fed into a further amplifier with a noise figure of 14 dB and a gain of 10 dB. If the amplified signal is then fed into a mixer whose noise figure is 25 dB, what is the overall noise factor? If $T_0 = 20$ K, what is the equivalent noise temperature overall? If the order of the mixing and amplifying were completely reversed, what would these figures be?

Friiss's formula can be used, considering the mixer as a unity-gain amplifier. Then $G_1 = 10^3$ (30 dB), $F_1 = 1.5$, $G_2 = 10$ (10 dB), $F_2 = 10^{1.4}$ (14 dB) = 25, $G_3 = 1$ and $F_3 = 10^{2.5}$ (25 dB) = 316. Substituting in Equation 24.73 leads to

$$F = 1.5 + (25 - 1)/10^3 + (316 - 1)/(10 \times 10^3) = 1.5555 \qquad (24.74)$$

To find the overall noise temperature we first must find the equivalent noise temperature of each stage from Equation 24.65, which leads to $T_{1eq} = (F_1 - 1)T_0 = 0.5T_0 = 10$ K, $T_{2eq} = 24T_0 = 480$ K and $T_{3eq} = 315T_0 = 6300$ K. Equation 24.73 then gives

$$T_{eq} = 10 + 480/10^3 + 6300/(10^3 \times 10) = 11.11 \text{ K} \qquad (24.75)$$

Reversing the order of the stages has a dramatic effect on the noise figure, since now $G_1 = 1$, $F_1 = 316$, $G_2 = 10$, $F_2 = 25$, $G_3 = 10^3$ and $F_3 = 1.5$. Using Equation 24.72 we find

$$F = 315 + 25/1 + 1.5/10 = 340.15 \qquad (24.76)$$

The equivalent noise temperatures are $T_{1eq} = 6300$ K, $T_{2eq} = 480$ K and $T_{3eq} = 10$ K, and Equation 24.73 gives

$$T_{eq} = 6300 + 480/1 + 10/10 = 6781 \text{ K} \qquad (24.77)$$

By placing the noisiest element (the mixer) before the amplifiers we have amplified all its noise as well as the signal and hugely degraded the noise figure.

24.7.8 Noise in AM and FM receivers

The noise power in a SSB receiver of bandwidth B_{SSB} and noise factor F will be $FkTB_{SSB}$. If the received power is proportional to the transmitted power, P_T, then the SNR is

$$SNR_{SSB} \propto P_T/FkTB_{SSB} \qquad (24.78)$$

In a full-carrier, double-sideband AM system the transmitted power by Equation 24.11 is $P_T = (1 + 0.5m^2)P_c$, where P_c is the carrier power. In this transmitted power the signal power is $0.5m^2 P_c$, which means

$$P_s = \frac{0.5m^2 P_T}{1 + 0.5m^2} \qquad (24.79)$$

Now the bandwidth of full AM transmission is twice that of SSB transmission, so the AM receiver's noise power will be $2FkTB_{SSB}$, and its SNR becomes

$$SNR_{AM} \propto \frac{0.5m^2 P_T}{(1+0.5m^2)2FkTB_{SSB}} \tag{24.80}$$

Dividing Equation 24.78 by 24.80 produces

$$\frac{SNR_{SSB}}{SNR_{AM}} = \frac{P_T}{FkTB_{SSB}} \times \frac{(1+0.5m^2)2FkTB_{SSB}}{0.5m^2 P_T} = 2 + \frac{4}{m^2} \tag{24.81}$$

assuming that F and P_T are the same for both. Since $m \leq 1$ Equation 24.81 implies that the SNR for SSB transmission is at least six times better than for full-carrier, double-sideband AM. This is a consequence of the tripled signal power and halved bandwidth of SSB.

In FM the picture is more complicated. Unlike AM, in which the effect of noise is the same at all frequencies, in FM the effects of a constant noise-to-carrier voltage ratio are different at different frequencies. In fact the noise-to-signal *voltage* ratio, $(1/SNR)^2$, increases linearly with frequency if the noise is uniformly distributed over all frequencies. This gives rise to the *noise triangle* shown in Figure 24.19, in which noise *voltage* is plotted against frequency.

For a modulation index, m_f, the noise power is

$$N_{FM} \propto 2 \int_0^{f_m} (f/m_f)^2 \, df = 2f_m^3/3m_f^2 \tag{24.82}$$

where f_m is the maximum frequency in the unmodulated signal (the *baseband*) and the factor 2 accounts for two sidebands. As m_f increases the bandwidth increases and the effect of the noise is reduced.

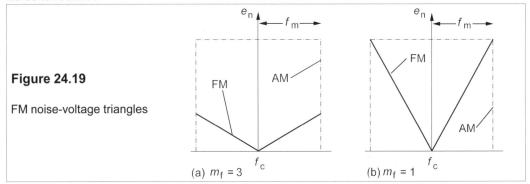

Figure 24.19

FM noise-voltage triangles

(a) $m_f = 3$ (b) $m_f = 1$

Signal power is proportional to f_m^2 and then the FM SNR is

$$SNR_{FM} \propto \frac{f_m^2}{2f_m^3/3m_f^2} = \frac{1.5A m_f^2}{f_m} \tag{24.83}$$

The constant of proportionality, A, is P_c/FkT, where P_c is the carrier power received. The AM noise in the same band is proportional to f_m, the *maximum* modulating frequency, so for SSB operation the SNR is

$$SNR_{SSB} = P_s/FkTf_m \tag{24.84}$$

where P_s is the AM signal power at the receiver. If the received power is the same for AM and FM, that is $P_s = P_c$, and the receivers have the same noise factor, F, the SNRs are related by

$$SNR_{FM} = 1.5m_f^2 SNR_{SSB} \qquad (24.85)$$

Considering an FM transmission where $f_m = 15$ kHz and δf, the maximum deviation, is 75 kHz, we have $m_f = \delta f/f_{max} = 5$, and substituting into Equation 24.85 gives an improvement in the SNR of 37.5 or 15.7 dB. Of course, the improved SNR is largely due to the increased bandwidth, $2(75 + 15) = 180$ kHz, by Carson's rule, compared to 15 kHz, 12 times as much and therefore contributing $10\log_{10}12 = 10.7$ dB. However, the residual 5 dB represents an additional advantage for FM over SSB AM. As a general rule

- ### *SNR can be traded for bandwidth and vice versa*

Pre-emphasis and de-emphasis

Because noise has a greater effect on the higher frequencies of FM signals it is possible to gain further in SNR by boosting the higher frequencies on transmission and reducing them by the same amount at the receiver. The boosting process is called *pre-emphasis* and the compensating reduction at the receiver is called *de-emphasis*. The circuits used are simple first-order RL or RC filters, shown in Figure 24.20.

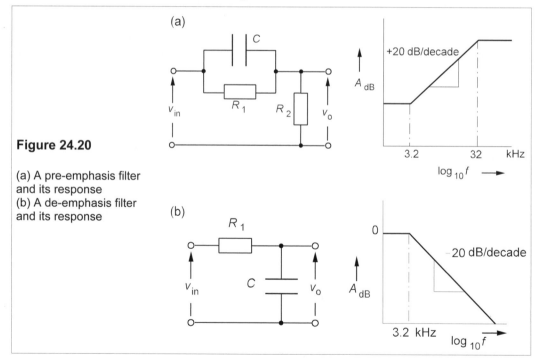

Figure 24.20

(a) A pre-emphasis filter and its response
(b) A de-emphasis filter and its response

The time constants for pre-emphasis and de-emphasis must be the same, and are standardised in Europe at 50 µs and in N America at 75 µs. These time-constants correspond to frequencies of $(2\pi \times 50 \times 10^{-6})^{-1} = 3.2$ kHz and $(2\pi \times 75 \times 10^{-6})^{-1} = 2.1$ kHz respectively. The pre-emphasis filter's upper cut-off must be high enough to pass the highest modulation frequency; 32 kHz will suffice for audio signals. Pre-emphasis/de-emphasis is the principle used in the well-known Dolby noise-reduction system: it results in an additional gain in discrimination against noise comparable to that of FM over AM.

Suggestions for further reading – see Chapter 26

Problems

1. In a full-carrier, double-sideband AM broadcasting system the unmodulated carrier current in the transmitting antenna is 12 A, which increases to 13.6 A when amplitude modulated. What is the modulation index? What is the signal power in the modulated wave if the antenna has a resistance of 75 Ω? What fraction of the total power is the signal power? If the carrier frequency is 1.273 MHz and the modulating frequency is 15.6 kHz, what is the antenna voltage as a function of time?
    [0.754, 3.07 kW, 0.221]

2. If the carrier in the previous problem is just overmodulated, what are the antenna current and the signal power? What is the ratio of signal power to carrier power? *[14.7 A, 5.4 kW, 1:2]*

3. The modulating signal in an FM signal is a sinewave of frequency 2.4 kHz and the maximum deviation is to be 6 kHz, what is the effective bandwidth of the signal, using (a) Carson's rule and (b) Table 24.1, neglecting components whose coefficients are < 0.1? If the modulating signal is doubled in amplitude what will the bandwidth be?
    [(a) 16.8 kHz (b) between 14.4 kHz and 19.2 kHz, or 16.8 kHz on average. 28.8 kHz]

4. An FM receiver has a bandwidth of 50 kHz and receives an FM signal whose maximum deviation is 20 kHz and whose modulation frequency is 5 kHz. If the total signal power at the receiver is 10 mW, what is the power received at the carrier frequency? What signal power resides in the sidebands? What signal power is lost in the receiver? Repeat the calculations for a modulating frequency of 10 kHz and the same amplitude and power. Repeat the calculations for a modulating frequency that is 20 kHz and reduced in amplitude by half.
    [1.6 mW, 8.4 mW, 0.106 mW; 0.48 mW, 9.52 mW, 0.34 mW; 8.84 mW, 1.16 mW, 0.018 mW]

5. A VCO uses a varactor diode which has a capacitance that follows Equation 24.31 with $n = -0.4$ and when its reverse bias is 5 V it has a capacitance of 20 pF. The series capacitance is 180 pF, the modulating frequency is 20 kHz and the carrier frequency is 120 MHz. If the modulating voltage must change by ± 25 mV, with a modulation index, $m_f = 7.5$, what must the bias voltage of the diode be? *[3.55 V]*

6. An oscilloscope displays a white-noise voltage of r.m.s. value 0.6 mV. If the bandwidth of the oscilloscope is 20 MHz and the temperature is 300 K, what is the equivalent noise resistance of the oscilloscope? *[1.09 MΩ]*

7. The power gains and equivalent noise temperatures of three amplifiers are $G_1 = 6$ dB, $T_{1eq} = 50$ K, $G_2 = 30$ dB, $T_{2eq} = 100$ K, $G_3 = 20$ dB, $T_{3eq} = 600$ K. If the amplifiers are cascaded, what is the maximum possible equivalent noise temperature of the combination? What is the minimum? What are the corresponding overall noise factors? (Take T_A as 300 K) *[601 K, 75.15 K, 3.0033, 1.2505]*

8. What is the thermal-noise voltage measured in a bandwidth of 200 kHz across a resistance of 10 kΩ at 450 K? What is the probability that the magnitude of the instantaneous noise voltage will be greater than 6 μV? What voltage has a 50% chance of being exceeded in magnitude? What is the probability that two successive instantaneous voltages are both greater than -5 μV? What is the probability that the instantaneous voltage lies between -3 μV and $+5$ μV? (Use Table 25.1 for erfc.)
    [7.05 μV, 0.395, 4.76 μV, 0.58, 42.45%]

9. Calculate the noise voltage in a FET which has a transconductance, g_m, of 1.3 mS, if the measurement's bandwidth is 1 MHz and the temperature is 350 K. If the drain current, I_D, is 1 mA when g_m is 1 mS, what will the thermal-noise voltage be at a drain current of 100 μA and a

temperature of 400 K? If the gate resistance is 1 MΩ, what is the thermal-noise voltage at 300 K in a 1 MHz bandwidth? If the input signal to the gate is 0.8 mV, what is the SNR, if thermal noise is the only noise source? *[3.86 μV, 8.36 μV, 0.129 mV, 15.85 dB]*

10. If the base spreading resistance, r_b, in a BJT is 75 Ω and the emitter current, I_E, is 0.1 mA, what is the overall noise voltage in a bandwidth of 3 MHz at 390 K? If the temperature were increased to 500 K and the thermal noise and shot noise are equal in contributing to the overall noise voltage, what is the emitter current? *[3.96 μV, 0.2875 mA]*

11. Examine Figure 24.16 and determine, on the basis of the proportion of voltage values exceeding E_n, whether the noise is truly gaussian. You must decide what to do with those values 'on the line'. Devise a rapid test to see if the mean voltage is approximately zero.

12. Calculate the values required for R_1, R_2 and C in a pre-emphasis filter with $\tau = 50$ μs and 0.5 ms, if the filter is to have an input impedance of 1 kΩ at the lower cut-off.

13. If the modulation frequency in an FM transmission is 5 kHz and the maximum deviation is to be less than 30 kHz, with all the power residing in the sidebands and none in the carrier, what must the maximum deviation and the bandwidth be? *[27.5 and 65 kHz]*

14. What are the flicker noise and the thermal noise in a 330kΩ metal-film resistor carrying a current of 0.1 mA at 410 K, measured in the band between 30 Hz and 100 kHz? What is the total noise voltage? *[11.6 μV, 27.3 μV, 29.7 μV]*

25 Digital communications

THERE IS effectively an infinite range of values available for the encoded signal in analogue modulation methods like AM and FM. The same can be said for the analogue pulse-modulation techniques, PAM, PWM and PPM. Digital modulation employs a small number of discrete pulse sizes, often just two, and consequently the signal amplitude must be encoded in some way. The significant words here are 'in some way', for that determines the capacity of the communications channel – how much information it can convey in a given time. A great advantage of digital communications is that the encoded message can be made as free from errors as desired and moreover, given the appropriate parameters of the channel, the error rate is predictable. It is inherently more difficult to maintain the relative levels of a continuously varying waveform than to recognise the presence or absence of a pulse. Thus signal distortion is cumulative in an analogue communications system, whereas in a digital system, no matter how many repeaters are used, the signal can be recovered with a guaranteed level of distortion. Digitising a signal also enables it to be encoded, so making it hard to intercept and decode: digital communications can be made secure fairly easily. The bandwidth available can also be used more efficiently by digitising signals and as the available wireless spectrum becomes ever more crowded this factor will prove decisive in shifting non-cable communications to a digital form.

25.1 Binary modulation

Just as analogue modulation can be achieved using either the amplitude, frequency or phase of a carrier wave, so in binary pulse modulation we can correspondingly use on-off keying or OOK (sometimes this is called amplitude-shift keying or ASK), frequency-shift keying (FSK) or phase-shift keying (PSK) as shown in Figure 25.1.

OOK simply switches the carrier frequency, f_0, on or off to encode a 1 or 0. FSK uses different frequencies: f_1 for a 0 and f_2 for a 1, keeping the carrier amplitude the same. PSK employs a single carrier frequency, f_0, and a 180° phase shift to change a 1 to a 0 and vice versa. A further type of keying is quadrature phase-shift keying (QPSK) in which the phase shift is 90°. OOK is the simplest to achieve and PSK the most difficult, so it is not surprising therefore to find, by way of compensation for the trouble, that for the same noise levels the pulse power required is more for OOK and for FSK than for PSK for a given error rate[1].

---

[1] Strictly speaking this is binary phase-shift keying (BPSK). By phase-shifting 90° more information can be sent as two bits encoded in a single pulse called a *dibit*. Differential phase-shift keying (DPSK) is sometimes used for high-data-rate telephone modems. The phase shift of one bit is made relative to the previous bit and no absolute phase reference is needed, unlike BPSK. Quadrature amplitude modulation (QAM) is a combination of OOK and PSK.

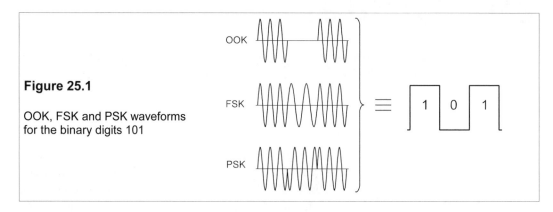

Figure 25.1

OOK, FSK and PSK waveforms
for the binary digits 101

25.1.1 Types of pulse

A constant-amplitude digital pulse can be one of many types, three of which are shown in
Figure 25.2.

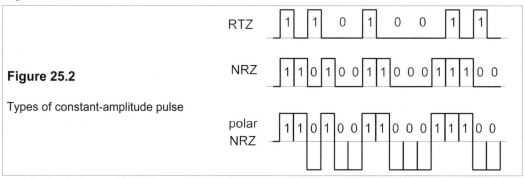

Figure 25.2

Types of constant-amplitude pulse

In each case only the pulse envelope is drawn. The first type uses only half the available time
to send a 1 (pulse) or zero (absence of pulse) because after every digit, whether 0 or 1, the pulse
returns to zero and is called a *return-to-zero* (RTZ) signal. The second type of signal has no
gaps between digits and is known as a *non-return-to-zero* (NRZ) pulse; it is also an on-off pulse
as the 0s are just pulse absences. The third type of pulse is a polar pulse in which the 1s are of
one polarity (+) and the 0s are of opposite polarity (−). Various combinations of these basic
types are possible, such as the bipolar signal in which successive 1s are coded in opposite
polarity. Mostly NRZ signals are used as they make more efficient use of the channel.

25.1.2 Error probabilities for OOK, FSK and PSK

With OOK, a 1 is signalled with a voltage of positive amplitude, A. Let us suppose the noise
is gaussian of zero mean with an r.m.s. noise voltage, σ, the probability that the signal is
misread as a 0 (which means its amplitude falls below $A/2$) is given by Equation 24.40, with
$E = A/2$ and $E_n = \sigma$, that is[2]

---

[2] σ is often called the r.m.s. noise, since it may be measured as a voltage or a current.

$$P_e = 0.5 \, \text{erfc} \, (A/\sigma 2\sqrt{2}) \tag{25.1}$$

The probability of misreading a 0 as a 1 (signal amplitude $> A/2$) is the same. In a message of more than a few bits a 1 and a 0 are equiprobable and Equation 25.1 gives the probability of error in any bit. Since $A^2/\sigma^2 = 2S/N$ (the average signal power is $A^2/2$ if a 1 and a 0 are equiprobable) we can rewrite Equation 25.1 in the form

$$P_e = 0.5 \, \text{erfc}\sqrt{S/4N} \tag{25.2}$$

The probability of error, $0.5\text{erfc}(x)$, is tabulated in Table 25.1.

Table 25.1 *Error probabilities, $P_e = 0.5\text{erfc}(x)$*

x	0.0	0.1	0.2	0.3	0.4	0.5	0.6	0.7	0.8	0.9
0	5.000E−1	4.438E−1	3.886E−1	3.357E−1	2.858E−1	2.398E−1	1.981E−1	1.611E−1	1.289E−1	1.015E−1
1	7.865E−2	5.900E−2	4.484E−2	3.300E−2	2.386E−2	1.695E−2	1.183E−2	8.105E−3	5.455E−3	3.605E−3
2	2.339E−3	1.490E−3	9.314E−4	5.716E−4	3.443E−4	2.035E−4	1.181E−4	6.717E−5	3.751E−5	2.055E−5
3	1.105E−5	5.824E−6	3.013E−6	1.529E−6	7.610E−7	3.716E−7	1.779E−7	8.358E−8	3.850E−8	1.740E−8

Note: For $x \geq 4$, $0.5\text{erfc}(x) \approx 0.5\exp(-x^2)/x\sqrt{\pi}$.

For arguments greater than about 4 the approximation given below the table is sufficiently accurate. For $P_e = 10^{-5}$, a typical maximum error rate in digital communications systems, we can estimate $x = 3.02$ from Table 25.1, that is $\sqrt{(S/4N)} = 3.02$ and $S/N = 36.5$ or $A/\sigma = \sqrt{(2 \times 36.5)} = 8.54$. In fibre-optical communications the maximum error rate is normally 10^{-9} for which $x = 4.244$ and the SNR for OOK is 72 or 18.6 dB. A probability of error of 10^{-9} is minuscule; it implies that a signal sent at 64 kbit/s (such as a digitised telephone call) would have an average of one bit error in over 4 hours. The rapid fall in error rate as the SNR rises above about 17 dB is said to be a *threshold effect*.

OOK, FSK and PSK differ in their requirements for detection and PSK in particular requires phase synchronization of transmitter and receiver. FSK requires two detection filters whose outputs are compared, and these filters must be matched to the respective signals. In the case of OOK the detector output is

$$V_o = \begin{pmatrix} A & (\rightarrow 1) \\ 0 & (\rightarrow 0) \end{pmatrix} + \sqrt{N} \tag{25.3}$$

where N is the noise power. In FSK the detector outputs are subtracted with the result

$$V_o = \begin{pmatrix} +A & (\rightarrow 1) \\ -A & (\rightarrow 0) \end{pmatrix} + \sqrt{[n_2(t) - n_1(t)]^2} \tag{25.4}$$

Both detectors are on all the time, so the random noise voltage, $n_1(t)$, of the f_1 (or 0) detector is subtracted from the random noise voltage (which may be positive or negative with zero mean), $n_2(t)$, of the f_2 (or 1) detector. The r.m.s. value is then taken, which is not $\sqrt{(N_1 - N_2)}$, the square root of the difference of the noise powers, but the square root of their *sum*, $\sqrt{(N_1 + N_2)}$, as may readily be seen by doing the calculation on two series of random numbers. Normally $N_1 = N_2 = N$ and then the output of the detector is

$$V_0 = \begin{pmatrix} +A & (\equiv 1) \\ -A & (\equiv 0) \end{pmatrix} + \sqrt{2N} \qquad (25.5)$$

For PSK only one detector is needed and the 180° phase shift can be converted to $-A$. The detector output is then

$$V_0 = \begin{pmatrix} +A & (\rightarrow 1) \\ -A & (\rightarrow 0) \end{pmatrix} + \sqrt{N} \qquad (25.6)$$

since the noise is the same as for OOK.

The *within-pulse* SNRs for each method of keying are

(1) OOK: $S/N = A^2/N$
(2) FSK: $S/N = (2A)^2/2N = 2A^2/N$
(3) PSK: $S/N = (2A)^2/N = 4A^2/N$

These equations imply that for a given amplitude of received pulse, the within-pulse SNR for OOK is half that for FSK and one quarter that for PSK.

Now the probability of detection may be calculated. For OOK the decision threshold (the voltage level at which a decision is made whether the detector output is a 1 or a 0) is $0.5A$, voltages above $0.5A$ being interpreted as a 1 and voltages below $0.5A$ as a 0. The noise voltage is σ ($= \sqrt{N}$) and the probability of an error is therefore

$$P_{e,OOK} = 0.5 \operatorname{erfc} \left(\frac{0.5A}{\sqrt{2}\sigma} \right) = 0.5 \operatorname{erfc} \left(\frac{A}{2\sqrt{2}\sigma} \right) \qquad (25.7)$$

And for FSK the decision threshold is 0, but the signal is polar, going from $+A$ to $-A$, which is effectively the same as having a decision threshold at A when the signal goes from 0 to $2A$. The noise voltage is $\sqrt{(2\sigma^2)} = \sigma\sqrt{2}$, so the probability of error is

$$P_{e,FSK} = 0.5 \operatorname{erfc} (A/2\sigma) \qquad (25.8)$$

Finally for PSK the decision threshold is again 0 with a polar signal of $\pm A$, but the noise voltage is σ and the probability of error becomes

$$P_{e,PSK} = 0.5 \operatorname{erfc} \left(\frac{A}{\sqrt{2}\sigma} \right) \qquad (25.9)$$

Combining these three equations into one, we have

$$P_e = 0.5 \operatorname{erfc} \left(\frac{A}{k\sigma} \right) \qquad (25.10)$$

where $k = \sqrt{2}$ for OOK, 2 for FSK and $2\sqrt{2}$ for PSK.

The error probabilities appear to give an advantage of 3 dB in signal power to PSK over FSK and the same advantage to FSK over OOK (for the same error probability); however, there are several things to consider. The *average* signal power needed for OOK is only half the *average* signal power needed for FSK or PSK, since the pulse is there for only half the time. This wipes out the 3dB advantage that FSK has over OOK; moreover, FSK requires two detectors and a greater bandwidth than either PSK or OOK. Provided synchronisation can be achieved, PSK offers a real advantage over both FSK and OOK.

Example 25.1

A fibre-optic communications system operates at 565 Mbit/s using OOK with an average laser power of −11 dBm (which means −11 dB relative to 1 mW) and an equivalent noise power at the receiver of −47 dBm. The transmission losses over 108 km of cable are found to be 29.6 dB, made up of splicing losses, insertion losses, recovery losses and fibre attenuation losses of 0.2 dB/km. Find the BER (bit-error rate) for 108 km of cable. Assuming the splicing, insertion and recovery losses are independent of cable length, what length of cable would result in an error rate of 1 bit/s?

The signal level at the receiver is −11 − 29.6 = −40.6 dB, so the SNR is −40.6 −(−47) = 6.4 dB, or 4.4. Equation 25.2 gives $P_e = 0.5\,\mathrm{erfc}\sqrt{(4.4/4)} = 0.5\,\mathrm{erfc}(1.049) = 0.0688$ (from Table 25.1). The BER is unacceptably high at 0.0688 for 108 km of cable.

The BER for an error of 1 bit/s is $1/(565 \times 10^6) = 1.77 \times 10^{-9}$. This is outside the values in Table 25.1, but using the approximation $0.5\,\mathrm{erfc}(x) = 0.5\exp(-x^2)/x\sqrt{\pi}$, we can find by trial and error $x = 4.178$ for $P_e = 1.77 \times 10^{-9}$. Then Equation 25.2 gives $x = \sqrt{(S/4N)}$, that is $S/4N = 4.178^2$, and the SNR = 70 or 18.5 dB. The signal level at the receiver must be 18.5 − 47 = −28.5 dBm for an error rate of 1 bit/s. Since the laser power is −11 dBm, this means the cable losses can be no more than 28.5 − 11 = 17.5 dB. The attenuation loss in 108 km is 108 × 0.2 = 21.6 dB, while the total losses are 29.6 dB, implying fixed losses of 8 dB. Subtracting these fixed losses from the required overall loss of 17.5 dB gives the attenuation loss as 9.5 dB and the cable length as 9.5/0.2 = 47.5 km.

25.1.3 *Pulse energy and spectrum noise level*

The error probability can be expressed in terms of pulse energy rather than average voltages. If the *spectrum noise level* (the noise power per Hz in the detector bandwidth, assuming all the noise is white) is n_0 and the detector bandwidth is B, then the noise power in the detector is Bn_0. The SNR is

$$S/N = A^2/Bn_0 \tag{25.11}$$

But for optimal detection the detector filter must be matched to the signal, that is to say the filter bandwidth must be the reciprocal of the pulse length, $B = 1/\tau$. If the pulse length is τ, the energy in a pulse, E, is $A^2\tau$ and Equation 25.11 then gives

$$\frac{A^2}{N} = \frac{A^2}{Bn_0} = \frac{A^2\tau}{n_0} = \frac{E}{n_0} \tag{25.12}$$

In terms of pulse amplitude and r.m.s. noise voltage this is

$$\frac{A}{\sqrt{N}} = \frac{A}{\sigma} = \sqrt{\frac{E}{n_0}} \tag{25.13}$$

Equation 25.10 remains the same with A/σ replaced by $\sqrt{(E/n_0)}$:

$$P_e = 0.5\,\mathrm{erfc}\,(x), \quad x = \begin{cases} \sqrt{E/8n_0}, & \mathrm{OOK} \\ \sqrt{E/4n_0}, & \mathrm{FSK} \\ \sqrt{E/2n_0}, & \mathrm{PSK} \end{cases} \tag{25.14}$$

Figure 25.3 shows a plot of P_e against x^2.

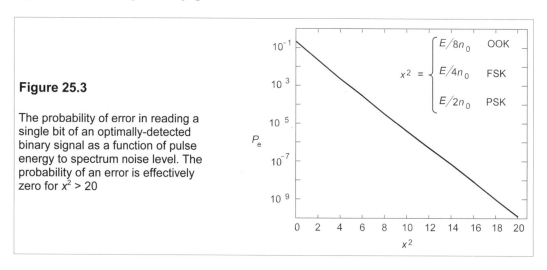

Figure 25.3

The probability of error in reading a single bit of an optimally-detected binary signal as a function of pulse energy to spectrum noise level. The probability of an error is effectively zero for $x^2 > 20$

$$x^2 = \begin{cases} E/8n_0 & \text{OOK} \\ E/4n_0 & \text{FSK} \\ E/2n_0 & \text{PSK} \end{cases}$$

25.1.4 *Transmission losses*

When a signal is broadcast through an effectively unbounded medium such as air or space as in most radio, television and microwave transmissions, it is subject to attenuation by *spherical spreading*. For example if a transmitter of signal power, P, is sending out power equally in all directions, at a distance of r the power per m² will be $P/4\pi r^2$, since the area of a sphere of radius r is $4\pi r^2$. If this power is intercepted by an antenna of effective area, A, the received signal power is $PA/4\pi r^2$. In addition there is usually attenuation by the medium of transmission which produces a loss which is constant when expressed in dB/m. An example will illustrate how these losses can be taken into account, using an unusual transmission medium: the sea.

Example 25.2

An underwater telephone uses OOK at a frequency of 70 kHz. The pulses are 1 ms long and are transmitted from an omnidirectional projector and received at an omnidirectional hydrophone which has an effective area of 10^{-3} m². The efficiencies of both projector (η_{Tx}) and hydrophone (η_{Rx}) are 65%. The acoustic signal is attenuated by absorption at a rate of 0.02 dB/m by sea water and also suffers spherical spreading loss. The transmitter pulse-power input is 16 W (that is on average the power into the projector will be 8 W) and the spectrum noise level is -142 dB relative to 1 W/Hz. Assuming the detection process is optimal, what will the bit-error rate (BER) be at 800 m? At 750 m?

We first need to find the signal level at the receiver. At a range, r, the absorption is αr, where α is the absorption coefficient, which is $0.02 \times 800 = 16$ dB. The spherical spreading results in a within-pulse power at the receiver of

$$P_{Rx}{'} = \eta_{Tx} P_{Tx} \times \frac{\eta_{Rx} A_{Rx}}{4\pi r^2} = \frac{(0.65)^2 \times 16 \times 10^{-3}}{4\pi r^2} = 5.38 \times 10^{-4} r^{-2} \qquad (25.15)$$

with no absorption. In dB relative to 1 W (or dB//1 W) this is

$$10\log_{10} P_{Rx}{'} = -44.7 - 20\log_{10} r$$

$$(25.16)$$

$$= -34.7 - 20 \times 2.845 = -91.6 \text{ dB//W}$$

The overall receiver power is found by subtracting the absorption losses from this: $P_{Rx} = -92.8 - 16$ = -108.8 dB//W (or -78.8 dBm).

Optimal detection implies that the detector bandwidth is $1/\tau = 1/10^{-3} = 1$ kHz, so the receiver noise power is Bn_0, or

$$N_{dB} = n_0(dB) + 10\log_{10}B$$

(25.17)

$$= -142 + 10\log_{10}1000 = -112 \text{ dB//W}$$

The pulse SNR is therefore $-108.8 - (-112) = +3.2$ dB, that is $A/\sqrt{N} = 10^{0.32} = 2.09$. From Equation 25.9 we find $P_e = 0.5$ erfc(2.09), which is about 1.57×10^{-3} by interpolation from Table 25.1, uncomfortably high.

Repeating the calculation for $r = 750$ m leads to SNR = 4.8 dB and $P_e = 10^{-5}$, a much more acceptable error rate. A relatively small change in distance has led to a much smaller probability of error: the telephone has an effective range (or a range threshold) of 750 m. Slight changes in noise levels due to ambient conditions can also reduce the range appreciably. Notice how small the power levels are: at the receiver the pulse power is only antilog($-108.8/10$) = 16 pW and the energy in a pulse is $16 \times 10^{-12} \times 10^{-3} = 16$ fJ! (f = femto = 10^{-15}.)

Example 25.3

Repeat Example 25.1 using PSK instead of OOK. How much energy is there in a bit when the BER is 10^{-5} with PSK?

From Equation 25.10 the BER is 0.5 erfc($A/\sigma\sqrt{2}$) = 0.5erfc$\sqrt{(S/2N)}$. Therefore in the first instance when the cable is 108 km long, $\sqrt{(S/2N)} = \sqrt{(4.4/2)} = 1.483$ and 0.5erfc(1.483) = 0.0181, still a high BER. The required SNR is 6 dB less for PSK, so the cable losses can be 6 dB more, or 12.5 dB, and the cable length is then 12.5/0.2 = 62.5 km for an error rate of 1 bit/s.

The BER is 10^{-5} when $x = \sqrt{(S/2N)} = 3.02$. The noise power is -47 dBm or 20 nW which leads to a signal power of $2 \times 20 \times 3.02^2 = 0.365$ µW. The energy per bit is then $0.365 \times 10^{-6}/(565 \times 10^{6}) = 0.646$ fJ.

25.2 Pulse-code modulation

The simplest way to send information is by representing 1 by a pulse and a zero by its absence, that is we send *bits* (Binary digITS). The receiver looks for the presence of a pulse of carrier within a certain time slot. If it detects a pulse it registers a 1 and if it does not it registers a zero. We can then decide that a certain number of sequential time slots should be a byte – a collection of bits treated as a unit.

When a byte is used to represent the sampled amplitude of a signal, which is called *pulse-code modulation* (PCM), the number of bits in that byte will determine the accuracy of the encoded information. For example, with three bits to a byte it can take one of 2^3, or 8 possible values. The unavoidable error in the encoding process is called *quantization noise*. PCM is the same as A/D conversion and demodulation is accomplished by a D/A converter. Figure 25.4 shows a PCM signal with 3 bits. Because the sampling frequency here is low, the quantization error is compounded by the sampling error.

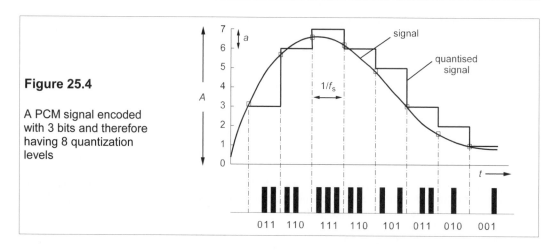

Figure 25.4

A PCM signal encoded with 3 bits and therefore having 8 quantization levels

25.2.1 PCM transmission bandwidth and bit rate

The transmission bandwidth required for PCM is proportional to the sampling frequency, f_s, and the number of bits required for each quantized level, b. The sampling frequency, by the Nyquist criterion (Equation 23.33) must be at least $2f_{max}$, where f_{max} is the highest frequency to be transmitted. The PCM bit rate, r (the number of bits transmitted per second), is therefore $2bf_{max}$. Let the time taken to send a bit be τ seconds to give a bit rate of $1/\tau$, then if the bits are sent as a square wave such as 1010... (as in Figure 25.5) the period of the square wave is 2τ. Its frequency is $1/2\tau = r/2$, which is the minimum bandwidth.

Figure 25.5

The bit rate is $1/\tau$ and the square wave's frequency is $1/2\tau$

In practice no receiver with this bandwidth could unscramble the digits and the actual bandwidth employed is approximately double, that is

$$B = r = 2bf_{max} \qquad (25.18)$$

Thus in telephony, where $f_{max} \approx 4$ kHz and $b = 8$ (256 quantization levels), the transmission bandwidth required is about $2 \times 8 \times 4 = 64$ kHz.

25.2.2 Quantization noise

Quantization noise is indistinguishable to the recipient of the message from other noise, though its characteristics and origin are quite different. If the maximum signal amplitude is A and there are m equally spaced quantization levels, then the step size of the sampled signal will be $a = A/(m - 1) = A/(2^b - 1)$, where b is the number of bits. Assuming that over a period of time the signal within a step can take on any value with equal probability, the maximum error will be ±0.5*a*; see Figure 25.6.

If the error is denoted by ε, then the mean square error is given by

$$\epsilon^2_{\text{rms}} = \frac{1}{a}\int_{-a/2}^{+a/2}\epsilon^2 d\epsilon = \frac{\left[\frac{1}{3}\epsilon^3\right]_{-a/2}^{+a/2}}{a} = \frac{a^2}{12} \tag{25.19}$$

which is a little greater than the time-average of *a*/4 for ε. ε² is the quantization noise, N_q.

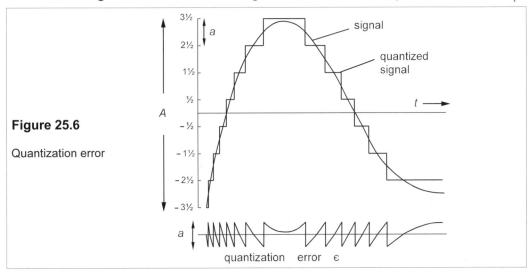

Figure 25.6

Quantization error

Assuming that all signal levels have equal likelihood a similar calculation for the mean signal level yields

$$S = [(m^2 - 1)a^2]/12 \tag{25.20}$$

(The signal is coded with polar pulses of amplitudes ±0.5*a*, ±1.5*a*, ±2.5*a* and ±3.5*a*. Had it been coded with non-polar pulses of amplitudes *a*, 2*a*,..., 7*a*, the mean signal level would have been nearly four times as great.) The quantization SNR for the polar signal is

$$S/N_q = S/\epsilon^2_{\text{rms}} = (m^2 - 1)a^2/a^2 = m^2 - 1 \tag{25.21}$$

In dB, $$\left(S/N_q\right)_{\text{dB}} = 10\log_{10}(m^2 - 1) = 10\log_{10}(2^{2b} - 1) \tag{24.22}$$

$$\approx 20n\log_{10}2 = 6b$$

For example if the quantization SNR in a PCM system is to be kept to below 40 dB, Equation 25.22 shows that $b \approx 40/6 = 7$ (the next highest integer) and $m = 2^7 = 128$ levels. An 'exact' calculation gives the same result. Interestingly, the quantization SNR for $b = 1$ (that is only two levels) is about 5 dB, yet telephonic speech is quite comprehensible with two-level PCM. Digital telephone systems normally use 256 levels, which corresponds to 8 bits and $SNR_q = 48$ dB, but for high-fidelity audio as many as 16 bits might be used (65,536 levels, $SNR_q = 96$ dB).

25.2.3 Delta modulation

In many cases the successive samples of an encoded signal are correlated, they are not very different and cannot be considered random. For example the successive samples of the sinewave shown in Figure 25.7 differ by one bit at most.

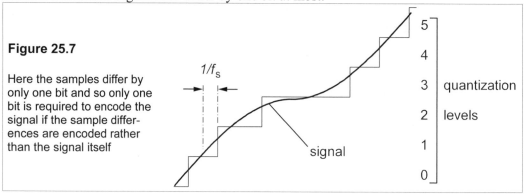

Figure 25.7

Here the samples differ by only one bit and so only one bit is required to encode the signal if the sample differences are encoded rather than the signal itself

If only the differences between samples were sent, there would be a saving in bits transmitted yet the information sent would be the same. Of course, if the signal were to change more rapidly with the same sampling rate, the number of bits required to send the differences would be little fewer than those required for the signal proper; but a suitable choice of sampling rate should alleviate the problem. The general form of difference modulation is known as *differential pulse-code modulation* (DPCM) and *delta modulation* (DM) is a special case of DPCM in which the number of bits used is the minimum – two.

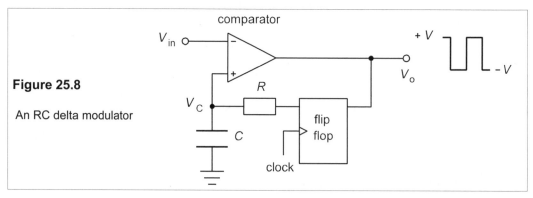

Figure 25.8

An RC delta modulator

Figure 25.8 shows a simple form of RC delta modulator. The comparator output voltage is either $+V$ or $-V$ according to whether the capacitor voltage, v_C, is respectively below or above the signal level, v_{in}. The flip-flop sends back samples of the output on receipt of clock pulses, causing the capacitor to charge or discharge exponentially with time constant RC as in Figure 25.9. The sampling frequency, f_s, is the clock frequency. The receiver is an identical RC network.

Both RC networks are low-pass filters whose frequency responses are

$$H(\omega) = \frac{V}{\sqrt{1+(\omega/\omega_c)^2}}$$

(25.23)

where $\omega_c \equiv 1/RC$. In order to follow the input waveform, $f_c \ (= \omega_c/2\pi)$, should normally be set equal to the input signal bandwidth, f_m, but in the case of speech waveforms, this turns out not to be so. The power spectrum of speech tends to fall off at 20 dB/decade like a low-pass filter and the best value for f_c is about 150 Hz. If f_c is set to 3 kHz for example, most of the time the capacitor is charged up to $\pm V$ and the quantization noise is increased. On the other hand, if f_c is made too small the overload noise will increase.

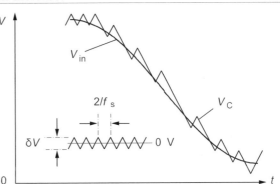

Figure 25.9

How the capacitor voltage changes with time. When the signal level is zero the idling waveform is a triangular wave of peak-to-peak amplitude δV_C and frequency $f_s/2$.

25.2.4 Channel capacity with PAM and PCM

We can calculate the maximum rate at which a PAM signal can be sent over a channel as a function of SNR and transmission bandwidth. The calculation is essentially the same as for a PCM signal since this is derived initially from a PAM signal. Suppose we code the pulses as in Figure 25.4 and the sampling rate (pulses/second) is the Nyquist rate of $2B$, where B is the transmission bandwidth. If there are m levels for each pulse, there are $\log_2 m$ bits of information per pulse, where $\log_2 2^b = b$. For example if there are 8 levels, they can be represented by $\log_2 8 = 3$ bits. Each pulse carries 3 bits of information: it is a 3-bit *symbol*[3]. If the sampling rate is $2B$ the maximum information rate in bits per second is

$$C = 2B\log_2 m = B\log_2 m^2$$

(25.24)

where C is the channel capacity. Now the average signal power, S, assuming each level is equally likely, is

$$S = \frac{2([a/2]^2 + [3a/2]^2 + ... + [(m-1)a/2]^2)}{m} = \frac{(m^2-1)a^2}{12}$$

(25.25)

From which we find

$$m^2 = 1 + 12S/a^2$$

(25.26)

---

[3] The information rate is sometimes given in *bauds* where 1 baud = 1 symbol/s. Baud rates are mostly used when sending ASCII (American Standard Code for Information Interchange) in which each symbol comprises 7 bits plus a start bit and a stop bit. Consequently there are $2^7 = 128$ possible characters in ASCII.

and substituting into Equation 25.24 produces

$$C = B\log_2(1 + 12S/a^2) \qquad (25.27)$$

Equation 25.27 suggests that, even in the event of restricted bandwidth, we can still increase the information rate without limit simply by increasing S and reducing a, in other words by making the signal amplitude and the number of bits per pulse larger and larger. Of course, we must pay a penalty in power, but that may be worthwhile, even though the increase in power must be *exponential* for a *linear* increase in capacity. The channel capacity may also be increased by decreasing the spacing between levels, a. However, there is a limit on how small the step size, a, can be made, which depends on the noise level. It is obvious that if the noise level is greater than the step size then the probability of misreading the information is nearly certain. Suppose that the r.m.s. noise voltage is σ and let $a = h\sigma$, where h is to be large enough to cause as few errors as we desire, then $a^2 = h^2\sigma^2$ and Equation 25.27 becomes

$$C = B\log_2\left(1 + \frac{12S}{h^2\sigma^2}\right) = B\log_2\left(1 + \frac{12S}{h^2N}\right) \qquad (25.28)$$

taking N as σ^2, the r.m.s. noise voltage squared.

The question is, what value of h results in an acceptable error rate? For a BER of 10^{-5}, it has been shown in Section 25.1.2 that the value of $h\,(= a/\sigma)$ required is about 8.5. Substituting $h = 8.5$ into Equation 25.28 gives the channel capacity as

$$C = B\log_2(1 + S/6N) \qquad (25.29)$$

At a lower BER the channel capacity is larger (see Problems 25.7 and 25.8). Had better encoding been used the signal capacity of the channel would be greater, as we shall now see.

25.3 The theoretical maximum channel capacity

In a series of important papers[4] published in 1948 and 1949, Shannon showed that above a certain rate of transmission, the errors in a message subject to band-limited white noise became unacceptable. Shannon's formula for the maximum channel capacity, C, is

$$C = B\log_2(1 + S/N) \qquad (25.30)$$

where C is in bits/s, B is the transmission bandwidth (not necessarily the modulation bandwidth) and $\log_2$ is the logarithm to base 2, which can be converted to base 10 or base e using

$$\log_2 x = \log_n x/\log_n 2 = \log_{10} x/\log_{10} 2 = 3.322\log_{10} x \qquad (25.31)$$

Shannon's formula does not indicate what method should be used to encode the information – a fruitful research area for many years – merely what the maximum possible value is.

The channel capacity for a PCM signal with a BER of 10^{-5} has already been shown to be

$$C = B\log_2(1 + S/6N) \qquad (25.32)$$

---

[4] C E Shannon, 'A mathematical theory of communication', *Bell System Technical Journal* **27** 379–423 and 623–656 (1948); C E Shannon, 'Communication in the presence of noise', *Proc. IRE* **37** 1–21 (1949).

implying that the signal power required is *six times* the theoretical value for an optimally-encoded transmission. Accordingly PCM is some way from the optimal encoding method and a good deal of effort has been expended at finding better ones.

Example 25.4

What is the maximum capacity in bit/s of a telephone line using PCM with a bandwidth of 3 kHz, a signal voltage of 5 V and an r.m.s. noise spectral density (σ_0) of 2 mV/$\sqrt{\text{Hz}}$? What is its maximum capacity (a) if the signal power is doubled and the bandwidth is kept constant and (b) if the bandwidth is doubled while the signal power is kept constant? The spectral noise density is the same for all three cases. What are the theoretical maximum capacities in each case?

In a bandwidth of 3 kHz the noise power will be

$$N = B\sigma_0^2 = 3000 \times (2 \times 10^{-3})^2 = 0.012 \text{ W} \tag{25.33}$$

(taking the voltage to be across a notional 1Ω resistance). Then $S = 5^2 = 25$ W and the SNR is $25/0.012 = 2083$ (33.2 dB). Equation 25.31 yields

$$C = 3 \log_2(1 + 2083/6) = 3 \times 3.322 \log_{10}348 = 25.3 \text{ kbit/s} \tag{25.34}$$

If the signal power is doubled while B and σ_0 remain constant, then the SNR is $2 \times 2083 = 4166$, and

$$C = 3 \times 3.322 \log_{10}(1 + 4166/6) = 28.3 \text{ kbit/s} \tag{25.35}$$

at best a marginal improvement. If the signal power stays the same and the bandwidth doubles, the noise power doubles to 0.024 W, so the SNR is halved to 1042 and

$$C = 2 \times 3 \times 3.322 \log_{10}(1 + 1042/6) = 44.7 \text{ kbit/s} \tag{25.36}$$

almost double the original capacity. The theoretical maximum capacities are

$$\begin{aligned} 3 \times 3.322 \log_{10}(1 + 2083) &= 33.1 \text{ kbit/s} \\ 3 \times 3.322 \log_{10}(1 + 4166) &= 36.1 \text{ kbit/s} \\ 2 \times 3 \times 3.322 \log_{10}(1 + 2083) &= 66.2 \text{ kbit/s} \end{aligned} \tag{25.37}$$

Note that increasing the bandwidth pays much better than increasing the signal strength, even though the noise increases proportionally. There is a limit to the improvement by this means because eventually the noise engendered by the wider bandwidth will be comparable to the signal. In the limit when $S = N = B\sigma_0^2$, then $B = S/\sigma_0^2$ and the channel capacity is

$$C_\infty = 2B\log_2(1 + S/N) = 2B\log_2 2 = 2B = 2S/\sigma_0^2 \tag{25.38}$$

where C_∞ is the asymptotic (unlimited bandwidth) channel capacity. In practice a SNR of unity would be impractical. Taking a more realistic limit of $S = 4N$ leads to $C_\infty = S/2\sigma_0^2$.

Suggestions for further reading - see Chapter 26

Problems

1. A digital communications system operates at a bit rate of 10 Mbit/s and a transmitter power of 3 kW. The transmission losses are 6 dB/km and the noise power in the receiver is 2 nW. What is the distance between transmitter and receiver (the range) that will result in a BER of 10^{-5} with OOK?

If the range is 18 km, what is the error rate in bit/s? What range would result in an error rate of 1 bit/s? (Assume average, not pulse, power in the calculations.) *[17.6 km, 2.91 kbit/s 17.35 km]*

2. The signal power from a radio source is subject to spherical spreading, but no other losses. At 1 km from the transmitter the signal power is 195 $\mu W/m^2$. A receiver and its antenna are placed 100 million km from this transmitter. If the effective antenna area is 38 m^2 and it has a gain of 76 dB, what is the SNR at the receiver if its noise power is 4.9 pW? What would the BER be with OOK? With PSK? The receiver is cooled from 300 K to 77 K to improve its SNR, as all the noise is thermal in origin. What will the BER be now with PSK? To what temperature would the receiver have to be cooled to produce a BER of 10^{-9} with PSK? (Use average power, not pulse power.) *[0.042, 7.2 × 10⁻³, 6.6 × 10⁻⁷, 50 K]*

3. A probe is to send back information from Neptune which is 4.5×10^9 km from earth. If the noise level in the receiver is 0.1 pW and the receiving antenna has an area of 100 m^2 with a gain of 80 dB, what transmitted pulse power is required to produce a BER of 10^{-3} with OOK? If the average transmitter power is 850 W and the minimum pulse length is 0.1 μs, what is the maximum bit rate? Colour television pictures are to be sent back from Neptune with 625 lines, each of 400 tri-colour pixel dots; what frame rate is possible? If delta modulation is used only 12% of the pixels need be changed from frame to frame. What frame rate is now possible?
    [97.4 kW, 175 kbit/s, 0.233 Hz, 1.94 Hz]

4. Repeat the previous problem using PSK instead of OOK. If the BER is to be 10^{-5} what is the required pulse power? (The picture is sent frame by frame with an appropriate interval between frames to keep the average power down to 850 W.)
    [24.3 kW, 350 kbit/s, 0.466 Hz, 3.88 Hz, 46.4 kW]

5. A communications system comprises a transmitter, a receiver and a channel which has a fixed loss of 6 dB and a further cable loss of 0.13 dB/m. If the receiver has an equivalent noise temperature of 2500 K and a bandwidth of 10 MHz, what is the maximum cable length if the transmitted power is 176 W and the maximum BER is 10^{-6} with PSK. What cable length is permitted if the equivalent noise temperature is 300 K? *[981 m, 1052 m]*

6. A sonar transmitter operates with a pulse length of 100 μs and a pulse power of 20 kW. The receiver uses an optimal detection process and has an area of 0.2 m^2. The spectrum noise level is -135 dB relative to 1 W/Hz. If the attenuation of sea-water is 8 dB/km and the overall gain of the system is 78 dB, what is the range at which the BER is 10^{-4} with OOK? (Take into account spherical spreading.) *[12.3 km]*

7. What is the channel capacity of a communications system with a spectrum noise of 0.8 $\mu V/\sqrt{Hz}$ and a signal level of 50 mV, using PAM with a BER of 10^{-5} and a bandwidth of 80 MHz? If the BER were to be reduced to 10^{-7} show that the channel capacity would be given by $C = 3.322 \, B \log_{10}(1 + S/9N)$. What is the channel capacity now with the same signal and noise levels? What does Shannon's formula give for the maximum capacity? What is the capacity if the bandwidth is unlimited and the SNR is 4? *[255 Mbit/s, 215 Mbit/s, 451 Mbit/s, 1.95 Gbit/s]*

8. A communications system operates with a signal and transmission bandwidth of 2 MHz and a SNR of 20 dB. What is the theoretical maximum information rate in bits per second? If PAM is used with a BER of 10^{-4}, what is the maximum information rate? If the bandwidth is tripled and the signal and spectrum noise levels are unchanged, what do these two rates become?
    [13.3 Mbit/s, 9 Mbit/s, 30.6 Mbit/s, 18.2 Mbit/s]

9. How many bits/sample are required in a PCM signal to keep the quantization SNR below 30 dB? If the SNR at the receiver is 25 dB (apart from quantization noise), what is the overall SNR with 5-bit encoding? What is the BER? If the sampling rate is 6 kHz, what is the theoretical minimum transmission bandwidth required? What is the information rate? What is the maximum information rate according to Shannon's formula (Equation 25.30)? And according to Equation 25.29? (Take the minimum theoretical bandwidth for these calculations.) Why the differences?
    [5, 23.8 dB, 3.3 × 10⁻²⁸, 15 kHz, 30 kbit/s, 119 kbit/s, 80 kbit/s]

26 Fibre-optic communications

OPTICAL FREQUENCIES cover the wavelength decade from about 2 µm to 200 nm, that is from near infra-red to near ultra-violet. Though desirable for communications because of their huge bandwidth, they languished until a transmission medium of low attenuation was found. Aerial communications become increasingly difficult as the frequency is raised from microwave to infra-red because of increasing absorption by molecules and particles in the atmosphere. In addition a coherent[1] source of light was essential as well as a means of modulating it. The coherent source was the laser, invented in 1958, and the low-loss medium was discovered in 1966. In that year, Kao and Hockham of STC Laboratories in England succeeded in sending light through an optical fibre, suggesting that transmission over several kilometres might be feasible with less attenuating fibres. By 1976 losses had been reduced to 0.5 dB/km in laboratory fibre specimens and the large-scale use of fibres for telephone trunk lines began. For some years all new trunk lines in the UK have been fibre-optic cables.

26.1 Pros and cons of fibre-optics

The advantages of using fibre-optical systems in communications are chiefly:

- Large bandwidth
- Immunity from electromagnetic interference
- Smaller size, weight and cost
- Long life: fibres cannot corrode like wires
- Higher security: fibres cannot be 'tapped' like wires
- Elimination of sparking and fire hazards

There are some disadvantages too:

- Fibres cannot be joined as easily as wires; however, splicing is becoming easier as time goes on
- Fibres need a strong protective sheath as they are quite easily broken

---

[1] Coherent means that phase relationships in the propagating wave are preserved, which does not happen with incoherent sources such as a light bulb. The light beam is spatially coherent if, at all points on a plane normal to the direction of propagation, the electric field varies identically with time. The light beam is temporally coherent if the electric field at a point varies uniformly with time.

- Remote power has to be sent by a separate wire

Wherever the required bandwidth is small it is easier to use a twisted pair or coaxial cable, but the extra cost of fibre-optic cabling is negligible when higher capacity is needed.

26.2 The transmitter

The elements of a fibre-optic communications system are the same as for any other communications system: a transmitter/modulator, a channel and a receiver/demodulator. The transmitter is usually a semiconductor laser diode (SLD) made of a III-V compound like gallium arsenide (GaAs – Ga from group III and As from group V of the periodic table). The SLD is similar to an edge-emitting LED in principle, but is operated at a higher current density that produces laser emission from a small region near the p-n junction. The wavelength of the light emitted is determined by the bandgap of the active region, which is also the minimum voltage across the diode:

$$\lambda = hc/E_g = 1.24/E_g \ \mu\text{m}, \quad (E_g \text{ in eV}) \tag{26.1}$$

where h is Planck's constant (6.63×10^{-34} Js), c is the speed of light in a vacuum and λ is the wavelength of the emitted light. Thus to obtain a wavelength of 1.55 μm (where silica-fibre attenuation is minimal) a bandgap of 0.8 eV is required. The bandgap of GaAs is 1.43 eV and that of InAs (indium arsenide) is 0.36 eV, so a compound such as $\text{In}_x\text{Ga}_{1-x}\text{As}$ with $x \approx 0.6$ would have roughly the required bandgap. However the SLDs are made up of epitaxial (meaning 'having the same crystallographic orientation') layers which require closely-matching lattice parameters (crystallographic unit cells of the same size), and a quaternary composition ($\text{In}_x\text{Ga}_{1-x}\text{As}_y\text{P}_{1-y}$) is necessary to achieve this.

Because the number of photons emitted while lasing is proportional to the number of electrons passing through the p-n junction, modulation is a straightforward matter of modulating the current. However, although analogue modulation is readily performed, in most fibre-optic applications digital modulation is used, often simply OOK.

Example 26.1

The semiconductor laser of Figure 26.1 operates with a current-density of 30 MA/m². Given that the wavelength of the light emitted is 1.3 μm, estimate the power consumed. If the laser's overall efficiency is 0.3% and a NRZ pulse is used to modulate it with OOK at 565 Mbit/s, how many photons are emitted per bit? The lasing channel is 0.5 μm thick, what is the optical power density at the channel exit?

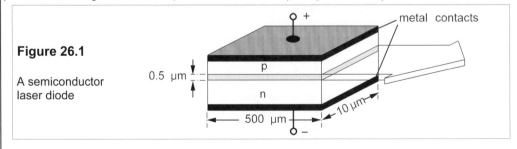

Figure 26.1

A semiconductor
laser diode

Since $I = JA$, we have $I = 3 \times 10^7 \times 500 \times 10^{-6} \times 10 \times 10^{-6} = 0.15$ A. We can estimate the power consumed from the bandgap energy and the current, using Equation 26.1. With $\lambda = 1.3$ µm, we find $E_g = 0.954$ eV. The voltage must be at least this (divided by the electronic charge), and so the power must be at least $0.15 \times 0.954 = 0.143$ W or 21.6 dBm. If the efficiency is 0.3%, then 0.3% of the electrons in the current flowing across the junction produce photons. A current of 0.15 A is an electron flow rate of $0.15/1.6 \times 10^{-19} = 9.4 \times 10^{17}$ electrons/s, resulting in an emission rate of $0.003 \times 9.4 \times 10^{17} = 2.8 \times 10^{15}$ photons/s, and that is $2.8 \times 10^{15}/565 \times 10^6 = 5 \times 10^6$ photons/bit. This beam passes out of an area of 10 µm by 0.5 µm (5×10^{-12} m²), so the optical power density is $0.003 \times 0.143/5 \times 10^{-12} = 86$ MW/m² or 86 W/mm² – a reasonably high intensity, though very small beer compared to the levels reached in high-energy, pulsed, ruby lasers.

At low current densities SLDs act just like LEDs, until a current-density threshold, J_{th}, is reached when the light output increases rapidly and laser action begins. Laser diodes are much more efficient than LEDs, but to improve the efficiency still further the lasing area can be compressed into a thin strip at the centre of the laser. The emitted light is confined to a narrow channel by making the refractive index of the material higher at the lasing junction. This can be achieved by using a different material at the junction from that either side of it. For example (Ga,Al)As has a lower refractive index (RI) and a higher bandgap than GaAs but can be doped in the same way to form a P-p-N or a P-n-N *heterojunction* (different materials either side of the junction), where upper-case letters refer to the higher-bandgap (lower RI) semiconductor. Besides vertical confinement, horizontal confinement of the active region is possible. In the *buried double-heterojunction (DH) laser* (see Figure 26.2a) the N-type (Ga,Al)As at the sides has a lower refractive index than the lasing channel between. The emitting area can be as small as 2 µm × 0.4 µm. The stripe-contact laser of Figure 26.2b is another way of obtaining horizontal confinement.

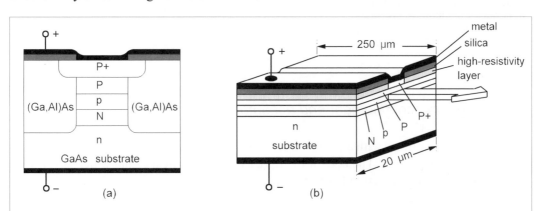

Figure 26.2 DH laser diodes (DHLDs) (a) A cross-section of a buried DHLD. The active region is the central p-type area (b) A stripe-contact DHLD, in which the lasing region is constricted by a high-resistivity layer beneath the insulating silica. The active layer is about 0.5 µm thick.

Unfortunately the light output of SLDs is very dependent on temperature, as well as current and it is necessary not only to control the temperature of the SLD but also to monitor its output and use feedback to control the intensity. The SLD can be mounted on a controlled Peltier-effect device which will keep its temperature constant to within ± 1 K. If the duty cycle is low, mounting the laser on a heat sink could suffice.

SLDs have another important advantage over LEDs and that is narrow *linewidth*. The laser does not emit light at a single wavelength but at a number of closely-spaced wavelengths occupying a band whose width, though narrow, is finite. This bandwidth is known as the line-width and is sometimes expressed as a Q-factor, defined by

$$Q = \lambda / \Delta\lambda \qquad (26.2)$$

where λ is the wavelength at the centre of the band and $\Delta\lambda$ the linewidth. For example, a GaAs SLD emitting at 800 nm might have a Q of 500, producing a linewidth of $800/500 = 1.6$ nm. Linewidth is important as it limits the bit-rate of transmission.

26.3 The channel: optical fibres

Low-loss optical fibres have undoubtedly been the most important advance in materials for communications in the last 20 years. The fibres are drawn from a furnace containing molten silica (SiO_2) with small amounts of additives such as GeO_2 to permit the control of refractive index. Though a fibre-optic cable superficially looks like a piece of coaxial cable, it contains only a very small-diameter fibre at its centre, the rest being made up of strength members and protective material as shown in Figure 26.3.

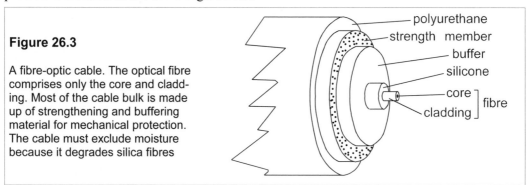

Figure 26.3

A fibre-optic cable. The optical fibre comprises only the core and cladding. Most of the cable bulk is made up of strengthening and buffering material for mechanical protection. The cable must exclude moisture because it degrades silica fibres

26.3.1 Light ray confinement and acceptance

Let us first consider fibres with a core of uniform RI, n_1, surrounded by a cladding of uniform RI, n_2, – a *step-index* fibre. The core of the fibre carries the optical signal which is confined there by total internal reflection from the cladding.

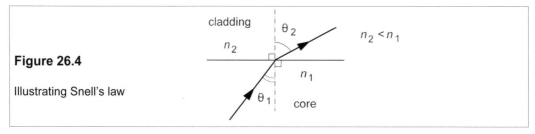

Figure 26.4

Illustrating Snell's law

In Figure 26.4, the incident ray in the core (RI $= n_1$) makes an angle θ_1 with the normal to the axis, while the refracted ray in the cladding (RI $= n_2$) makes an angle θ_2 with it. Snell's law states that

$$n_1 \sin \theta_1 = n_2 \sin \theta_2 \tag{26.3}$$

For the incident ray to be reflected rather than refracted, $\theta_2 \geq 90°$, that is, using Equation 26.3 with $\sin \theta_2 = 1$,

$$\sin \theta_1 \geq n_2/n_1, \quad \text{therefore} \quad \theta_1 \geq \sin^{-1}(n_2/n_1) \tag{26.4}$$

The critical angle, $\theta_c = \sin^{-1}(n_2/n_1)$, and unless the cladding has a *lower* refractive index than the core, total internal reflection is impossible. In a typical fibre, $n_1 \approx 1.54$ and $n_2 \approx 1.51$ giving a critical angle of 79°. The cladding absorbs rays that are not totally internally reflected and they will not propagate. Similar calculations can be made to discover that a ray may only enter an optical fibre if it lies within a cone having a semiangle less than a certain value known as the *acceptance angle*, α_{max}.

Consider a ray entering a fibre from the air at an angle α to the axis as in Figure 26.5.

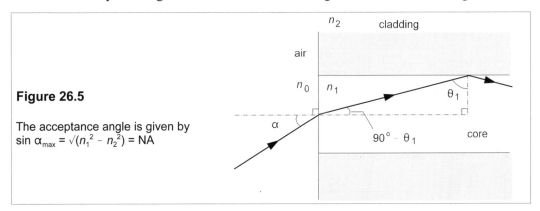

Figure 26.5

The acceptance angle is given by $\sin \alpha_{max} = \sqrt{(n_1^2 - n_2^2)} = \text{NA}$

On entering the fibre the ray is refracted by Snell's law and then

$$n_0 \sin \alpha = n_1 \sin (90° - \theta_1) \quad \Rightarrow \quad \sin \alpha = n_1 \cos \theta_1 \tag{26.5}$$

since the RI of air, n_0, is 1. When $\theta_1 < \theta_c$, the ray is not reflected but is absorbed rapidly by the cladding, but when $\theta_1 \geq \theta_c$ the ray is reflected totally by the cladding and will propagate: it will be accepted by the fibre. The minimum value of θ_1 is θ_c and this angle gives the maximum value of α:

$$\sin \alpha_{max} = n_1 \cos \theta_c \tag{26.6}$$

But the critical angle is given by Equation 26.4 so that the *numerical aperture*, $\sin \alpha_{max}$, is

$$\text{NA} = \sin \alpha_{max} = n_1 \sqrt{1 - \sin^2 \theta_c} = \sqrt{n_1^2 - n_2^2} \tag{26.7}$$

Because $n_1 \approx n_2$ the acceptance angle is small – only 18° when $n_1 = 1.54$ and $n_2 = 1.51$. Lasers and fibres must be aligned precisely to ensure most of the light goes into the fibre.

26.3.2 Monomodal and multimodal fibres

Light energy propagates down the fibre and its electric and magnetic fields obey Maxwell's equations subject to boundary conditions imposed by the fibre, which is a circular waveguide. Each solution of Maxwell's equations corresponds to a *mode of propagation* – mode for short. When the number of modes, N, is large it can be shown that it is given by

$$N \approx (\pi d/\lambda)^2 (n_1^2 - n_2^2)/2 \approx (n_1 \pi d/\lambda)^2 \Delta \tag{26.8}$$

where $\Delta \equiv (n_1 - n_2)/n_1$, d is the core diameter and λ is the vacuum wavelength.

Example 26.2

A fibre has a core diameter of 20 μm and RI = 1.515 with a cladding of RI = 1.495. It carries pulses of light whose vacuum wavelength is 1.55 μm. How many modes are there, approximately? What would the diameter have to be approximately to reduce the number to just one?
  Here, $\Delta = (1.515 - 1.495)/1.515 = 0.0132$ and then by Equation 26.8 we find

$$N \approx \left(\frac{1.515 \times \pi \times 20 \times 10^{-6}}{1.55 \times 10^{-6}} \right)^2 \times 0.0132 \approx 50 \tag{26.9}$$

Putting $N = 1$ (though this is hardly a 'large' number) into Equation 26.9 yields

$$1 \approx \left(\frac{1.515 \times \pi d}{1.55 \times 10^{-6}} \right)^2 \times 0.0132 \tag{26.10}$$

from which $d \approx 2.8$ μm.

Thus, by making the core diameter very small it is possible to exclude all but one mode and make a single-mode (monomodal) fibre. The exact formula for this diameter is

$$d < \frac{0.77\lambda}{\sqrt{n_1^2 - n_2^2}} \tag{26.11}$$

and with the values of RI from Example 26.2 we find $d = 4.9$ μm, roughly twice that given by the formula for large numbers of modes.

26.3.3 *Intermodal dispersion*

The light energy passes down the fibre with different speeds according to which mode is taken, but roughly speaking the fastest mode travels at a speed of $v_2 = c/n_2$ and the slowest mode at a speed of $v_1 = c/n_1$. The difference in time of arrival of the first part of a light pulse travelling though a fibre of length, L, will thus be

$$\Delta \tau = \frac{L}{v_1} - \frac{L}{v_2} = \frac{Ln_1}{c} - \frac{Ln_2}{c} = \frac{(n_1 - n_2)L}{c} \tag{26.12}$$

Expressed per unit length (usually 1 km) of fibre, $\Delta \tau$ is the *intermodal dispersion* (sometimes called *modal dispersion*). In practice the effects of inter-modal dispersion are mitigated by the presence of inhomogeneities and small bends in the fibre. These have the effect of transferring energy from faster modes to slower modes, so that the dispersion is reduced and becomes proportional to $\sqrt{L}$ instead of L, when L is greater than about 1 km.
  This can be written

$$\Delta \tau = \Delta \tau_1 \sqrt{L}, \quad L > 1 \text{ km} \tag{26.13}$$

where $\Delta\tau_1$ is the dispersion for 1 km of fibre and L is in km.

Because of the spreading in the pulse, the maximum bit rate is reduced (apart from other limitations which there may be). The leading edge of the pulse will be spread by about $\Delta\tau$ as will the trailing edge, so the best that one might hope for would be a bit rate of $1/(2\Delta\tau)$. However, this estimate is a little too optimistic and in practice the maximum bit rate is about

$$R_{max} = 0.25/\Delta\tau \tag{26.14}$$

Example 26.3

For the fibre in Example 26.2 calculate the intermodal dispersion/km and the intermodal dispersion of a fibre 100 m long and one 50 km long.
   Taking $L = 1$ km, Equation 26.12 gives

$$\Delta\tau_1 = \frac{(1.515 - 1.495) \times 1000}{3 \times 10^8} = 67 \text{ ns} \tag{26.15}$$

Thus a fibre 100 m long will have an intermodal dispersion of $67 \times 0.1 = 6.7$ ns and one 50 km long a dispersion of $67 \times \sqrt{50} = 474$ ns.

26.3.4 Graded-index fibres

Inter-modal dispersion can be reduced to nil by using a monomodal fibre, but the problem with these is that the very tiny core makes it much more difficult to align the transmitter and receiver with the fibre. An alternative is to change the RI gradually from n_1 at the core's centre to n_2 at the core's circumference, that is to use a *graded-index* fibre. The manner of the grading determines the precise amount of inter-modal dispersion. The variation in RI across the core causes rays that travel furthest to do so in material of lower RI and at greater speed so minimising the dispersion.

One can express the dependence of RI with radius in the form

$$\begin{aligned} n(r) &= n_1\sqrt{1 - 2(r/a)^\alpha \Delta}, \quad &r < a \\ &= n_2, \quad &r \geq a \end{aligned} \tag{26.16}$$

where a is the core radius and α is the *grading index*. The reader may verify that when $r = a$, Equation 26.16 gives $n(a) = n_2$, as is required. When $\alpha = 1$, the grading is linear, and $\alpha = \infty$ corresponds to a step-index fibre. Figure 26.6 shows the RI profile for various values of the grading index.

If N_s is the number of modes in a step-index fibre of core radius, a and RI, n_1, the number of propagating modes in a graded-index fibre is given by

$$N = \frac{N_s \alpha}{\alpha + 2} \tag{26.17}$$

The dispersion in a graded-index fibre is found to have a minimum when

$$\alpha = \alpha_{opt} = 2(1 - \Delta) \tag{26.18}$$

which is just a little less than 2. In this case the optimal dispersion is

$$\Delta\tau_{min} = Ln_1\Delta^2/8c \qquad (26.19)$$

– very much less than in a step-index fibre. (Substitution of the figures for the fibre in Example 26.2 leads to $\Delta\tau_{min} = 0.11$ ns/km compared to 67 ns/km.)

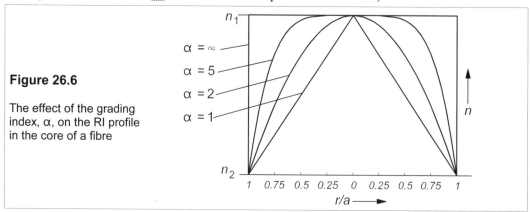

Figure 26.6

The effect of the grading index, α, on the RI profile in the core of a fibre

Unfortunately the dispersion minimum is very sharp and small deviations in α from its optimal value cause large increases in dispersion. Near the minimum the dispersion is

$$\Delta\tau = \frac{L(n_1 - n_2)|\alpha - \alpha_{opt}|}{c(\alpha + 2)} \qquad (26.20)$$

For example, if $\alpha_{opt} = 1.97$ and $\alpha = 2$ while $n_1 - n_2 = 0.02$, $\Delta\tau = 0.5$ ns/km, about $5\Delta\tau_{min}$. However, dispersion still occurs even in single-mode fibres, and arises from two different causes: waveguide dispersion and material dispersion, sometimes lumped together and confusingly called *intramodal dispersion*: it is better to call this *within-mode dispersion*.

26.3.5 *Within-mode dispersion*

Material dispersion is the change in RI with wavelength that is responsible for splitting white light into its component colours as it passes through a glass prism. In silica fibres between wavelengths of 800 nm (0.8 μm) and 1600 nm (1.6 μm) it is roughly given by

$$D_{mat} = \Delta\tau/L\Delta\lambda \approx -0.2\lambda + 260 \text{ ps/nm/km} \quad (\lambda \text{ in nm}) \qquad (26.21)$$

where $\Delta\lambda$ is the linewidth of the source (laser diode or LED) in nm. Equation 26.21 shows that the material dispersion is zero in silica fibres for wavelengths around 1.3 μm.

Waveguide dispersion is given by

$$D_{wg} = \frac{\Delta\tau}{L\Delta\lambda} = \frac{10^6\lambda}{cn_1\pi^2d^2} \text{ ps/km/nm} \qquad (26.22)$$

where $c = 3 \times 10^8$ m/s, n_1 is the core RI, d is the core diameter in m and λ is the wavelength in m. Waveguide dispersion is only important in single-mode fibres.

Waveguide and material dispersions are additive and, because D_{mat} and D_{wg} are opposite in sign at wavelengths greater than 1.3 μm in silica, complete cancellation of within-mode dispersion occurs at about 1.5 μm. However, intermodal dispersion is different and cannot be added directly to within-mode dispersion. Instead, like noise, we must add the squares of the dispersions/km and take the square root:

$$\sigma_{tot} = \sqrt{\sigma_{mod}^2 + (\sigma_{mat} + \sigma_{wg})^2} \qquad (26.23)$$

where $\sigma_{mod} = \Delta\tau/L$, as given by Equation 26.12 for a step-index fibre, $\sigma_{wg} = D_{wg}\Delta\lambda$ and $\sigma_{mat} = D_{mat}\Delta\lambda$.

Example 26.4

A step-index fibre has a core diameter of 12 μm with $n_1 = 1.51$ and a cladding with $n_2 = 1.47$. It is to be used in conjunction with a SLD producing light of wavelength 1.55 μm and a linewidth of 2 nm. What is the (inter)modal dispersion/km? What is the within-mode dispersion/km? Calculate the overall dispersion/km. What diameter core would be required for single-mode operation and what would be the overall dispersion/km for this? At what wavelength is the dispersion zero for the single-mode fibre?

The first thing is to check whether the fibre is multi-mode. We find from Equation 26.8 that $N_s = 35$, so it is. Then from Equation 26.12, $\Delta\tau = 1000(1.51 - 1.47)/3 \times 10^8 = 133$ ns/km. This is the modal dispersion/km, D_{mod}.

Equation 26.21 gives the material dispersion:

$$D_{mat} = -0.2 \times 1550 + 260 = -50 \text{ ps/km/nm} \qquad (26.24)$$

And Equation 26.22 the waveguide dispersion:

$$D_{wg} = \frac{10^6 \times 1.55 \times 10^{-6}}{3 \times 10^8 \times 1.51 \times (\pi \times 12 \times 10^{-6})^2} = 2.4 \text{ ps/km/nm} \qquad (26.25)$$

The within-mode dispersion is then $-50 + 2 = -48$ ps/km/nm and the within-mode dispersion per km is $-48 \times 2 = -96$ ps/km, as the linewidth is 2 nm. The total dispersion, σ_{tot}, is $\sqrt{(133^2 + 0.096^2)} = 133$ ns/km. The contribution of the within-mode dispersion is utterly swamped by the intermodal dispersion.

For single-mode operation the core diameter is given by Equation 26.11 as 3.5 μm. The intermodal dispersion is of course zero and the material dispersion is unchanged at -50 ps/km/nm. The waveguide dispersion is increased, however, to 28 ps/km/nm to give a total within-mode dispersion of $-50 + 28 = -22$ ps/km/nm. Multiplying by the source linewidth gives $\sigma = 22 \times 2 = 44$ ps/km; the minus sign is irrelevant.

The dispersion will be zero for the single-mode fibre when the material dispersion is negative and equal in magnitude to the waveguide dispersion. According to Equation 26.22 the waveguide dispersion is $10^6\lambda/cn\pi^2d^2$ in ps/km/nm. With λ in nm and the values substituted for the other variables, the waveguide dispersion becomes 0.0183λ ps/km/nm, and cancellation of the material dispersion occurs when

$$0.0183\lambda = 0.2\lambda - 260 \qquad (26.26)$$

that is when $\lambda = 1430$ nm = 1.43 μm.

Example 26.5

If the 12μm diameter fibre in Example 26.4 is a graded-index fibre with $\alpha = 1.95$, what is the dispersion? What is the maximum bit rate over a fibre 50 km long?

We calculate Δ as $(1.51 - 1.47)/1.51 = 0.0265$, and $\alpha_{opt} = 2(1 - 0.0265) = 1.947$. The grading index is effectively optimal and so the dispersion is given by Equation 26.19:

$$\frac{\Delta\tau}{L} = \frac{n_1\Delta^2}{8c} = \frac{1.51 \times 0.0265^2}{8 \times 3 \times 10^8} = 0.44 \text{ ps/m} \qquad (26.27)$$

which is 0.44 ns/km. There is still within-mode dispersion as before at 96 ps/km for a total dispersion of $\sqrt{(0.44^2 + 0.096^2)} = 0.45$ ns/km. Once again, even with optimal grading, the modal dispersion predominates.

To find the maximum bit rate we must recalculate the modal dispersion as the fibre is 'long'. This becomes $\sqrt{50} \times 0.44 = 3.11$ ns. However, the within-mode dispersion is now $50 \times 0.096 = 4.8$ ns, being linearly dependent on fibre length. The total dispersion is $\sqrt{(3.11^2 + 4.8^2)} = 5.7$ ns. Within-mode dispersion is important in long graded-index fibres. The maximum bit rate is $1/4\Delta\tau_{tot} = 1/(4 \times 5.7 \times 10^{-9})$ = 44 Mbit/s.

26.3.6 Absorption in silica fibres

Absorption in silica fibres is a product of the intrinsic absorption of silica plus the absorption of impurities, particularly water. Figure 26.7 shows the absorption in silica fibres as a function of wavelength. Intrinsic absorption comes from two sources: Rayleigh scattering which goes as λ^{-4}, and the fundamental absorption of the silicon-oxygen bonds. These produce a minimum in intrinsic absorption at about 1.6 μm, but there is a pronounced peak at 1.4 μm caused by water. Rayleigh scattering is caused by small inhomogeneities in the silica, which are a fundamental part of its make-up and cannot be eradicated. (Microscopic particles in the atmosphere cause Rayleigh scattering of sunlight and make the sky blue and the sunset red.) The Si-O absorption peak is at a wavelength of about 10 μm, but it is so huge that its tail extends down to 1.6 μm. The only way to reduce this absorption is to change to another material.

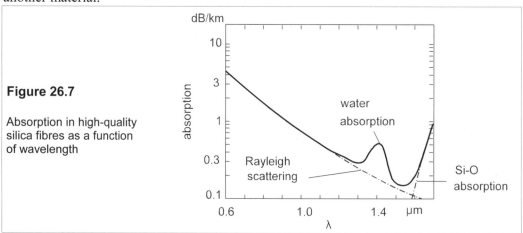

Figure 26.7

Absorption in high-quality silica fibres as a function of wavelength

26.3.7 Optical-fibre splices and connectors

Optical fibres can be spliced by melting them together with an electric arc after carefully aligning them under a microscope. Patent, non-fusion, permanent splicing kits are available, but the cost is high (£400). A less-permanent connection can be made with a demountable butt-joint as shown in Figure 26.8. It is essential that the fibres are accurately aligned, and that the ends are square and close together to avoid too much loss. There is an inevitable reflection loss of about 0.3 dB because of the air-fibre RI mismatch, which can be reduced by enclosing the joint in plastic of an RI approximating to that of the fibre. 'Dry' connectors that can be mated with suitably equipped fibre-optic cables are now quite cheap (about £5 each), provided that the insertion loss of 0.8 dB can be tolerated. It must be said, however, that the cables are not cheap – a 100m length costs about £60 – and are damaged by excessive bending. Usually a minimum bend radius of 30 to 50 mm is stipulated.

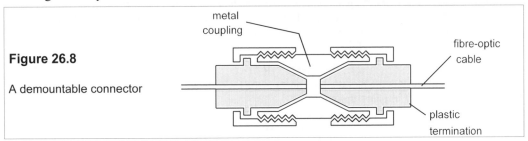

Figure 26.8

A demountable connector

26.4 Optoelectronic signal detectors

Light which reaches the receiving end of an optical fibre has to be detected and demodulated, a process accomplished with photodiodes. Before discussing particular types of photodiode we shall examine photoconduction in semiconductors in general.

26.4.1 *Photoconduction in semiconductors*

When a semiconducting material absorbs a sufficiently energetic photon, an electron in the valence band is excited into the conduction band, leaving a hole behind. The formation of this electron-hole pair leads to an increase in the conductivity of the diode. Now it is clearly necessary for the photon to have at least the gap energy to promote an electron in this way, and photons of lower energy are not absorbed by the material. If the gap is direct the *absorption edge* is quite sharp and occurs at a wavelength given by

$$\lambda_g = hc/E_g = 1.24/E_g \quad (\lambda_g \text{ in } \mu\text{m}, E_g \text{ in eV}) \tag{26.28}$$

The process is the reverse of photoemission in LEDs and SLDs.

As light passes through the detector material the intensity fall as photons are absorbed. The absorption coefficient, α, is defined by

$$I = I_0\exp(-\alpha x) \tag{26.29}$$

where I_0 is the intensity of the incident light and x is the distance in metres from the illuminated surface. It is important in photodiodes to strike a good balance between α and x, so that the light is nearly all absorbed by the active layer of the device, but the charge carriers do not have far to travel, which would lead to a slower response. A fast response is essential if high bit-transmission rates are to be achieved. Indirect-gap materials like silicon have lower absorption coefficients than direct-gap ones like (In,Ga)As, and so silicon detectors require much thicker active layers than those made from (In,Ga)As.

Example 26.6

A silicon photodetector has an active layer 50 µm thick. If the operating voltage is 60 V and 90% of this is dropped across the active layer, what is the transit time for an electron crossing the active layer, if the mobility of electrons in the device is 0.14 m²/V/s? If the absorption coefficient is 6×10^4/m, what percentage of photons is absorbed?

The electron's speed is given by

$$v = \mu\mathscr{E} = \mu V / x \qquad (26.30)$$

where µ is the mobility, $\mathscr{E}$ is the electric field and x is the layer thickness. The voltage across the active layer is $0.9 \times 60 = 54$ V, making the speed

$$v = \frac{0.14 \times 54}{50 \times 10^{-6}} = 151 \times 10^3 \text{ m/s} \qquad (26.31)$$

And the transit time is $x/v = 50 \times 10^{-6}/151000 = 330$ ps.

The total absorption is $\alpha x = 6 \times 10^5 \times 50 \times 10^{-6} = 3$, and the intensity of the light passing out of the active layer is $I = I_0\exp(-3) = 0.05I_0$, that is 5% of the incident light is lost by transmission through the active layer and 95% is absorbed. If the device were made thinner to reduce the response time, the fraction of photons absorbed would fall.

Silicon has a bandgap of about 1.1 eV, corresponding to a photon wavelength of 1.13 µm, resulting in its being substantially transparent at the frequently-used wavelengths of 1.3 and 1.55 µm. For absorption at these wavelengths a material must be used with a bandgap of under 0.8 eV. Germanium has the right bandgap (0.66 eV), but it suffers from an unacceptably high leakage current, about 10000 times silicon's. (In,Ga)As diodes have bandgaps in the right region and much lower dark currents than germanium.

There have been principally two types of photodiode used as fibre-optic detectors: the *avalanche photodiode* and the *PIN diode*. Both of these are used in reverse bias where the dark current is small and the photocurrent relatively large.

26.4.2 The PIN diode

The PIN diode is made from a thin p+ layer on top of a thick, almost-intrinsic (actually lightly-doped n-type) layer on top of an n+ layer, whence the name. Figure 26.9a shows a silicon PIN diode in cross-section.

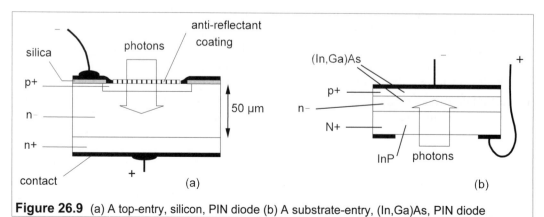

Figure 26.9 (a) A top-entry, silicon, PIN diode (b) A substrate-entry, (In,Ga)As, PIN diode

An antireflectant coating is needed since the reflection coefficient for normally-incident light at an interface is given by

$$R = [(n_1 - n_2)/(n_1 + n_2)]^2$$

Thus if $n_1 = 1$ (air) and $n_2 = 3.4$ (for Ga(As,P) for example), then $R = 0.31$ and 31% of the incident light is reflected. With an antireflectant coating of intermediate RI this can be reduced to about 18%. (See Problem 26.3).

The diode operates under reverse bias and the depletion region embraces the whole of the intrinsic layer, across which all the voltage is dropped. Photons incident on the thin p+ layer pass through and are nearly all absorbed in the intrinsic layer. Because the carrier density in the intrinsic layer is low the photo-generated holes do not recombine but drift to the negative terminal where they combine with electrons flowing as photocurrent. Electrons likewise drift towards the positive terminal and exit as photocurrent. Any photo-generated electrons in the p+ layer soon recombine with the majority holes there. The conversion efficiency of PIN diodes is close to 100% – almost all photons result in an electron appearing in the external circuit.

The depletion layer has a width given by

$$d = \sqrt{\frac{2\varepsilon_0\varepsilon_r(V_b + V_r)}{qN_D}} \tag{26.32}$$

where ϵ_r is the relative permittivity (12 for silicon), V_b is the built-in potential (about 0.8 V for silicon), V_r is the reverse bias, q is the electronic charge and N_D the dopant density in the I-layer. Thus if $N_D = 10^{19}$ per m^3 and the I-layer is 50 μm thick, we find that V_r has to be at least 18 V if the depletion region is to fill the I-layer.

PIN diodes made from (In,Ga)As consist of hetero-epitaxial layers on an InP (N+ doped) substrate as in Figure 26.9b. InP has a bandgap of 1.35 eV and a cut-off wavelength of 900 nm, so that infra-red light of 1.3 or 1.55 μm wavelength passes through and is absorbed in the I-(In,Ga)As layer. The absorption coefficient of the (In,Ga)As I-layer is about ten times higher than silicon's so it is only one tenth as thick, 5 μm instead of 50 μm. The dopant density in the I-layer is about 10^{21} m^{-3} and the relative permittivity is 14, resulting in a bias voltage of 15 V, using Equation 26.32. Carrier transit times in the device are about ten times less than in silicon which would lead to faster devices if the capacitance could be reduced too.

26.4.3 The avalanche photodiode or APD

Avalanche photodiodes operate in reverse bias, but close to the breakdown region of the device. As a result the photo-generated carriers can gain sufficient energy from the bias field to produce more carriers when they collide with the lattice. By producing such carrier multiplication an APD can in theory detect fainter light signals than a PIN diode. The bias voltage must be very carefully maintained to keep the unilluminated device on the edge of avalanching. Figure 26.10 shows the structure of a silicon APD. The I-layer is actually lightly doped with boron (about 10^{20} atoms/m^3, denoted p−) and the depletion region embraces the whole of it. Most light absorption takes place here and the high electric field causes avalanche multiplication of the photo-generated carriers. The n-type guard ring ensures that the central n+/p junction breaks down before the n/p junction at the edge. A disadvantage of the silicon

APD is that the operating voltage is about 200 V. As with PIN diodes, so too with APDs, the use of III-V compounds produces a thinner active layer and a reduced operating voltage of about 50 V. The current multiplication factor, M, is from 100 to 1000 for silicon APDs and about 10 to 20 for APDs made of III-V compounds. Thus an APD should be a factor of M^2 more sensitive than a PIN diode. However, the statistics of the avalanche process are such that the device is intrinsically noisier than a PIN diode by a factor of 5 or 10. Then, when there is no source of noise but light quanta (the so-called *quantum noise level*), the APD SNR is smaller for a given signal level by this factor of 5 or 10. But if the post-detector amplifiers are the primary source of noise, the APD has a superior SNR by a factor of M^2. In other words, APDs are best when the receiver is noisy.

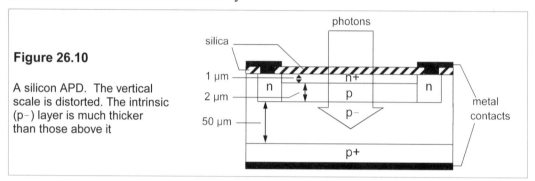

Figure 26.10

A silicon APD. The vertical scale is distorted. The intrinsic (p–) layer is much thicker than those above it

26.5 **The ultimate performance of optical receivers**

The most common digital fibre-optical systems use simple OOK and envelope detection and, if the statistics of photon arrivals at the detector are known, we can work out the detection limit for optical-fibre communications. The emission of photons in a pulse is found to follow a Poisson distribution and their arrival at the detector is likewise a Poisson process. The Poisson distribution is such that if there are on average m photons in a time interval, T, the probability that there are n photons in the same time interval is

$$P(n) = (m^n/n!)\exp(-m) \tag{26.33}$$

With OOK we will take any non-zero photon count within a time interval of T ($= 1/B$, where B is the bit rate) to signal a 1 and any zero count to signal a zero. The probability of getting a zero count when a pulse is present is just

$$P(0) = (m^0/0!)\exp(-m) = \exp(-m) \tag{26.34}$$

In an ideal system no photons are sent when a 0 is signalled and therefore the probability of error for 0s is zero. On average there will be equal numbers of 1s and 0s, so the overall probability of error is

$$P_e = 0.5P(0) = 0.5\exp(-m) \tag{26.35}$$

For a BER (bit-error rate) of 10^{-9}, or $P_e = 10^{-9}$, we find $m = 20$. Since there are as many 1s as 0s in a message of reasonable length, the smallest possible average signal level for OOK at a BER of 10^{-9} is 10 photons/bit. This is the *quantum limit* for direct detection and the assumption is that the receiver (detector plus amplifiers) is ideal.

Example 26.7

A fibre-optic link operates at 500 Mbit/s and 1.3 µm. Find the signal power required for a BER of 10^{-9} if OOK is used and the receiver is noise free and uses direct detection. The noise current in the receiver is 6 nA; what must the signal power be for the same BER assuming that each photon entering the detector generates one electron and that 20% of incident photons are reflected?

The energy per photon at 1.3 µm is

$$E = \frac{hc}{\lambda} = \frac{6.626 \times 10^{-34} \times 3 \times 10^8}{1.3 \times 10^{-6}} = 1.53 \times 10^{-19} \text{ J} \tag{26.36}$$

Direct detection in the quantum limit requires 10 photons/bit. The average energy per bit is therefore $10 \times 1.53 \times 10^{-18} = 1.53 \times 10^{-17}$ J. And then if the bit rate is 500 Mbit/s the optical power is $5 \times 10^8 \times 1.53 \times 10^{-17} = 7.65$ nW.

For a BER of 10^{-9} a SNR of 72 is required with OOK (see Section 24.1.2, Chapter 24), so the signal current needed is therefore $\sqrt{72}$ times the noise current or $\sqrt{72} \times 6 = 51$ nA. This is $51 \times 10^{-9}/1.6 \times 10^{-19} = 3.2 \times 10^{11}$ electrons/s. There must be the same number of photons in the detector and the number incident onto the detector must be $1.25 \times 3.2 \times 10^{11} = 4 \times 10^{11}$ each second or $4 \times 10^{11}/5 \times 10^8 = 800$ photons/bit, a factor of 80 higher than the quantum limit. To get down to the quantum limit the noise current in the receiver must be reduced to around 75 pA, roughly 10 times less than the present state of the art.

Suggestions for further reading

The following deal with electronic communications in general:

Introduction to communications by M P Godwin (Butterworth-Heinemann, 2nd ed., 2002). Cheap and concise.
Digital and analog communications systems by L W Couch (Prentice Hall, 6th ed., 2001)
Electronic communications system: fundamentals through advanced by W Tomasi (Prentice Hall, 4th ed., 2001) Covers a lot. Not cheap.
Communications systems and networks by R Horak (M&T Books, 2nd ed., 2000). Inexpensive, not over mathematical.
Introduction to digital communications by R E Ziemer and P Peterson (Prentice Hall, 2nd ed., 2000)

The following deal with fibre-optical communications:

Introduction to fibre optics by J Crisp (Butterworth-Heinemann, 2nd ed., 2001). Short and not too dear.
Fibre optic communications systems by G P Agrawal (J Wiley, 3rd ed., 2002). Comprehensive, expensive.
Understanding optical communications by H J R Dutton (Prentice Hall, 1999). Long.
Fibre optics by R J Hoss and E A Lacy (Prentice hall, 2nd ed., 1997). Simple and brief.

Problems

1. Show that the Q-factor of a semiconductor laser is given by $Q = E_g/\Delta E_g$, and that if the Q-factor of a GaAs laser is 1000, this is equivalent to a bandgap variation, ΔE_g, of 1.43 meV. If the bandgap variation in $In_xGa_{1-x}As$, with $x = 0.6$, is caused by inhomogeneities in composition, that is variations in x, how much does x have to vary to produce a Q-factor of 1000? What is the linewidth when $x = 0$ and $Q = 1000$? (Assume a linear relationship between bandgap and x.) *[$\pm 4 \times 10^{-4}$, 0.87 nm]*

2. Estimate the current density in a laser diode of electrode area 100×20 μm^2, operating with an overall efficiency of 5% at bit rate of 100 Mbit/s, if there are 10^7 photons/bit. If the wavelength of the light emitted is 0.8 μm, what is the electrical power input and the optical power output? How many photons are there in a 1 and in a 0? (The modulation is OOK with NRZ pulses.)
[*1.6 MA/m^2, 5 mW, 0.25 mW, 2×10^7, 0*]

3. What is the reflection coefficient for normally-incident light at an air/GaP interface if the RI of GaP is 3.4? What is the reflection coefficient for an air/silicon oxynitride interface if the oxynitride has a RI of 1.5? What will be the overall reflection coefficient of an air/oxynitride/GaP combination? Show that the RI of the oxynitride must be √3.4 for minimum reflection. What is R then?
[*0.298,0.184, 0.168*]

4. If the acceptance angle for an step-index optical fibre is 12° when the cladding has a refractive index of 1.52, what is the refractive index of the core? What is the critical angle at the core-cladding interface? If the core is 25 μm in diameter, what is the maximum distance that a critically-angled ray can travel between reflections? How many reflections is this per m? If the absorption on reflection is 0.0001 dB, what is the attenuation per m for the critically-angled ray?
[*1.534, 82.2°, 183 μm, 5480, 0.55 dB*]

5. In the fibre of the previous problem, how many modes are there if the light has a wavelength of 0.5 μm? Estimate approximately the diameter of the core for it to be monomodal. What does the exact formula give for this diameter? [*527, 1.1 μm, 1.86 μm*]

6. What is the intermodal dispersion per km for a step-index optical fibre with core RI of 1.54 and cladding RI of 1.53? What is the maximum bit rate for this dispersion in fibres of length 250 m and 2.5 km? [*33.3 ns/km, 30 Mbit/s, 4.75 Mbit/s*]

7. In a graded index fibre the core RI, $n_1 = 1.54$ and the cladding RI, $n_2 = 1.51$, while the grading index, α, is 2. Estimate the dispersion per km. What is the ratio between this figure and the dispersion of an optimally-graded fibre? [*0.97 ns/km, 4:1*]

8. What is the material dispersion in a graded-index silica fibre at a wavelength of 1 μm if the Q-factor of the laser source is 600? If the intermodal dispersion is 0.2 ns/km what is the total dispersion in fibres of length 1 km and 10 km? What is the maximum bit rate for these fibres?
[*0.1 ns/km, 0.224 ns, 1.18 ns, 1.1 Gbit/s, 212 Mbit/s*]

9. What is the minimum number of photons/bit on average if the BER using OOK is not more than 10^{-6}? If a fibre-optic system operates with OOK at 200 Mbit/s and a wavelength of 0.8 μm, what is the signal power needed in the quantum limit and a BER of 10^{-6} at most? If it uses a receiver with a noise current of 10 nA, a quantum efficiency of 10% and 40% of the incident light is reflected, what signal power is needed? How many photons/bit does that represent?
[*7, 0.35 nW, 1.74 μW, 3.5×10^4*]

10. If the voltage across the active layer in a PIN diode is 30 V and the layer is 35 μm thick, what is the transit time of an electron if its mobility is 0.2 m^2/V/s? What absorption coefficient is required to cause 50% of incident photons to be absorbed, if the reflection coefficient is 20%? What must the doping density be in the active layer if the refractive index, n, is 2.5 and $V_b = 1.5$ V? (Take $\epsilon_r = n^2$ and assume the depletion layer penetrates the whole active layer.)
[*0.2 ns, 2.8×10^4 m^{-1}, 1.78×10^{19} m^{-3}*]

27   Telephony

OTHER THAN by direct voice, telephones are the most important means of personal communication in the developed countries; and in N America, where local calls are not directly paid for or are exceedingly cheap, telephone conversation ousts direct speech into second place. The immediacy of the telephone is its great advantage over other forms of communication and the coverage of the global network is virtually complete. Thirty years ago transatlantic calls were rarely made, today they are commonplace. Mobile telephones ('cellphones' in N America and elsewhere) are now almost indispensable, particularly in those parts of the world where telephone lines are non-existent through theft or under-investment. Spectacular use of the mobile telephone was seen in the recent war in Afghanistan where television reports were made routinely over mobile videophones. Yet is not much more than 30 years since the last manual exchange in England closed [1].

In 2001 came the third generation of mobile communications [2], and spectacular sums were paid for 3G licences, particularly in the UK. The 3G system will give access to voice, fax, video and Internet communications throughout the world, transforming the telephone into an all-purpose communications centre.

Telecommunications have made such a huge impact on communications because each of several billion telephones worldwide can be connected to any other by means of elaborate switching systems or exchanges. At first these were hand operated with jack plugs to connect incoming calls to the subscriber lines. Though the operators were extraordinarily dexterous they were slow and expensive compared to automatic switching systems and they had crucial defects which led sooner to their replacement than the expense of their labour would have done. [3] Nowadays all new exchanges use stored-program control (SPC) electronic switching, and digital technology is being progressively extended down to the subscriber.

---

[1] It operated until 1969 in Hastings, E Sussex; sometimes several minutes elapsed before the operator could answer. Yet the first stored-program-control (SPC) exchange, the No. 1 ESS (electronic switching system), was installed at Succasunna, N. J., USA in 1965.

[2] The first generation was analogue voice only, the second digital voice and some data such as fax and text messages, the third digital with Internet and video, making the telephone a mobile communications centre.

[3] The story goes that Mr Strowger ran an undertakers in Kansas City where a rival establishment was owned by a man whose wife was the operator at the telephone exchange. Mr Strowger noticed that he received very few business calls by telephone and discovered that his rival's wife was diverting his calls to her husband. He thereupon devised his automatic switch in 1889. This used pulses from the subscriber to operate solenoids to move a wiper around a set of horizontal contacts. Further pulses caused the wiper to be moved up to select a second contact and so on according to the number of digits in the called number. Thus the Strowger switch ingeniously used two-dimensional motion to set up a connection. Until the dial was invented in 1895, the Strowger switch was rather error prone. Strowger switches were unchallenged until the invention of crossbar switching in Sweden in 1919, though crossbar switches were not used in the USA until the 1930s and were never used in the UK. The operational details of both switches are now largely of historical interest.

27.1 **Signalling**

The stages in a telephone call require a signalling system. In pre-digital days the signals were sent over a channel which was associated with that call. In a digital network this is not necessary or desirable and signals are sent over a channel common to many calls (known as *common-channel signalling*). Signalling can be divided up into ten stages:

1. Lifting the handset closes a switch to power the telephone, simultaneously sending an 'off-hook' signal to tell the local exchange a connection is wanted.
2. The call is detected at the subscriber's line-termination unit (SLTU) and the SLTU code is translated into a directory number.
3. Storage is allocated for the call record and equipment for processing the call. Then a dial tone is sent.
4. The subscriber dials or keys in – 'dialling' henceforth means either – a number which is sent as a signal also and stored at the exchange.
5. The control system examines the digits dialled and decides a route for the call. Local calls – calls to an SLTU on the same exchange – can take only one path and if the called party's line is busy a 'busy' tone is sent to the caller. Calls that require other exchanges must use one of the outgoing circuits. If a circuit is free for a call it is seized by the control system and cannot be used for other calls. In SPC exchanges seizure just means changing the status flag in the memory allocated to that circuit. If no free circuit can be found a busy tone is sent.
6. Circuits having been allocated to the call, the control system sets up a switch path to connect them.
7. When a connection has been successfully completed a ringing current is sent to the called telephone and a ringing tone to the caller.
8. The removal of the called party's handset initiates charging at the appropriate tariff and call-timing, while the ringing and ringing tones are discontinued.
9. While in progress the control system monitors the call for a 'clear' signal.
10. Replacing the handset by either party to the call results in a clear signal which terminates the call, stops call-timing and frees the circuits for further use.

Signalling is an important part of making a call and can be divided into subscriber signalling and inter-exchange signalling. Subscriber signalling used only the DC loop in the past – dialling consisted of breaking the DC loop a number of times to sent each digit but multi-frequency (MF) AC signalling is now used in push-button telephones. Pressing a button on the handset sends a combination of one each of four high-frequency and four low-frequency tones to the exchange. The tones are so chosen that voice frequencies cannot be mistaken for dial tones. The problem with signalling used to be that it required a channel dedicated to it for each and every call, and was thus known as *channel-associated signalling*, but most of the time the signal channel was idle, thereby wasting valuable capacity. Even with PCM the allocation to signalling of TS16 in the 32-slot PCM frame means that the signalling is still associated with 30 specific channels.

Controllers in SPC exchanges could, however, communicate directly just as any other computers could and so telephone calls could be sent as discrete pieces called *messages* or *packets* which contained five fields: origin, destination, message number, data and error check.

The use of packets frees the telephone network from channel dedication and makes much more efficient use of the system. For example, speech consists mostly of silence, which can be sent very efficiently. These savings could also be extended to signalling and lead to common-channel signalling in which the signals associated with a call were sent as messages between controllers, and signal channels could be separated from speech channels. Signalling standards – such as CCITT signalling system No. 7, which is now in use nationally and internationally – can now be made which are independent of the physical form of the network[4]. Signals could also now be made to carry much more information about the network as well as calls and so lead to efficient network management. It is, for example, now possible to monitor the state of every SPC exchange in the UK every five minutes and re-route or block traffic to stop the spread of congestion due to emergency conditions. Network congestion occurs for example when many callers simultaneously try to call a single number, their call attempts taking up equipment and blocking the exchange for potentially-successful calls. When one exchange is knocked out the congestion can snowball and cause widespread failure of communications[5].

27.2 **Transmission systems**

Subscribers' telephones are connected to the local exchange by twisted pairs of 600Ω wires known as subscriber loops, though it is now often the practice to concentrate subscribers' lines remotely and feed a multiplexed PCM signal down a much smaller number of twisted pairs to the local exchange. Power for the telephone is sent through these wires as medium-voltage DC (–48 V in the UK) from the exchange, though the ringing signal is sent as medium-voltage AC: between 50 V and 100 V r.m.s. To keep down costs no amplification is used on these lines. Since the attenuation is less at voice frequencies when inductance is added to the line, long subscriber loops are loaded with 88 mH of inductance every 1.829 km (6000 ft) and can span 10 km without amplification. Loading reduces the speed of propagation tenfold – from 2.2×10^8 m/s for unloaded lines to 2.2×10^7 for loaded. A subscriber's 2-wire loop can carry only one analogue signal at a time if it is to remain intelligible, so transmission and reception occur alternately, not simultaneously. At the handset the 2-wire circuit must be split into separate transmitting and receiving circuits (a 4-wire circuit). At the exchange (unless it is an old one) the 2-wire circuit must also be converted into 4-wire form so that the transmission path can be separated from the reception path and amplified as necessary.

Coaxial cables were used for many trunk lines since they are far less prone to electromagnetic interference (EMI), offer a greater bandwidth and suffer less attenuation than twisted pairs. Microwave links are especially popular in N America for trunks as larger distances can be spanned more economically than with cables. The distance between transmitter and repeater antennas is normally about 40–50 km. Microwaves are now used worldwide for cellphones Though they are not susceptible to crosstalk and EMI so much as copper cables, microwave

---

[4] The Comité Consultatif International Télégraphique et Téléphonique was set up in 1956 to formulate international technical standards in telephony.

[5] A famous example occurred in 1960 during a conference on network management in the USA. Hurricane Donna struck Florida at the time and although the conference was far away, it suffered a telephone blackout when the increased calls from worried subscribers in Florida overloaded a tandem exchange and then spread congestion over a great part of the USA.

links suffer from enhanced attenuation during rainstorms, and they require a clear path with no obstructing hills or buildings between stations. As numbers of microwave transmitters have grown in the larger cities, interference has required stricter licensing and control.

Since 1980 all new trunk lines in the UK and elsewhere have been made from silica-glass fibres operating with short infra-red wavelengths to minimise attenuation. Many thousands of telephone channels can be sent down a single fibre and the attenuation is so low that repeaterless connection between London and Birmingham (160 km) is possible.

Table 27.1 *Some telephone cables between Europe and USA*

Cable ID	Year	Repeaters	Channels
TAT1	1956	118	42
TAT2	1959	114	48
TAT3	1963	182	138
TAT4	1965	186	138
TAT5	1970	361	845
TAT6	1976	694	4000
TAT7	1983	662	4246
TAT8[A]	1988	109	7560
PTAT1	1989	c.100	40k
TAT9	1991	51	80k
PTAT2	1992	50	40k
TAT10	1992	c.40	113k
TAT11	1993	37	113k
TAT12/13[B]	1995/6	150	600k
TAT14[C]	2001	278	10M

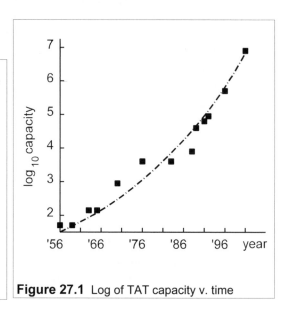

Figure 27.1 Log of TAT capacity v. time

Notes: [A] First fibre-optic cable [B] First to use optical amplifiers [C] First to use WDM

The history of submarine communications cables, dating back to the electric telegraph of 1839, is one of heroic engineering that cannot, alas, be covered here. The first transatlantic cable to carry speech (TAT1) rather than telegraphy was not laid until 1956, prior to which radio telephony was used, and was subject to fading and other interference. TAT1 was 6775 km long, had 118 repeaters, roughly one every 50 km, and a capacity of only 42 voice channels, despite which it was a great success. Table 27.1 shows the progression of telephone cables between Europe and N America, which seems to have been little affected by the advent of communications satellites, and though in 1990 only half the traffic was carried by cable, now all is. Cable-borne conversation is more acceptable than satellite-borne because of the unavoidably long delay, about half a second, satellites introduce between speaking and hearing a response. Submarine telephone cables were made of coaxial cable and the attenuation of the signal varied with frequency, requiring periodic equalising and amplification using repeaters. The number of these used per km increased steadily as the power consumption of solid-state devices fell, but in 1988 the first fibre-optic transatlantic link (TAT-8) was introduced, and the number of repeaters fell sharply. The superior performance of optical fibres ensured that all suboceanic cables would thenceforth be optical.

TAT12/13 broke new ground by using erbium-doped optical amplifiers in its flexible equalisers, which meant that the cable could be laid continuously without stopping the ship.

Then the success of wave-division multiplexing (WDM) in 1995 meant that the next cable, TAT14, could provide *ten times* the existing transatlantic cable capacity, about 640 Gbit/s, or 1000 CD ROM's worth per second. The enormous growth in Internet traffic is expected to utilise much of this capacity soon. Driven by demand for new services, competitive prices and profiting from new technology, TAT cable capacity – shown in Figure 27.1 – has grown faster and faster with time.

27.3 **Multiplexing**

It is clearly necessary to combine individual telephone channels into one large channel to make efficient use of trunk lines. In the past the analogue signals were frequency-division multiplexed (FDM), but with digital signals time-division multiplexing (TDM) is used, as illustrated in Figure 27.2.

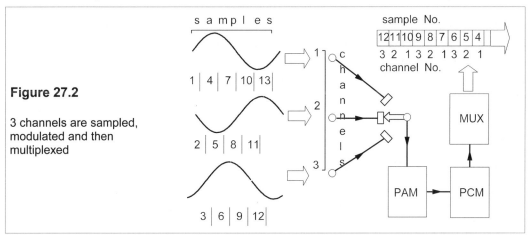

Figure 27.2

3 channels are sampled, modulated and then multiplexed

In FDM each channel of a group, usually twelve, is allocated a 4kHz bandwidth. Groups can be combined into 60-channel supergroups, 300-channel mastergroups or 900-channel supermaster groups giving a total bandwidth of 3.872 MHz. With TDM, samples from a number of channels are interleaved for transmission over a TDM highway. The samples, each comprising eight PCM bits, from a number of channels, usually 30, are contained within a time interval called a frame. The order of the samples within each frame is the same. For 3G mobile telephones code-division multiplexing (CDM) will be used. In this each telephone uses all the available frequencies according to a code specific to each handset, and the capacity of the available bandwidth will be further increased.

Figure 27.2 shows TDM applied to three channels, as an illustration, though in practice there would be thirty channels, or 24 in N America. The CCITT standard for 30-channel PCM requires a frame size of 125 µs, equivalent to 8 kHz, subdivided into 32 time slots: 30 for the telephone channels, one for signalling and one for frame alignment. Each of the time slots contains an 8-bit PCM signal sample, corresponding to 256 quantization levels in the speech sample, and each bit is separated by an equal space from the next, so the time/bit is only $125/32/16 = 0.244$ µs. 30-channel multiplexed PCM thus operates at $8 \times 32 \times 8 = 2048$ kbit/s or about 2 Mbit/s. The layout of a frame is shown in Figure 27.3.

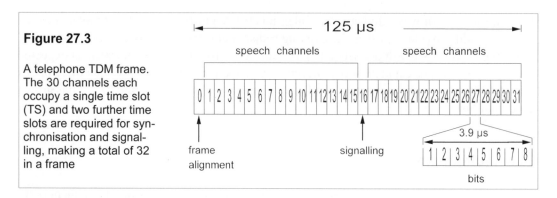

Figure 27.3

A telephone TDM frame. The 30 channels each occupy a single time slot (TS) and two further time slots are required for synchronisation and signalling, making a total of 32 in a frame

27.4 **Companding**

Speech is now sent over trunk telephone lines as a PCM signal, but in practice linear spacing of quantization levels leads to considerable loss of information from speech because low-level signals are inadequately catered for and high-level signals saturate amplifiers. The solution is to use a logarithmic scheme for the spacing of the quantization levels, as in Figure 27.4.

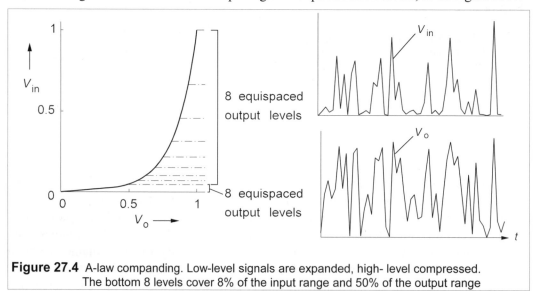

Figure 27.4 A-law companding. Low-level signals are expanded, high- level compressed. The bottom 8 levels cover 8% of the input range and 50% of the output range

In Europe the analogue signal is compressed using the formula known as the *A-law*, which is

$$v_o = \frac{A|v_{in}|}{1 + \ln A}, \quad 0 \le |v_{in}| \le 1/A$$

$$= \frac{1 + \ln(A|v_{in}|)}{1 + \ln A}, \quad 1/A \le |v_{in}| \le 1$$

(27.1)

where A is the compression factor (the CCITT recommendation is $A = 87.6$, see Problem

27.1). In N America and Japan, the μ-law is used which is similar to the A-law, but sufficiently different to cause compatibility problems with equipment. At the receiving end of the line the reverse process of signal expansion takes place to restore the original as faithfully as possible. Thus the process is known as *companding* (COMpressing and exPANDING). The A-law is plotted in Figure 27.4, together with an example of an A-law-compressed signal. The quantization SNR is greatly reduced for low-level signals.

27.5 **Telephone exchanges**

The need for some central point where calls can be connected from one line to another is clear from Figure 27.5, where four telephones are connected individually to each other.

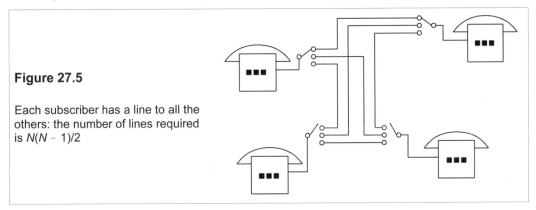

Figure 27.5

Each subscriber has a line to all the others: the number of lines required is $N(N-1)/2$

Each subscriber performs his own switching and a total of six interconnecting lines is required. For N subscribers the number lines required is $N(N-1)/2$, so that 100 subscribers need 4950 lines. By carrying out the switching at a central point or exchange, the number of lines is reduced to one per subscriber. Calls are made not only locally, however, and local exchanges (known as central offices in N America) are connected by other exchanges and so on up to the highest level, the international gateway exchange, forming a hierarchy. Groups of local exchanges within a large urban area, such as those in London, Birmingham or Manchester, are interconnected by means of tandem exchanges forming a star network, though in general, depending on the level of traffic between centres, local exchanges are connected in both mesh and star as in Figure 27.6.

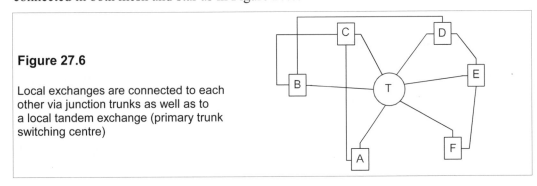

Figure 27.6

Local exchanges are connected to each other via junction trunks as well as to a local tandem exchange (primary trunk switching centre)

The network shown has twice as many connections as strictly required, that is it exhibits redundancy. Redundancy is important as it allows emergencies to be catered for, such as a surge in traffic or the loss of inter-exchange links. All of the local exchanges in Figure 27.6 have at least two links and if for example the traffic between exchanges D and E overloads their direct link (called a *junction* in the UK, a *junction circuit* by the CCITT and an *inter-office trunk* in N America), the overflow would be automatically routed through the local tandem exchange, T (primary trunk-switching centre in CCITT parlance).

In turn the local tandems are linked to regional tandem exchanges (secondary trunk-switching centres), national exchanges (tertiary trunk-switching centres) and international gateway exchanges via trunk lines (*toll* lines in N America) as shown in Figure 27.7. Connecting links between exchanges are expensive and are only constructed after a detailed analysis of present traffic and an exhaustive forecast of future traffic (the telephone network must serve the needs of its customers for many years ahead). In the past this data was gathered by sampling the exchange links, but with SPC exchanges traffic information can be provided at any time.

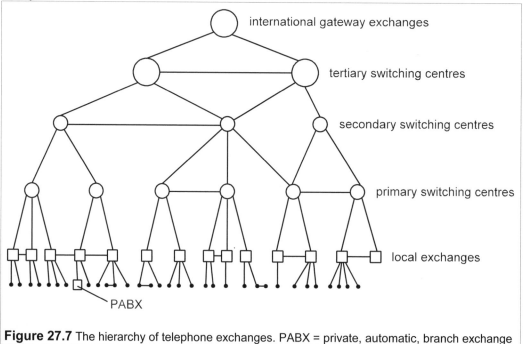

Figure 27.7 The hierarchy of telephone exchanges. PABX = private, automatic, branch exchange

27.6 Telephone-traffic theory

In 1909 Erlang published a theory of telephone traffic which is still used. The amount of traffic is measured in erlangs (E), defined by

$$A = CH \qquad (27.2)$$

where C is the number of calls per hour and H is the average duration of a call in hours. If there is only one line and $C = 1$ and $H = 1$, that is there is only one call per hour which lasts for one hour, then the line is fully occupied. A single line or connection cannot carry more than 1 E of traffic: if a line carries traffic of 1 E then it is 100% occupied, if it carries 0.5 E it is 50% occupied.

The time of arrival and duration of a telephone call at an exchange are both random if the number of subscribers is large compared to the number of exchange connections (*M*) available. If those calls are cleared which cannot be connected because all the exchange connections are busy ('lost-calls-cleared', the usual method used in telephone exchanges; the alternative would be to hold them) then Erlang showed that the probability that *k* connections are busy out of a total of *M* is

$$P_k = \frac{A^k/k!}{\sum_{n=0}^{M} A^n/n!}$$ (27.3)

Clearly, if *k* = *M*, the incoming call will find all the lines busy and will be lost. This probability is

$$B = \frac{A^M/M!}{\sum_{n=0}^{M} A^n/n!}$$ (27.4)

B, the probability that all lines are busy, is known as the *grade of service* (GOS). Telephone systems are designed to give a certain GOS *at peak times* at a future date, often twenty years after the equipment was installed. The standard GOS has been 0.002, that is under normal conditions at peak times, 0.2% of incoming calls will be lost. However, overload conditions must also be met by a system, a typical condition being that the GOS should not get worse than 1% if the overload is 20%.

Table 27.2 gives the traffic (A) in erlangs for values of *M* ranging from 1 to 100 at two grades of service (*B*), 0.002 and 0.01. In queuing theory, *M* is the number of *servers*, which might be for example the number of trunks or the number of exchange switches available. As the number of servers increases, the average occupancy of those servers rises for the same GOS. For instance when *M* = 5 and GOS = 1%, *A* = 1.361, that is only 1.361 × 0.99/5 = 0.2695 E per server. While when *M* = 50, *A* = 37.9 for the same GOS, an occupancy rate of 0.75 E per server: the 0.99 takes account of the fact that 1% of calls are lost. It therefore pays to concentrate traffic so that expensive equipment, such as trunks, are more fully utilised. However, with more servers the GOS is much more sensitive to traffic changes: a 20% increase in traffic, from 0.9 to 1.08 E, with five servers causes a change in GOS from 0.2% to 0.42%, whereas when *M* = 50, a 20% increase in traffic, from 33.9 to 40.7 E, causes the GOS to change from 0.2% to 2.2%. In the former case the increase in lost calls is 110% and in the latter case it is 1000%.

Traffic from residential lines is rather sparse (typically < 0.03 E in the peak hour) and exchanges usually use a concentrator (which may be remote from the exchange to save wiring cost) before the switching is done. Figure 27.8 shows one such scheme in which 30 incoming lines are concentrated into six and five groups of six are connected to a 30 × 30 switch (900 connections) and then to an expander and so to the outgoing lines. Fully connecting 150 incoming with 150 outgoing lines would require 150^2 = 22500 connections. If 30 lines can be taken to be a large number then the Erlang formula can be used to estimate the blocking probability of the concentrator. The incoming traffic will be about $30 \times 0.03 = 0.9$ E, so putting $A = 0.9$ and $M = 6$ into Equation 27.4 gives $B = 1.5 \times 10^{-4}$, an acceptably low figure.

Table 27.2 *Traffic capacity, A, for GOS = 1% and 0.2% and M = 1–100*

					GOS = 1%					
M	1	2	3	4	5	6	7	8	9	10
0	0.01	0.153	0.455	0.869	1.36	1.91	2.50	3.13	3.78	4.46
10	5.16	5.88	6.61	7.35	8.11	8.88	9.65	10.4	11.2	12.0
20	12.8	13.7	14.5	15.3	16.1	17.0	17.8	18.6	19.5	20.3
30	21.2	22.0	22.9	23.8	24.6	25.5	26.4	27.3	28.1	29.0
40	29.9	30.8	31.7	32.5	33.4	34.3	35.2	36.1	37.0	37.9
50	38.8	39.7	40.6	41.5	42.4	43.3	44.2	45.1	46.0	46.9
60	47.9	48.8	49.7	50.6	51.5	52.4	53.4	54.3	55.2	56.1
70	57.0	58.0	58.9	59.8	60.7	61.7	62.6	63.5	64.4	65.4
80	66.3	67.2	68.2	69.1	70.0	70.9	71.9	72.8	73.7	74.7
90	75.6	76.5	77.5	78.4	79.3	80.3	81.2	82.1	83.1	84.0

					GOS = 0.2%					
M	1	2	3	4	5	6	7	8	9	10
0	0.002	0.065	0.249	0.535	0.900	1.33	1.80	2.31	2.85	3.43
10	4.02	4.63	5.27	5.92	6.58	7.26	7.95	8.64	9.35	10.1
20	10.8	11.5	12.3	13.0	13.8	14.5	15.3	16.1	16.9	17.7
30	18.4	19.2	20.0	20.8	21.6	22.4	23.2	24.0	24.9	25.7
40	26.5	27.3	28.1	28.9	29.7	30.5	31.4	32.2	33.0	33.9
50	34.7	35.6	36.4	37.2	38.1	38.9	39.8	40.7	41.5	42.4
60	43.3	44.1	45.0	45.8	46.7	47.6	48.4	49.3	50.1	51.0
70	51.9	52.7	53.6	54.4	55.3	56.2	57.1	58.0	58.8	59.7
80	60.6	61.5	62.4	63.2	64.1	65.0	65.9	66.8	67.7	68.6
90	69.4	70.3	71.2	72.1	73.0	73.9	74.8	75.7	76.6	77.5

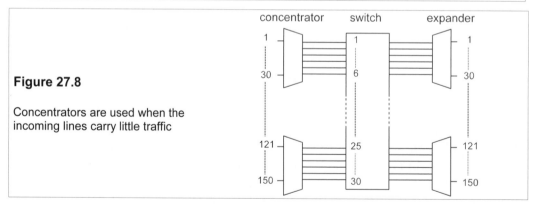

Figure 27.8

Concentrators are used when the incoming lines carry little traffic

Example 27.1

Traffic of 2 E from a large number of subscribers is served by 6 lines. What is the lost traffic? If the same traffic is served by 3 lines and the overflow by another 3, what will be the lost traffic and overall GOS assuming the statistics of the overflow are the same as the rest of the traffic ? What happens in each case if the traffic increases by 20%? What is the average occupancy of the servers in all cases?

(The assumption regarding the statistics of the overload traffic is by no means obviously valid[6].)
Using Equation 27.4 with $A = 2$ and $M = 6$ yields

$$B = \frac{2^6/6!}{1 + 2 + 2^2/2! + 2^3/3! + 2^4/4! + 2^5/5! + 2^6/6!} = 0.0121 \tag{27.5}$$

B is the GOS and the probability of blocking. The lost traffic is $AB = 0.0242$ E. The average server occupancy is $(2 - 0.0242)/6 = 0.3293$.
In the second case ($M = 3 + 3$) the probability of blocking at the first stage is

$$B_1 = \frac{2^3/3!}{1 + 2 + 2^2/2! + 2^3/3!} = 0.2105 \tag{27.6}$$

The overflow traffic is then $AB_1 = 2 \times 0.2105 = 0.421$ E. The occupancy of the first group of servers is $(2 - 0.421)/3 = 0.5263$. This traffic is fed into one line and the probability of its being blocked is

$$B_2 = \frac{0.421^3/3!}{1 + 0.421 + 0.421^2/2! + 0.421^3/3!} = 8.17 \times 10^{-3} \tag{27.7}$$

The lost traffic is then $AB_1B_2 = 0.421 \times 0.00817 = 0.00344$ E and the overall GOS is $0.00344/2 = 0.00172$. The average occupancy of the second group is $(0.421 - 0.00344)/3 = 0.1392$. A considerable improvement in the GOS has been produced by splitting the servers in this way.
When the traffic increases by 20% to 2.4 E, the same calculations give first $B = 0.0244$, the lost traffic is 0.0585 E and the average occupancy is 0.39. Then for 3 + 3 servers we find $B_1 = 0.2684$ and the overflow traffic is $AB_1 = 0.644$ E. Feeding this into 3 lines gives $B_2 = 0.0235$ and the lost traffic, $AB_1B_2 = 0.0151$ E and the overall GOS = $0.0151/2.4 = 0.0063$. The first group's average occupancy is 0.585 and the second's is 0.21. These results are summarised in Table 27.3. Notice that the improvement produced by splitting the servers in this way is reduced as the occupancy increases.

Table 27.3 *The solutions for Example 27.1*

A	M	Occupancy	GOS	Lost Traffic
2	6	0.3293	0.0121	0.0242
	3 + 3	0.5263, 0.1392	0.00172	0.00344
2.4	6	0.39	0.0244	0.0585
	3 + 3	0.585, 0.21	0.0063	0.0151

27.7 SPC exchanges

All local and trunk exchanges manufactured after about 1985 are almost fully digital, apart from a few analogue devices used in the subscriber-line interfaces. The advantages of these SPC exchanges over the older electromechanical ones are very substantial:

- Economy – cheaper to build, cheaper to install, cheaper to run

---

[6] The overload traffic will be more peaky than the normal traffic, so the assumption is not valid, but it does give a qualitative picture. The theory of overload traffic is too specialised to be discussed here, but the interested reader is referred to the *Bell System Technical Journal* **35**, 421 (1956).

- Flexibility – software is changed, not hardware
- Maintainability – self-diagnosis of faults
- Rapidity – calls can be set up as fast as the number is dialled
- Reliability – no moving parts in the switches
- Quality – lower noise levels than with analogue switching
- Quantity – there is much less space used; SPC exchanges are far smaller

As an example, the changing of a subscriber's number requires only a program change and not an alteration to the wiring. A great deal of statistical data can be generated as required and sent automatically wherever it is needed, which is very helpful in network management and planning.

The essence of an exchange is setting up calls. The subscribers' lines are monitored in groups of around 500 and when a call request is signalled a microprocessor (usually called a remote processor or RP) decodes the digits and send them as a single message to dual central processors (CPU) that set up the call path. Two CPUs are used and any disagreement triggers a checking routine to discover the faulty CPU which can then be sidelined for repair. Many RPs are used to build in a huge amount of redundancy and remove from the CPU as much of the workload as possible apart from path setting. The No. 1 ESS (electronic switching system) had no microprocessors (they were not available until the mid 1970s) so the CPU had perform many more functions. Programming for the No. 1 ESS was therefore one of the most formidable software tasks ever undertaken. Bell Labs eventually developed a new operating system (UNIX) and a new high-level language (C) in the early 1970s to cope with the problem. UNIX is now written in C and is installed in the No. 5 ESS. The structure of C was designed to reduce programming errors and make them easier to trace.

The CPU handles all the calls at an exchange and is therefore time-shared by them. By working extremely fast it gives the illusion that it processes them simultaneously. Its main function is to supervise the switching of the call path through the central switching region of Figure 27.8, which is designed to carry all the calls even when every incoming and outgoing line is busy. To do this a new switching concept was required: the digital time-switch as opposed to the conventional digital space-switch. Digital space-switching of multiplexed PCM signals merely involves the transference of the contents of one PCM system's time slot, say No. 3 (TS3), to TS3 of another PCM system, which is fairly easily accomplished if they occur simultaneously, but only if both are free. If we could select any TS on the PCM signal for transference of the incoming data, there would be much less probability of blocking. This selection of a different TS, achieved by suitable electronic delays, is the essence of digital time-switching. A practical digital switching system incorporates both space and time-switching.

Figure 27.9 shows a digital TDM space-switch. The connection memories activate the appropriate crosspoints at the right time so that the contents of TS1 of incoming PCM system A are placed in TS1 of outgoing PCM system D, an action which is repeated every 125 μs if the multiplexing scheme is like that in Figure 27.9. Each time slot of incoming PCM system A can be associated with any one of the corresponding time slots of outgoing systems A-E. Altogether there are $5 \times 32 (= 160)$ incoming time slots which can be connected to 160 outgoing time slots through $5 \times 5 = 25$ crosspoints. The five crosspoints connected to each outgoing system are controlled by connection memories containing 32 addresses which operate the crosspoints at the appropriate time by crosspoint-decoding logic. If analogue space-switching had been used it would have required $160 \times 160 = 25\,600$ crosspoints instead of 25!

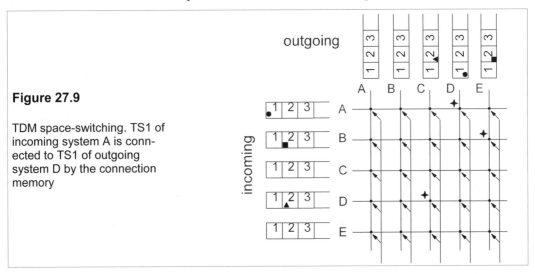

Figure 27.9

TDM space-switching. TS1 of incoming system A is conn-ected to TS1 of outgoing system D by the connection memory

27.8 The integrated-services digital network (ISDN)

Telephony has been very much dominated by analogue voice communications. The acquisition of digital transmission capability was mainly used to improve the quality and capacity of trunk-line communications. However, after digitising, speech loses its unique characteristics and to a digital network it is a stream of bits like any other. The distinction between voice and data communications is lost. The all-digital telephone network is much more flexible and versatile than a hybrid network and has been given the name *integrated-services digital network*. The CCITT published its Red Book recommendations for ISDNs in 1985 and in the same year British Telecom ran a pilot ISDN. An ISDN is one which crucially includes the subscriber's apparatus, whereas the integrated digital network (IDN) stopped short at the local exchange. The telephone network, when fully digitised, can be expected to perform a great many services other than its primary one of voice communication; indeed it is possible that one day the non-voice services will be the more important. Some of the present and future non-voice services are:

- TEXT   Facsimile
  Electronic mail
  Text transmission – telex and teletex
  Viewdata
  Electronic shopping by videotex

- DATA   Electronic funds transfer
  Computer interconnection
  Telemetry (gas, electricity and water)
  Telecontrol and monitoring of the 'intelligent' home

- VISUAL   Cable television
             Slow-scan televison
             Videophone

27.9 **Mobile telephony**

Mobile telephones are just radiophones, but what a difference between today's and yesterday's! The mobile telephone handset contains more sophisticated technology per unit volume than any other consumer product, and at an astonishingly low price. Mobile handsets in parts of London have reached staggering densities and yet they all use a very limited band of frequencies without many problems. The cell is the means by which small bandwidth can accommodate many calls. The transmitted power is kept to less than one or two watts, depending on the frequency band and the range of the transmissions is restricted to about 30 km at most in a macrocell. Signals are sensed by neighbouring cells and when the signal strength is greater in a neighbouring cell the signal is switched to it. If the traffic becomes denser and the available channels are fully utilised the cell size and transmitter power must be reduced. Microcells cover roughly a square kilometre and picocells can be used to cover just a single office building. If there are C channels available to a mobile telephone network and it has T transmitters (= cells), then it can accommodate a maximum of CT simultaneous calls.

The handset contains a keyboard, a microphone, a speaker, an LCD display, an aerial (antenna), a PCB containing a number of ICs, and a battery to power it. Everything has been done to minimise power consumption for two reasons: first to conserve the battery and second to minimise UHF radiation. The batteries are themselves products of far greater energy density than obtained a few years ago, and the microphone, aerial, keyboard, display and speaker have all been improved and miniaturised. But the chief technological foundation of the mobile telephone's success is the integrated circuit.

One of the ICs in the handset controls the power of the signal using *adaptive power control* (APC) so that the signal strength is adjusted according to conditions and in general is reduced by d^2 where d is the distance from the cell's base station. APC also reduces power to very low levels when the handset is not transmitting. In the UK the average power transmitted by a handset is limited to ¼ W at 800 MHz and ⅛ W at 1.8 GHz.

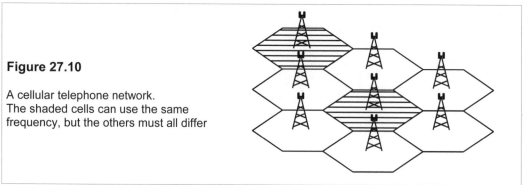

Figure 27.10

A cellular telephone network.
The shaded cells can use the same
frequency, but the others must all differ

Bell Telephone devised the first scheme for mobile telephones (cellphones in the USA) about thirty year ago. Figure 27.10 shows the arrangement of base stations. The signals were analogue and the system used frequency-division multiplexing (FDM). Adjacent cells used differing frequencies, so that on a hexagonal grid seven frequency bands were required. Bandwidth was used inefficiently and the signals were subject to interference and interception and handsets were easily cloned, that is their identity was stolen and used in other handsets, the original subscriber being stuck with the bill for all the clones' calls.

This first-generation system was superseded by a second-generation system which uses time-division multiplexing (TDM) and frequency-shift keying of the digitally-encoded signals. Voice compression is also used to increase the capacity. This can be done with no great loss in quality since small samples of speech can be expanded at the receiving end using predictive techniques which restore the signal almost completely. Typically the compression ratio is 10:1. The use of available bandwidth is several times better than with analogue signals and there was greater resistance to interference and interception, so that cloning became much more difficult. The third generation of mobile communications has just begun and will be vastly more capable than the previous two, in particular it will have excellent video and multi-media capability.

There is some discussion as to whether it is better for the 3G mobile communications to use the Global System for Mobile Communications (GSM) favoured in Europe and using advanced TDM, or the N American system called *code-division multiplexing* (CDM) or *code-division multiple access* (CDMA), also called *spread spectrum*[7]. In this each handset is assigned a unique encoding sequence which is passed on to the receiver of the message. All frequencies may be used by all handsets at once (hence 'spread spectrum') and the available channels are increased to several times that of TDM. Because the encryption sequence is lengthy and pseudo-random the transmissions are essentially secure, they are also much less subject to noise, narrowband interference and multipath[8] effects. The technical advantages of CDM may overcome the present preference for GSM in the longer term.

Suggestions for further reading

Probably the best source of information on telephony is the *Bell System Technical Journal* (now the *AT&T Technical Journal*). Many articles are written in an informative, not-too-technical style. Numerous important papers have been published in the journal.

The essential guide to telecommunications by A Z Dodd (Prentice Hall, 3rd ed., 2001)
Newnes telecommunications pocket book by S Winder (Butterworth-Heinemann, 3rd ed., 2001). Inexpensive.
Digital telephony by J C Bellamy (John Wiley, 3rd ed., 2000). Extensive.
The Irwin handbook of telecommunications by J H Green (McGraw-Hill, 4th ed., 2000). Expensive.
Telecommunications by W Hioki (Prentice Hall, 4th ed., 2000). Not over mathematical.
Cellular radio by R C V Macario (Palgrave, 2nd ed., 1997). Covers the basics and not too expensive.
Introduction to digital mobile communication by Y Akaiwa (Wiley, 1997). Mathematical, expensive.
Telecommunications, switching, traffic and networks by J E Flood (Longman, 1994). A commendably concise text, but needs a little updating.

---

[7] The original patent, US 2,292,387, was taken out by Hedy Lamarr (1914–2000), a Hollywood film star with far greater claims to fame, and a composer associate of hers, George Antheil, in 1942. It was designed to prevent jamming, but was too far in advance of available technology. When this patent had expired in 1957 the Sylvania Co. coined the term 'spread spectrum' for the idea, and eventually it was used to secure US communications during the Cuban missile crisis of 1962 and subsequently.

[8] Interference caused by reflections from buildings especially, responsible for 'ghosts' in television reception.

Problems

1. If a signal, $v_{in} = \sin 2000t$ V, is to be companded using the A-law of Equation 27.1, at what time does the signal begin to be compressed? What is its magnitude at that time? What is the companded signal's magnitude after (a) 1 μs and (b) 10 μs? Sketch the graph of the companded signal for $0 \leq t \leq 785$ μs. Why has A been chosen as 87.6? *[t = 5.7 μs, 0.1826 V, 32 mV, 0.285 V]*

2. A local exchange has 2000 subscriber lines each with an average peak traffic of 0.03 E. How many switching circuits are required to give a GOS of (a) 0.2% and (b) 1%? What concentration ratio do these correspond to? If the average peak traffic increases by 20%, what will the GOS be for case (a) and (b)? (You will have to write a small computer or calculator program to find the GOS from Equation 27.4) *[81, 75, 24.7, 26.7, 2.97%, 6.5%]*

3. If traffic of 17.8 E from a large number of subscribers is served by 27 lines, what is the GOS and the lost traffic? If the lines are split into two groups, with 22 lines in the first group and 5 in the second, which services the overflow calls, what is the lost traffic, assuming the overflow traffic can be treated statistically the same as the rest? *[1%, 0.178 E, 0.0049 E]*

4. In the telephone network of Figure P27.4, the traffic on the lines between exchanges is shown in erlangs. If the lines between exchanges C and T are inoperable, what will the GOS be on the lines between C and D, and between D and T, if all the traffic is automatically transferred and there are 8 lines serving the traffic between C and D and 6 serving that between D and T? What is the overall GOS and lost traffic for calls that originate at C and go through T? *[7%, 11.72%, 17.9%, 0.537 E]*

5. In the network of Figure P27.4 there are 8 lines serving the link between A and T. What will the GOS be if the lines between A and B are inoperable and there are enough lines between B and T not to affect the GOS? How many more lines would be needed between A and T to provide a 1% GOS in that event? *[12.2%, 13]*

6. In the fully-connected network of Figure P27.6a, the exchange positions are indicated by coordinates in units of 10 km. The traffic flows in erlangs are given next to the connecting lines. If the network of Figure P27.6a is to be replaced by a star network as in Figure 27.6b, write down the traffic flows in Figure 27.6b and then calculate the best place to put exchange T. Assume that the best place to put T is where ΣAd is a minimum, A being the traffic between exchanges and d the distance between exchanges – like a centre-of-mass calculation. What is the percentage change in ΣAd? *[(2.5,1.86), +26.5%]*

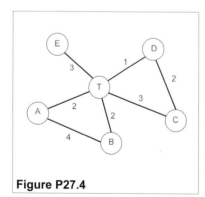

Figure P27.4

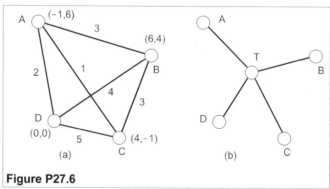

Figure P27.6

7. If a mobile telephone transmits at a pulse power of 1 W when it is 10 km from its base sation, what is the average transmitted power level assuming 20% idle time, 40% receiving time and OOK without voice compression? Using FSK? What voice compression ratio is needed to make the transmission's average power legal when the handset is 20 km from its base station, assuming it uses FSK at 1.8 GHz? *[0.2 W, 0.4 W, 12.8:1]*

28 Electromagnetic compatibility

ELECTRICAL appliances, machines, instruments and systems must work properly in their environment, implying that they must not only be insensitive to electromagnetic interference (EMI), but also not interfere with the working of other electrical equipment. Fitness for the electromagnetic environment is called *electromagnetic compatibility* (EMC). As the number of electrical products of all kinds is rising exponentially and as new electrical applications are found almost daily EMC has assumed major importance in design. Many standards for EMC have been issued, a few of the EU ones being given in Table 28.1. The standards are of three types: type A, Generic Product Standards; type B, Product Family Standards; and type C, Specific Product Standards. To find what applies to a particular product you first look for type C, then if not found, type B and if none, type A. Examples of type A are EN50081 and EN50082, of type B EN55014, EN55022 and EN55024, and of type C EN50199 and EN50263.

Table 28.1 *Some CENELEC[1] standards for EMC*

EN50065-1	Mains signalling on low-voltage installations, from 3 kHz to 148.5 kHz
EN50081-1	Generic emission standard for residential, commercial and light-industrial environments
EN50081-2	Generic emission standard for industrial environments
EN50082-1	Generic immunity standard for residential, commercial and light-industrial environments
EN50082-2	Generic immunity standard for industrial environments.
EN50083-2	Cable distribution systems for television and sound signals
EN50090-2-2	Home and building electronic systems
EN50091-2	Uninterruptible power supplies (UPS)
EN50130-4	Immunity for fire, intruder and social alarm systems
EN50199	EMC product standard for arc-welding equipment
EN50263	EMC product standard for relays and other protection equipment
EN55011	Emissions from industrial, scientific and medical equipment
EN55013	Sound and TV broadcast receivers and associated equipment
EN55014	Household appliances, portable tools etc.
EN55015	Lighting and similar equipment
EN55020	Sound and TV receivers and associated equipment
EN55022	Emissions from equipment used in information technology (ITE)
EN55024	Immunity standard for ITE
EN55103-1	Emissions from audio, video and entertainment-lighting controls
EN55103-2	Immunity of audio, video and entertainment-lighting controls
EN55104	Immunity of household appliances, portable tools etc.
EN60521	Watt-hour meters
EN60601-1-2	Medical electronic equipment
EN61000-3-2	Mains harmonics
EN61000-3-3	Mains flicker

---

[1] Comité Européen de Normalisation Electrotechnique, rue Bréderode 2, B-1000, Brussels, Belgium

Of the utmost consequence for all manufacturers of electrical goods is the European Community's Directive[2] on EMC, which took full effect on January 1st, 1996. From this date all electrical products sold in the EC, including imports, must not cause 'excessive' EMI, nor be 'unduly' affected by it. Products which conform to the standards laid down by CENELEC will be given the CE mark of conformity. Nine CENELEC standards are listed in Table 28.1 and more are to be issued concerning disturbances caused by equipment connected to the mains supply. Anyone selling a product in breach of the standards will be committing a criminal offence: several successful prosections have already been made in the UK. In addition Trading Standards officers have the power to seize and destroy any non-complying products.

28.1 **Sources of electromagnetic interference**

EMI requires a source, a transmission path and a susceptor – an apparatus, device or equipment that suffers the interference. The transmission can be by conduction through wires, metal tracks and components or by radiation through the air. Many sources produce EMI which takes both routes. Common EMI sources are

1. Power-electronic devices which switch rapidly at mains frequency and produce conducted interference with very large voltage spikes having significant harmonics to 150 kHz and beyond. Examples are motor controllers and rectifiers/inverters in power supplies.
2. Fluorescent lamps produce conducted and radiated EMI with a peak at 1 MHz because of their gaseous electrical discharges.
3. Desk-top computers produce large amounts of EMI at their clock frequencies. Early models in unscreened plastic cases radiate especially strongly.
4. DC motors with brushes produce wide-spectrum EMI by arcing and rapid current reversal.
5. Motor vehicles can be a powerful source of radiated EMI not only from the ignition system but also from alternators and ancillary electrical equipment.
6. Electrical driers and heaters can produce conducted broadband EMI because of rapid on-off switching. The precise nature depends on how the current is controlled.
7. Atmospheric noise arises from lightning discharges and is relatively low level except during a local thunderstorm. Cosmic emissions from space are slightly more intense, especially during sunspot activity. Most affected are frequencies in the tens of MHz range.
8. Radio and TV transmitters. Radiated frequencies are well defined and narrow band.
9. Household appliances and portable power tools.
10. Other devices and modules inside the equipment housing, such as flip-flops, timers and multiplexers. These produce interference which is enhanced by their close proximity to the susceptor and because they are within the primary EMI shield. Sharp edges to digital pulses cause interference at very high frequencies ($\approx 1/2\pi t_r$, where t_r is the rise time).

The solution to an EMI problem may be to suppress it at source, or to break the transmission path or to protect the susceptor; often there is no choice but the last. Often very simple measures will do, such as moving an amplifier from one end of a cable to the other, as in Figure 28.1, where even though the noise introduced through the cable is only 20 dB, the effect is very marked.

---

[2] Council Directive 89/336/EEC (3 May 1989) as amended by Directive 92/31/EEC (28 April 1992).

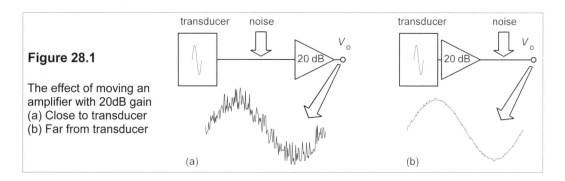

Figure 28.1

The effect of moving an
amplifier with 20dB gain
(a) Close to transducer
(b) Far from transducer

It is generally accepted that EMC has to be considered at the inception of the design of any electrical product, as the consequence of trying to add on EMC at the prototype stage may be large and avoidable expense. The principal methods of reducing EMI in order of increasing cost are grounding, shielding and filtering. We shall discuss these next.

28.2 Conductor shielding

Shielding takes two forms: conductor shielding and shielding by enclosing the sensitive device or equipment within a box which reflects or absorbs the incident radiation. Conductor shielding is intimately connected with grounding.

28.2.1 Capacitive coupling

Conductors pick up EMI by capacitive or inductive coupling. First we shall consider capacitive coupling (sometimes called electric-field coupling). The two conductors in Figure 28.2a which are coupled to ground by stray capacitances C_1 and C_2 and to each other only by the stray capacitance, C_{12}. The resistance to ground of conductor 2 is R_2. A voltage source, $\mathbf{E_s}$, connected to conductor 1 causes a voltage, $\mathbf{E_n}$, to be induced in conductor 2 whose magnitude is calculated with the aid of the equivalent circuit of Figure 28.2b.

The impedance to ground of conductor 1 is not in the equation for $\mathbf{E_n}$, which is

$$\mathbf{E_n} = \frac{\omega C_{12} R_2 \mathbf{E_s}}{R_2 \omega (C_{12} + C_2) - j} \tag{28.1}$$

Now C_{12} is given by

$$C_{12} \approx \frac{\pi \varepsilon_r \varepsilon_0 l}{\ln(2x/d)} \tag{28.2}$$

where x is the centre-to-centre separation of the conductors, d is their diameter and l their length. In air $\varepsilon_r = 1$ and Equation 28.2 gives $C_{12} \approx 10$ to 40 pF/m. Typically the stray capacitances are a few pF and $\omega R_2 (C_{12} + C_2) \ll 1$ at normal frequencies, when Equation 28.1 reduces to

$$E_n = \omega R_2 C_{12} E_s \tag{28.3}$$

At high frequencies, or if R_2 is large, Equation 28.1 becomes

$$E_n = \frac{C_{12}E_s}{C_{12} + C_2} \tag{28.4}$$

This is the limiting case when the noise voltage is determined by the capacitor-divider circuit and is independent of frequency.

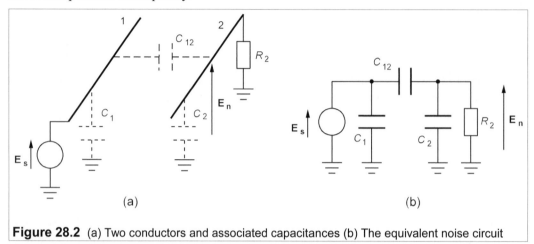

(a)   (b)

Figure 28.2 (a) Two conductors and associated capacitances (b) The equivalent noise circuit

The effects of capacitive coupling are much reduced by screening the conductors. Suppose a cylindrical metal screen is placed about conductor 2 (as in a coaxial cable) and earthed at one point as in Figure 28.3a. Additional stray capacitances now exist between conductor 1 and the screen, C_{1S}, between the screen and conductor 2, C_{2S}, and between screen and ground, C_S. Conductor 2 will not be covered by the screen at its ends, and will therefore still have a stray capacitance, C_{12}', between it and conductor 1 and another to ground, C_2' (though both are much smaller than without the screen). The equivalent circuit of Figure 28.3a is that of Figure 28.3b. Figure 28.3b can be rearranged in the form of Figure 28.3c. But this circuit is identical to that of Figure 28.3b, apart from having C_{2S} in parallel with C_2'. Equations 28.3 and 28.4 become

$$E_n = \omega C_{12}'R_2E_s \tag{28.5}$$

And,

$$E_n = \frac{C_{12}'E_s}{C_{12}' + C_{2S} + C_2'} \approx \frac{C_{12}'E_s}{C_{2S}} \tag{28.6}$$

However as $C_{12}' \ll C_{12}$, and as $C_{2S} \gg C_{12}' \approx C_2'$, the noise voltages are much reduced in each case, typically by a factor of 100, or 40 dB. (See Problem 28.4.) If a braided screen is used as in a coaxial cable, its effectiveness is reduced at high frequencies because of the gaps in the braid which allow electric fields to penetrate. (See Problem 28.1.)

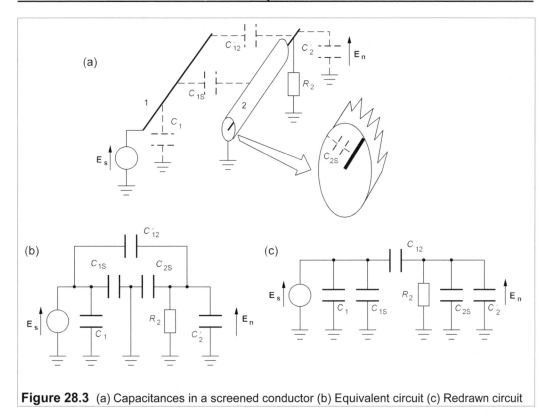

Figure 28.3 (a) Capacitances in a screened conductor (b) Equivalent circuit (c) Redrawn circuit

28.2.2 *Inductive coupling*

Inductive coupling (sometimes called magnetic coupling) occurs when magnetic flux from a noise source links with a conductor as in Figure 28.4. The degree of coupling depends on the area of the flux enclosed by the susceptible conductor, since the induced e.m.f. is

$$E_n = \omega BA \cos\theta \qquad\qquad (28.7)$$

where θ is the angle between **B** and the normal to A. Current is returned through a *ground loop* which tends to have maximal area.

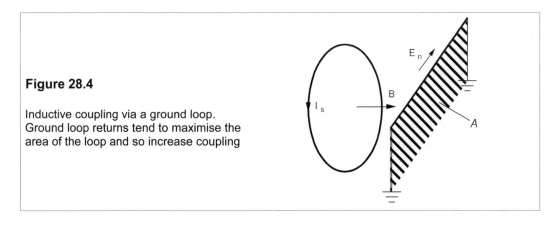

Figure 28.4

Inductive coupling via a ground loop.
Ground loop returns tend to maximise the
area of the loop and so increase coupling

Unfortunately, shielding with a non-magnetic metal does not by itself help because the magnetic flux will penetrate the shield. The only way to reduce inductively-coupled noise is to reduce the effective area of the current loop in the susceptor circuit. If the susceptor circuit is grounded at one end, returning the current through the shield, which is itself grounded at one end, will do this. If the susceptor circuit is grounded at both ends, so must the shield be. Figure 28.5 shows these configurations. In the case of Figure 28.5c, some current returns through the ground loop and the effective area is greater than that shown.

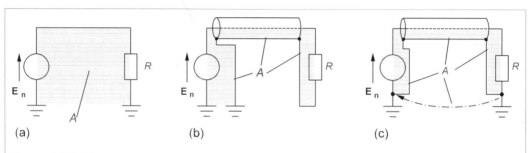

Figure 28.5 (a) Full ground loop, large area (b) Single ground and screen return, minimal loop area
(c) Some current returns through the ground loop, increasing the effective loop area

Tests have been made on various susceptor-circuit grounding configurations, some of which are shown in Figure 28.6. The circuits were arranged as coils in close proximity to a radiating coil and the voltage was measured across the 1 MΩ resistor. The reference circuit is grounded at both ends with a shield connected only to one ground. It can be seen that a twisted pair is a simple and efficient way to reduce loop area when the susceptor circuit is grounded at one end as in Figure 28.6c.

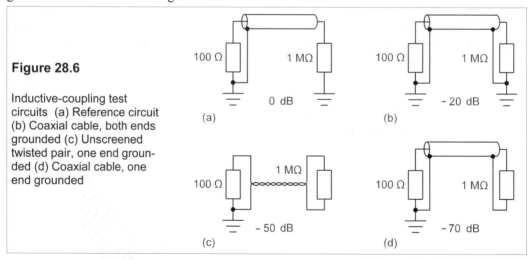

Figure 28.6

Inductive-coupling test circuits (a) Reference circuit (b) Coaxial cable, both ends grounded (c) Unscreened twisted pair, one end grounded (d) Coaxial cable, one end grounded

Experimental tests have shown that at about 100 kHz the reduction in inductive coupling is around 50 dB for 'large' twists (20/m) and about 75 dB for 'small' twists (60/m). The effectiveness of a single ground is reduced as the length of cable is increased. Roughly speaking, cables over λ/10 are 'long' and should be grounded every twentieth of a wavelength (which is only 0.15 m at 100 MHz). Twisted pairs are used when the signal frequencies are below 100 kHz and coaxial cables up to 200 MHz. They are more effective against inductive coupling at low frequencies than coaxial cables.

28.3 **Grounding**

We have already seen that proper grounding is essential to minimise inductive coupling. There are three types of ground shown in Figure 28.7: the series-connected ground (often called a *common ground*), the parallel-connected (or *separate*) ground and the multipoint ground. The latter (Figure 28.7c) is used at high frequencies (say > 30 MHz) when the separation between ground points should be about λ/20.

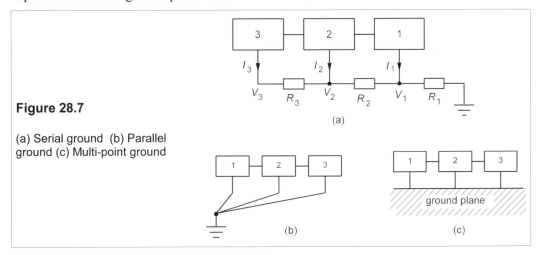

Figure 28.7

(a) Serial ground (b) Parallel ground (c) Multi-point ground

Ground connections should always be as short as possible and of low resistance. The serial ground is easiest and cheapest, but introduces additional resistance and sums the ground currents as in Figure 28.7a. There may be nothing wrong in using a serial ground for parts of a system that are not much affected by noise and a parallel ground for the critical subsystems.

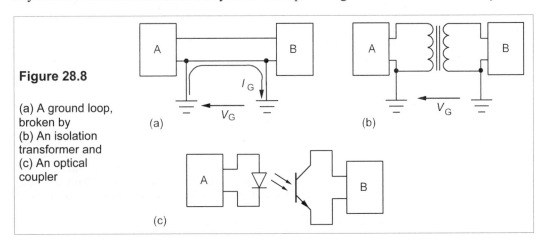

Figure 28.8

(a) A ground loop, broken by
(b) An isolation transformer and
(c) An optical coupler

Noisy equipment and that drawing high currents (motors, heaters, relays and power supplies) should be given their own ground connection. The parallel ground of Figure 28.7b uses much extra conductor, but is necessary to separate noisy and high-current subsystems from low-noise, low-current subsystems.

Where an offending ground cannot be removed, the solution is to break the circuit with an isolation transformer, as in Figure 28.8b, or by optical coupling as in Figure 28.8c. If DC continuity is required this solution cannot be used, but a *longitudinal* or *neutralising* transformer can be used instead as in Figure 28.9b.

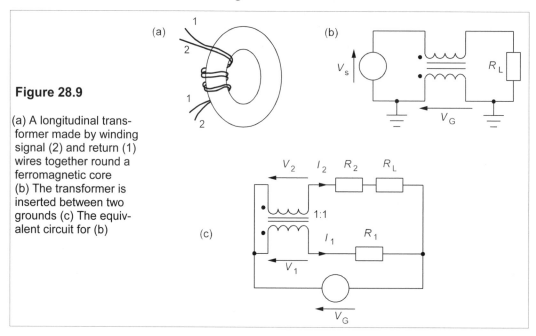

Figure 28.9

(a) A longitudinal transformer made by winding signal (2) and return (1) wires together round a ferromagnetic core
(b) The transformer is inserted between two grounds (c) The equivalent circuit for (b)

In a longitudinal transformer the primary winding is inserted into one line while an identical secondary goes into the other. Figure 28.9c shows the equivalent circuit considering the ground voltage source, V_G, by itself. The ground voltage causes a primary current, I_1, which produces voltages, V_1, across the primary, and V_2 across the secondary winding. As the return wire has resistance, $R_1 \approx 0$, then $V_G \approx V_1$, and as the transformer windings are identical, $V_2 \approx V_1 \approx V_G$. This means that V_G is almost cancelled by V_2 in the signal line and cannot appear across the load, R_L. The signal, however, is differential mode and remains unaffected by the transformer. A longitudinal transformer can easily be made by winding both lines around a single ferromagnetic toroid as in Figure 28.9a.

Example 28.1

A neutralising transformer has equal primary and secondary inductances of 2 mH and the line resistances are both 1 Ω. If the ground-to-ground voltage between two subsystems is 40 mV at 3 kHz, and the load resistance is 50 Ω, what is the noise voltage in the load? If the signal voltage at 3 kHz is 56 mV, what is the SNR?

We assume the transformer is ideal so that $V_1 = V_2$ in Figure 28.9b. Using Kirchhoff's voltage law in the upper mesh gives

$$V_1 + I_1R_1 = V_2 + I_2(R_2 + R_L) \quad \Rightarrow \quad I_1R_1 = I_2(R_2 + R_L) \tag{28.8}$$

After putting in the values for R_1, R_2 and R_L, we find $I_2 = 0.0196I_1$. We find I_1 by using Kirchhoff's voltage law on the lower mesh of Figure 28.9c, which produces

$$\mathbf{V_G} = \mathbf{V_1} + \mathbf{I_1}R_1 = j\omega L\mathbf{I_1} + \mathbf{I_1}R_1 \tag{28.9}$$

(The voltage due to the mutual inductance, $j\omega MI_2$, can be overlooked as I_2 is small). Thus

$$I_1 = \frac{V_G}{\sqrt{R_1^2 + \omega^2 L^2}} = \frac{0.04}{\sqrt{1 + 37.7^2}} = 1.06 \text{ mA} \tag{28.10}$$

Thus $I_2 = 0.0196 \times 1.06$ mA $= 20.8$ μA. The noise voltage is $I_2R_L = 1.04$ mV, so the SNR is 20 $\log_{10}(56/1.04) = 34.6$ dB. With no neutralising transformer, all the noise voltage would appear in the load and the SNR would be 2.9 dB: the SNR is higher by over 30 dB.

28.4 **Shielding with sheet conductors**

Electromagnetic radiation can be prevented from causing interference by excluding it from equipment with an electrically-conducting box. This screening works both ways: it protects the instrument from external sources of EMI and it reduces the EMI emitted. Many instruments and appliances have an outer case made of non-conducting material such as plastic which can be coated internally with a metal-loaded paint to provide an effective shield. Electromagnetic radiation incident on a sheet of conducting material is partly absorbed and partly reflected. First we shall discuss absorption.

28.4.1 *Absorption losses in sheet conductors*

An e.m. wave incident upon a conducting sheet of material will decline in strength as it penetrates through the surface. The e.m. field decays exponentially with depth so that $F(x) = F_0\exp(-x/\delta)$, where $F(x)$ is the magnitude of the electric or magnetic field at a depth, x, beneath the surface, F_0 is the field strength at the surface and δ is the *skin depth*. Roughly speaking, the field penetrates a conductor only to a depth, δ, which is given by

$$\delta = \sqrt{\frac{2}{\mu\sigma\omega}} \tag{28.11}$$

For example, copper has a magnetic permeability, $\mu = \mu_0 = 4\pi \times 10^{-7}$ H/m, a conductivity, $\sigma = 58$ MS/m, so that at 50 Hz the skin depth in copper, δ_{Cu}, is 9 mm and at 50 MHz it is only 9 μm. The radiation is absorbed by the metal and the amount of absorption is

$$-20\log_{10}[F(x)/F_0] = -20\log_{10}[\exp(-x/\delta)]$$

$$= -20\log_{10}e\ln[\exp(-x/\delta)] \tag{28.12}$$

$$= 8.686x/\delta$$

In other words, the absorption is 8.686 dB times the number of skin-depths. At 50 kHz the skin depth in copper is 0.29 mm, so a 2mm thick sheet would have an absorption of $8.686 \times 2/0.29 = 60$ dB.

Ferromagnetic metals have large values of μ_r and hence smaller skin depths. For example, soft iron has $\mu_r = 1000$ and $\sigma = 10$ MS/m, which means that at any frequency its skin depth will be $\delta_{Fe} = \sqrt{(5.8/1000)}\delta_{Cu} = 0.076\delta_{Cu}$. The absorption losses for soft iron in dB are $1/0.076 = 13$ times greater than those in a similar thickness of copper. At high frequencies the absorption is great, but at low frequencies it is small. However, a substantial amount of the incident radiation may still be reflected.

28.4.2 Reflection of electromagnetic waves by a plane conductor

The amount of incident e.m. radiation that is reflected by a plane of conducting material depends on the relative impedances of the wave (Z_w) and the conductor surface (Z_c). The transmission coefficient is

$$T = \frac{4Z_w Z_c}{(Z_w + Z_c)^2} \tag{28.13}$$

But for metals $Z_w \gg Z_c$, in which case Equation 28.13 becomes

$$T \approx 4Z_c/Z_w \tag{28.14}$$

For radiation in the far field (which is the region more than $\lambda/2\pi$ from the source), the wave impedance is the free-space impedance, $Z_0 = \sqrt{(\mu_0/\epsilon_0)} = 377$ Ω. The conductor's surface impedance is given by

$$Z_c = \sqrt{\omega\mu/\sigma} \tag{28.15}$$

Substituting this into Equation 28.14 and also using $Z_w = 377$ Ω, we find

$$T = 4Z_c/Z_w = 0.0106\sqrt{\omega\mu/\sigma} \tag{28.16}$$

The reflection loss in dB is

$$R = -20\log_{10}T = -20\log_{10}(0.0106\sqrt{\omega\mu/\sigma})$$
$$= 39.5 - 10\log_{10}(\omega\mu/\sigma) \tag{28.17}$$

This loss is independent of the thickness of the shield, provided it is not too thin, and is far more than the absorption losses for shields of normal thickness.

If the previously quoted values of μ and σ are substituted into Equation 27.17, it can be written

$$R_{Cu} = 168 - 10\log_{10}f \tag{28.18}$$

At 50 Hz, therefore, the reflection loss for a copper shield is 151 dB, which is far in excess of the absorption losses at low frequencies for shields of normal thickness. The reflection loss for iron is some 40 dB less than for copper. The absorption loss goes up with frequency, while the reflection loss goes down, resulting in a frequency where the sum of these losses is a minimum for a given shield thickness (see Problem 28.5).

28.4.3 Near-field radiation

Many sources are closer than $\lambda/2\pi$ to the susceptor, which means the radiation is not plane wave but near field in character. The radiation field's impedance then depends on the type of source. High-impedance (or electric) sources such as a dipole antenna emit e.m. radiation which has a higher E/H ratio than the plane-wave value of 377 Ω and thus a higher wave impedance, Z_w. Equation 28.14 shows that in this case the fraction of the incident radiation that is transmitted into a shield is even less than in the case of a plane wave. One can therefore assume that shields will reflect electric fields almost totally. On the other hand, low-impedance (or magnetic) sources such as a loop antenna emit radiation with a lower E/H ratio than a plane wave. In these transmission through a shield is greater and the reflection loss is smaller than given by Equation 28.17. The impedance of a magnetic-field point source is given by

$$Z_M = \mu_0 \omega r \tag{28.19}$$

where r is the distance from the point source. In this case the reflection loss is

$$R_M = -20\log_{10}(4Z_c/Z_M) = -20\log_{10}\left(\frac{4\sqrt{\omega\mu/\sigma}}{\omega\mu_0 r}\right) \tag{28.20}$$

$$= 20\log_{10}(0.25r\sqrt{\omega\mu_0\sigma/\mu_r})$$

Substituting $\sigma = 58$ MS/m and $\mu_r = 1$ for a copper shield yields

$$R_M(\text{Cu}) = 14.6 + 10\log_{10}r^2 f \text{ dB} \tag{28.21}$$

In the case of a 50Hz source at 1 m, Equation 28.21 gives $R_M = 31.6$ dB, compared to 151 dB for a 50Hz plane wave.

Another problem with using thin shields against magnetic fields is multiple reflections, which reduce the reflection loss so that for thin shields ($< 0.1\delta$ thick) the reflection loss is 15 dB less than that given by Equation 28.20. For low-frequency magnetic fields, shields are likely to be ineffective. A steel or iron shield would perform better, considering both absorption and reflection, than copper at frequencies below 100 kHz. (See Problem 28.8.)

28.4.4 Shield integrity

By and large conductive shields are effective screens for high-frequency radiated EMI and against low-frequency electric fields, provided the shield integrity is maintained. Unfortunately it is impossible in practice to do so because wires have to be brought in and access to components may be required from time to time. Breaks in the shield's integrity are more likely to result in excessive EMI than deficient reflectivity or absorption of the shield. Leakage through gaps or holes in the shield depends on their maximum linear dimensions and not on area, provided the largest dimension is greater than 0.01λ. Seams, covers, doors and lids are therefore much more serious sources of leaks than round holes. The gaps in the shield can be considered to be waveguides which attenuate radiation whose half wavelength is much greater than L, the largest slot or gap dimension. The *cut-off* frequency can be found from

$$\lambda_c/2 = L = c/2f_c \rightarrow f_c = c/2L \tag{28.22}$$

In fact f_c is the frequency of maximum radiation through a slot or gap, but frequencies well below f_c will be attenuated. If $f \ll f_c$ the attenuation is

$$S \approx 30t/L \text{ dB}$$
(28.23)

where t is the thickness of the shield.

Example 28.2

An EMI shield is 1 mm thick and has an overlapping, spot-welded seam with welds every 10 mm. Between welds the gap is 0.1 mm wide and the overlap is 15 mm, as in Figure 28.10. What is the cut-off frequency and what is the attenuation of the seam at 300 MHz? What are the reflection and the attenuation losses at 300 MHz if the shield is copper? [Assume far-field radiation.]

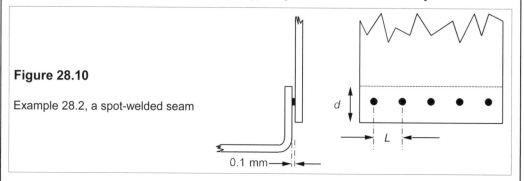

Figure 28.10

Example 28.2, a spot-welded seam

0.1 mm

The welded seam is effectively a slot with L = 10 mm and t = 15 mm. Using Equation 28.22, f_c = $3 \times 10^8/(2 \times 0.01)$ = 15 GHz. Thus at 300 MHz we can take the seam to be an attenuating waveguide. Now the large dimension has to be > 0.01λ or

$$L > 0.01\lambda = 0.01c/f \;\rightarrow\; f_{min} = 0.01c/L = 300 \text{ MHz}$$
(28.24)

That is to say, frequencies well below 15 GHz will be attenuated, and those below 300 MHz will not radiate through the slot. The attenuation at 300 MHz is given by Equation 28.23 as S = 30 × 15/10 = 45 dB. Note that if the shield did not have overlap at the seam the attenuation at 300 MHz would be negligible.

The reflection losses are given by Equation 28.17, with σ = 58 MS/m and μ = μ_0, that is R = 39.5 – 44 ≈ 0 dB. The skin depth is δ = $\sqrt{(2/\mu\sigma\omega)}$ = 3.8 μm, so the shield is 263δ thick and the absorption is 263 × 8.686 = 2284 dB. The shield's performance at 300 MHz is wholly dependent on the quality of the seam.

Seams can be welded continuously and should not be a problem, but it is impossible to avoid leakage at doors, lids and inspections covers. A conducting gasket can be used to minimize the leakage but it must be placed inside screw holes or these will leak EMI. Conducting gaskets can be made from wire mesh or conducting rubber. Note too that coaxial cables are made from wire braid which will leak at high frequencies because the braiding is criss-crossed at a length of about 1 mm, which is the effective the slot length, L. If the braid is poor or damaged, holes may occur which will destroy its integrity. The braid must also be properly connected to a BNC connector to ensure shield continuity, in particular the shield around the whole of the circumference of the wire must be joined to the connector so that a uniform current distribution is achieved at the termination.

28.5 **EMI filtering**

Filtering is essential to limit the bandwidth of outgoing EMI and to attenuate incoming EMI. In many respects EMI filtering is like any other, but in a few particulars it is special. Firstly there are statutory requirements to be met by equipment which puts EMI onto the public supply. Secondly the power supplies in equipments often serve several devices or units which are thereby interconnected. Any EMI put onto power lines will be coupled into all of these.

28.5.1 *Power supplies and power lines*

Sensitive units may need a dedicated power supply, rather than a filtered one. Power supplies should have a low internal impedance because fluctuations in one load will immediately couple into any load through this. A power supply connected to a noisy unit may also radiate considerable EMI from its outgoing lines. In general a filter put into a power line without much thought except to cut-offs and roll-offs will reflect much of the unwanted EMI back which can cause worse problems elsewhere. Power-line filters should therefore be designed as far as possible to absorb the unwanted EMI; thus RC or RLC filters are preferable to LC filters. If an RLC filter is used it should be well damped (see Chapter 4, Section 4.3.1) so as not to have much resonant response. Wound inductors in RL and RLC filters may be strong sources of magnetic interference. Ferrite beads or rods coated with a thin metal layer make reasonably effective and cheap lossy filters. A π-filter is best used for high-impedance, and a T-filter for low-impedance, sources and loads to minimise loading effects.

The power leads should be of low characteristic impedance because at high frequencies an instantaneous change in the line current, δI_L, will become an instantaneous (noise) voltage of $Z_0 \delta I_L$, where Z_0 is the characteristic impedance of the power line, given by

$$Z_0 = \sqrt{\frac{L_L}{C_L}} \qquad (28.25)$$

where L_L and C_L are the line inductance and capacitance. For parallel round wires as in Figure 28.11b, where the separation, $x \geq 2d$ (d is the wire diameter), the characteristic impedance is

$$Z_0 \approx \frac{377}{\pi\sqrt{\varepsilon_r}} = \frac{120}{\sqrt{\varepsilon_r}} \ \Omega \qquad (28.26)$$

For polythene, with $\varepsilon_r = 2$, Equation 28.26 gives $Z_0 = 85 \ \Omega$: a fairly large value.

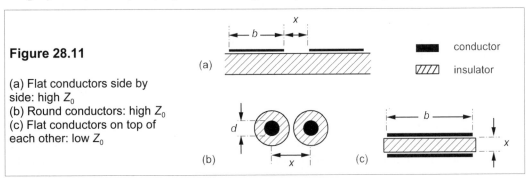

Figure 28.11

(a) Flat conductors side by side: high Z_0
(b) Round conductors: high Z_0
(c) Flat conductors on top of each other: low Z_0

Instead of round wires, thin rectangular conductors can be placed on top of each other as in Figure 28.11c. For these

$$Z_0 = 377x/b\sqrt{\epsilon_r} \quad \Omega \tag{28.27}$$

where x is the separation ($x \gg t$, the conductor thickness) and b is the width of the conductor. When $x = 0.05b$ and $\epsilon_r = 5$ (as in PVC), Equation 28.27 gives $Z_0 = 8\ \Omega$. PCB conductors are often placed side by side as in Figure 28.11a, in which case

$$Z_0 = (377/\pi\sqrt{\epsilon_r})\ln[\pi(x/b + 1)] = 120\epsilon_r^{-1/2}\ln[\pi(x/b + 1)]\ \Omega \tag{28.28}$$

The relative permittivity here is the substrate's (glass-fibre-reinforced epoxy for a PCB). Equation 28.28 gives $Z_0 = 110\ \Omega$ when $x/b = 1$ and $\epsilon_r = 4$ (PCB material). Only when the configuration of conductors is like Figure 28.11c is the impedance low.

28.5.2 Components

The components used in filters are not ideal and each on its own will exhibit self resonance at a higher or lower frequency. We shall consider these in turn, starting with one which is not usually considered to be a 'component' at all: the conductor.

Conductors

Though often overlooked, conductors are essential components which ideally have no resistance, capacitance or inductance. In practice they are not ideal and have resistance, inductance and capacitance. The resistance of a wire at high frequencies is often far more than its DC resistance because of the skin effect. Where the smallest linear dimension is more than twice the skin depth, δ, the conductor resistance is given by

$$R_{AC} = \frac{A}{\text{perimeter} \times \delta} \times R_{DC} \tag{28.29}$$

where the perimeter is that of the cross-section, A, normal to the current flow. Thus for a round wire the AC:DC resistance ratio is $\pi r^2/2\pi r\delta = r/2\delta$, while for a thin rectangular conductor of thickness, t, it will be $t/2\delta$. Flat, rectangular conductors make much better use of copper than do round wires.

The inductance of a conductor depends on its length, diameter and its distance from a return wire or a ground plane, but is approximately 1 µH per m of length. The capacitance of a conductor likewise will depend on its size and position. For parallel round conductors, rectangular side-by-side conductors or conductors above a ground plane it is roughly $70\epsilon_r$ pF per m of length, where ϵ_r is the relative permittivity of the medium separating the conductor from its return path.

Capacitors

In addition to capacitance, capacitors have series inductance and resistance, and parallel resistance too. The inductance depends on the type of construction as well as the lead length.

Leads add about 1 μH/m and dominate the inductance when their total length exceeds about 100 mm. Thus a 10nF capacitor with two 50mm leads will have an induc-tance of about 0.1 μH and be self-resonant at 5 MHz, a fairly modest frequency. Above this frequency the capacitor behaves like an inductance! Shortening its leads as far as poss-ible will make a capacitor self-resonant at the highest possible frequency. Large capacitances are often supplied by electrolytic capacitors which have relatively large inductances and therefore cannot be used above about 10 kHz. Where high-frequency performance is needed too a polycarbonate capacitor of lower value can be used in parallel. Note that high-quality, cylindrical capacitors, which are wound from metal foil and plastic film, have a band round the body nearer the wire attached to the foil. This wire should be the one that is grounded. For the best high-frequency performance, polystyrene capacitors are preferred. However, for effective line filtering in the 100MHz range, lead-through capacitors, costing about £2 each can be used in conjunction with a metal case or shield, as in Figure 28.12a. These capacitors have very low inductance and hence are self resonant only at GHz frequencies. Their construction is shown in Figure 28.12c and the circuit symbol in Figure 28.12d.

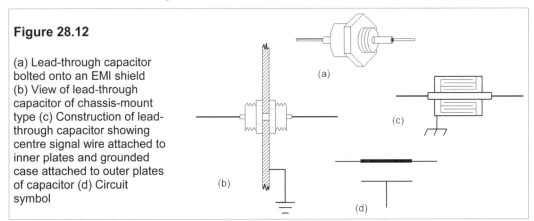

Figure 28.12

(a) Lead-through capacitor bolted onto an EMI shield
(b) View of lead-through capacitor of chassis-mount type (c) Construction of lead-through capacitor showing centre signal wire attached to inner plates and grounded case attached to outer plates of capacitor (d) Circuit symbol

Ferrite beads

Most people are conscious of the non-ideality of inductors and we shall say little here about 'conventional' inductors which have severe frequency limitations when iron-cored (a few kHz). Ferrite beads costing only a few pence apiece are, however, available which can be slipped over a conductor to provide a lossy inductor. At low frequencies, it offers small resis-tance or reactance, but at high frequencies the ferrite bead can act as a resistance in series with the conductor and damp out resonances. An example will show this.

Example 28.3

A round wire 0.5 mm in diameter and 50 mm long feeds a load inside a shielded enclosure via a 10nF, lead-through capacitor. What is the resonant frequency of the line and capacitor if the load impedance is very large? What is the Q of the circuit? What noise will be coupled into the load at this frequency if E_n = 2 mV? A ferrite bead of effective resistance 50 Ω is threaded onto the line to damp the resonance, what then is the noise voltage across the load?

The line has an inductance of 0.05 μH as it is 0.05 m long. Its capacitance is insignificant compared to the 10nF lead-through capacitor's. The circuit is a series resonant circuit, as shown in Figure 28.13

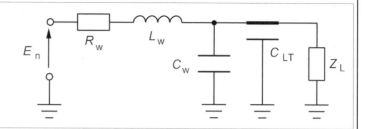

Figure 28.13

The equivalent circuit for Example 28.3. The resistance and inductance are due to the conductor only

The resonant frequency is

$$\omega_0 = (LC_{LT})^{-1/2} = (10 \times 10^{-9} \times 0.05 \times 10^{-6})^{-1/2} = 44.7 \text{ Mrad/s} \qquad (28.30)$$

which is 7.1 MHz. The line is a copper conductor, so its skin depth at 44.7 Mrad/s is

$$\delta = \sqrt{\frac{2}{\mu_0 \sigma \omega}} = \sqrt{\frac{2}{4\pi \times 10^{-7} \times 58 \times 10^6 \times 44.7 \times 10^6}} = 24.8 \text{ } \mu\text{m} \qquad (28.31)$$

The conductor is round and its radius is 0.25 mm, so its DC resistance is

$$R_{DC} = l/\sigma A = 0.05 \times [58 \times 10^6 \times \pi (2.5 \times 10^{-4})^2]^{-1} = 4.4 \text{ m}\Omega \qquad (28.32)$$

Thus the AC resistance is

$$R_{AC} = R_{DC} r/2\delta = R_{DC} \times 2.5 \times 10^{-4}/2 \times 24.8 \times 10^{-6} = 5.04 R_{DC} \qquad (28.33)$$

which is 0.023 Ω. The load is large compared to the shunt capacitance and can be ignored. The circuit's Q is $\omega_0 L/R = 44.7 \times 0.05/0.023 = 97$, a fairly high value. For this high-Q circuit the noise voltage across the reactive elements at resonance is $QE_n = 97 \times 2 = 194$ mV. This will be the noise voltage across the capacitor and also the load. Threading the line with a resistive ferrite bead will increase the series resistance to 50 Ω and so the resistance will increase by a factor of 50/0.023 = 2174 times, and the noise voltage across the capacitor will drop to 194/2174 = 0.089 mV. Inductive beads can be used similarly to make a low-pass filter out of a power line. If more resistance or inductance is needed, more beads can be threaded onto the line.

28.6 Legal requirements for EMC: EN 55014

The legal regulations for EMC are complex as can be seen from the variety of standards that have to be implemented. As an example of these we shall look at EN 55014 covering the conducted or radiated radio interference requirements for all domestic appliances. The spectrum covered is from 150 kHz to 300 MHz. This range is split into two parts, the lower part being considered to be conducted interference and the upper part to be radiated interference. In the lower range, from 150 kHz to 30 MHz, terminal voltages are measured across an artificial mains V-network (an impedance equivalent to 50 Ω resistance in parallel with a 50 μH inductance). From 30 to 300 MHz, the radiated power is measured. Noise is defined as continuous or intermittent and different standards apply. Figure 28.14 shows the conducted noise limits in dB(μV) against log$f$.

The radiated noise limit is 45 dB(pW) at 30 MHz, increasing linearly with frequency to 55 dB(pW) at 300 MHz, though in practice measurements are made only at six 'preferred' frequencies within the range. Because the standard requires that EMI from sources other than the appliance under test must be at least 20 dB below the lowest voltage that it is desired to measure, the tests are likely to require a screened enclosure (or Faraday cage). The standard recognises that this condition 'may be difficult to achieve'. The test procedure for the upper frequency range assumes that the main source of radiation is from the mains power lead and therefore the measurement method used is to place an absorber (ferrite rings) round that lead and measure the absorbed power by substitution.

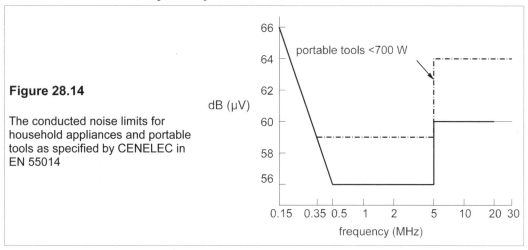

Figure 28.14

The conducted noise limits for household appliances and portable tools as specified by CENELEC in EN 55014

Suggestions for further reading

A very useful bi-monthly periodical is *Approval*, which contains up-to-date information on the latest standards and directives as well as other articles on EMC.

EMC for product designers by T Williams (Newnes, 3rd ed., 2001). Good and not overlong.
Engineering EMC: principles, measurements, technologies and computer models by V Prasad Kodali (Wiley/IEEE, 2nd ed.., 2001)
EMI troubleshooting techniques by M Mardiguian (Tab books, 1999)
Grounding and shielding in instrumentation by R Morrison (Wiley, 4th ed., 1998)
Introduction to EMC by J Scott and T Van Zyl (Newnes, 1997). Short.
EMC analysis methods and computational models by F M Tesche, M Ianoz and T Karlsson (Wiley, 1996). Large and expensive.
Introduction to EMC by C R Paul (Wiley, 1992). Very accessible.

Problems

1. A coaxial cable has a braided outer conductor (or woven shield) which is 0.4 mm thick. If the braid has an effective slot length of 1 mm and effective width of 0.1 mm, at what frequency, f_c, is the waveguide transmission greatest? Below what frequency is the waveguide transmission negligible? What is the waveguide attenuation at a frequency of $0.1f_c$? What do these three values become if the braid is damaged so that the slot length is 3 mm and the slot width 1 mm?
   [150 GHz, 3 GHz, 12 dB, 50 GHz, 1 GHz, 4 dB]
2. At what frequency is the skin depth 0.4 mm in copper at 300 K? What is the attenuation at this

frequency? What are the reflection losses in a plane copper sheet at this frequency? Recalculate these figures for a steel sheet with a conductivity of 2 MS/m and a relative permeability of 300. *[27.3 kHz, 8.7 dB, 124 dB, 2.64 kHz, 8.7 dB, 94.5 dB]*

3. Two round wires, each 250 mm long, are placed side by side as close together as possible. If the conductors are 1 mm in diameter and the insulation is 1 mm thick and of relative permittivity 2.3, calculate the capacitance between the wires. If now one wire has an alternating voltage of 2.5 V across it, and the capacitances to ground can be neglected, at what frequency will the noise voltage in the other wire be 1 V, given that its resistance to ground is 10 kΩ? Approximately what will the noise voltage be at 100 kHz and at 10 MHz, if the capacitance between each wire and ground is 2 pF? *[8.92 pF, 779 kHz, 0.14 V, 2.04 V]*

4. The second wire in the previous problem is screened over 90% of its length by a grounded conductor of negligible thickness. Calculate the capacitance between this screen and the inner wire using the formula

$$C_{2S} = \frac{2\pi\varepsilon_r\varepsilon_0 l}{\ln(d_2/d_1)}$$

where d_1 is the wire diameter and d_2 is the screen diameter. What is the capacitance between the wires now? What is the maximum noise voltage? At what frequency is the noise power half of the maximum? What are the noise voltages at 100 kHz and 10 MHz? *[26.2 pF, 0.892 pF, 85 mV, 607 kHz, 14 mV, 85 mV]*

5. Show that the sum of absorption and reflection losses for a sheet conductor are at a minimum when the shield thickness is equal to the skin depth.

6. Power is fed into a load via parallel, round, insulated wires with a conductor diameter of 1 mm. If the wires are insulated with a material of relative permittivity 2 and thickness 1 mm, what will the noise voltage be if the current in the load changes instantaneously by 1 mA and the separation of the wires is as small as possible? If the power lines are made of thin conductors of rectangular cross-section, 5 mm width, which are placed on top of each other and separated by 0.5mm thick PVC, what will the noise voltage be if the load current changes instantaneously by 1 mA? Take ϵ_r = 5 for PVC. Now suppose that these rectangular power lines are placed side by side on a PCB with ϵ_r = 4, separated by a 5mm gap; what will the noise voltage in them be if the load current changes instantaneously by 1 mA? *[85 mV, 17 mV, 110 mV]*

7. A power line comprises a thin copper strip 0.5 mm thick, 5 mm wide and 100 mm long which is stuck to a PCB with ϵ_r = 4. If the line's inductance is 0.1 μH and the ground plane is on the other side of the PCB whose thickness is 1.3 mm, what is the self-resonant frequency of the line? Take the capacitance to be given by

$$C = \varepsilon A / t$$

where t is the PCB thickness and A is the area of the power conductor. What are the AC resistance and the Q-factor of the line at this frequency? What is the induced voltage in the line at this frequency if the noise voltage is 0.2 mV? The conductivity of copper is 58 MS/m. *[136 MHz, 30.5 mΩ, 2810, 0.562 V]*

8. A piece of equipment suffers EMI from a low-impedance 50Hz source 0.2 m away such that the induced voltage in a critical component is 34 mV. It is decided to reduce the EMI with a copper shield of thickness 1 mm. Is this a 'thin' shield? What will the induced voltage be? If the shield is made from steel with μ_r = 400 and σ = 6 MS/m and the same thickness, what will the induced voltage be now? How thick would these shields need to be to reduce the induced voltage to 1 mV? Is it better to shield the source or the susceptor? *[25 mV, 17 mV, 14 mm, 5.1 mm]*

29 Measurements and instruments

ALL ENGINEERS make measurements and take instrumental readings from time to time. Often the accuracy of these observations is of less importance than their change with time, nevertheless we must know whether the impressions gained from our instruments are reliable or not. This chapter is a guide to what to look for in an instrument and what to expect from a measurement. We shall limit our discussion to a few variables, though these are of the greatest importance to more than electrical engineers, namely, frequency, voltage, current, resistance, capacitance, inductance, magnetic field, temperature and displacement. Before we examine any instruments, however, we need to consider the meaning of certain words associated with measurements in general.

29.1 Accuracy

Most of us have some idea of what we mean by 'accuracy', but it is confused in many minds with precision, resolution, reproducibility and other terms associated with measurements. Let us suppose some apparatus has been set up to make a measurement of voltage. The measuring equipment is in good working order and performs as well as it should. Nevertheless when the apparatus is dismantled and reassembled[1] and another measurement of the same quantity is taken, a series of readings is obtained as in Table 29.1.

Table 29.1 *Experimental data*

Reading No.	1	2	3	4	5	6	7	8	9	10	11	12	13	14	15
Voltage (mV)	88	85	95	94	92	91	87	89	93	86	94	90	92	92	94
Running mean	88	86.5	89.3	90.5	90.8	90.8	90.3	90.1	90.4	90.0	90.4	90.3	90.5	90.6	90.8

The apparatus does not seem to be capable of much *precision*, that is the readings vary over a range from 87 to 98 mV in a random way; but the mean or average of them settles down after ten or a dozen readings to about 91 mV. The mean is defined by

$$\mu = \frac{1}{n} \sum_{i=1}^{n} x_i \tag{29.1}$$

that is, we add up all the readings and divide by the total number of readings, *n*. The most

---

[1] It is essential that the apparatus be disassembled: merely disconnecting the measuring instrument is unlikely to give any idea of the inherent variability of the system.

useful measure of variability is the *variance*, which is defined by

$$\sigma^2 = \frac{1}{n-1} \sum_{i=1}^{n} (x_i - \mu)^2 \qquad (29.2)$$

where σ is the *standard deviation* of the mean[2] and works out to 3.2 mV for this data. That is with this apparatus it is 66% probable that a single reading would lie within 3.2 mV of the mean of 90.8 mV. The standard deviation gives the most likely error in a single reading.

But the standard deviation does not tell us the limits of accuracy for the mean value of 90.8 mV. This is the *standard error* of the mean and is defined by

$$s = \sigma / \sqrt{n} \qquad (29.3)$$

For the data given in Table 29.1, the standard error is $3.2/\sqrt{15} = 0.8$ mV, and the experimental results can be summarised as

$$V = 90.8 \pm 0.8 \text{ mV} \qquad (29.4)$$

The probability is 66% that the true value lies between 89.4 and 91.2 mV. The standard error gives the accuracy in the mean resulting from a number of readings.

There are thus two ways of improving the accuracy of an experiment:

1. Improve the accuracy or precision of the apparatus
2. Take more readings with the same apparatus

Taking more readings can be time-consuming, since to reduce the standard error of the mean by a factor of ten requires a hundred times as many readings, or 1300 in the above case. On the other hand, it may be that buying better apparatus would enable us to achieve the same goal but at greater expense, though more quickly.

Our instrumental readings are limited by the *resolution* of the instruments we use. Resolution means the smallest division that can be read. Thus if a digital voltmeter (DVM) reads a voltage of 1.018 V, when the true value is 1.000 V, the reading has an accuracy of 1.8%. If we change the voltage slightly, to say 1.001 V, the voltmeter reading may change to 1.019 V, so the accuracy is the same, but in both cases the resolution is very much better, being 1 mV, or 0.1%. Without an independent means of checking the results we should be misled into believing that the reading was accurate to about 0.1%, whereas the meter has a *systematic* error of +1.8%. A systematic error biases all the readings in the same way and can prove very hard to discover, let alone eliminate.

On the other hand, we might employ a moving-coil meter to measure the same voltage and find that it reads 1.00 V when the true value is 1.001 V, and does so no matter how many times the experimental arrangement is set up. We should then say that the accuracy is limited by the instrument's low resolution of 10 mV, while the reproducibility is ±0%, but only because the resolution is low.

Professional measurement engineers apart, most people take just a single reading and rely on their instrument, or rather, rely on their knowledge of it and how it is used and maintained. They assume that the data in the manufacturer's handbook will give them a reasonable idea of the error in their (single) reading. If a measurement is at all important, however, it *must* be

---

[2] We divide by $n - 1$ rather than n since we have estimated μ and have thereby lost a *degree of freedom*.

repeated and the instrument must be calibrated against a standard.

Numerical values should be written so that their accuracy is implied by the number of significant figures used. Thus if a voltage is measured to an accuracy of 1%, it is misleading to write it down as 12.13 V (four significant figures), since that implies an accuracy of ±0.1%, and it should have been rendered as 12.1 V (three significant figures). Official bodies are prone to quote far too many figures in their published statistics. When the census figures are published, for example, they are usually given to an implied accuracy of ±1 person, yet the Bureau of the Census in the USA for example admitted to an error of −3% (census figures are invariably under-recorded), which would be about 8 million people! In fact a census is very unlikely to be more accurate than about 1%, and three significant figures are sufficient.

29.1.1 The overall error

Often a quantity is found from a formula such as

$$y = F(a,b,c) \tag{29.5}$$

and we should like to know how to combine the errors associated with our measurements of a, b and c to find the error in y. If the respective errors are s_a, s_b and s_c, then the error in y is

$$s_y^2 = f_a^2 s_a^2 + f_b^2 s_b^2 + f_c^2 s_c^2 \tag{29.6}$$

where $f_a = \partial F/\partial a$, the partial derivative of F with respect to a, etc.

When $F(a,b,c)$ is a product of powers such as

$$F(a,b,c) = a^u b^v c^w \tag{29.7}$$

then $f_a = uF/a$, etc., in which case Equation 29.6 becomes

$$s_y^2 = u^2 s_a^2 + v^2 s_b^2 + w^2 s_c^2 \tag{29.8}$$

where s_y ($= s_y/y$) is now the relative error in y, etc.

Example 29.1

The current through a resistance has been measured and found to be 2.13 ± 0.02 A, and then the resistance has been measured and found to be 33 ± 0.5 Ω, what is the power consumption and what is its likely error?

The power is given by $P = I^2 R$ and the partial derivatives are

$$f_I = \partial P/\partial I = 2IR , \qquad f_R = \partial P/\partial R = I^2 \tag{29.9}$$

which work out to $f_I = 2 \times 2.13 \times 33 = 140.58$ V and $f_R = 2.13^2 = 4.5369$ A². Using Equation 29.6 then gives

$$s_P = \sqrt{(140.58 \times 0.02)^2 + (4.5369 \times 0.5)^2} = 3.613 \text{ W} \tag{29.10}$$

The relative errors are $s_I = 0.02/2.13 = 0.94\%$ and $s_R = 0.5/33 = 1.52\%$. And we could use Equation 29.8, with $u = 2$ and $v = 1$, to find

$$s_P = \sqrt{(2 \times 0.94)^2 + (1.51)^2} = 2.41\% \tag{29.11}$$

> The error in the power is 2.41%. The power is $2.13^2 \times 33 = 149.7$ W, so the relative error from Equation 29.12 is $3.61/149.7 = 0.0241 = 2.41\%$, agreeing with that found in Equation 29.13. The result can be written either as $P = 149.7 \pm 3.6$ W or as $P = 149.7$ W $\pm 2.41\%$. Little information is lost if these are rendered as $P = 150 \pm 4$ W or $P = 150$ W $\pm 2.4\%$.

29.1.2 The role of individual judgement

It happens sometimes that a series of measurements yields, say, a straight line graph which when plotted has one point which lies far from the line that would be drawn through the rest. What does one do about that point? It may be that including it in the results would lead to a substantial reorientation of the line, which could well be misleading, yet not to include it is a breach of professional etiquette. The best course of action might be to acknowledge the outlier by including it on the graph, but to ignore it when calculating the line of best fit, and make this clear when presenting the results. The observer is the person in the best position to decide on the most appropriate course of action.

29.2 Standards and transducers for measurements

Standards used in industry are working standards that are from time to time compared to standards kept in National Standards Laboratories such as the National Physical Laboratory in England. These serve to calibrate instruments which are used everywhere, but they can be supplemented by non-instrumental standards that are relatively cheap and easily maintained. There are also other non-standard means for measuring quantities which use various transducers[3]. We shall list these briefly below.

29.2.1 Voltage

The primary standard maintained at the National Physical Laboratory (NPL) in the UK, and in its equivalents elsewhere, is the Josephson junction. This consists of two superconducting thin films separated by an insulating layer about 1 nm thick. When a direct voltage is applied to the junction the current flowing across it oscillates at a frequency given by $f = 2Vq/h$ where q is the magnitude of the electronic charge ($\approx 1.601 \times 10^{-19}$ C) and h is Planck's constant ($\approx 6.626 \times 10^{-34}$ Js). The frequency can be measured very accurately and hence the voltage also (to about 1 part in 10^8). This primary standard is used to calibrate Zener diodes as a secondary standard. The Zeners are temperature compensated with a forward-biased series diode and can be kept in a temperature-controlled enclosure (to ± 10 mK). When provided with a constant-current source these can give voltages which are stable to about 1 in 10^6 over the range 2.4–12 V. 'Millivolt' sources are cheap and reasonably accurate (to about 0.1%), using bandgap-reference diodes and precision resistors to give reference voltages in the range from 1 mV to 1.200 V.

---

[3] A transducer is any device which converts one physical quantity into another, usually for the purposes of measurement or qualitative display.

29.2.2 *Current and resistance*

Standards for current are not generally available and they are measured by means of standard resistors and voltmeters. A standard resistor is fairly expensive compared to precision resistors. It is made from manganin wire having a low temperature coefficient of resistance, wound non-inductively with a large spacing between turns to minimise capacitance. The resistor is kept in an oil-filled can provided with a thermometer, so that the resistor's temperature is both known and uniform. A certificate giving its resistance at various temperatures should be provided. Typically, the temperature coefficient of resistance is about +10 ppm/K. Four terminals are provided, the outer two supply current and the inner two are used for voltage measurement. The accuracy is about 0.001% if used at rated current.

29.2.3 *Capacitance and inductance*

Standard capacitors and inductors can be made whose capacitance and inductance may be calculated from their geometry. Standard air-dielectric capacitors are available up to 1 nF and mica-dielectric capacitors up to 1 μF. The accuracy with which these can be made is astonishingly high, long term accuracy being about 1 in 10^5. Inductors are more difficult to make with the same precision and accuracies are restricted to about 1 in 10^3. If inductors are to be used for AC measurements, the wire size must be small compared to the skin depth at the operating frequency.

29.2.4 *Frequency*

Quartz crystals are cheap and highly accurate standards for frequency (and hence time). If a temperature-controlled enclosure is provided an accuracy of 1 part in 10^7 is achievable and maintainable for years if required. Frequency counters are often accurate to 1 part in $10^6 \pm 1$ count. The determination of frequency is always the most accurate routine measurement available.

29.2.5 *Temperature*

The mercury-in-glass thermometer is cheap (under £5) and accurate to about 0.25 K. Though it has a slow response time, it is a useful means of calibrating more convenient sensors such as thermistors. The standard temperature sensor is, however, the platinum resistance thermometer, which covers an enormous range (from about 20 K to 800 K) with an accuracy of about 10 mK if its resistance can be accurately measured. Its temperature coefficient is about 0.4%/K at 300 K and it is very cheap in thin-film form (about £3). Thin-film platinum thermometers are mounted on alumina which is an insulator with high thermal conductivity. They are readily encapsulated in plastic to provide electrical insulation and may then be immersed in all sorts of fluids. They suffer from bulkiness and hence slow response time just like mercury-in-glass thermometers, but they are the best standard for calibrating thermistors and thermocouples. To achieve a resolution of 10 mK with a thermometer of 100 Ω at 0°C requires a resistance measurement to ±1 mΩ or better.

Thermistors are very convenient and highly sensitive resistance thermometers which can have response times of 0.1 s or less. Their resistances obey the equation

$$R = A\exp(B/T) \qquad (29.12)$$

where R is in Ω and T in K. The resolution can be very high; for example (see Problem 29.6), a 10 kΩ (at 20°C) thermistor has a temperature coefficient of resistance of about 30 Ω/K at room temperature and temperature changes of 0.1 mK are detectable. Though they can be fragile they are reasonably cheap – about £4 apiece. They must be calibrated by the user if any accuracy at all is desired.

Thermocouples are also handy thermometers since they are small and normally respond as fast as small thermistors. There are various standard types such as chromel/alumel (type K), and platinum/platinum-rhodium (type R), for which standard tables are available. The latter are much more expensive than the former. A thermocouple produces an e.m.f. which depends on the junction temperature and junction materials, but its change with temperature is not great (10 μV/K for type R and 40 μV/K for type K). Cold-junctions or cold-junction compensation must also be provided. The great advantage of these thermometers is that they can be used up to 1800°C (type R) or higher with special materials.

29.2.6 Magnetic field

Magnetic field intensities can be measured to about 1% with a Hall-effect probe. For precise measurement an electronic integrating fluxmeter and search coil can be used (see Problem 13.4). If the coil is made with sufficient precision and the integrator is made with low-leakage components then an accuracy of about 0.1% is attainable. However, by using nuclear magnetic resonance (NMR), which requires a small (about 1 cm^3) sample with a coil wound round it to detect the frequency of proton resonance, it is possible to achieve an accuracy of about 1 part in 10^6.

29.2.7 Displacement

Displacement can be measured with linear voltage differential transformers (LVDTs), which can measure displacements of up to 20 mm with a resolution of 0.01% and linearity. They are priced at about £50 and require additional electronics to provide excitation (a few kHz) and readout. They can be checked with dial gauges (resolution about 10 μm) or travelling microscopes (resolution about 1 μm).

29.3 Instruments

Whatever we are measuring and with whatever instrument, it is necessary to ask, 'Under what conditions is the stated accuracy obtained?' The most important considerations are the frequency and the effect of the instrument on the circuit, component or experimental set-up being investigated. Provided the instrument is properly maintained and used within its limitations, the accuracy given in the manufacturer's handbook can usually be relied on. In most cases calibration can be obtained at modest cost from suppliers or manufacturers, or at independent laboratories.

29.3.1 *The moving-coil meter*[4]

These have largely had their day and been supplanted, especially in the case of multimeters, by instruments based on op amps and analogue-to-digital converters; however moving-coil meters are still often found and are more convenient for indicating slowly-changing voltages or currents than are digital panel meters. The movement (known as the *d'Arsonval* movement, see Figure 29.1) of a coil of N turns and area A, wound onto a permeable soft-iron cylinder which is free to rotate within the pole-pieces of a permanent magnet. The torque exerted on the coil when a current, I, flows through it is given by

$$T = BANI \qquad\qquad (29.13)$$

where B is the magnetic flux density in the cylinder. This torque is opposed by a helical spring attached to the cylinder, which exerts a restoring torque of $K\theta$, K being the spring constant and θ the angle of rotation. Thus equating torques leads to

$$\theta = BANI/K \propto I \qquad\qquad (29.14)$$

since for a given instrument B, A, N and K are constant.

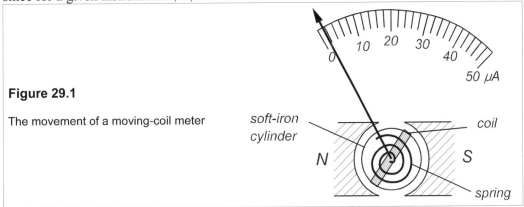

Figure 29.1

The movement of a moving-coil meter

soft-iron cylinder

coil

N S

spring

Multimeters typically require a direct current of about 50 μA for a full-scale deflection and have resistances of about 20 kΩ/V which means 50 kΩ on the most sensitive range (2.5V full-scale), rising to 1 MΩ on the 50V range. Voltage ranges are changed by switching in different resistances in series with the coil. This relatively low resistance can result in serious loading effects when measuring high-resistance voltage sources. Currents are measured by placing the meter in the circuit in series and employing shunt resistances of a suitable size. The instrument can then measure currents down to a 50μA full-scale deflection. On this range, however, the instrument presents maximum resistance, namely that of the coil, which may be 200 Ω.

Resistance can be measured with the so-called series circuit of Figure 29.2. Before each measurement the terminals are short circuited and R_p trimmed to give a reading of 0 Ω. The meter deflection is given by

---

[4] Moving-iron instruments work on the same principles as moving-coil instruments, but the coil is static and induces poles on two pieces of soft iron, one static and the other attached to a pointer. The neighbouring induced poles on the two pieces of are of opposite polarities, which causes repulsion. However, the repulsive force is proportional to the square of the current, resulting in a non-linear scale.

$$I_{\mathrm{M}} = \frac{V}{(R_{\mathrm{x}} + R_{\mathrm{s}})(1 + r) + R_{\mathrm{M}}} \tag{29.15}$$

where R_{M} is the meter resistance and $r \equiv R_{\mathrm{M}}/R_{\mathrm{p}}$. In terms of the full-scale deflection, R_{F}, this is

$$I_{\mathrm{M}} = \frac{[(1 + r)R_{\mathrm{s}} + R_{\mathrm{M}}]I_{\mathrm{F}}}{(1 + r)(R_{\mathrm{x}} + R_{\mathrm{s}}) + R_{\mathrm{M}}} \tag{29.16}$$

R_{s} is adjusted to give a suitable half-scale deflection, which can be derived from Equation 29.18 by setting $I_{\mathrm{M}} = 0.5I_{\mathrm{F}}$. The ohmmeter is most accurate at this point (see Problem 29.7). The resistance scale is inevitably non-linear.

AC measurements by multimeters are made by rectifying the input current with a full-wave, or half-wave, rectifier. In some meters the voltage drop across the rectifying diodes is compensated for by using a step-up transformer, which causes the input resistance to be rather small: about 500 Ω on the 2.5V range and only 2 kΩ/V on the higher ranges. The bandwidth of the instrument is limited to about 20 kHz. The meter deflection is proportional to the time-average of the input and the scaling resistors are adjusted to give r.m.s. values of sinusoidal input waveforms, in consequence the meter reads incorrectly all other waveforms (square waves are indicated too high by a factor of $\pi/2\sqrt{2}$, for example).

The moving-coil multimeter can be improved by using high-impedance amplifiers on its inputs, which may extend its range of application somewhat, but the bandwidth is not improved by this means, nor is the resolution. In any case, moving-coil instruments with small scales cannot be expected to give better resolution or accuracy than about 1%.

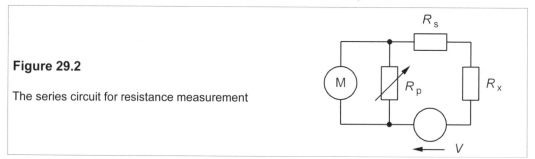

Figure 29.2

The series circuit for resistance measurement

29.3.2 *The electrodynamic wattmeter*

The dynamometer movement used in the low-frequency (usually line frequency) wattmeter was invented by the one of the Siemens brothers[5] as long ago as 1843. The load current is passed through two series current coils which are coaxial and between which a high-resistance voltage coil is free to rotate, as in Figure 29.3.

The field created by the current coils exerts a torque on the voltage coil, as this carries a small current. Thus the coil deflection is given by

---

[5] There were four Siemens brothers whose contribution to many branches of technology was enormous. They were born in Hanover between 1816 and 1829. Karl Wilhelm Siemens (1823–83) later became a British citizen and was knighted, being known as Sir William Siemens. He is commemorated by a window in Westminster Abbey.

$$\theta = Ki_{L}i_{v} = \frac{Ki_{L}v_{L}}{R_{v} + R} = ki_{L}v_{L} = kp_{L} \qquad (29.17)$$

where K, k are constants and i_{v} is given by $v_{L}/(R_{v} + R)$, R_{v} being the resistance of the voltage coil. The deflection would be proportional to the instantaneous power if the movement were capable of rapid response. In practice the movement cannot respond to line frequencies or

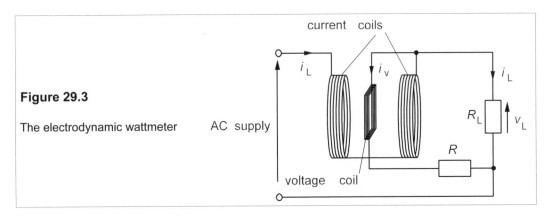

Figure 29.3

The electrodynamic wattmeter

above and so the voltage coil reading is a time average and so is the indicated power, regardless of the waveform. The meter scale can therefore be calibrated to read watts. The current coils have little reactance up to a few kHz, the frequency limit of the instrument.

29.3.3 *The digital multimeter*

With the advent of cheap op amps and LEDs, the digital multimeter soon became ubiquitous. More measurement functions have been added to them over the years, so that not only are DC and AC voltages and currents and resistance available, but also capacitance, frequency, conductivity, decibels and even temperature are available on many meters.

Since their first appearance the power consumption of portable DMMs has been greatly reduced by using passive, liquid-crystal displays and low-power CMOS ICs. Hand-held meters now contain most of the features formerly found only in bench meters, and are very popular as a result. The absence of a moving-coil has resulted in greater robustness, the absence of the needle and scale has led to greater resolution, and the use of precision resistors and low-drift ICs has resulted in improved accuracy. For modest cost the performance available is quite remarkable, as Table 29.2 shows. The more expensive meters often have an RS 232 or IEE 488 interface so that data can be output and some can be programmed to perform a range of measurements at intervals over an extended time. Calibration traceable to national standards is usually provided for a fee for any multimeter in the table.

Table 29.2 *Digital multimeters*

Make	Dig.	Acc.	Res.	Freq.	Price	Comments†
Wavetek DM7	3½	0.8%	1000	-	£20	hand-held (HH)
Isotech 63N	3½	0.5%	100	-	£40	hand-held, diode test
Isotech IDM101	3¾	0.7%	100	1000	£99	hand held, *C*, *f*, memory
Wavetek 2030	4½	0.1%	10	2000	£280	hand-held, trms. *C*, *f*
Fluke 8060A	4½	0.1%	10	100	£400	hand-held, trms, diode test, *f*
Wavetek DM85XT	4½	0.05%	10	200	£129	hand-held, trms
Thurlby 1705	4½	0.04%	10	200	£300	bench, trms, *C*, *f*,
Fluke 187	5	0.025%	1	1000	£254	HH, trms, *C*, *f*, *T*, duty cycle, PC i/f
Metrix M56B	5	0.025%	1	500	£178	HH, trms , *P*, *f*, mma, duty cycle
Thurlby 1906	5½	0.012%	1	20	£395	bench, trms, mem., data, mma
HP 34401A	6½	0.0035%	0.1	20	£795	bench

Notes: '3½ digits' implies a full-scale reading of 199.9 mV on the most sensitive voltage range. Accuracy and resolution (in µV) are for DC voltage only. Frequency is max. in kHz. Nearly all meters have bargraphs. † trms = true r.m.s., *P, C, f, T* = power, capacitance, frequency and/or temperature (using type K thermocouple) also measured, mma = minimum, maximum and average readings stored.

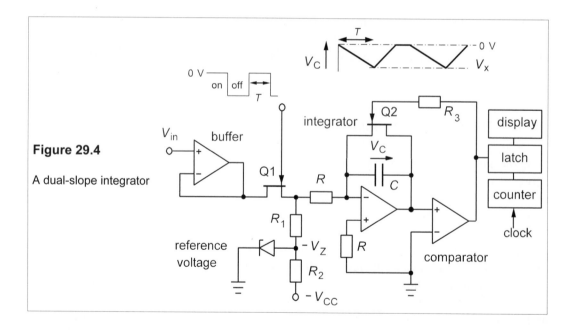

Figure 29.4

A dual-slope integrator

Digital voltmeters frequently employ the technique of *dual-slope integration*, which can be performed by a circuit such as Figure 29.4 shows. The input voltage is buffered to give an input resistance of the order of 10 MΩ, then fed to an integrator via Q_1, which is switched on for a time, T. The capacitor charges up to a voltage

$$V_x = \frac{-1}{RC}\int V_{in}\,dt = \frac{-V_{in}T}{RC}$$

(29.18)

Q_1 is then switched off and the capacitor discharges at a rate set by the reference voltage, $-V_Z$, and the resistors R and R_1. The discharging current is

$$I_{ref} = -V_Z/(R + R_1) \tag{29.19}$$

So the time taken to discharge from V_x to zero is

$$T_x = CV_x/I_{ref} = CV_x(R + R_1)/V_{ref} \tag{29.20}$$

Substituting for V_x from Equation 29.18 leads to

$$T_x = \frac{CV_{in}(R + R_1)T}{RCV_{ref}} = \frac{(R + R_1)T}{RV_{ref}}V_{in} \tag{29.21}$$

The discharging time is proportional to the input voltage, and independent of C and the clock frequency, provided it is constant. R_1 is usually factory-adjusted to allow some leeway in V_Z, which varies a little from Zener to Zener.

When the capacitor voltage, V_C is non-zero, the comparator puts out a negative voltage that keeps Q2 switched off, but when the capacitor is fully discharged the comparator puts out a positive voltage and turns Q2 on, keeping the capacitor voltage at zero. At the same time the counter output is latched and displayed. At the end of $2T$ seconds Q1 is switched on again and the cycle repeated. Since T is determined by the number of counts registered, it is not important that the clock frequency be adjusted accurately. For example suppose $V_{ref} = 5$ V, $T = 2000$ counts, $R = 10$ kΩ, $R_1 = 15$ kΩ and $T_x = 1234$ counts, then from Equation 29.21, $V_{in} = 1.234$ V. The 3½-digit display is normally updated every 0.5 s, so a suitable clock frequency would be 8 kHz. By using higher clock frequencies, higher-precision resistors and a temperature-compensated Zener or bandgap-reference diode, together with a 5½-digit display, the resolu-tion can be improved a hundredfold with at least a tenfold improvement in accuracy. To maintain accuracies of 0.01% or better, it is necessary to have the instrument recalibrated at least every 3 months.

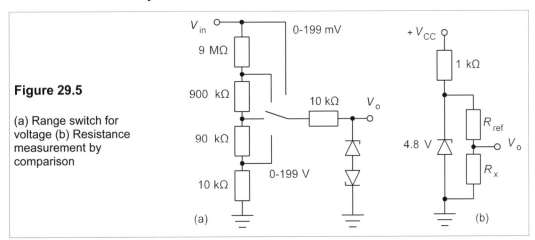

Figure 29.5

(a) Range switch for voltage (b) Resistance measurement by comparison

Currents can be measured by employing a shunt resistor to generate an input voltage to the circuit of Figure 29.4, and a wide measurement range is readily achieved by adding a range-switching op amp after the buffer. But in practice this is not done because DVM ICs (and panel meters which include the display) are available and range switching is accomplished with a potential-divider resistor network on the input such as in Figure 29.5a, which gives an input resistance of 10 MΩ. It is inadvisable to employ resistors of higher value than about 10 MΩ as surface leakage currents then become significant. Resistance can be measured with the circuit of Figure 29.5b, which is accurate to about 0.2% if a pair of 2.4V bandgap-reference diodes is used. In this circuit $V_o = V_Z R_x/(R_x + R_{ref})$, so if $V_Z = 4.8$ V, then $R_{ref} = 2.8$ kΩ for the 2kΩ range, $V_o = 28$ kΩ for the 20kΩ range and a range switch is required. R_x can be looked up in a table if V_o is read and a microcontroller is used in the DMM. Resistances up to about 20 MΩ can be accommodated, but if a 200Ω range is to be used the series 1kΩ resistor, which holds the Zener current around 5 mA, must be switched out.

AC measurements in a DMM are accomplished by rectification, just as in a moving-coil multimeters, but usually a 'precision' (meaning the forward-voltage drop of the diodes is not subtracted from the output) rectifier is used such as that of Figure 29.6.

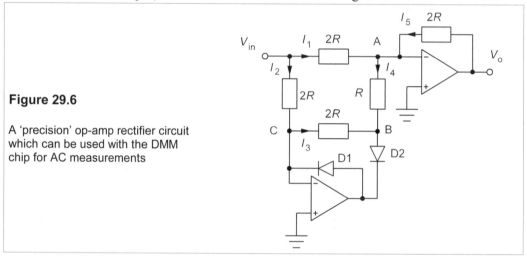

Figure 29.6

A 'precision' op-amp rectifier circuit which can be used with the DMM chip for AC measurements

First note that points A and C are virtual grounds and at a potential of 0 V, making $I_1 = I_2 = V_{in}/2R$. If D2 is reverse biased, no current flows through it and so I_3 and I_4 are both zero. I_2 must then be negative (and so must be V_{in}) and flow through D1, which is forward biased. But $I_2 = V_{in}/2R$, implying that V_{in} is negative. Also $I_1 + I_5 = 0$, and $I_5 = V_o/2R$, while $I_1 = V_{in}/2R$, meaning that $V_o = -V_{in}$. When D2 is forward biased, $I_3 = -V_B/2R$, where V_B is the potential at B, and $I_4 = -V_B/R = 2I_3$. This means V_B is negative and D1 is reverse biased. As D1 then conducts no current, $I_2 = I_3 = I_1 = V_{in}/2R$. Now $I_4 = 2I_3 = 2I_1$, and using Kirchhoff's current law at B leads to $I_5 = I_4 - I_1 = I_1$. Since $I_5 = V_o/2R$ and $I_1 = V_{in}/2R$, we see that when V_{in} is positive, $V_o = V_{in}$ and the circuit is indeed a precision rectifier.

29.3.4 *The cathode-ray oscilloscope (CRO)*

The CRO has considerable advantages over multimeters as a test and diagnostic instrument, and even as a measuring instrument in some cases, because it gives an immediate picture of a waveform as a function of time, making distortion and noise immediately recognisable, which is not the case with a multimeter. Though often considered to be of low accuracy (c. ±5%), when properly used and calibrated, a CRO can approach an accuracy of ±1%;

moreover calibration with a sine-wave source and an ordinary DVM is relatively quick. The input impedance of a cheap CRO is typically 1 MΩ in parallel with 20 pF, giving a bandwidth of about 10 MHz. Not only can voltage waveforms be examined, but also phase relationships by the use of a dual-beam oscilloscope with two input channels. Waveforms can also be added and subtracted and X-Y plots can be made readily, even on the cheapest instruments. Of course, far better performance, and more facilities, can be obtained for a larger capital outlay.

The heart of a CRO is an evacuated cathode-ray tube with an electron 'gun' at one end and a flat screen at the other, with accelerating, focussing and deflecting plates in between, as shown in Figure 29.7. The vacuum in the tube must be high enough that very few electrons collide with gas molecules as they travel from cathode to screen.

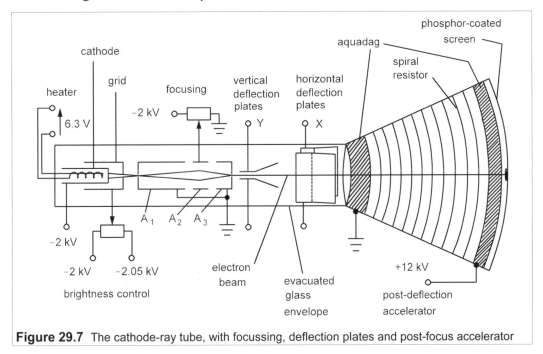

Figure 29.7 The cathode-ray tube, with focussing, deflection plates and post-focus accelerator

Electrons are drawn off the heated, oxide coated, nickel cathode by the large potential difference between it and the pre-focus accelerator (anode, A_1). They pass through a tiny hole (about 0.2 mm diameter) in the nickel can, which is the grid, and whose potential can be slightly adjusted by the brightness control to give a beam with more or fewer electrons. The beam converges then diverges before entering the pre-focus accelerator. After this the beam enters a focussing electrode (A_2) whose potential is adjusted by the focus-control knob, by which the beam can be brought to a spot of minimal diameter on the screen. Anode A_3 provides further acceleration and focussing. After this the beam of constant-velocity electrons is deflected by the horizontal (or X) and vertical (or Y) deflection plates. The horizontal plates are usually connected to a time base which is a saw-tooth or repetitive ramp voltage. The input signal is connected via suitable amplifiers to the Y-plates. The electron velocity after deflection is not high enough to 'write' on the screen and a post-deflection accelerator is used in which a high positive voltage (such as +12 kV) is applied to a spiral resistor wound onto a diverging cone inside the tube. The spiral is made from a conducting graphite paste called 'aquadag' which is baked onto the cone. The beam strikes a glass screen coated with an electroluminescent substance called a phosphor. The phosphor emits light when struck by energetic electrons and continues to do so for a short time afterwards (the *persistence*).

The electron has a very high charge-to-weight ratio which means that it is very rapidly accelerated by the potentials applied. The energy acquired when the electron rises in potential by V volts is qV, where q is the electronic charge. The kinetic energy of the electron is $\frac{1}{2}mv^2$, so that

$$qV = 0.5mv^2, \quad \Rightarrow \quad v = \sqrt{2qV/m} \tag{29.22}$$

Substituting for $V = 14$ kV, $m = 9.1 \times 10^{-31}$ kg and $q = 1.6 \times 10^{-19}$ C, leads to $v = 7 \times 10^7$ m/s ($\approx 25\%$ of the speed of light). The transit time from cathode to screen is only about 10 ns! The ease and speed of writing onto a screen by an electron beam, more than any other factor, has led to the continued use of the CRO long after the demise of most vacuum-tube devices.

Oscilloscope controls

This section refers to the inexpensive, basic two-channel CRO, the most obvious part of which is the screen, divided into a grid (or *graticule*) of 10 mm $\times$ 10 mm squares, usually eight vertically by ten horizontally. These are the 'divisions' referred to on the front panel by the multi-range knobs which control the timebase ('TIME/DIV', from 0.2 μs/cm to 100 ms/cm, in a 1–2–5 sequence) and the voltage gain of the channel 1 and channel 2 amplifiers ('VOLTS/DIV', from 5 mV/cm to 5 V/cm). By using the fine divisions on the centre lines it is possible to measure to 0.1 division or 1 mm, and thus the smallest interval of time that can be measured is about 20 ns, and the smallest voltage about 500 μV. The inputs to the channels are via female BNC bulkhead connectors on the front panel and these require the appropriate male BNC connector on the input cable, which is almost always 50Ω coaxial cable. The outer (screen) of the coaxial cable is grounded when connected to the CRO as can easily be checked with an ohmmeter connected between the earth pin of the CRO plug and the screen connector of the cable. The inner wire of the coaxial cable (the *signal* lead) should be connected to whatever voltage is to be measured, and this voltage is conveyed via amplifiers to the Y-deflection plates. The X-plates are connected to the timebase ramp and this deflects the beam from left to right across the screen. The beam can be adjusted by the 'FOCUS' control to give the sharpest possible line, and the line can be made horizontal with the 'ROTATE' control.

Often the *trace*, that is the $V(t)$ graph displayed on the screen, is not visible at first. Assuming that the instrument is on and warmed up (only a few seconds needed), the best way to find the trace is to ground the input (there is a switch next to the channel input labelled 'GD-AC-DC'), set the trigger AT/NORM control to 'AT' (automatic) and turn the appropriate '↕ POS.' or 'Y-POS.' (vertical position) knob up and down until a horizontal line can be seen. If it cannot be seen try turning the 'INTENS.' (intensity) control up, it may be too faint a trace. Set the trace in the centre of the screen and then revert to DC coupling. If the trace disappears it may be that the timebase is too fast – reduce it – or the gain is too high – reduce that too. If the 'X-Y' setting has been selected only a spot will appear, which can quickly burn out the phosphor, so switch this off at once.

Triggering is needed to start the voltage-time trace at the same point of the waveform (assuming it is periodic), so that successive traces overwrite. If the triggering point were random successive traces produced by the time base would be written all over the screen. If the trace is not triggering properly, for the trigger setting is often one that causes difficulty, the waveform may not be displayed at all. Check first that you are triggering off the channel being used and not on the other channel or on the external trigger ('EXT'). The trigger can be set to 'AC', 'DC','LF', 'HF'. There is usually a 'TV' setting for TV video signals and the LINE setting triggers off the main supply. It is best to select AC unless the triggering is from a very slow edge or a pulse of changing duty cycle, when DC should be used along with 'NORMAL' and a manually-adjusted trigger level. If the mode is AT (automatic) there should be a trace displayed. If you want to select your own trigger level, then you must switch out of AT to 'NORMAL'. Then you must turn the trigger 'LEVEL' knob until the trace appears. If you now turn the '↔ POS.' or 'X-POS.' (horizontal position) knob until you can see the beginning of the trace on the left, you will find that turning the trigger level control causes the triggering point to move up and down the positive slope of first bit of the voltage waveform. Switching the '+/–' slope control from '+' to '–' will cause triggering off the negative slope part instead of the positive slope of the waveform.

Each of the three range-switching knobs (CH1, CH2 and the timebase) has a calibration knob in its centre. In order to read times and voltages correctly these knobs must be set to 'CAL'. As you can see by trying, the calibration knob alters the amplification factor to any desired value between range settings. To measure a voltage accurately with respect to ground, you have to ground the trace by setting the input coupling to GND and turn the vertical position knob until the horizontal line lies along a horizontal line of the graticule. Then positive voltages are read upwards and negative downwards. With sinewave one normally reads peak-to-peak (really peak-to-trough) voltages which *must be divided by $2\sqrt{2}$* if you want the r.m.s. value. Should there be a large DC component on a waveform of interest (such as ripple on a rectifier) AC coupling can be used (by selecting with the GD-AC-DC switch), which inserts a capacitance of about 20 nF in series with the signal to cut out the DC component.

There is probably a probe-calibration point (or two next to each other) on the front panel marked 'CAL' '0.2 V' '2 V', which will supply $2V_{pp}$ and 200 mV$_{pp}$ square waves at 1 kHz. These are used to set up accurately CRO probes. Probes are useful for clipping onto or just touching points in a circuit for test purposes and they eliminate the circuit-loading effects of the input capacitance (see Problem 2.9). They may be 1×, 10× or 100× probes and will have a screw-adjustable capacitor, which can be set up with the calibration square wave. If this square wave is distorted when connected via the probe to one of the CRO inputs, the probe's adjustment screw can be turned to restore the squareness. Sometimes a probe can only be used with a particular CRO, so be warned!

The X-Y mode can be selected either with the TIME/DIV knob or with an 'X-Y' switch, and it enables the input of one channel to be displayed as a function of the other. Some CROs have CH1 as X and CH2 as Y, but others vice versa. With this setting one can, for example display current-voltage graphs for diodes and transistors.

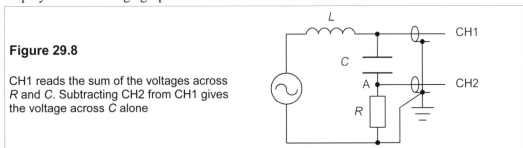

Figure 29.8

CH1 reads the sum of the voltages across *R* and *C*. Subtracting CH2 from CH1 gives the voltage across *C* alone

The ability to subtract signals from each other can be useful when one cannot ground a component as in the example of Figure 29.8. As set up the voltage displayed on CH1 will be the current (proportional to R) and the voltage displayed by CH2 will be the sum of the resistor and the capacitor voltage. To display the capacitor voltage alone is not possible because placing the screen of the input at point A will short-circuit the current probe. However, by subtracting CH2 from CH1 the capacitor voltage can be displayed. The channel VOLTS/DIV settings must be the same, though. Subtraction is achieved by using the 'ADD' switch and inverting CH2.

Phase differences between signals of the same frequency on CH1 and CH2 can be measured by making the traces the same amplitude with the CAL knob of one of the channels and then measuring the difference in time between the zero crossings of the two traces (this is the most accurate method). Phase differences can also be measured by using the X-Y setting to generate *Lissajous*[6] figures. When the frequencies of CH1 and CH2 are the same, the figures are ellipses in general as in Figures 29.9a-e.

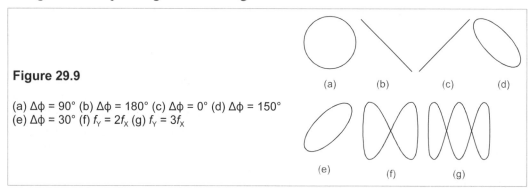

Figure 29.9

(a) $\Delta\phi = 90°$ (b) $\Delta\phi = 180°$ (c) $\Delta\phi = 0°$ (d) $\Delta\phi = 150°$
(e) $\Delta\phi = 30°$ (f) $f_Y = 2f_X$ (g) $f_Y = 3f_X$

To measure the phase angle between two waveforms the maximum height, $2Y_2$, and the Y-axis intercept, $2Y_1$, are measured as in Figure 29.10 and the phase angle is $\sin^{-1}(Y_1/Y_2)$. The ellipse must be centred about the Y-axis. There is some ambiguity in phase as the Lissajous figures for phase differences of $\Delta\phi$ and $360° - \Delta\phi$ are the same. When the frequencies are different on CH1 and CH2, the figures are as in Figure 29.9f and 29.9g. The frequency ratio is given by dividing the number of peaks on top by the number on the right-hand side.

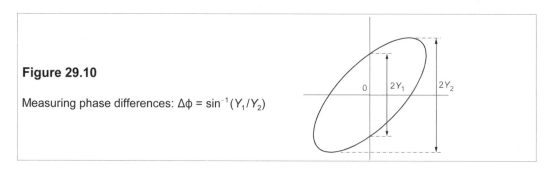

Figure 29.10

Measuring phase differences: $\Delta\phi = \sin^{-1}(Y_1/Y_2)$

29.3.5 True r.m.s. meters

True r.m.s. meters now use an analogue-to-digital converter (see Section 10.6) and a customised ASIC (application-specific IC) to turn the digital samples into an r.m.s. reading. The A/D must have sufficient bits to give the required accuracy and there is usually some restriction on the *crest factor* (the ratio of peak to r.m.s value of a waveform) to achieve the stated accuracy. In some instruments the functions are controlled and the readings are processed by microcontrollers, and in time perhaps all multimeters will be microcontroller based, particularly when 16-bit and 32-bit devices become cheaper.

29.3.6 Automatic bridges

We shall not deal with manual bridges in this chapter as the circuits have been dealt with elsewhere (in Chapter 2 for example), and in any case their use is declining as microprocessor-based instruments of high accuracy become more and more available, though admittedly only over a restricted frequency range. Automatic bridges can be bought for about £700 and will measure *R, L, C, Q* and tanδ at one or two fixed frequencies such as 1 kHz with a basic accuracy of ±0.25% + 1 digit, falling off at both low and high values. Their great advantage lies in the rapidity with which the measurement can be made and the instruments are well suited to manufacturing process control and may be used by unskilled personnel with a little training. The instrument measures the voltage across and the current through the component under test by means of sinusoidal excitation at a precisely-known frequency. Two square waves differing in phase by 90° are generated at the same time as the exciting sinewave and the measured voltage and current are compared to these references. Thus the in-phase and quadrature components of both current and voltage can be determined and hence the value of the resistance and reactance of the component. The sign of the reactance indicates whether the component is inductive or capacitive and its value is calculated from the reactance and the test frequency.

Suggestions for further reading

Measurement and instrumentation principles by A S Morris (Newnes, 3rd ed., 2001) Very good value.
Elements of instrumentation and measurements by J J Carr (Prentice Hall, 3rd ed., 1998). Comprehensive but expensive.
Electronic instrument design by K R Fowler (OUP USA, 1996). The whole process.
Handbook of modern sensors by J Fraden (Am. Inst. of Physics, 2nd ed., 1996). Expensive, thorough.
Sensors and transducers by M J Usher (Palgrave, 2nd ed., 1996). Short, covers most of the field.
Principles of engineering instrumentation by D C Ramsay (Arnold, 1996). Clear and concise.

Problems

1. A voltage of 3.63 ± 0.03 V is added to a voltage of 6.31 V ± 0.3%. What is the overall accuracy in volts and as a percentage? What will the percentage accuracy be if the first voltage is subtracted from the second? *[0.05 V, 0.5%. 1.9%]*
2. In an experimental reading of the same quantity the following data were obtained in μV:

    198.4, 198.9, 199.6, 200.1, 198.5, 198.8, 199.3, 198.9, 199.3, 199.6, 199.2, 199.9

What is the best estimate of the true value? Find the standard deviation and the standard error of the mean. How many more readings would be required to reduce the standard error of the mean to 0.1 µV? If the next (thirteenth) reading was 202.1 µV would you include it in the analysis or not? *[199.2 µV, 0.51 µV, 0.15 µV, 14]*

3. A metre ruler is made with only cm scale divisions. How many measurements would it take to measure a length of 300 mm to an accuracy of 0.5%? If an oscilloscope is used to measure a peak-to-peak voltage of 45 mV using the 10 mV/div scale, how many measurements would be needed to achieve an accuracy of 0.1 dB? (Assume the CRO can be read to no better than 0.2 div, but is otherwise accurate.) *[45, 15]*

4. A voltage, V_x, is calculated with the formula

$$V_x = \frac{VR_1}{R_2 + R_3 + R_4}$$

The error in resistance is ±0.1% and that in V is ±0.2%. What is the error in V_x? *[0.245%]*

5. A voltage, V_x, is calculated with the formula of the previous problem using the following data: $V = 3.573 \pm 0.005$ V, $R_1 = 5.562$ kΩ, $R_2 = 90$ kΩ, $R_3 = 9$ kΩ and $R_4 = 1$ kΩ. This time all the resistances are in error by ±10 Ω. What is V_x and its uncertainty? *[198.73 ± 0.46 mV]*

6. A platinum resistance thermometer is used to measure the temperature of a liquid. The resistance of the thermometer is given by the formula $R_T = R_0(1 + \alpha T)$, where $R_0 = 100.00 \pm 0.02$ Ω, $\alpha = 0.00388 \pm 0.00001$, T is the temperature to be found in Celsius and $R_T = 115.33 \pm 0.05$ Ω. What is T and its error? What is the largest source of error? *[39.51 ± 0.17 °C. The error in R_T]*

7. A thermistor's resistance is known to obey the equation $R_T = R_\infty \exp(B/T)$, where T is the absolute temperature in K, $R_\infty = 0.05652 \pm 0.00001$ Ω and $B = 3333.0 \pm 0.1$ K. If R_T is found to be 5.238 ± 0.001 kΩ, what is the temperature, T, and its error? If the errors in R_∞ and B are made negligible, how accurately must R_T be measured to achieve an error of 1 mK at a temperature of 0°C? (Take 0°C to be 273.15 K.) *[291.43 ± 0.01 K. To ±0.5 Ω or ±0.0044%]*

8. The error in reading a moving-coil ohmmeter is s% of the full-scale deflection. Show that the percentage error at a fraction, α, of full-scale deflection is $s/\alpha(1 - \alpha)$. Hence show that the minimum error occurs at half full scale. If $s = 1$%, what are the percentage errors at 20%, 50% and 80% of full scale? *[6.25%, 4% and 6.25%]*

9. What are the crest factors for (a) a square wave alternating between $+V_m$ and $-V_m$ (b) a triangular wave alternating between $+V_m$ and $-V_m$ (c) the sine wave $V_m \sin \omega t$ and (d) a voltage spike? *[(a) 1 (b) √3 (c) √2 (d) ∞]*

10. What is the smallest number of bits required by the A/D converter in a 5½-digit AC voltmeter if it must achieve an accuracy of ±1 digit in AC voltage measurements? If a DMM uses a 16-bit A/D, what is its AC accuracy at best? *[18 bits, 0.002%]*

11. What is the minimum meter resistance required to read the voltage from a source of resistance 1 kΩ to an accuracy of 0.2%? An AC voltage source of internal resistance 10 kΩ is connected to an oscilloscope via 0.75 m of coaxial cable of capacitance 110 pF/m. If the input impedance of the CRO is 20 pF in parallel with 1 MΩ, at what frequency will the CRO's reading be down by 0.5 dB from its true value? *[499 kΩ, 49.6 kHz]*

12. An electronic fluxmeter (see Problem 13.4) has a circular coil of 20 ± 0.02 turns and is 100 ± 0.1 mm in diameter. It has a resistance of 12 Ω, neglected in calculating a magnetic field which produces a voltage change of 5.379 ± 0.002 V from its integrator. If the input resistor, R, is 22 kΩ ± 0.1% and the capacitor, C, is 2.252 ± 0.003 µF what is the magnetic field strength and its error? Which error contributes most to the overall error? *[1.350 MA/m, –3 kA/m +4.5 kA/m; the coil diameter]*

Index